Student Solutions Man

Beginning and Intermediate Algebra
A Guided Approach

SEVENTH EDITION

Rosemary M. Karr
Collin College

Marilyn B. Massey
Collin College

R. David Gustafson
Rock Valley College

Prepared by

Michael G. Welden
Mt. San Jacinto College

CENGAGE
Learning·

Australia • Brazil • Mexico • Singapore • United Kingdom • United States

For product information and technology assistance, contact us at **Cengage Learning Customer & Sales Support, 1-800-354-9706**.

For permission to use material from this text or product, submit all requests online at **www.cengage.com/permissions** Further permissions questions can be emailed to **permissionrequest@cengage.com**.

ISBN-13: 978-1-285-84642-2
ISBN-10: 1-285-84642-7

Cengage Learning
200 First Stamford Place, 4th Floor
Stamford, CT 06902
USA

Cengage Learning is a leading provider of customized learning solutions with office locations around the globe, including Singapore, the United Kingdom, Australia, Mexico, Brazil, and Japan. Locate your local office at: **www.cengage.com/global**.

Cengage Learning products are represented in Canada by Nelson Education, Ltd.

To learn more about Cengage Learning Solutions, visit **www.cengage.com**.

Purchase any of our products at your local college store or at our preferred online store **www.cengagebrain.com**.

Printed in the United States of America
1 2 3 4 5 6 7 18 17 16 15 14

Contents

Chapter 1	Real Numbers and Their Basic Properties	1
Chapter 2	Equations and Inequalities	23
Chapter 3	Graphing; Writing Equations of Lines; Functions	76
Chapter 4	Polynomials	113
Chapter 5	Factoring Polynomials	143
Chapter 6	Rational Expressions and Equations; Proportion and Variation	169
Chapter 7	Transitioning to Intermediate Algebra	208
Chapter 8	Solving Systems of Linear Equations and Inequalities	256
Chapter 9	Radicals and Rational Exponents	316
Chapter 10	Quadratic and Other Nonlinear Functions and Inequalities	350
Chapter 11	Algebra, Composition, and Inverses of Functions; Exponential and Logarithmic Functions	398
Chapter 12	Conic Sections, Systems of Equations and Inequalities, and More Graphing	433
Chapter 13	Miscellaneous Topics	463
Appendix 1	Measurement Conversions	490
Appendix 2	Symmetries of Graphs	491

Preface

This manual contains detailed solutions to the odd exercises of the text *Beginning and Intermediate Algebra*, seventh edition, by Karr, Massey, and Gustafson. It also contains solutions to all of the exercises in the Chapter Review and Chapter Test sections of the text.

Many of the exercises in the text may be solved using more than one method, but it is not feasible to list all possible solutions in this manual. Also, some of the exercises may have been solved in this manual using a method that differs slightly from that presented in the text. There are a few exercises in the text whose solutions may vary from person to person. Some of these solutions may not have been included in this manual. For the solution to an exercise like this, the notation "answers may vary" has been included.

Please remember that only reading a solution does not teach you how to solve a problem. To repeat a commonly used phrase, mathematics is not a spectator sport. You MUST make an honest attempt to solve each exercise in the text without using this manual first. This manual should be viewed more or less as a last resort. Above all, DO NOT simply copy the solution from this manual onto your own paper. Doing so will not help you learn how to do the exercise, nor will it help you to do better on quizzes or tests.

This book is dedicated to my nieces Hannah and Mairin Welden.

May your study of this material be successful and rewarding.

Michael G. Welden

SECTION 1.1

Exercises 1.1 (page 11)

1-9. Answers may vary.

11. $-|-7| = -(+7) = -7$

13. set **15.** whole **17.** integers **19.** subset

21. rational **23.** real **25.** natural, prime **27.** odd

29. $<$ **31.** variables **33.** 7 **35.** parenthesis, open

37. distance, 6

39. natural: $1, 2, 6, 9$

41. positive integers: $1, 2, 6, 9$

43. integers: $-3, -1, 0, 1, 2, 6, 9$

45. real: $-3, -\frac{1}{2}, -1, 0, 1, 2, \frac{5}{3}, \sqrt{7}, 3.25, 6, 9$

47. odd integers: $-3, -1, 1, 9$

49. composite: $6, 9$

51. $7 \boxed{<} 10$

53. $9 \boxed{\phantom{<}} 2 + 5$
$9 \boxed{>} 7$

55. $-6 \boxed{>} -8$

57. $5 + 7 \boxed{\phantom{<}} 10$
$12 \boxed{>} 10$

59.

4 is greater than 2. 4 is to the right of 2.

61.

11 is greater than 6. 11 is to the right of 6.

63.

-2 is greater than -5.
-2 is to the right of -5.

65.

8 is greater than 0. 8 is to the right of 0.

67.

69.

71.

73.

75. $|36| = 36$

77. $|0| = 0$

79. $-|-23| = -(+23) = -23$

81. $|12 - 4| = |8| = 8$

1

SECTION 1.1

83. $6 + 3 = 9$
9: natural, odd, composite, whole

85. $15 - 15 = 0$
0: even, whole

87. $3 \cdot 8 = 24$
24: natural, even, composite, whole

89. $24 \div 8 = 3$
3: natural, odd, prime, whole

91. $5 + 6 \boxed{\phantom{<}} 13 - 1$
$11 \boxed{<} 12$

93. $4 \cdot 3 \boxed{} 3 \cdot 4$
$12 \boxed{=} 12$

95. $0 \div 6 \boxed{\phantom{<}} 1$
$0 \boxed{<} 1$

97. $45 \div 9 \boxed{} 36 \div 12$
$5 \boxed{>} 3$

99. $3 + 2 + 5 \boxed{} 5 + 2 + 3$
$10 \boxed{=} 10$

101. $9 > 4$

103. $8 \leq 8$

105. $3 + 4 = 7$

107. $\sqrt{2} \approx 1.41$

109. $3 \leq 7 \Rightarrow \boxed{7 \geq 3}$

111. $6 > 0 \Rightarrow \boxed{0 < 6}$

113. $3 + 8 > 8 \Rightarrow \boxed{8 < 3 + 8}$

115. $6 - 2 < 10 - 4 \Rightarrow \boxed{10 - 4 > 6 - 2}$

117. $2 \cdot 3 < 3 \cdot 4 \Rightarrow \boxed{3 \cdot 4 > 2 \cdot 3}$

119. $\dfrac{12}{4} < \dfrac{24}{6} \Rightarrow \boxed{\dfrac{24}{6} > \dfrac{12}{4}}$

121.

123.

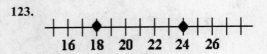

125.

![number line with bracket at -5 and parenthesis at 4]
$-5 \qquad 4$

127. $|21 - 19| = |2| = 2$

129. If you think you have the greatest natural number, just add 1 to it to get a greater natural number.

131. The absolute value of a positive number (or 0) is equal to the positive number (or 0). The absolute value of a negative number is equal to the opposite of the negative number.

133. Answers may vary.

Exercises 1.2   (page 25)

1. $3 = 1 \cdot 3, 6 = 2 \cdot 3$
largest common factor: 3

3. $12 = 2 \cdot 6, 18 = 3 \cdot 6$
largest common factor: 6

5. $\dfrac{3}{4} \cdot \dfrac{1}{2} = \dfrac{3 \cdot 1}{4 \cdot 2} = \dfrac{3}{8}$

7. $\dfrac{3}{4} \div \dfrac{4}{3} = \dfrac{3}{4} \cdot \dfrac{3}{4} = \dfrac{3 \cdot 3}{4 \cdot 4} = \dfrac{9}{16}$

9. $\dfrac{4}{9} + \dfrac{7}{9} = \dfrac{4 + 7}{9} = \dfrac{11}{9}$

2

SECTION 1.2

11. $\dfrac{2}{3} - \dfrac{1}{2} = \dfrac{2 \cdot \mathbf{2}}{3 \cdot \mathbf{2}} - \dfrac{1 \cdot \mathbf{3}}{2 \cdot \mathbf{3}} = \dfrac{4}{6} - \dfrac{3}{6}$

$\qquad\qquad = \dfrac{4-3}{6} = \dfrac{1}{6}$

13.

$$
\begin{array}{r}
5 \;.\; 1 \\
+ \;\; 0 \;.\; 6 \; 2 \\
\hline
5 \;.\; 7 \; 2
\end{array}
$$

15.

$$
\begin{array}{r}
0 \;.\; 2 \\
\times \;\; 2 \;.\; 5 \\
\hline
1 \; 0 \\
4 \; 0 \\
\hline
5 \; 0
\end{array}
$$

Put two digits to the right of the decimal point. Answer: 0.5

17. The digit in the 2nd decimal place is 6. The next digit to the right is 5. Since this digit is 5 or more, round up. Change the 6 in the 2nd decimal place to 7, and delete all digits to the right. 5.17

19. true

21. false; 21 has factors of 3 and 7.

23. false; -5 is to the left of -2.

25. true; $|-9| = 9$, so $9 \le |-9|$.

27. $3 + 7 \;\boxed{=}\; 10$

29. $|-2| = 2$, so $|-2| \;\boxed{=}\; 2$

31. numerator

33. undefined

35. prime

37. improper

39. 1

41. multiply

43. numerators, denominator

45. least common denominator, equivalent

47. terminating, 2

49. divisor, dividend, quotient

51. $24 = 4 \cdot 6 = 2 \cdot 2 \cdot 2 \cdot 3$

53. $48 = 8 \cdot 6 = 2 \cdot 4 \cdot 2 \cdot 3$
$\qquad = 2 \cdot 2 \cdot 2 \cdot 2 \cdot 3$

55. $\dfrac{6}{12} = \dfrac{1 \cdot \overset{1}{\cancel{6}}}{2 \cdot \underset{1}{\cancel{6}}} = \dfrac{1}{2}$

57. $\dfrac{15}{20} = \dfrac{3 \cdot \overset{1}{\cancel{5}}}{4 \cdot \underset{1}{\cancel{5}}} = \dfrac{3}{4}$

59. $\dfrac{27}{18} = \dfrac{3 \cdot \overset{1}{\cancel{9}}}{2 \cdot \underset{1}{\cancel{9}}} = \dfrac{3}{2}$

61. $\dfrac{72}{64} = \dfrac{9 \cdot \overset{1}{\cancel{8}}}{8 \cdot \underset{1}{\cancel{8}}} = \dfrac{9}{8}$

63. $\dfrac{1}{3} \cdot \dfrac{2}{5} = \dfrac{1 \cdot 2}{3 \cdot 5} = \dfrac{2}{15}$

65. $\dfrac{4}{3} \cdot \dfrac{6}{5} = \dfrac{4 \cdot 6}{3 \cdot 5} = \dfrac{4 \cdot 2 \cdot \overset{1}{\cancel{3}}}{\underset{1}{\cancel{3}} \cdot 5} = \dfrac{8}{5}$

67. $12 \cdot \dfrac{5}{6} = \dfrac{12}{1} \cdot \dfrac{5}{6} = \dfrac{12 \cdot 5}{1 \cdot 6} = \dfrac{2 \cdot \overset{1}{\cancel{6}} \cdot 5}{1 \cdot \underset{1}{\cancel{6}}}$

$\qquad\qquad = \dfrac{10}{1} = 10$

69. $\dfrac{10}{21} \cdot 14 = \dfrac{10}{21} \cdot \dfrac{14}{1} = \dfrac{10 \cdot 14}{21 \cdot 1} = \dfrac{10 \cdot 2 \cdot \overset{1}{\cancel{7}}}{3 \cdot \underset{1}{\cancel{7}}}$

$\qquad\qquad\qquad = \dfrac{20}{3}$

71. $\dfrac{2}{5} \div \dfrac{3}{2} = \dfrac{2}{5} \cdot \dfrac{2}{3} = \dfrac{2 \cdot 2}{5 \cdot 3} = \dfrac{4}{15}$

73. $\dfrac{3}{4} \div \dfrac{6}{5} = \dfrac{3}{4} \cdot \dfrac{5}{6} = \dfrac{3 \cdot 5}{4 \cdot 6} = \dfrac{\overset{1}{\cancel{3}} \cdot 5}{4 \cdot 2 \cdot \underset{1}{\cancel{3}}} = \dfrac{5}{8}$

75. $9 \div \dfrac{3}{8} = \dfrac{9}{1} \div \dfrac{3}{8} = \dfrac{9}{1} \cdot \dfrac{8}{3} = \dfrac{9 \cdot 8}{1 \cdot 3} = \dfrac{3 \cdot \overset{1}{\cancel{3}} \cdot 8}{1 \cdot \underset{1}{\cancel{3}}} = \dfrac{24}{1} = 24$

77. $\dfrac{54}{20} \div 3 = \dfrac{54}{20} \div \dfrac{3}{1} = \dfrac{54}{20} \cdot \dfrac{1}{3} = \dfrac{54 \cdot 1}{20 \cdot 3} = \dfrac{\overset{1}{\cancel{2}} \cdot \overset{1}{\cancel{3}} \cdot 9}{10 \cdot \underset{1}{\cancel{2}} \cdot \underset{1}{\cancel{3}}} = \dfrac{9}{10}$

79. $\dfrac{3}{5} + \dfrac{3}{5} = \dfrac{3+3}{5} = \dfrac{6}{5}$

81. $\dfrac{5}{17} - \dfrac{3}{17} = \dfrac{5-3}{17} = \dfrac{2}{17}$

83. $\dfrac{1}{42} + \dfrac{1}{6} = \dfrac{1}{42} + \dfrac{1 \cdot \mathbf{7}}{6 \cdot \mathbf{7}} = \dfrac{1}{42} + \dfrac{7}{42} = \dfrac{1+7}{42} = \dfrac{8}{42} = \dfrac{4 \cdot \overset{1}{\cancel{2}}}{21 \cdot \underset{1}{\cancel{2}}} = \dfrac{4}{21}$

85. $\dfrac{7}{10} - \dfrac{1}{14} = \dfrac{7 \cdot \mathbf{7}}{10 \cdot \mathbf{7}} - \dfrac{1 \cdot \mathbf{5}}{14 \cdot \mathbf{5}} = \dfrac{49}{70} - \dfrac{5}{70} = \dfrac{49-5}{70} = \dfrac{44}{70} = \dfrac{22 \cdot \overset{1}{\cancel{2}}}{35 \cdot \underset{1}{\cancel{2}}} = \dfrac{22}{35}$

87. $4\dfrac{3}{5} + \dfrac{3}{5} = \left(4 + \dfrac{3}{5}\right) + \dfrac{3}{5} = \left(\dfrac{20}{5} + \dfrac{3}{5}\right) + \dfrac{3}{5} = \dfrac{23}{5} + \dfrac{3}{5} = \dfrac{26}{5} = 5\dfrac{1}{5}$

89. $3\dfrac{1}{3} - 1\dfrac{2}{3} = \left(3 + \dfrac{1}{3}\right) - \left(1 + \dfrac{2}{3}\right) = \left(\dfrac{9}{3} + \dfrac{1}{3}\right) - \left(\dfrac{3}{3} + \dfrac{2}{3}\right) = \dfrac{10}{3} - \dfrac{5}{3} = \dfrac{5}{3} = 1\dfrac{2}{3}$

91. $3\dfrac{3}{4} - 2\dfrac{1}{2} = \left(3 + \dfrac{3}{4}\right) - \left(2 + \dfrac{1}{2}\right) = \left(\dfrac{12}{4} + \dfrac{3}{4}\right) - \left(\dfrac{8}{4} + \dfrac{2}{4}\right) = \dfrac{15}{4} - \dfrac{10}{4} = \dfrac{5}{4} = 1\dfrac{1}{4}$

93. $8\dfrac{2}{9} - 7\dfrac{2}{3} = \left(8 + \dfrac{2}{9}\right) - \left(7 + \dfrac{2}{3}\right) = \left(\dfrac{72}{9} + \dfrac{2}{9}\right) - \left(\dfrac{63}{9} + \dfrac{6}{9}\right) = \dfrac{74}{9} - \dfrac{69}{9} = \dfrac{5}{9}$

95.

```
      0.  6
  5 | 3.  0
      3   0
      ─────
          0
```

$\dfrac{3}{5} = 0.6$, terminating

97.

```
       0.  4  0  9  0  9
  22 | 9.  0  0  0  0  0
       8   8
       ──────
           2   0
               0
           ────────
           2   0  0
           1   9  8
           ─────────
               2   0
                   0
               ─────────
               2   0  0
               1   9  8
               ─────────
                   2
```

$\dfrac{9}{22} = 0.4\overline{09}$

repeating

4

99.

$$
\begin{array}{r}
\;\;\overset{1}{}\\
4\;\;3\;.\;5\;\;4\\
+\;\;3\;\;1\;\;5\;.\;7\\
\hline
3\;\;5\;\;9\;.\;2\;\;4
\end{array}
$$

101.

$$
\begin{array}{r}
\overset{6}{}\;\;\overset{\overset{11}{1}}{}\;\;\overset{13}{}\\
6\;\;7\;.\;2\;\;3\;\;5\\
-\;\;2\;\;2\;.\;4\;\;5\\
\hline
4\;\;4\;.\;7\;\;8\;\;5
\end{array}
$$

103.

$$
\begin{array}{r}
7\;.\;2\\
\times\;\;\;\;1\;\;5\;.\;6\\
\hline
4\;\;3\;\;2\\
3\;\;6\;\;0\\
7\;\;2\\
\hline
1\;\;1\;\;2\;\;3\;\;2
\end{array}
$$

Put two digits to the right of the decimal point. Answer: 112.32

105.

$$0.23\overline{)1.0465}$$

Move decimal points 2 places right.

$$
\begin{array}{r}
4.\;5\;\;5\\
2\;3.\;\overline{)1\;0\;4.\;6\;\;5}\\
9\;2\\
\hline
1\;\;2\;\;6\\
1\;\;1\;\;5\\
\hline
1\;\;1\;\;5\\
1\;\;1\;\;5\\
\hline
0
\end{array}
$$

107. The digit in the 2nd decimal place is 5. The next digit to the right is 8. Since this digit is 5 or more, round up. Change the 5 in the 2nd decimal place to 6, and delete all digits to the right. 496.26

The digit in the 3rd decimal place is 8. The next digit to the right is 3. Since this digit is less than 5, round down. Leave the 8 in the 3rd decimal place, and delete all digits to the right. 496.258

109. The digit in the 2nd decimal place is 9. The next digit to the right is 8. Since this digit is 5 or more, round up. Change the 9 in the 2nd decimal place to to 0, increase the digit in the 1st decimal place from 3 to 4, and delete all digits to the right. 6,025.40

The digit in the 3rd decimal place is 8. The next digit to the right is 2. Since this digit is less than 5, round down. Leave the 8 in the 3rd decimal place, and delete all digits to the right. 6,025.398

111. $\dfrac{5}{12} \cdot \dfrac{18}{5} = \dfrac{5 \cdot 18}{12 \cdot 5} = \dfrac{\overset{1}{\cancel{5}} \cdot 3 \cdot \overset{1}{\cancel{6}}}{2 \cdot \underset{1}{\cancel{6}} \cdot \underset{1}{\cancel{5}}} = \dfrac{3}{2}$

113. $\dfrac{17}{34} \cdot \dfrac{3}{6} = \dfrac{17 \cdot 3}{34 \cdot 6} = \dfrac{\overset{1}{\cancel{17}} \cdot \overset{1}{\cancel{3}}}{2 \cdot \underset{1}{\cancel{17}} \cdot 2 \cdot \underset{1}{\cancel{3}}} = \dfrac{1}{4}$

115. $\dfrac{2}{13} \div \dfrac{8}{13} = \dfrac{2}{13} \cdot \dfrac{13}{8} = \dfrac{2 \cdot 13}{13 \cdot 8} = \dfrac{\overset{1}{\cancel{2}} \cdot \overset{1}{\cancel{13}}}{\underset{1}{\cancel{13}} \cdot 4 \cdot \underset{1}{\cancel{2}}}$
$= \dfrac{1}{4}$

117. $\dfrac{21}{35} \div \dfrac{3}{14} = \dfrac{21}{35} \cdot \dfrac{14}{3} = \dfrac{21 \cdot 14}{35 \cdot 3}$
$= \dfrac{7 \cdot \overset{1}{\cancel{3}} \cdot 14}{5 \cdot 7 \cdot \underset{1}{\cancel{3}}} = \dfrac{14}{5}$

119. $\dfrac{3}{5} + \dfrac{2}{3} = \dfrac{3 \cdot \mathbf{3}}{5 \cdot \mathbf{3}} + \dfrac{2 \cdot \mathbf{5}}{3 \cdot \mathbf{5}} = \dfrac{9}{15} + \dfrac{10}{15}$
$= \dfrac{9 + 10}{15} = \dfrac{19}{15}$

121. $\dfrac{9}{4} - \dfrac{5}{6} = \dfrac{9 \cdot \mathbf{3}}{4 \cdot \mathbf{3}} - \dfrac{5 \cdot \mathbf{2}}{6 \cdot \mathbf{2}} = \dfrac{27}{12} - \dfrac{10}{12}$
$= \dfrac{27 - 10}{12} = \dfrac{17}{12}$

123. $3 - \dfrac{3}{4} = \dfrac{3}{1} - \dfrac{3}{4} = \dfrac{3 \cdot \mathbf{4}}{1 \cdot \mathbf{4}} - \dfrac{3}{4} = \dfrac{12}{4} - \dfrac{3}{4} = \dfrac{12 - 3}{4} = \dfrac{9}{4}$

125. $\dfrac{17}{3} + 4 = \dfrac{17}{3} + \dfrac{4}{1} = \dfrac{17}{3} + \dfrac{4 \cdot \mathbf{3}}{1 \cdot \mathbf{3}} = \dfrac{17}{3} + \dfrac{12}{3} = \dfrac{17 + 12}{3} = \dfrac{29}{3}$

Problems 127-133 are to be solved using a calculator. The keystrokes needed to solve each problem using a TI-84 graphing calculator appear in each solution. There may be other solutions. Keystrokes for other calculators may be slightly different.

127. $\boxed{4}\ \boxed{7}\ \boxed{4}\ \boxed{.}\ \boxed{8}\ \boxed{1}\ \boxed{+}\ \boxed{2}\ \boxed{3}\ \boxed{.}\ \boxed{4}\ \boxed{5}\ \boxed{3}\ \boxed{2}\ \boxed{\text{ENTER}}\ \{498.2632\} \Rightarrow 498.26$

129. $\boxed{2}\ \boxed{5}\ \boxed{.}\ \boxed{2}\ \boxed{5}\ \boxed{\times}\ \boxed{1}\ \boxed{3}\ \boxed{2}\ \boxed{.}\ \boxed{1}\ \boxed{7}\ \boxed{9}\ \boxed{\text{ENTER}}\ \{3337.51975\} \Rightarrow 3,337.52$

131. $\boxed{4}\ \boxed{.}\ \boxed{5}\ \boxed{6}\ \boxed{9}\ \boxed{4}\ \boxed{3}\ \boxed{2}\ \boxed{3}\ \boxed{\div}\ \boxed{.}\ \boxed{4}\ \boxed{5}\ \boxed{6}\ \boxed{\text{ENTER}}\ \{10.02068487\} \Rightarrow 10.02$

133. $\boxed{5}\ \boxed{5}\ \boxed{.}\ \boxed{7}\ \boxed{7}\ \boxed{4}\ \boxed{3}\ \boxed{-}\ \boxed{.}\ \boxed{5}\ \boxed{6}\ \boxed{8}\ \boxed{2}\ \boxed{4}\ \boxed{5}\ \boxed{\text{ENTER}}$
$\{55.206185\} \Rightarrow 55.21$

135. $43\dfrac{1}{2} - 12\dfrac{1}{3} = 43 + \dfrac{1}{2} - 12 - \dfrac{1}{3} = \dfrac{258}{6} + \dfrac{3}{6} - \dfrac{72}{6} - \dfrac{2}{6} = \dfrac{187}{6} = 31\dfrac{1}{6}$ acres

137. $15 \cdot 4\dfrac{1}{3} = \dfrac{15}{1} \cdot \dfrac{13}{3} = \dfrac{15 \cdot 13}{1 \cdot 3} = \dfrac{5 \cdot \overset{1}{\cancel{3}} \cdot 13}{1 \cdot \underset{1}{\cancel{3}}} = \dfrac{65}{1} = 65$ yd

139. $187.75 - 46.8 - 72.5 = \$68.45$ million   **141.** 34% of $36,000 = 0.34(36,000) = \$12,240$

143. $0.23(17,500) = 4,025$ defective $\Rightarrow 17,500 - 4025 = 13,475$ acceptable units

145. $0.12(18,700,000) = 2,244,000$ increase $\Rightarrow$ sales $= 18,700,000 + 2,244,000 = \$20,944,000$

147. # gallons $= 16,275.3 \div 25.5 = 638.24705882 \Rightarrow$ cost $= 638.24705882(3.45) \approx \$2,201.95$

149. Area $=$ length $\cdot$ width $= (253.5 \text{ ft})(178.5 \text{ ft}) = 45,249.75 \text{ ft}^2$
Drums of sealer $= 45,249.75 \div 4,000 \approx 11.3 \Rightarrow$ needs 12 drums; Cost $= 12(97.50) = \$1,170$

151. Standard $= 37.50(2,530) = \$94,875$; High-capacity $= 57.35(1,670) = \$95,774.50$
The high-capacity order will produce the greater profit.

153. Silage per cow $= 0.57(12,000) = 6840$ pounds; $30(6,840) = 205,200$ lb of silage

155. Regular $= 1,730 + 36(107.75) = 1,730 + 3,879 = 5,609$
High $= 4,170 + 36(57.50) = 4,170 + 2,070 = 6,240$
The high-efficiency furnace will be more expensive after 3 years.

157-161. Answers may vary.

6

163. No. Each proper fraction is less than 1. When a number is multiplied by a number less than 1, the result is smaller than the original number.

Exercises 1.3 (page 36)

1. $2 \cdot 2 \cdot 2 \cdot 2 \cdot 2 = 32$

3. $4 \cdot 4 \cdot 4 = 64$

5. $\frac{2}{3} \cdot \frac{2}{3} \cdot \frac{2}{3} = \frac{8}{27}$

7. base: y

9. base: $4x$

11.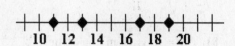

13. 17 is a prime number.

15. exponent

17. grouping

19. perimeter, circumference

21. $P = 4s$, units

23. $P = 2l + 2w$, units

25. $P = a + b + c$, units

27. $P = a + b + c + d$, units

29. $C = \pi D$, or $C = 2\pi r$, units

31. $V = lwh$, cubic units

33. $V = \frac{1}{3}Bh$, cubic units

35. $V = \frac{4}{3}\pi r^3$, cubic units

37. $6^2 = 6 \cdot 6 = 36$

39. $\left(-\frac{1}{5}\right)^4 = \left(-\frac{1}{5}\right)\left(-\frac{1}{5}\right)\left(-\frac{1}{5}\right)\left(-\frac{1}{5}\right)$
$= \frac{1}{625}$

41. $x^3 = x \cdot x \cdot x$

43. $8z^4 = 8 \cdot z \cdot z \cdot z \cdot z$

45. $(4x)^3 = 4x \cdot 4x \cdot 4x$

47. $3(6y)^2 = 3 \cdot 6y \cdot 6y$

49. $4(3^2) = 4 \cdot 9 = 36$

51. $(2 \cdot 5)^4 = 10^4 = 10{,}000$

53. $5(4)^2 = 5 \cdot 16 = 80$

55. $(3 \cdot 2)^3 = 6^3 = 216$

57. $3 \cdot 5 - 4 = 15 - 4 = 11$

59. $3(5 - 4) = 3(1) = 3$

61. $2 + 3 \cdot 5 - 4 = 2 + 15 - 4 = 17 - 4 = 13$

63. $48 \div (4 + 2) = 48 \div 6 = 8$

65. $3^2 + 2(1 + 4) - 2 = 9 + 2(5) - 2 = 9 + 10 - 2 = 19 - 2 = 17$

67. $\frac{3}{5} \cdot \frac{10}{3} + \frac{1}{2} \cdot 12 = \frac{3}{5} \cdot \frac{10}{3} + \frac{1}{2} \cdot \frac{12}{1} = \frac{\cancel{3}}{\cancel{5}_1} \cdot \frac{2 \cdot \cancel{5}}{\cancel{3}_1} + \frac{1}{\cancel{2}_1} \cdot \frac{6 \cdot \cancel{2}}{1} = \frac{2}{1} + \frac{6}{1} = 2 + 6 = 8$

69. $\left[\frac{1}{3} - \left(\frac{1}{2}\right)^2\right]^2 = \left[\frac{1}{3} - \frac{1}{4}\right]^2 = \left[\frac{1 \cdot 4}{3 \cdot 4} - \frac{1 \cdot 3}{4 \cdot 3}\right]^2 = \left[\frac{4}{12} - \frac{3}{12}\right]^2 = \left[\frac{1}{12}\right]^2 = \frac{1}{144}$

71. $\dfrac{(3+5)^2+2}{2(8-5)}=\dfrac{8^2+2}{2(3)}=\dfrac{64+2}{6}=\dfrac{66}{6}$
$=11$

73. $\dfrac{(5-3)^2+2}{4^2-(8+2)}=\dfrac{2^2+2}{16-10}=\dfrac{4+2}{6}=\dfrac{6}{6}=1$

75. $\dfrac{3\cdot7-5(3\cdot4-11)}{4(3+2)-3^2+5}=\dfrac{3\cdot7-5(12-11)}{4(5)-3^2+5}=\dfrac{3\cdot7-5(1)}{4(5)-9+5}=\dfrac{21-5}{20-9+5}=\dfrac{16}{16}=1$

77. $P=4s=4(5\text{ in.})=20\text{ in.}$

79. $P=a+b+c=3\text{ m}+5\text{ m}+7\text{ m}$
$=15\text{ m}$

81. $A=s^2=(6\text{ m})^2=36\text{ m}^2$

83. $A=bh=(5\text{ ft})(11\text{ ft})=55\text{ ft}^2$

85. $C=2\pi r\approx2\left(\dfrac{22}{7}\right)(14\text{ m})=\dfrac{2}{1}\cdot\dfrac{22}{7}\cdot\dfrac{14}{1}\text{ m}=\dfrac{2\cdot22\cdot2\cdot\overset{1}{\cancel{7}}}{\underset{1}{\cancel{7}}}\text{ m}=88\text{ m}$

87. $A=\pi r^2\approx\dfrac{22}{7}(21\text{ ft})^2=\dfrac{22}{7}\left(441\text{ ft}^2\right)=\dfrac{22}{7}\cdot\dfrac{441}{1}\text{ ft}^2=\dfrac{22}{\underset{1}{\cancel{7}}}\cdot\dfrac{63\cdot\overset{1}{\cancel{7}}}{1}\text{ ft}^2=1{,}386\text{ ft}^2$

89. $V=\tfrac{1}{3}Bh=\tfrac{1}{3}(3\text{ cm})^2(2\text{ cm})=\tfrac{1}{3}(9\text{ cm}^2)(2\text{ cm})=(3\text{ cm}^2)(2\text{ cm})=6\text{ cm}^3$

91. $V=\tfrac{4}{3}\pi r^3\approx\dfrac{4}{3}\cdot\dfrac{22}{7}(6\text{ m})^3=\dfrac{88}{21}\left(216\text{ m}^3\right)=\dfrac{88}{21}\cdot\dfrac{216}{1}\text{ m}^3\approx905\text{ m}^3$

93. Cylinder: $V=Bh=\pi(4\text{ cm})^2(14\text{ cm})\approx\dfrac{22}{7}\cdot\dfrac{16}{1}\text{ cm}^2\cdot\dfrac{14}{1}\text{ cm}=\dfrac{22\cdot16\cdot2\cdot\overset{1}{\cancel{7}}}{\underset{1}{\cancel{7}}\cdot1\cdot1}\text{ cm}^3=704\text{ cm}^3$

Cone: $V=\tfrac{1}{3}Bh=\tfrac{1}{3}\pi(4\text{ cm})^2(21\text{ cm})\approx\dfrac{1}{3}\cdot\dfrac{22}{7}\cdot\dfrac{16}{1}\text{ cm}^2\cdot\dfrac{21}{1}\text{ cm}=\dfrac{22\cdot16\cdot\overset{1}{\cancel{3}}\cdot\overset{1}{\cancel{7}}}{\underset{1}{\cancel{3}}\cdot\underset{1}{\cancel{7}}\cdot1\cdot1}\text{ cm}^3=352\text{ cm}^3$

Total $=704\text{ cm}^3+352\text{ cm}^3=1{,}056\text{ cm}^3$

95. $6^2=6\cdot6=36$

97. $2+4^2=2+16=18$

99. $(2+4)^2=(6)^2=36$

101. $(7+9)\div(2\cdot4)=16\div8=22$

103. $(5+7)\div3\cdot4=12\div3\cdot4=4\cdot4=16$

105. $24\div4\cdot3+3=6\cdot3+3=18+3=21$

107. $6^2-(8-3)^2=6^2-5^2=36-25=11$

109. $(2\cdot3-4)^3=(6-4)^3=2^3=8$

111. $\dfrac{2[4+2(3-1)]}{3[3(2\cdot3-4)]}=\dfrac{2[4+2(2)]}{3[3(6-4)]}=\dfrac{2[4+4]}{3[3(2)]}=\dfrac{2[8]}{3[6]}=\dfrac{16}{18}=\dfrac{8}{9}$

SECTION 1.3

Problems 113-115 are to be solved using a calculator. The keystrokes needed to solve each problem using a TI-84 graphing calculator appear in each solution. There may be other solutions. Keystrokes for other calculators may be slightly different.

113. $\boxed{7}\ \boxed{.}\ \boxed{9}\ \boxed{\wedge}\ \boxed{3}\ \boxed{\text{ENTER}}$
$\{493.039\}$

115. $\boxed{2}\ \boxed{5}\ \boxed{.}\ \boxed{3}\ \boxed{\wedge}\ \boxed{2}\ \boxed{\text{ENTER}}$
$\{640.09\}$

117. $39 = (3 \cdot 8) + (5 \cdot 3)$

119. $87 = (3 \cdot 8 + 5) \cdot 3$

121. $V = \frac{4}{3}\pi r^3 = \frac{4}{3}\pi (21.35\,\text{ft})^3 \approx 40{,}764.51\,\text{ft}^3$

123. $P = 4s = 4\left(30\frac{2}{5}\,\text{m}\right) = 4\left(30 + \frac{2}{5}\,\text{m}\right) = 4\left(\frac{150}{5} + \frac{2}{5}\,\text{m}\right) = \frac{4}{1}\left(\frac{152}{5}\,\text{m}\right)$
$$= \frac{608}{5}\,\text{m} = 121\frac{3}{5}\,\text{m}$$

125. $V = lwh = (40\,\text{ft})(40\,\text{ft})(9\,\text{ft}) = 14{,}400\,\text{ft}^3$; Per student $= 14{,}400\,\text{ft}^3 \div 30 = 480\,\text{ft}^3$ per student

127. $f = \dfrac{rs}{(r+s)(n-1)} = \dfrac{(8)(12)}{(8+12)(1.6-1)} = \dfrac{96}{(20)(0.6)} = \dfrac{96}{12} = 8$

129. Answers may vary.

131. Increasing powers produce larger numbers.

Exercises 1.4 (page 45)

1. $2 + 3 = +(2 + 3) = 5$

3. $-4 + 7 = +(7 - 4) = 3$

5. $6 - 2 = 4$

7. $-5 - (-7) = -5 + (+7) = +(7 - 5) = 2$

9. $5 + 3(7 - 2) = 5 + 3(5) = 5 + 15 = 20$

11. $5 + 3(7) - 2 = 5 + 21 - 2 = 26 - 2 = 24$

13. arrows

15. unlike

17. subtract, greater

19. add, opposite

21. $5 + 9 = +(5 + 9) = 14$

23. $(-7) + (-2) = -(7 + 2)$
$= -9$

25. $\dfrac{1}{5} + \left(+\dfrac{1}{7}\right) = \dfrac{7}{35} + \left(+\dfrac{5}{35}\right) = +\left(\dfrac{7}{35} + \dfrac{5}{35}\right) = \dfrac{12}{35}$

27. $44.902 + 33.098 = +(44.902 + 33.098) = 78$

29. $7 + (-3) = +(7 - 3) = 4$

31. $(-0.4) + 0.9 = +(0.9 - 0.4) = 0.5$

33. $\dfrac{2}{3} + \left(-\dfrac{1}{4}\right) = \dfrac{8}{12} + \left(-\dfrac{3}{12}\right) = +\left(\dfrac{8}{12} - \dfrac{3}{12}\right) = +\dfrac{5}{12}$

35. $73.82 + (-108.4) = -(108.4 - 73.82) = -34.58$

37. $5 + [4 + (-2)] = 5 + [2] = 7$

39. $-2 + (-4 + 5) = -2 + 1 = -1$

41. $(-7 + 5) + 2 = -2 + 2 = 0$

43. $-9 + [-6 + (-4)] = -9 + [-10] = -19$

45.
$$\begin{array}{r} 5 \\ + \underline{-4} \\ 1 \end{array}$$

47.
$$\begin{array}{r} -1.3 \\ + \underline{3.5} \\ 2.2 \end{array}$$

49. $8 - 4 = 8 + (-4) = 4$

51. $8 - (-4) = 8 + (+4) = 12$

53. $0 - (-5) = 0 + (+5) = 5$

55. $\dfrac{5}{3} - \dfrac{7}{6} = \dfrac{10}{6} - \dfrac{7}{6} = \dfrac{10}{6} + \left(-\dfrac{7}{6}\right)$
$$= \dfrac{3}{6} = \dfrac{1}{2}$$

57.
$$\begin{array}{r} 8 \\ - \underline{4} \\ \end{array} \quad \Rightarrow \quad \begin{array}{r} 8 \\ + \underline{-4} \\ 4 \end{array}$$

59.
$$\begin{array}{r} -10 \\ - \underline{-3} \\ \end{array} \quad \Rightarrow \quad \begin{array}{r} -10 \\ + \underline{3} \\ -7 \end{array}$$

61. $5 - [(-2) - 4] = 5 - [(-2) + (-4)] = 5 - [-6] = 5 + [+6] = 11$

63. $4 - [(-3) - 5] = 4 - [(-3) + (-5)] = 4 - [-8] = 4 + [+8] = 12$

65. $\dfrac{5 - (-4)}{3 - (-6)} = \dfrac{5 + (+4)}{3 + (+6)} = \dfrac{9}{9} = 1$

67. $\dfrac{-6 - (-3)}{5 + (-8)} = \dfrac{-6 + (+3)}{-3} = \dfrac{-6 + 3}{-3} = \dfrac{-3}{-3}$
$$= 1$$

Problems 69-71 are to be solved using a calculator. The keystrokes needed to solve each problem using a TI-84 graphing calculator appear in each solution. There may be other solutions. Keystrokes for other calculators may be slightly different.

69. $\boxed{4}\ \boxed{.}\ \boxed{2}\ \boxed{6}\ \boxed{-}\ \boxed{6}\ \boxed{.}\ \boxed{3}\ \boxed{4}\ \boxed{+}\ \boxed{.}\ \boxed{5}\ \boxed{6}\ \boxed{\text{ENTER}}$
$\{-1.52\} \Rightarrow -1.52$

71. $\boxed{2}\ \boxed{.}\ \boxed{3}\ \boxed{4}\ \boxed{x^2}\ \boxed{-}\ \boxed{3}\ \boxed{.}\ \boxed{4}\ \boxed{7}\ \boxed{x^2}\ \boxed{-}\ \boxed{.}\ \boxed{7}\ \boxed{2}\ \boxed{x^2}\ \boxed{\text{ENTER}}$
$\{-7.0837\} \Rightarrow -7.08$

73. $\left(\dfrac{5}{2} - 3\right) - \left(\dfrac{3}{2} - 5\right) = \left(\dfrac{5}{2} - \dfrac{6}{2}\right) - \left(\dfrac{3}{2} - \dfrac{10}{2}\right) = \left(-\dfrac{1}{2}\right) - \left(-\dfrac{7}{2}\right) = \left(-\dfrac{1}{2}\right) + \left(+\dfrac{7}{2}\right)$
$$= \dfrac{6}{2} = 3$$

75. $(5.2 - 2.5) - (5.25 - 5) = [5.2 + (-2.5)] - [5.25 + (-5)] = 2.7 - [0.25] = 2.7 + (-0.25)$
$$= 2.45$$

77. $4 + (-12) = -(12 - 4) = -8$

79. $[-4 + (-3)] + [2 + (-2)] = [-7] + [0] = -7$

81. $-4 + (-3 + 2) + (-3) = -4 + (-1) + (-3) = -5 + (-3) = -8$

SECTION 1.4

83. $-|8 + (-4)| + 7 = -|4| + 7 = -4 + 7 = 3$ **85.** $-5.2 + |-2.5 + (-4)| = -5.2 + |-6.5|$
$$= -5.2 + 6.5 = 1.3$$

87. $-3\dfrac{1}{2} - 5\dfrac{1}{4} = -\dfrac{7}{2} - \dfrac{21}{4} = -\dfrac{14}{4} - \dfrac{21}{4} = -\dfrac{14}{4} + \left(-\dfrac{21}{4}\right) = -\dfrac{35}{4} = -8\dfrac{3}{4}$

89. $-6.7 - (-2.5) = -6.7 + (+2.5) = -4.2$ **91.** $\dfrac{-4-2}{-[2+(-3)]} = \dfrac{-4+(-2)}{-[-1]} = \dfrac{-6}{+1} = -6$

93. $\left(\dfrac{3}{4} - \dfrac{4}{5}\right) - \left(\dfrac{2}{3} + \dfrac{1}{4}\right) = \left(\dfrac{15}{20} - \dfrac{16}{20}\right) - \left(\dfrac{8}{12} + \dfrac{3}{12}\right) = \left(-\dfrac{1}{20}\right) - \left(\dfrac{11}{12}\right) = \left(-\dfrac{3}{60}\right) - \left(\dfrac{55}{60}\right)$
$$= -\left(\dfrac{3}{60} + \dfrac{55}{60}\right)$$
$$= -\dfrac{58}{60} = -\dfrac{29}{30}$$

95. $(-735) + (+500) = -235$
She still owes \$235.

97. $(+13) + (-4) = +9$

99. $(-14) + 10 = -4°$

101. $1700 - (-300) = 2000$ years

103. $(-2,300) + (1,750) + (1,875) = +1,325$ m

105. $32,000 - 28,000 = 4,000$ ft

107. $+32 - (+27) = 5°$

109. $12,153 - 23 + 57 = 12,187$

111. $500 \cdot 2 - 300 = 1000 - 300 = 700$ shares

113. $437.45 + 25.17 + 37.93 + 45.26 - 17.13 - 83.44 - 22.58 = \$422.66$

115. $115,000 - 78 - 446 - 216 - 7,612.32 - 23,445.11 + 223 = \$83,425.57$

117. Answers may vary.

119. The answers agree if the two numbers have the same sign. The answers do not agree if the numbers have opposite signs.

Exercises 1.5 (page 52)

1. $1(3) = 3$

3. $2(3)(4) = 6(4) = 24$

5. $\dfrac{12}{6} = 2$

7. $\dfrac{3(6)}{2} = \dfrac{18}{2} = 9$

9. $12 \div 4(3) = 3(3) = 9$

11. $30 \cdot 37\dfrac{1}{2} = \dfrac{30}{1} \cdot \dfrac{75}{2} = \dfrac{15 \cdot \overset{1}{\cancel{2}} \cdot 75}{1 \cdot \underset{1}{\cancel{2}}} = 1,125$ lb   **13.** $3^3 - 8(3)^2 = 27 - 8(9) = 27 - 72 = -45$

11

15. positive   **17.** positive   **19.** positive   **21.** a

23. 0   **25.** $(+4)(+9) = 36$ **27.** $(-8)(-7) = 56$ **29.** $(-10)(+9) = -90$

31. $(-32)(-14) = 448$ **33.** $(-2)(3)(4) = (-6)(4) = -24$

35. $(-5)^2 = (-5)(-5) = 25$ **37.** $(-4)^3 = (-4)(-4)(-4) = (+16)(-4) = -64$

39. $(-3)(5)(-6) = (-15)(-6) = 90$ **41.** $2 + (-1)(-3) = 2 + 3 = 5$

43. $(-1 + 2)(-3) = 1(-3) = -3$ **45.** $[-1 - (-3)][-1 + (-3)] = [-1 + 3][-4]$
$$= [2][-4] = -8$$

47. $2(-1)^2 - 3(-2)^2 = 2(1) - 3(4) = 2 - 12 = -10$

49. $\left(\dfrac{2}{3}\right)(-36) = -\dfrac{2}{3} \cdot \dfrac{36}{1} = -\dfrac{72}{3} = -24$ **51.** $\left(-\dfrac{20}{3}\right)\left(-\dfrac{3}{5}\right) = +\dfrac{20}{3} \cdot \dfrac{3}{5} = \dfrac{60}{15} = 4$

53. $\dfrac{80}{-20} = -4$ **55.** $\dfrac{-110}{-55} = 2$ **57.** $\dfrac{-120}{30} = -4$ **59.** $\dfrac{320}{-16} = -20$

61. $\dfrac{-3(6)}{-(-2)} = \dfrac{-18}{2} = -9$ **63.** $\dfrac{(-2)^3(10)}{-(-5)} = \dfrac{(-8)(10)}{5} = \dfrac{-80}{5} = -16$

65. $\dfrac{18 - 20}{-2} = \dfrac{-2}{-2} = 1$ **67.** $\dfrac{-3(-2)(-4)}{-4 - 2(-5)} = \dfrac{6(-4)}{-4 + 10} = \dfrac{-24}{6} = -4$

69. $\dfrac{6 - 3(2)^2}{-1(7 - 4)} = \dfrac{6 - 3(4)}{-1(3)} = \dfrac{6 - 12}{-3} = \dfrac{-6}{-3} = 2$

71. $\dfrac{-4(5)(2) + 2(-10)(3)}{-2(-4) - 8} = \dfrac{-20(2) + (-20)(3)}{8 - 8} = \dfrac{-40 + (-60)}{0} \Rightarrow$ undefined

Problems 73-75 are to be solved using a calculator. The keystrokes needed to solve each problem using a TI-84 graphing calculator appear in each solution. There may be other solutions. Keystrokes for other calculators may be slightly different.

73. ((−) 6 + 4 × (−) 3) ÷ (4 − 6) ENTER {9}

75. (4 × ((−) 6) x^2 × (−) 3 + 4 x^2 × (−) 6) ÷ (2 × (−) 6 − 2 × (−) 3) ENTER {88}

77. $(-4)\left(\dfrac{-3}{4}\right) = +\dfrac{4}{1} \cdot \dfrac{3}{4} = \dfrac{12}{4} = 3$ **79.** $(-1)(2^3) = (-1)(8) = -8$

81. $(-2)(-2)(-2)(-3)(-4) = (+4)(-2)(-3)(-4) = (-8)(-3)(-4) = (+24)(-4) = -96$

83. $(2)(-5)(-6)(-7) = (-10)(-6)(-7) = (+60)(-7) = -420$

85. $(-7)^2 = (-7)(-7) = 49$

87. $-(-3)^2 = -(-3)(-3) = -(+9) = -9$

89. $(-1)^2[2 - (-3)] = (-1)(-1)[2 + (+3)]$
$\qquad = 1[5] = 5$

91. $(-3)(-1) - (-3)(2) = 3 - (-6) = 3 + 6$
$\qquad = 9$

93. $(-1)^3(-2)^2 + (-3)^2 = (-1)(-1)(-1)(-2)(-2) + (-3)(-3) = -4 + 9 = 5$

95. $\dfrac{4 + (-12)}{(-2)^2 - 4} = \dfrac{-8}{4 - 4} = \dfrac{-8}{0} \Rightarrow$ undefined

97. $\dfrac{(-2)(5)(4)}{-3 + 1} = \dfrac{-40}{-2} = 20$

99. $\dfrac{1}{2} - \dfrac{2}{3} - \dfrac{3}{4} = \dfrac{6}{12} + \left(-\dfrac{8}{12}\right) + \left(-\dfrac{9}{12}\right)$
$\qquad = -\dfrac{11}{12}$

101. $\dfrac{1}{2} - \dfrac{2}{3} = \dfrac{3}{6} + \left(-\dfrac{4}{6}\right) = -\dfrac{1}{6}$

103. $\left(\dfrac{1}{2} - \dfrac{2}{3}\right)\left(\dfrac{1}{2} + \dfrac{2}{3}\right) = \left(\dfrac{3}{6} - \dfrac{4}{6}\right)\left(\dfrac{3}{6} + \dfrac{4}{6}\right) = \left(-\dfrac{1}{6}\right)\left(\dfrac{7}{6}\right) = -\dfrac{7}{36}$

105. $\left(\dfrac{1}{4} - \dfrac{2}{3}\right)\left(\dfrac{3}{4} - \dfrac{1}{3}\right) = \left(\dfrac{3}{12} - \dfrac{8}{12}\right)\left(\dfrac{9}{12} - \dfrac{4}{12}\right) = \left(-\dfrac{5}{12}\right)\left(\dfrac{5}{12}\right) = -\dfrac{25}{144}$

107. $(+425)(-12) = -\$5{,}100$

109. $\dfrac{-18}{-3} = +6$

111. **a.** $75(-32) = -\$2400$ **b.** $57(-17) = -\$969$ **c.** $87(-12) = -\$1044$
  d. $(-2400) + (-969) + (-1044) = -\$4413$

113. $\dfrac{(+26) + (+35) + (+17) + (-25) + (-31) + (-12) + (-24)}{7} = \dfrac{-14}{7} = -2$ per day

115. $613.50(18) = \$11{,}043 \Rightarrow$ enough $\$$

117. Answers may vary.

119. If the quotient is undefined, then the denominator must equal 0, and the product of the two numbers is 0.

121. If x^5 is negative, then x must be negative.

Exercises 1.6 (page 59)

1. sum

3. product

5. quotient

7. difference

9. $0.14 \cdot 3{,}800 = 532$

11. $\dfrac{-4 + (7 - 9)}{(-9 - 7) + 4} = \dfrac{-4 + (-2)}{-16 + 4} = \dfrac{-6}{-12} = \dfrac{1}{2}$

13. sum

15. multiplication

17. algebraic

19. term, coefficient

SECTION 1.6

21. $x + y$ **23.** $x - 3$ **25.** $(2x)y$ **27.** $3xy$

29. $\dfrac{y}{x}$ **31.** $\dfrac{3z}{4x}$ **33.** $x + y = (-2) + 5 = 3$

35. $4xyz = 4(-2)(5)(-3) = -8(5)(-3)$
$= -40(-3) = 120$

37. $\dfrac{x^2 y}{z-1} = \dfrac{(-2)^2(5)}{-3-1} = \dfrac{4(5)}{-4} = \dfrac{20}{-4} = -5$

39. $\dfrac{4z^2 y}{3(x-z)} = \dfrac{4(-3)^2(5)}{3[-2-(-3)]} = \dfrac{4(9)(5)}{3[1]} = \dfrac{36(5)}{3} = \dfrac{180}{3} = 60$

41. $\dfrac{x(y+z)-25}{(x+z)^2-y^2} = \dfrac{-2[5+(-3)]-25}{[-2+(-3)]^2-5^2} = \dfrac{-2[2]-25}{[-5]^2-25} = \dfrac{-4-25}{25-25} = \dfrac{-29}{0} \Rightarrow$ undefined

43. $\dfrac{3(x+z^2)+4}{y(x-z)} = \dfrac{3[(-2)+(-3)^2]+4}{5[(-2)-(-3)]} = \dfrac{3(-2+9)+4}{5(1)} = \dfrac{3(7)+4}{5} = \dfrac{21+4}{5} = \dfrac{25}{5} = 5$

45. $-7c$: 1 term; coef. $= -7$ **47.** $-xy - 5z + 8$: 3 terms; coef. $= -1$

49. $-3xy + yz - zw + 5$: 4 terms; coef. $= -3$ **51.** $9abc - 5ab - c$: 3 terms; coef. $= 9$

53. $5x - 4y + 3z + 2$: 4 terms; coef. $= 5$

55. $z + \dfrac{x}{y}$ **57.** $z - xy$ **59.** $\dfrac{xy}{x+z}$ **61.** $\dfrac{x-4}{3y}$

Several answers are possible for problems 63-73. Only one possible answer is listed for each problem.

---

63. the sum of y and 4

65. the product of x, y and the sum of x and y

67. the quotient obtained when the sum of x and 2 is divided by z

69. the quotient obtained when y is divided by z

71. the product of 2, x and y

73. the quotient obtained when 5 is divided by the sum of x and y

---

75. $x + z = 8 + 2 = 10$ **77.** $y - z = 4 - 2 = 2$

79. $yz - 3 = (4)(2) - 3 = 5$ **81.** $\dfrac{xy}{z} = \dfrac{(8)(4)}{2} = 16$

83. 3rd term: $19x$; factors: $19, x$ **85.** x is common to the 1st and 3rd terms.

87. 1st term: $3xyz$; factors: $3, x, y, z$ **89.** 3rd term: $17xz$; factors: $17, x, z$

14

91. coefficients: 5, 1 and 8

93. x and y are common to the 1st and 3rd terms.

95. coefficients: 3, 1 and 25; $3 \cdot 1 \cdot 25 = 75$

97. x and y are common to the 1st and 3rd terms.

99. $c + 6$

101. a. $(h - 20)$ ft   **b.** $(c + 20)$ ft

103. \$35,000n

105. $(500 - x)$ in.

107. $\$(3d + 5)$

109. $\dfrac{N(N-1)}{2} = \dfrac{10,000(10,000-1)}{2} = \dfrac{10,000(9,999)}{2} = \dfrac{99,990,000}{2} = 49,995,000$ comparisons

111-113. Answers may vary.

115. $37x \Rightarrow 37(2x)$
$37(2x) = 2(37x)$
$37x$ is doubled.

Exercises 1.7 (page 67)

Several answers are possible for problems 1-5. Only one possible answer is listed for each problem.

1. $2(xy) = (2x)y$

3. $2(x + y) = 2x + 2y$

5. $5 - 3 = 2; 3 - 5 = -2$

7. $x + y^2 \geq z$

9. $|x| \boxed{\geq} 0$

11. positive

13. real

15. $a + b = b + \underline{\,a\,}$

17. $(a + b) + c = a + \underline{(b + c)}$

19. $a(b + c) = ab + \underline{\,ac\,}$

21. $a \cdot 1 = \underline{\,a\,}$

23. element, multiplication

25. $a, \dfrac{1}{a}$, multiplicative

27. $x + y = 12 + (-2) = 10$

29. $xy = 12(-2) = -24$

31. $x^2 = 12^2 = 144$

33. $\dfrac{x}{y^2} = \dfrac{12}{(-2)^2} = \dfrac{12}{4} = 3$

35. $x + y = 5 + 7 = 12$
$y + x = 7 + 5 = 12$

37. $3x + 2y = 3(5) + 2(7) = 15 + 14 = 29$
$2y + 3x = 2(7) + 3(5) = 14 + 15 = 29$

39. $x(x + y) = 5(5 + 7) = 5(12) = 60$
$(x + y)x = (5 + 7)5 = (12)5 = 60$

41. $x^2(yz^2) = 5^2[7(-1)^2] = 25[7(1)] = 25[7] = 175$
$(x^2y)z^2 = [5^2(7)](-1)^2 = [25(7)](1) = [175](1) = 175$

43. $3(x + 5) = 3x + 15$

45. $5(z - 4) = 5z - 20$

47. $-2(3x + y) = -6x - 2y$

49. $x(x + 3) = x \cdot x + x \cdot 3 = x^2 + 3x$

51. $-x(a + b) = (-x)a + (-x)b = -ax - bx$

53. $-4(x^2 + x + 2) = (-4)x^2 + (-4)x + (-4)2 = -4x^2 - 4x - 8$

55. additive inverse: -5

multiplicative inverse: $\frac{1}{5}$

57. additive inverse: $-\frac{1}{3}$

multiplicative inverse: 3

59. additive inverse: 0

multiplicative inverse: none

61. additive inverse: $\frac{2}{3}$

multiplicative inverse: $-\frac{3}{2}$

63. additive inverse: 0.2

multiplicative inverse: -5

65. additive inverse: $-\frac{5}{4}$

multiplicative inverse: $\frac{4}{5}$

67. $8(x + 2) = 8x + 8(2) = 8x + 16$

69. $xy^3 = y^3 x$

71. $(x + y)z = (y + x)z$

73. $(xy)z = x(yz)$

75. $(x + y) + z = [2 + (-3)] + 1 = -1 + 1 = \boxed{0}$; $x + (y + z) = 2 + (-3 + 1) = 2 + (-2) = \boxed{0}$

77. $(xz)y = [2(1)](-3) = [2](-3) = \boxed{-6}$; $x(yz) = 2[-3(1)] = 2[-3] = \boxed{-6}$

79. $-6(a + 4) = -6a + (-6)(4) = -6a - 24$

81. $-3x(x - a) = -3x \cdot x + (-3x)(-a)$
$= -3x^2 + 3ax$

83. commutative property of addition

85. commutative property of multiplication

87. distributive property

89. commutative property of addition

91. multiplication identity property

93. additive inverse property

95. addition identity property

97. **Answers may vary.**

99. Closure for addition would not be true (odd number plus odd number equals even number).
Closure for multiplication would be true (odd number times odd number equals odd number).
There would be no additive identity (0 is an even number).
There would be a multiplicative identity, 1 (1 is an odd number).

Chapter 1 Review (page 71)

1. natural: $1, 2, 3, 4, 5$

2. prime: $2, 3, 5$

3. odd, natural: $1, 3, 5$

4. composite: 4

5. integers: $-6, 0, 5$

6. rational: $-6, -\frac{2}{3}, 0, 2.6, 5$

7. prime: 5

8. real: $-6, -\frac{2}{3}, 0, \sqrt{2}, 2.6, \pi, 5$

9. even integers: $-6, 0$

10. odd integers: 5

11. irrational: $\sqrt{2}, \pi$

12. negative numbers: $-6, -\frac{2}{3}$

13. $-3 \boxed{\phantom{<}} 5 - 5$

$-3 \boxed{<} 0$

14. $\dfrac{12}{4} \boxed{\phantom{<}} 7$

$3 \boxed{<} 7$

15. $\dfrac{36}{4} \boxed{} -2$

$9 \boxed{>} -2$

16. $2 - 2 \boxed{} 8 - \dfrac{24}{3}$

$0 \boxed{} 8 - 8$

$0 \boxed{=} 0$

17. $-(-9) = +9$

18. $-(12 - 4) = -(8) = -8$

19.

20.

21.

22.

23. $|29 - 24| = |5| = 5$

24. $|-25| = 25$

25. $\dfrac{45}{27} = \dfrac{5 \cdot \overset{1}{\cancel{9}}}{3 \cdot \underset{1}{\cancel{9}}} = \dfrac{5}{3}$

26. $\dfrac{48}{18} = \dfrac{8 \cdot \overset{1}{\cancel{6}}}{3 \cdot \underset{1}{\cancel{6}}} = \dfrac{8}{3}$

27. $\dfrac{31}{15} \cdot \dfrac{10}{62} = \dfrac{\overset{1}{\cancel{31}} \cdot \overset{1}{\cancel{2}} \cdot \overset{1}{\cancel{5}}}{3 \cdot \underset{1}{\cancel{5}} \cdot \underset{1}{\cancel{2}} \cdot \underset{1}{\cancel{31}}} = \dfrac{1}{3}$

28. $\dfrac{25}{36} \cdot \dfrac{12}{15} \cdot \dfrac{3}{5} = \dfrac{\overset{1}{\cancel{5}} \cdot \overset{1}{\cancel{5}} \cdot \overset{1}{\cancel{12}} \cdot \overset{1}{\cancel{3}}}{3 \cdot \underset{1}{\cancel{12}} \cdot 3 \cdot \underset{1}{\cancel{5}} \cdot \underset{1}{\cancel{5}}} = \dfrac{1}{3}$

29. $\dfrac{18}{21} \div \dfrac{6}{7} = \dfrac{18}{21} \cdot \dfrac{7}{6} = \dfrac{\overset{1}{\cancel{3}} \cdot \overset{1}{\cancel{6}} \cdot \overset{1}{\cancel{7}}}{\underset{1}{\cancel{3}} \cdot \underset{1}{\cancel{7}} \cdot \underset{1}{\cancel{6}}} = \dfrac{1}{1} = 1$

30. $\dfrac{14}{24} \div \dfrac{7}{12} \div \dfrac{2}{5} = \dfrac{14}{24} \cdot \dfrac{12}{7} \cdot \dfrac{5}{2}$

$= \dfrac{\overset{1}{\cancel{2}} \cdot \overset{1}{\cancel{7}} \cdot \overset{1}{\cancel{12}} \cdot 5}{2 \cdot \underset{1}{\cancel{12}} \cdot \underset{1}{\cancel{7}} \cdot 2} = \dfrac{5}{2}$

31. $\dfrac{7}{12} + \dfrac{9}{12} = \dfrac{7 + 9}{12} = \dfrac{16}{12} = \dfrac{4 \cdot \overset{1}{\cancel{4}}}{3 \cdot \underset{1}{\cancel{4}}} = \dfrac{4}{3}$

32. $\dfrac{13}{24} - \dfrac{5}{24} = \dfrac{13 - 5}{24} = \dfrac{8}{24} = \dfrac{\overset{1}{\cancel{8}}}{3 \cdot \underset{1}{\cancel{8}}} = \dfrac{1}{3}$

33. $\dfrac{1}{5} + \dfrac{1}{4} = \dfrac{1 \cdot \mathbf{4}}{5 \cdot \mathbf{4}} + \dfrac{1 \cdot \mathbf{5}}{4 \cdot \mathbf{5}} = \dfrac{4}{20} + \dfrac{5}{20} = \dfrac{9}{20}$

34. $\dfrac{5}{7} + \dfrac{4}{9} = \dfrac{5 \cdot \mathbf{9}}{7 \cdot \mathbf{9}} + \dfrac{4 \cdot \mathbf{7}}{9 \cdot \mathbf{7}} = \dfrac{45}{63} + \dfrac{28}{63}$

$= \dfrac{45 + 28}{63} = \dfrac{73}{63}$

35. $\dfrac{2}{3} - \dfrac{1}{7} = \dfrac{2 \cdot \mathbf{7}}{3 \cdot \mathbf{7}} - \dfrac{1 \cdot \mathbf{3}}{7 \cdot \mathbf{3}} = \dfrac{14}{21} - \dfrac{3}{21}$

$= \dfrac{14 - 3}{21} = \dfrac{11}{21}$

36. $\dfrac{4}{5} - \dfrac{2}{3} = \dfrac{4 \cdot \mathbf{3}}{5 \cdot \mathbf{3}} - \dfrac{2 \cdot \mathbf{5}}{3 \cdot \mathbf{5}} = \dfrac{12}{15} - \dfrac{10}{15}$

$= \dfrac{12 - 10}{15} = \dfrac{2}{15}$

37. $3\dfrac{2}{3} + 5\dfrac{1}{4} = \dfrac{11}{3} + \dfrac{21}{4} = \dfrac{11 \cdot \mathbf{4}}{3 \cdot \mathbf{4}} + \dfrac{21 \cdot \mathbf{3}}{4 \cdot \mathbf{3}} = \dfrac{44}{12} + \dfrac{63}{12} = \dfrac{44 + 63}{12} = \dfrac{107}{12} = 8\dfrac{11}{12}$

38. $7\dfrac{5}{12} - 4\dfrac{1}{2} = \dfrac{89}{12} - \dfrac{9}{2} = \dfrac{89}{12} - \dfrac{9 \cdot \mathbf{6}}{2 \cdot \mathbf{6}} = \dfrac{89}{12} - \dfrac{54}{12} = \dfrac{89 - 54}{12} = \dfrac{35}{12} = 2\dfrac{11}{12}$

39. $48.29 + 31.9 = 80.19$

40. $36.85 - 15.86 = 20.99$

41. $4.32 \cdot 1.5 = 6.48$

42. $21.83 \div 5.9 = 3.7$

43. $2.7(4.92 - 3.18) = 2.7(1.74) \approx 4.70$

44. $\dfrac{3.3 + 2.5}{0.22} = \dfrac{5.8}{0.22} \approx 26.36$

45. $\dfrac{12.5}{14.7 - 11.2} = \dfrac{12.5}{3.5} \approx 3.57$

46. $(3 - 0.7)(3.63 - 2) = (2.3)(1.63) \approx 3.75$

47. $17\dfrac{1}{2} + 15\dfrac{3}{4} = 17 + \dfrac{2}{4} + 15 + \dfrac{3}{4} = 32 + \dfrac{5}{4} = 32 + 1\dfrac{1}{4} = 33\dfrac{1}{4}$

$100 - 33\dfrac{1}{4} = 100 - 33 - \dfrac{1}{4} = 67 - \dfrac{1}{4} = 66\dfrac{3}{4}$ acres left

48. avg. $= \dfrac{5.2 + 4.7 + 9.5 + 8}{4} = \dfrac{27.4}{4}$

$\qquad = 6.85$ hours

49. $0.20(425) = 85$

50. Front/Back: $2(2.7 + 2.7 + 4.2) = 2(9.6) = 19.2$ ft $\qquad$ TOTAL $= 19.2 + 13.2 + 7.8 = 40.2$ ft

Top/Bottom: $2(1.2 + 1.2 + 4.2) = 2(6.6) = 13.2$ ft

Sides: $2(1.2 + 2.7) = 2(3.9) = 7.8$ ft

51. $3^4 = 3 \cdot 3 \cdot 3 \cdot 3 = 81$

52. $\left(\dfrac{2}{3}\right)^2 = \dfrac{2}{3} \cdot \dfrac{2}{3} = \dfrac{4}{9}$

53. $(0.5)^2 = (0.5)(0.5) = 0.25$

54. $5^2 + 2^3 = 5 \cdot 5 + 2 \cdot 2 \cdot 2 = 25 + 8 = 33$

55. $3^2 + 4^2 = 9 + 16 = 25$

56. $(3 + 4)^2 = 7^2 = 49$

57. $A = \dfrac{1}{2}bh = \dfrac{1}{2}\left(6\dfrac{1}{2} \text{ ft}\right)(7 \text{ ft}) = \dfrac{1}{2} \cdot \dfrac{13}{2} \cdot \dfrac{7}{1} \text{ ft}^2 = \dfrac{91}{4} \text{ ft}^2 = 22\dfrac{3}{4} \text{ ft}^2$

58. $V = Bh = \pi r^2 h = \pi\left(\dfrac{32.1}{2} \text{ ft}\right)^2 (18.7 \text{ ft}) = \pi(257.6025 \text{ ft}^2)(18.7 \text{ ft}) \approx 15{,}133.6 \text{ ft}^3$

59. $7 + 3^3 = 7 + 27 = 34$

60. $6 + 2 \cdot 4 = 6 + 8 = 14$

61. $5 + 6 \div 2 = 5 + 3 = 8$

62. $(8 + 6) \div 2 = 14 \div 2 = 7$

63. $5^3 - \dfrac{81}{3} = 125 - 27 = 98$

64. $(5-2)^2 + 5^2 + 2^2 = 3^2 + 5^2 + 2^2$
$$= 9 + 25 + 4 = 38$$

65. $\dfrac{4 \cdot 3 + 3^4}{31} = \dfrac{12 + 81}{31} = \dfrac{93}{31} = 3$

66. $\dfrac{4}{3} \cdot \dfrac{9}{2} + \dfrac{1}{2} \cdot 18 = \dfrac{2 \cdot \overset{1}{\cancel{2}} \cdot 3 \cdot \overset{1}{\cancel{3}}}{\underset{1}{\cancel{3}} \cdot \underset{1}{\cancel{2}}} + \dfrac{1 \cdot 9 \cdot \overset{1}{\cancel{2}}}{\underset{1}{\cancel{2}} \cdot 1}$
$$= \dfrac{6}{1} + \dfrac{9}{1} = 15$$

67. $8^2 - 6 = 64 - 6 = 58$

68. $(8-6)^2 = 2^2 = 4$

69. $\dfrac{10+2}{10-6} = \dfrac{12}{4} = 3$

70. $\dfrac{6(8)-12}{4+8} = \dfrac{48-12}{12} = \dfrac{36}{12} = 3$

71. $2^2 + 2(3)^2 = 4 + 2(9) = 4 + 18 = 22$

72. $\dfrac{2^2+3}{2^3-1} = \dfrac{4+3}{8-1} = \dfrac{7}{7} = 1$

73. $(+15) + (+9) = +(15+9) = 24$

74. $(-17) + (-16) = -(17+16) = -33$

75. $(-2.7) + (-3.8) = -(2.7+3.8) = -6.5$

76. $\dfrac{1}{2} + \left(-\dfrac{1}{6}\right) = \dfrac{3}{6} + \left(-\dfrac{1}{6}\right) = +\left(\dfrac{3}{6} - \dfrac{1}{6}\right)$
$$= \dfrac{2}{6} = \dfrac{1}{3}$$

77. $(+12) + (-24) = -(24-12) = -12$

78. $(-44) + (+60) = +(60-44) = 16$

79. $3.7 + (-2.5) = +(3.7-2.5) = 1.2$

80. $-5.6 + (+2.06) = -(5.6-2.06)$
$$= -3.54$$

81. $15 - (-4) = 15 + (+4) = 19$

82. $-8 - (-15) = -8 + (+15) = 7$

83. $[-5 + (-5)] - (-5) = [-10] + (+5) = -5$

84. $1 - [5 - (-3)] = 1 - [5 + (+3)]$
$$= 1 - [8] = -7$$

85. $-\dfrac{7}{10} - \left(-\dfrac{2}{5}\right) = -\dfrac{7}{10} + \dfrac{2}{5} = -\dfrac{7}{10} + \dfrac{4}{10} = -\dfrac{3}{10}$

86. $\dfrac{2}{3} - \left(\dfrac{1}{3} - \dfrac{2}{3}\right) = \dfrac{2}{3} - \left(-\dfrac{1}{3}\right) = \dfrac{2}{3} + \dfrac{1}{3} = \dfrac{3}{3} = 1$

87. $\left|\dfrac{3}{7} - \left(-\dfrac{4}{7}\right)\right| = \left|\dfrac{3}{7} + \dfrac{4}{7}\right| = \left|\dfrac{7}{7}\right| = |1| = 1$

88. $\dfrac{3}{7} - \left|-\dfrac{4}{7}\right| = \dfrac{3}{7} - \left(+\dfrac{4}{7}\right) = \dfrac{3}{7} + \left(-\dfrac{4}{7}\right) = -\dfrac{1}{7}$

89. $(+5)(+8) = 40$

90. $(-5)(-12) = 60$

91. $\left(-\dfrac{3}{14}\right)\left(-\dfrac{7}{6}\right) = +\dfrac{3}{14} \cdot \dfrac{7}{6}$

$$= \dfrac{\overset{1}{\cancel{3}} \cdot \overset{1}{\cancel{7}}}{2 \cdot \underset{1}{\cancel{7}} \cdot 2 \cdot \underset{1}{\cancel{3}}} = \dfrac{1}{4}$$

92. $(3.75)(0.37) = 1.3875$

93. $5(-7) = -35$

94. $(-15)(7) = -105$

95. $\left(-\dfrac{1}{2}\right)\left(\dfrac{4}{3}\right) = -\dfrac{1}{2} \cdot \dfrac{4}{3} = -\dfrac{1 \cdot 2 \cdot \overset{1}{\cancel{2}}}{\underset{1}{\cancel{2}} \cdot 3} = -\dfrac{2}{3}$

96. $(2.1)(-8.2) = -17.22$

97. $\dfrac{+36}{+12} = 3$

98. $\dfrac{-14}{-2} = 7$

99. $\dfrac{(-2)(-7)}{4} = \dfrac{+14}{4} = +\dfrac{7 \cdot \overset{1}{\cancel{2}}}{2 \cdot \underset{1}{\cancel{2}}} = \dfrac{7}{2}$

100. $\dfrac{-22.5}{-3.75} = 6$

101. $\dfrac{(-2)(-9)}{-3} = \dfrac{+18}{-3} = -6$

102. $\dfrac{(-6)(12)}{-4} = \dfrac{-72}{-4} = 18$

103. $\left(\dfrac{-10}{2}\right)^2 - (-1)^3 = (-5)^2 - (-1)^3$

$$= 25 - (-1)$$
$$= 25 + 1 = 26$$

104. $\dfrac{[-3 + (-4)]^2}{10 + (-3)} = \dfrac{[-7]^2}{7} = \dfrac{49}{7} = 7$

105. $\left(\dfrac{-3 + (-3)}{3}\right)\left(\dfrac{-15}{5}\right) = \left(\dfrac{-6}{3}\right)\left(\dfrac{-15}{5}\right)$

$$= (-2)(-3) = 6$$

106. $\dfrac{-2 - (-8)}{5 + (-1)} = \dfrac{-2 + (+8)}{4} = \dfrac{6}{4} = \dfrac{3}{2}$

107. xz

108. $x + 2y$

109. $2(x + y)$

110. $x - yz$

111. the product of 5, x and z

112. 5 decreased by the product of y and z

113. 4 less than the product of x and y

114. the quotient obtained when the sum of x, y and z is divided by twice their product

115. $x + z = 2 + (-1) = 1$

116. $x + y + z = 2 + (-3) + (-1) = -1 + (-1)$
$$= -2$$

117. $5x + (y - z) = 5(2) + [-3 - (-1)]$
$$= 10 + [-3 + 1]$$
$$= 10 + (-2) = 8$$

118. $z^2 - y = (-1)^2 - (-3) = 1 - (-3)$
$$= 1 + 3 = 4$$

20

119.
$$x - (y - z) = 2 - [-3 - (-1)]$$
$$= 2 - [-3 + (+1)]$$
$$= 2 - [-2] = 2 + (+2) = 4$$

120.
$$(x - y) - z = [2 - (-3)] - (-1)$$
$$= [2 + (+3)] + (+1)$$
$$= 5 + 1 = 6$$

121. $yz = (-3)(-1) = 3$

122. $xyz = (2)(-3)(-1) = -6(-1) = 6$

123.
$$(x + y)(y + z) = [2 + (-3)][-3 + (-1)]$$
$$= [-1][-4] = 4$$

124.
$$\frac{3(x - y)}{x + (y - z)} = \frac{3[2 - (-3)]}{2 + [-3 - (-1)]}$$
$$= \frac{3[5]}{2 + [-2]} = \frac{15}{0} : \text{undefined}$$

125.
$$y^2 z + x = (-3)^2(-1) + 2$$
$$= 9(-1) + 2 = -9 + 2 = -7$$

126.
$$yz^3 + (xy)^2 = (-3)(-1)^3 + [2(-3)]^2$$
$$= (-3)(-1) + [-6]^2$$
$$= 3 + 36 = 39$$

127.
$$\frac{2y^2}{3x - 6} = \frac{2(-3)^2}{3(2) - 6} = \frac{2(9)}{6 - 6} = \frac{18}{0}$$
undefined

128.
$$\frac{|xy|}{3z} = \frac{|2(-3)|}{3(-1)} = \frac{|-6|}{-3} = \frac{6}{-3} = -2$$

129. three terms

130. 7

131. 1

132. $2 + 4 + 3 = 9$

133. closure property of addition

134. commutative property of multiplication

135. associative property of addition

136. distributive property

137. commutative property of addition

138. associative property of multiplication

139. commutative property of addition

140. multiplicative identity property

141. additive inverse property

142. additive identity property

Chapter 1 Test (page 77)

1. $31, 37, 41, 43, 47$

2. 2

3.

4.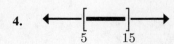

5. $-|-17| = -(+17) = -17$

6. $-|9| + |-9| = -(+9) + (+9) = -9 + 9 = 0$

7. $3(4-2)\;\boxed{}\;-2(2-5)$
$\quad\;\;3(2)\;\boxed{}\;-2(-3)$
$\qquad\quad 6\;\boxed{=}\;6$

8. $1+4\cdot3\;\boxed{\phantom{<}}\;-2(-7)$
$\quad\;\; 1+12\;\boxed{\phantom{<}}\;+14$
$\qquad\quad 13\;\boxed{<}\;14$

9. $25\%\text{ of }136\;\boxed{}\;\dfrac{1}{2}\text{ of }66$
$\quad\; 0.25(136)\;\boxed{}\;\dfrac{1}{2}(66)$
$\qquad\quad 34\;\boxed{>}\;33$

10. $-8.5\;\boxed{}\;-|-8.5|$
$\quad -8.5\;\boxed{}\;-(+8.5)$
$\quad -8.5\;\boxed{=}\;-8.5$

11. $\dfrac{26}{40}=\dfrac{13\cdot\overset{1}{\cancel{2}}}{20\cdot\underset{1}{\cancel{2}}}=\dfrac{13}{20}$

12. $\dfrac{9}{11}\cdot\dfrac{44}{45}=\dfrac{\overset{1}{\cancel{9}}\cdot4\cdot\overset{1}{\cancel{11}}}{\underset{1}{\cancel{11}}\cdot\underset{1}{\cancel{9}}\cdot5}=\dfrac{4}{5}$

13. $\dfrac{14}{21}\div\dfrac{28}{9}=\dfrac{14}{21}\cdot\dfrac{9}{28}=\dfrac{\overset{1}{\cancel{14}}\cdot\overset{1}{\cancel{3}}\cdot3}{3\cdot7\cdot2\cdot\underset{1}{\cancel{14}}}=\dfrac{3}{14}$

14. $\dfrac{24}{16}+3=\dfrac{3\cdot\overset{1}{\cancel{8}}}{2\cdot\underset{1}{\cancel{8}}}+\dfrac{3}{1}=\dfrac{3}{2}+\dfrac{3\cdot\mathbf{2}}{1\cdot\mathbf{2}}=\dfrac{3}{2}+\dfrac{6}{2}=\dfrac{3+6}{2}=\dfrac{9}{2}\;\left(\text{or }4\dfrac{1}{2}\right)$

15. $\dfrac{17-5}{36}-\dfrac{2(13-5)}{12}=\dfrac{12}{36}-\dfrac{2(8)}{12}=\dfrac{12}{36}-\dfrac{16}{12}=\dfrac{\overset{1}{\cancel{12}}}{3\cdot\underset{1}{\cancel{12}}}-\dfrac{4\cdot\overset{1}{\cancel{4}}}{3\cdot\underset{1}{\cancel{4}}}=\dfrac{1}{3}-\dfrac{4}{3}=\dfrac{1-4}{3}=\dfrac{-3}{3}=-1$

16. $\dfrac{|-7-(-6)|}{-7-|-6|}=\dfrac{|-7+(+6)|}{-7-(+6)}=\dfrac{|-1|}{-7+(-6)}=\dfrac{1}{-13}=-\dfrac{1}{13}$

17. $0.13(256)=33.28\approx33.3$

18. $A=lw=(18.9\text{ ft})(21.25\text{ ft})=401.625\text{ ft}^2$
$\qquad\qquad\qquad\qquad\qquad\quad\approx401.63\text{ ft}^2$

19. $A=\dfrac{1}{2}bh=\dfrac{1}{2}(16\text{ cm})(8\text{ cm})=\dfrac{1}{2}(128\text{ cm}^2)$
$\qquad\qquad\qquad\qquad\qquad\qquad=64\text{ cm}^2$

20. $V=Bh=\pi r^2h=\pi(7\text{ in.})^2(10\text{ in.})$
$\qquad\qquad\quad=\pi(49\text{ in.}^2)(10\text{ in.})$
$\qquad\qquad\quad=\pi(490\text{ in.}^3)\approx1{,}539\text{ in.}^3$

21. $xy+z=(-2)(3)+4=-6+4=-2$

22. $x(y+z)=-2(3+4)=-2(7)=-14$

23. $\dfrac{z+4y}{2x}=\dfrac{4+4(3)}{2(-2)}=\dfrac{4+12}{-4}=\dfrac{16}{-4}=-4$

24. $|x^3-z|=|(-2)^3-4|=|-8-4|=|-12|$
$\qquad\qquad\qquad\qquad\qquad\qquad\quad=12$

25. $x^3+y^2+z=(-2)^3+(3)^2+4$
$\qquad\qquad\quad=-8+9+4=5$

26. $|x|-3|y|-4|z|=|-2|-3|3|-4|4|$
$\qquad\qquad\qquad\quad=2-3(3)-4(4)$
$\qquad\qquad\qquad\quad=2-9-16=-23$

27. $\dfrac{xy}{x+y}$

28. $5y-(x+y)$

29. $x(12 + 12) + y(7 + 7) = 24x + 14y$

30. $\$(12a + 8b)$

31. -5

32. 4 terms

33. $3(x + 2) = 3x + 6$

34. $-p(r - t) = (-p)r + (-p)(-t) = -pr + pt$

35. 0

36. 5

37. commutative property of multiplication

38. distributive property

39. commutative property of addition

40. multiplicative inverse property

Exercises 2.1 (page 91)

1. $x - 5 = 15$
addition

3. $w + 5 = 7$
subtraction

5. $-8x = -24$
division

7. $\dfrac{x}{5} = 2$
multiplication

9. $\dfrac{4}{5} + \dfrac{2}{3} = \dfrac{4 \cdot 3}{5 \cdot 3} + \dfrac{2 \cdot 5}{3 \cdot 5} = \dfrac{12}{15} + \dfrac{10}{15} = \dfrac{22}{15}$

11. $\dfrac{5}{9} \div \dfrac{3}{5} = \dfrac{5}{9} \cdot \dfrac{5}{3} = \dfrac{25}{27}$

13. $3 + 5 \cdot 6 = 3 + 30 = 33$

15. $3 + 4^3(-5) = 3 + 64(-5) = 3 + (-320)$
$= -317$

17. equation, expression

19. equivalent

21. equal

23. equal

25. regular price

27. 100

29. $x = -4$
equation

31. $6x + 7$
expression

33. $x^2 + 2x = 3$
equation

35. $3(x - 4)$
expression

37.
$x + 3 = 6$
$3 + 3 \stackrel{?}{=} 6$
$6 = 6$
3 is a solution.

39.
$2y - 5 = y$
$2(4) - 5 \stackrel{?}{=} 4$
$8 - 5 \stackrel{?}{=} 4$
$3 \neq 4$
4 is not a solution.

41.
$\dfrac{y}{7} = 4$
$\dfrac{28}{7} \stackrel{?}{=} 4$
$4 = 4$
28 is a solution.

43.
$\dfrac{x}{5} = x$
$\dfrac{0}{5} \stackrel{?}{=} 0$
$0 = 0$
0 is a solution.

45.
$3k + 5 = 5k - 1$
$3(3) + 5 \stackrel{?}{=} 5(3) - 1$
$9 + 5 \stackrel{?}{=} 15 - 1$
$14 = 14$
3 is a solution.

47.
$\dfrac{5 + x}{10} - x = \dfrac{1}{2}$
$\dfrac{5 + 0}{10} - 0 \stackrel{?}{=} \dfrac{1}{2}$
$\dfrac{5}{10} - 0 \stackrel{?}{=} \dfrac{1}{2}$
$\dfrac{1}{2} = \dfrac{1}{2}$
0 is a solution.

49.
$$a - 6 = 9$$
$$a - 6 + \mathbf{6} = 9 + \mathbf{6}$$
$$a = 15$$

$$a - 6 = 9$$
$$15 - 6 \overset{?}{=} 9$$
$$9 = 9$$

51.
$$b - 5 = -19$$
$$b - 5 + \mathbf{5} = -19 + \mathbf{5}$$
$$b = -14$$

$$b - 5 = -19$$
$$-14 - 5 \overset{?}{=} -19$$
$$-19 = -19$$

53.
$$4 = c - 9$$
$$4 + \mathbf{9} = c - 9 + \mathbf{9}$$
$$13 = c$$

$$4 = c - 9$$
$$4 \overset{?}{=} 13 - 9$$
$$4 = 4$$

55.
$$r - \frac{1}{5} = \frac{3}{10}$$
$$r - \frac{1}{5} + \frac{\mathbf{1}}{\mathbf{5}} = \frac{3}{10} + \frac{\mathbf{1}}{\mathbf{5}}$$
$$r = \frac{3}{10} + \frac{2}{10} = \frac{5}{10} = \frac{1}{2}$$

Check: $\quad r - \dfrac{1}{5} = \dfrac{3}{10}$
$$\frac{1}{2} - \frac{1}{5} \overset{?}{=} \frac{3}{10}$$
$$\frac{5}{10} - \frac{2}{10} \overset{?}{=} \frac{3}{10}$$
$$\frac{3}{10} = \frac{3}{10}$$

57.
$$y + 8 = 11$$
$$y + 8 - \mathbf{8} = 11 - \mathbf{8}$$
$$y = 3$$

$$y + 8 = 11$$
$$3 + 8 \overset{?}{=} 11$$
$$11 = 11$$

59.
$$a + 9 = -12$$
$$a + 9 - \mathbf{9} = -12 - \mathbf{9}$$
$$a = -21$$

$$a + 9 = -12$$
$$-21 + 9 \overset{?}{=} -12$$
$$-12 = -12$$

61.
$$41 = 45 + q$$
$$41 - \mathbf{45} = 45 - \mathbf{45} + q$$
$$-4 = q$$

$$41 = 45 + q$$
$$41 \overset{?}{=} 45 + (-4)$$
$$41 = 41$$

63.
$$k + \frac{2}{3} = \frac{1}{5}$$
$$k + \frac{2}{3} - \frac{\mathbf{2}}{\mathbf{3}} = \frac{1}{5} - \frac{\mathbf{2}}{\mathbf{3}}$$
$$k = \frac{3}{15} - \frac{10}{15} = -\frac{7}{15}$$

$$k + \frac{2}{3} = \frac{1}{5}$$
$$-\frac{7}{15} + \frac{2}{3} \overset{?}{=} \frac{1}{5}$$
$$-\frac{7}{15} + \frac{10}{15} \overset{?}{=} \frac{1}{5}$$
$$\frac{3}{15} \overset{?}{=} \frac{1}{5}$$
$$\frac{1}{5} = \frac{1}{5}$$

65.
$$\frac{x}{6} = 3$$
$$\mathbf{6} \cdot \frac{x}{6} = \mathbf{6} \cdot \mathbf{3}$$
$$x = 18$$

$$\frac{x}{6} = 3$$
$$\frac{18}{6} \overset{?}{=} 3$$
$$3 = 3$$

67.
$$\frac{b}{3} = 5$$
$$\mathbf{3} \cdot \frac{b}{3} = \mathbf{3} \cdot 5$$
$$b = 15$$

$$\frac{b}{3} = 5$$
$$\frac{15}{3} \overset{?}{=} 5$$
$$5 = 5$$

69.
$$\frac{a}{3} = \frac{1}{9}$$
$$3 \cdot \frac{a}{3} = 3 \cdot \frac{1}{9}$$
$$a = \frac{3}{1} \cdot \frac{1}{9}$$
$$a = \frac{3}{9}$$
$$a = \frac{1}{3}$$

$$\frac{a}{3} = \frac{1}{9}$$
$$\frac{\frac{1}{3}}{3} \stackrel{?}{=} \frac{1}{9}$$
$$\frac{\frac{1}{3}}{\frac{3}{1}} \stackrel{?}{=} \frac{1}{9}$$
$$\frac{1}{3} \div \frac{3}{1} \stackrel{?}{=} \frac{1}{9}$$
$$\frac{1}{3} \cdot \frac{1}{3} \stackrel{?}{=} \frac{1}{9}$$
$$\frac{1}{9} = \frac{1}{9}$$

71.
$$\frac{u}{5} = -\frac{3}{10}$$
$$5 \cdot \frac{u}{5} = 5\left(-\frac{3}{10}\right)$$
$$u = \frac{5}{1}\left(-\frac{3}{10}\right)$$
$$u = -\frac{15}{10}$$
$$u = -\frac{3}{2}$$

$$\frac{u}{5} = -\frac{3}{10}$$
$$\frac{-\frac{3}{2}}{5} \stackrel{?}{=} -\frac{3}{10}$$
$$\frac{-\frac{3}{2}}{\frac{5}{1}} \stackrel{?}{=} -\frac{3}{10}$$
$$-\frac{3}{2} \div \frac{5}{1} \stackrel{?}{=} -\frac{3}{10}$$
$$-\frac{3}{2} \cdot \frac{1}{5} \stackrel{?}{=} -\frac{3}{10}$$
$$-\frac{3}{10} = -\frac{3}{10}$$

73.
$$7x = 28$$
$$\frac{7x}{7} = \frac{28}{7}$$
$$x = 4$$

$$7x = 28$$
$$7(4) \stackrel{?}{=} 28$$
$$28 = 28$$

75.
$$11x = -121$$
$$\frac{11x}{11} = \frac{-121}{11}$$
$$x = -11$$

$$11x = -121$$
$$11(-11) \stackrel{?}{=} -121$$
$$-121 = -121$$

77.
$$-4x = 36$$
$$\frac{-4x}{-4} = \frac{36}{-4}$$
$$x = -9$$

$$-4x = 36$$
$$-4(-9) \stackrel{?}{=} 36$$
$$36 = 36$$

79.
$$4w = 108$$
$$\frac{4w}{4} = \frac{108}{4}$$
$$w = 27$$

$$4w = 108$$
$$4(27) \stackrel{?}{=} 108$$
$$108 = 108$$

81.
$$5x = \frac{5}{8}$$
$$\frac{1}{5} \cdot 5x = \frac{1}{5} \cdot \frac{5}{8}$$
$$x = \frac{1}{8}$$

$$5x = \frac{5}{8}$$
$$5 \cdot \frac{1}{8} \stackrel{?}{=} \frac{5}{8}$$
$$\frac{5}{8} = \frac{5}{8}$$

83.
$$\frac{1}{7}w = 14$$
$$7 \cdot \frac{1}{7}w = 7 \cdot 14$$
$$w = 98$$

$$\frac{1}{7}w = 14$$
$$\frac{1}{7}(98) \stackrel{?}{=} 14$$
$$14 = 14$$

85.
$$-1.2w = -102$$
$$\frac{-1.2w}{-1.2} = \frac{-102}{-1.2}$$
$$w = 85$$

$$-1.2w = -102$$
$$-1.2(85) \stackrel{?}{=} -102$$
$$-102 = -102$$

87.
$$0.25x = 1228$$
$$\frac{0.25x}{0.25} = \frac{1228}{0.25}$$
$$x = 4912$$

$$0.25x = 1228$$
$$0.25(4912) \stackrel{?}{=} 1228$$
$$1228 = 1228$$

89. Let r = the regular price. Then

$$\boxed{\text{Sale price}} = \boxed{\text{Regular price}} - \boxed{\text{Markdown}}$$
$$7995 = r - 1350$$
$$7995 + 1350 = r - 1350 + 1350$$
$$9345 = r$$

The regular price is \$9,345.

91. Let w = the wholesale price. Then

$$\boxed{\text{Retail price}} = \boxed{\text{Wholesale price}} + \boxed{\text{Markup}}$$
$$175 = w + 85$$
$$175 - 85 = w + 85 - 85$$
$$90 = w$$

The wholesale price is \$90.

93.
$$rb = a$$
$$40\% \cdot 200 = a$$
$$0.40(200) = a$$
$$80 = a$$

95.
$$rb = a$$
$$50\% \cdot 38 = a$$
$$0.50(38) = a$$
$$19 = a$$

97.
$$rb = a$$
$$35\% \cdot b = 182$$
$$0.35b = 182$$
$$\frac{0.35b}{0.35} = \frac{182}{0.35}$$
$$b = 520$$

99.
$$rb = a$$
$$15\% \cdot b = 48$$
$$0.15b = 48$$
$$\frac{0.15b}{0.15} = \frac{48}{0.15}$$
$$b = 320$$

101.
$$rb = a$$
$$28\% \cdot b = 42$$
$$0.28b = 42$$
$$\frac{0.28b}{0.28} = \frac{42}{0.28}$$
$$b = 150$$

103.
$$rb = a$$
$$r(357.5) = 71.5$$
$$\frac{r(357.5)}{357.5} = \frac{71.5}{357.5}$$
$$r = 0.20$$
$$r = 20\%$$

105.
$$p + 0.27 = 3.57$$
$$p + 0.27 - \mathbf{0.27} = 3.57 - \mathbf{0.27}$$
$$p = 3.3$$

Check: $p + 0.27 = 3.57$
$$3.3 + 0.27 \overset{?}{=} 3.57$$
$$3.57 = 3.57$$

107.
$$\frac{x}{15} = -4$$
$$15 \cdot \frac{x}{15} = 15 \cdot (-4)$$
$$x = -60$$

$$\frac{x}{15} = -4$$
$$\frac{-60}{15} \overset{?}{=} -4$$
$$-4 = -4$$

109.
$$-57 = b - 29$$
$$-57 + \mathbf{29} = b - 29 + \mathbf{29}$$
$$-28 = b$$

Check: $-57 = b - 29$
$$-57 \overset{?}{=} -28 - 29$$
$$-57 = -57$$

111.
$$y - 2.63 = -8.21$$
$$y - 2.63 + \mathbf{2.63} = -8.21 + \mathbf{2.63}$$
$$y = -5.58$$

Check: $y - 2.63 = -8.21$
$$-5.58 - 2.63 \overset{?}{=} -8.21$$
$$-8.21 = -8.21$$

113.
$$\frac{y}{-3} = -\frac{5}{6}$$
$$-\mathbf{6} \cdot \frac{y}{-3} = -\mathbf{6} \cdot \left(-\frac{5}{6}\right)$$
$$2y = 5$$
$$\frac{2y}{2} = \frac{5}{2}$$
$$y = \frac{5}{2}$$

$$\frac{y}{-3} = -\frac{5}{6}$$
$$\frac{\frac{5}{2}}{-3} \overset{?}{=} -\frac{5}{6}$$
$$\frac{\frac{5}{2} \cdot 2}{-3 \cdot 2} \overset{?}{=} -\frac{5}{6}$$
$$\frac{5}{-6} = -\frac{5}{6}$$

115.
$$-18 + y = 18$$
$$-18 + \mathbf{18} + y = 18 + \mathbf{18}$$
$$y = 36$$

Check: $-18 + y = 18$
$$-18 + 36 \overset{?}{=} 18$$
$$18 = 18$$

117.
$$-3 = \frac{x}{11}$$
$$11 \cdot (-3) = \mathbf{11} \cdot \frac{x}{11}$$
$$-33 = x$$

$$-3 = \frac{x}{11}$$
$$-3 \overset{?}{=} \frac{-33}{11}$$
$$-3 = -3$$

119.
$$b + 7 = \frac{20}{3}$$
$$b + 7 - \mathbf{7} = \frac{20}{3} - \mathbf{7}$$
$$b = \frac{20}{3} - \frac{21}{3} = -\frac{1}{3}$$

Check: $\qquad b + 7 = \frac{20}{3}$
$$-\frac{1}{3} + 7 \overset{?}{=} \frac{20}{3}$$
$$-\frac{1}{3} + \frac{21}{3} \overset{?}{=} \frac{20}{3}$$
$$\frac{20}{3} = \frac{20}{3}$$

121.
$$3x = -\frac{1}{4} \qquad\qquad 3x = -\frac{1}{4}$$
$$\frac{1}{3} \cdot 3x = \frac{1}{3}\left(-\frac{1}{4}\right) \qquad 3\left(-\frac{1}{12}\right) \overset{?}{=} -\frac{1}{4}$$
$$x = -\frac{1}{12} \qquad\qquad -\frac{3}{12} \overset{?}{=} -\frac{1}{4}$$
$$-\frac{1}{4} = -\frac{1}{4}$$

123.
$$-\frac{3}{5} = x - \frac{2}{5} \qquad\qquad \text{Check:} \quad -\frac{3}{5} = x - \frac{2}{5}$$
$$-\frac{3}{5} + \mathbf{\frac{2}{5}} = x - \frac{2}{5} + \mathbf{\frac{2}{5}} \qquad\qquad -\frac{3}{5} \overset{?}{=} -\frac{1}{5} - \frac{2}{5}$$
$$-\frac{1}{5} = x \qquad\qquad\qquad -\frac{3}{5} = -\frac{3}{5}$$

125.
$$\frac{1}{7}x = \frac{5}{7} \qquad\qquad \frac{1}{7}x = \frac{5}{7}$$
$$\mathbf{7} \cdot \frac{1}{7}x = \mathbf{7} \cdot \frac{5}{7} \qquad \frac{1}{7}(5) = \frac{5}{7}$$
$$x = 5$$

127.
$$-27w = 81 \qquad\qquad -27w = 81$$
$$\frac{-27w}{-27} = \frac{81}{-27} \qquad -27(-3) \overset{?}{=} 81$$
$$w = -3 \qquad\qquad 81 = 81$$

129.
$$18x = -9 \qquad\qquad 18x = -9$$
$$\frac{18x}{18} = \frac{-9}{18} \qquad 18\left(-\frac{1}{2}\right) \overset{?}{=} -9$$
$$x = -\frac{1}{2} \qquad\qquad -9 = -9$$

131.
$$rb = a$$
$$r(8) = 0.48$$
$$\frac{r(8)}{8} = \frac{0.48}{8}$$
$$r = 0.06$$
$$r = 6\%$$

133.
$$rb = a$$
$$r(17) = 34$$
$$\frac{r(17)}{17} = \frac{34}{17}$$
$$r = 2.00$$
$$r = 200\%$$

135. Let $b =$ the selling price
$$rb = a$$
$$0.05b = 13.50$$
$$\frac{0.05b}{0.05} = \frac{13.50}{0.05}$$
$$b = 270$$
The selling price is \$270.

137. Let $r =$ the percentage for sales tax.
$$rb = a$$
$$r(12) = 0.72$$
$$\frac{r(12)}{12} = \frac{0.72}{12}$$
$$r = 0.06$$
$$r = 6\%$$
The sales tax is computed at a rate of 6%.

139.
$$A = p + i$$
$$5010 = 4750 + i$$
$$5010 - \mathbf{4750} = 4750 - \mathbf{4750} + i$$
$$260 = i$$

The deposit earned $260 in interest.

141. Let $x =$ the original number in the audience.

$$\frac{1}{3} \cdot \boxed{\begin{array}{c}\text{Original} \\ \text{audience}\end{array}} = \boxed{\begin{array}{c}\text{Number} \\ \text{who left}\end{array}}$$

$$\frac{1}{3}x = 78$$
$$\mathbf{3} \cdot \frac{1}{3}x = \mathbf{3} \cdot 78$$
$$x = 234$$

There were originally 234 in the audience.

143. Let $r =$ the percentage not pleased.
$$a = 9200 - 4140 = 5060$$
$$rb = a$$
$$r(9200) = 5060$$
$$\frac{r(9200)}{9200} = \frac{5060}{9200}$$
$$r = 0.55$$
$$r = 55\%$$

55% of those surveyed were not pleased.

145. Let $b =$ the number employed at the factory
$$rb = a$$
$$0.90(b) = 2484$$
$$\frac{0.90b}{0.90} = \frac{2484}{0.90}$$
$$b = 2760$$

There are a total of 2,760 employees.

147. Let $x =$ the original number of shares owned.

$$1.5 \cdot \boxed{\begin{array}{c}\text{Original} \\ \text{\# of shares}\end{array}} = \boxed{\begin{array}{c}\text{\# of shares} \\ \text{after split}\end{array}}$$

$$1.5x = 555$$
$$\frac{1.5x}{1.5} = \frac{555}{1.5}$$
$$x = 370$$

The shareholder owned 370 shares.

149. Let $c =$ the original cost.

$$\boxed{\begin{array}{c}\text{Original} \\ \text{cost}\end{array}} - \boxed{\text{Depreciation}} = \boxed{\begin{array}{c}\text{New} \\ \text{cost}\end{array}}$$

$$c - 7500 = 10250$$
$$c - 7500 + \mathbf{7500} = 10250 + \mathbf{7500}$$
$$c = 17750$$

The original cost was $17,750.

151. Let $t =$ the tax paid.

$$\boxed{\text{Cost}} = \boxed{\text{Price}} + \boxed{\text{Tax}}$$

$$39.32 = 37.10 + t$$
$$39.32 - \mathbf{37.10} = 37.10 - \mathbf{37.10} + t$$
$$2.22 = t$$

The tax was $2.22.

153. Let $b =$ the cost of the brush. Then

$$\boxed{\begin{array}{c}\text{Primer} \\ \text{cost}\end{array}} + \boxed{\begin{array}{c}\text{Paint} \\ \text{cost}\end{array}} + \boxed{\begin{array}{c}\text{Brush} \\ \text{cost}\end{array}} = \boxed{\begin{array}{c}\text{Total} \\ \text{cost}\end{array}}$$

$$10.99 + 14.50 + b = 30.44$$
$$25.49 + b = 30.44$$
$$25.49 - \mathbf{25.49} + b = 30.44 - \mathbf{25.49}$$
$$b = 4.95$$

The cost of the brush was $4.95.

155. Let $c =$ the condominium price. Then

$$\boxed{\begin{array}{c}\text{Condo} \\ \text{price}\end{array}} = \boxed{\begin{array}{c}\text{House} \\ \text{price}\end{array}} - 57595$$

$$c = 202744 - 57595$$
$$c = 145149$$

The price of the condominium is $145,149.

157. Answers may vary.

159.
$$A_{\text{circle}} = A_{\text{square}} \, 4$$
$$\pi r^2 = s^2$$
$$\pi(4.5)^2 = 8^2$$
$$\pi(20.25) = 64$$
$$\frac{\pi(20.25)}{20.25} = \frac{64}{20.25}$$
$$\pi \approx 3.16$$

Exercises 2.2 (page 100)

1. add 9

3. add 3

5. multiply by 3

7.
$$7x - 7 = 14$$
$$7x - 7 + 7 = 14 + 7$$
$$7x = 21$$
$$\frac{7x}{7} = \frac{21}{7}$$
$$x = 3$$

9. $P = 2l + 2w = 2(8.5\,\text{cm}) + 2(16.5\,\text{cm})$
$$= 17\,\text{cm} + 33\,\text{cm} = 50\,\text{cm}$$

11. $A = \frac{1}{2}h(b+d) = \frac{1}{2}(8.5\,\text{in.})(6.7\,\text{in.} + 12.2\,\text{in.}) = \frac{1}{2}(8.5\,\text{in.})(18.9\,\text{in.}) = \frac{1}{2}(160.65\,\text{in.}^2) = 80.325\,\text{in.}^2$

13. cost

15. percent

17. percent of increase

19.
$$5x - 1 = 4$$
$$5x - 1 + 1 = 4 + 1$$
$$5x = 5$$
$$\frac{5x}{5} = \frac{5}{5}$$
$$x = 1$$

$$5x - 1 = 4$$
$$5(1) - 1 \overset{?}{=} 4$$
$$5 - 1 \overset{?}{=} 4$$
$$4 = 4$$

21.
$$-6x + 2 = 14$$
$$-6x + 2 - 2 = 14 - 2$$
$$-6x = 12$$
$$\frac{-6x}{-6} = \frac{12}{-6}$$
$$x = -2$$

$$-6x + 2 = 14$$
$$-6(-2) + 2 \overset{?}{=} 14$$
$$12 + 2 \overset{?}{=} 14$$
$$14 = 14$$

23.
$$6x + 2 = -4$$
$$6x + 2 - 2 = -4 - 2$$
$$6x = -6$$
$$\frac{6x}{6} = \frac{-6}{6}$$
$$x = -1$$

$$6x + 2 = -4$$
$$6(-1) + 2 \overset{?}{=} -4$$
$$-6 + 2 \overset{?}{=} -4$$
$$-4 = -4$$

25.
$$3x - 8 = 1$$
$$3x - 8 + 8 = 1 + 8$$
$$3x = 9$$
$$\frac{3x}{3} = \frac{9}{3}$$
$$x = 3$$

$$3x - 8 = 1$$
$$3(3) - 8 \overset{?}{=} 1$$
$$9 - 8 \overset{?}{=} 1$$
$$1 = 1$$

27.
$$4x - 7 = 5$$
$$4x - 7 + 7 = 5 + 7$$
$$4x = 12$$
$$\frac{4x}{4} = \frac{12}{4}$$
$$x = 3$$

$$4x - 7 = 5$$
$$4(3) - 7 \overset{?}{=} 5$$
$$12 - 7 \overset{?}{=} 5$$
$$5 = 5$$

29.
$$-3x - 6 = 12$$
$$-3x - 6 + 6 = 12 + 6$$
$$-3x = 18$$
$$\frac{-3x}{-3} = \frac{18}{-3}$$
$$x = -6$$

$$-3x - 6 = 12$$
$$-3(-6) - 6 \overset{?}{=} 12$$
$$18 - 6 \overset{?}{=} 12$$
$$12 = 12$$

31.
$$-2x - 8 = -2$$
$$-2x - 8 + 8 = -2 + 8$$
$$-2x = 6$$
$$\frac{-2x}{-2} = \frac{6}{-2}$$
$$x = -3$$

$$-2x - 8 = -2$$
$$-2(-3) - 8 \stackrel{?}{=} -2$$
$$6 - 8 \stackrel{?}{=} -2$$
$$-2 = -2$$

33.
$$\frac{z}{9} + 5 = -1$$
$$\frac{z}{9} + 5 - 5 = -1 - 5$$
$$\frac{z}{9} = -6$$
$$9 \cdot \frac{z}{9} = 9(-6)$$
$$z = -54$$

$$\frac{z}{9} + 5 = -1$$
$$\frac{-54}{9} + 5 \stackrel{?}{=} -1$$
$$-6 + 5 \stackrel{?}{=} -1$$
$$-1 = -1$$

35.
$$\frac{x}{4} + 7 = 3$$
$$\frac{x}{4} + 7 - 7 = 3 - 7$$
$$\frac{x}{4} = -4$$
$$4 \cdot \frac{x}{4} = 4(-4)$$
$$x = -16$$

$$\frac{x}{4} + 7 = 3$$
$$\frac{-16}{4} + 7 \stackrel{?}{=} 3$$
$$-4 + 7 \stackrel{?}{=} 3$$
$$3 = 3$$

37.
$$\frac{x}{3} - 10 = -1$$
$$\frac{x}{3} - 10 + 10 = -1 + 10$$
$$\frac{x}{3} = 9$$
$$3 \cdot \frac{x}{3} = 3 \cdot 9$$
$$x = 27$$

$$\frac{x}{3} - 10 = -1$$
$$\frac{27}{3} - 10 \stackrel{?}{=} -1$$
$$9 - 10 \stackrel{?}{=} -1$$
$$-1 = -1$$

39.
$$\frac{p}{11} + 9 = 6$$
$$\frac{p}{11} + 9 - 9 = 6 - 9$$
$$\frac{p}{11} = -3$$
$$11 \cdot \frac{p}{11} = 11(-3)$$
$$p = -33$$

$$\frac{p}{11} + 9 = 6$$
$$\frac{-33}{11} + 9 \stackrel{?}{=} 6$$
$$-3 + 9 \stackrel{?}{=} 6$$
$$6 = 6$$

41.
$$\frac{b+5}{3} = 11$$
$$3 \cdot \frac{b+5}{3} = 3 \cdot 11$$
$$b + 5 = 33$$
$$b + 5 - 5 = 33 - 5$$
$$b = 28$$

$$\frac{b+5}{3} = 11$$
$$\frac{28+5}{3} \stackrel{?}{=} 11$$
$$\frac{33}{3} \stackrel{?}{=} 11$$
$$11 = 11$$

43.
$$\frac{x+5}{2} = 4$$
$$2 \cdot \frac{x+5}{2} = 2 \cdot 4$$
$$x + 5 = 8$$
$$x + 5 - 5 = 8 - 5$$
$$x = 3$$

$$\frac{x+5}{2} = 4$$
$$\frac{3+5}{2} \stackrel{?}{=} 4$$
$$\frac{8}{2} \stackrel{?}{=} 4$$
$$4 = 4$$

45.
$$\frac{3x - 12}{2} = 9$$
$$2 \cdot \frac{3x - 12}{2} = 2 \cdot 9$$
$$3x - 12 = 18$$
$$3x - 12 + 12 = 18 + 12$$
$$3x = 30$$
$$\frac{3x}{3} = \frac{30}{3}$$
$$x = 10$$

$$\frac{3x - 12}{2} = 9$$
$$\frac{3(10) - 12}{2} \stackrel{?}{=} 9$$
$$\frac{30 - 12}{2} \stackrel{?}{=} 9$$
$$\frac{18}{2} \stackrel{?}{=} 9$$
$$9 = 9$$

47.

$$\frac{4k-1}{5}=3$$

$$5\cdot\frac{4k-1}{5}=5\cdot 3$$

$$4k-1=15$$

$$4k-1+1=15+1$$

$$4k=16$$

$$\frac{4k}{4}=\frac{16}{4}$$

$$k=4$$

$$\frac{4k-1}{5}=3$$

$$\frac{4(4)-1}{5}\overset{?}{=}3$$

$$\frac{16-1}{5}\overset{?}{=}3$$

$$\frac{15}{5}\overset{?}{=}3$$

$$3=3$$

49.

$$\frac{k}{5}-\frac{1}{2}=\frac{3}{2}$$

$$\frac{k}{5}-\frac{1}{2}+\frac{1}{2}=\frac{3}{2}+\frac{1}{2}$$

$$\frac{k}{5}=\frac{4}{2}$$

$$\frac{k}{5}=2$$

$$5\cdot\frac{k}{5}=5\cdot 2$$

$$k=10$$

$$\frac{k}{5}-\frac{1}{2}=\frac{3}{2}$$

$$\frac{10}{5}-\frac{1}{2}\overset{?}{=}\frac{3}{2}$$

$$2-\frac{1}{2}\overset{?}{=}\frac{3}{2}$$

$$\frac{4}{2}-\frac{1}{2}\overset{?}{=}\frac{3}{2}$$

$$\frac{3}{2}=\frac{3}{2}$$

51.

$$\frac{w}{16}+\frac{5}{4}=1$$

$$\frac{w}{16}+\frac{5}{4}-\frac{5}{4}=1-\frac{5}{4}$$

$$\frac{w}{16}=\frac{4}{4}-\frac{5}{4}$$

$$\frac{w}{16}=-\frac{1}{4}$$

$$16\cdot\frac{w}{16}=16\left(-\frac{1}{4}\right)$$

$$w=-4$$

$$\frac{w}{16}+\frac{5}{4}=1$$

$$\frac{-4}{16}+\frac{5}{4}\overset{?}{=}1$$

$$-\frac{1}{4}+\frac{5}{4}\overset{?}{=}1$$

$$\frac{4}{4}\overset{?}{=}1$$

$$1=1$$

53.

$$\frac{3x}{2}-6=9$$

$$\frac{3x}{2}-6+6=9+6$$

$$\frac{3x}{2}=15$$

$$2\cdot\frac{3x}{2}=2\cdot 15$$

$$3x=30$$

$$\frac{3x}{3}=\frac{30}{3}$$

$$x=10$$

$$\frac{3x}{2}-6=9$$

$$\frac{3(10)}{2}-6\overset{?}{=}9$$

$$\frac{30}{2}-6\overset{?}{=}9$$

$$15-6\overset{?}{=}9$$

$$9=9$$

55.

$$\frac{9y}{2}+3=-15$$

$$\frac{9y}{2}+3-3=-15-3$$

$$\frac{9y}{2}=-18$$

$$2\cdot\frac{9y}{2}=2(-18)$$

$$9y=-36$$

$$\frac{9y}{9}=\frac{-36}{9}$$

$$y=-4$$

$$\frac{9y}{2}+3=-15$$

$$\frac{9(-4)}{2}+3\overset{?}{=}-15$$

$$\frac{-36}{2}+3\overset{?}{=}-15$$

$$-18+3\overset{?}{=}-15$$

$$-15=-15$$

57.

$$43p+72=158$$

$$43p+72-72=158-72$$

$$43p=86$$

$$\frac{43p}{43}=\frac{86}{43}$$

$$p=2$$

$$43p+72=158$$

$$43(2)+72\overset{?}{=}158$$

$$86+72\overset{?}{=}158$$

$$158=158$$

59.

$$-47 - 21n = 58$$
$$-47 + 47 - 21n = 58 + 47$$
$$-21n = 105$$
$$\frac{-21n}{-21} = \frac{105}{-21}$$
$$n = -5$$

$$-47 - 21n = 58$$
$$-47 - 21(-5) \stackrel{?}{=} 58$$
$$-47 + 105 \stackrel{?}{=} 58$$
$$58 = 58$$

61.

$$2y - \frac{5}{3} = \frac{4}{3}$$
$$2y - \frac{5}{3} + \frac{5}{3} = \frac{4}{3} + \frac{5}{3}$$
$$2y = \frac{9}{3}$$
$$2y = 3$$
$$\frac{2y}{2} = \frac{3}{2}$$
$$y = \frac{3}{2}$$

$$2y - \frac{5}{3} = \frac{4}{3}$$
$$2\left(\frac{3}{2}\right) - \frac{5}{3} \stackrel{?}{=} \frac{4}{3}$$
$$3 - \frac{5}{3} \stackrel{?}{=} \frac{4}{3}$$
$$\frac{9}{3} - \frac{5}{3} \stackrel{?}{=} \frac{4}{3}$$
$$\frac{4}{3} = \frac{4}{3}$$

63.

$$-0.4y - 12 = -20$$
$$-0.4y - 12 + 12 = -20 + 12$$
$$-0.4y = -8$$
$$\frac{-0.4y}{-0.4} = \frac{-8}{-0.4}$$
$$y = 20$$

$$-0.4y - 12 = -20$$
$$-0.4(20) - 12 \stackrel{?}{=} -20$$
$$-8 - 12 \stackrel{?}{=} -20$$
$$-20 = -20$$

65.

$$\frac{2x}{3} + \frac{1}{2} = 3$$
$$\frac{2x}{3} + \frac{1}{2} - \frac{1}{2} = 3 - \frac{1}{2}$$
$$\frac{2x}{3} = \frac{5}{2}$$
$$3 \cdot \frac{2x}{3} = 3 \cdot \frac{5}{2}$$
$$2x = \frac{15}{2}$$
$$\frac{1}{2} \cdot 2x = \frac{1}{2} \cdot \frac{15}{2}$$
$$x = \frac{15}{4}$$

$$\frac{2x}{3} + \frac{1}{2} = 3$$
$$\frac{2\left(\frac{15}{4}\right)}{3} + \frac{1}{2} \stackrel{?}{=} 3$$
$$\frac{\frac{15}{2}}{3} + \frac{1}{2} \stackrel{?}{=} 3$$
$$\frac{15}{6} + \frac{1}{2} \stackrel{?}{=} 3$$
$$\frac{5}{2} + \frac{1}{2} \stackrel{?}{=} 3$$
$$\frac{6}{2} \stackrel{?}{=} 3$$
$$3 = 3$$

67.

$$\frac{3x}{4} - \frac{2}{5} = 2$$
$$\frac{3x}{4} - \frac{2}{5} + \frac{2}{5} = 2 + \frac{2}{5}$$
$$\frac{3x}{4} = \frac{12}{5}$$
$$4 \cdot \frac{3x}{4} = 4 \cdot \frac{12}{5}$$
$$3x = \frac{48}{5}$$
$$\frac{1}{3} \cdot 3x = \frac{1}{3} \cdot \frac{48}{5}$$
$$x = \frac{16}{5}$$

$$\frac{3x}{4} - \frac{2}{5} = 2$$
$$\frac{3\left(\frac{16}{5}\right)}{4} - \frac{2}{5} \stackrel{?}{=} 2$$
$$\frac{\frac{48}{5}}{4} - \frac{2}{5} \stackrel{?}{=} 2$$
$$\frac{48}{20} - \frac{2}{5} \stackrel{?}{=} 2$$
$$\frac{12}{5} - \frac{2}{5} \stackrel{?}{=} 2$$
$$\frac{10}{5} \stackrel{?}{=} 2$$
$$2 = 2$$

69.
$$\frac{u-4}{7} = 1$$
$$7 \cdot \frac{u-4}{7} = 7 \cdot 1$$
$$u - 4 = 7$$
$$u - 4 + 4 = 7 + 4$$
$$u = 11$$

$$\frac{u-4}{7} = 1$$
$$\frac{11-4}{7} \overset{?}{=} 1$$
$$\frac{7}{7} \overset{?}{=} 1$$
$$1 = 1$$

71.
$$\frac{x-5}{3} = -4$$
$$3 \cdot \frac{x-5}{3} = 3(-4)$$
$$x - 5 = -12$$
$$x - 5 + 5 = -12 + 5$$
$$x = -7$$

$$\frac{x-5}{3} = -4$$
$$\frac{-7-5}{3} \overset{?}{=} -4$$
$$\frac{-12}{3} \overset{?}{=} -4$$
$$-4 = -4$$

73.
$$\frac{3z+2}{17} = 0$$
$$17 \cdot \frac{3z+2}{17} = 17 \cdot 0$$
$$3z + 2 = 0$$
$$3z + 2 - 2 = 0 - 2$$
$$3z = -2$$
$$\frac{3z}{3} = \frac{-2}{3}$$
$$z = -\frac{2}{3}$$

$$\frac{3z+2}{17} = 0$$
$$\frac{3\left(-\frac{2}{3}\right)+2}{17} \overset{?}{=} 0$$
$$\frac{-2+2}{17} \overset{?}{=} 0$$
$$\frac{0}{17} \overset{?}{=} 0$$
$$0 = 0$$

75.
$$\frac{17k-28}{21} + \frac{4}{3} = 0$$
$$\frac{17k-28}{21} + \frac{4}{3} - \frac{4}{3} = 0 - \frac{4}{3}$$
$$\frac{17k-28}{21} = -\frac{4}{3}$$
$$21 \cdot \frac{17k-28}{21} = 21\left(-\frac{4}{3}\right)$$
$$17k - 28 = -28$$
$$17k - 28 + 28 = -28 + 28$$
$$17k = 0$$
$$\frac{17k}{17} = \frac{0}{17}$$
$$k = 0$$

$$\frac{17k-28}{21} + \frac{4}{3} = 0$$
$$\frac{17 \cdot 0 - 28}{21} + \frac{4}{3} \overset{?}{=} 0$$
$$\frac{0-28}{21} + \frac{4}{3} \overset{?}{=} 0$$
$$-\frac{4}{3} + \frac{4}{3} \overset{?}{=} 0$$
$$0 = 0$$

77.
$$-\frac{x}{3} - \frac{1}{2} = -\frac{5}{2}$$
$$-\frac{x}{3} - \frac{1}{2} + \frac{1}{2} = -\frac{5}{2} + \frac{1}{2}$$
$$-\frac{x}{3} = -\frac{4}{2}$$
$$-\frac{x}{3} = -2$$
$$-3\left(-\frac{x}{3}\right) = -3(-2)$$
$$x = 6$$

$$-\frac{x}{3} - \frac{1}{2} = -\frac{5}{2}$$
$$-\frac{6}{3} - \frac{1}{2} \overset{?}{=} -\frac{5}{2}$$
$$-2 - \frac{1}{2} \overset{?}{=} -\frac{5}{2}$$
$$-\frac{4}{2} - \frac{1}{2} \overset{?}{=} -\frac{5}{2}$$
$$-\frac{5}{2} = -\frac{5}{2}$$

79.
$$\frac{10-3w}{9} = \frac{2}{3}$$
$$9 \cdot \frac{10-3w}{9} = 9 \cdot \frac{2}{3}$$
$$10 - 3w = 6$$
$$10 - 10 - 3w = 6 - 10$$
$$-3w = -4$$
$$\frac{-3w}{-3} = \frac{-4}{-3}$$
$$w = \frac{4}{3}$$

$$\frac{10-3w}{9} = \frac{2}{3}$$
$$\frac{10-3\left(\frac{4}{3}\right)}{9} \overset{?}{=} \frac{2}{3}$$
$$\frac{10-4}{9} \overset{?}{=} \frac{2}{3}$$
$$\frac{6}{9} \overset{?}{=} \frac{2}{3}$$
$$\frac{2}{3} = \frac{2}{3}$$

81. Let $x =$ her former rent.

Then $2x - 200 =$ the new rent.

$$\boxed{\text{The new rent}} = 450$$
$$2x - 200 = 450$$
$$2x - 200 + 200 = 450 + 200$$
$$2x = 650$$
$$\frac{2x}{2} = \frac{650}{2}$$
$$x = 325$$

Her former rent was \$325.

83. Let $x =$ the number of days.

Then $20 + 14x =$ the total cost.

$$\boxed{\text{The total cost}} = 104$$
$$20 + 14x = 104$$
$$20 - 20 + 14x = 104 - 20$$
$$14x = 84$$
$$\frac{14x}{14} = \frac{84}{14}$$
$$x = 6$$

The owner was gone for 6 days.

85. Let $x =$ the regular price. Then $0.80x =$ the sale price.

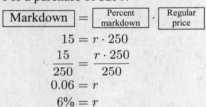

$$\boxed{\text{Final price}} = 0.90 \cdot \boxed{\text{Sale price}}$$
$$36 = 0.90(0.80x)$$
$$36 = 0.72x$$
$$\frac{36}{0.72} = \frac{0.72x}{0.72}$$
$$50 = x \Rightarrow \text{The original price was } \$50.$$

87. For a purchase of \$100:

$$\boxed{\text{Markdown}} = \boxed{\text{Percent markdown}} \cdot \boxed{\text{Regular price}}$$
$$15 = r \cdot 100$$
$$\frac{15}{100} = \frac{r \cdot 100}{100}$$
$$0.15 = r$$
$$15\% = r$$

For a purchase of \$250:

$$\boxed{\text{Markdown}} = \boxed{\text{Percent markdown}} \cdot \boxed{\text{Regular price}}$$
$$15 = r \cdot 250$$
$$\frac{15}{250} = \frac{r \cdot 250}{250}$$
$$0.06 = r$$
$$6\% = r$$

The range of the percent discount is from 6% to 15%.

89. Let $x =$ the original number.

Then $3x - 6 =$ the other number.

$$\boxed{\text{The other number}} = 9$$
$$3x - 6 = 9$$
$$3x - 6 + 6 = 9 + 6$$
$$3x = 15$$
$$\frac{3x}{3} = \frac{15}{3}$$
$$x = 5$$

The original number is 5.

91. Let $x =$ the original number.

Then $\dfrac{x + 7}{2} =$ the other number.

$$\boxed{\text{The other number}} = 5$$
$$\frac{x + 7}{2} = 5$$
$$2 \cdot \frac{x + 7}{2} = 2 \cdot 5$$
$$x + 7 = 10$$
$$x + 7 - 7 = 10 - 7$$
$$x = 3$$

The original number is 3.

93. Let $x =$ the # of minutes (after the 1st). Then $0.85 + 0.27x =$ the total cost.

$$\boxed{\text{The total cost}} = 8.68$$
$$0.85 + 0.27x = 8.68$$
$$0.85 - 0.85 + 0.27x = 8.68 - 0.85$$
$$0.27x = 7.83$$
$$\frac{0.27x}{0.27} = \frac{7.83}{0.27}$$
$$x = 29$$

She can talk for 29 minutes **after the first minute**, for a total of 30 minutes.

95. Let $x =$ the money from ticket sales.
Then $1500 + 0.20x =$ the total income.

$$\boxed{\text{The total income}} = 2980$$
$$1500 + 0.20x = 2980$$
$$1500 - 1500 + 0.20x = 2980 - 1500$$
$$0.20x = 1480$$
$$\frac{0.20x}{0.20} = \frac{1480}{0.20}$$
$$x = 7400$$

The total ticket sales were $7,400.

97. Let $x =$ the score on the fifth exam.

$$\boxed{\text{Average score}} = 90$$
$$\frac{85 + 80 + 95 + 78 + x}{5} = 90$$
$$\frac{338 + x}{5} = 90$$
$$5 \cdot \frac{338 + x}{5} = 5 \cdot 90$$
$$338 + x = 450$$
$$338 - 338 + x = 450 - 338$$
$$x = 112$$

It is impossible to receive an A.

99. Answers may vary.

101.
$$\frac{7x + \#}{22} = \frac{1}{2}$$
$$\frac{7(1) + \#}{22} = \frac{1}{2}$$
$$\frac{7 + \#}{22} = \frac{1}{2}$$
$$22 \cdot \frac{7 + \#}{22} = 22 \cdot \frac{1}{2}$$
$$7 + \# = 11$$
$$7 - 7 + \# = 11 - 7$$
$$\# = 4$$

The original equation was $\dfrac{7x + 4}{22} = \dfrac{1}{2}$.

Exercises 2.3 (page 107)

1. $5x - 2$: expression

3. $3x + 7 = -1$: equation

5. $6 - 4x = 7$: equation

7. $2x + 5 = 2x + 5$: identity

9. $6x - 2 = 6x + 4$: contradiction

11. $x^2 z(y^3 - z) = (-3)^2(0)\left[(-5)^3 - 0\right] = 0$

13. $\dfrac{x - y^2}{2y - 1 + x} = \dfrac{-3 - (-5)^2}{2(-5) - 1 + (-3)} = \dfrac{-3 - (+25)}{-10 - 1 + (-3)} = \dfrac{-28}{-14} = 2$

15. $\dfrac{6}{7} - \dfrac{5}{8} = \dfrac{6 \cdot \mathbf{8}}{7 \cdot \mathbf{8}} - \dfrac{5 \cdot \mathbf{7}}{8 \cdot \mathbf{7}} = \dfrac{48}{56} - \dfrac{35}{56}$

$= \dfrac{48 - 35}{56} = \dfrac{13}{56}$

17. $\dfrac{6}{7} \div \dfrac{5}{8} = \dfrac{6}{7} \cdot \dfrac{8}{5} = \dfrac{48}{35}$

19. variables, like, unlike, coefficient

21. identity, contradiction

23. $8x + 12x = (8 + 12)x = 20x$

25. $8x^2 - 5x^2 = (8 - 5)x^2 = 3x^2$

27. $9x + 3y \Rightarrow$ unlike terms

29. $4(x + 3) - 2x = 4 \cdot x + 4 \cdot 3 - 2x$

$= 4x + 12 - 2x = 2x + 12$

31. $5(z - 3) + 2z = 5 \cdot z - 5 \cdot 3 + 2z$

$= 5z - 15 + 2z = 7z - 15$

33. $12(x + 11) - 11 = 12 \cdot x + 12 \cdot 11 - 11$

$= 12x + 132 - 11$

$= 12x + 121$

35. $6(y - 2) - 3(y + 1) = 6 \cdot y + 6(-2) + (-3) \cdot y + (-3)(1) = 6y - 12 - 3y - 3 = 3y - 15$

37. $5x - 2(y - x) + 4y = 5x + (-2) \cdot y + (-2)(-x) + 4y = 5x - 2y + 2x + 4y = 7x + 2y$

39. $8(x + 5) + 6(7 - x) = 0$

$8x + 40 + 42 - 6x = 0$

$2x + 82 = 0$

$2x + 82 - 82 = 0 - 82$

$2x = -82$

$\dfrac{2x}{2} = \dfrac{-82}{2}$

$x = -41$

$8(x + 5) + 6(7 - x) = 0$

$8(-41 + 5) + 6[7 - (-41)] \stackrel{?}{=} 0$

$8(-36) + 6[7 + 41] \stackrel{?}{=} 0$

$-288 + 6(48) \stackrel{?}{=} 0$

$-288 + 288 \stackrel{?}{=} 0$

$0 = 0$

41. $12x - 4(5 + x) = 4$

$12x - 20 - 4x = 4$

$8x - 20 = 4$

$8x - 20 + 20 = 4 + 20$

$8x = 24$

$\dfrac{8x}{8} = \dfrac{24}{8}$

$x = 3$

$12x - 4(5 + x) = 4$

$12(3) - 4(5 + 3) \stackrel{?}{=} 4$

$36 - 4(8) \stackrel{?}{=} 4$

$36 - 32 \stackrel{?}{=} 4$

$4 = 4$

43.
$$3x + 2 = 2x$$
$$3x - 3x + 2 = 2x - 3x$$
$$2 = -x$$
$$-1(2) = -1(-x)$$
$$-2 = x$$

$$3x + 2 = 2x$$
$$3(-2) + 2 \overset{?}{=} 2(-2)$$
$$-6 + 2 \overset{?}{=} -4$$
$$-4 = -4$$

45.
$$5x - 3 = 4x$$
$$5x - 5x - 3 = 4x - 5x$$
$$-3 = -x$$
$$-1(-3) = -1(-x)$$
$$3 = x$$

$$5x - 3 = 4x$$
$$5(3) - 3 \overset{?}{=} 4(3)$$
$$15 - 3 \overset{?}{=} 12$$
$$12 = 12$$

47.
$$9y - 3 = 6y$$
$$9y - 9y - 3 = 6y - 9y$$
$$-3 = -3y$$
$$\frac{-3}{-3} = \frac{-3y}{-3}$$
$$1 = y$$

$$9y - 3 = 6y$$
$$9(1) - 3 \overset{?}{=} 6(1)$$
$$9 - 3 \overset{?}{=} 6$$
$$6 = 6$$

49.
$$10y - 10 = 5y$$
$$10y - 10y - 10 = 5y - 10y$$
$$-10 = -5y$$
$$\frac{-10}{-5} = \frac{-5y}{-5}$$
$$2 = y$$

$$10y - 10 = 5y$$
$$10(2) - 10 \overset{?}{=} 5(2)$$
$$20 - 10 \overset{?}{=} 10$$
$$10 = 10$$

51.
$$3(a + 2) = 4a$$
$$3a + 6 = 4a$$
$$3a - 3a + 6 = 4a - 3a$$
$$6 = a$$

$$3(a + 2) = 4a$$
$$3(6 + 2) \overset{?}{=} 4(6)$$
$$3(8) \overset{?}{=} 24$$
$$24 = 24$$

53.
$$6(b + 4) = 8b$$
$$6b + 24 = 8b$$
$$6b - 6b + 24 = 8b - 6b$$
$$24 = 2b$$
$$\frac{24}{2} = \frac{2b}{2}$$
$$12 = b$$

$$6(b + 4) = 8b$$
$$6(12 + 4) \overset{?}{=} 8(12)$$
$$6(16) \overset{?}{=} 96$$
$$96 = 96$$

55.
$$2 + 3(x - 5) = 4(x - 1)$$
$$2 + 3x - 15 = 4x - 4$$
$$3x - 13 = 4x - 4$$
$$3x - 3x - 13 = 4x - 3x - 4$$
$$-13 = x - 4$$
$$-13 + 4 = x - 4 + 4$$
$$-9 = x$$

$$2 + 3(x - 5) = 4(x - 1)$$
$$2 + 3(-9 - 5) \overset{?}{=} 4(-9 - 1)$$
$$2 + 3(-14) \overset{?}{=} 4(-10)$$
$$2 + (-42) \overset{?}{=} -40$$
$$-40 = -40$$

57.

$$3(a + 2) = 2(a - 7)$$
$$3a + 6 = 2a - 14$$
$$3a - 2a + 6 = 2a - 2a - 14$$
$$a + 6 = -14$$
$$a + 6 - 6 = -14 - 6$$
$$a = -20$$

$$3(a + 2) = 2(a - 7)$$
$$3(-20 + 2) \stackrel{?}{=} 2(-20 - 7)$$
$$3(-18) \stackrel{?}{=} 2(-27)$$
$$-54 = -54$$

59.

$$\frac{3(t - 7)}{2} = t - 6$$
$$2 \cdot \frac{3(t - 7)}{2} = 2(t - 6)$$
$$3(t - 7) = 2(t - 6)$$
$$3t - 21 = 2t - 12$$
$$3t - 2t - 21 = 2t - 2t - 12$$
$$t - 21 = -12$$
$$t - 21 + 21 = -12 + 21$$
$$t = 9$$

$$\frac{3(t - 7)}{2} = t - 6$$
$$\frac{3(9 - 7)}{2} \stackrel{?}{=} 9 - 6$$
$$\frac{3(2)}{2} \stackrel{?}{=} 9 - 6$$
$$\frac{6}{2} \stackrel{?}{=} 3$$
$$3 = 3$$

61.

$$\frac{2(t - 1)}{6} - 2 = \frac{t + 2}{6}$$
$$6\left[\frac{2(t - 1)}{6} - 2\right] = 6 \cdot \frac{t + 2}{6}$$
$$6 \cdot \frac{2(t - 1)}{6} - 6 \cdot 2 = t + 2$$
$$2(t - 1) - 12 = t + 2$$
$$2t - 2 - 12 = t + 2$$
$$2t - 14 = t + 2$$
$$2t - t - 14 = t - t + 2$$
$$t - 14 = 2$$
$$t - 14 + 14 = 2 + 14$$
$$t = 16$$

$$\frac{2(t - 1)}{6} - 2 = \frac{t + 2}{6}$$
$$\frac{2(16 - 1)}{6} - 2 \stackrel{?}{=} \frac{16 + 2}{6}$$
$$\frac{2(15)}{6} - 2 \stackrel{?}{=} \frac{18}{6}$$
$$\frac{30}{6} - 2 \stackrel{?}{=} 3$$
$$5 - 2 \stackrel{?}{=} 3$$
$$3 = 3$$

63.

$$3.1(x - 2) = 1.3x + 2.8$$
$$3.1x - 6.2 = 1.3x + 2.8$$
$$3.1x - 1.3x - 6.2 = 1.3x - 1.3x + 2.8$$
$$1.8x - 6.2 = 2.8$$
$$1.8x - 6.2 + 6.2 = 2.8 + 6.2$$
$$1.8x = 9.0$$
$$\frac{1.8x}{1.8} = \frac{9.0}{1.8}$$
$$x = 5$$

$$3.1(x - 2) = 1.3x + 2.8$$
$$3.1(5 - 2) \stackrel{?}{=} 1.3(5) + 2.8$$
$$3.1(3) \stackrel{?}{=} 6.5 + 2.8$$
$$9.3 = 9.3$$

SECTION 2.3

65.

$$2.7(y + 1) = 0.3(3y + 33)$$
$$2.7y + 2.7 = 0.9y + 9.9$$
$$2.7y - 0.9y + 2.7 = 0.9y - 0.9y + 9.9$$
$$1.8y + 2.7 = 9.9$$
$$1.8y + 2.7 - 2.7 = 9.9 - 2.7$$
$$1.8y = 7.2$$
$$\frac{1.8y}{1.8} = \frac{7.2}{1.8}$$
$$y = 4$$

$$2.7(y + 1) = 0.3(3y + 33)$$
$$2.7(4 + 1) = 0.3[3(4) + 33]$$
$$2.7(5) \stackrel{?}{=} 0.3(12 + 33)$$
$$13.5 \stackrel{?}{=} 0.3(45)$$
$$13.5 = 13.5$$

67.

$$7x + 5(3 - x) = 2(x + 5) + 5$$
$$7x + 15 - 5x = 2x + 10 + 5$$
$$2x + 15 = 2x + 15$$
$$2x - 2x + 15 = 2x - 2x + 15$$
$$15 = 15$$

Identity, $\mathbb{R}$

69.

$$2(s + 2) = 2(s + 1) + 3$$
$$2s + 4 = 2s + 2 + 3$$
$$2s + 4 = 2s + 5$$
$$2s - 2s + 4 = 2s - 2s + 5$$
$$4 \neq 5$$

Contradiction, $\emptyset$

71.

$$\frac{5(x + 3)}{3} - x = \frac{2(x + 8)}{3}$$
$$3\left[\frac{5(x + 3)}{3} - x\right] = 3 \cdot \frac{2(x + 8)}{3}$$
$$3 \cdot \frac{5(x + 3)}{3} - 3 \cdot x = 2(x + 8)$$
$$5(x + 3) - 3x = 2x + 16$$
$$5x + 15 - 3x = 2x + 16$$
$$2x + 15 = 2x + 16$$
$$2x - 2x + 15 = 2x - 2x + 16$$
$$15 \neq 16$$

Contradiction, $\emptyset$

73.

$$x + 7 = \frac{2x + 6}{2} + 4$$
$$2(x + 7) = 2\left[\frac{2x + 6}{2} + 4\right]$$
$$2x + 14 = 2 \cdot \frac{2x + 6}{2} + 2 \cdot 4$$
$$2x + 14 = 2x + 6 + 8$$
$$2x + 14 = 2x + 14$$
$$2x - 2x + 14 = 2x - 2x + 14$$
$$14 = 14$$

Identity, $\mathbb{R}$

75. Expression:

$$2(x - y) - (x + y) + y = 2(x - y) - 1(x + y) + y = 2 \cdot x + 2(-y) + (-1) \cdot x + (-1) \cdot y + y$$
$$= 2x - 2y - x - y + y$$
$$= x - 2y$$

77.   Equation:

$$\frac{4(2x - 10)}{3} = 2(x - 4)$$
$$3 \cdot \frac{4(2x - 10)}{3} = 3 \cdot 2(x - 4)$$
$$4(2x - 10) = 6(x - 4)$$
$$8x - 40 = 6x - 24$$
$$8x - 6x - 40 = 6x - 6x - 24$$

$$8x - 6x - 40 = 6x - 6x - 24$$
$$2x - 40 = -24$$
$$2x - 40 + 40 = -24 + 40$$
$$2x = 16$$
$$\frac{2x}{2} = \frac{16}{2}$$
$$x = 8$$

79. Expression:

$$2\left(4x + \frac{9}{2}\right) - 3\left(x + \frac{2}{3}\right) = 2 \cdot 4x + 2 \cdot \frac{9}{2} + (-3) \cdot x + (-3) \cdot \frac{2}{3} = 8x + 9 - 3x - 2 = 5x + 7$$

81.   Equation:

$$\frac{8(5 - q)}{5} = -2q$$

$$5 \cdot \frac{8(5 - q)}{5} = 5(-2q)$$

$$8(5 - q) = -10q$$

$$40 - 8q = -10q$$

$$40 - 8q + 8q = -10q + 8q$$

$$40 = -2q$$

$$\frac{40}{-2} = \frac{-2q}{-2}$$

$$-20 = q$$

83.   Equation:

$$\frac{3x + 14}{2} = x - 2 + \frac{x + 18}{2}$$

$$2 \cdot \frac{3x + 14}{2} = 2\left[x - 2 + \frac{x + 18}{2}\right]$$

$$3x + 14 = 2x - 2 \cdot 2 + 2 \cdot \frac{x + 18}{2}$$

$$3x + 14 = 2x - 4 + x + 18$$

$$3x + 14 = 3x + 14$$

$$3x - 3x + 14 = 3x - 3x + 14$$

$$14 = 14$$

Identity, $\mathbb{R}$

85.   Equation:

$$6 - 5r = 7r$$

$$6 - 5r + 5r = 7r + 5r$$

$$6 = 12r$$

$$\frac{6}{12} = \frac{12r}{12}$$

$$\frac{1}{2} = r$$

87.   Equation:

$$22 - 3r = 8r$$

$$22 - 3r + 3r = 8r + 3r$$

$$22 = 11r$$

$$\frac{22}{11} = \frac{11r}{11}$$

$$2 = r$$

89. Expression:

$$8(x + 3) - 3x = 8 \cdot x + 8 \cdot 3 - 3x$$
$$= 8x + 24 - 3x$$
$$= 5x + 24$$

91.   Equation:

$$19.1x - 4(x + 0.3) = -46.5$$

$$19.1x - 4x - 1.2 = -46.5$$

$$15.1x - 1.2 = -46.5$$

$$15.1x - 1.2 + 1.2 = -46.5 + 1.2$$

$$15.1x = -45.3$$

$$\frac{15.1x}{15.1} = \frac{-45.3}{15.1}$$

$$x = -3$$

93. Expression: $3.2(m + 1.3) - 2.5(m - 7.2) = 3.2m + 4.16 - 2.5m + 18 = 0.7m + 22.16$

95.   Equation:

$$14.3(x + 2) + 13.7(x - 3) = 15.5 \qquad 28x - 12.5 + 12.5 = 15.5 + 12.5$$

$$14.3x + 28.6 + 13.7x - 41.1 = 15.5 \qquad 28x = 28$$

$$28.0x - 12.5 = 15.5 \qquad \frac{28x}{28} = \frac{28}{28}$$

$$28x - 12.5 + 12.5 = 15.5 + 12.5 \qquad x = 1$$

97. Equation:
$$10x + 3(2 - x) = 5(x + 2) - 4$$
$$10x + 6 - 3x = 5x + 10 - 4$$
$$7x + 6 = 5x + 6$$
$$7x - 5x + 6 = 5x - 5x + 6$$
$$2x + 6 = 6$$
$$2x + 6 - 6 = 6 - 6$$
$$2x = 0$$
$$\frac{2x}{2} = \frac{0}{2}$$
$$x = 0$$

99.
$$\frac{3.7(2.3x - 2.7)}{1.5} = 5.2(x - 1.2)$$
$$1.5 \cdot \frac{3.7(2.3x - 2.7)}{1.5} = 1.5(5.2)(x - 1.2)$$
$$3.7(2.3x - 2.7) = 7.8(x - 1.2)$$
$$8.51x - 9.99 = 7.8x - 9.36$$
$$8.51x - 7.8x - 9.99 = 7.8x - 7.8x - 9.36$$
$$0.71x - 9.99 = -9.36$$
$$0.71x - 9.99 + 9.99 = -9.36 + 9.99$$
$$0.71x = 0.63$$
$$\frac{0.71x}{0.71} = \frac{0.63}{0.71}$$
$$x \approx 0.887 \approx 0.9$$

101. They are like terms because the variable parts are exactly the same.

103. $7xxy^3 = 7x^2y^3$; $5x^2yyy = 5x^2y^3$
They are like terms because the variable parts are exactly the same.

105. Let $x =$ the number.
$$x = 2x$$
$$x - x = 2x - x$$
$$0 = x \Rightarrow \text{The number is 0.}$$

Exercises 2.4 (page 113)

1.
$$ab + c = 0$$
$$ab + c - c = 0 - c$$
$$ab = -c$$
$$\frac{ab}{b} = \frac{-c}{b}$$
$$a = -\frac{c}{b}$$

3.
$$a = \frac{b}{c}$$
$$ac = \frac{b}{c} \cdot c$$
$$ac = b, \text{ or } b = ac$$

41

5. $7x - 4y - 4x = 7x - 4x - 4y$
$$= (7-4)x - 4y = 3x - 4y$$

7. $\dfrac{2}{3}(a+3) - \dfrac{5}{3}(6+a) = \dfrac{2}{3}a + \dfrac{2}{3}\cdot 3 - \dfrac{5}{3}\cdot 6 - \dfrac{5}{3}a = \dfrac{2}{3}a + 2 - 10 - \dfrac{5}{3}a = \left(\dfrac{2}{3} - \dfrac{5}{3}\right)a - 8$

$$= -\dfrac{3}{3}a - 8$$
$$= -a - 8$$

9. literal

11. isolate

13. subtract

15. $E = IR$
$$\dfrac{E}{R} = \dfrac{IR}{R}$$
$$\dfrac{E}{R} = I, \text{ or } I = \dfrac{E}{R}$$

17. $V = lwh$
$$\dfrac{V}{lh} = \dfrac{lwh}{lh}$$
$$\dfrac{V}{lh} = w, \text{ or } w = \dfrac{V}{lh}$$

19. $x = y + 12$
$$x - 12 = y + 12 - 12$$
$$x - 12 = y, \text{ or } y = x - 12$$

21. $V = \dfrac{1}{3}Bh$
$$3V = 3 \cdot \dfrac{1}{3}Bh$$
$$3V = Bh$$
$$\dfrac{3V}{B} = \dfrac{Bh}{B}$$
$$\dfrac{3V}{B} = h, \text{ or } h = \dfrac{3V}{B}$$

23. $V = \dfrac{1}{3}\pi r^2 h$
$$3 \cdot V = 3 \cdot \dfrac{1}{3}\pi r^2 h$$
$$3V = \pi r^2 h$$
$$\dfrac{3V}{\pi r^2} = \dfrac{\pi r^2 h}{\pi r^2}$$
$$\dfrac{3V}{\pi r^2} = h, \text{ or } h = \dfrac{3V}{\pi r^2}$$

25. $y = \dfrac{1}{2}(x+2)$
$$2y = 2 \cdot \dfrac{1}{2}(x+2)$$
$$2y = x + 2$$
$$2y - 2 = x + 2 - 2$$
$$2y - 2 = x, \text{ or } x = 2y - 2$$

27. $A = \dfrac{5}{2}(B+3)$
$$\dfrac{2}{5}A = \dfrac{2}{5} \cdot \dfrac{5}{2}(B+3)$$
$$\dfrac{2}{5}A = B + 3$$
$$\dfrac{2}{5}A - 3 = B + 3 - 3$$
$$\dfrac{2}{5}A - 3 = B, \text{ or } B = \dfrac{2}{5}A - 3$$

29. $p = \dfrac{h}{2}(q+r)$
$$\dfrac{2}{h} \cdot p = \dfrac{2}{h} \cdot \dfrac{h}{2}(q+r)$$
$$\dfrac{2p}{h} = q + r$$
$$\dfrac{2p}{h} - r = q + r - r$$
$$\dfrac{2p}{h} - r = q, \text{ or } q = \dfrac{2p}{h} - r$$

42

31.
$$G = 2b(r - 1)$$
$$\frac{G}{2b} = \frac{2b(r - 1)}{2b}$$
$$\frac{G}{2b} = r - 1$$
$$\frac{G}{2b} + 1 = r - 1 + 1$$
$$\frac{G}{2b} + 1 = r, \text{ or } r = \frac{G}{2b} + 1$$

33.
$$d = rt \qquad t = \frac{d}{r}$$
$$\frac{d}{r} = \frac{rt}{r} \qquad t = \frac{455}{65}$$
$$\frac{d}{r} = t \qquad t = 7$$

35.
$$P = a + b + c \qquad b = P - a - c$$
$$P - a = a - a + b + c \qquad b = 37 - 15 - 6$$
$$P - a = b + c \qquad b = 22 - 6$$
$$P - a - c = b + c - c \qquad b = 16$$
$$P - a - c = b$$

37.
$$3x + 2y = 5$$
$$3x - 3x + 2y = 5 - 3x$$
$$2y = 5 - 3x$$
$$\frac{2y}{2} = \frac{5 - 3x}{2}$$
$$y = \frac{5 - 3x}{2}$$

39.
$$C = \pi d$$
$$\frac{C}{\pi} = \frac{\pi d}{\pi}$$
$$\frac{C}{\pi} = d, \text{ or } d = \frac{C}{\pi}$$

41.
$$P = 2l + 2w$$
$$P - 2l = 2l - 2l + 2w$$
$$P - 2l = 2w$$
$$\frac{P - 2l}{2} = \frac{2w}{2}$$
$$\frac{P - 2l}{2} = w, \text{ or } w = \frac{P - 2l}{2}$$

43.
$$A = P + Prt$$
$$A - P = P - P + Prt$$
$$A - P = Prt$$
$$\frac{A - P}{Pr} = \frac{Prt}{Pr}$$
$$\frac{A - P}{Pr} = t, \text{ or } t = \frac{A - P}{Pr}$$

45.
$$K = \frac{wv^2}{2g}$$
$$2g \cdot K = 2g \cdot \frac{wv^2}{2g}$$
$$2gK = wv^2$$
$$\frac{2gK}{v^2} = \frac{wv^2}{v^2}$$
$$\frac{2gK}{v^2} = w, \text{ or } w = \frac{2gK}{v^2}$$

47.
$$K = \frac{wv^2}{2g}$$
$$2g \cdot K = 2g \cdot \frac{wv^2}{2g}$$
$$2gK = wv^2$$
$$\frac{2gK}{2K} = \frac{wv^2}{2K}$$
$$g = \frac{wv^2}{2K}$$

49.
$$F = \frac{GMm}{d^2}$$
$$d^2 \cdot F = d^2 \cdot \frac{GMm}{d^2}$$
$$d^2 F = GMm$$
$$\frac{d^2 F}{Gm} = \frac{GMm}{Gm}$$
$$\frac{d^2 F}{Gm} = M, \text{ or } M = \frac{d^2 F}{Gm}$$

51.
$$i = prt \qquad p = \frac{i}{rt}$$
$$\frac{i}{rt} = \frac{prt}{rt} \qquad p = \frac{90}{0.03(4)}$$
$$\frac{i}{rt} = p \qquad p = \frac{90}{0.12} = 750$$

53.
$$K = \frac{1}{2}h(a+b) \qquad h = \frac{2K}{a+b}$$
$$2 \cdot K = 2 \cdot \frac{1}{2}h(a+b) \qquad h = \frac{2(48)}{7+5}$$
$$2K = h(a+b) \qquad h = \frac{96}{12} = 8$$
$$\frac{2K}{a+b} = \frac{h(a+b)}{a+b}$$
$$\frac{2K}{a+b} = h$$

55.
$$V = \frac{1}{3}\pi r^2 h \qquad h = \frac{3V}{\pi r^2}$$
$$3 \cdot V = 3 \cdot \frac{1}{3}\pi r^2 h \qquad h = \frac{3(36\pi)}{\pi(6)^2}$$
$$3V = \pi r^2 h$$
$$\frac{3V}{\pi r^2} = \frac{\pi r^2 h}{\pi r^2} \qquad h = \frac{108\pi}{36\pi} = 3 \text{ inches}$$
$$\frac{3V}{\pi r^2} = h$$

57.
$$E = IR \qquad I = \frac{E}{R}$$
$$\frac{E}{R} = \frac{IR}{R} \qquad I = \frac{48}{12}$$
$$\frac{E}{R} = I \qquad I = 4 \text{ amperes}$$

59.
$$P = I^2 R \qquad R = \frac{P}{I^2}$$
$$\frac{P}{I^2} = \frac{I^2 R}{I^2} \qquad R = \frac{2700}{14^2}$$
$$\frac{P}{I^2} = R \qquad R = \frac{2700}{196} = 13.78 \text{ ohms}$$

61.
$$F = \frac{GMm}{d^2}$$
$$d^2 \cdot F = d^2 \cdot \frac{GMm}{d^2}$$
$$d^2 F = GMm$$
$$\frac{Fd^2}{GM} = \frac{GMm}{GM}$$
$$\frac{Fd^2}{GM} = m, \text{ or } m = \frac{Fd^2}{GM}$$

63.
$$L = 2D + 3.25(r+R)$$
$$L - 3.25(r+R) = 2D + 3.25(r+R) - 3.25(r+R)$$
$$L - 3.25(r+R) = 2D$$
$$\frac{L - 3.25(r+R)}{2} = \frac{2D}{2}$$
$$\frac{L - 3.25r - 3.25R}{2} = D$$

$$D = \frac{L - 3.25r - 3.25R}{2}$$
$$D = \frac{25 - 3.25(1) - 3.25(3)}{2}$$
$$D = \frac{25 - 3.25 - 9.75}{2} = \frac{12}{2} = 6 \text{ ft}$$

65.
$$C = 0.15(T - C)$$
$$C = 0.15T - 0.15C$$
$$C + 0.15C = 0.15T - 0.15C + 0.15C$$
$$1.15C = 0.15T$$
$$\frac{1.15C}{1.15} = \frac{0.15T}{1.15}$$
$$C \approx 0.1304T$$

The maximum contribution is about 13% of taxable income.

67. **Answers may vary.**

69. $E = mc^2$
$$E = 1(300,000)^2$$
$$E = 90,000,000,000 \text{ joules}$$

Exercises 2.5 (page 122)

1. $12 - x$

3. $I = Pr = (18000 - x)(0.07)$
$$= 0.07(18000 - x)$$

5. $V = \frac{1}{3}Bh = \frac{1}{3}s^2h = \frac{1}{3}(10\,\text{cm})^2(6\,\text{cm}) = \frac{1}{3}(100\,\text{cm}^2)(6\,\text{cm}) = \frac{1}{3}(600\,\text{cm}^3) = 200\,\text{cm}^3$

7. $5(x - 2) + 3(x + 4) = 5 \cdot x + 5(-2) + 3 \cdot x + 3 \cdot 4 = 5x - 10 + 3x + 12 = 8x + 2$

9. $\frac{1}{2}(x + 1) - \frac{1}{2}(x + 4) = \frac{1}{2} \cdot x + \frac{1}{2} \cdot 1 - \frac{1}{2} \cdot x - \frac{1}{2} \cdot 4 = \frac{1}{2}x + \frac{1}{2} - \frac{1}{2}x - \frac{4}{2} = -\frac{3}{2}$

11. $A = P + Prt$
$$= 1200 + 1200(0.08)(3)$$
$$= 1200 + 288 = \$1488$$

13. $2l + 2w$

15. vertex

17. degrees

19. straight

21. supplementary

23. Let x = the length of one part.
Then $2x$ = the length of the other part.

$$\boxed{\text{Length of first part}} + \boxed{\text{Length of second part}} = 12$$
$$x + 2x = 12$$
$$3x = 12$$
$$\frac{3x}{3} = \frac{12}{3}$$
$$x = 4$$

The parts are 4 feet and 8 feet long.

25.
$$\boxed{\text{Sum of 3 lengths}} = 30$$
$$x + 10 + 2x + x = 30$$
$$4x + 10 = 30$$
$$4x + 10 - 10 = 30 - 10$$
$$4x = 20$$
$$\frac{4x}{4} = \frac{20}{4}$$
$$x = 5$$

The sections are 15, 10, and 5 feet long.

27. $\boxed{\text{Sum of 3 lengths}} = 24$

$x + x + 4 + x + 2 = 24$

$3x + 6 = 24$

$3x + 6 - 6 = 24 - 6$

$3x = 18$

$\dfrac{3x}{3} = \dfrac{18}{3}$

$x = 6$

The sections are 6, 8, and 10 feet long.

29. Let $x =$ the number of hardcover books.
Then $14x =$ the number of paperbacks.

$\boxed{\text{Number of hardcover books}} + \boxed{\text{Number of paperbacks}} = 210{,}000$

$x + 14x = 210{,}000$

$15x = 210{,}000$

$\dfrac{15x}{15} = \dfrac{210{,}000}{15}$

$x = 14{,}000$

There were 14,000 hardcover books and 196,000 paperbacks sold.

31. $x + 40° = 50°$

$x + 40° - 40° = 50° - 40°$

$x = 10°$

33. $x + 12° = 59°$

$x + 12° - 12° = 59° - 12°$

$x = 47°$

35. $x + 21° = 180°$

$x + 21° - 21° = 180° - 21°$

$x = 159°$

37. $x + 93° = 180°$

$x + 93° - 93° = 180° - 93°$

$x = 87°$

39. Let $x =$ the measure of the complement.

$x + 46° = 90°$

$x + 46° - 46° = 90° - 46°$

$x = 44°$

41. Let $x =$ the measure of the complement.

$x + 40° = 90°$

$x + 40° - 40° = 90° - 40°$

$x = 50°$

Let $x =$ the measure of the supplement of the complement.

$x + 50° = 180°$

$x + 50° - 50° = 180° - 50°$

$x = 130°$

43. $\boxed{\text{Perimeter}} = 90$

$2(w + 7) + 2w = 90$

$2w + 14 + 2w = 90$

$4w + 14 = 90$

$4w + 14 - 14 = 90 - 14$

$4w = 76$

$\dfrac{4w}{4} = \dfrac{76}{4}$

$w = 19$

The dimensions are 19 cm by 26 cm.

45. Let $w =$ the width of the picture.
Then $2w - 6 =$ the length of the picture.

$\boxed{\text{Perimeter}} = 60$

$2w + 2(2w - 6) = 60$

$2w + 4w - 12 = 60$

$6w - 12 = 60$

$6w - 12 + 12 = 60 + 12$

$6w = 72$

$\dfrac{6w}{6} = \dfrac{72}{6}$

$w = 12$

The dimensions are 12 in. by 18 in.

47. $\boxed{\text{Sum of three sides}} = 57$

$x + x + x = 57$

$3x = 57$

$\dfrac{3x}{3} = \dfrac{57}{3}$

$x = 19$

Each side is 19 feet long.

49. Let $a =$ the measure of the vertex angle.

Then $4a =$ the measure of the other angles.

$\boxed{\text{Sum of angle measures}} = 180$

$a + 4a + 4a = 180$

$9a = 180$

$\dfrac{9a}{9} = \dfrac{180}{9}$

$a = 20$

The vertex angle measures $20°$.

51. Let $x =$ amount invested at 5%.

$I = PRT$

$300 = x(0.05)(1)$

$\dfrac{300}{0.05} = \dfrac{0.05x}{0.05}$

$6000 = x$

He invested $6,000.

53. Let $x =$ amount in 9% fund. Then $24000 - x =$ amount in 14% fund.

$\boxed{\text{Interest at 9\%}} + \boxed{\text{Interest at 14\%}} = \boxed{\text{Total interest}}$

$0.09x + 0.14(24000 - x) = 3135$

$9x + 14(24000 - x) = 313500$

$9x + 336000 - 14x = 313500$

$-5x + 336000 = 313500$

$-5x + 336000 - 336000 = 313500 - 336000$

$-5x = -22500$

$\dfrac{-5x}{-5} = \dfrac{-22500}{-5}$

$x = 4500 \Rightarrow$ $4,500 was invested at 9%, and $19,500 was invested at 14%.

55. Let $x =$ amount invested in each account.

$\boxed{\text{Interest at 8\%}} + \boxed{\text{Interest at 11\%}} = \boxed{\text{Total interest}}$

$0.08x + 0.11x = 712.50$

$8x + 11x = 71250$

$19x = 71250$

$\dfrac{19x}{19} = \dfrac{71250}{19}$

$x = 3750 \Rightarrow$ $3,750 is invested in each account.

57. Let $x =$ amount invested at 7%.

$$\boxed{\begin{array}{c}\text{Interest}\\\text{at 6\%}\end{array}} + \boxed{\begin{array}{c}\text{Interest}\\\text{at 7\%}\end{array}} = \boxed{\begin{array}{c}\text{Total}\\\text{interest}\end{array}}$$

$$0.06(15000) + 0.07x = 1250$$
$$6(15000) + 7x = 125000$$
$$90000 + 7x = 125000$$
$$90000 - 90000 + 7x = 125000 - 90000$$
$$7x = 35000$$
$$\frac{7x}{7} = \frac{35000}{7}$$
$$x = 5000$$

$5,000 should be invested at 7%.

59. Let $r =$ the fund rate and $r + 0.01 =$ the CD rate.
The client invests $21,000 in CDs, and $10,500 in the fund.

$$\boxed{\begin{array}{c}\text{CD}\\\text{interest}\end{array}} = \boxed{\begin{array}{c}\text{Fund}\\\text{interest}\end{array}} + 840$$

$$(r + 0.01)21000 = r(10500) + 840$$
$$21000r + 210 = 10500r + 840$$
$$21000r - 10500r + 210 = 10500r - 10500r + 840$$
$$10500r + 210 = 840$$
$$10500r + 210 - 210 = 840 - 210$$
$$10500r = 630$$
$$\frac{10500r}{10500} = \frac{630}{10500}$$
$$r = 0.06 \Rightarrow \text{The rates are 6\% and 7\%.}$$

61. Answers may vary.

63. Pairs of vertical angles have equal measures.

Exercises 2.6 (page 130)

1. $d = rt = 60h$ mi

3. $0.25(20) = 5$ oz

5. $5 + 2(-7) = 5 + (-14) = -9$

7. $3^3 - 5^2 = 3 \cdot 3 \cdot 3 - 5 \cdot 5 = 27 - 25 = 2$

9.
$$-6x + 4 = -8$$
$$-6x + 4 - 4 = -8 - 4$$
$$-6x = -12$$
$$\frac{-6x}{-6} = \frac{-12}{-6}$$
$$x = 2$$

11.
$$\frac{2}{3}p + 1 = 5$$
$$\frac{2}{3}p + 1 - 1 = 5 - 1$$
$$\frac{2}{3}p = 4$$
$$\frac{3}{2} \cdot \frac{2}{3}p = \frac{3}{2} \cdot 4$$
$$p = \frac{12}{2}$$
$$p = 6$$

SECTION 2.6

13. $d = rt$

15. $v = pn$

17. Let t = time for cars to meet.

	r	t	d
Car 1 (A to B)	50	t	$50t$
Car 2 (B to A)	55	t	$55t$

$$\boxed{\text{Distance for car 1}} + \boxed{\text{Distance for car 2}} = 315$$
$$50t + 55t = 315$$
$$105t = 315$$
$$\frac{105t}{105} = \frac{315}{105}$$
$$t = 3$$

The cars meet after 3 hours.

19. Let t = days for crews to meet.

	r	t	d
Crew 1	1.5	t	$1.5t$
Crew 2	1.2	t	$1.2t$

$$\boxed{\text{Distance for crew 1}} + \boxed{\text{Distance for crew 2}} = 9.45$$
$$1.5t + 1.2t = 9.45$$
$$2.7t = 9.45$$
$$\frac{2.7t}{2.7} = \frac{9.45}{2.7}$$
$$t = 3.5$$

The crews meet after 3.5 days.

21. Let t = time for cars to be 715 miles apart.

	r	t	d
Car 1 (going east)	60	t	$60t$
Car 2 (going west)	50	t	$50t$

$$\boxed{\text{Distance 1}} + \boxed{\text{Distance 2}} = 715$$
$$60t + 50t = 715$$
$$110t = 715$$
$$\frac{110t}{110} = \frac{715}{110}$$
$$t = 6.5$$

They will be 715 miles apart after 6.5 hours.

23. Let t = time for boys to be 2 miles apart.

	r	t	d
Boy 1	3	t	$3t$
Boy 2	4	t	$4t$

$$\boxed{\text{Distance 1}} + \boxed{\text{Distance 2}} = 2$$
$$3t + 4t = 2$$
$$7t = 2$$
$$\frac{7t}{7} = \frac{2}{7}$$
$$t = \frac{2}{7}$$

They will lose contact after $\frac{2}{7}$ hour, or approximately 17 minutes.

25.

	r	t	d
Car	60 mph	t	$60t$
Bus	50 mph	$t + 2$	$50(t + 2)$

$$\boxed{\text{Car distance}} = \boxed{\text{Bus distance}}$$
$$60t = 50(t + 2)$$
$$60t = 50t + 100$$
$$60t - 50t = 50t - 50t + 100$$
$$10t = 100$$
$$\frac{10t}{10} = \frac{100}{10}$$
$$t = 10$$

The car will overtake the bus after 10 hours.

27. Let t = time for cars to be 82.5 miles apart.

	r	t	d
Car 1	42	t	$42t$
Car 2	53	t	$53t$

$$\boxed{\text{Distance 2}} - \boxed{\text{Distance 1}} = 82.5$$
$$53t - 42t = 82.5$$
$$11t = 82.5$$
$$\frac{11t}{11} = \frac{82.5}{11}$$
$$t = 7.5$$

They will be 82.5 miles apart after 7.5 hours.

29. Let r = rate of slow train. Then
$r + 20$ = rate of fast train.

	r	t	d
Slow train	r	3	$3r$
Fast train	$r + 20$	3	$3(r + 20)$

$$\boxed{\text{Slow dist.}} + \boxed{\text{Fast dist.}} = 330$$
$$3r + 3(r + 20) = 330$$
$$3r + 3r + 60 = 330$$
$$6r + 60 = 330$$
$$6r + 60 - 60 = 330 - 60$$
$$6r = 270$$
$$\frac{6r}{6} = \frac{270}{6}$$
$$r = 45$$

The rates are 45 mph and 65 mph.

31. Let t = slower time.
Then $5 - t$ = faster time.

	r	t	d
1st part	40	t	$40t$
2nd part	50	$5 - t$	$50(5 - t)$

$$\boxed{\text{1st dist.}} + \boxed{\text{2nd dist.}} = 210$$
$$40t + 50(5 - t) = 210$$
$$40t + 250 - 50t = 210$$
$$-10t + 250 = 210$$
$$-10t + 250 - 250 = 210 - 250$$
$$-10t = -40$$
$$\frac{-10t}{-10} = \frac{-40}{-10}$$
$$t = 4$$

The car averaged 40 mph for 4 hours.

33. Let T = total number of liters of solution.
12% of the total = liters of acid
$$0.12T = 0.3$$
$$\frac{0.12T}{0.12} = \frac{0.3}{0.12}$$
$$T = 2.5$$

There are 2.5 liters of the solution.

35. Let x = gallons of \$3.35 fuel.

$$\boxed{\substack{\text{Value of} \\ \text{\$3.35 fuel}}} + \boxed{\substack{\text{Value of} \\ \text{\$3.85 fuel}}} = \boxed{\substack{\text{Value of} \\ \text{mixture}}}$$
$$3.35x + 3.85(20) = 3.55(20 + x)$$
$$335x + 385(20) = 355(20 + x)$$
$$335x + 7700 = 7100 + 355x$$
$$335x - 335x + 7700 = 7100 + 355x - 335x$$
$$7700 = 7100 + 20x$$
$$7700 - 7100 = 7100 - 7100 + 20x$$
$$600 = 20x$$
$$\frac{600}{20} = \frac{20x}{20}$$
$$30 = x \Rightarrow 30 \text{ gallons of the \$3.35 fuel should be used.}$$

37. Let x = gallons of 3% solution used.

$$\boxed{\text{Salt in 3\% solution}} + \boxed{\text{Salt in 7\% solution}} = \boxed{\text{Salt in mixture}}$$

$$0.03x + 0.07(50) = 0.05(x + 50)$$
$$3x + 7(50) = 5(x + 50)$$
$$3x + 350 = 5x + 250$$
$$3x - 5x + 350 = 5x - 5x + 250$$
$$-2x + 350 = 250$$
$$-2x + 350 - 350 = 250 - 350$$
$$-2x = -100$$
$$\frac{-2x}{-2} = \frac{-100}{-2}$$
$$x = 50$$

50 gallons of the 3% mixture should be used.

39. Let x = ounces of water (0%) added.

$$\boxed{\text{Amt. in 10\% sol.}} + \boxed{\text{Amt. in water}} = \boxed{\text{Amt. in 8\% sol.}}$$

$$0.10(30) + 0(x) = 0.08(30 + x)$$
$$10(30) + 0 = 8(30 + x)$$
$$300 = 240 + 8x$$
$$300 - 240 = 240 - 240 + 8x$$
$$60 = 8x$$
$$\frac{60}{8} = \frac{8x}{8}$$
$$7.5 = x$$

7.5 ounces of water should be added.

41. Let x = pounds of lemon drops. Then $100 - x$ = pounds of jelly beans.

$$\boxed{\text{Value of lemon drops}} + \boxed{\text{Value of jelly beans}} = \boxed{\text{Value of mixture}}$$

$$1.90x + 1.20(100 - x) = 1.48(100)$$
$$190x + 120(100 - x) = 148(100)$$
$$190x + 12000 - 120x = 14800$$
$$70x + 12000 = 14800$$
$$70x + 12000 - 12000 = 14800 - 12000$$
$$70x = 2800$$
$$\frac{70x}{70} = \frac{2800}{70}$$
$$x = 40 \Rightarrow 40 \text{ lb of lemon drops and 60 lb of jelly beans should be used.}$$

43. Let c = cost of cashews. Then $c - 0.30$ = cost of peanuts. 20 pounds of each are used.

$$\boxed{\text{Value of cashews}} + \boxed{\text{Value of peanuts}} = \boxed{\text{Value of mixture}}$$

$$20c + 20(c - 0.30) = 1.05(40)$$
$$20c + 20c - 6 = 42$$
$$40c - 6 = 42$$
$$40c - 6 + 6 = 42 + 6$$
$$40c = 48$$
$$\frac{40c}{40} = \frac{48}{40}$$
$$c = 1.20$$

A bag of cashews is worth $1.20.

45. Let x = pounds of regular coffee used.

$$\boxed{\text{Value of regular}} + \boxed{\text{Value of gourmet}} = \boxed{\text{Value of mixture}}$$

$$7(x) + 12(30) = 9(x + 30)$$
$$7x + 360 = 9x + 270$$
$$7x - 9x + 360 = 9x - 9x + 270$$
$$-2x + 360 = 270$$
$$-2x + 360 - 360 = 270 - 360$$
$$-2x = -90$$
$$\frac{-2x}{-2} = \frac{-90}{-2}$$
$$x = 45$$

45 pounds of regular coffee should be used.

47. Let $c =$ cost of hazelnut beans per pound.

$$\boxed{\begin{array}{c}\text{Value of}\\\text{chocolate}\end{array}} + \boxed{\begin{array}{c}\text{Value of}\\\text{hazelnut}\end{array}} = \boxed{\begin{array}{c}\text{Value of}\\\text{mixture}\end{array}}$$

$$7(2) + c(5) = 6(7)$$
$$14 + 5c = 42$$
$$14 - 14 + 5c = 42 - 14$$
$$5c = 28$$
$$c = \tfrac{28}{5} = \$5.60 \Rightarrow \text{The hazelnut beans cost \$5.60 per pound.}$$

49. Answers may vary.   **51. Answers may vary.**

53. Let x and $x + 2$ represent the integers.

$$x + x + 2 = 16$$
$$2x + 2 = 16$$
$$2x = 14$$
$$x = 7$$

This says that the integers are 7 and 9, but these are not **even** integers. The equation has a solution, but not the problem itself.

55. You cannot mix a 10% solution and a 20% solution and end up with a solution that has a greater concentration (30%) than either of the original solutions.

Exercises 2.7 (page 137)

1. $2x < 4$
same

3. $-3x \le -6$
reverses

5. $2x - 5 < 7$
same

7. $3x^2 - 2(y^2 - x^2) = 3x^2 + (-2)y^2 - (-2)x^2 = 3x^2 - 2y^2 + 2x^2 = 5x^2 - 2y^2$

9. $\frac{1}{3}(x + 6) - \frac{4}{3}(x - 9) = \frac{1}{3}x + \frac{1}{3} \cdot 6 - \frac{4}{3}x - \frac{4}{3}(-9) = -\frac{3}{3}x + 2 + 12 = -x + 14$

11. is less than, is greater than   **13.** double inequality   **15.** inequality

17.
$$x + 5 > 8$$
$$x + 5 - 5 > 8 - 5$$
$$x > 3$$

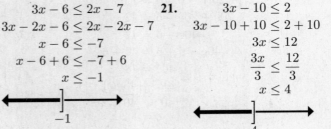

19.
$$3x - 6 \le 2x - 7$$
$$3x - 2x - 6 \le 2x - 2x - 7$$
$$x - 6 \le -7$$
$$x - 6 + 6 \le -7 + 6$$
$$x \le -1$$

21.
$$3x - 10 \le 2$$
$$3x - 10 + 10 \le 2 + 10$$
$$3x \le 12$$
$$\frac{3x}{3} \le \frac{12}{3}$$
$$x \le 4$$

23.
$$6x - 2 > 3x - 11$$
$$6x - 3x - 2 > 3x - 3x - 11$$
$$3x - 2 > -11$$
$$3x - 2 + 2 > -11 + 2$$
$$3x > -9$$
$$\frac{3x}{3} > \frac{-9}{3}$$
$$x > -3$$

⟵————(———⟶
     -3

25.
$$9x + 4 > 4x - 1$$
$$9x - 4x + 4 > 4x - 4x - 1$$
$$5x + 4 > -1$$
$$5x + 4 - 4 > -1 - 4$$
$$5x > -5$$
$$\frac{5x}{5} > \frac{-5}{5}$$
$$x > -1$$

⟵————(———⟶
     -1

27.
$$\frac{5}{2}(7x - 15) + x \geq \frac{13}{2}x - \frac{3}{2}$$
$$2\left[\frac{5}{2}(7x - 15) + x\right] \geq 2\left[\frac{13}{2}x - \frac{3}{2}\right]$$
$$2 \cdot \frac{5}{2}(7x - 15) + 2x \geq 2 \cdot \frac{13}{2}x - 2 \cdot \frac{3}{2}$$
$$5(7x - 15) + 2x \geq 13x - 3$$
$$35x - 75 + 2x \geq 13x - 3$$
$$37x - 75 \geq 13x - 3$$
$$37x - 13x - 75 \geq 13x - 13x - 3$$
$$24x - 75 \geq -3$$
$$24x - 75 + 75 \geq -3 + 75$$
$$24x \geq 72$$
$$\frac{24x}{24} \geq \frac{72}{24}$$
$$x \geq 3$$

⟵————[———⟶
     3

29.
$$-x - 3 \leq 7$$
$$-x - 3 + 3 \leq 7 + 3$$
$$-x \leq 10$$
$$-1(-x) \geq -1(10)$$
$$x \geq -10$$

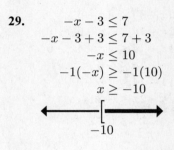

⟵————[———⟶
     -10

31.
$$-3x - 5 < 4$$
$$-3x - 5 + 5 < 4 + 5$$
$$-3x < 9$$
$$\frac{-3x}{-3} > \frac{9}{-3}$$
$$x > -3$$

⟵————(———⟶
     -3

33.
$$-5x + 17 > 37$$
$$-5x + 17 - 17 > 37 - 17$$
$$-5x > 20$$
$$\frac{-5x}{-5} < \frac{20}{-5}$$
$$x < -4$$

⟵————)———⟶
     -4

35.
$$-3x - 7 > -1$$
$$-3x - 7 + 7 > -1 + 7$$
$$-3x > 6$$
$$\frac{-3x}{-3} < \frac{6}{-3}$$
$$x < -2$$

37.
$$9 - 2x > 24 - 7x$$
$$9 - 2x + 7x > 24 - 7x + 7x$$
$$9 + 5x > 24$$
$$9 - 9 + 5x > 24 - 9$$
$$5x > 15$$
$$\frac{5x}{5} > \frac{15}{5}$$
$$x > 3$$

39.
$$2(x + 7) \leq 4x - 6$$
$$2x + 14 \leq 4x - 6$$
$$2x - 4x + 14 \leq 4x - 4x - 6$$
$$-2x + 14 \leq -6$$
$$-2x + 14 - 14 \leq -6 - 14$$
$$-2x \leq -20$$
$$\frac{-2x}{-2} \geq \frac{-20}{-2}$$
$$x \geq 10$$

41.
$$2 < \quad x - 5 \quad < 5$$
$$2 + 5 < x - 5 + 5 < 5 + 5$$
$$7 < \quad x \quad < 10$$

43.
$$-4 < \quad 2(x + 1) \quad \leq 12$$
$$-4 < \quad 2x + 2 \quad \leq 12$$
$$-4 - 2 < 2x + 2 - 2 \leq 12 - 2$$
$$-6 < \quad 2x \quad \leq 10$$
$$\frac{-6}{2} < \quad \frac{2x}{2} \quad \leq \frac{10}{2}$$
$$-3 < \quad x \quad \leq 5$$

45.
$$0 \leq \quad x + 10 \quad \leq 10$$
$$0 - 10 \leq x + 10 - 10 \leq 10 - 10$$
$$-10 \leq \quad x \quad \leq 0$$

47.
$$-6 < \quad 3(x + 2) \quad < 9$$
$$-6 < \quad 3x + 6 \quad < 9$$
$$-6 - 6 < 3x + 6 - 6 < 9 - 6$$
$$-12 < \quad 3x \quad < 3$$
$$\frac{-12}{3} < \quad \frac{3x}{3} \quad < \frac{3}{3}$$
$$-4 < \quad x \quad < 1$$

49.
$$8 + x > 7$$
$$8 - 8 + x > 7 - 8$$
$$x > -1$$

54

51.
$$7 - x \le 3x - 1$$
$$7 - x - 3x \le 3x - 3x - 1$$
$$7 - 4x \le -1$$
$$7 - 7 - 4x \le -1 - 7$$
$$-4x \le -8$$
$$\frac{-4x}{-4} \ge \frac{-8}{-4}$$
$$x \ge 2$$

53.
$$8(5 - x) \le 10(8 - x)$$
$$40 - 8x \le 80 - 10x$$
$$40 - 8x + 10x \le 80 - 10x + 10x$$
$$40 + 2x \le 80$$
$$40 - 40 + 2x \le 80 - 40$$
$$2x \le 40$$
$$\frac{2x}{2} \le \frac{40}{2}$$
$$x \le 20$$

55.
$$\frac{3x - 3}{2} < 2x + 2$$
$$2 \cdot \frac{3x - 3}{2} < 2(2x + 2)$$
$$3x - 3 < 4x + 4$$
$$3x - 4x - 3 < 4x - 4x + 4$$
$$-x - 3 < 4$$
$$-x - 3 + 3 < 4 + 3$$
$$-x < 7$$
$$-1(-x) > -1(7)$$
$$x > -7$$

57.
$$\frac{2(x + 5)}{3} \le 3x - 6$$
$$3 \cdot \frac{2(x + 5)}{3} \le 3(3x - 6)$$
$$2(x + 5) \le 9x - 18$$
$$2x + 10 \le 9x - 18$$
$$2x - 9x + 10 \le 9x - 9x - 18$$
$$-7x + 10 \le -18$$
$$-7x + 10 - 10 \le -18 - 10$$
$$-7x \le -28$$
$$\frac{-7x}{-7} \ge \frac{-28}{-7}$$
$$x \ge 4$$

59.
$$9 < -3x < 15$$
$$\frac{9}{-3} > \frac{-3x}{-3} > \frac{15}{-3}$$
$$-3 > x > -5$$
$$-5 < x < -3$$

61.
$$-3 \le \frac{x}{2} \le 5$$
$$2(-3) \le 2 \cdot \frac{x}{2} \le 2(5)$$
$$-6 \le x \le 10$$

63.
$$3 \le 2x - 1 < 5$$
$$3 + 1 \le 2x - 1 + 1 < 5 + 1$$
$$4 \le 2x < 6$$
$$\frac{4}{2} \le \frac{2x}{2} < \frac{6}{2}$$
$$2 \le x < 3$$

65.
$$0 < 10 - 5x \le 15$$
$$0 - 10 < 10 - 10 - 5x \le 15 - 10$$
$$-10 < -5x \le 5$$
$$\frac{-10}{-5} > \frac{-5x}{-5} \ge \frac{5}{-5}$$
$$2 > x \ge -1$$
$$-1 \le x < 2$$

67.
$$-4 < \frac{x-2}{2} < 6$$
$$2(-4) < 2 \cdot \frac{x-2}{2} < 2(6)$$
$$-8 < x - 2 < 12$$
$$-8 + 2 < x - 2 + 2 < 12 + 2$$
$$-6 < x < 14$$

$$\longleftarrow \!\!(\underset{-6}{\quad}\underset{14}{\quad})\!\!\longrightarrow$$

69. Let s = score on last exam.
$$\text{Average score} \geq 80$$
$$\frac{68 + 75 + 79 + s}{4} \geq 80$$
$$4 \cdot \frac{68 + 75 + 79 + s}{4} \geq 4(80)$$
$$68 + 75 + 79 + s \geq 320$$
$$222 + s \geq 320$$
$$222 - 222 + s \geq 320 - 222$$
$$s \geq 98$$
Her last test score must be at least 98%.

71. Let s = length of each side.
$$\text{Perimeter} \geq 68$$
$$4s \geq 68$$
$$\frac{4s}{4} \geq \frac{68}{4}$$
$$s \geq 17$$
Each side must be at least 17 cm long.

73. Let r = rating of third model.
$$\text{Average rating} \geq 21$$
$$\frac{17 + 19 + r}{3} \geq 21$$
$$3 \cdot \frac{17 + 19 + r}{3} \geq 3(21)$$
$$17 + 19 + r \geq 63$$
$$36 + r \geq 63$$
$$36 - 36 + r \geq 63 - 36$$
$$r \geq 27$$
It must have a rating of at least 27 mpg.

75.
$$470 \text{ ft} \leq \text{range in feet} \leq 13{,}143 \text{ ft}$$
$$\frac{470 \text{ ft}}{5280} \leq \text{range in miles} \leq \frac{13143 \text{ ft}}{5280}$$
$$0.1 \text{ miles} \leq \text{range in miles} \leq 2.5 \text{ miles}$$
The range is from 0.1 miles to 2.5 miles.

77.
$$17500 \text{ ft} < \text{range in feet} < 21700 \text{ ft}$$
$$\frac{17500 \text{ ft}}{5280} < \text{range in miles} < \frac{21700 \text{ ft}}{5280}$$
$$3.3 \text{ miles} < \text{range in miles} < 4.1 \text{ miles}$$
The range is between 3.3 miles and 4.1 miles.

79.
$$23° < C < 26°$$
$$23° < \frac{5}{9}(F - 32) < 26°$$
$$\frac{9}{5}(23°) < \frac{9}{5} \cdot \frac{5}{9}(F - 32) < \frac{9}{5}(26°)$$
$$41.4° < F - 32 < 46.8°$$
$$41.4° + 32 < F - 32 + 32 < 46.8° + 32$$
$$73.4° < F < 78.8°$$
The Fahrenheit temperature is between 73.4° and 78.8°.

81.
$$5.9 \text{ in.} < r < 6.1 \text{ in.}$$
$$2\pi(5.9 \text{ in.}) < 2\pi \cdot r < 2\pi(6.1 \text{ in})$$
$$2(3.14)(5.9 \text{ in.}) < 2\pi r < 2(3.14)(6.1 \text{ in.})$$
$$37.052 \text{ in.} < C < 38.308 \text{ in.}$$
The circumference can vary between about 37.052 inches and 38.308 inches.

83.
$$150 \text{ lb} < \text{ range in lbs } < 190 \text{ lb}$$
$$\frac{150 \text{ lb}}{2.2} < \text{ range in kg } < \frac{190 \text{ lb}}{2.2}$$
$$68.18 \text{ kg} < \text{ range in kg } < 86.36 \text{ kg}$$
The weight is between 68.18 and 86.36 kg.

85. Let $w =$ width. Then $2w - 3 =$ length.
$$24 < \quad \text{perimeter} \quad < 48$$
$$24 < 2w + 2(2w - 3) < 48$$
$$24 < \quad 2w + 4w - 6 \quad < 48$$
$$24 < \quad 6w - 6 \quad < 48$$
$$24 + 6 < \quad 6w - 6 + 6 \quad < 48 + 6$$
$$30 < \quad 6w \quad < 54$$
$$\frac{30}{6} < \quad \frac{6w}{6} \quad < \frac{54}{6}$$
$$5 < \quad w \quad < 9$$
The width could be between 5 and 9 feet.

87. **Answers may vary.**

89. This is not correct because x could represent a negative number, and then multiplying both sides of the equation would require the inequality to change from $<$ to $>$.

Chapter 2 Review (page 142)

1. $3(x + 4)$
expression

2. $x + 4 = 3$
equation

3.
$$6 - 4x = 2$$
$$6 - 4(-1) \stackrel{?}{=} 2$$
$$6 + 4 \stackrel{?}{=} 2$$
$$10 \neq 2$$
-1 is not a solution.

4.
$$2(x + 3) = 3(x - 4)$$
$$2(18 + 3) \stackrel{?}{=} 3(18 - 4)$$
$$2(21) \stackrel{?}{=} 3(14)$$
$$42 = 42$$
18 is a solution.

5.
$$x - 6 = -7 \qquad x - 6 = -7$$
$$x - 6 + 6 = -7 + 6 \qquad -1 - 6 \stackrel{?}{=} -7$$
$$x = -1 \qquad -7 = -7$$

6.
$$-y - 3 = 10 \qquad -y - 3 = 10$$
$$-y - 3 + 3 = 10 + 3 \qquad -(-13) - 3 \stackrel{?}{=} 10$$
$$-y = 13 \qquad 13 - 3 \stackrel{?}{=} 10$$
$$-1(-y) = -1(13) \qquad 10 = 10$$
$$y = -13$$

7.
$$p + 5 = 9 \qquad p + 5 = 9$$
$$p + 5 - 5 = 9 - 5 \qquad 4 + 5 \stackrel{?}{=} 9$$
$$p = 4 \qquad 9 = 9$$

8.
$$c + 9 = -6 \qquad c + 9 = -6$$
$$c + 9 - 9 = -6 - 9 \qquad -15 + 9 \stackrel{?}{=} -6$$
$$c = -15 \qquad -6 = -6$$

9.
$$p + \frac{1}{2} = -\frac{1}{2} \qquad\qquad p + \frac{1}{2} = -\frac{1}{2}$$
$$p + \frac{1}{2} - \frac{1}{2} = -\frac{1}{2} - \frac{1}{2} \qquad -1 + \frac{1}{2} \overset{?}{=} -\frac{1}{2}$$
$$p = \frac{-2}{2} \qquad\qquad \frac{-2}{2} + \frac{1}{2} \overset{?}{=} -\frac{1}{2}$$
$$p = -1 \qquad\qquad\qquad -\frac{1}{2} = -\frac{1}{2}$$

10.
$$x + \frac{5}{7} = \frac{5}{7} \qquad\qquad x + \frac{5}{7} = \frac{5}{7}$$
$$x + \frac{5}{7} - \frac{5}{7} = \frac{5}{7} - \frac{5}{7} \qquad 0 + \frac{5}{7} \overset{?}{=} \frac{5}{7}$$
$$x = 0 \qquad\qquad\qquad \frac{5}{7} = \frac{5}{7}$$

11.
$$z + \frac{2}{3} = \frac{1}{3} \qquad\qquad z + \frac{2}{3} = \frac{1}{3}$$
$$z + \frac{2}{3} - \frac{2}{3} = \frac{1}{3} - \frac{2}{3} \qquad -\frac{1}{3} + \frac{2}{3} \overset{?}{=} \frac{1}{3}$$
$$z = -\frac{1}{3} \qquad\qquad\qquad \frac{1}{3} = \frac{1}{3}$$

12.
$$b - \frac{1}{4} = -\frac{3}{4} \qquad\qquad b - \frac{1}{4} = -\frac{3}{4}$$
$$b - \frac{1}{4} + \frac{1}{4} = -\frac{3}{4} + \frac{1}{4} \qquad -\frac{1}{2} - \frac{1}{4} \overset{?}{=} -\frac{3}{4}$$
$$b = -\frac{2}{4} \qquad\qquad -\frac{2}{4} - \frac{1}{4} \overset{?}{=} -\frac{3}{4}$$
$$b = -\frac{1}{2} \qquad\qquad\qquad -\frac{3}{4} = -\frac{3}{4}$$

13. Let r = the regular price. Then

$$\boxed{\text{Sale price}} = \boxed{\text{Regular price}} - \boxed{\text{Markdown}}$$

$$69.95 = r - 35.45$$
$$69.95 + 35.45 = r - 35.45 + 35.45$$
$$105.40 = r$$

The regular price is \$105.40.

14. Let w = the wholesale cost. Then

$$\boxed{\text{Retail price}} = \boxed{\text{Wholesale cost}} + \boxed{\text{Markup}}$$

$$212.95 = w + 115.25$$
$$212.95 - 115.25 = w + 115.25 - 115.25$$
$$97.70 = w$$

The wholesale cost is \$97.70.

15.
$$3x = 15 \qquad\qquad 3x = 15$$
$$\frac{3x}{3} = \frac{15}{3} \qquad 3(5) \overset{?}{=} 15$$
$$x = 5 \qquad\qquad 15 = 15$$

16.
$$8r = -16 \qquad\qquad 8r = -16$$
$$\frac{8r}{8} = \frac{-16}{8} \qquad 8(-2) \overset{?}{=} -16$$
$$r = -2 \qquad\qquad -16 = -16$$

17.
$$-10z = 5 \qquad\qquad -10z = 5$$
$$\frac{-10z}{-10} = \frac{5}{-10} \qquad -10\left(-\frac{1}{2}\right) \overset{?}{=} 5$$
$$z = -\frac{1}{2} \qquad\qquad 5 = 5$$

18.
$$14q = 21 \qquad\qquad 14q = 21$$
$$\frac{14q}{14} = \frac{21}{14} \qquad 14\left(\frac{3}{2}\right) \overset{?}{=} 21$$
$$q = \frac{3}{2} \qquad\qquad 21 = 21$$

19.
$$\frac{y}{3} = 6 \qquad\qquad \frac{y}{3} = 6$$
$$3 \cdot \frac{y}{3} = 3 \cdot 6 \qquad \frac{18}{3} \overset{?}{=} 6$$
$$y = 18 \qquad\qquad 6 = 6$$

20.
$$\frac{w}{7} = -5 \qquad\qquad \frac{w}{7} = -5$$
$$7 \cdot \frac{w}{7} = 7(-5) \qquad \frac{-35}{7} \overset{?}{=} -5$$
$$w = -35 \qquad\qquad -5 = -5$$

CHAPTER 2 REVIEW

21.
$$\frac{a}{-7} = \frac{1}{14} \qquad \frac{a}{-7} = \frac{1}{14}$$
$$-7 \cdot \frac{a}{-7} = -7\left(\frac{1}{14}\right) \qquad \frac{-\frac{1}{2}}{-7} \overset{?}{=} \frac{1}{14}$$
$$a = -\frac{7}{14} \qquad \frac{2\left(-\frac{1}{2}\right)}{2(-7)} \overset{?}{=} \frac{1}{14}$$
$$a = -\frac{1}{2} \qquad \frac{-1}{-14} = \frac{1}{14}$$

22.
$$\frac{p}{12} = \frac{1}{2} \qquad \frac{p}{12} = \frac{1}{2}$$
$$12 \cdot \frac{p}{12} = 12\left(\frac{1}{2}\right) \qquad \frac{6}{12} \overset{?}{=} \frac{1}{2}$$
$$p = \frac{12}{2} \qquad \frac{1}{2} = \frac{1}{2}$$
$$p = 6$$

23.
$$rb = a$$
$$35\% \cdot 700 = a$$
$$0.35(700) = a$$
$$245 = a$$

24.
$$rb = a$$
$$72\% \cdot b = 936$$
$$0.72b = 936$$
$$\frac{0.72b}{0.72} = \frac{936}{0.72}$$
$$b = 1{,}300$$

25.
$$rb = a$$
$$r \cdot 2300 = 851$$
$$\frac{r \cdot 2300}{2300} = \frac{851}{2300}$$
$$r = 0.37$$
$$r = 37\%$$

26.
$$rb = a$$
$$r \cdot 576 = 72$$
$$\frac{r \cdot 576}{576} = \frac{72}{576}$$
$$r = 0.125$$
$$r = 12.5\%$$

27.
$$5y + 6 = 21 \qquad 5y + 6 = 21$$
$$5y + 6 - 6 = 21 - 6 \qquad 5(3) + 6 \overset{?}{=} 21$$
$$5y = 15 \qquad 15 + 6 \overset{?}{=} 21$$
$$\frac{5y}{5} = \frac{15}{5} \qquad 21 = 21$$
$$y = 3$$

28.
$$5y - 9 = 1 \qquad 5y - 9 = 1$$
$$5y - 9 + 9 = 1 + 9 \qquad 5(2) - 9 \overset{?}{=} 1$$
$$5y = 10 \qquad 10 - 9 \overset{?}{=} 1$$
$$\frac{5y}{5} = \frac{10}{5} \qquad 1 = 1$$
$$y = 2$$

29.
$$-12z + 4 = -8 \qquad -12z + 4 = -8$$
$$-12z + 4 - 4 = -8 - 4 \qquad -12(1) + 4 \overset{?}{=} -8$$
$$-12z = -12 \qquad -12 + 4 \overset{?}{=} -8$$
$$\frac{-12z}{-12} = \frac{-12}{-12} \qquad -8 = -8$$
$$z = 1$$

30.
$$17z + 3 = 20 \qquad 17z + 3 = 20$$
$$17z + 3 - 3 = 20 - 3 \qquad 17(1) + 3 \overset{?}{=} 20$$
$$17z = 17 \qquad 17 + 3 \overset{?}{=} 20$$
$$\frac{17z}{17} = \frac{17}{17} \qquad 20 = 20$$
$$z = 1$$

31.
$$13 - 13p = 0 \qquad 13 - 13p = 0$$
$$13 - 13 - 13p = 0 - 13 \qquad 13 - 13(1) \overset{?}{=} 0$$
$$-13p = -13 \qquad 13 - 13 \overset{?}{=} 0$$
$$\frac{-13p}{-13} = \frac{-13}{-13} \qquad 0 = 0$$
$$p = 1$$

32.
$$10 + 7p = -4 \qquad 10 + 7p = -4$$
$$10 - 10 + 7p = -4 - 10 \qquad 10 + 7(-2) \overset{?}{=} -4$$
$$7p = -14 \qquad 10 + (-14) \overset{?}{=} -4$$
$$\frac{7p}{7} = \frac{-14}{7} \qquad -4 = -4$$
$$p = -2$$

33.

$$23a - 43 = 3$$
$$23a - 43 + 43 = 3 + 43$$
$$23a = 46$$
$$\frac{23a}{23} = \frac{46}{23}$$
$$a = 2$$

$$23a - 43 = 3$$
$$23(2) - 43 \stackrel{?}{=} 3$$
$$46 - 43 \stackrel{?}{=} 3$$
$$3 = 3$$

34.

$$84 - 21a = -63$$
$$84 - 84 - 21a = -63 - 84$$
$$-21a = -147$$
$$\frac{-21a}{-21} = \frac{-147}{-21}$$
$$a = 7$$

$$84 - 21a = -63$$
$$84 - 21(7) \stackrel{?}{=} -63$$
$$84 - 147 \stackrel{?}{=} -63$$
$$-63 = -63$$

35.

$$3x + 7 = 1$$
$$3x + 7 - 7 = 1 - 7$$
$$3x = -6$$
$$\frac{3x}{3} = \frac{-6}{3}$$
$$x = -2$$

$$3x + 7 = 1$$
$$3(-2) + 7 \stackrel{?}{=} 1$$
$$-6 + 7 \stackrel{?}{=} 1$$
$$1 = 1$$

36.

$$7 - 9x = 16$$
$$7 - 7 - 9x = 16 - 7$$
$$-9x = 9$$
$$\frac{-9x}{-9} = \frac{9}{-9}$$
$$x = -1$$

$$7 - 9x = 16$$
$$7 - 9(-1) \stackrel{?}{=} 16$$
$$7 + 9 \stackrel{?}{=} 16$$
$$16 = 16$$

37.

$$\frac{b + 3}{4} = 2$$
$$4 \cdot \frac{b + 3}{4} = 4 \cdot 2$$
$$b + 3 = 8$$
$$b + 3 - 3 = 8 - 3$$
$$b = 5$$

$$\frac{b + 3}{4} = 2$$
$$\frac{5 + 3}{4} \stackrel{?}{=} 2$$
$$\frac{8}{4} \stackrel{?}{=} 2$$
$$2 = 2$$

38.

$$\frac{b - 7}{2} = -2$$
$$2 \cdot \frac{b - 7}{2} = 2(-2)$$
$$b - 7 = -4$$
$$b - 7 + 7 = -4 + 7$$
$$b = 3$$

$$\frac{b - 7}{2} = -2$$
$$\frac{3 - 7}{2} \stackrel{?}{=} -2$$
$$\frac{-4}{2} \stackrel{?}{=} -2$$
$$-2 = -2$$

39.

$$\frac{3y - 2}{4} = -5$$
$$4 \cdot \frac{3y - 2}{4} = 4(-5)$$
$$3y - 2 = -20$$
$$3y - 2 + 2 = -20 + 2$$
$$3y = -18$$
$$\frac{3y}{3} = \frac{-18}{3}$$
$$y = -6$$

$$\frac{3y - 2}{4} = -5$$
$$\frac{3(-6) - 2}{4} = -5$$
$$\frac{-18 - 2}{4} \stackrel{?}{=} -5$$
$$\frac{-20}{4} \stackrel{?}{=} -5$$
$$-5 = -5$$

40.

$$\frac{3x + 10}{2} = -1 \qquad\qquad \frac{3x + 10}{2} = -1$$

$$2 \cdot \frac{3x + 10}{2} = 2(-1) \qquad \frac{3(-4) + 10}{2} \overset{?}{=} -1$$

$$3x + 10 = -2 \qquad\qquad \frac{-12 + 10}{2} \overset{?}{=} -1$$

$$3x + 10 - 10 = -2 - 10$$

$$3x = -12 \qquad\qquad \frac{-2}{2} \overset{?}{=} -1$$

$$\frac{3x}{3} = \frac{-12}{3} \qquad\qquad -1 = -1$$

$$x = -4$$

41.

$$\frac{x}{2} + 7 = 11 \qquad \frac{x}{2} + 7 = 11$$

$$\frac{x}{2} + 7 - 7 = 11 - 7 \qquad \frac{8}{2} + 7 \overset{?}{=} 11$$

$$\frac{x}{2} = 4 \qquad 4 + 7 \overset{?}{=} 11$$

$$2 \cdot \frac{x}{2} = 2 \cdot 4 \qquad 11 = 11$$

$$x = 8$$

42.

$$\frac{r}{3} - 3 = 7 \qquad \frac{r}{3} - 3 = 7$$

$$\frac{r}{3} - 3 + 3 = 7 + 3 \qquad \frac{30}{3} - 3 \overset{?}{=} 7$$

$$\frac{r}{3} = 10 \qquad 10 - 3 \overset{?}{=} 7$$

$$3 \cdot \frac{r}{3} = 3 \cdot 10 \qquad 7 = 7$$

$$r = 30$$

43.

$$\frac{2x}{3} - 4 = 6 \qquad\qquad \frac{2x}{3} - 4 = 6$$

$$\frac{2x}{3} - 4 + 4 = 6 + 4 \qquad \frac{2(15)}{3} - 4 \overset{?}{=} 6$$

$$\frac{2x}{3} = 10 \qquad\qquad \frac{30}{3} - 4 \overset{?}{=} 6$$

$$3 \cdot \frac{2x}{3} = 3(10) \qquad 10 - 4 \overset{?}{=} 6$$

$$2x = 30 \qquad\qquad 6 = 6$$

$$\frac{2x}{2} = \frac{30}{2}$$

$$x = 15$$

44.

$$\frac{y}{4} - \frac{6}{5} = -\frac{1}{5} \qquad \frac{y}{4} - \frac{6}{5} = -\frac{1}{5}$$

$$\frac{y}{4} - \frac{6}{5} + \frac{6}{5} = -\frac{1}{5} + \frac{6}{5} \qquad \frac{4}{4} - \frac{6}{5} \overset{?}{=} -\frac{1}{5}$$

$$\frac{y}{4} = \frac{5}{5} \qquad 1 - \frac{6}{5} \overset{?}{=} -\frac{1}{5}$$

$$4 \cdot \frac{y}{4} = 4(1) \qquad \frac{5}{5} - \frac{6}{5} \overset{?}{=} -\frac{1}{5}$$

$$y = 4 \qquad -\frac{1}{5} = -\frac{1}{5}$$

45.

$$\frac{a}{2} + \frac{3}{4} = 6 \qquad\qquad \frac{a}{2} + \frac{3}{4} = 6$$

$$\frac{a}{2} + \frac{3}{4} - \frac{3}{4} = \frac{24}{4} - \frac{3}{4} \qquad \frac{\frac{21}{2}}{2} + \frac{3}{4} \overset{?}{=} 6$$

$$\frac{a}{2} = \frac{21}{4} \qquad \frac{2 \cdot \frac{21}{2}}{2 \cdot 2} + \frac{3}{4} \overset{?}{=} 6$$

$$2 \cdot \frac{a}{2} = 2 \cdot \frac{21}{4} \qquad \frac{21}{4} + \frac{3}{4} \overset{?}{=} 6$$

$$a = \frac{21}{2} \qquad\qquad \frac{24}{4} \overset{?}{=} 6$$

$$6 = 6$$

46.

$$\frac{x}{8} - 2.3 = 3.2 \qquad \frac{x}{8} - 2.3 = 3.2$$

$$\frac{x}{8} - 2.3 + 2.3 = 3.2 + 2.3 \qquad \frac{44}{8} - 2.3 \overset{?}{=} 3.2$$

$$\frac{x}{8} = 5.5 \qquad 5.5 - 2.3 \overset{?}{=} 3.2$$

$$8 \cdot \frac{x}{8} = 8(5.5) \qquad 3.2 = 3.2$$

$$x = 44$$

47. Let $x =$ the regular price.

$$\boxed{\text{Sale price}} = \boxed{\text{Regular price}} - \boxed{\text{Markdown}}$$

$$240 = x - 0.25x$$
$$240 = 0.75x$$
$$\frac{240}{0.75} = \frac{0.75x}{0.75}$$
$$320 = x$$

The regular price is $320.

48. Let $r =$ the sales tax rate.

$$\boxed{\text{Total price}} = \boxed{\text{Price before tax}} + \boxed{\text{Sales tax}}$$

$$40.47 = 38 + 38r$$
$$40.47 - 38 = 38 - 38 + 38r$$
$$2.47 = 38r$$
$$\frac{2.47}{38} = \frac{38r}{38}$$
$$0.065 = r$$

The sales tax rate is 6.5%.

49. Let $r =$ the % increase.

$$\boxed{\text{New price}} = \boxed{\text{Original price}} + \boxed{\text{Increase}}$$

$$1100 = 560 + 560r$$
$$1100 - 560 = 560 - 560 + 560r$$
$$540 = 560r$$
$$\frac{540}{560} = \frac{560r}{560}$$
$$0.964 = r$$

The percent increase is 96.4%.

50. Let $r =$ the % discount.

$$\boxed{\text{Markdown}} = \boxed{\text{Percent markdown}} \cdot \boxed{\text{Regular price}}$$

$$470 - 221.84 = r \cdot 470$$
$$248.16 = 470r$$
$$\frac{248.16}{470} = \frac{470r}{470}$$
$$0.528 = r$$

The percent markdown is 52.8%.

51. $5x + 9x = (5 + 9)x = 14x$

52. $7a + 12a = (7 + 12)a = 19a$

53. $18b - 13b = (18 - 13)b = 5b$

54. $21x - 23x = (21 - 23)x = -2x$

55. $5y - 7y = (5 - 7)y = -2y$

56. $19x - 19 \Rightarrow$ unlike terms

57. $7(x + 2) + 2(x - 7) = 7x + 14 + 2x - 14$
$$= 7x + 2x + 14 - 14$$
$$= (7 + 2)x + 0 = 9x$$

58. $2(3 - x) + x - 6x = 6 - 2x + x - 6x$
$$= 6 + (-2 + 1 - 6)x$$
$$= 6 - 7x$$

59.
$$10y - 14 = 3y$$
$$10y - 10y - 14 = 3y - 10y$$
$$-14 = -7y$$
$$\frac{-14}{-7} = \frac{-7y}{-7}$$
$$2 = y$$

$$10y - 14 = 3y$$
$$10(2) - 14 \overset{?}{=} 3(2)$$
$$20 - 14 \overset{?}{=} 6$$
$$6 = 6$$

60.
$$6(a + 4) = 3a$$
$$6a + 24 = 3a$$
$$6a - 6a + 24 = 3a - 6a$$
$$24 = -3a$$
$$\frac{24}{-3} = \frac{-3a}{-3}$$
$$-8 = a$$

$$6(a + 4) = 3a$$
$$6(-8 + 4) \overset{?}{=} 3(-8)$$
$$6(-4) \overset{?}{=} -24$$
$$-24 = -24$$

61.

$$2x - 19 = 2 - x$$
$$2x + x - 19 = 2 - x + x$$
$$3x - 19 = 2$$
$$3x - 19 + 19 = 2 + 19$$
$$3x = 21$$
$$\frac{3x}{3} = \frac{21}{3}$$
$$x = 7$$

$$2x - 19 = 2 - x$$
$$2(7) - 19 \overset{?}{=} 2 - 7$$
$$14 - 19 \overset{?}{=} -5$$
$$-5 = -5$$

62.

$$5b - 19 = 2b + 20$$
$$5b - 2b - 19 = 2b - 2b + 20$$
$$3b - 19 = 20$$
$$3b - 19 + 19 = 20 + 19$$
$$3b = 39$$
$$\frac{3b}{3} = \frac{39}{3}$$
$$b = 13$$

$$5b - 19 = 2b + 20$$
$$5(13) - 19 \overset{?}{=} 2(13) + 20$$
$$65 - 19 \overset{?}{=} 26 + 20$$
$$46 = 46$$

63.

$$3x + 20 = 5 - 2x$$
$$3x + 2x + 20 = 5 - 2x + 2x$$
$$5x + 20 = 5$$
$$5x + 20 - 20 = 5 - 20$$
$$5x = -15$$
$$\frac{5x}{5} = \frac{-15}{5}$$
$$x = -3$$

$$3x + 20 = 5 - 2x$$
$$3(-3) + 20 \overset{?}{=} 5 - 2(-3)$$
$$-9 + 20 \overset{?}{=} 5 + 6$$
$$11 = 11$$

64.

$$0.9x + 10 = 0.7x + 1.8$$
$$0.9x - 0.7x + 10 = 0.7x - 0.7x + 1.8$$
$$0.2x + 10 = 1.8$$
$$0.2x + 10 - 10 = 1.8 - 10$$
$$0.2x = -8.2$$
$$\frac{0.2x}{0.2} = \frac{-8.2}{0.2}$$
$$x = -41$$

$$0.9x + 10 = 0.7x + 1.8$$
$$0.9(-41) + 10 \overset{?}{=} 0.7(-41) + 1.8$$
$$-36.9 + 10 \overset{?}{=} -28.7 + 1.8$$
$$-26.9 = -26.9$$

65.

$$10(p - 3) = 3(p + 11)$$
$$10p - 30 = 3p + 33$$
$$10p - 3p - 30 = 3p - 3p + 33$$
$$7p - 30 = 33$$
$$7p - 30 + 30 = 33 + 30$$
$$7p = 63$$
$$\frac{7p}{7} = \frac{63}{7}$$
$$p = 9$$

$$10(p - 3) = 3(p + 11)$$
$$10(9 - 3) \overset{?}{=} 3(9 + 11)$$
$$10(6) \overset{?}{=} 3(20)$$
$$60 = 60$$

63

66.

$$2(5x - 7) = 2(x - 35)$$
$$10x - 14 = 2x - 70$$
$$10x - 2x - 14 = 2x - 2x - 70$$
$$8x - 14 = -70$$
$$8x - 14 + 14 = -70 + 14$$
$$8x = -56$$
$$\frac{8x}{8} = \frac{-56}{8}$$
$$x = -7$$

$$2(5x - 7) = 2(x - 35)$$
$$2[5(-7) - 7] \stackrel{?}{=} 2(-7 - 35)$$
$$2[-35 - 7] \stackrel{?}{=} 2(-42)$$
$$2(-42) \stackrel{?}{=} 2(-42)$$
$$-84 = -84$$

67.

$$\frac{3u - 6}{5} = 3$$
$$5 \cdot \frac{3u - 6}{5} = 5(3)$$
$$3u - 6 = 15$$
$$3u - 6 + 6 = 15 + 6$$
$$3u = 21$$
$$\frac{3u}{3} = \frac{21}{3}$$
$$u = 7$$

$$\frac{3u - 6}{5} = 3$$
$$\frac{3(7) - 6}{5} \stackrel{?}{=} 3$$
$$\frac{21 - 6}{5} \stackrel{?}{=} 3$$
$$\frac{15}{5} \stackrel{?}{=} 3$$
$$3 = 3$$

68.

$$\frac{5v - 35}{3} = -5$$
$$3 \cdot \frac{5v - 35}{3} = 3(-5)$$
$$5v - 35 = -15$$
$$5v - 35 + 35 = -15 + 35$$
$$5v = 20$$
$$\frac{5v}{5} = \frac{20}{5}$$
$$v = 4$$

$$\frac{5v - 35}{3} = -5$$
$$\frac{5(4) - 35}{3} \stackrel{?}{=} -5$$
$$\frac{20 - 35}{3} \stackrel{?}{=} -5$$
$$\frac{-15}{3} \stackrel{?}{=} -5$$
$$-5 = -5$$

69.

$$\frac{2(b + 4)}{3} = b - 2$$
$$3 \cdot \frac{2b + 8}{3} = 3(b - 2)$$
$$2b + 8 = 3b - 6$$
$$2b - 3b + 8 = 3b - 3b - 6$$
$$-b + 8 = -6$$
$$-b + 8 - 8 = -6 - 8$$
$$-b = -14$$
$$-1(-b) = -1(-14)$$
$$b = 14$$

$$\frac{2(b + 4)}{3} = b - 2$$
$$\frac{2(14 + 4)}{3} \stackrel{?}{=} 14 - 2$$
$$\frac{2(18)}{3} \stackrel{?}{=} 12$$
$$\frac{36}{3} \stackrel{?}{=} 12$$
$$12 = 12$$

64

70.
$$\frac{3(x-1)}{6} - 5 = \frac{2(x+3)}{6}$$
$$6 \cdot \frac{3x-3}{6} - 6(5) = 6 \cdot \frac{2x+6}{6}$$
$$3x - 3 - 30 = 2x + 6$$
$$3x - 33 = 2x + 6$$
$$3x - 2x - 33 = 6$$
$$x - 33 = 6$$
$$x - 33 + 33 = 6 + 33$$
$$x = 39$$

$$\frac{3(x-1)}{6} - 5 = \frac{2(x+3)}{6}$$
$$\frac{3(39-1)}{6} - 5 \overset{?}{=} \frac{2(39+3)}{6}$$
$$\frac{3(38)}{6} - 5 \overset{?}{=} \frac{2(42)}{6}$$
$$\frac{114}{6} - 5 \overset{?}{=} \frac{84}{6}$$
$$19 - 5 = 14$$
$$14 = 14$$

71.
$$4x - 2 = x + 3(x+1) - 5$$
$$4x - 2 = x + 3x + 3 - 5$$
$$4x - 2 = 4x - 2$$
$$4x - 4x - 2 = 4x - 4x - 2$$
$$-2 = -2$$
Identity, $\mathbb{R}$

72.
$$-3(a+1) - a = -4a + 3$$
$$-3a - 3 - a = -4a + 3$$
$$-4a - 3 = -4a + 3$$
$$-4a + 4a - 3 = -4a + 4a + 3$$
$$-3 \neq 3$$
Contradiction, $\emptyset$

73.
$$2(x-1) + 4 = 4(1+x) - (2x+2)$$
$$2x - 2 + 4 = 4 + 4x - 2x - 2$$
$$2x + 2 = 2x + 2$$
$$2x - 2x + 2 = 2x - 2x + 2$$
$$2 = 2$$
Identity, $\mathbb{R}$

74.
$$3(2x+1) + 3 = 9(x+2) + 9 - 3x$$
$$6x + 3 + 3 = 9x + 18 + 9 - 3x$$
$$6x + 6 = 6x + 27$$
$$6x - 6x + 6 = 6x - 6x + 27$$
$$6 \neq 27$$
Contradiction, $\emptyset$

75.
$$E = IR$$
$$\frac{E}{I} = \frac{IR}{I}$$
$$\frac{E}{I} = R, \text{ or } R = \frac{E}{I}$$

76.
$$i = prt$$
$$\frac{i}{pr} = \frac{prt}{pr}$$
$$\frac{i}{pr} = t, \text{ or } t = \frac{i}{pr}$$

77.
$$P = I^2 R$$
$$\frac{P}{I^2} = \frac{I^2 R}{I^2}$$
$$\frac{P}{I^2} = R, \text{ or } R = \frac{P}{I^2}$$

78.
$$V = \frac{1}{3}Bh$$
$$3V = 3 \cdot \frac{1}{3}Bh$$
$$3V = Bh$$
$$\frac{3V}{h} = \frac{Bh}{h}$$
$$\frac{3V}{h} = B, \text{ or } B = \frac{3V}{h}$$

79.
$$p = a + b + c$$
$$p - a = a - a + b + c$$
$$p - a - b = b - b + c$$
$$p - a - b = c,$$
$$\text{or } c = p - a - b$$

80.
$$y = mx + b$$
$$y - b = mx + b - b$$
$$y - b = mx$$
$$\frac{y-b}{x} = \frac{mx}{x}$$
$$\frac{y-b}{x} = m, \text{ or } m = \frac{y-b}{x}$$

81.
$$V = \pi r^2 h$$
$$\frac{V}{\pi r^2} = \frac{\pi r^2 h}{\pi r^2}$$
$$\frac{V}{\pi r^2} = h, \text{ or } h = \frac{V}{\pi r^2}$$

82.
$$A = \frac{3}{2}(B + 4)$$
$$\frac{2}{3}A = \frac{2}{3} \cdot \frac{3}{2}(B + 4)$$
$$\frac{2}{3}A = B + 4$$
$$\frac{2}{3}A - 4 = B + 4 - 4$$
$$\frac{2}{3}A - 4 = B, \text{ or } B = \frac{2}{3}A - 4$$

83.
$$F = \frac{GMm}{d^2}$$
$$d^2 \cdot F = d^2 \cdot \frac{GMm}{d^2}$$
$$d^2 F = GMm$$
$$\frac{d^2 F}{Mm} = \frac{GMm}{Mm}$$
$$\frac{d^2 F}{Mm} = G, \text{ or } G = \frac{d^2 F}{Mm}$$

84.
$$P = \frac{RT}{mV}$$
$$mV \cdot P = mV \cdot \frac{RT}{mV}$$
$$mVP = RT$$
$$\frac{mVP}{VP} = \frac{RT}{VP}$$
$$m = \frac{RT}{VP}$$

85. Let x = the length of one part. Then $2x - 7$ = the length of the other part.

$$\boxed{\begin{array}{c}\text{Length of}\\\text{first part}\end{array}} + \boxed{\begin{array}{c}\text{Length of}\\\text{second part}\end{array}} = 8$$
$$x + 2x - 7 = 8$$
$$3x - 7 = 8$$
$$3x - 7 + 7 = 8 + 7$$
$$3x = 15$$
$$\frac{3x}{3} = \frac{15}{3}$$
$$x = 5 \Rightarrow \text{One part should be cut 5 feet long.}$$

86.
$$x + 47° = 62°$$
$$x + 47° - 47° = 62° - 47°$$
$$x = 15°$$

87.
$$x + 135° = 180°$$
$$x + 135° - 135° = 180° - 135°$$
$$x = 45°$$

88. Let x = the measure of the complement.
$$x + 12° = 90°$$
$$x + 12° - 12° = 90° - 12°$$
$$x = 78°$$

89. Let x = the measure of the supplement.
$$x + 75° = 180°$$
$$x + 75° - 75° = 180° - 75°$$
$$x = 105°$$

66

CHAPTER 2 REVIEW

90. Let w = the width of the picture. Then $2w + 3$ = the length of the picture.

$$\boxed{\text{Perimeter}} = 84$$
$$2w + 2(2w + 3) = 84$$
$$2w + 4w + 6 = 84$$
$$6w + 6 = 84$$
$$6w + 6 - 6 = 84 - 6$$
$$6w = 78$$
$$\frac{6w}{6} = \frac{78}{6}$$
$$w = 13 \Rightarrow \text{The width is 13 inches.}$$

91. Let x = amount in 7% CD. Then $27000 - x$ = amount in 9% fund.

$$\boxed{\begin{array}{c}\text{Interest}\\\text{at 7%}\end{array}} + \boxed{\begin{array}{c}\text{Interest}\\\text{at 9%}\end{array}} = \boxed{\begin{array}{c}\text{Total}\\\text{interest}\end{array}}$$
$$0.07x + 0.09(27000 - x) = 2110$$
$$7x + 9(27000 - x) = 211000$$
$$7x + 243000 - 9x = 211000$$
$$-2x + 243000 = 211000$$
$$-2x + 243000 - 243000 = 211000 - 243000$$
$$-2x = -32000$$
$$\frac{-2x}{-2} = \frac{-32000}{-2}$$
$$x = 16000 \Rightarrow \$16,000 \text{ is invested at 7%, and } \$11,000 \text{ is invested at 9%.}$$

92. Let t = time for friends to meet.

	r	t	d
Walk	4	t	$4t$
Bike	10	t	$10t$

$$\boxed{\begin{array}{c}\text{Distance}\\\text{walked}\end{array}} + \boxed{\begin{array}{c}\text{Distance}\\\text{biked}\end{array}} = 7$$
$$4t + 10t = 7$$
$$14t = 7$$
$$\frac{14t}{14} = \frac{7}{14}$$
$$t = \frac{1}{2}$$

They meet after $\frac{1}{2}$ hour (or 30 minutes).

93. Let t = time to be 90 miles apart.

	r	t	d
1st Team	20	t	$20t$
2nd Team	25	t	$25t$

$$\boxed{\begin{array}{c}\text{Distance}\\\text{of 1st Team}\end{array}} + \boxed{\begin{array}{c}\text{Distance}\\\text{of 2nd Team}\end{array}} = 90$$
$$20t + 25t = 90$$
$$45t = 90$$
$$\frac{45t}{45} = \frac{90}{45}$$
$$t = 2$$

They will lose contact in 2 hours.

94. Let t = time needed to be 75 miles apart.

	r	t	d
Bus	60	t	$60t$
Truck	50	t	$50t$

$$\boxed{\begin{array}{c}\text{Bus}\\\text{Distance}\end{array}} - \boxed{\begin{array}{c}\text{Truck}\\\text{Distance}\end{array}} = 90$$

$$60t - 50t = 90$$
$$10t = 90$$
$$\frac{10t}{10} = \frac{90}{10}$$
$$t = 9$$

They will be 90 miles apart after 9 hours.

95. Let x = liters of 1% butterfat milk used.

$$\boxed{\begin{array}{c}\text{Butterfat in}\\\text{4\% milk}\end{array}} + \boxed{\begin{array}{c}\text{Butterfat in}\\\text{1\% milk}\end{array}} = \boxed{\begin{array}{c}\text{Butterfat in}\\\text{2\% mixture}\end{array}}$$

$$0.04(12) + 0.01x = 0.02(12 + x)$$
$$4(12) + 1x = 2(12 + x)$$
$$48 + x = 24 + 2x$$
$$x - 2x + 48 = 24 + 2x - 2x$$
$$-x + 48 = 24$$
$$-x + 48 - 48 = 24 - 48$$
$$-x = -24$$
$$x = 24$$

24 liters of the 1% milk should be used.

96. Let x = liters of 12% solution used.

$$\boxed{\begin{array}{c}\text{Acid in 6\%}\\\text{solution}\end{array}} + \boxed{\begin{array}{c}\text{Acid in 12\%}\\\text{solution}\end{array}} = \boxed{\begin{array}{c}\text{Acid in 8\%}\\\text{solution}\end{array}}$$

$$0.06(2) + 0.12x = 0.08(2 + x)$$
$$0.12 + 0.12x = 0.16 + 0.08x$$
$$12 + 12x = 16 + 8x$$
$$12 + 12x - 8x = 16 + 8x - 8x$$
$$12 + 4x = 16$$
$$12 - 12 + 4x = 16 - 12$$
$$4x = 4$$
$$\frac{4x}{4} = \frac{4}{4}$$
$$x = 1$$

1 liter of the 12% solution should be used.

97. Let x = pounds of 90¢ candy.

Then $20 - x$ = pounds of \$1.50 candy.

$$\boxed{\begin{array}{c}\text{Value of}\\\text{90¢ candy}\end{array}} + \boxed{\begin{array}{c}\text{Value of}\\\text{\$1.50 candy}\end{array}} = \boxed{\begin{array}{c}\text{Value of}\\\text{mixture}\end{array}}$$

$$0.90x + 1.50(20 - x) = 1.20(20)$$
$$90x + 150(20 - x) = 120(20)$$
$$90x + 3000 - 150x = 2400$$
$$-60x + 3000 = 2400$$
$$-60x + 3000 - 3000 = 2400 - 3000$$
$$-60x = -600$$
$$\frac{-60x}{-60} = \frac{-600}{-60}$$
$$x = 10$$

10 lb of each should be used.

98.
$$3x + 2 < 5$$
$$3x + 2 - 2 < 5 - 2$$
$$3x < 3$$
$$\frac{3x}{3} < \frac{3}{3}$$
$$x < 1$$

99.
$$-5x - 8 > 7$$
$$-5x - 8 + 8 > 7 + 8$$
$$-5x > 15$$
$$\frac{-5x}{-5} < \frac{15}{-5}$$
$$x < -3$$

100.
$$5x - 3 \geq 2x + 9$$
$$5x - 2x - 3 \geq 2x - 2x + 9$$
$$3x - 3 \geq 9$$
$$3x - 3 + 3 \geq 9 + 3$$
$$3x \geq 12$$
$$\frac{3x}{3} \geq \frac{12}{3}$$
$$x \geq 4$$

101.
$$7x + 1 \leq 8x - 5$$
$$7x - 8x + 1 \leq 8x - 8x - 5$$
$$-x + 1 \leq -5$$
$$-x + 1 - 1 \leq -5 - 1$$
$$-x \leq -6$$
$$-(-x) \geq -(-6)$$
$$x \geq 6$$

102.
$$5(3 - x) \leq 3(x - 3)$$
$$15 - 5x \leq 3x - 9$$
$$15 - 5x - 3x \leq 3x - 3x - 9$$
$$15 - 8x \leq -9$$
$$15 - 15 - 8x \leq -9 - 15$$
$$-8x \leq -24$$
$$\frac{-8x}{-8} \geq \frac{-24}{-8}$$
$$x \geq 3$$

103.
$$8(2 - x) > 4 - 2x$$
$$16 - 8x > 4 - 2x$$
$$16 - 8x + 8x > 4 - 2x + 8x$$
$$16 > 4 + 6x$$
$$16 - 4 > 4 - 4 + 6x$$
$$12 > 6x$$
$$\frac{12}{6} > \frac{6x}{6}$$
$$2 > x, \text{ or } x < 2$$

104.
$$8 < \quad x + 2 \quad < 13$$
$$8 - 2 < x + 2 - 2 < 13 - 2$$
$$6 < \quad x \quad < 11$$

105.
$$4 \leq \quad 2 - 2x \quad < 8$$
$$4 - 2 \leq 2 - 2 - 2x < 8 - 2$$
$$2 \leq \quad -2x \quad < 6$$
$$\frac{2}{-2} \geq \quad \frac{-2x}{-2} \quad > \frac{6}{-2}$$
$$-1 \geq \quad x \quad > -3$$
$$-3 < \quad x \quad \leq -1$$

106. Let l = length. Then $l - 6$ = width.

$$0 \text{ ft} < \quad \text{perimeter} \quad \leq 68 \text{ ft}$$
$$0 \text{ ft} < 2l + 2(l - 6) \leq 68 \text{ ft}$$
$$0 \text{ ft} < 2l + 2l - 12 \quad \leq 68 \text{ ft}$$
$$0 \text{ ft} < \quad 4l - 12 \quad \leq 68 \text{ ft}$$
$$0 + 12 \text{ ft} < 4l - 12 + 12 \leq 68 + 12 \text{ ft}$$
$$12 \text{ ft} < \quad 4l \quad \leq 80 \text{ ft}$$
$$\frac{12}{4} \text{ ft} < \quad \frac{4l}{4} \quad \leq \frac{80}{4} \text{ ft}$$
$$3 \text{ ft} < \quad l \quad \leq 20 \text{ ft}$$

If $l = 3$, then the width is negative, which is impossible. To keep the width positive, we must have $l - 6 > 0$, which implies $l > 6$. Thus the solution is $6 \text{ ft} < l \leq 20 \text{ ft}$.

Chapter 2 Test (page 146)

1.
$$4x + 5 = -3$$
$$4(-2) + 5 \stackrel{?}{=} -3$$
$$-8 + 5 \stackrel{?}{=} -3$$
$$-3 = -3$$
-2 is a solution.

2.
$$3(x + 2) = 2x + 4$$
$$3(2 + 2) \stackrel{?}{=} 2(2) + 4$$
$$3(4) \stackrel{?}{=} 4 + 4$$
$$12 \neq 8$$
2 is not a solution.

3.
$$x + 17 = -19$$
$$x + 17 - 17 = -19 - 17$$
$$x = -36$$

4.
$$a - 15 = 32$$
$$a - 15 + 15 = 32 + 15$$
$$a = 47$$

5.
$$12x = -144$$
$$\frac{12x}{12} = \frac{-144}{12}$$
$$x = -12$$

6.
$$\frac{x}{7} = -1$$
$$7 \cdot \frac{x}{7} = 7(-1)$$
$$x = -7$$

7.
$$8x + 2 = -14$$
$$8x + 2 - 2 = -14 - 2$$
$$8x = -16$$
$$\frac{8x}{8} = \frac{-16}{8}$$
$$x = -2$$

8.
$$3 = 5 - 2x$$
$$3 - 5 = 5 - 5 - 2x$$
$$-2 = -2x$$
$$\frac{-2}{-2} = \frac{-2x}{-2}$$
$$1 = x$$

9.
$$\frac{2x - 5}{3} = 3$$
$$3 \cdot \frac{2x - 5}{3} = 3(3)$$
$$2x - 5 = 9$$
$$2x - 5 + 5 = 9 + 5$$
$$2x = 14$$
$$\frac{2x}{2} = \frac{14}{2}$$
$$x = 7$$

10.
$$23 - 5(x + 10) = -12$$
$$23 - 5x - 50 = -12$$
$$-5x - 27 = -12$$
$$-5x - 27 + 27 = -12 + 27$$
$$-5x = 15$$
$$\frac{-5x}{-5} = \frac{15}{-5}$$
$$x = -3$$

11. $x + 5(x - 3) = x + 5x - 15 = 6x - 15$

12. $3x - 5(2 - x) = 3x - 10 + 5x = 8x - 10$

13. $-3(x + 3) + 3(x - 3) = -3x - 9 + 3x - 9 = -3x + 3x - 9 - 9 = -18$

14. $-4(2x - 5) - 7(4x + 1) = -8x + 20 - 28x - 7 = -36x + 13$

15.
$$8x + 6 = 2(4x + 3)$$
$$8x + 6 = 8x + 6$$
$$8x - 8x + 6 = 8x - 8x + 6$$
$$6 = 6$$
Identity, $\mathbb{R}$

16.
$$2(x + 6) = 2(x - 2)$$
$$2x + 12 = 2x - 4$$
$$2x - 2x + 12 = 2x - 2x - 4$$
$$12 \neq -4$$
Contradiction, $\emptyset$

70

17.
$$\frac{3x - 18}{2} = 6x$$
$$2 \cdot \frac{3x - 18}{2} = 2(6x)$$
$$3x - 18 = 12x$$
$$3x - 3x - 18 = 12x - 3x$$
$$-18 = 9x$$
$$\frac{-18}{9} = \frac{9x}{9}$$
$$-2 = x$$

18.
$$\frac{7}{8}(x - 4) = 5x - \frac{7}{2}$$
$$8 \cdot \frac{7}{8}(x - 4) = 8\left(5x - \frac{7}{2}\right)$$
$$7(x - 4) = 8(5x) - 8 \cdot \frac{7}{2}$$
$$7x - 28 = 40x - 28$$
$$7x - 40x - 28 = 40x - 40x - 28$$
$$-33x - 28 = -28$$
$$-33x - 28 + 28 = -28 + 28$$
$$-33x = 0$$
$$x = 0$$

19.
$$d = rt$$
$$\frac{d}{r} = \frac{rt}{r}$$
$$\frac{d}{r} = t, \text{ or } t = \frac{d}{r}$$

20.
$$S = 2\pi rh + 2\pi r^2$$
$$S - 2\pi r^2 = 2\pi rh + 2\pi r^2 - 2\pi r^2$$
$$S - 2\pi r^2 = 2\pi rh$$
$$\frac{S - 2\pi r^2}{2\pi r} = \frac{2\pi rh}{2\pi r}$$
$$\frac{S - 2\pi r^2}{2\pi r} = h, \text{ or } h = \frac{S - 2\pi r^2}{2\pi r}$$

21.
$$A = 2\pi rh$$
$$\frac{A}{2\pi r} = \frac{2\pi rh}{2\pi r}$$
$$\frac{A}{2\pi r} = h, \text{ or } h = \frac{A}{2\pi r}$$

22.
$$x + 2y = 5$$
$$x - x + 2y = 5 - x$$
$$2y = -x + 5$$
$$\frac{2y}{2} = \frac{-x + 5}{2}$$
$$y = \frac{-x + 5}{2}$$

23.
$$x + 45 = 120$$
$$x + 45 - 45 = 120 - 45$$
$$x = 75°$$

24. Let $x =$ the measure of the supplement.
$$x + 79 = 180$$
$$x + 79 - 79 = 180 - 79$$
$$x = 101°$$

25. Let $x =$ amount at 6%. Then $10000 - x =$ amount at 5%.

$$\boxed{\begin{array}{c}\text{Interest}\\\text{at 6\%}\end{array}} + \boxed{\begin{array}{c}\text{Interest}\\\text{at 5\%}\end{array}} = \boxed{\begin{array}{c}\text{Total}\\\text{interest}\end{array}}$$

$$0.06x + 0.05(10000 - x) = 560$$
$$6x + 5(10000 - x) = 56000$$
$$6x + 50000 - 5x = 56000$$
$$x + 50000 = 56000$$
$$x + 50000 - 50000 = 56000 - 50000$$
$$x = 6000 \Rightarrow \$6,000 \text{ is invested at 6\%, and } \$4,000 \text{ is invested at 5\%.}$$

26. Let t = time to meet.

	r	t	d
Car	65	t	$65t$
Truck	55	t	$55t$

$$\boxed{\text{Car Dist}} + \boxed{\text{Truck Dist}} = 72$$
$$65t + 55t = 72$$
$$120t = 72$$
$$\frac{120t}{120} = \frac{72}{120}$$
$$t = \frac{3}{5}$$

They meet after $\frac{3}{5}$ of an hour,
or $\frac{3}{5}(60) = 36$ minutes.

27. Let x = liters of water used.

$$\boxed{\substack{\text{Amt. in}\\\text{10\% sol.}}} + \boxed{\substack{\text{Amt. in}\\\text{water}}} = \boxed{\substack{\text{Amt. in}\\\text{8\% sol.}}}$$
$$0.10(30) + 0(x) = 0.08(30 + x)$$
$$10(30) + 0 = 8(30 + x)$$
$$300 = 240 + 8x$$
$$300 - 240 = 8x$$
$$60 = 8x$$
$$\frac{60}{8} = \frac{8x}{8}$$
$$\frac{15}{2} = x$$

$7\frac{1}{2}$ liters of water should be used.

28. Let x = pounds of peanuts used.

$$\boxed{\substack{\text{Value of}\\\text{cashews}}} + \boxed{\substack{\text{Value of}\\\text{peanuts}}} = \boxed{\substack{\text{Value of}\\\text{mixture}}}$$
$$6(20) + 3x = 4(20 + x)$$
$$120 + 3x = 80 + 4x$$
$$120 + 3x - 4x = 80 + 4x - 4x$$
$$-x + 120 = 80$$
$$-x + 120 - 120 = 80 - 120$$
$$-x = -40$$
$$x = 40; \quad 40 \text{ pounds of peanuts should be used.}$$

29.
$$8x - 20 \geq 4$$
$$8x - 20 + 20 \geq 4 + 20$$
$$8x \geq 24$$
$$\frac{8x}{8} \geq \frac{24}{8}$$
$$x \geq 3$$

30.
$$x - 2(x + 7) > 14$$
$$x - 2x - 14 > 14$$
$$-x - 14 > 14$$
$$-x - 14 + 14 > 14 + 14$$
$$-x > 28$$
$$-(-x) < -(28)$$
$$x < -28$$

31.
$$-4 \leq \ 2(x+1) \ < 10$$
$$-4 \leq \ \ 2x + 2 \ \ < 10$$
$$-4 - 2 \leq 2x + 2 - 2 < 10 - 2$$
$$-6 \leq \ \ 2x \ \ < 8$$
$$\frac{-6}{2} \leq \ \ \frac{2x}{2} \ \ < \frac{8}{2}$$
$$-3 \leq \ \ x \ \ < 4$$

32.
$$-10 < \ 2(4-x) \ \leq 12$$
$$-10 < \ \ 8 - 2x \ \ \leq 12$$
$$-10 - 8 < 8 - 8 - 2x \leq 12 - 8$$
$$-18 < \ \ -2x \ \ \leq 4$$
$$\frac{-18}{-2} > \ \ \frac{-2x}{-2} \ \ \geq \frac{4}{-2}$$
$$9 > \ \ x \ \ \geq -2$$
$$-2 \leq \ \ x \ \ < 9$$

Cumulative Review Exercises (page 147)

1. $\dfrac{27}{9} = 3$: integer, rational number, real number

2. $-0.25 = -\dfrac{1}{4}$: rational number, real number

3.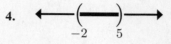

4.

5. $\dfrac{|-3| - |3|}{|-3 - 3|} = \dfrac{(+3) - (+3)}{|-6|} = \dfrac{0}{+6} = 0$

6. $\dfrac{14}{15} \cdot \dfrac{3}{4} = \dfrac{7 \cdot \overset{1}{\cancel{2}}}{5 \cdot \underset{1}{\cancel{3}}} \cdot \dfrac{\overset{1}{\cancel{3}}}{\underset{2}{\cancel{4}}} = \dfrac{7}{10}$

7. $2\dfrac{3}{5} + 5\dfrac{1}{2} = \dfrac{13}{5} + \dfrac{11}{2} = \dfrac{13 \cdot 2}{5 \cdot 2} + \dfrac{11 \cdot 5}{2 \cdot 5} = \dfrac{26}{10} + \dfrac{55}{10} = \dfrac{81}{10} = 8\dfrac{1}{10}$

8. $35.7 - 0.05 = 35.65$

9. $(2z - 3x)y = [2(0) - 3(-5)]3 = [0 + 15]3$
$$= 15(3) = 45$$

10. $\dfrac{x - 3y + |z|}{2 - x} = \dfrac{-5 - 3(3) + |0|}{2 - (-5)} = \dfrac{-5 - 9 + 0}{2 + 5} = \dfrac{-14}{7} = -2$

11. $x^2 - y^2 + z^2 = (-5)^2 - (3)^2 + 0^2$
$$= 25 - 9 + 0 = 16$$

12. $\dfrac{x}{y} + \dfrac{y + 2}{3 - z} = \dfrac{-5}{3} + \dfrac{3 + 2}{3 - 0} = -\dfrac{5}{3} + \dfrac{5}{3} = 0$

13. $rb = a$
$0.045(220) = a$
$9.9 = a$

14. $rb = a$
$0.32b = 1688$
$\dfrac{0.32b}{0.32} = \dfrac{1688}{0.32}$
$b = 5{,}275$

15. 2nd term: $5x^2y$; coefficient: 5

16. 3rd term: $37y$; factors: $37, y$

17. $4x + 7y - 9x = -5x + 7y$

18. $3(x - 7) + 2(8 - x) = 3x - 21 + 16 - 2x$
$$= x - 5$$

19. $2x^2y^3 - 4x^2y^3 = -2x^2y^3$

20. $3(5 - x) - 6(x + 2) = 15 - 3x - 6x - 12 = -9x + 3$

CUMULATIVE REVIEW EXERCISES

21.
$$2(x-7)+5=3x$$
$$2x-14+5=3x$$
$$2x-9=3x$$
$$2x-2x-9=3x-2x$$
$$-9=x$$

22.
$$\frac{x-5}{3}-5=7$$
$$\frac{x-5}{3}-5+5=7+5$$
$$\frac{x-5}{3}=12$$
$$3 \cdot \frac{x-5}{3}=3(12)$$
$$x-5=36$$
$$x-5+5=36+5$$
$$x=41$$

23.
$$\frac{2x-1}{5}=\frac{1}{2}$$
$$10 \cdot \frac{2x-1}{5}=10 \cdot \frac{1}{2}$$
$$2(2x-1)=5$$
$$4x-2=5$$
$$4x-2+2=5+2$$
$$4x=7$$
$$\frac{4x}{4}=\frac{7}{4}$$
$$x=\frac{7}{4}$$

24.
$$2(a-3)-3(a-2)=-a$$
$$2a-6-3a+6=-a$$
$$-a=-a$$
$$-a+a=-a+a$$
$$0=0$$
Identity, $\mathbb{R}$

25.
$$A=\frac{1}{2}h(b+B)$$
$$2A=2 \cdot \frac{1}{2}h(b+B)$$
$$2A=h(b+B)$$
$$\frac{2A}{b+B}=\frac{h(b+B)}{b+B}$$
$$\frac{2A}{b+B}=h, \text{ or } h=\frac{2A}{b+B}$$

26.
$$y=mx+b$$
$$y-b=mx+b-b$$
$$y-b=mx$$
$$\frac{y-b}{m}=\frac{mx}{m}$$
$$\frac{y-b}{m}=x, \text{ or } x=\frac{y-b}{m}$$

27. Let x = Dealer's invoice.

$$\boxed{\text{Price}} = \boxed{\begin{array}{c}\text{Dealer's}\\\text{invoice}\end{array}} + \boxed{\text{Markup}}$$
$$23499=x+0.03x$$
$$23499=1.03x$$
$$\frac{23499}{1.03}=\frac{1.03x}{1.03}$$
$$22814.56=x$$
The dealer's invoice was \$22,814.56.

28. Let x = original sofa price.

$$\boxed{\text{Price}} = \boxed{\begin{array}{c}\text{New sofa}\\\text{price}\end{array}} + \boxed{\begin{array}{c}\text{New chair}\\\text{price}\end{array}}$$
$$780=x-0.35x+300-0.35(300)$$
$$780=0.65x+300-105$$
$$780=0.65x+195$$
$$780-195=0.65x+195-195$$
$$585=0.65x$$
$$\frac{585}{0.65}=\frac{0.65x}{0.65}$$
$$900=x$$
The original price of the sofa was \$900.

74

CUMULATIVE REVIEW EXERCISES

29. Let x = original car price.

$$24618.65 = x + 0.085x$$
$$24618.65 = 1.085x$$
$$\frac{24618.65}{1.085} = \frac{1.085x}{1.085}$$
$$22690 = x$$

The original price was $22,690.

30. Let x and $3x$ = the unknown amounts.

$$\boxed{\text{Cement}} + \boxed{\text{Gravel}} = \boxed{\text{Concrete}}$$
$$x + 3x = 500$$
$$4x = 500$$
$$\frac{4x}{4} = \frac{500}{4}$$
$$x = 125 \text{ lbs cement}$$

31. Let x and $2x$ = the lengths.

$$\text{Total length} = 35$$
$$14 + x + 2x = 35$$
$$14 + 3x = 35$$
$$14 - 14 + 3x = 35 - 14$$
$$3x = 21$$
$$\frac{3x}{3} = \frac{21}{3}$$
$$x = 7 \text{ ft}$$

The section will not span the doorway.

32. Let x = the length of one panel.

$$\boxed{\text{Total length}} = 18$$
$$x + x + 3.4 = 18$$
$$2x + 3.4 = 18$$
$$2x + 3.4 - 3.4 = 18 - 3.4$$
$$2x = 14.6$$
$$\frac{2x}{2} = \frac{14.6}{2}$$
$$x = 7.3$$

The lengths are 7.3 and 10.7 feet.

33. Let n = number of kwh used.

$$52.50 + 0.28n = 203.70$$
$$52.50 - 52.50 + 0.28n = 203.70 - 52.50$$
$$0.28n = 151.20$$
$$\frac{0.28n}{0.28} = \frac{151.20}{0.28}$$
$$n = 540$$

540 kwh were used that month.

34. Let n = number of feet required.

$$47.75 + 2.05n = 427$$
$$47.75 - 47.75 + 2.05n = 427 - 47.75$$
$$2.05n = 379.25$$
$$\frac{2.05n}{2.05} = \frac{379.25}{2.05}$$
$$n = 185$$

185 ft of gutters were required.

35. $4^2 - 5^2 = 16 - 25 = -9$

36. $(4 - 5)^2 = (-1)^2 = 1$

37. $5(4^3 - 2^3) = 5(64 - 8) = 5(56) = 280$

38. $-2(5^4 - 7^3) = -2(625 - 343)$
$$= -2(282) = -564$$

39.
$$8(4 + x) > 10(6 + x)$$
$$32 + 8x > 60 + 10x$$
$$32 + 8x - 10x > 60 + 10x - 10x$$
$$32 - 2x > 60$$
$$32 - 32 - 2x > 60 - 32$$
$$-2x > 28$$
$$\frac{-2x}{-2} < \frac{28}{-2}$$
$$x < -14$$

40.
$$-9 < 3(x + 2) \le 3$$
$$-9 < 3x + 6 \le 3$$
$$-9 - 6 < 3x + 6 - 6 \le 3 - 6$$
$$-15 < 3x \le -3$$
$$\frac{-15}{3} < \frac{3x}{3} \le \frac{-3}{3}$$
$$-5 < x \le -1$$

75

Exercises 3.1 (page 158)

1. III

3. IV

5. $-5 - 5(-4) = -5 - (-20) = -5 + 20$
$= 15$

7. opposite of -8: 8

9. $-4x + 7 = -21$
$-4x + 7 - 7 = -21 - 7$
$-4x = -28$
$\dfrac{-4x}{-4} = \dfrac{-28}{-4}$
$x = 7$

11. $(x+1)(x+y)^2 = (-2+1)[-2+(-5)]^2$
$= (-1)(-7)^2$
$= -1(+49) = -49$

13. ordered pair

15. origin

17. rectangular, Cartesian

19. coordinates

21. no

23. origin, left, up

25. II

27.

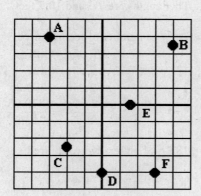

29.

x	y
2	0
-2	2
4	-1
-4	3
0	1

31. The point $(-10, 60)$ indicates that 10 minutes before the workout started, her heart rate was 60 beats per minute.

33. The point with an x-coordinate of 30 is $(30, 150)$, so her heart rate was 150 beats per minute one half-hour after starting.

35. The points on the graph with a y-coordinate of 100 have x-coordinates of approximately 5 and 50, so her heart rate was 100 beats per minute after about 5 and 50 minutes.

37. Before the workout, her heart rate was 60 beats per minute. After the workout, her heart rate was about 70 beats per minute, or about 10 beats per minute higher.

39. To find the charge for a 1-day rental, find the y-coordinate of the point with an x-coordinate of 1. This is the point $(1, 2)$. The charge will be $2.

41. To find the charge for a 5-day rental, find the y-coordinate of the point with an x-coordinate of 5. This is the point $(5, 7)$. The charge will be $7.

43. $A(-1, 4)$, $B(2, -2)$, $C(-3, 0)$, $D(1, -4)$, $E(4, 3)$, $F(-4, 4)$

45. To find the cost for a 1-oz. letter, find the y-coordinate of the point with an x-coordinate of 1. This is the point $(1, 45)$. Cost $= 45\cent$

To find the cost for a $2\frac{1}{2}$-oz. letter, find the y-coordinate of the point with an x-coordinate of $2\frac{1}{2}$. This is the point $\left(2\frac{1}{2}, 85\right)$. Cost $= 85\cent$

47. To find the cost for a 0.75-oz. letter, find the y-coordinate of the point with an x-coordinate of 0.75. This is the point $(0.75, 45)$, with a cost of $45\cent$. To find the cost for a 2.75-oz. letter, find the y-coordinate of the point with an x-coordinate of 2.75. This is the point $(2.75, 85)$, with a cost of $85\cent$. The difference is $85 - 45 = 40\cent$.

49.

City	Ordered Pair
Carbondale	$\left(5\frac{1}{2}, J\right)$
Champaign	$\left(6\frac{1}{2}, D\right)$
Chicago	$(8, B)$
Peoria	$\left(5\frac{1}{2}, C\right)$
Rockford	$\left(5\frac{1}{2}, A\right)$

51. **a.** The highest point on the graph for the 60° angle is $(-2, 7)$, for a height of 7 ft. The highest point on the graph for the 30° angle is $(0, 3)$, for a height of 3 ft. The 60° angle results in a height 4 ft greater.

b. The farthest point on the graph for the 60° angle is $(3, 0)$, for a final location of 3 ft. The farthest point on the graph for the 30° angle is $(7, 0)$, for a final location of 7 ft. The 30° angle results in a distance 4 ft farther.

53.

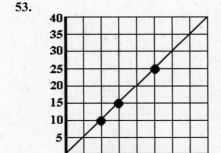

a. $(6, 30) \Rightarrow 30$ miles
b. $(8, 40) \Rightarrow 8$ gallons
c. $(6.5, 32.5) \Rightarrow 32.5$ miles

55.

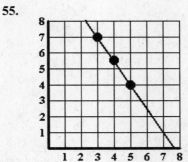

a. A 3-year old car is worth $7000.
b. $(7, 1) \Rightarrow \$1000$
c. $(6, 2500) \Rightarrow 6$ years

57-61. Answers may vary.

Exercises 3.2 (page 171)

1. 2

3. $y = 3x + 2$
$y = 3(1) + 2 = 3 + 2 = 5$

5. Vertical lines have no y-intercepts. (except $x = 0$)

7. $\dfrac{2}{3}x = -12$
$\dfrac{3}{2} \cdot \dfrac{2}{3}x = \dfrac{3}{2}\left(-\dfrac{12}{1}\right)$
$x = -18$

9. expression

11. $rb = a$
$0.005(250) = a$
$1.25 = a$

13. $-2.5 - (-2.6) = -2.5 + 2.6 = 0.1$

15. linear, two

17. independent, dependent

19. linear

21. y-intercept

23. $x - 2y = -4$
$4 - 2(4) \overset{?}{=} -4$
$4 - 8 \overset{?}{=} -4$
$-4 = -4 \Rightarrow (4, 4)$ is a solution.

25. $y = \dfrac{2}{3}x + 5$
$12 \overset{?}{=} \dfrac{2}{3}(6) + 5$
$12 \overset{?}{=} 4 + 5$
$12 \neq 9 \Rightarrow (6, 12)$ is not a solution.

27. $y = x - 3$

$x = 0$	$x = 1$	$x = -2$	$x = -4$
$y = 0 - 3$	$y = 1 - 3$	$y = -2 - 3$	$y = -4 - 3$
$y = -3$	$y = -2$	$y = -5$	$y = -7$

x	y	(x, y)
0	-3	$(0, -3)$
1	-2	$(1, -2)$
-2	-5	$(-2, -5)$
-4	-7	$(-4, -7)$

29. $y = -2x$

$x = 0$	$x = 1$	$x = 3$	$x = -2$
$y = -2(0)$	$y = -2(1)$	$y = -2(3)$	$y = -2(-2)$
$y = 0$	$y = -2$	$y = -6$	$y = 4$

x	y	(x, y)
0	0	$(0, 0)$
1	-2	$(1, -2)$
3	-6	$(3, -6)$
-2	4	$(-2, 4)$

31.
$$y = 2x$$

$x = 0$	$x = 1$	$x = -1$
$y = 2(0)$	$y = 2(1)$	$y = 2(-1)$
$y = 0$	$y = 2$	$y = -2$

x	y
0	0
1	2
-1	-2

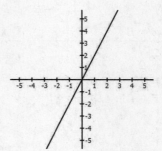

33.
$$y = 2x - 1$$

$x = 0$	$x = 1$	$x = -1$
$y = 2(0) - 1$	$y = 2(1) - 1$	$y = 2(-1) - 1$
$y = 0 - 1$	$y = 2 - 1$	$y = -2 - 1$
$y = -1$	$y = 1$	$y = -3$

x	y
0	-1
1	1
-1	-3

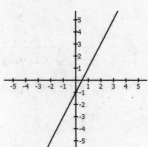

35.
$$y = 1.2x - 2$$

$x = 0$	$x = 5$	$x = -5$
$y = 1.2(0) - 2$	$y = 1.2(5) - 2$	$y = 1.2(-5) - 2$
$y = 0 - 2$	$y = 6 - 2$	$y = -6 - 2$
$y = -2$	$y = 4$	$y = -8$

x	y
0	-2
5	4
-5	-8

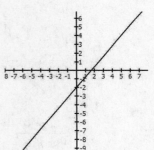

37.
$$y = 2.5x - 5$$

$x = -4$	$x = -2$
$y = 2.5(-4) - 5$	$y = 2.5(-2) - 5$
$y = -10 - 5$	$y = -5 - 5$
$y = -15$	$y = -10$

$x = 0$	$x = 4$
$y = 2.5(0) - 5$	$y = 2.5(4) - 5$
$y = 0 - 5$	$y = 10 - 5$
$y = -5$	$y = 5$

x	y
-4	-15
-2	-10
0	-5
4	5

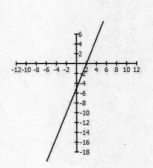

79

39.

$$y = \frac{x}{2} - 2$$

$x = 0$	$x = 2$	$x = -2$
$y = \dfrac{0}{2} - 2$	$y = \dfrac{2}{2} - 2$	$y = \dfrac{-2}{2} - 2$
$y = 0 - 2$	$y = 1 - 2$	$y = -1 - 2$
$y = -2$	$y = -1$	$y = -3$

x	y
0	-2
2	-1
-2	-3

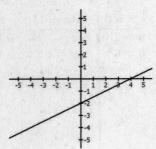

41.

$$y - 3 = -\frac{1}{2}(2x + 4)$$

$x = 0$	$x = 1$
$y - 3 = -\dfrac{1}{2}(2(0) + 4)$	$y - 3 = -\dfrac{1}{2}(2(1) + 4)$
$y - 3 = -\dfrac{1}{2}(4)$	$y - 3 = -\dfrac{1}{2}(6)$
$y - 3 = -2$	$y - 3 = -3$
$y = 1$	$y = 0$

$x = -1$
$y - 3 = -\dfrac{1}{2}(2(-1) + 4)$
$y - 3 = -\dfrac{1}{2}(2)$
$y - 3 = -1$
$y = 2$

x	y
0	1
1	0
-1	2

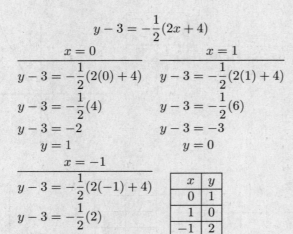

43.

$$x + y = 7$$

$x = 0$	$y = 0$	$x = 2$
$0 + y = 7$	$x + 0 = 7$	$2 + y = 7$
$y = 7$	$x = 7$	$y = 5$
$(0, 7)$	$(7, 0)$	$(2, 5)$

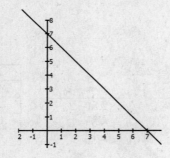

45.

$$2x + 3y = 12$$

$x = 0$	$y = 0$	$x = 3$
$2(0) + 3y = 12$	$2x + 3(0) = 12$	$2(3) + 3y = 12$
$0 + 3y = 12$	$2x + 0 = 12$	$6 + 3y = 12$
$3y = 12$	$2x = 12$	$3y = 6$
$y = 4$	$x = 6$	$y = 2$
$(0, 4)$	$(6, 0)$	$(3, 2)$

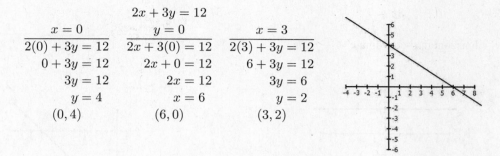

47.
$$y = -5$$
horizontal, y-coordinate $= -5$

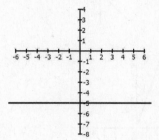

49.
$$x = 5$$
vertical, x-coordinate $= 5$

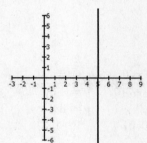

51.
$$y = -3x - 1$$
$$3x + y = -1$$

$x = 0$	$y = 0$	$x = 1$
$3(0) + y = -1$	$3x + 0 = -1$	$y = -3x - 1$
$0 + y = -1$	$3x = -1$	$y = -3(1) - 1$
$y = -1$	$x = -\frac{1}{3}$	$y = -4$
$(0, -1)$	$\left(-\frac{1}{3}, 0\right)$	$(1, -4)$

53.
$$x - y = -2$$

$x = 0$	$y = 0$	$x = 2$
$0 - y = -2$	$x - 0 = -2$	$2 - y = -2$
$-y = -2$	$x = -2$	$-y = -4$
$y = 2$	$(-2, 0)$	$y = 4$
$(0, 2)$		$(2, 4)$

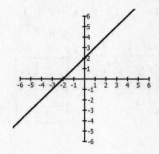

81

55.
$$3y = 7$$
$$y = \frac{7}{3}$$
horizontal, y-coordinate $= \frac{7}{3} = 2\frac{1}{3}$

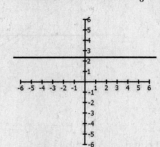

57.
$$x - y = 7$$

$x = 0$	$y = 0$	$x = 2$
$0 - y = 7$	$x - 0 = 7$	$2 - y = 7$
$-y = 7$	$x = 7$	$-y = 5$
$y = -7$	$(7, 0)$	$y = -5$
$(0, -7)$		$(2, -5)$

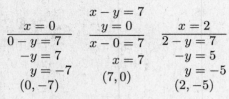

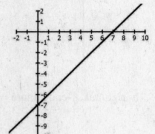

59.
$$y = -3x$$

$x = 0$	$x = -1$	$x = 1$
$y = -3(0)$	$y = -3(-1)$	$y = -3(1)$
$y = 0$	$y = 3$	$y = -3$
$(0, 0)$	$(-1, 3)$	$(1, -3)$

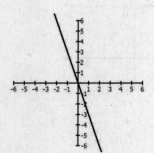

61.
$$y + 2 = \frac{3}{4}(4x + 8)$$

$x = 0$	$x = 1$
$y + 2 = \frac{3}{4}(4(0) + 8)$	$y + 2 = \frac{3}{4}(4(1) + 8)$
$y + 2 = \frac{3}{4}(8)$	$y + 2 = \frac{3}{4}(12)$
$y + 2 = 6$	$y + 2 = 9$
$y = 4$	$y = 7$

$$x = -1$$
$$y + 2 = \frac{3}{4}(4(-1) + 8)$$
$$y + 2 = \frac{3}{4}(4)$$
$$y + 2 = 3$$
$$y = 1$$

x	y
0	4
1	7
-1	1

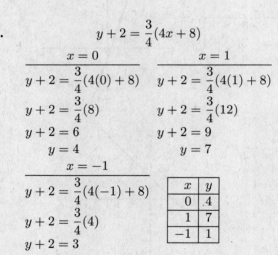

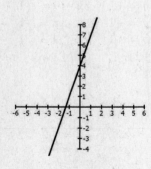

82

63. **a.** $c = 50 + 25u$

b.

u	c	(u, c)
4	$50 + 25(4) = 150$	$(4, 150)$
8	$50 + 25(8) = 250$	$(8, 250)$
14	$50 + 25(14) = 400$	$(14, 400)$

c. The service fee is \$50.

d. cost for 18 units = \$500

cost for 12 units = \$350

Total cost = \$850

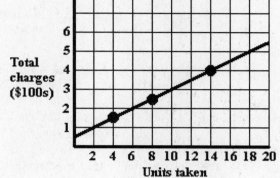

65. **a.**

r	h	(r, h)
7	$3.9(7) + 28.9 = 56.2$	$(7, 56.2)$
8.5	$3.9(8.5) + 28.9 = 62.1$	$(8.5, 62.1)$
9	$3.9(9) + 28.9 = 64.0$	$(9, 64.0)$

b. ...taller the woman is.

c. 68 inches tall

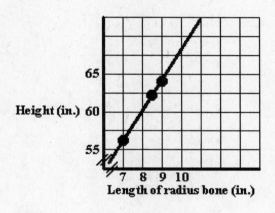

67-71. Answers may vary.

73. $x_M = \dfrac{a + c}{2} = \dfrac{5 + 7}{2} = \dfrac{12}{2} = 6$

$y_M = \dfrac{b + d}{2} = \dfrac{3 + 9}{2} = \dfrac{12}{2} = 6$

$M(6, 6)$

75. $x_M = \dfrac{a + c}{2} = \dfrac{2 + (-3)}{2} = \dfrac{-1}{2} = -\dfrac{1}{2}$

$y_M = \dfrac{b + d}{2} = \dfrac{-7 + 12}{2} = \dfrac{5}{2}$

$M\left(-\dfrac{1}{2}, \dfrac{5}{2}\right)$

77. $x_M = \dfrac{a + c}{2} = \dfrac{4 + 10}{2} = \dfrac{14}{2} = 7$

$y_M = \dfrac{b + d}{2} = \dfrac{6 + 6}{2} = \dfrac{12}{2} = 6$

$M(7, 6)$

79. $x_M = \dfrac{a + c}{2} = \dfrac{x + x - 1}{2} = \dfrac{2x - 1}{2}$

$y_M = \dfrac{b + d}{2} = \dfrac{3 + (-4)}{2} = \dfrac{-1}{2} = -\dfrac{1}{2}$

$M\left(\dfrac{2x - 1}{2}, -\dfrac{1}{2}\right)$

Exercises 3.3 (page 185)

1. $\dfrac{3-7}{-6-4} = \dfrac{-4}{-10} = \dfrac{2}{5}$

3. $\dfrac{9-(-8)}{-2-(-2)} = \dfrac{9+8}{-2+2} = \dfrac{17}{0}$: undefined

5. $4(a-3)+2a = 4a-12+2a = 6a-12$

7. $4z-6(z+w)+2w = 4z-6z-6w+2w$
$= -2z-4w$

9. $3(a-b)-2(a+b) = 3a-3b-2a-2b$
$= a-5b$

11. y; x

13. rise; run; rise; run   **15.** hypotenuse

17. perpendicular   **19.** increasing; decreasing

21.

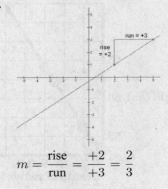

$m = \dfrac{\text{rise}}{\text{run}} = \dfrac{+2}{+3} = \dfrac{2}{3}$

23.

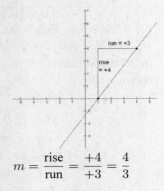

$m = \dfrac{\text{rise}}{\text{run}} = \dfrac{+4}{+3} = \dfrac{4}{3}$

25.

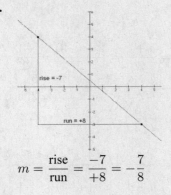

$m = \dfrac{\text{rise}}{\text{run}} = \dfrac{-7}{+8} = -\dfrac{7}{8}$

27.

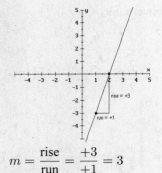

$m = \dfrac{\text{rise}}{\text{run}} = \dfrac{+3}{+1} = 3$

29. $m = \dfrac{y_2 - y_1}{x_2 - x_1} = \dfrac{9-0}{3-0} = \dfrac{9}{3} = 3$

31. $m = \dfrac{y_2 - y_1}{x_2 - x_1} = \dfrac{-3-(-5)}{4-2} = \dfrac{2}{2} = 1$

33. $m = \dfrac{y_2 - y_1}{x_2 - x_1} = \dfrac{2-(-1)}{-6-3} = \dfrac{3}{-9} = -\dfrac{1}{3}$

35. $m = \dfrac{y_2 - y_1}{x_2 - x_1} = \dfrac{-11-15}{-6-4} = \dfrac{-26}{-10} = \dfrac{13}{5}$

84

37. Pick two x-coordinates at random and find the corresponding y-coordinates. For example, use $x = 0$ and $x = 4$.

x	0	4
y	6	0

$$m = \frac{y_2 - y_1}{x_2 - x_1} = \frac{0 - 6}{4 - 0} = \frac{-6}{4} = -\frac{3}{2}$$

39. Pick two x-coordinates at random and find the corresponding y-coordinates. For example, use $x = 0$ and $x = 5$.

x	0	5
y	2	4

$$m = \frac{y_2 - y_1}{x_2 - x_1} = \frac{4 - 2}{5 - 0} = \frac{2}{5}$$

41. Pick two x-coordinates at random and find the corresponding y-coordinates. For example, use $x = 4$ and $x = 8$.

x	4	8
y	0	2

$$m = \frac{y_2 - y_1}{x_2 - x_1} = \frac{2 - 0}{8 - 4} = \frac{2}{4} = \frac{1}{2}$$

43. Pick two x-coordinates at random and find the corresponding y-coordinates. For example, use $x = 0$ and $x = 1$.

x	0	1
y	-4	1

$$m = \frac{y_2 - y_1}{x_2 - x_1} = \frac{1 - (-4)}{1 - 0} = \frac{5}{1} = 5$$

45. horizontal line; $m = 0$

47. $m = \dfrac{y_2 - y_1}{x_2 - x_1} = \dfrac{-2 - (-5)}{-7 - (-7)} = \dfrac{3}{0}$

There is no defined slope.

49. $m = \dfrac{y_2 - y_1}{x_2 - x_1} = \dfrac{-2 - (-2)}{-4 - 6} = \dfrac{0}{-10} = 0$

51. $4x - 2 = 12$

$4x = 14$

$x = \dfrac{14}{4} = \dfrac{7}{2}$; vertical, undefined slope

53. $2(y + 1) = 3y$

$2y + 2 = y$

$y = -2$; horizontal, $m = 0$

55. $x - y = \dfrac{7 - 5y}{5}$

$5(x - y) = 5 \cdot \dfrac{7 - 5y}{5}$

$5x - 5y = 7 - 5y$

$5x = 7$

$x = \dfrac{7}{5}$; vertical, no defined slope

57. $3 \neq -\dfrac{1}{3}; 3\left(-\dfrac{1}{3}\right) = -1$

perpendicular

59. $4 \neq 0.25; 4(0.25) \neq -1$

neither

61. $\dfrac{2}{3} = \dfrac{4}{6}$; parallel

63. $m_1 = 0 \Rightarrow$ horizontal

m_2 is undefined $\Rightarrow$ vertical; perpendicular

65. $m = \dfrac{y_2 - y_1}{x_2 - x_1} = \dfrac{2 - (-1)}{-3 - 2} = \dfrac{3}{-5} = -\dfrac{3}{5}$

67. $m = \dfrac{y_2 - y_1}{x_2 - x_1} = \dfrac{2 - (-3)}{-3 - 2} = \dfrac{5}{-5} = -1$

SECTION 3.3

69. negative   **71.** positive   **73.** undefined

75. PQ: $m = \dfrac{y_2 - y_1}{x_2 - x_1} = \dfrac{8-4}{4-(-2)} = \dfrac{4}{6} = \dfrac{2}{3}$ PR: $m = \dfrac{y_2 - y_1}{x_2 - x_1} = \dfrac{12-4}{8-(-2)} = \dfrac{8}{10} = \dfrac{4}{5}$
not on same line

77. PQ: $m = \dfrac{y_2 - y_1}{x_2 - x_1} = \dfrac{0-10}{-6-(-4)} = \dfrac{-10}{-2} = 5$ PR: $m = \dfrac{y_2 - y_1}{x_2 - x_1} = \dfrac{5-10}{-1-(-4)} = \dfrac{-5}{3}$
not on same line

79. PQ: $m = \dfrac{y_2 - y_1}{x_2 - x_1} = \dfrac{8-4}{0-(-2)} = \dfrac{4}{2} = 2$ PR: $m = \dfrac{y_2 - y_1}{x_2 - x_1} = \dfrac{12-4}{2-(-2)} = \dfrac{8}{4} = 2$
on same line

81. $m = \dfrac{y_2 - y_1}{x_2 - x_1} = \dfrac{1-(-3)}{-4-(-2)} = \dfrac{4}{-2} = -2$ **83.** $m = \dfrac{y_2 - y_1}{x_2 - x_1} = \dfrac{5-1}{6-(-2)} = \dfrac{4}{8} = \dfrac{1}{2}$
parallel   perpendicular

85. $m = \dfrac{y_2 - y_1}{x_2 - x_1} = \dfrac{0-(-3)}{-2-4} = \dfrac{3}{-6} = -\dfrac{1}{2}$ **87.** x-axis: $y = 0$, $m = 0$
neither

89. $m = \dfrac{\text{rise}}{\text{run}} = \dfrac{24}{5280} = \dfrac{1}{220}$ **91.** $m = \dfrac{\text{rise}}{\text{run}} = \dfrac{3}{15} = \dfrac{1}{5}$

93. Let $x =$ the number of years the program has been offered, and let $y =$ the enrollment. Then two points are $(1, 12)$ and $(5, 26)$. Find the slope:
$m = \dfrac{y_2 - y_1}{x_2 - x_1} = \dfrac{26-12}{5-1} = \dfrac{14}{4} = 3.5$
The enrollment is growing by 3.5 students per year.

95. Let $x =$ the year, and let $y =$ the price. Then two points are $(-10, 6700)$ and $(-3, 2200)$. Find the slope:
$m = \dfrac{y_2 - y_1}{x_2 - x_1} = \dfrac{2200-6700}{-3-(-10)} = \dfrac{-4500}{7}$
≈ -642.86
The price has decreased by about \$642.86 per year.

97. Answers may vary.

99. Slope from $(3, a)$ to $(5, 7)$: $m = \dfrac{y_2 - y_1}{x_2 - x_1} = \dfrac{7-a}{5-3} = \dfrac{7-a}{2}$

Slope from $(5, 7)$ to $(7, 10)$: $m = \dfrac{y_2 - y_1}{x_2 - x_1} = \dfrac{10-7}{7-5} = \dfrac{3}{2}$

Set slopes equal: $\dfrac{7-a}{2} = \dfrac{3}{2}$

$$2 \cdot \dfrac{7-a}{2} = 2 \cdot \dfrac{3}{2}$$
$$7 - a = 3$$
$$4 = a$$

Exercises 3.4 (page 192)

1. $y - 5 = 2(x - 3)$
$y - y_1 = m(x - x_1)$
$m = 2$; point: $(3, 5)$

3. $y - (-5) = \dfrac{7}{8}[x - (-4)]$
$y - y_1 = m(x - x_1)$
$m = \frac{7}{8}$; point: $(-4, -5)$

5. $3(x + 2) + x = 5x$
$3x + 6 + x = 5x$
$4x + 6 = 5x$
$6 = x$

7. $\dfrac{5(2 - x)}{3} - 1 = x + 5$
$3\left[\dfrac{5(2 - x)}{3} - 1\right] = 3(x + 5)$
$5(2 - x) - 3 = 3x + 15$
$10 - 5x - 3 = 3x + 15$
$-5x + 7 = 3x + 15$
$-8x = 8$
$x = -1$

9. $rb = a$
$0.53(15000) = a$
$7950 = a$
About 7,950 of the ads are likely to be read.

11. $y - y_1 = m(x - x_1)$

13. $(1, 2); 2; 3$

15. $y - y_1 = m(x - x_1)$
$y - 0 = 3(x - 0)$

17. $y - y_1 = m(x - x_1)$
$y - (-2) = -7(x - (-1))$

19. $y - y_1 = m(x - x_1)$
$y - 3 = 2(x - (-5))$

21. $y - y_1 = m(x - x_1)$
$y - 5 = -\dfrac{6}{7}(x - 6)$

23. $y - y_1 = m(x - x_1)$
$y - (-3) = -5(x - 2)$
$y + 3 = -5x + 10$
$y = -5x + 7$

25. $y - y_1 = m(x - x_1)$
$y - 7 = 5(x - 0)$
$y - 7 = 5x$
$y = 5x + 7$

27. $y - y_1 = m(x - x_1)$
$y - 0 = -3(x - 2)$
$y = -3x + 6$

29. $y - y_1 = m(x - x_1)$
$y - (-2) = \dfrac{1}{3}(x - 6)$
$y + 2 = \dfrac{1}{3}x - 2$
$y = \dfrac{1}{3}x - 4$

31. Use the points $(-1, 3)$ and $(2, 5)$:
$m = \dfrac{y_2 - y_1}{x_2 - x_1} = \dfrac{5 - 3}{2 - (-1)} = \dfrac{2}{3}$
$y - y_1 = m(x - x_1)$
$y - 3 = \dfrac{2}{3}(x - (-1))$
$y - 3 = \dfrac{2}{3}(x + 1)$
$y - 3 = \dfrac{2}{3}x + \dfrac{2}{3}$
$y = \dfrac{2}{3}x + \dfrac{2}{3} + \dfrac{3}{1}$
$y = \dfrac{2}{3}x + \dfrac{2}{3} + \dfrac{9}{3}$
$y = \frac{2}{3}x + \frac{11}{3}$

SECTION 3.4

33. Use the points $(-2, 0)$ and $(0, -1)$: $m = \dfrac{y_2 - y_1}{x_2 - x_1} = \dfrac{-1 - 0}{0 - (-2)} = \dfrac{-1}{2} = -\dfrac{1}{2}$

$$y - y_1 = m(x - x_1)$$
$$y - 0 = -\frac{1}{2}(x - (-2))$$
$$y = -\frac{1}{2}(x + 2)$$
$$y = -\frac{1}{2}x - 1$$

35. $(x_1, y_1) = (1, 3);\ m = 2 = \dfrac{+2}{+1} = \dfrac{\Delta y}{\Delta x}$

Start at $(1, 3)$ and go up 2 and right 1.

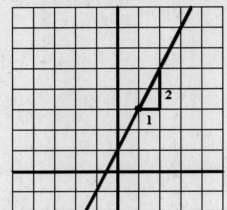

37. $(x_1, y_1) = (2, 4);\ m = -\dfrac{2}{3} = \dfrac{-2}{+3} = \dfrac{\Delta y}{\Delta x}$

Start at $(2, 4)$ and go down 2 and right 3.

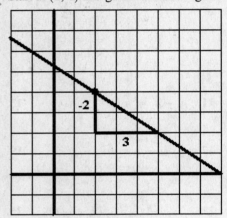

39.
$$y - y_1 = m(x - x_1)$$
$$y - (-8) = 0.5(x - (-1))$$
$$y + 8 = 0.5(x + 1)$$

41.
$$y - y_1 = m(x - x_1)$$
$$y - 2 = -4(x - (-3))$$
$$y - 2 = -4(x + 3)$$

43. $m = \dfrac{y_2 - y_1}{x_2 - x_1} = \dfrac{4 - 0}{4 - 0} = \dfrac{4}{4} = 1$
$$y - y_1 = m(x - x_1)$$
$$y - 0 = 1(x - 0)$$
$$y = x$$

45. $m = \dfrac{y_2 - y_1}{x_2 - x_1} = \dfrac{-3 - 4}{0 - 3} = \dfrac{-7}{-3} = \dfrac{7}{3}$
$$y - y_1 = m(x - x_1)$$
$$y - 4 = \frac{7}{3}(x - 3)$$
$$y - 4 = \frac{7}{3}x - 7$$
$$y = \frac{7}{3}x - 3$$

47. $(x_1, y_1) = (-2, 1); m = -\dfrac{1}{2} = \dfrac{-1}{+2} = \dfrac{\Delta y}{\Delta x}$

Start at $(-2, 1)$ and go down 1 and right 2.

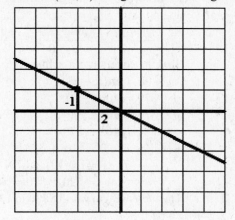

49. $(x_1, y_1) = (2, -2); m = 3 = \dfrac{+3}{+1} = \dfrac{\Delta y}{\Delta x}$

Start at $(2, -2)$ and go up 3 and right 1.

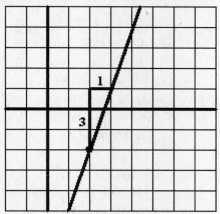

51.
$$y - y_1 = m(x - x_1)$$
$$y - (-1) = -\frac{2}{3}(x - 3)$$
$$y + 1 = -\frac{2}{3}x + 2$$
$$y = -\frac{2}{3}x + 1$$

53.
$$y - y_1 = m(x - x_1)$$
$$y - 8 = 0.5(x - 10)$$
$$y - 8 = 0.5x - 5$$
$$y = 0.5x + 3$$

55.
$$m = \frac{y_2 - y_1}{x_2 - x_1} = \frac{-4 - 2}{-3 - 1} = \frac{-6}{-4} = \frac{3}{2}$$
$$y - y_1 = m(x - x_1)$$
$$y - 2 = \frac{3}{2}(x - 1)$$
$$y - 2 = \frac{3}{2}x - \frac{3}{2}$$
$$y = \frac{3}{2}x - \frac{3}{2} + \frac{2}{1}$$
$$y = \frac{3}{2}x - \frac{3}{2} + \frac{4}{2}$$
$$y = \frac{3}{2}x + \frac{1}{2}$$

57. Let $x =$ the # of kilowatt-hours, and let $y =$ the bill. Then two points are known: $(1450, 200)$ and $(1850, 250)$. Find the slope:
$$m = \frac{y_2 - y_1}{x_2 - x_1} = \frac{250 - 200}{1850 - 1450} = \frac{50}{400} = \frac{1}{8}$$
Find the equation of the line:
$$y - y_1 = m(x - x_1)$$
$$y - 200 = \tfrac{1}{8}(x - 1450)$$
$$y - 200 = \tfrac{1}{8}x - 181.25$$
$$y = \tfrac{1}{8}x + 18.75$$
Let $x = 1500$ and find y:
$$y = \tfrac{1}{8}x + 18.75$$
$$= \tfrac{1}{8}(1500) + 18.75$$
$$= 187.50 + 18.75 = 206.25$$
Her bill will be \$206.25.

59. Let $x =$ the number of years owned, and let $y =$ the value. Then two points are known: $(0, 370)$ and $(2, 450)$. Find the slope:
$$m = \frac{y_2 - y_1}{x_2 - x_1} = \frac{450 - 370}{2 - 0} = \frac{80}{2} = 40$$
Find the equation of the line:
$$y - y_1 = m(x - x_1)$$
$$y - 370 = 40(x - 0)$$
$$y - 370 = 40x$$
$$y = 40x + 370$$
Let $x = 13$: $\quad y = 40x + 370$
$$= 40(13) + 370$$
$$= 520 + 370 = \$890$$
The table will be worth \$890 in 13 years.

61. Let $x =$ the number of hours, and let $y =$ the total charge.
$$m = \frac{y_2 - y_1}{x_2 - x_1} = \frac{105 - 70}{4 - 2}$$
$$= \frac{35}{2} = \$17.50 \text{ per hr}$$

63. Let $x =$ the year after purchase, and let $y =$ the value. Then two points are known: $(3, 147700)$ and $(10, 172200)$.
$$m = \frac{y_2 - y_1}{x_2 - x_1} = \frac{172200 - 147700}{10 - 3}$$
$$= \frac{24500}{7} = 3500$$
Find the equation of the line:
$$y - y_1 = m(x - x_1)$$
$$y - 147700 = 3500(x - 3)$$
$$y - 147700 = 3500x - 10500$$
$$y = 3500x + 137200$$
Let $x = 0$:
$$y = 3500(0) + 137200 = 137200$$
The original price was \$137,200.

65. **Answers may vary.**

67. The slope of a vertical line is undefined, so the equation cannot be written in point-slope form.

Exercises 3.5 (page 201)

1. $2x + 4y = 12$
$4y = -2x + 12$
$\dfrac{4y}{4} = \dfrac{-2x}{4} + \dfrac{12}{4}$
$y = -\dfrac{1}{2}x + 3$

3. $2 = \dfrac{3}{5}(10) + b$
$2 = 6 + b$
$-4 = b$

5. $2x + 3 = 7$
$2x + 3 - 3 = 7 - 3$
$2x = 4$
$\dfrac{2x}{2} = \dfrac{4}{2}$
$x = 2$

7. $3(y - 2) = y + 1$
$3y - 6 = y + 1$
$3y - y - 6 = y - y + 1$
$2y - 6 = 1$
$2y - 6 + 6 = 1 + 6$
$2y = 7$
$\frac{2y}{2} = \frac{7}{2}$
$y = \frac{7}{2}$

9. $y = mx + b; -3; (0, 7)$

11. reciprocals

13. $y = 7x - 5$
$y = mx + b; m = 7, (0, -5)$

15. $y = -\frac{2}{5}x + 6$
$y = mx + b; m = -\frac{2}{5}, (0, 6)$

17. $3x - 2y = 8$
$-2y = -3x + 8$
$y = \dfrac{-3x}{-2} + \dfrac{8}{-2}$
$y = \frac{3}{2}x - 4$
$y = mx + b$
$m = \frac{3}{2}, (0, -4)$

19. $-2x - 6y = 5$
$-6y = 2x + 5$
$y = \dfrac{2x}{-6} + \dfrac{5}{-6}$
$y = -\frac{1}{3}x - \frac{5}{6}$
$y = mx + b$
$m = -\frac{1}{3}, \left(0, -\frac{5}{6}\right)$

21. $y = mx + b$
$y = 12x + b$
$0 = 12(0) + b$
$0 = 0 + b$
$0 = b$
$y = 12x + 0 \Rightarrow y = 12x$

23. $y = mx + b$
$y = -5x + b$
$-4 = -5(0) + b$
$-4 = 0 + b$
$-4 = b$
$y = -5x - 4$

25.
$$y = mx + b$$
$$y = -7x + b$$
$$5 = -7(7) + b$$
$$5 = -49 + b$$
$$54 = b$$
$$y = -7x + 54$$

27.
$$y = mx + b$$
$$y = 0x + b$$
$$-5 = 0(3) + b$$
$$-5 = 0 + b$$
$$-5 = b$$
$$y = 0x + (-5)$$
$$y = -5$$

29.
$$m = \frac{y_2 - y_1}{x_2 - x_1} = \frac{-11 - (-5)}{3 - 1} = \frac{-6}{2} = -3$$
$$y = mx + b$$
$$-5 = -3(1) + b$$
$$-5 = -3 + b$$
$$-2 = b \Rightarrow y = -3x - 2$$

31.
$$m = \frac{y_2 - y_1}{x_2 - x_1} = \frac{2 - 10}{8 - (-8)} = \frac{-8}{16} = -\frac{1}{2}$$
$$y = mx + b$$
$$10 = -\frac{1}{2}(-8) + b$$
$$10 = 4 + b$$
$$6 = b \Rightarrow y = -\frac{1}{2}x + 6$$

33.
$$x - y = 1$$
$$-y = -x + 1$$
$$y = x - 1$$
$$m = 1 = \frac{+1}{+1}, (0, -1)$$

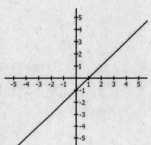

35.
$$2x = 3y - 6$$
$$2x + 6 = 3y$$
$$\frac{2x + 6}{3} = y$$
$$\frac{2}{3}x + \frac{6}{3} = y$$
$$\frac{2}{3}x + 2 = y$$
$$m = \frac{2}{3} = \frac{+2}{+3}, (0, 2)$$

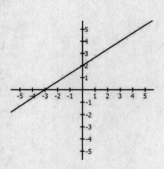

92

SECTION 3.5

37.
$$3y = -2x + 18$$
$$y = \frac{-2x + 18}{3}$$
$$y = -\frac{2}{3}x + \frac{18}{3}$$
$$y = -\frac{2}{3}x + 6$$
$$m = -\frac{2}{3} = \frac{-2}{+3}, \ (0, 6)$$

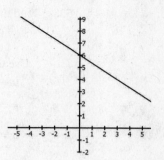

39.
$$y = -6x + 3 \quad y = -6x - 4$$
$$m = -6 \qquad m = -6$$
$$\text{parallel}$$

41.
$$x + y = 7 \qquad y = x - 3$$
$$y = -x + 7 \qquad m = 1$$
$$m = -1$$
$$\text{perpendicular}$$

43.
$$y = 3x + 9 \quad 2y = 6x - 10$$
$$m = 3 \qquad y = 3x - 5$$
$$m = 3$$
$$\text{parallel}$$

45.
$$x = 3y + 9 \quad y = -3x + 2$$
$$x - 9 = 3y \qquad m = -3$$
$$\frac{1}{3}x - 3 = y$$
$$m = \frac{1}{3}$$
$$\text{perpendicular}$$

47. Find the slope of the given line:
$$y = 4x - 9 \Rightarrow m = 4$$
Use the parallel slope.
$$y = mx + b$$
$$y = 4x + b$$
$$0 = 4(0) + b$$
$$0 = 0 + b$$
$$0 = b$$
$$y = 4x + 0 \Rightarrow y = 4x$$

49. Find the slope of the given line:
$$4x - y = 7$$
$$4x - 7 = y \Rightarrow m = 4$$
Use the parallel slope.
$$y = mx + b$$
$$y = 4x + b$$
$$5 = 4(2) + b$$
$$5 = 8 + b$$
$$-3 = b$$
$$y = 4x - 3$$

93

51. Find the slope of the given line:

$$y = 4x - 9 \Rightarrow m = 4$$

Use the perpendicular slope.

$$y = mx + b$$

$$y = -\frac{1}{4}x + b$$

$$0 = -\frac{1}{4}(0) + b$$

$$0 = 0 + b$$

$$0 = b$$

$$y = -\frac{1}{4}x + 0 \Rightarrow y = -\frac{1}{4}x$$

53. Find the slope of the given line:

$$4x - y = 7$$

$$4x - 7 = y \Rightarrow m = 4$$

Use the perpendicular slope.

$$y = mx + b$$

$$y = -\frac{1}{4}x + b$$

$$5 = -\frac{1}{4}(2) + b$$

$$5 = -\frac{1}{2} + b$$

$$\frac{10}{2} = -\frac{1}{2} + b$$

$$\frac{11}{2} = b$$

$$y = -\frac{1}{4}x + \frac{11}{2}$$

55.
$$7x = 2y - 4$$

$$-2y = -7x - 4$$

$$y = \frac{-7x - 4}{-2}$$

$$y = \frac{-7}{-2}x - \frac{4}{-2}$$

$$y = \frac{7}{2}x + 2$$

$$m = \frac{7}{2}, (0, 2)$$

57.
$$y = mx + b$$

$$y = \frac{2}{3}x + b$$

$$4 = \frac{2}{3}(-3) + b$$

$$4 = -2 + b$$

$$6 = b$$

$$y = \frac{2}{3}x + 6$$

59.
$$y = mx + b$$

$$y = -\frac{4}{3}x + b$$

$$-2 = -\frac{4}{3}(6) + b$$

$$-2 = -8 + b$$

$$6 = b$$

$$y = -\frac{4}{3}x + 6$$

61.
$$m = \frac{y_2 - y_1}{x_2 - x_1} = \frac{-9 - (-1)}{-3 - 3} = \frac{-8}{-6} = \frac{4}{3}$$

$$y = mx + b$$

$$-1 = \frac{4}{3}(3) + b$$

$$-1 = 4 + b$$

$$-5 = b \Rightarrow y = \frac{4}{3}x - 5$$

63. Find the slope of the given line: Use the parallel slope.

$$x = \frac{5}{4}y - 2$$
$$4x = 5y - 8$$
$$4x + 8 = 5y$$
$$\frac{4}{5}x + \frac{8}{5} = y \Rightarrow m = \frac{4}{5}$$

$$y = mx + b$$
$$y = \frac{4}{5}x + b$$
$$-2 = \frac{4}{5}(4) + b$$
$$-2 = \frac{16}{5} + b$$
$$-\frac{10}{5} = \frac{16}{5} + b$$
$$-\frac{26}{5} = b$$
$$y = \frac{4}{5}x - \frac{26}{5}$$

65. Find the slope of the given line:

$$x = -\frac{3}{4}y + 5$$
$$4x = -3y + 20$$
$$3y = -4x + 20$$
$$y = -\frac{4}{3}x + \frac{20}{3} \Rightarrow m = -\frac{4}{3}$$

Use the perpendicular slope.

$$y = mx + b$$
$$y = \frac{3}{4}x + b$$
$$-5 = \frac{3}{4}(1) + b$$
$$-5 = \frac{3}{4} + b$$
$$-\frac{20}{4} = \frac{3}{4} + b$$
$$-\frac{23}{4} = b$$
$$y = \frac{3}{4}x - \frac{23}{4}$$

67. The line $y = 5$ is horizontal. A perpendicular line will be vertical. Find the vertical line through $(-2, 7)$. $\boxed{x = -2}$

69. The line $x = 8$ is vertical. A parallel line will also be vertical. Find the vertical line through $(5, 2)$. $\boxed{x = 5}$

71. $\quad y = 8 \qquad x = 4$

horizontal $\quad$ vertical

perpendicular

73. $\quad 3x = y - 2 \quad 3(y - 3) + x = 0$

$3x + 2 = y \qquad 3y - 9 + x = 0$

$m = 3 \qquad\qquad 3y = -x + 9$

$y = -\frac{1}{3}x + 3$

$m = -\frac{1}{3}$

perpendicular

95

75.
$$4x + 5y = 20 \qquad\qquad 5x - 4y = 20$$
$$5y = -4x + 20 \qquad -4y = -5x + 20$$
$$y = -\tfrac{4}{5}x + 4 \qquad\quad y = \tfrac{5}{4}x - 5$$
$$m_1 = -\tfrac{4}{5} \qquad\qquad m_2 = \tfrac{5}{4}$$

perpendicular

77.
$$2x + 3y = 12 \qquad\qquad 6x + 9y = 32$$
$$3y = -2x + 12 \qquad\quad 9y = -6x + 32$$
$$y = -\tfrac{2}{3}x + 4 \qquad\quad y = \frac{-6}{9}x + \frac{32}{9}$$
$$m_1 = -\tfrac{2}{3} \qquad\qquad\qquad y = -\tfrac{2}{3}x + \tfrac{32}{9}$$
$$m_2 = -\tfrac{2}{3}$$

parallel

79. Let $x =$ the year of operation, and let $y =$ the value. Then two points are known: $(0, 24300)$ and $(7, 1900)$. Find the slope:
$$m = \frac{y_2 - y_1}{x_2 - x_1} = \frac{1900 - 24300}{7 - 0}$$
$$= \frac{-22400}{7} = -3200$$

Find the equation of the line:
$$y = mx + b$$
$$24300 = -3200(0) + b$$
$$24300 = 0 + b$$
$$24300 = b \Rightarrow y = -3200x + 24300$$

81. Let $x =$ the year of appreciation, and let $y =$ the value. Then two points are known: $(0, 450000)$ and $(12, 900000)$.
$$m = \frac{y_2 - y_1}{x_2 - x_1} = \frac{900000 - 450000}{12 - 0}$$
$$= \frac{450000}{12} = 37500$$

Find the equation of the line:
$$y = mx + b$$
$$450000 = 37500(0) + b$$
$$450000 = 0 + b$$
$$450000 = b \Rightarrow y = 37500x + 450000$$

83. Let $x =$ the year of depreciation, and let $y =$ the value. Then two points are known: $(0, 1900)$ and $(3, 1190)$. Find the slope:
$$m = \frac{y_2 - y_1}{x_2 - x_1} = \frac{1190 - 1900}{3 - 0} = \frac{-710}{3}$$

Find the equation of the line:
$$y = mx + b$$
$$1900 = -\tfrac{710}{3}(0) + b$$
$$1900 = 0 + b$$
$$1900 = b \Rightarrow y = -\tfrac{710}{3}x + 1900$$

85. Let $x =$ the age, and let $y =$ the value. Then two points are known: $(0, 1050)$ and $(8, 90)$. Find the slope:
$$m = \frac{y_2 - y_1}{x_2 - x_1} = \frac{90 - 1050}{8 - 0} = \frac{-960}{8}$$
$$= -120$$

The depreciation rate is $120 per year.

87. Let $x =$ the number of copies (in 100's), and let $y =$ total charge. Then two points are known: $(7, 375)$ and $(10, 525)$

$$m = \frac{y_2 - y_1}{x_2 - x_1} = \frac{525 - 375}{10 - 7} = \frac{150}{3} = 50$$

$y = mx + b$
$375 = 50(7) + b$
$375 = 350 + b$
$25 = b$
$y = 50x + 25 \Rightarrow$ The setup cost is \$25.

89. Let $x =$ the year $(0 = 1993)$, and let $y =$ the avg. pension. Then 2 pts are known: $(0, 22176)$ and $(9, 42144)$. Find the slope:

$$m = \frac{y_2 - y_1}{x_2 - x_1} = \frac{42144 - 22176}{9 - 0}$$
$$= \frac{19968}{9} = 2218.6667$$

Find the equation of the line:
$$y = mx + b$$
$22176 = 2218.6667(0) + b$
$22176 = 0 + b$
$22176 = b \Rightarrow y = 2218.6667x + 22176$
Let $x = 19$:
$y = 2218.6667x + 22176$
$= 2218.6667(19) + 22176$
$= 42154.67 + 22176 = 64330.67 \approx$ \$64,331

91. $0.80(465000) = \boxed{372000}$
Let $x =$ the # of yrs, and let $y =$ the value. Then 2 points are known: $(0, 465000)$ and $(40, 372000)$. Find the slope:

$$m = \frac{y_2 - y_1}{x_2 - x_1} = \frac{372000 - 465000}{40 - 0}$$
$$= \frac{-93000}{40} = \boxed{-2325}$$

Find the equation of the line:
$$y = mx + b$$
$465000 = -2325(0) + b$
$465000 = 0 + b$
$465000 = b \Rightarrow y = -2325x + 465000$
$\boxed{y = -2325x + 465000}$

93. **Answers may vary.**

95. $Ax + By = C$
$$By = -Ax + C$$
$$y = \frac{-Ax + C}{B}$$
$$y = -\frac{A}{B}x + \frac{C}{B}$$

97. The slope of a vertical line is undefined, so the equation cannot be written in slope-intercept form.

99. $a < 0, b > 0$

101-105. **Answers may vary.**

Exercises 3.6 (page 209)

1. $y = 2(0) + 1 = 0 + 1 = 1$

3. $y = 2(-1) + 1 = -2 + 1 = -1$

5. $4x = 3(x + 2)$
$4x = 3x + 6$
$x = 6$

7. $5(2 - a) = 3(a + 6)$
$10 - 5a = 3a + 18$
$-8a = 8$
$a = -1$

9. relation **11.** input; function **13.** range **15.** independent

17. cannot

19. domain = $\{-3, 1, 3\}$
range = $\{-1, 2, 7\}$

21. domain = $\{-3, 0, 2, 4\}$
range = $\{-8, 0, 5, 7\}$

23. Each value of x is paired with only one value of y. FUNCTION

25. $y = 2$ and $y = -2$ are both paired with $x = -4$. NOT A FUNCTION

27. $f(3) = -3(3) = -9$ $f(x) = 3$
$f(0) = -3(0) = 0$ $-3x = 3$
$f(-1) = -3(-1) = 3$ $x = -1$

29. $f(3) = 2(3) - 3 = 6 - 3 = 3$
$f(0) = 2(0) - 3 = 0 - 3 = -3$
$f(-1) = 2(-1) - 3 = -2 - 3 = -5$
$f(x) = 3$
$2x - 3 = 3$
$2x = 6$
$x = 3$

31. $f(3) = 7 + 5(3) = 7 + 15 = 22$
$f(0) = 7 + 5(0) = 7 + 0 = 7$
$f(-1) = 7 + 5(-1) = 7 - 5 = 2$
$f(x) = 3$
$7 + 5x = 3$
$5x = -4$
$x = -\frac{4}{5}$

33. $f(3) = 9 - 2(3) = 9 - 6 = 3$
$f(0) = 9 - 2(0) = 9 - 0 = 9$
$f(-1) = 9 - 2(-1) = 9 + 2 = 11$
$f(x) = 3$
$9 - 2x = 3$
$-2x = -6$
$x = 3$

35. $f(3) = \frac{1}{2}(3) + \frac{3}{2} = \frac{3}{2} + \frac{3}{2} = \frac{6}{2} = 3$
$f(0) = \frac{1}{2}(0) + \frac{3}{2} = 0 + \frac{3}{2} = \frac{3}{2}$
$f(-1) = \frac{1}{2}(-1) + \frac{3}{2} = -\frac{1}{2} + \frac{3}{2} = \frac{2}{2} = 1$

$f(x) = 3$
$\frac{1}{2}x + \frac{3}{2} = 3$
$\frac{1}{2}x + \frac{3}{2} = \frac{6}{2}$
$\frac{1}{2}x = \frac{3}{2}$
$2 \cdot \frac{1}{2}x = 2 \cdot \frac{3}{2}$
$x = 3$

37.
$$y = x + 2$$
Pick some values for x and find y for each:

x	-2	0	2
y	0	2	4

domain $= \mathbb{R}$; range $= \mathbb{R}$

39.
$$y = f(x) = \tfrac{1}{2}|x|$$

x	-2	-1	0	1	2
y	1	$\frac{1}{2}$	0	$\frac{1}{2}$	1

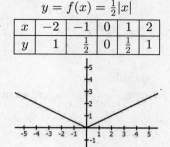

domain $= \mathbb{R}$;
range $= \{y | y$ is a real number and $y \geq 0\}$

41. function

43. not a function

45.
$f(1) = 1^2 + 1 = 1 + 1 = 2$
$f(-2) = (-2)^2 + 1 = 4 + 1 = 5$
$f(3) = 3^2 + 1 = 9 + 1 = 10$

47.
$f(1) = 1^3 - 1 = 1 - 1 = 0$
$f(-2) = (-2)^3 - 1 = -8 - 1 = -9$
$f(3) = 3^3 - 1 = 27 - 1 = 26$

49.
$f(1) = (1 - 1)^2 = (0)^2 = 0$
$f(-2) = (-2 - 1)^2 = (-3)^2 = 9$
$f(3) = (3 - 1)^2 = 2^2 = 4$

51.
$f(1) = 3(1)^2 - 2(1) = 3(1) - 2 = 3 - 2 = 1$
$f(-2) = 3(-2)^2 - 2(-2) = 3(4) + 4 = 12 + 4 = 16$
$f(3) = 3(3)^2 - 2(3) = 3(9) - 6 = 27 - 6 = 21$

53.
$f(1) = \tfrac{1}{2}(1)^2 - 2(1) + 3 = 0.5(1) - 2 + 3 = 0.5 - 2 + 3 = 1.5$
$f(-2) = \tfrac{1}{2}(-2)^2 - 2(-2) + 3 = 0.5(4) + 4 + 3 = 2 + 4 + 3 = 9$
$f(3) = \tfrac{1}{2}(3)^2 - 2(3) + 3 = 0.5(9) - 6 + 3 = 4.5 - 6 + 3 = 1.5$

55.
$$y = -\tfrac{1}{2}x + 2$$
Pick some values for x and find y for each:

x	-2	0	2
y	3	2	1

domain $= \mathbb{R}$; range $= \mathbb{R}$

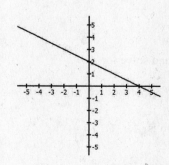

57. $y = f(x) = |x| - 1$

x	-2	-1	0	1	2
y	1	0	-1	0	1

domain $= \mathbb{R}$;
range $= \{y | y$ is a real number and $y \geq -1\}$

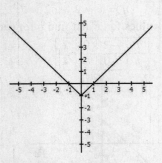

59. $\begin{aligned} C &= 0.10n + 12 = 0.10(20) + 12 \\ &= 2 + 12 = \$14 \end{aligned}$

61. $\begin{aligned} C &= 0.10n + 12 = 0.10(100) + 12 \\ &= 10 + 12 = \$22 \end{aligned}$

63. $\begin{aligned} C &= 0.08n + 17 = 0.08(500) + 17 \\ &= 40 + 17 = \$57 \end{aligned}$

65. $\begin{aligned} C &= 0.08n + 17 = 0.08(800) + 17 \\ &= 64 + 17 = \$81 \end{aligned}$

67.   Answers may vary.

69. $f(x) + g(x) = (2x + 1) + (x) = 3x + 1;\ g(x) + f(x) = (x) + (2x + 1) = 3x + 1$
Thus, $f(x) + g(x) = g(x) + f(x)$.

71. $f(x) \cdot g(x) = (2x + 1)(x) = 2x^2 + x;\ g(x) \cdot f(x) = (x)(2x + 1) = 2x^2 + x$
Thus, $f(x) \cdot g(x) = g(x) \cdot f(x)$.

Exercises 3.7   (page 216)

1. $(0, 0)$: below

3. $(6, 1)$: above

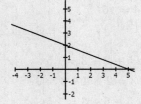

5. $(0, 0)$: left

7. $(0, 0)$: above

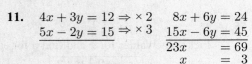

9. $\begin{aligned} x + y &= 3 \\ x - y &= 5 \\ \hline 2x\ \ \ &= 8 \\ x\ \ \ &= 4 \end{aligned}$

Substitute and solve for y:
$x + y = 3$
$4 + y = 3$
$\ \ \ \ y = -1$
The solution is $(4, -1)$.

11. $\begin{array}{ll} 4x + 3y = 12 \Rightarrow \times 2 & 8x + 6y = 24 \\ 5x - 2y = 15 \Rightarrow \times 3 & \underline{15x - 6y = 45} \\ & 23x\ \ \ \ \ \ = 69 \\ & x\ \ \ \ \ \ = 3 \end{array}$

Substitute and solve for y:
$4x + 3y = 12$
$4(3) + 3y = 12$
$\ \ \ \ 3y = 0$
$\ \ \ \ y = 0$
The solution is $(3, 0)$.

100

13. linear **15.** edge **17.** dashed

19.
$$y \geq x$$

Boundary   Test point: $(1, 2)$

(solid) $y \geq x$

$y = x$ $2 \overset{?}{\geq} 1$

x	y
0	0
2	2

$2 \geq 1$

same half-plane

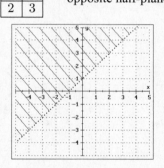

21.
$$y \geq -2$$

The boundary will be a solid line.

Shade all points above $y = -2$.

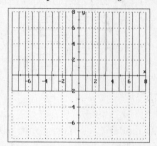

23.
$$y > x + 1$$

Boundary   Test point: $(0, 0)$

(dashed) $y > x + 1$

$y = x + 1$ $0 \overset{?}{>} 0 + 1$

x	y
0	1
2	3

$0 \not> 1$

opposite half-plane

25.
$$x < 4$$

The boundary will be a dashed line.

Shade all points left of $x = 4$.

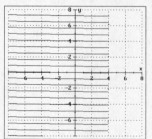

101

27.
$$-2 \leq x < 0$$
Shade points between $x = -2$ (solid) and $x = 0$ (dashed).

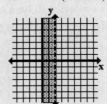

29.
$$y < -2 \text{ or } y > 3$$
Shade points below $y = -2$ (dashed) and points above $y = 3$ (dashed).

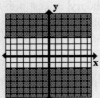

31.
$$y \geq 1 - \frac{3}{2}x$$

Boundary (solid)

$y = 1 - \frac{3}{2}x$

x	y
0	1
2	-2

Test point: $(0, 0)$

$y \geq 1 - \frac{3}{2}x$

$0 \overset{?}{\geq} 1 - \frac{3}{2}(0)$

$0 \not\geq 1$

opposite half-plane

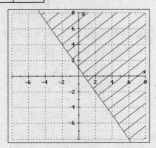

33.
$$0.5x + y > 1.5 + x$$
$$y > 0.5x + 1.5$$

Boundary (dashed)

$y = 0.5x + 1.5$

x	y
1	2
3	3

Test point: $(0, 0)$

$y > 0.5x + 1.5$

$0 \overset{?}{>} 0.5(0) + 1.5$

$0 \not> 1.5$

opposite half-plane

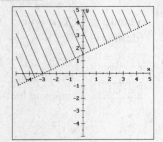

35.
$$y < \frac{1}{3}x - 1$$

Boundary (dashed)

$y = \frac{1}{3}x - 1$

x	y
0	-1
3	0

Test point: $(0, 0)$

$y < \frac{1}{3}x - 1$

$0 \overset{?}{<} \frac{1}{3}(0) - 1$

$0 \not< -1$

opposite half-plane

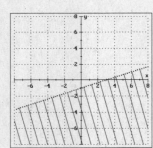

37.
$$3x \geq -y + 3$$

Boundary   Test point: $(0, 0)$
(solid) $3x \geq -y + 3$

$3x = -y + 3$ $\overset{?}{3(0) \geq -(0) + 3}$

x	y
0	3
1	0

$0 \ngeq 3$
opposite half-plane

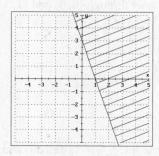

39. $y < 0.27x - 1$

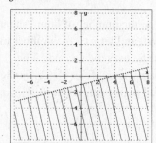

41. $y \geq -2.37x + 1.5$

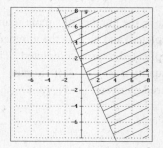

43. The boundary line is $x = 3$. Since points with x-coordinates less than 3 are shaded, and since the line is solid, the inequality is $\boxed{x \leq 3}$.

45. The boundary line goes through $(2, 0)$ and $(0, 3)$. Use point-slope form to find the equation:

$$m = \frac{\Delta y}{\Delta x} = \frac{3 - 0}{0 - 2} = \frac{3}{-2} = -\frac{3}{2}$$

$$y - y_1 = m(x - x_1)$$

$$y - 0 = -\frac{3}{2}(x - 2)$$

$$y = -\frac{3}{2}x + 3$$

Pick a point which is shaded, like $(2, 1)$, and substitute its coordinates for x and y:

$$y \ \square \ -\tfrac{3}{2}x + 3$$
$$1 \ \square \ -\tfrac{3}{2}(2) + 3$$
$$1 > -3 + 3$$

Since the line is dashed, do not include the line:

$$\boxed{y > -\tfrac{3}{2}x + 3}$$

47. The points between the lines $x = -2$ and $x = 3$ are shaded. Both lines are solid, so they should be included. $\boxed{-2 \leq x \leq 3}$

49. The boundary line goes through $(0,0)$ and $(1,1)$. Use point-slope form to find the equation:

$$m = \frac{\Delta y}{\Delta x} = \frac{1-0}{1-0} = \frac{1}{1} = 1$$
$$y - y_1 = m(x - x_1)$$
$$y - 0 = 1(x - 0)$$
$$y = x$$

Pick a point which is shaded, and substitute the coordinates for x and y:

$$y \ \boxed{\phantom{<}} \ x$$
$$0 < 1$$

Since the line is solid, include the line:

$$\boxed{y \leq x}$$

51. The points above the line $y = -1$ (dashed) and below the line $y = -3$ (solid) are shaded.

$$\boxed{y > -1 \text{ or } y \leq -3}$$

53. Let $x =$ the number of simple returns completed, and let $y =$ the number of complicated returns completed. The inequality is $x + 3y \leq 9$. Some ordered pairs are: $(1,1), (2,1), (2,2)$

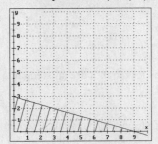

55. Let $x =$ the number of hours she uses the first, and let $y =$ the number of hours she uses the second. The inequality is $6x + 7y \leq 42$. Some ordered pairs are: $(2,2), (3,3), (5,1)$

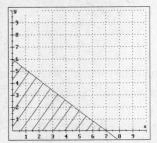

57. Let $x =$ the number of shares of Traffico, and let $y =$ the number of shares of Cleanco. The inequality is $50x + 60y \leq 6000$. Some ordered pairs are: $(40,20), (60,40), (80,20)$

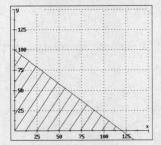

59. **Answers may vary.**

61. An inequality such as $x + (-x) < 10$ is an identity.

104

Chapter 3 Review (page 222)

1-6.

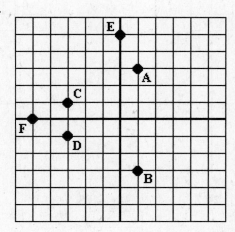

7. $(3, 1)$

8. $(-4, 5)$

9. $(-3, -4)$

10. $(2, -3)$

11. $(0, 0)$

12. $(0, 4)$

13. $(-5, 0)$

14. $(0, -3)$

15.
$$3x - 4y = 12$$
$$3(0) - 4(-3) \overset{?}{=} 12$$
$$0 + 12 \overset{?}{=} 12$$
$$12 = 12$$
$(0, -3)$ is a solution.

16.
$$3x - 4y = 12$$
$$3\left(\frac{4}{3}\right) - 4(2) \overset{?}{=} 12$$
$$4 - 8 \overset{?}{=} 12$$
$$-4 \neq 12$$
$\left(\frac{4}{3}, 2\right)$ is not a solution.

17. $y = x - 5$

x	y
0	-5
2	-3

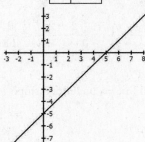

18. $y = 2x + 1$

x	y
0	1
2	5

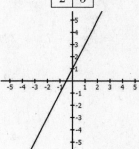

19. $y = \dfrac{x}{2} + 2$

x	y
0	2
2	3

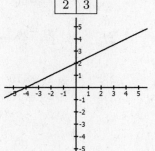

20. $y = 3$

x	y
0	3
2	3

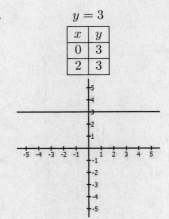

21. $x + y = 4$

x	y
0	4
4	0

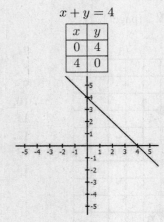

22. $x - y = -3$

x	y
0	3
-3	0

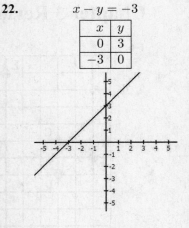

23. $3x + 5y = 15$

x	y
0	3
5	0

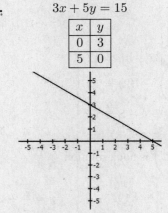

24. $7x - 4y = 28$

x	y
0	-7
4	0

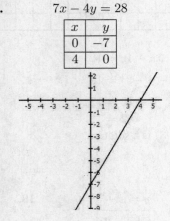

25. $m = \dfrac{y_2 - y_1}{x_2 - x_1} = \dfrac{2 - 8}{7 - 3} = \dfrac{-6}{4} = -\dfrac{3}{2}$

26. $m = \dfrac{y_2 - y_1}{x_2 - x_1} = \dfrac{-2 - 3}{3 - (-1)} = \dfrac{-5}{4} = -\dfrac{5}{4}$

27. $m = \dfrac{y_2 - y_1}{x_2 - x_1} = \dfrac{9 - (-5)}{-4 - (-2)} = \dfrac{14}{-2} = -7$

28. $m = \dfrac{y_2 - y_1}{x_2 - x_1} = \dfrac{2 - 2}{3 - (-8)} = \dfrac{0}{11} = 0$

29. $5x - 2y = 10$

$-2y = -5x + 10$

$y = \dfrac{-5x}{-2} + \dfrac{10}{-2}$

$y = \dfrac{5}{2}x - 5 \Rightarrow m = \dfrac{5}{2}$

30. $x = -4$

vertical line $\Rightarrow$ undefined slope

31. positive

32. 0

33. undefined

34. negative

35. perpendicular

36. parallel

37. neither

38. neither

39. perpendicular

40. parallel

41. $m = \dfrac{\text{rise}}{\text{run}} = \dfrac{4}{12} = \dfrac{1}{3}$

42. $m = \dfrac{y_2 - y_1}{x_2 - x_1} = \dfrac{66000 - 25000}{3 - 1}$

$= \dfrac{41000}{2} = \$20{,}500 \text{ per year}$

43. $y - y_1 = m(x - x_1)$
$y - 3 = 5(x - (-2))$
$y - 3 = 5(x + 2)$
$y - 3 = 5x + 10$
$y = 5x + 13$

44. $y - y_1 = m(x - x_1)$
$y - \dfrac{2}{3} = -\dfrac{1}{3}(x - 1)$
$y - \dfrac{2}{3} = -\dfrac{1}{3}x + \dfrac{1}{3}$
$y = -\dfrac{1}{3}x + 1$

45. $y - y_1 = m(x - x_1)$
$y - (-2) = \dfrac{1}{9}(x - (-27))$
$y + 2 = \dfrac{1}{9}x + 3$
$y = \dfrac{1}{9}x + 1$

46. $y - y_1 = m(x - x_1)$
$y - \left(-\dfrac{1}{5}\right) = -\dfrac{3}{5}(x - 1)$
$y + \dfrac{1}{5} = -\dfrac{3}{5}x + \dfrac{3}{5}$
$y = -\dfrac{3}{5}x + \dfrac{2}{5}$

47. $y = -\dfrac{x}{3} + 6$
$y = -\dfrac{1}{3}x + 6$
$m = -\dfrac{1}{3}, \ (0, 6)$

48. $3x - 6y = 9$
$-6y = -3x + 9$
$y = \dfrac{-3x}{-6} + \dfrac{9}{-6}$
$y = \dfrac{1}{2}x - \dfrac{3}{2}$
$m = \dfrac{1}{2}, \ \left(0, -\dfrac{3}{2}\right)$

49. $2x + 5y = 1$
$5y = -2x + 1$
$y = \dfrac{-2x}{5} + \dfrac{1}{5}$
$y = -\dfrac{2}{5}x + \dfrac{1}{5}$
$m = -\dfrac{2}{5}, \ \left(0, \dfrac{1}{5}\right)$

50. $x + 3y = 1$
$3y = -x + 1$
$y = \dfrac{-x + 1}{3}$
$y = -\dfrac{1}{3}x + \dfrac{1}{3}$
$m = -\dfrac{1}{3}, \ \left(0, \dfrac{1}{3}\right)$

51. $y = mx + b$
$y = -3x + 2$

52. $y = mx + b$
$y = 0x + (-7)$
$y = -7$

53. $y = mx + b$
$y = 7x + 0$
$y = 7x$

54. $y = mx + b$
$y = \dfrac{1}{2}x + \left(-\dfrac{3}{2}\right)$
$y = \dfrac{1}{2}x - \dfrac{3}{2}$

55. $(-1, 3), m = -2$

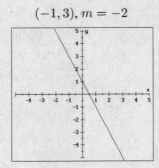

56. $(1, -2), m = 2$

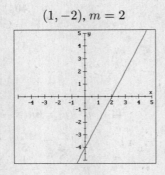

57. $\left(0, \dfrac{1}{2}\right), m = \dfrac{3}{2}$

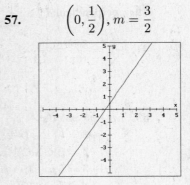

58. $(-3, 0), m = -\dfrac{5}{2}$

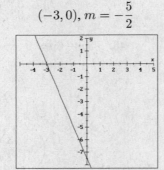

59. $y = 3x \qquad x = 3y$

   $m = 3 \qquad \dfrac{1}{3}x = y$

   $\qquad\qquad\quad m = \dfrac{1}{3}$

   neither

60. $3x = y \qquad x = -3y$

   $m = 3 \qquad -\dfrac{1}{3}x = y$

   $\qquad\qquad\quad m = -\dfrac{1}{3}$

   perpendicular

61. $x + 2y = y - x \qquad 2x + y = 3$

   $\qquad y = -2x \qquad\quad y = -2x + 3$

   $\qquad m = -2 \qquad\qquad m = -2$

   parallel

62. $3x + 2y = 7 \qquad 2x - 3y = 8$

   $\quad 2y = -3x + 7 \qquad -3y = -2x + 8$

   $\quad y = -\dfrac{3}{2}x + \dfrac{7}{2} \qquad y = \dfrac{2}{3}x - \dfrac{8}{3}$

   $\quad m = -\dfrac{3}{2} \qquad\qquad m = \dfrac{2}{3}$

   perpendicular

63.   Find the slope of the given line:

   $y = 7x - 18 \Rightarrow m = 7$

   Use the parallel slope.

   $y = mx + b$

   $y = 7x + b$

   $5 = 7(2) + b$

   $5 = 14 + b$

   $-9 = b$

   $y = 7x - 9$

64.   Find the slope of the given line:

   $3x + 2y = 7$

   $2y = -3x + 7$

   $y = -\dfrac{3}{2}x + \dfrac{7}{2} \Rightarrow m = -\dfrac{3}{2}$

   Use the parallel slope.

   $y = mx + b$

   $y = -\dfrac{3}{2}x + b$

   $5 = -\dfrac{3}{2}(-3) + b$

   $\dfrac{10}{2} = \dfrac{9}{2} + b$

   $\dfrac{1}{2} = b$

   $y = -\dfrac{3}{2}x + \dfrac{1}{2}$

65.   Find the slope of the given line: Use the perpendicular slope.

$$2x - 5y = 12$$

$$2x - 12 = 5y$$

$$\frac{2}{5}x - \frac{12}{5} = y \Rightarrow m = \frac{2}{5}$$

$$y = mx + b$$

$$y = -\frac{5}{2}x + b$$

$$0 = -\frac{5}{2}(0) + b$$

$$0 = 0 + b$$

$$0 = b$$

$$y = -\frac{5}{2}x$$

66.   Find the slope of the given line: Use the perpendicular slope.

$$y = \frac{x}{3} + 17 \Rightarrow m = \frac{1}{3}$$

$$y = mx + b$$

$$y = -3x + b$$

$$-4 = -3(0) + b$$

$$-4 = 0 + b$$

$$-4 = b$$

$$y = -3x - 4$$

67. $m = \dfrac{y_2 - y_1}{x_2 - x_1} = \dfrac{-1 - 5}{1 - (-2)} = \dfrac{-6}{3} = -2$ **68.** $m = \dfrac{y_2 - y_1}{x_2 - x_1} = \dfrac{4 - 8}{2 - 10} = \dfrac{-4}{-8} = \dfrac{1}{2}$

$$y = mx + b$$

$$5 = -2(-2) + b$$

$$5 = 4 + b$$

$$1 = b \Rightarrow y = -2x + 1$$

$$y = mx + b$$

$$8 = \frac{1}{2}(10) + b$$

$$8 = 5 + b$$

$$3 = b \Rightarrow y = \frac{1}{2}x + 3$$

69.   Let $x =$ the year of operation, and let $y =$ the value. Then two points are known:

$(0, 2700)$ and $(5, 200)$. Find the slope: $m = \dfrac{y_2 - y_1}{x_2 - x_1} = \dfrac{200 - 2700}{5 - 0} = \dfrac{-2500}{5} = -500$

Find the equation of the line: $y - y_1 = m(x - x_1)$

$$y - 2700 = -500(x - 0)$$

$$y - 2700 = -500x$$

$$y = -500x + 2700$$

Let $x = 3$: $y = -500x + 2700 = -500(3) + 2700 = -1500 + 2700 = \$1200$

70.   domain $= \{-3, 0\}$; range $= \{-2, -1, 5\}$

71. $y = \frac{x}{2} - 1 = \frac{1}{2}x - 1$

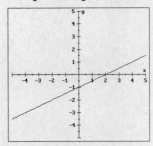

y is a function of x.

72. $f(0) = -2(0) + 5 = 0 + 5 = 5$

73. $f(-5) = -2(-5) + 5 = 10 + 5 = 15$

74. $f\left(-\frac{1}{2}\right) = -2\left(-\frac{1}{2}\right) + 5 = 1 + 5 = 6$

75. $f(6) = -2(6) + 5 = -12 + 5 = -7$

76.

$$y = f(x) = |x| - 3$$

x	-2	-1	0	1	2
y	-1	-2	-3	-2	-1

domain $= \mathbb{R}$

range $= \{y | y \text{ is a real number and } y \geq -3\}$

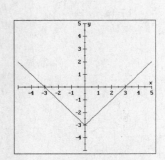

77. not a function

78. function

79.

$$2x + 3y > 6$$

Boundary (dashed)

$2x + 3y = 6$

x	y
0	2
3	0

Test point: $(0, 0)$

$2x + 3y > 6$

$2(0) + 3(0) \overset{?}{>} 6$

$0 \not> 6$

opposite half-plane

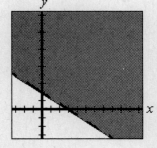

80.

$$y \leq 4 - x$$

Boundary (solid)

$y = 4 - x$

x	y
0	4
2	2

Test point: $(0, 0)$

$y \leq 4 - x$

$0 \overset{?}{\leq} 4 - 0$

$0 \leq 4$

same half-plane

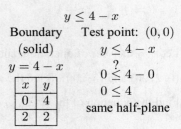

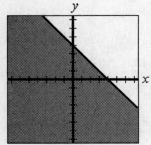

81.
$$-2 < x < 4$$
Shade points between $x = -2$ and $x = 4$.

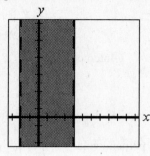

82.
$$y \leq -2 \text{ or } y > 1$$
Shade points below $y = -2$ or above $y = 1$.

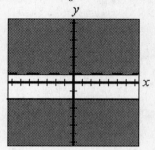

Chapter 3 Test   (page 227)

1. $y = \dfrac{x}{2} + 1$

x	y
0	1
2	2

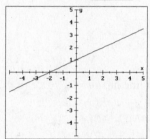

2. $2(x+1) - y = 4$
$2x + 2 - y = 4$
$2x - y = 2$

x	y
0	-2
1	0

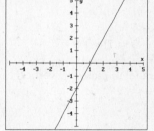

3. $x = 1$

x	y
1	2
1	0

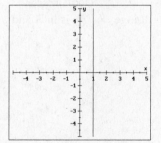

4. $2y = 8$
$y = 4$

x	y
0	4
2	4

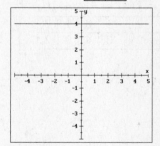

5. $m = \dfrac{y_2 - y_1}{x_2 - x_1} = \dfrac{8 - 3}{5 - (-1)} = \dfrac{5}{6}$

6. $m = \dfrac{y_2 - y_1}{x_2 - x_1} = \dfrac{-1 - 3}{3 - (-1)} = \dfrac{-4}{4} = -1$

7.
$$3x - 4y = 5$$
$$-4y = -3x + 5$$
$$\frac{-4y}{-4} = \frac{-3x}{-4} + \frac{5}{-4}$$
$$y = \frac{3}{4}x - \frac{5}{4}$$
$$m = \frac{3}{4}$$

8.
$$2y - 7(x + 5) = 7$$
$$2y - 7x - 35 = 7$$
$$2y = 7x + 42$$
$$y = \frac{7}{2}x + 21$$
y-intercept: $(0, 21)$

9. horizontal line $\Rightarrow m = 0$

10. vertical line $\Rightarrow m$ is undefined.

11. equal   **12.** -1

13. perpendicular   **14.** parallel

15. $m = \dfrac{\text{rise}}{\text{run}} = \dfrac{3}{12} = \dfrac{1}{4}$

16.
$$m = \frac{y_2 - y_1}{x_2 - x_1} = \frac{100000 - 50000}{5 - 2}$$
$$= \frac{50000}{3} \approx \$16{,}666.67$$

17. 2

18. $-\frac{1}{2}$

19.
$$y - y_1 = m(x - x_1)$$
$$y - 5 = 7(x - (-2))$$
$$y - 5 = 7(x + 2)$$

20.
$$y = mx + b$$
$$y = \frac{3}{4}x - 5$$

21. Parallel to y-axis $\Rightarrow$ vertical
$$x = -7$$

22. Given line has $m = \frac{1}{3}$. Use $m = -3$.
$$y - y_1 = m(x - x_1)$$
$$y - (-5) = -3(x - 3)$$
$$y + 5 = -3x + 9$$
$$y = -3x + 4$$

23. $f(2) = 3(2) - 2 = 6 - 2 = 4$

24. $f(-3) = 3(-3) - 2 = -9 - 2 = -11$

25. $y = f(x) = -|x| + 4$

x	-2	-1	0	1	2
y	2	3	4	3	2

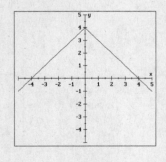

Domain $= \mathbb{R}$

Range $= \{y | y$ is in $\mathbb{R}$ and $y \leq 4\}$

26. not a function

27. function

112

28.
$$3x + 2y \geq 6$$

Boundary   Test point: $(0,0)$
(solid) $3x + 2y \geq 6$
$3x + 2y = 6$ $\overset{?}{3(0) + 2(0) \geq 6}$

x	y
0	3
2	0

$0 \ngeq 6$

opposite half-plane

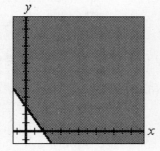

29.
$$-2 \leq y < 5$$

Shade points between $y = -2$ and $y = 5$.

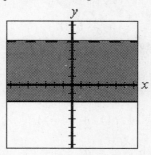

30. The boundary line goes through $(0,1)$ and $(1,0)$. Use point-slope form to find the equation:

$$m = \frac{\Delta y}{\Delta x} = \frac{0-1}{1-0} = \frac{-1}{1} = -1$$
$$y - y_1 = m(x - x_1)$$
$$y - 1 = -1(x - 0)$$
$$y = -x + 1$$

Pick a point which is shaded, like $(0,0)$ and substitute the coordinates for x and y:

$y \;\boxed{\phantom{<}}\; -x + 1$
$0 < -0 + 1$

Since the line is dashed, do not include the line:

$$\boxed{y < -x + 1}$$

Exercises 4.1 (page 237)

1. $4 \cdot 4 \cdot 4 = 64$

3. $(-7)(-7)(-7) = -343$

5. $6^2 = 6 \cdot 6 = 36$

7. $3(4)^2 = 3(16) = 48$

9.

+⟨─◆─⟩◆⟨─◆─⟩◆⟨─◆─⟩+
-3 -1 1 3

11. the product of 3 and the sum of x and y

13. $|2x| + 3$

15. base; -5; 3

17. $(4y)(4y)(4y)$

19. $y \cdot y \cdot y \cdot y \cdot y$

21. $x^n y^n$

23. a^{bc}

25. Answers may vary.

27. Base: x
Exponent: 4

29. Base: 7
Exponent: 2

31. Base: $2y$
Exponent: 3

33. Base: x
Exponent: 4

35. Base: x; Exponent: 1

37.   Base: x; Exponent: 3

39. $-5^2 = -(5 \cdot 5) = -25$

41. $2^2 + 3^2 = 4 + 9 = 13$

43. $-3(6^2 - 2^3) = -3(36 - 8) = -3(28)$
$$= -84$$

45. $(-4)^3 + (-3)^2 = -64 + 9 = -55$

47. $5^3 = 5 \cdot 5 \cdot 5$

49. $-5x^6 = -5 \cdot x \cdot x \cdot x \cdot x \cdot x \cdot x$

51. $-2y^4 = -2 \cdot y \cdot y \cdot y \cdot y$

53. $(3t)^5 = (3t)(3t)(3t)(3t)(3t)$

55. $2 \cdot 2 \cdot 2 = 2^3$

57. $x \cdot x \cdot x \cdot x = x^4$

59. $(2x)(2x)(2x) = (2x)^3$

61. $-4 \cdot t \cdot t \cdot t \cdot t = -4t^4$

63. $x^4 x^3 = x^{4+3} = x^7$

65. $x^5 x^5 = x^{5+5} = x^{10}$

67. $a^3 a^4 a^5 = a^{3+4+5} = a^{12}$

69. $y^3(y^2 y^4) = y^3 y^{2+4} = y^3 y^6 = y^{3+6} = y^9$

71. $4x^2(3x^5) = 4 \cdot 3x^2 x^5 = 12x^{2+5} = 12x^7$

73. $(5x^4)(-2x) = 5(-2)x^4 x^1 = -10x^{4+1}$
$$= -10x^5$$

75. $(3^2)^4 = 3^{2 \cdot 4} = 3^8$

77. $(y^5)^3 = y^{5 \cdot 3} = y^{15}$

79. $(x^2 x^3)^5 = (x^{2+3})^5 = (x^5)^5 = x^{5 \cdot 5} = x^{25}$

81. $(a^2 a^7)^3 = (a^{2+7})^3 = (a^9)^3 = a^{9 \cdot 3} = a^{27}$

83. $(x^5)^2 (x^7)^3 = x^{5 \cdot 2} x^{7 \cdot 3} = x^{10} x^{21} = x^{10+21} = x^{31}$

85. $(r^3 r^2)^4 (r^3 r^5)^2 = (r^{3+2})^4 (r^{3+5})^2 = (r^5)^4 (r^8)^2 = r^{5 \cdot 4} r^{8 \cdot 2} = r^{20} r^{16} = r^{20+16} = r^{36}$

87. $(xy)^3 = x^3 y^3$

89. $(r^3 s^2)^2 = (r^3)^2 (s^2)^2 = r^{3 \cdot 2} s^{2 \cdot 2} = r^6 s^4$

91. $(4ab^2)^2 = 4^2 a^2 (b^2)^2 = 16a^2 b^{2 \cdot 2}$
$$= 16a^2 b^4$$

93. $(-2r^2 s^3 t)^3 = (-2)^3 (r^2)^3 (s^3)^3 t^3$
$$= -8r^{2 \cdot 3} s^{3 \cdot 3} t^3 = -8r^6 s^9 t^3$$

95. $\left(\dfrac{a}{b}\right)^3 = \dfrac{a^3}{b^3}$

97. $\left(\dfrac{2x}{3y^2}\right)^4 = \dfrac{(2x)^4}{(3y^2)^4} = \dfrac{2^4 \cdot x^4}{3^4 \cdot y^{2 \cdot 4}} = \dfrac{16x^4}{81y^8}$

99. $\dfrac{x^5}{x^3} = x^{5-3} = x^2$

101. $\dfrac{y^3 y^4}{yy^2} = \dfrac{y^{3+4}}{y^{1+2}} = \dfrac{y^7}{y^3} = y^{7-3} = y^4$

103. $\dfrac{12a^2 a^3 a^4}{4(a^4)^2} = \dfrac{12a^{2+3+4}}{4a^{4 \cdot 2}} = \dfrac{12a^9}{4a^8} = \dfrac{12}{4}a^{9-8} = 3a^1 = 3a$

105. $\dfrac{(ab^2)^3}{(ab)^2} = \dfrac{a^3 (b^2)^3}{a^2 b^2} = \dfrac{a^3 b^{2 \cdot 3}}{a^2 b^2} = \dfrac{a^3 b^6}{a^2 b^2} = a^{3-2} b^{6-2} = a^1 b^4 = ab^4$

107. $tt^2 = t^{1+2} = t^3$

109. $6x^3(-x^2)(-x^4) = 6(-1)(-1)x^3 x^2 x^4$
$$= 6x^{3+2+4} = 6x^9$$

111. $(-2a^5)^3 = (-2)^3 \cdot a^{5\cdot3} = -8a^{15}$

113. $(3zz^2z^3)^5 = (3z^{1+2+3})^5 = (3z^6)^5 = 3^5(z^6)^5 = 243z^{6\cdot5} = 243z^{30}$

115. $(s^3)^3(s^2)^2(s^5)^4 = s^{3\cdot3}s^{2\cdot2}s^{5\cdot4} = s^9s^4s^{20} = s^{9+4+20} = s^{33}$

117. $\left(\dfrac{-2a}{b}\right)^5 = \dfrac{(-2a)^5}{b^5} = \dfrac{(-2)^5a^5}{b^5} = \dfrac{-32a^5}{b^5}$ **119.** $\left(\dfrac{b^2}{3a}\right)^3 = \dfrac{(b^2)^3}{(3a)^3} = \dfrac{b^{2\cdot3}}{3^3a^3} = \dfrac{b^6}{27a^3}$

121. $\dfrac{17(x^4y^3)^8}{34(x^5y^2)^4} = \dfrac{17(x^4)^8(y^3)^8}{34(x^5)^4(y^2)^4} = \dfrac{17x^{4\cdot8}y^{3\cdot8}}{34x^{5\cdot4}y^{2\cdot4}} = \dfrac{17x^{32}y^{24}}{34x^{20}y^8} = \dfrac{17}{34}x^{32-20}y^{24-8} = \dfrac{1}{2}x^{12}y^{16} = \dfrac{x^{12}y^{16}}{2}$

123. $\left(\dfrac{y^3y}{2yy^2}\right)^3 = \left(\dfrac{y^{3+1}}{2y^{1+2}}\right)^3 = \left(\dfrac{y^4}{2y^3}\right)^3 = \left(\dfrac{1}{2}y^{4-3}\right)^3 = \left(\dfrac{1}{2}y^1\right)^3 = \left(\dfrac{1}{2}\right)^3(y^1)^3 = \dfrac{1}{8}y^{1\cdot3} = \dfrac{y^3}{8}$

125. $\left(\dfrac{-2r^3r^3}{3r^4r}\right)^3 = \left(\dfrac{-2r^{3+3}}{3r^{4+1}}\right)^3 = \left(\dfrac{-2r^6}{3r^5}\right)^3 = \left(\dfrac{-2}{3}r^{6-5}\right)^3 = \dfrac{(-2)^3}{3^3}(r^1)^3 = \dfrac{-8}{27}r^{1\cdot3} = -\dfrac{8r^3}{27}$

127. $\dfrac{20(r^4s^3)^4}{6(rs^3)^3} = \dfrac{20(r^4)^4(s^3)^4}{6r^3(s^3)^3} = \dfrac{20r^{4\cdot4}s^{3\cdot4}}{6r^3s^{3\cdot3}} = \dfrac{20r^{16}s^{12}}{6r^3s^9} = \dfrac{20}{6}r^{16-3}s^{12-9} = \dfrac{10}{3}r^{13}s^3 = \dfrac{10r^{13}s^3}{3}$

129.

Bounce	1	2	3	4
Height	$\frac{1}{2}(32)$	$\frac{1}{2}\left[\frac{1}{2}(32)\right] = 32\left(\frac{1}{2}\right)^2$	$\frac{1}{2}\left[32\left(\frac{1}{2}\right)^2\right] = 32\left(\frac{1}{2}\right)^3$	$\frac{1}{2}\left[32\left(\frac{1}{2}\right)^3\right] = 32\left(\frac{1}{2}\right)^4$

$$32\left(\dfrac{1}{2}\right)^4 = 32\left(\dfrac{1^4}{2^4}\right) = 32\left(\dfrac{1}{16}\right) = \dfrac{32}{16} = 2 \text{ ft}$$

131.

Years	7	14	21	28
Value	2,000	4,000	8,000	16,000

133. $A = P(1+r)^t = 8000(1+0.06)^{30} = 8000(1.06)^{30} \approx 8000(5.743491) \approx \$45,947.93$

135. **Answers may vary.**   **137.** No. $2^3 = 8,\ 3^2 = 9 \Rightarrow 2^3 \neq 3^2$

Exercises 4.2   (page 242)

1. x **3.** $(5x)$

5. $\dfrac{3 \cdot a \cdot a}{3 \cdot a \cdot a \cdot a} = \dfrac{\not{3} \cdot \not{a} \cdot \not{a}}{\not{3} \cdot \not{a} \cdot \not{a} \cdot a} = \dfrac{1}{a}$ **7.** $\dfrac{a \cdot a \cdot b}{a \cdot b \cdot b} = \dfrac{\not{a} \cdot a \cdot \not{b}}{\not{a} \cdot \not{b} \cdot b} = \dfrac{a}{b}$

9. $\dfrac{3a^2 + 4b + 8}{a + 2b^2} = \dfrac{3(-2)^2 + 4(3) + 8}{-2 + 2(3)^2} = \dfrac{3(4) + 12 + 8}{-2 + 2(9)} = \dfrac{12 + 12 + 8}{-2 + 18} = \dfrac{32}{16} = 2$

11. $5\left(x - \dfrac{1}{2}\right) = \dfrac{7}{2}$

$5x - \dfrac{5}{2} = \dfrac{7}{2}$

$5x - \dfrac{5}{2} + \dfrac{5}{2} = \dfrac{7}{2} + \dfrac{5}{2}$

$5x = \dfrac{12}{2}$

$5x = 6$

$\dfrac{5x}{5} = \dfrac{6}{5}$

$x = \dfrac{6}{5}$

13. $P = L + \dfrac{s}{f}i$

$P - L = \dfrac{s}{f}i$

$f(P - L) = f \cdot \dfrac{s}{f}i$

$f(P - L) = si$

$\dfrac{f(P - L)}{i} = \dfrac{si}{i}$

$\dfrac{f(P - L)}{i} = s, \text{ or } s = \dfrac{f(P - L)}{i}$

15. $1, \dfrac{1}{x^n}$

17. $\dfrac{1}{8^2}$

19. $9^0 = 1$

21. $5x^0 = 5 \cdot x^0 = 5 \cdot 1 = 5$

23. $\left(\dfrac{a^2 b^3}{ab^4}\right)^0 = 1$

25. $\dfrac{8^y}{8^y} = 8^{y-y} = 8^0 = 1$

27. $(-x)^0 = 1$

29. $\dfrac{x^0 - 5x^0}{2x^0} = \dfrac{1 - 5 \cdot 1}{2 \cdot 1} = \dfrac{1 - 5}{2} = \dfrac{-4}{2} = -2$

31. $5^{-4} = \dfrac{1}{5^4} = \dfrac{1}{625}$

33. $a^{-5} = \dfrac{1}{a^5}$

35. $(2y)^{-4} = \dfrac{1}{(2y)^4} = \dfrac{1}{2^4 y^4} = \dfrac{1}{16y^4}$

37. $(-5p)^{-3} = \dfrac{1}{(-5p)^3} = \dfrac{1}{(-5)^3 p^3} = -\dfrac{1}{125p^3}$

39. $\left(y^2 y^4\right)^{-2} = \left(y^6\right)^{-2} = y^{-12} = \dfrac{1}{y^{12}}$

41. $-5x^{-4} = -5 \cdot x^{-4} = -5 \cdot \dfrac{1}{x^4} = -\dfrac{5}{x^4}$

43. $\dfrac{y^4}{y^5} = y^{4-5} = y^{-1} = \dfrac{1}{y^1} = \dfrac{1}{y}$

45. $\dfrac{a^4}{a^9} = a^{4-9} = a^{-5} = \dfrac{1}{a^5}$

47. $\dfrac{x^{-2} x^{-3}}{x^{-10}} = \dfrac{x^{-2+(-3)}}{x^{-10}} = \dfrac{x^{-5}}{x^{-10}} = x^{-5-(-10)}$

$= x^5$

49. $\dfrac{15a^3 b^8}{3a^4 b^4} = \dfrac{15}{3} a^{3-4} b^{8-4} = 5a^{-1} b^4 = \dfrac{5b^4}{a}$

51. $\left(a^{-6}\right)^4 = a^{-6 \cdot 4} = a^{-24} = \dfrac{1}{a^{24}}$

53. $\left(-\dfrac{b^5}{b^{-3}}\right)^{-4} = \left(-b^{5-(-3)}\right)^{-4} = \left(-b^8\right)^{-4} = (-b)^{-32} = \dfrac{1}{(-b)^{32}} = \dfrac{1}{b^{32}}$

55. $x^{2m} x^m = x^{2m+m} = x^{3m}$

57. $\dfrac{x^{3n}}{x^{6n}} = x^{3n-6n} = x^{-3n} = \dfrac{1}{x^{3n}}$

59. $y^{3m+2}y^{-m} = y^{3m+2+(-m)} = y^{2m+2}$

61. $\left(x^{n+4}\right)^3 = x^{(n+4)\cdot 3} = x^{3n+12}$

63. $u^{2m}v^{3n}u^{3m}v^{-3n} = u^{2m+3m}v^{3n+(-3n)}$
$= u^{5m}v^0 = u^{5m}$

65. $\left(y^{2-n}\right)^{-4} = y^{(2-n)(-4)} = y^{-8+4n} = y^{4n-8}$

67. $2^5 \cdot 2^{-2} = 2^{5+(-2)} = 2^3 = 8$

69. $5^{-2} \cdot 5^5 \cdot 5^{-3} = 5^{-2+5+(-3)} = 5^0 = 1$

71. $\dfrac{8^5 \cdot 8^{-3}}{8} = \dfrac{8^{5+(-3)}}{8^1} = \dfrac{8^2}{8^1} = 8^{2-1}$
$= 8^1 = 8$

73. $\dfrac{2^5 \cdot 2^7}{2^6 \cdot 2^{-3}} = \dfrac{2^{5+7}}{2^{6+(-3)}} = \dfrac{2^{12}}{2^3} = 2^{12-3}$
$= 2^9 = 512$

75. $a^{-9} = \dfrac{1}{a^9}$

77. $\dfrac{y^{3m}}{y^{2m}} = y^{3m-2m} = y^m$

79. $(4t)^{-3} = \dfrac{1}{(4t)^3} = \dfrac{1}{4^3 t^3} = \dfrac{1}{64t^3}$

81. $\left(ab^2\right)^{-3} = \dfrac{1}{\left(ab^2\right)^3} = \dfrac{1}{a^3 \left(b^2\right)^3} = \dfrac{1}{a^3 b^6}$

83. $\left(x^2 y\right)^{-2} = \dfrac{1}{\left(x^2 y\right)^2} = \dfrac{1}{\left(x^2\right)^2 y^2} = \dfrac{1}{x^4 y^2}$

85. $\dfrac{b^0 b^3}{b^{-3} b^4} = \dfrac{b^3}{b^1} = b^{3-1} = b^2$

87. $\left(m^3 n^4\right)^{-3} = \dfrac{1}{\left(m^3 n^4\right)^3} = \dfrac{1}{\left(m^3\right)^3 \left(n^4\right)^3} = \dfrac{1}{m^9 n^{12}}$

89. $\dfrac{x^{12} x^{-7}}{x^3 x^4} = \dfrac{x^5}{x^7} = x^{5-7} = x^{-2} = \dfrac{1}{x^2}$

91. $\left(ab^2\right)^{-2} = \dfrac{1}{\left(ab^2\right)^2} = \dfrac{1}{a^2 \left(b^2\right)^2} = \dfrac{1}{a^2 b^4}$

93. $\left(-2x^3 y^{-2}\right)^{-5} = (-2)^{-5}\left(x^3\right)^{-5}\left(y^{-2}\right)^{-5} = \dfrac{1}{(-2)^5} x^{-15} y^{10} = -\dfrac{y^{10}}{32x^{15}}$

95. $\left(a^{-2}b^{-3}\right)^{-4} = \left(a^{-2}\right)^{-4}\left(b^{-3}\right)^{-4} = a^8 b^{12}$

97. $\left(\dfrac{b^5}{b^{-2}}\right)^{-2} = \left(b^{5-(-2)}\right)^{-2} = \left(b^7\right)^{-2}$
$= b^{-14} = \dfrac{1}{b^{14}}$

99. $\left(\dfrac{6a^2 b^3}{2ab^2}\right)^{-2} = \left(\dfrac{6}{2} a^{2-1} b^{3-2}\right)^{-2} = \left(3a^1 b^1\right)^{-2} = \dfrac{1}{(3ab)^2} = \dfrac{1}{3^2 a^2 b^2} = \dfrac{1}{9a^2 b^2}$

101. $\left(\dfrac{18a^2 b^3 c^{-4}}{3a^{-1} b^2 c}\right)^{-3} = \left(\dfrac{18}{3} a^{2-(-1)} b^{3-2} c^{-4-1}\right)^{-3} = \left(6a^3 b^1 c^{-5}\right)^{-3} = 6^{-3}\left(a^3\right)^{-3}\left(b^1\right)^{-3}\left(c^{-5}\right)^{-3}$
$= \dfrac{1}{6^3} a^{-9} b^{-3} c^{15} = \dfrac{c^{15}}{216a^9 b^3}$

103. $\left(\dfrac{-3r^4 r^{-3}}{r^{-3} r^7}\right)^3 = \left(\dfrac{-3r^1}{r^4}\right)^3 = \left(-3r^{1-4}\right)^3 = \left(-3r^{-3}\right)^3 = (-3)^3 \left(r^{-3}\right)^3 = -27r^{-9} = -\dfrac{27}{r^9}$

105. $\left(\dfrac{14u^{-2}v^3}{21u^{-3}v}\right)^4 = \left(\dfrac{14}{21}u^{-2-(-3)}v^{3-1}\right)^4 = \left(\dfrac{2}{3}u^1v^2\right)^4 = \left(\dfrac{2}{3}\right)^4(u^1)^4(v^2)^4 = \dfrac{16}{81}u^4v^8 = \dfrac{16u^4v^8}{81}$

107. $\dfrac{x^{3n}}{x^{6n}} = x^{3n-6n} = x^{-3n} = \dfrac{1}{x^{3n}}$

Problems 109-111 are to be solved using a calculator. The keystrokes needed to solve each problem using a TI-84 graphing calculator appear in each solution. There may be other solutions. Keystrokes for other calculators may be slightly different.

109. $P = A(1+i)^{-n} = 100,000(1+0.07)^{-40}$

$\boxed{1}\boxed{0}\boxed{0}\boxed{0}\boxed{0}\boxed{0}\boxed{(}\boxed{1}\boxed{+}\boxed{.}\boxed{0}\boxed{7}\boxed{)}\boxed{\wedge}\boxed{(-)}\boxed{4}\boxed{0}\boxed{=}$

The original deposit should be $6,678.04.

111. $P = A(1+i)^{-n} = 1,000,000(1+0.04)^{-60}$

$\boxed{1}\boxed{0}\boxed{0}\boxed{0}\boxed{0}\boxed{0}\boxed{0}\boxed{(}\boxed{1}\boxed{+}\boxed{.}\boxed{0}\boxed{4}\boxed{)}\boxed{\wedge}\boxed{(-)}\boxed{6}\boxed{0}\boxed{=}$

The original deposit should be $95,060.40.

113. Answers may vary.

115. If $x > 1$, then x raised to a negative power is less than x. If $x = 1$, then x raised to a negative power is equal to x. If $0 < x < 1$, then x raised to a negative power is greater than x.

Exercises 4.3 (page 249)

1. 39,000,000: use 3.9

3. 83700: use 8.37

5. 0.0000000001052: use 1.052

7. $-3a^{33} = -3(-1)^{33} = -3(-1) = 3$

9. commutative property of multiplication

11. $6(x-5)+9 = 3x$
$6x-30+9 = 3x$
$6x-21 = 3x$
$-21 = -3x$
$7 = x$

13. scientific notation

15. $450,000 = 4.5 \times 10^5$

17. $1,700,000 = 1.7 \times 10^6$

19. $0.0059 = 5.9 \times 10^{-3}$

21. $0.00000275 = 2.75 \times 10^{-6}$

23. $42.5 \times 10^2 = 4.25 \times 10^1 \times 10^2$
$= 4.25 \times 10^3$

25. $0.37 \times 10^{-4} = 3.7 \times 10^{-1} \times 10^{-4}$
$= 3.7 \times 10^{-5}$

27. $2.3 \times 10^2 = 230$

29. $8.12 \times 10^5 = 812,000$

31. $1.15 \times 10^{-3} = 0.00115$

33. $9.76 \times 10^{-4} = 0.000976$

SECTION 4.3

35. $(3.4 \times 10^2)(2.1 \times 10^3) = (3.4)(2.1) \times 10^2 \times 10^3 = 7.14 \times 10^5 = 714,000$

37. $\dfrac{9.3 \times 10^2}{3.1 \times 10^{-2}} = \dfrac{9.3}{3.1} \times \dfrac{10^2}{10^{-2}} = 3 \times 10^4 = 30,000$

39. $\dfrac{96,000}{(12,000)(0.00004)} = \dfrac{9.6 \times 10^4}{(1.2 \times 10^4)(4 \times 10^{-5})} = \dfrac{9.6 \times 10^4}{4.8 \times 10^{-1}} = \dfrac{9.6}{4.8} \times \dfrac{10^4}{10^{-1}} = 2 \times 10^5 = 200,000$

41. $\dfrac{2,475}{(132,000,000)(0.25)} = \dfrac{2.475 \times 10^3}{(1.32 \times 10^8)(2.5 \times 10^{-1})} = \dfrac{2.475 \times 10^3}{3.3 \times 10^7} = \dfrac{2.475}{3.3} \times \dfrac{10^3}{10^7} = 0.75 \times 10^{-4}$
$= 0.000075$

43. $0.0000051 = 5.1 \times 10^{-6}$ **45.** $863,000,000 = 8.63 \times 10^8$

47. $\dfrac{2.4 \times 10^2}{6 \times 10^{23}} = \dfrac{2.4}{6} \times \dfrac{10^2}{10^{23}} = 0.4 \times 10^{-21} = 4 \times 10^{-1} \times 10^{-21} = 4 \times 10^{-22}$

49. $37 \times 10^7 = 370,000,000$ **51.** $0.32 \times 10^{-4} = 0.000032$

53. $3.72 \times 10^2 = 372$, so 3.72×10^2 is larger.

55. $3.72 \times 10^3 = 3,720$; $4.72 \times 10^3 = 4,720$
4.72×10^3 is larger.

57. $3.72 \times 10^{-1} = 0.372$
$4.72 \times 10^{-2} = 0.0472$; 3.72×10^{-1} is larger.

59. $25,700,000,000,000 = 2.57 \times 10^{13}$ mi   **61.** $1.14 \times 10^8 = 114,000,000$ mi

63. $0.00622 = 6.22 \times 10^{-3}$ mi

65. 3.6×10^7 mi $\times 5280$ ft/mi $= 19008 \times 10^7$ ft $= 1.9008 \times 10^{11}$ ft

67. $\$1,645.12$ billion $= \$1,645,120,000,000 = \$1.64512 \times 10^{12}$

69. $\dfrac{3.3 \times 10^4 \text{ cm}}{\text{sec}} = \dfrac{3.3 \times 10^4 \text{ cm}}{\text{sec}} \cdot \dfrac{1 \text{ m}}{100 \text{ cm}} \cdot \dfrac{1 \text{ km}}{1,000 \text{ m}} = \dfrac{3.3 \times 10^4}{10^5}$ km/sec $= 3.3 \times 10^{-1}$ km/sec

71. x-rays, visible light, infrared

73. 1.5×10^{-4} in.; $25,000,000,000,000 = 2.5 \times 10^{13}$

75. Answers may vary.   **77. Answers may vary.**

Exercises 4.4 (page 259)

1. $5 \cdot a \cdot a + 2 \cdot b \cdot b \cdot b = 5a^2 + 2b^3$ **3.** $4 \cdot a \cdot a \cdot a \cdot b \cdot b = 4a^3b^2$

5. $x \cdot x \cdot x + 3 \cdot x \cdot x + x = x^3 + 3x^2 + x$ **7.** $a \cdot a \cdot a + b \cdot b \cdot b = a^3 + b^3$

119

9.
$$5(u - 5) + 9 = 2(u + 4)$$
$$5u - 25 + 9 = 2u + 8$$
$$5u - 16 = 2u + 8$$
$$3u - 16 = 8$$
$$3u = 24$$
$$u = 8$$

11.
$$-4(3y + 2) \leq 28$$
$$-12y - 8 \leq 28$$
$$-12y \leq 36$$
$$\frac{-12y}{-12} \geq \frac{36}{-12}$$
$$y \geq -3$$

13. $\left(x^3 x^5\right)^4 = \left(x^8\right)^4 = x^{32}$

15. $\left(\dfrac{y^2 y^5}{y^4}\right)^3 = \left(\dfrac{y^7}{y^4}\right)^3 = \left(y^3\right)^3 = y^9$

17. algebraic

19. monomial; binomial; trinomial

21. sum

23. polynomial

25. cubic

27. descending; 5

29. function

31. yes

33. no; fractional exponent on x

35. binomial

37. trinomial

39. monomial

41. binomial

43. 7th

45. 3rd

47. 8th
$(3 + 5 = 8)$

49. 6th
$(3 + 1 + 2 = 6)$

51. $-x^2 - 4 = -(0)^2 - 4 = -0 - 4 = -4$

53. $-x^2 - 4 = -(-1)^2 - 4 = -1 - 4 = -5$

55.

x	$x^2 - 3$
-2	$(-2)^2 - 3 = 4 - 3 = 1$
-1	$(-1)^2 - 3 = 1 - 3 = -2$
0	$(0)^2 - 3 = 0 - 3 = -3$
1	$(1)^2 - 3 = 1 - 3 = -2$
2	$(2)^2 - 3 = 4 - 3 = 1$

57.

x	$x^3 + 2$
-2	$(-2)^3 + 2 = -8 + 2 = -6$
-1	$(-1)^3 + 2 = -1 + 2 = 1$
0	$(0)^3 + 2 = 0 + 2 = 2$
1	$(1)^3 + 2 = 1 + 2 = 3$
2	$(2)^3 + 2 = 8 + 2 = 10$

59. $f(-3) = 5(-3) + 1 = -15 + 1 = -14$

61. $f\left(-\dfrac{1}{2}\right) = 5\left(-\dfrac{1}{2}\right) + 1 = -\dfrac{5}{2} + \dfrac{2}{2} = -\dfrac{3}{2}$

63. $f(x) = x^2 - 1$

x	$f(x)$
-2	3
-1	0
0	-1
1	0
2	3

D: $(-\infty, \infty)$
R: $[-1, \infty)$

65. $f(x) = x^3 + 2$

x	$f(x)$
-2	-6
-1	1
0	2
1	3
2	10

D: $(-\infty, \infty)$
R: $(-\infty, \infty)$

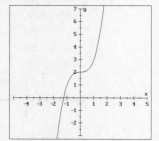

67. $5x - 3 = 5(2) - 3 = 10 - 3 = 7$

69. $5x - 3 = 5(-1) - 3 = -5 - 3 = -8$

71. trinomial

73. none of these (4 terms)

75. binomial

For exercises 77-83, answers may vary. The answers below are only examples of correct answers.

77. $x + 2$

79. $x^2 + 2x + 1$

81. $x^3 + 4$

83. 5

85. 12th

87. 0th

89. $f(5) = (5)^2 - 2(5) + 3 = 25 - 10 + 3 = 18$

91. $f(-2) = (-2)^2 - 2(-2) + 3$
$= 4 + 4 + 3 = 11$

93. $f(0.5) = (0.5)^2 - 2(0.5) + 3$
$= 0.25 - 1 + 3 = 2.25$

95. $h = -16t^2 + 64t$
$h = -16(2)^2 + 64(2)$
$h = -16(4) + 128$
$h = -64 + 128 = 64$ ft

97. $f(d) = -0.08d^2 + 100d$
$f(815) = -0.08(815)^2 + 100(815)$
$= -0.08(664225) + 81500$
$= -53138 + 81500$
$= \$28,362$

99. $d = 0.04v^2 + 0.9v$
$d = 0.04(30)^2 + 0.9(30)$
$d = 0.04(900) + 27$
$d = 36 + 27 = 63$ ft

101. **Answers may vary.**

103. There are many possible polynomials. One is $2x - 2$.

Exercises 4.5 (page 265)

1. $x(7 + 2) = x(9) = 9x$

3. $a(16 - 4) = a(12) = 12a$

5. $6x^2, 6x$: unlike terms

7. $4x^3, 5x^3$: like terms

9. $ab + cd = (3)(-2) + (-1)(2)$
$= -6 + (-2) = -8$

11. $a(b - c) = 3[-2 - (-1)] = 3(-2 + 1)$
$= 3(-1) = -3$

13. $-4(2x - 9) \geq 12$

$-8x + 36 \geq 12$

$-8x \geq -24$

$\dfrac{-8x}{-8} \leq \dfrac{-24}{-8}$

$x \leq 3$

15. monomial

17. coefficients; variables

19. like terms

21. like terms; $3y + 4y = 7y$

23. unlike terms

25. like terms

$3x^3 + 4x^3 + 6x^3 = 13x^3$

27. like terms; $-5x^3y^2 + 13x^3y^2 = 8x^3y^2$

29. like terms

$15x^4y^2 + \left(-9x^4y^2\right) + 4x^4y^2 = 10x^4y^2$

31. unlike terms

33. $4y + 5y = 9y$

35. $15x^2 + 10x^2 = 25x^2$

37. $-7t^6 + 3(t^2)^3 = -7t^6 + 3t^6 = -4t^6$

39. $26x^2y^4 + 3x^2y^4 = 29x^2y^4$

41. $-18a - 3a = -21a$

43. $32u^3 - 16u^3 = 16u^3$

45. $18x^5y^2 - 11x^5y^2 = 7x^5y^2$

47. $-14ab^3 - 6ab^3 = -20ab^3$

49. $(3x + 7) + (4x - 3) = 3x + 7 + 4x - 3 = 7x + 4$

51. $(6y^2 - 2y + 5) + (2y^2 + 5y - 8) = 6y^2 - 2y + 5 + 2y^2 + 5y - 8 = 8y^2 + 3y - 3$

53.
$$\begin{array}{r} 3x^2 + 4x + 5 \\ +\ \underline{2x^2 - 3x + 6} \\ 5x^2 + x + 11 \end{array}$$

55.
$$\begin{array}{r} 2x^3 - 3x^2 + 4x - 7 \\ +\ \underline{-9x^3 - 4x^2 - 5x + 6} \\ -7x^3 - 7x^2 - x - 1 \end{array}$$

57. $(4a + 3) - (2a - 4) = 4a + 3 - 2a + 4 = 2a + 7$

59. $(2a^2 - 6a + 3) - (-3a^2 - 4a + 5) = 2a^2 - 6a + 3 + 3a^2 + 4a - 5 = 5a^2 - 2a - 2$

61.
$$\begin{array}{r} 3x^2 + 4x - 5 \\ -\ \underline{-2x^2 - 2x + 3} \\ 3x^2 + 4x - 5 \\ +\ \underline{2x^2 + 2x - 3} \\ 5x^2 + 6x - 8 \end{array}$$

63.
$$\begin{array}{r} 4x^3 + 4x^2 - 3x + 10 \\ -\ \underline{5x^3 - 2x^2 - 4x - 4} \\ 4x^3 + 4x^2 - 3x + 10 \\ +\ \underline{-5x^3 + 2x^2 + 4x + 4} \\ -x^3 + 6x^2 + x + 14 \end{array}$$

65. $(-3x - 7y) - (8x + 2y) = -3x - 7y - 8x - 2y = -11x - 9y$

67. $(2x^2 - 3x + 1) - (4x^2 - 3x + 2) = 2x^2 - 3x + 1 - 4x^2 + 3x - 2 = -2x^2 - 1$

69. $2(x+3) + 4(x-2) = 2x + 6 + 4x - 8 = 6x - 2$

71. $2(x^2 - 5x - 4) - 3(x^2 - 5x - 4) + 6(x^2 - 5x - 4)$
$$= 2x^2 - 10x - 8 - 3x^2 + 15x + 12 + 6x^2 - 30x - 24 = 5x^2 - 25x - 20$$

73. $3rst + 4rst + 7rst = 14rst$ 　　　　　　**75.** $-4a^2bc + 5a^2bc - 7a^2bc = -6a^2bc$

77. $-3x^3y^6 + 2(xy^2)^3 - (3x)^3 y^6 = -3x^3y^6 + 2x^3y^6 - 27x^3y^6 = -28x^3y^6$

79. $5x^5y^{10} - (2xy^2)^5 + (3x)^5 y^{10} = 5x^5y^{10} - 32x^5y^{10} + 243x^5y^{10} = 216x^5y^{10}$

81. $-8(x-y) + 11(x-y) = -8x + 8y + 11x - 11y = 3x - 3y$

83. $(-3z^2 - 4z + 7) + (2z^2 + 2z - 1) - (2z^2 - 3z + 7)$
$$= -3z^2 - 4z + 7 + 2z^2 + 2z - 1 - 2z^2 + 3z - 7 = -3z^2 + z - 1$$

85.
$$\begin{array}{r} -6x^3z - 4x^2z^2 + 7z^3 \\ + \quad -7x^3z + 9x^2z^2 - 21z^3 \\ \hline -13x^3z + 5x^2z^2 - 14z^3 \end{array}$$

87.
$$\begin{array}{r} 25x^3 - 45x^2z + 31xz^2 \\ - \quad 12x^3 + 27x^2z - 17xz^2 \\ \hline 25x^3 - 45x^2z + 31xz^2 \\ + \quad -12x^3 - 27x^2z + 17xz^2 \\ \hline 13x^3 - 72x^2z + 48xz^2 \end{array}$$

89. $3(xy^2 + y^2) - 2(xy^2 - 4y^2 + y^3) + 2(y^3 + y^2) = 3xy^2 + 3y^2 - 2xy^2 + 8y^2 - 2y^3 + 2y^3 + 2y^2$
$$= xy^2 + 13y^2$$

91. $[(2x^2 - 3x + 4) + (3x^2 - 2)] + (x^2 + x - 3)$
$$= [2x^2 - 3x + 4 + 3x^2 - 2] + x^2 + x - 3$$
$$= 5x^2 - 3x + 2 + x^2 + x - 3 = 6x^2 - 2x - 1$$

93. $[(2z^2 + 3z - 7) + (-4z^3 - 2z - 3)] - (-3z^3 - 4z + 7)$
$$= [2z^2 + 3z - 7 - 4z^3 - 2z - 3] + 3z^3 + 4z - 7$$
$$= -4z^3 + 2z^2 + z - 10 + 3z^3 + 4z - 7 = -z^3 + 2z^2 + 5z - 17$$

95. $y = 900x + 105{,}000; \; y = 900(10) + 105{,}000 = 9{,}000 + 105{,}000 = \$114{,}000$

97. $y = (900x + 105{,}000) + (1{,}000x + 120{,}000) \Rightarrow y = 1{,}900x + 225{,}000$

99a. $y = -1{,}100x + 6{,}600$ 　　　　　　**99b.** $y = -1{,}700x + 9{,}200$

101. Answers may vary.

103. $P(x + h) + P(x) = [3(x + h) - 5] + [3x - 5] = 3x + 3h - 5 + 3x - 5 = 6x + 3h - 10$

105. $P(x) - Q(x) = (x^{23} + 5x^2 + 73) - (x^{23} + 4x^2 + 73) = x^{23} + 5x^2 + 73 - x^{23} - 4x^2 - 73 = x^2$
$P(7) - Q(7) = 7^2 = 49$

Exercises 4.6 (page 275)

1. $3x^2(5x) = 3 \cdot 5 \cdot x^2 \cdot x = 15x^3$

3. $-4x(2y) = -4 \cdot 2 \cdot x \cdot y = -8xy$

5. $-5(x - 4) = -5(x) - (-5)(4) = -5x + 20$

7. $2(4x^2 - 9) = 2(4x^2) - (2)(9) = 8x^2 - 18$

9. commutative property of addition

11. commutative property of multiplication

13.
$$\frac{5}{3}(5y + 6) - 10 = 0$$
$$3\left[\frac{5}{3}(5y + 6) - 10\right] = 3(0)$$
$$3 \cdot \frac{5}{3}(5y + 6) - 3 \cdot 10 = 0$$
$$5(5y + 6) - 30 = 0$$
$$25y + 30 - 30 = 0$$
$$25y = 0$$
$$y = 0$$

15. monomial

17. special products

19. $(2x)(3x) = 6x^2$

21. $(5)(3x) = 15x$

23. $(3x^2)(4x^3) = (3)(4)x^2x^3 = 12x^5$

25. $(-5t^3)(2t^4) = (-5)(2)t^3t^4 = -10t^7$

27. $(2x^2y^3)(3x^3y^2) = (2)(3)x^2x^3y^3y^2$
$$= 6x^5y^5$$

29. $(3b^2)(-2b)(4b^3) = (3)(-2)(4)b^2bb^3$
$$= -24b^6$$

31. $3(x + 4) = 3(x) + 3(4) = 3x + 12$

33. $-4(t + 7) = -4(t) + (-4)(7) = -4t - 28$

35. $3x(x - 2) = 3x(x) - 3x(2) = 3x^2 - 6x$

37. $-2x^2(3x^2 - x) = -2x^2(3x^2) - (-2x^2)(x)$
$$= -6x^4 + 2x^3$$

39. $3xy(x + y) = 3xy(x) + 3xy(y) = 3x^2y + 3xy^2$

41. $-6x^2(2x^2 + 3x - 5) = -6x^2(2x^2) + (-6x^2)(3x) - (-6x^2)(5) = -12x^4 - 18x^3 + 30x^2$

43. $\frac{1}{4}x^2(8x^5 - 4) = \frac{1}{4}x^2(8x^5) - \frac{1}{4}x^2(4) = 2x^7 - x^2$

45. $-\frac{2}{3}r^2t^2(9r - 3t) = -\frac{2}{3}r^2t^2(9r) - \left(-\frac{2}{3}r^2t^2\right)(3t) = -6r^3t^2 + 2r^2t^3$

47. $(a + 4)(a + 5) = a(a) + a(5) + 4(a) + 4(5) = a^2 + 5a + 4a + 20 = a^2 + 9a + 20$

49. $(3x - 2)(x + 4) = 3x(x) + 3x(4) + (-2)(x) + (-2)(4) = 3x^2 + 12x - 2x - 8 = 3x^2 + 10x - 8$

51. $(4a - 2)(2a - 3) = 4a(2a) + 4a(-3) - 2(2a) - 2(-3) = 8a^2 - 12a - 4a + 6 = 8a^2 - 16a + 6$

53. $(3x - 5)(2x + 1) = 3x(2x) + 3x(1) + (-5)(2x) + (-5)(1) = 6x^2 + 3x - 10x - 5$
$$= 6x^2 - 7x - 5$$

55. $(2s + 3t)(3s - t) = 2s(3s) - 2s(t) + 3t(3s) - 3t(t) = 6s^2 - 2st + 9st - 3t^2 = 6s^2 + 7st - 3t^2$

57. $(u + v)(u + 2t) = u(u) + u(2t) + v(u) + v(2t) = u^2 + 2tu + uv + 2tv$

59. $2(x - 4)(x + 1) = 2[x(x) + x(1) + (-4)(x) + (-4)(1)] = 2[x^2 + x - 4x - 4]$
$$= 2[x^2 - 3x - 4] = 2x^2 - 6x - 8$$

61. $3a(a + b)(a - b) = 3a[a(a) - a(b) + b(a) - b(b)] = 3a[a^2 - ab + ab - b^2]$
$$= 3a[a^2 - b^2] = 3a^3 - 3ab^2$$

63. $(3x - 2y)(x + y) = 3x(x) + 3x(y) - 2y(x) - 2y(y) = 3x^2 + 3xy - 2xy - 2y^2 = 3x^2 + xy - 2y^2$

65. $(2x - 3)(x + 1) - 5x(x + 2) = [2x(x) + 2x(1) - 3(x) - 3(1)] - [5x(x) + 5x(2)]$
$$= [2x^2 + 2x - 3x - 3] - [5x^2 + 10x]$$
$$= [2x^2 - x - 3] - [5x^2 + 10x]$$
$$= 2x^2 - x - 3 - 5x^2 - 10x = -3x^2 - 11x - 3$$

67. $(x + 5)^2 = (x + 5)(x + 5) = x(x) + x(5) + 5(x) + 5(5) = x^2 + 5x + 5x + 25 = x^2 + 10x + 25$

69. $(x - 4)^2 = (x - 4)(x - 4) = x(x) + x(-4) - 4(x) - 4(-4) = x^2 - 4x - 4x + 16 = x^2 - 8x + 16$

71. $(4t + 3)^2 = (4t + 3)(4t + 3) = 4t(4t) + 4t(3) + 3(4t) + 3(3) = 16t^2 + 12t + 12t + 9$
$$= 16t^2 + 24t + 9$$

73. $(x - 2y)^2 = (x - 2y)(x - 2y) = x(x) - x(2y) + (-2y)(x) - (-2y)(2y)$
$$= x^2 - 2xy - 2xy + 4y^2 = x^2 - 4xy + 4y^2$$

75. $(r + 4)(r - 4) = r(r) - r(4) + 4(r) - 4(4) = r^2 - 4r + 4r - 16 = r^2 - 16$

77. $(4x + 5)(4x - 5) = 4x(4x) - 4x(5) + 5(4x) - 5(5) = 16x^2 - 20x + 20x - 25 = 16x^2 - 25$

79. $(2x + 3)(x^2 + 4x - 1) = (2x + 3)(x^2) + (2x + 3)(4x) + (2x + 3)(-1)$
$$= 2x(x^2) + 3(x^2) + 2x(4x) + 3(4x) + 2x(-1) + 3(-1)$$
$$= 2x^3 + 3x^2 + 8x^2 + 12x - 2x - 3 = 2x^3 + 11x^2 + 10x - 3$$

81. $(4t + 3)(t^2 + 2t + 3) = 4t(t^2) + 4t(2t) + 4t(3) + 3(t^2) + 3(2t) + 3(3)$
$$= 4t^3 + 8t^2 + 12t + 3t^2 + 6t + 9 = 4t^3 + 11t^2 + 18t + 9$$

83.
$$\begin{array}{r} 4x + 3 \\ x + 2 \\ \hline 4x^2 + 3x \\ 8x + 6 \\ \hline 4x^2 + 11x + 6 \end{array}$$

85.
$$\begin{array}{r} 4x - 2y \\ 3x + 5y \\ \hline 12x^2 - 6xy \\ 20xy - 10y^2 \\ \hline 12x^2 + 14xy - 10y^2 \end{array}$$

87. $(s-4)(s+1) = s^2 + 5$

$s^2 + s - 4s - 4 = s^2 + 5$

$s^2 - 3s - 4 = s^2 + 5$

$-3s - 4 = 5$

$-3s = 9$

$s = -3$

89. $z(z+2) = (z+4)(z-4)$

$z^2 + 2z = z^2 - 4z + 4z - 16$

$z^2 + 2z = z^2 - 16$

$2z = -16$

$z = -8$

91. $(x+4)(x-4) = (x-2)(x+6)$

$x^2 - 4x + 4x - 16 = x^2 + 6x - 2x - 12$

$x^2 - 16 = x^2 + 4x - 12$

$-16 = 4x - 12$

$-4 = 4x$

$-1 = x$

93. $(a-3)^2 = (a+3)^2$

$(a-3)(a-3) = (a+3)(a+3)$

$a^2 - 3a - 3a + 9 = a^2 + 3a + 3a + 9$

$a^2 - 6a + 9 = a^2 + 6a + 9$

$-6a + 9 = 6a + 9$

$-6a = 6a$

$12a = 0$

$a = 0$

95. $(x^2 y^5)(x^2 z^5)(-3y^2 z^3) = (1)(1)(-3)x^2 x^2 y^5 y^2 z^5 x^3 = -3x^4 y^7 z^8$

97. $(x+3)(2x-3) = x(2x) - x(3) + 3(2x) - 3(3) = 2x^2 - 3x + 6x - 9 = 2x^2 + 3x - 9$

99. $(t-3)(t-3) = t(t) - t(3) + (-3)(t) - (-3)(3) = t^2 - 3t - 3t + 9 = t^2 - 6t + 9$

101. $(-2r - 3s)(2r + 7s) = (-2r)(2r) + (-2r)(7s) + (-3s)(2r) + (-3s)(7s)$

$= -4r^2 - 14rs - 6rs - 21s^2 = -4r^2 - 20rs - 21s^2$

103. $(3x - 2)^2 = (3x - 2)(3x - 2) = 3x(3x) + 3x(-2) - 2(3x) - 2(-2) = 9x^2 - 6x - 6x + 4$

$= 9x^2 - 12x + 4$

105. $(3x - y)(x^2 + 3xy - y^2)$

$= 3x(x^2) + 3x(3xy) - 3x(y^2) + (-y)(x^2) + (-y)(3xy) - (-y)(y^2)$

$= 3x^3 + 9x^2 y - 3xy^2 - x^2 y - 3xy^2 + y^3 = 3x^3 + 8x^2 y - 6xy^2 + y^3$

107. $3xy(x + y) - 2x(xy - x) = [3xy(x) + 3xy(y)] - [(2x)(xy) - (2x)(x)]$

$= [3x^2 y + 3xy^2] - [2x^2 y - 2x^2]$

$= 3x^2 y + 3xy^2 - 2x^2 y + 2x^2 = x^2 y + 3xy^2 + 2x^2$

109. $(2x - 1)(2x + 1) = x(4x + 1)$
$$4x^2 + 2x - 2x - 1 = 4x^2 + x$$
$$4x^2 - 1 = 4x^2 + x$$
$$-1 = x$$

111.
$$(x + 2)^2 = (x - 2)^2$$
$$(x + 2)(x + 2) = (x - 2)(x - 2)$$
$$x^2 + 2x + 2x + 4 = x^2 - 2x - 2x + 4$$
$$x^2 + 4x + 4 = x^2 - 4x + 4$$
$$4x + 4 = -4x + 4$$
$$8x + 4 = 4$$
$$8x = 0$$
$$x = 0$$

113. $(x + y)(x - y) + x(x + y) = [x(x) - x(y) + y(x) - y(y)] + [x(x) + x(y)]$
$$= [x^2 - xy + xy - y^2] + [x^2 + xy]$$
$$= [x^2 - y^2] + [x^2 + xy] = x^2 - y^2 + x^2 + xy = 2x^2 + xy - y^2$$

115. $(3x + 4)(2x - 2) - (2x + 1)(x + 3)$
$$= [3x(2x) - 3x(2) + 4(2x) - 4(2)] - [2x(x) + 2x(3) + 1(x) + 1(3)]$$
$$= [6x^2 - 6x + 8x - 8] - [2x^2 + 6x + x + 3]$$
$$= [6x^2 + 2x - 8] - [2x^2 + 7x + 3] = 6x^2 + 2x - 8 - 2x^2 - 7x - 3 = 4x^2 - 5x - 11$$

117. $3y(y + 2) = 3(y + 1)(y - 1)$
$$3y^2 + 6y = 3(y^2 - y + y - 1)$$
$$3y^2 + 6y = 3(y^2 - 1)$$
$$3y^2 + 6y = 3y^2 - 3$$
$$6y = -3$$
$$y = -\frac{3}{6} = -\frac{1}{2}$$

119. Let $r =$ the smaller radius.
Then $r + 3 =$ the larger radius.

$$\boxed{\text{Larger area}} - \boxed{\text{Smaller area}} = 15\pi$$
$$\pi(r + 3)^2 - \pi r^2 = 15\pi$$
$$\pi(r + 3)(r + 3) - \pi r^2 = 15\pi$$
$$\pi(r^2 + 6r + 9) - \pi r^2 = 15\pi$$
$$\pi r^2 + 6\pi r + 9\pi - \pi r^2 = 15\pi$$
$$6\pi r + 9\pi = 15\pi$$
$$6\pi r = 6\pi$$
$$\frac{6\pi r}{6\pi} = \frac{6\pi}{6\pi}$$
$$r = 1$$

The larger radius $= 1 + 3 = 4$ m.

121. Let $s =$ the side of the softball field. Then $s + 30 =$ the side of the baseball field.

$$\boxed{\text{Larger area}} - \boxed{\text{Smaller area}} = 4500$$
$$(s + 30)^2 - s^2 = 4500$$
$$(s + 30)(s + 30) - s^2 = 4500$$
$$s^2 + 30s + 30s + 900 - s^2 = 4500$$
$$60s + 900 = 4500$$
$$60s = 3600$$
$$s = 60$$

The baseball field has a side of
$60 + 30 = 90$ feet.

123. Answers may vary.

125. Answers may vary.

Exercises 4.7 (page 281)

1. $\dfrac{3}{21} = \dfrac{\cancel{3} \cdot 1}{\cancel{3} \cdot 7} = \dfrac{1}{7}$

3. $-\dfrac{64}{72} = -\dfrac{\cancel{8} \cdot 8}{\cancel{8} \cdot 9} = -\dfrac{8}{9}$

5. $\dfrac{70}{420} = \dfrac{\cancel{70} \cdot 1}{\cancel{70} \cdot 6} = \dfrac{1}{6}$

7. $\dfrac{8{,}423}{-8{,}423} = \dfrac{\cancel{8423} \cdot 1}{-\cancel{8423} \cdot 1} = -1$

9. binomial

11. none of these

13. 2

15. polynomial

17. 1

19. $\dfrac{a}{b}$

21. $\dfrac{xy}{yz} = \dfrac{x}{z}$

23. $\dfrac{r^3 s^2}{r s^3} = r^{3-1} s^{2-3} = r^2 s^{-1} = \dfrac{r^2}{s}$

25. $\dfrac{8x^3 y^2}{4xy^3} = \dfrac{8}{4} x^{3-1} y^{2-3} = 2x^2 y^{-1} = \dfrac{2x^2}{y}$

27. $\dfrac{12u^5 v}{-4u^2 v^3} = \dfrac{12}{-4} u^{5-2} v^{1-3} = -3u^3 v^{-2}$
$= -\dfrac{3u^3}{v^2}$

29. $\dfrac{6x + 9y}{3xy} = \dfrac{6x}{3xy} + \dfrac{9y}{3xy} = \dfrac{2}{y} + \dfrac{3}{x}$

31. $\dfrac{xy + 6}{3y} = \dfrac{xy}{3y} + \dfrac{6}{3y} = \dfrac{x}{3} + \dfrac{2}{y}$

33. $\dfrac{5x - 10y}{25xy} = \dfrac{5x}{25xy} - \dfrac{10y}{25xy} = \dfrac{1}{5y} - \dfrac{2}{5x}$

35. $\dfrac{3x^2 + 6y^3}{3x^2 y^2} = \dfrac{3x^2}{3x^2 y^2} + \dfrac{6y^3}{3x^2 y^2} = \dfrac{1}{y^2} + \dfrac{2y}{x^2}$

37. $\dfrac{4x - 2y + 8z}{4xy} = \dfrac{4x}{4xy} - \dfrac{2y}{4xy} + \dfrac{8z}{4xy} = \dfrac{1}{y} - \dfrac{1}{2x} + \dfrac{2z}{xy}$

39. $\dfrac{12x^3 y^2 - 8x^2 y - 4x}{4xy} = \dfrac{12x^3 y^2}{4xy} - \dfrac{8x^2 y}{4xy} - \dfrac{4x}{4xy} = 3x^2 y - 2x - \dfrac{1}{y}$

41. $\dfrac{-25x^2 y + 30xy^2 - 5xy}{-5xy} = \dfrac{-25x^2 y}{-5xy} + \dfrac{30xy^2}{-5xy} - \dfrac{5xy}{-5xy} = 5x - 6y + 1$

43. $\dfrac{15a^3 b^2 - 10a^2 b^3}{5a^2 b^2} = \dfrac{15a^3 b^2}{5a^2 b^2} - \dfrac{10a^2 b^3}{5a^2 b^2} = 3a - 2b$

45. $\dfrac{5x(4x - 2y)}{2y} = \dfrac{20x^2 - 10xy}{2y} = \dfrac{20x^2}{2y} - \dfrac{10xy}{2y} = \dfrac{10x^2}{y} - 5x$

47. $\dfrac{(-2x)^3 + (3x^2)^2}{6x^2} = \dfrac{-8x^3 + 9x^4}{6x^2} = \dfrac{-8x^3}{6x^2} + \dfrac{9x^4}{6x^2} = -\dfrac{4x}{3} + \dfrac{3x^2}{2}$

49. $\dfrac{4x^2y^2 - 2(x^2y^2 + xy)}{2xy} = \dfrac{4x^2y^2 - 2x^2y^2 - 2xy}{2xy} = \dfrac{2x^2y^2 - 2xy}{2xy} = \dfrac{2x^2y^2}{2xy} - \dfrac{2xy}{2xy} = xy - 1$

51. $\dfrac{(a+b)^2 - (a-b)^2}{2ab} = \dfrac{(a+b)(a+b) - (a-b)(a-b)}{2ab}$

$= \dfrac{[a^2 + ab + ab + b^2] - [a^2 - ab - ab + b^2]}{2ab}$

$= \dfrac{[a^2 + 2ab + b^2] - [a^2 - 2ab + b^2]}{2ab}$

$= \dfrac{a^2 + 2ab + b^2 - a^2 + 2ab - b^2}{2ab} = \dfrac{4ab}{2ab} = 2$

53. $l = \dfrac{P - 2w}{2}$

$l = \dfrac{P}{2} - \dfrac{2w}{2}$

$l = \dfrac{P}{2} - w$

They are the same.

55. $C = \dfrac{0.15x + 12}{x}$

$C = \dfrac{0.15x}{x} + \dfrac{12}{x}$

$C = 0.15 + \dfrac{12}{x}$

They are the same.

57. $\dfrac{120}{160} = \dfrac{40(3)}{40(4)} = \dfrac{3}{4}$

59. $\dfrac{5880}{2660} = \dfrac{140(42)}{140(19)} = \dfrac{42}{19}$

61. $\dfrac{-16r^3y^2}{-4r^2y^4} = \dfrac{-16}{-4}r^{3-2}y^{2-4} = 4ry^{-2}$

$= \dfrac{4r}{y^2}$

63. $\dfrac{-65rs^2t}{15r^2s^3t} = \dfrac{-65}{15}r^{1-2}s^{2-3}t^{1-1}$

$= -\dfrac{13}{3}r^{-1}s^{-1}t^0 = -\dfrac{13}{3rs}$

65. $\dfrac{x^2x^3}{xy^6} = \dfrac{x^5}{xy^6} = \dfrac{x^{5-1}}{y^6} = \dfrac{x^4}{y^6}$

67. $\dfrac{(a^3b^4)^3}{ab^4} = \dfrac{a^9b^{12}}{ab^4} = a^{9-1}b^{12-4} = a^8b^8$

69. $\dfrac{12a + 2b}{6ab} = \dfrac{12a}{6ab} + \dfrac{2b}{6ab} = \dfrac{2}{b} + \dfrac{1}{3a}$

71. $\dfrac{16x - 8y}{24xy} = \dfrac{16x}{24xy} - \dfrac{8y}{24xy} = \dfrac{2}{3y} - \dfrac{1}{3x}$

73. $\dfrac{2x(8x - 3y)}{4x} = \dfrac{16x^2 - 6xy}{4x} = \dfrac{16x^2}{4x} - \dfrac{6xy}{4x} = 4x - \dfrac{3y}{2}$

75. $\dfrac{8x^3 + 16x^2 - 4x}{4x} = \dfrac{8x^3}{4x} + \dfrac{16x^2}{4x} - \dfrac{4x}{4x} = 2x^2 + 4x - 1$

77. $\dfrac{-(3x^3y^4)^3}{-(9x^4y^5)^2} = \dfrac{-27x^9y^{12}}{-81x^8y^{10}} = \dfrac{-27}{-81}x^{9-8}y^{12-10} = \dfrac{1}{3}xy^2 = \dfrac{xy^2}{3}$

79. $\dfrac{(a^2a^3)^4}{(a^4)^3} = \dfrac{(a^5)^4}{a^{12}} = \dfrac{a^{20}}{a^{12}} = a^{20-12} = a^8$

81. $\dfrac{(3x-y)(2x-3y)}{6xy} = \dfrac{6x^2 - 9xy - 2xy + 3y^2}{6xy} = \dfrac{6x^2 - 11xy + 3y^2}{6xy} = \dfrac{6x^2}{6xy} - \dfrac{11xy}{6xy} + \dfrac{3y^2}{6xy}$

$$= \dfrac{x}{y} - \dfrac{11}{6} + \dfrac{y}{2x}$$

83. **Answers may vary.**

85. $\dfrac{x^{500} - x^{499}}{x^{499}} = \dfrac{x^{500}}{x^{499}} - \dfrac{x^{499}}{x^{499}} = x - 1$

Let $x = 501$: $\quad x - 1 = 501 - 1 = 500$

Exercises 4.8 (page 287)

1.
$$\begin{array}{r} 2\ 4 \\ 16\,\overline{)3\ 8\ 4} \\ \underline{3\ 2} \\ 6\ 4 \\ \underline{6\ 4} \\ 0 \end{array}$$

3.
$$\begin{array}{r} 1\ 4\ \frac{5}{19} \\ 19\,\overline{)2\ 7\ 1} \\ \underline{1\ 9} \\ 8\ 1 \\ \underline{7\ 6} \\ 5 \end{array}$$

5. $\begin{aligned} x + 2 &\neq 0 \\ x &\neq -2 \end{aligned}$

7. $\begin{aligned} 2x - 7 &\neq 0 \\ 2x &\neq 7 \\ x &\neq \tfrac{7}{2} \end{aligned}$

9. $21, 22, 24, 25, 26, 27, 28$

11. $|a - b| = |-2 - 3| = |-5| = 5$

13. $-|a^2 - b^2| = -\left|(-2)^2 - 3^2\right| = -|4 - 9| = -|-5| = -(+5) = -5$

15. $4(3x^2 - 5x + 2) + 3(2x^2 + 6x - 3) = 12x^2 - 20x + 8 + 6x^2 + 18x - 9 = 18x^2 - 2x - 1$

17. divisor; dividend

19. remainder

21. $2x^3 + 8x^2 + 3x - 5$

23. $5x^4 + 6x^3 - 4x^2 + 7x$

25. $0x^3$ and $0x$

27.
$$\begin{array}{r} x + \ \ 4 \\ x + 1\,\overline{)x^2 + 5x + 4} \\ \underline{x^2 + \ \ x} \\ 4x + 4 \\ \underline{4x + 4} \\ 0 \end{array}$$

29.
$$\begin{array}{r} x + \ \ 2 \\ x + 5\,\overline{)x^2 + 7x + 10} \\ \underline{x^2 + 5x} \\ 2x + 10 \\ \underline{2x + 10} \\ 0 \end{array}$$

31.
$$\begin{array}{r} x - \ \ 3 \\ x - 2\,\overline{)x^2 - 5x + 6} \\ \underline{x^2 - 2x} \\ -3x + 6 \\ \underline{-3x + 6} \\ 0 \end{array}$$

33.
$$\begin{array}{r} a - \ \ 4 \\ a - 7\,\overline{)a^2 - 11a + 28} \\ \underline{a^2 - \ \ 7a} \\ -4a + 28 \\ \underline{-4a + 28} \\ 0 \end{array}$$

35.
$$\begin{array}{r} 2a - \ \ 1 \\ 4a - 3\,\overline{)8a^2 - 10a + 3} \\ \underline{8a^2 - \ \ 6a} \\ -4a + 3 \\ \underline{-4a + 3} \\ 0 \end{array}$$

37.
$$\begin{array}{r} b + \ \ 3 \\ 3b + 2\,\overline{)3b^2 + 11b + 6} \\ \underline{3b^2 + \ \ 2b} \\ 9b + 6 \\ \underline{9b + 6} \\ 0 \end{array}$$

39.
$$
\begin{array}{r}
x + 1 + \frac{-1}{2x+3} \\
2x+3 \overline{\smash{\big)}\ 2x^2 + 5x + 2} \\
\underline{2x^2 + 3x} \\
2x + 2 \\
\underline{2x + 3} \\
-1
\end{array}
$$

41.
$$
\begin{array}{r}
2x + 2 + \frac{-3}{2x+1} \\
2x+1 \overline{\smash{\big)}\ 4x^2 + 6x - 1} \\
\underline{4x^2 + 2x} \\
4x - 1 \\
\underline{4x + 2} \\
-3
\end{array}
$$

43.
$$
\begin{array}{r}
a + b \\
a+b \overline{\smash{\big)}\ a^2 + 2ab + b^2} \\
\underline{a^2 + ab} \\
ab + b^2 \\
\underline{ab + b^2} \\
0
\end{array}
$$

45.
$$
\begin{array}{r}
2x - y \\
x+2y \overline{\smash{\big)}\ 2x^2 + 3xy - 2y^2} \\
\underline{2x^2 + 4xy} \\
- xy - 2y^2 \\
\underline{- xy - 2y^2} \\
0
\end{array}
$$

47.
$$
\begin{array}{r}
x - 3y \\
2x-y \overline{\smash{\big)}\ 2x^2 - 7xy + 3y^2} \\
\underline{2x^2 - xy} \\
- 6xy + 3y^2 \\
\underline{- 6xy + 3y^2} \\
0
\end{array}
$$

49.
$$
\begin{array}{r}
4a + b \\
3a-b \overline{\smash{\big)}\ 12a^2 - ab - b^2} \\
\underline{12a^2 - 4ab} \\
3ab - b^2 \\
\underline{3ab - b^2} \\
0
\end{array}
$$

51.
$$
\begin{array}{r}
2x + 1 \\
5x+3 \overline{\smash{\big)}\ 10x^2 + 11x + 3} \\
\underline{10x^2 + 6x} \\
5x + 3 \\
\underline{5x + 3} \\
0
\end{array}
$$

53.
$$
\begin{array}{r}
x - 7 \\
2x+4 \overline{\smash{\big)}\ 2x^2 - 10x - 28} \\
\underline{2x^2 + 4x} \\
- 14x - 28 \\
\underline{- 14x - 28} \\
0
\end{array}
$$

55.
$$
\begin{array}{r}
x + 4 \\
x-4 \overline{\smash{\big)}\ x^2 + 0x - 16} \\
\underline{x^2 - 4x} \\
4x - 16 \\
\underline{4x - 16} \\
0
\end{array}
$$

57.
$$
\begin{array}{r}
2x - 3 \\
2x+3 \overline{\smash{\big)}\ 4x^2 + 0x - 9} \\
\underline{4x^2 + 6x} \\
- 6x - 9 \\
\underline{- 6x - 9} \\
0
\end{array}
$$

59.
$$
\begin{array}{r}
x^2 + 2x + 4 \\
x-2 \overline{\smash{\big)}\ x^3 + 0x^2 + 0x - 8} \\
\underline{x^3 - 2x^2} \\
2x^2 + 0x \\
\underline{2x^2 - 4x} \\
4x - 8 \\
\underline{4x - 8} \\
0
\end{array}
$$

61.
$$
\begin{array}{r}
x^2 + 2x - 1 \\
2x+3 \overline{\smash{\big)}\ 2x^3 + 7x^2 + 4x - 3} \\
\underline{2x^3 + 3x^2} \\
4x^2 + 4x \\
\underline{4x^2 + 6x} \\
- 2x - 3 \\
\underline{- 2x - 3} \\
0
\end{array}
$$

63.
$$
\begin{array}{r}
x^2 + 2x + 1 \\
x+1 \overline{\smash{\big)}\ x^3 + 3x^2 + 3x + 1} \\
\underline{x^3 + x^2} \\
2x^2 + 3x \\
\underline{2x^2 + 2x} \\
x + 1 \\
\underline{x + 1} \\
0
\end{array}
$$

65.
$$
\begin{array}{r}
5x^2 - x + 4 + \frac{16}{3x-4} \\
3x-4 \overline{\smash{\big)}\ 15x^3 - 23x^2 + 16x + 0} \\
\underline{15x^3 - 20x^2} \\
- 3x^2 + 16x \\
\underline{- 3x^2 + 4x} \\
12x + 0 \\
\underline{12x - 16} \\
16
\end{array}
$$

67.
$$
\begin{array}{r}
2x^2 + 2x + 1 \\
3x+2 \overline{\smash{\big)}\ 6x^3 + 10x^2 + 7x + 2} \\
\underline{6x^3 + 4x^2} \\
6x^2 + 7x \\
\underline{6x^2 + 4x} \\
3x + 2 \\
\underline{3x + 2} \\
0
\end{array}
$$

69.
$$
\begin{array}{r}
x^2 + 2x - 1 + \frac{-1}{2x+3} \\
2x+3 \overline{\smash{\big)}\ 2x^3 + 7x^2 + 4x - 4} \\
\underline{2x^3 + 3x^2} \\
4x^2 + 4x \\
\underline{4x^2 + 6x} \\
- 2x - 4 \\
\underline{- 2x - 3} \\
- 1
\end{array}
$$

71.
$$
\begin{array}{r}
2x^2 + 8x + 14 + \frac{31}{x-2} \\
x - 2 \overline{\smash{\big)}\, 2x^3 + 4x^2 - 2x + 3} \\
\underline{2x^3 - 4x^2} \\
8x^2 - 2x \\
\underline{8x^2 - 16x} \\
14x + 3 \\
\underline{14x - 28} \\
31
\end{array}
$$

73.
$$
\begin{array}{r}
3x + 2y \\
2x - y \overline{\smash{\big)}\, 6x^2 + xy - 2y^2} \\
\underline{6x^2 - 3xy} \\
4xy - 2y^2 \\
\underline{4xy - 2y^2} \\
0
\end{array}
$$

75. Answers may vary.

77. $x^2 - 2x$ is added to the dividend when it should be subtracted.

79.
$$
\begin{array}{r}
a^2 - 3a + 10 + \frac{-30}{a+3} \\
a + 3 \overline{\smash{\big)}\, a^3 + 0a^2 + a + 0} \\
\underline{a^3 + 3a^2} \\
-3a^2 + a \\
\underline{-3a^2 - 9a} \\
10a + 0 \\
\underline{10a + 30} \\
-30
\end{array}
$$

Exercises 4.9 (page 294)

1.
$$
\begin{array}{r}
2x - 5 \\
x + 4 \overline{\smash{\big)}\, 2x^2 + 3x - 18} \\
\underline{2x^2 + 8x} \\
-5x - 18 \\
\underline{-5x - 20} \\
2
\end{array}
$$
remainder $= 2$

3. $2(-4)^2 + 3(-4) - 18 = 2(16) - 12 - 18$
$$= 32 - 12 - 18 = 2$$

5. $f(-1) = 3(-1)^2 + 2(-1) - 1 = 0$

7. $f(2a) = 3(2a)^2 + 2(2a) - 1$
$$= 12a^2 + 4a - 1$$

9. $5(x^2 - 3x + 2) + 4(3x^2 - x + 1) = 5x^2 - 15x + 10 + 12x^2 - 4x + 4 = 17x^2 - 19x + 14$

11. $x - r$

13. $P(r)$

15.
$$
\begin{array}{r}
9 \,\underline{\big|\, 1 \quad -13 \quad 30} \\
9 \quad -36 \\
\overline{1 \quad -4 \quad -6}
\end{array}
$$
$\Rightarrow \boxed{x - 4 + \frac{-6}{x-9}}$

17.
$$
\begin{array}{r}
-5 \,\underline{\big|\, 1 \quad 4 \quad -10} \\
-5 \quad 5 \\
\overline{1 \quad -1 \quad -5}
\end{array}
$$
$\Rightarrow \boxed{x - 1 + \frac{-5}{x+5}}$

19.
$$
\begin{array}{r}
2 \,\underline{\big|\, 2 \quad 0 \quad -5 \quad -6} \\
4 \quad 8 \quad 6 \\
\overline{2 \quad 4 \quad 3 \quad 0}
\end{array}
$$
$\Rightarrow \boxed{2x^2 + 4x + 3}$

21.
$$
\begin{array}{r}
-2 \,\underline{\big|\, 4 \quad 0 \quad -2 \quad 36} \\
-8 \quad 16 \quad -28 \\
\overline{4 \quad -8 \quad 14 \quad 8}
\end{array}
$$
$\Rightarrow \boxed{4x^2 - 8x + 14 + \frac{8}{x+2}}$

23.

$$\begin{array}{r|rrrr} -1 & 6 & 5 & 0 & 4 \\ & & -6 & 1 & -1 \\ \hline & 6 & -1 & 1 & 3 \end{array}$$

$\Rightarrow \boxed{6x^2 - x + 1 + \frac{3}{x+1}}$

25.

$$\begin{array}{r|rrrr} -2 & 3 & 4 & 0 & 8 \\ & & -6 & 4 & -8 \\ \hline & 3 & -2 & 4 & 0 \end{array}$$

$\Rightarrow \boxed{3x^2 - 2x + 4}$

27. $P(1) = 2(1)^3 - 4(1)^2 + 2(1) - 1 = \boxed{-1}$

$$\begin{array}{r|rrrr} 1 & 2 & -4 & 2 & -1 \\ & & 2 & -2 & 0 \\ \hline & 2 & -2 & 0 & \boxed{-1} \end{array}$$

29. $P(-3) = 2(-3)^3 - 4(-3)^2 + 2(-3) - 1$

$= \boxed{-97}$

$$\begin{array}{r|rrrr} -3 & 2 & -4 & 2 & -1 \\ & & -6 & 30 & -96 \\ \hline & 2 & -10 & 32 & \boxed{-97} \end{array}$$

31. $P(3) = 2(3)^3 - 4(3)^2 + 2(3) - 1 = \boxed{23}$

$$\begin{array}{r|rrrr} 3 & 2 & -4 & 2 & -1 \\ & & 6 & 6 & 24 \\ \hline & 2 & 2 & 8 & \boxed{23} \end{array}$$

33. $P(-5) = 2(-5)^3 - 4(-5)^2 + 2(-5) - 1$

$= \boxed{-361}$

$$\begin{array}{r|rrrr} -5 & 2 & -4 & 2 & -1 \\ & & -10 & 70 & -360 \\ \hline & 2 & -14 & 72 & \boxed{-361} \end{array}$$

35.

$$\begin{array}{r|rrrr} 2 & 1 & -4 & 1 & -2 \\ & & 2 & -4 & -6 \\ \hline & 1 & -2 & -3 & \boxed{-8} \end{array}$$

37.

$$\begin{array}{r|rrrrr} -2 & 1 & -2 & 1 & -3 & 2 \\ & & -2 & 8 & -18 & 42 \\ \hline & 1 & -4 & 9 & -21 & \boxed{44} \end{array}$$

39.

$$\begin{array}{r|rrrr} 3 & 1 & -3 & 5 & -15 \\ & & 3 & 0 & 15 \\ \hline & 1 & 0 & 5 & \boxed{0} \end{array}$$

factor; $P(x) = (x - 3)(x^2 + 5)$

41.

$$\begin{array}{r|rrrr} 6 & 2 & -8 & 1 & -36 \\ & & 12 & 24 & 150 \\ \hline & 2 & 4 & 25 & \boxed{114} \end{array}$$

not a factor

43.

$$\begin{array}{r|rrrr} 3 & 3 & -10 & 5 & -6 \\ & & 9 & -3 & 6 \\ \hline & 3 & -1 & 2 & 0 \end{array}$$

$\Rightarrow \boxed{3x^2 - x + 2}$

45.

$$\begin{array}{r|rrr} -2 & 1 & -5 & 14 \\ & & -2 & 14 \\ \hline & 1 & -7 & 28 \end{array}$$

$\Rightarrow \boxed{x - 7 + \frac{28}{x+2}}$

47.

$$\begin{array}{r|rrr} -4 & 1 & 6 & 8 \\ & & -4 & -8 \\ \hline & 1 & 2 & 0 \end{array}$$

$\Rightarrow \boxed{x + 2}$

49.

$$\begin{array}{r|rrr} -6 & 1 & 13 & 42 \\ & & -6 & -42 \\ \hline & 1 & 7 & 0 \end{array}$$

$\Rightarrow \boxed{x + 7}$

51.

$$\begin{array}{r|rrrr} -2 & 1 & 1 & 0 & 1 \\ & & -2 & 2 & -4 \\ \hline & 1 & -1 & 2 & \boxed{-3} \end{array}$$

53.

$$\begin{array}{r|rrrr} \frac{1}{2} & 2 & -1 & 4 & -5 \\ & & 1 & 0 & 2 \\ \hline & 2 & 0 & 4 & \boxed{-3} \end{array}$$

55.

$$\begin{array}{r|rrrr} 0 & 7 & -5 & -8 & 0 \\ & & 0 & 0 & 0 \\ \hline & 7 & -5 & -8 & \boxed{0} \end{array}$$

factor; $P(x) = x\left(7x^2 - 5x - 8\right)$

57.

$$\begin{array}{r|rrrr} 5 & 1 & 0 & 0 & -125 \\ & & 5 & 25 & 125 \\ \hline & 1 & 5 & 25 & \boxed{0} \end{array}$$

factor; $P(x) = (x - 5)\left(x^2 + 5x + 25\right)$

59.

$$\begin{array}{r|rrrrrr} -\frac{1}{2} & 3 & 0 & 0 & 0 & 0 & 1 \\ & & -\frac{3}{2} & \frac{3}{4} & -\frac{3}{8} & \frac{3}{16} & -\frac{3}{32} \\ \hline & 3 & -\frac{3}{2} & \frac{3}{4} & -\frac{3}{8} & \frac{3}{16} & \boxed{\frac{29}{32}} \end{array}$$

61. $Q(-1) = (-1)^4 - 3(-1)^3 + 2(-1)^2 + (-1) - 3 = \boxed{2}$

$$\begin{array}{r|rrrrr} -1 & 1 & -3 & 2 & 1 & -3 \\ & & -1 & 4 & -6 & 5 \\ \hline & 1 & -4 & 6 & -5 & \boxed{2} \end{array}$$

63. $Q(2) = (2)^4 - 3(2)^3 + 2(2)^2 + (2) - 3$
$$= \boxed{-1}$$

$$\begin{array}{r|rrrrr} 2 & 1 & -3 & 2 & 1 & -3 \\ & & 2 & -2 & 0 & 2 \\ \hline & 1 & -1 & 0 & 1 & \boxed{-1} \end{array}$$

65.

$$\begin{array}{r|rrrrrrr} 2 & 1 & 0 & 0 & 0 & 0 & 0 & 0 \\ & & 2 & 4 & 8 & 16 & 32 & 64 \\ \hline & 1 & 2 & 4 & 8 & 16 & 32 & \boxed{64} \end{array}$$

67. Answers may vary.

69. remainder $= P(1) = 1^{100} - 1^{99} + 1^{98} - 1^{97} + \cdots + 1^2 - 1 + 1 = 1$

Chapter 4 Review (page 297)

1. $(-3x)^4 = (-3x)(-3x)(-3x)(-3x)$

2. $\left(\frac{1}{2}pq\right)^3 = \left(\frac{1}{2}pq\right)\left(\frac{1}{2}pq\right)\left(\frac{1}{2}pq\right)$

3. $5^3 = 5 \cdot 5 \cdot 5 = 125$

4. $3^5 = 3 \cdot 3 \cdot 3 \cdot 3 \cdot 3 = 243$

5. $(-6)^2 = (-6)(-6) = 36$

6. $-6^2 = -1 \cdot 6 \cdot 6 = -36$

7. $3^2 + 2^2 = 9 + 4 = 13$

8. $(3 + 2)^2 = (5)^2 = 25$

9. $x^4 x^6 = x^{4+6} = x^{10}$

10. $x^2 x^7 = x^{2+7} = x^9$

11. $\left(y^7\right)^3 = y^{7 \cdot 3} = y^{21}$

12. $\left(x^{21}\right)^2 = x^{21 \cdot 2} = x^{42}$

13. $(ab)^3 = a^3 b^3$

14. $(3x)^4 = 3^4 x^4 = 81x^4$

15. $b^3 b^4 b^5 = b^{3+4+5} = b^{12}$

16. $-z^2(z^3 y^2) = -z^{2+3} y^2 = -y^2 z^5$

17. $(16s)^2 s = 16^2 s^2 s = 256s^3$

18. $-3y(y^5) = -3y^6$

19. $(2x^4 y^2)^3 = 2^3 (x^4)^3 (y^2)^3 = 8x^{12} y^6$

20. $(5x^3 y)^2 = 5^2 (x^3)^2 y^2 = 25x^6 y^2$

21. $\dfrac{x^7}{x^3} = x^{7-3} = x^4$

22. $\left(\dfrac{x^2 y}{xy^2}\right)^2 = \dfrac{x^4 y^2}{x^2 y^4} = x^2 y^{-2} = \dfrac{x^2}{y^2}$

23. $\dfrac{8(y^2 x)^2}{4(yx^2)^2} = \dfrac{8y^4 x^2}{4y^2 x^4} = \dfrac{8}{4} y^2 x^{-2} = \dfrac{2y^2}{x^2}$

24. $\dfrac{(5y^2 z^3)^3}{25(yz)^5} = \dfrac{125 y^6 z^9}{25 y^5 z^5} = 5yz^4$

25. $x^0 = 1$

26. $(3x^2 y^2)^0 = 1$

27. $(3x^0)^2 = (3 \cdot 1)^2 = 3^2 = 9$

28. $(3x^2 y^0)^2 = (3x^2 \cdot 1)^2 = (3x^2)^2 = 9x^4$

29. $x^{-3} = \dfrac{1}{x^3}$

30. $x^{-2} x^3 = x^1 = x$

31. $y^4 y^{-3} = y^1 = y$

32. $\dfrac{x^3}{x^{-7}} = x^{3-(-7)} = x^{10}$

33. $(x^{-3} x^4)^{-2} = (x^1)^{-2} = x^{-2} = \dfrac{1}{x^2}$

34. $(a^{-2} b)^{-3} = a^6 b^{-3} = \dfrac{a^6}{b^3}$

35. $\left(\dfrac{x^2}{x}\right)^{-5} = (x)^{-5} = \dfrac{1}{x^5}$

36. $\left(\dfrac{15z^4}{5z^3}\right)^{-2} = (3z)^{-2} = \dfrac{1}{(3z)^2} = \dfrac{1}{9z^2}$

37. $728 = 7.28 \times 10^2$

38. $6{,}230 = 6.23 \times 10^3$

39. $0.0275 = 2.75 \times 10^{-2}$

40. $0.00942 = 9.42 \times 10^{-3}$

41. $7.73 = 7.73 \times 10^0$

42. $753 \times 10^3 = 7.53 \times 10^2 \times 10^3$
$= 7.53 \times 10^5$

43. $0.018 \times 10^{-2} = 1.8 \times 10^{-2} \times 10^{-2}$
$= 1.8 \times 10^{-4}$

44. $600 \times 10^2 = 6.00 \times 10^2 \times 10^2$
$= 6 \times 10^4$

45. $3.87 \times 10^4 = 38{,}700$

46. $7.98 \times 10^{-5} = 0.0000798$

47. $2.68 \times 10^0 = 2.68$

48. $5.76 \times 10^1 = 57.6$

49. $739 \times 10^{-2} = 7.39$

50. $0.437 \times 10^{-3} = 0.000437$

51. $\dfrac{(0.00012)(0.00004)}{0.00000016} = \dfrac{(1.2 \times 10^{-4})(4 \times 10^{-5})}{1.6 \times 10^{-7}} = \dfrac{4.8 \times 10^{-9}}{1.6 \times 10^{-7}} = 3 \times 10^{-2} = 0.03$

52. $\dfrac{(4,800)(20,000)}{600,000} = \dfrac{(4.8 \times 10^3)(2 \times 10^4)}{6 \times 10^5} = \dfrac{9.6 \times 10^7}{6 \times 10^5} = 1.6 \times 10^2 = 160$

53. $29x^8$: monomial, degree $= 8$

54. $5^3x + x^2$: binomial, degree $= 2$

55. $-3x^5 + x - 1$: trinomial, degree $= 5$

56. $9xy + 21x^3y^2$: binomial, degree $= 3 + 2 = 5$

57. $3x + 2 = 3(7) + 2 = 21 + 2 = 23$

58. $3x + 2 = 3(-4) + 2 = -12 + 2 = -10$

59. $3x + 2 = 3(-2) + 2 = -6 + 2 = -4$

60. $3x + 2 = 3\left(\frac{2}{3}\right) + 2 = 2 + 2 = 4$

61. $5x^4 - x = 5(3)^4 - 3 = 5(81) - 3$
$ = 405 - 3 = 402$

62. $5x^4 - x = 5(0)^4 - 0 = 5(0) - 0$
$ = 0 - 0 = 0$

63. $5x^4 - x = 5(-2)^4 - (-2) = 5(16) + 2$
$ = 80 + 2 = 82$

64. $5x^4 - x = 5(-0.3)^4 - (-0.3)$
$ = 5(0.0081) + 0.3$
$ = 0.0405 + 3 = 0.3405$

65. $f(0) = 0^2 - 4 = 0 - 4 = -4$

66. $f(-4) = (-4)^2 - 4 = 16 - 4 = 12$

67. $f(-2) = (-2)^2 - 4 = 4 - 4 = 0$

68. $f\left(\frac{1}{2}\right) = \left(\frac{1}{2}\right)^2 - 4 = \frac{1}{4} - 4 = -\frac{15}{4}$

69. $f(x) = x^2 - 5$

x	-2	-1	0	1	2
$f(x)$	-1	-4	-5	-4	-1

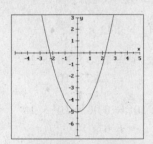

Domain $= \mathbb{R}$, Range $= [-5, \infty)$

70. $f(x) = x^3 - 2$

x	-2	-1	0	1	2
$f(x)$	-10	-3	-2	-1	6

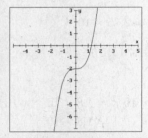

Domain $= \mathbb{R}$, Range $= \mathbb{R}$

71. $6x - 4x + x = 3x$

72. $5x + 4y$: not like terms, so not possible

73. $(xy)^2 + 3x^2y^2 = x^2y^2 + 3x^2y^2 = 4x^2y^2$

74. $-2x^2yz + 3yx^2z = -2x^2yz + 3x^2yz$
$ = x^2yz$

75. $(3x^2 + 2x) + (5x^2 - 8x) = 3x^2 + 2x + 5x^2 - 8x = 8x^2 - 6x$

76. $(7a^2 + 2a - 5) - (3a^2 - 2a + 1) = 7a^2 + 2a - 5 - 3a^2 + 2a - 1 = 4a^2 + 4a - 6$

77. $3(9x^2 + 3x + 7) - 2(11x^2 - 5x + 9) = 27x^2 + 9x + 21 - 22x^2 + 10x - 18 = 5x^2 + 19x + 3$

78. $4(4x^3 + 2x^2 - 3x - 8) - 5(2x^3 - 3x + 8) = 16x^3 + 8x^2 - 12x - 32 - 10x^3 + 15x - 40$
$$= 6x^3 + 8x^2 + 3x - 72$$

79. $(4x^3 y^5)(3x^2 y) = 12x^5 y^6$

80. $(xyz^3)(x^3 z)^2 = xyz^3 x^6 z^2 = x^7 y z^5$

81. $4(2x + 3) = 8x + 12$

82. $3(2x + 4) = 6x + 12$

83. $x^2(3x^2 - 5) = 3x^4 - 5x^2$

84. $2y^2(y^2 + 5y) = 2y^4 + 10y^3$

85. $-x^2 y(y^2 - xy) = -x^2 y^3 + x^3 y^2$

86. $-3xy(xy - x) = -3x^2 y^2 + 3x^2 y$

87. $(x + 5)(x + 4) = x^2 + 4x + 5x + 20$
$$= x^2 + 9x + 20$$

88. $(2x + 1)(x - 1) = 2x^2 - 2x + x - 1$
$$= 2x^2 - x - 1$$

89. $(3a - 3)(2a + 2) = 6a^2 + 6a - 6a - 6$
$$= 6a^2 - 6$$

90. $6(a - 1)(a + 1) = 6(a^2 + a - a - 1)$
$$= 6(a^2 - 1) = 6a^2 - 6$$

91. $(a - b)(2a + b) = 2a^2 + ab - 2ab - b^2$
$$= 2a^2 - ab - b^2$$

92. $(3x - y)(2x + y) = 6x^2 + 3xy - 2xy - y^2$
$$= 6x^2 + xy - y^2$$

93. $(x + 6)(x + 6) = x^2 + 6x + 6x + 36$
$$= x^2 + 12x + 36$$

94. $(x + 5)(x - 5) = x^2 - 5x + 5x - 25$
$$= x^2 - 25$$

95. $(y - 7)(y + 7) = y^2 + 7y - 7y - 49$
$$= y^2 - 49$$

96. $(x + 4)^2 = (x + 4)(x + 4)$
$$= x^2 + 4x + 4x + 16$$
$$= x^2 + 8x + 16$$

97. $(x - 3)^2 = (x - 3)(x - 3) = x^2 - 3x - 3x + 9 = x^2 - 6x + 9$

98. $(y - 2)^2 = (y - 2)(y - 2) = y^2 - 2y - 2y + 4 = y^2 - 4y + 4$

99. $(3y + 2)^2 = (3y + 2)(3y + 2) = 9y^2 + 6y + 6y + 4 = 9y^2 + 12y + 4$

100. $(y^2 + 1)(y^2 - 1) = y^4 - y^2 + y^2 - 1 = y^4 - 1$

101. $(3x + 1)(x^2 + 2x + 1) = 3x^3 + 6x^2 + 3x + x^2 + 2x + 1 = 3x^3 + 7x^2 + 5x + 1$

102. $(2a - 3)(4a^2 + 6a + 9) = 8a^3 + 12a^2 + 18a - 12a^2 - 18a - 27 = 8a^3 - 27$

103. $x^2 + 3 = x(x+3)$
$\quad\ \ x^2 + 3 = x^2 + 3x$
$\qquad\quad 3 = 3x$
$\qquad\quad 1 = x$

104. $x^2 + x = (x+1)(x+2)$
$\quad\ x^2 + x = x^2 + 2x + x + 2$
$\quad\ x^2 + x = x^2 + 3x + 2$
$\qquad\quad x = 3x + 2$
$\qquad -2x = 2$
$\qquad\quad x = -1$

105. $\qquad (x+2)(x-5) = (x-4)(x-1)$
$\quad x^2 - 5x + 2x - 10 = x^2 - x - 4x + 4$
$\qquad\ x^2 - 3x - 10 = x^2 - 5x + 4$
$\qquad\quad -3x - 10 = -5x + 4$
$\qquad\qquad -3x = -5x + 14$
$\qquad\qquad\ 2x = 14$
$\qquad\qquad\ x = 7$

106. $\qquad (x-1)(x-2) = (x-3)(x+1)$
$\quad x^2 - 2x - x + 2 = x^2 + x - 3x - 3$
$\qquad\ x^2 - 3x + 2 = x^2 - 2x - 3$
$\qquad\quad -3x + 2 = -2x - 3$
$\qquad\qquad -3x = -2x - 5$
$\qquad\qquad\ -x = -5$
$\qquad\qquad\ x = 5$

107. $x^2 + x(x+2) = x(2x+1) + 1$
$\ x^2 + x^2 + 2x = 2x^2 + x + 1$
$\qquad 2x^2 + 2x = 2x^2 + x + 1$
$\qquad\qquad 2x = x + 1$
$\qquad\qquad\ x = 1$

108. $(x+5)(3x+1) = x^2 + (2x-1)(x-5)$
$\quad\ 3x^2 + 16x + 5 = 3x^2 - 11x + 5$
$\qquad\quad 16x + 5 = -11x + 5$
$\qquad\qquad 16x = -11x$
$\qquad\qquad 27x = 0$
$\qquad\qquad\ x = 0$

109. $\dfrac{3x+6y}{2xy} = \dfrac{3x}{2xy} + \dfrac{6y}{2xy} = \dfrac{3}{2y} + \dfrac{3}{x}$

110. $\dfrac{21x^2y^2 - 7xy}{7xy} = \dfrac{21x^2y^2}{7xy} - \dfrac{7xy}{7xy} = 3xy - 1$

111. $\dfrac{15a^2bc + 20ab^2c - 25abc^2}{-5abc} = \dfrac{15a^2bc}{-5abc} + \dfrac{20ab^2c}{-5abc} - \dfrac{25abc^2}{-5abc} = -3a - 4b + 5c$

112. $\dfrac{(x+y)^2 + (x-y)^2}{-2xy} = \dfrac{(x+y)(x+y) + (x-y)(x-y)}{-2xy}$

$\qquad\qquad\qquad\ = \dfrac{x^2 + xy + xy + y^2 + x^2 - xy - xy + y^2}{-2xy}$

$\qquad\qquad\qquad\ = \dfrac{2x^2 + 2y^2}{-2xy} = \dfrac{2x^2}{-2xy} + \dfrac{2y^2}{-2xy} = -\dfrac{x}{y} - \dfrac{y}{x}$

113.
$$
\begin{array}{r}
x + \ \ 1 + \frac{3}{x+2} \\
x+2\ \overline{\smash{\big)}\ x^2 + 3x + \quad 5} \\
\underline{x^2 + 2x} \\
x + \quad 5 \\
\underline{x + \quad 2} \\
3
\end{array}
$$

114.
$$
\begin{array}{r}
x - \quad 5 \\
x-1\ \overline{\smash{\big)}\ x^2 - 6x + 5} \\
\underline{x^2 - \quad x} \\
-5x + 5 \\
\underline{-5x + 5} \\
0
\end{array}
$$

115.
$$x - 4 \overline{\smash{\big)}\ 3x^2 - 11x - 4}$$
with quotient $3x + 1$
$$\underline{3x^2 - 12x}$$
$$x - 4$$
$$\underline{x - 4}$$
$$0$$

116.
quotient $x + 5 + \frac{3}{3x-1}$
$$3x - 1 \overline{\smash{\big)}\ 3x^2 + 14x - 2}$$
$$\underline{3x^2 - x}$$
$$15x - 2$$
$$\underline{15x - 5}$$
$$3$$

117.
quotient $3x^2 + 2x + 1 + \frac{2}{2x-1}$
$$2x - 1 \overline{\smash{\big)}\ 6x^3 + x^2 + 0x + 1}$$
$$\underline{6x^3 - 3x^2}$$
$$4x^2 + 0x$$
$$\underline{4x^2 - 2x}$$
$$2x + 1$$
$$\underline{2x - 1}$$
$$2$$

118.
quotient $3x^2 - x - 4$
$$3x + 1 \overline{\smash{\big)}\ 9x^3 + 0x^2 - 13x - 4}$$
$$\underline{9x^3 + 3x^2}$$
$$-3x^2 - 13x$$
$$\underline{-3x^2 - x}$$
$$-12x - 4$$
$$\underline{-12x - 4}$$
$$0$$

119.
$$\begin{array}{r|rrrr} 3 & 1 & -7 & 9 & 9 \\ & & 3 & -12 & -9 \\ \hline & 1 & -4 & -3 & \boxed{0} \end{array}$$

factor; $P(x) = (x - 3)\left(x^2 - 4x - 3\right)$

120.
$$\begin{array}{r|rrrr} -5 & 1 & 4 & -5 & 5 \\ & & -5 & 5 & 0 \\ \hline & 1 & -1 & 0 & \boxed{5} \end{array} \Rightarrow \text{not a factor}$$

Chapter 4 Test (page 302)

1. $2xxxyyyy = 2x^3y^4$ **2.** $3^2 + 5^3 = 9 + 125 = 134$ **3.** $y^3(y^5y) = y^3y^6 = y^9$

4. $\left(-3b^2\right)\left(2b^3\right)\left(-b^2\right) = (-3)(2)(-1)b^2b^3b^2 = 6b^7$

5. $\left(2x^3\right)^5\left(x^2\right)^3 = 32x^{15}x^6 = 32x^{21}$ **6.** $\left(2rr^2r^3\right)^3 = \left(2r^6\right)^3 = 8r^{18}$

7. $-7x^0 = -7(1) = -7$ **8.** $5y^{-6}y^3 = 5y^{-3} = \dfrac{5}{y^3}$ **9.** $\dfrac{y^2}{yy^{-2}} = \dfrac{y^2}{y^{-1}} = y^3$

10. $\left(\dfrac{a^2b^{-1}}{4a^3b^{-2}}\right)^{-3} = \left(\dfrac{1}{4}a^{-1}b^1\right)^{-3} = \left(\dfrac{1}{4}\right)^{-3}a^3b^{-3} = \dfrac{64a^3}{b^3}$

11. $540{,}000 = 5.4 \times 10^5$ **12.** $0.0025 = 2.5 \times 10^{-3}$ **13.** $7.4 \times 10^3 = 7{,}400$

14. $6.7 \times 10^{-4} = 0.00067$ **15.** $3x^2 + 2$: binomial

16. $2 + 3 + 5 = 10$th degree **17.** $x^2 + x - 2 = (-2)^2 + (-2) - 2$
$$= 4 - 2 - 2 = 0$$

18. $f(x) = x^2 + 2$ 　　　　　　　Domain $= \mathbb{R}$, Range $= [2, \infty)$

x	$f(x)$
-2	6
-1	3
0	2
1	3
2	6

19. $-6(x - y) + 2(x + y) - 3(x + 2y) = -6x + 6y + 2x + 2y - 3x - 6y = -7x + 2y$

20. $-2\left(x^2 + 3x - 1\right) - 3\left(x^2 - x + 2\right) + 5\left(x^2 + 2\right) = -2x^2 - 6x + 2 - 3x^2 + 3x - 6 + 5x^2 + 10$
$$= -3x + 6$$

21.
$$\begin{array}{r} 3x^3 + 4x^2 - x - 7 \\ + 2x^3 - 2x^2 + 3x + 2 \\ \hline 5x^3 + 2x^2 + 2x - 5 \end{array}$$

22.
$$\begin{array}{r} 2x^2 - 7x + 3 \\ - 3x^2 - 2x - 1 \\ \hline \end{array} \Rightarrow \begin{array}{r} 2x^2 - 7x + 3 \\ + -3x^2 + 2x + 1 \\ \hline -x^2 - 5x + 4 \end{array}$$

23. $(-2x^3)(2x^2y) = -4x^5y$ 　　　　　　**24.** $(-5x^4)(4x^5y) = -20x^9y$

25. $(2x - 5)(3x + 4) = 6x^2 + 8x - 15x - 20 = 6x^2 - 7x - 20$

26. $(2x - 3)(x^2 - 2x + 4) = 2x^3 - 4x^2 + 8x - 3x^2 + 6x - 12 = 2x^3 - 7x^2 + 14x - 12$

27. $\dfrac{8x^2y^3z^4}{16x^3y^2z^4} = \dfrac{1}{2}x^{-1}y^1z^0 = \dfrac{y}{2x}$

28. $\dfrac{6a^2 - 12b^2}{24ab} = \dfrac{6a^2}{24ab} - \dfrac{12b^2}{24ab}$
$$= \dfrac{a}{4b} - \dfrac{b}{2a}$$

29.
$$\begin{array}{r} x - 2 \\ 2x + 3 \overline{)\,2x^2 - x - 6\,} \\ \underline{2x^2 + 3x} \\ -4x - 6 \\ \underline{-4x - 6} \\ 0 \end{array}$$

30.
$$(a + 2)^2 = (a - 3)^2$$
$$(a + 2)(a + 2) = (a - 3)(a - 3)$$
$$a^2 + 4a + 4 = a^2 - 6a + 9$$
$$4a + 4 = -6a + 9$$
$$10a = 5$$
$$a = \frac{5}{10} = \frac{1}{2}$$

31.
$$\begin{array}{r} x^2 - 5x + 10 \\ x + 1 \overline{)\,x^3 - 4x^2 + 5x + 3\,} \\ \underline{x^3 + x^2} \\ -5x^2 + 5x \\ \underline{-5x^2 - 5x} \\ 10x + 3 \\ \underline{10x + 10} \\ - 7 \end{array}$$

remainder $= -7$

32.
$$\begin{array}{r|rrrr} 2 & 4 & 3 & 2 & -1 \\ & & 8 & 22 & 48 \\ \hline & 4 & 11 & 24 & \boxed{47} \end{array}$$

Cumulative Review Exercises (page 303)

1. $4 + 5 \cdot 2 = 4 + 10 = 14$

2. $3(-5)^2 - 4 = 3 \cdot 25 - 4 = 75 - 4 = 71$

3. $\dfrac{3x - y}{xy} = \dfrac{3(2) - (-5)}{2(-5)} = \dfrac{6 + 5}{-10} = -\dfrac{11}{10}$

4. $\dfrac{x^2 - y^2}{x + y} = \dfrac{2^2 - (-5)^2}{2 + (-5)} = \dfrac{4 - 25}{-3} = \dfrac{-21}{-3} = 7$

5. $\dfrac{4}{5}x + 6 = 18$

$\dfrac{4}{5}x = 12$

$5 \cdot \dfrac{4}{5}x = 5(12)$

$4x = 60$

$x = 15$

6. $x - 2 = \dfrac{x + 2}{3}$

$3(x - 2) = 3 \cdot \dfrac{x + 2}{3}$

$3x - 6 = x + 2$

$2x - 6 = 2$

$2x = 8$

$x = 4$

7. $2(5x + 2) = 3(3x - 2)$

$10x + 4 = 9x - 6$

$x + 4 = -6$

$x = -10$

8. $4(y + 1) = -2(4 - y)$

$4y + 4 = -8 + 2y$

$2y + 4 = -8$

$2y = -12$

$y = -6$

9. $3x - 4 > 2$

$3x > 6$

$x > 2$

10. $7x - 9 < 5$

$7x < 14$

$x < 2$

11. $-2 < -x + 3 < 5$

$-5 < \quad -x \quad < 2$

$\dfrac{-5}{-1} > \dfrac{-x}{-1} > \dfrac{2}{-1}$

$5 > \quad x \quad > -2$

$-2 < \quad x \quad < 5$

12. $0 \leq \dfrac{4 - x}{3} \leq 2$

$3(0) \leq 3 \cdot \dfrac{4 - x}{3} \leq 3(2)$

$0 \leq \quad 4 - x \quad \leq 6$

$-4 \leq \quad -x \quad \leq 2$

$\dfrac{-4}{-1} \geq \dfrac{-x}{-1} \geq \dfrac{2}{-1}$

$4 \geq \quad x \quad \geq -2$

$-2 \leq \quad x \quad \leq 4$

13. $A = p + prt$

$A - p = p - p + prt$

$A - p = prt$

$\dfrac{A - p}{pt} = \dfrac{prt}{pt}$

$\dfrac{A - p}{pt} = r,$ or $r = \dfrac{A - p}{pt}$

14. $A = \dfrac{1}{2}bh$

$2A = 2 \cdot \dfrac{1}{2}bh$

$2A = bh$

$\dfrac{2A}{b} = \dfrac{bh}{b}$

$\dfrac{2A}{b} = h,$ or $h = \dfrac{2A}{b}$

15.
$$3x - 4y = 12$$

$x = 0$	$y = 0$
$3(0) - 4y = 12$	$3x - 4(0) = 12$
$0 - 4y = 12$	$3x - 0 = 12$
$-4y = 12$	$3x = 12$
$y = -3$	$x = 4$

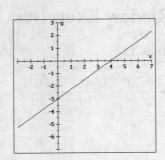

16.
$$y - 2 = \frac{1}{2}(x - 4)$$

$x = 0$	$x = 4$
$y - 2 = \frac{1}{2}(0 - 4)$	$y - 2 = \frac{1}{2}(4 - 4)$
$y - 2 = \frac{1}{2}(-4)$	$y - 2 = \frac{1}{2}(0)$
$y - 2 = -2$	$y - 2 = 0$
$y = 0$	$y = 2$

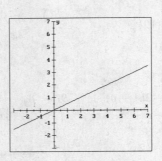

17. $f(4) = 5(4) - 2 = 20 - 2 = 18$

18. $f(-1) = 5(-1) - 2 = -5 - 2 = -7$

19. $f(-2) = 5(-2) - 2 = -10 - 2 = -12$

20. $f\left(\frac{1}{5}\right) = 5\left(\frac{1}{5}\right) - 2 = 1 - 2 = -1$

21. $y^4(y^2 y^8) = y^4 y^{10} = y^{14}$

22. $\dfrac{x^5 y^3}{x^4 y^4} = x^{5-4} y^{3-4} = x^1 y^{-1} = \dfrac{x}{y}$

23. $\dfrac{a^4 b^{-3}}{a^{-3} b^3} = a^{4-(-3)} b^{-3-3} = a^7 b^{-6} = \dfrac{a^7}{b^6}$

24. $\left(\dfrac{-x^{-2} y^3}{x^{-3} y^2}\right)^2 = \left(-x^{-2-(-3)} y^{3-2}\right)^2$
$$= \left(-x^1 y^1\right)^2 = x^2 y^2$$

25. $(3x^2 + 2x - 7) - (2x^2 - 2x + 7) = 3x^2 + 2x - 7 - 2x^2 + 2x - 7 = x^2 + 4x - 14$

26. $(4x - 5)(3x + 2) = 12x^2 + 8x - 15x - 10 = 12x^2 - 7x - 10$

27. $(x - 2)(x^2 + 2x + 4) = x^3 + 2x^2 + 4x - 2x^2 - 4x - 8 = x^3 - 8$

28.
$$\require{enclose}\begin{array}{r} 2x + 1 \\ x - 3 \enclose{longdiv}{2x^2 - 5x - 3} \\ \underline{2x^2 - 6x } \\ x - 3 \\ \underline{x - 3} \\ 0 \end{array}$$

29. $(1.6 \times 10^2)(3 \times 10^{16}) = 4.8 \times 10^{18}$ m

142

SECTION 5.1

30.
$$A = 2lw + 2wd + 2ld$$
$$202 = 2(9)(5) + 2(5)d + 2(9)d$$
$$202 = 90 + 28d$$
$$112 = 28d \Rightarrow d = 4 \text{ units}$$

31.
$$A = \pi(R + r)(R - r)$$
$$= \pi(17 + 3)(17 - 3)$$
$$= \pi(20)(14) = 280\pi \approx 879.6 \text{ sq. units}$$

32. Let $r =$ the regular price. Then an employee can purchase the TV for $0.75r$.
Purchase price + Sales tax = Total price
$$0.75r + 0.08(0.75r) = 414.72$$
$$0.81r = 414.72$$
$$\frac{0.81r}{0.81} = \frac{414.72}{0.81}$$
$$r = 512 \Rightarrow \text{The regular price is } \$512.$$

Exercise 5.1   (page 313)

1. $3 = 3; 6 = 2 \cdot 3; 9 = 3 \cdot 3$
GCF $= 3$

3. $4 = 2^2; 16 = 2^4; 32 = 2^5$
GCF $= 2^2 = 4$

5. $4 = 2^2; 6 = 2 \cdot 3; 10 = 2 \cdot 5$
GCF $= 2$

7. $12 = 2^2 \cdot 3; 18 = 2 \cdot 3^2; 24 = 2^3 \cdot 3$
GCF $= 2 \cdot 3 = 6$

9.
$$3x - 2(x + 1) = 5$$
$$3x - 2x - 2 = 5$$
$$x = 7$$

11.
$$\frac{2x - 7}{5} = 3$$
$$5 \cdot \frac{2x - 7}{5} = 5(3)$$
$$2x - 7 = 15$$
$$2x = 22$$
$$x = 11$$

13. prime-factored

15. largest (or greatest)

17. grouping

19. $12 = 2 \cdot 6 = 2 \cdot (2 \cdot 3) = 2^2 \cdot 3$

21. $40 = 4 \cdot 10 = (2 \cdot 2) \cdot (2 \cdot 5) = 2^3 \cdot 5$

23. $225 = 15 \cdot 15 = (3 \cdot 5) \cdot (3 \cdot 5) = 3^2 \cdot 5^2$

25.
$$288 = 24 \cdot 12 = (4 \cdot 6) \cdot (4 \cdot 3)$$
$$= \left(2^2 \cdot 2 \cdot 3\right) \cdot \left(2^2 \cdot 3\right)$$
$$= 2^5 \cdot 3^2$$

27. $3x = 3 \cdot x; 6xy = 2 \cdot 3 \cdot x \cdot y$
GCF $= 3x$

29. $5x^2 = 5 \cdot x \cdot x; 10x = 2 \cdot 5 \cdot x$
GCF $= 5x$

31. $4ab = 2 \cdot 2 \cdot a \cdot b; 18b = 2 \cdot 3 \cdot 3 \cdot b$
GCF $= 2b$

33. $6x^2y^2 = 2 \cdot 3 \cdot x \cdot x \cdot y \cdot y; 12xyz = 2 \cdot 2 \cdot 3 \cdot x \cdot y \cdot z; 18xy^2z^3 = 2 \cdot 3 \cdot 3 \cdot x \cdot y \cdot y \cdot z \cdot z \cdot z$
GCF $= 2 \cdot 3 \cdot xy = 6xy$

143

35. $9a + 15 = 3(\boxed{3a} + 5)$ **37.** $4a + 12 = \boxed{4}(a + 3)$

39. $8x + 12 = \boxed{4}(2x + 3)$ **41.** $4y^2 + 8y - 2xy = 2y(2y + \boxed{4} - \boxed{x})$

43. $3x^2 - 6xy + 9xy^2 = \boxed{3x}(\boxed{x} - 2y + 3y^2)$

45. $r^4 + r^2 = r^2(\boxed{r^2} + 1)$ **47.** $3x + 6 = 3(x + 2)$ **49.** $4x - 8 = 4(x - 2)$

51. $6x^2 - 9x = 3x(2x - 3)$ **53.** $4b^3 - 10b^2 = 2b^2(2b - 5)$ **55.** $t^3 + 2t^2 = t^2(t + 2)$

57. $10x^2y^3 + 15xy^4 = 5xy^3(2x + 3y)$ **59.** $a^3b^3z^3 - a^2b^3z^2 = a^2b^3z^2(az - 1)$

61. $24x^2y^3z^4 + 8xy^2z^3 = 8xy^2z^3(3xyz + 1)$ **63.** $3x + 3y - 6z = 3(x + y - 2z)$

65. $ab + ac - ad = a(b + c - d)$ **67.** $rs - rt + ru = r(s - t + u)$

69. $a^2b^2x^2 + a^3b^2x^2 - a^3b^3x^3 = a^2b^2x^2(1 + a - abx)$

71. $-x - 2 = -(x + 2)$ **73.** $-a - b = -(a + b)$ **75.** $-2x + 5y = -(2x - 5y)$

77. $-3ab - 5ac + 9bc = -(3ab + 5ac - 9bc)$ **79.** $-3x^2y - 6xy^2 = -3xy(x + 2y)$

81. $-4a^2b^2c^2 + 14a^2b^2c - 10ab^2c^2 = -2ab^2c(2ac - 7a + 5c)$

83. $a(x + y) + b(x + y) = (x + y)\boxed{(a + b)}$ **85.** $p(m - n) - q(m - n) = \boxed{(m - n)}(p - q)$

87. $3(r - 2s) - x(r - 2s) = \boxed{(r - 2s)}(3 - x)$

89. $(x + 3)(x + 1) - y(x + 1) = (x + 1)\boxed{(x + 3 - y)}$

91. $x(y + 1) - 5(y + 1) = (y + 1)(x - 5)$

93. $(3x - y)(x^2 - 2) + (x^2 - 2) = (3x - y)(x^2 - 2) + 1(x^2 - 2) = (x^2 - 2)(3x - y + 1)$

95. $(x + y)^2 + b(x + y) = (x + y)(x + y) + b(x + y) = (x + y)(x + y + b)$

97. $(x - 3)^2 + (x - 3) = (x - 3)(x - 3) + 1(x - 3) = (x - 3)(x - 3 + 1) = (x - 3)(x - 2)$

99. $5a(2a - 1) - 10b(2a - 1) = (2a - 1)(5a - 10b) = 5(2a - 1)(a - 2b)$

101. $3x(c - 3d) + 6y(c - 3d) = (c - 3d)(3x + 6y) = 3(c - 3d)(x + 2y)$

103. $2x + 2y + ax + ay = 2(x + y) + a(x + y)$ **105.** $9p - 9q + mp - mq = 9(p - q) + m(p - q)$
$\qquad\qquad\qquad\quad = (x + y)(2 + a)$ $= (p - q)(9 + m)$

107. $9x^3 + 3x^2 + 12x + 4 = 3x^2(3x + 1) + 4(3x + 1) = (3x + 1)(3x^2 + 4)$

109. $8a^3 - 2a^2 - 4a + 1 = 2a^2(4a - 1) - 1(4a - 1) = (4a - 1)(2a^2 - 1)$

111. $ax + bx - a - b = x(a + b) - 1(a + b) = (a + b)(x - 1)$

113. $x(a - b) + y(b - a) = x(a - b) - y(a - b) = (a - b)(x - y)$

115. $r^4 + r^2 = r^2(r^2 + 1)$ **117.** $12uvw^3 - 18uv^2w^2 = 6uvw^2(2w - 3v)$

119. $-14a^6b^6 + 49a^2b^3 - 21ab = -7ab(2a^5b^5 - 7ab^2 + 3)$

121. $3x(a + b + c) - 2y(a + b + c) = (a + b + c)(3x - 2y)$

123. $14x^2y(r + 2s - t) - 21xy(r + 2s - t) = (r + 2s - t)(14x^2y - 21xy) = 7xy(r + 2s - t)(2x - 3)$

125. $3tv - 9tw + uv - 3uw = 3t(v - 3w) + u(v - 3w) = (v - 3w)(3t + u)$

127. $-4abc - 4ac^2 + 2bc + 2c^2 = -2c[2ab + 2ac - b - c] = -2c[2a(b + c) - 1(b + c)]$
$$= -2c(b + c)(2a - 1)$$

129. $ax^3 + bx^3 + 2ax^2y + 2bx^2y = x^2[ax + bx + 2ay + 2by] = x^2[x(a + b) + 2y(a + b)]$
$$= x^2(a + b)(x + 2y)$$

131. $y^3 - 3y^2 - 5y + 15 = y^2(y - 3) - 5(y - 3) = (y - 3)(y^2 - 5)$

133. $2r - bs - 2s + br = 2r - 2s + br - bs = 2(r - s) + b(r - s) = (r - s)(2 + b)$

135. $ar^2 - brs + ars - br^2 = r[ar - bs + as - br] = r[ar - br + as - bs] = r[r(a - b) + s(a - b)]$
$$= r(a - b)(r + s)$$

137. Answers may vary.   **139.** Answers may vary.

Exercise 5.2   (page 318)

1. $x^2 - 9 = x^2 - 3^2 = (x + 3)(x - 3)$ **3.** $z^2 - 4 = z^2 - 2^2 = (z + 2)(z - 2)$

5. $25 - t^2 = 5^2 - t^2 = (5 + t)(5 - t)$ **7.** $81 - y^2 = 9^2 - y^2 = (9 + y)(9 - y)$

9. $4m^2 - 9n^2 = (2m)^2 - (3n)^2$ **11.** $25y^6 - 64x^4 = (5y^3)^2 - (8x^2)^2$
$$= (2m + 3n)(2m - 3n)$$ $$= (5y^3 + 8x^2)(5y^3 - 8x^2)$$

13. $\dfrac{p}{w} + \dfrac{v^2}{2g} + h = k$

$\dfrac{p}{w} = k - h - \dfrac{v^2}{2g}$

$w \cdot \dfrac{p}{w} = w\left(k - h - \dfrac{v^2}{2g}\right)$

$p = w\left(k - h - \dfrac{v^2}{2g}\right)$

15. difference of two squares

17. prime

19. $x^2 - 36 = x^2 - 6^2 = (x + 6)(x - 6)$

21. $y^2 - 49 = y^2 - 7^2 = (y + 7)(y - 7)$

23. $4y^2 - 49 = (2y)^2 - 7^2 = (2y + 7)(2y - 7)$

25. $25x^4 - 81 = \left(5x^2\right)^2 - 9^2$
$= \left(5x^2 + 9\right)\left(5x^2 - 9\right)$

27. $9x^2 - y^2 = (3x)^2 - y^2 = (3x + y)(3x - y)$

29. $25t^2 - 36u^2 = (5t)^2 - (6u)^2$
$= (5t + 6u)(5t - 6u)$

31. $100a^2 - 49b^2 = (10a)^2 - (7b)^2$
$= (10a + 7b)(10a - 7b)$

33. $x^4 - 9y^2 = \left(x^2\right)^2 - (3y)^2$
$= \left(x^2 + 3y\right)\left(x^2 - 3y\right)$

35. $8x^2 - 32y^2 = 8[x^2 - 4y^2] = 8\left[x^2 - (2y)^2\right] = 8(x + 2y)(x - 2y)$

37. $2a^2 - 8y^2 = 2[a^2 - 4y^2] = 2\left[a^2 - (2y)^2\right] = 2(a + 2y)(a - 2y)$

39. $100x^2 - 16y^2 = 4[25x^2 - 4y^2] = 4\left[(5x)^2 - (2y)^2\right] = 4(5x + 2y)(5x - 2y)$

41. $x^3 - xy^2 = x(x^2 - y^2) = x(x + y)(x - y)$

43. $4a^2x - 9b^2x = x[4a^2 - 9b^2] = x\left[(2a)^2 - (3b)^2\right] = x(2a + 3b)(2a - 3b)$

45. $3m^3 - 3mn^2 = 3m\left(m^2 - n^2\right)$
$= 3m(m + n)(m - n)$

47. $a^4 - 16 = \left(a^2 + 4\right)\left(a^2 - 4\right)$
$= \left(a^2 + 4\right)(a + 2)(a - 2)$

49. $a^4 - b^4 = (a^2 + b^2)(a^2 - b^2) = (a^2 + b^2)(a + b)(a - b)$

51. $2x^4 - 2y^4 = 2(x^4 - y^4) = 2(x^2 + y^2)(x^2 - y^2) = 2(x^2 + y^2)(x + y)(x - y)$

53. $a^4b - b^5 = b(a^4 - b^4) = b(a^2 + b^2)(a^2 - b^2) = b(a^2 + b^2)(a + b)(a - b)$

55. $2x^4y - 512y^5 = 2y(x^4 - 256y^4) = 2y(x^2 + 16y^2)(x^2 - 16y^2) = 2y(x^2 + 16y^2)(x + 4y)(x - 4y)$

57. $a^3 - 9a + 3a^2 - 27 = a\left(a^2 - 9\right) + 3\left(a^2 - 9\right) = \left(a^2 - 9\right)(a + 3)$
$= (a + 3)(a - 3)(a + 3) = (a + 3)^2(a - 3)$

59. $49y^2 - 225z^4 = (7y)^2 - (15z^2)^2 = (7y + 15z^2)(7y - 15z^2)$

61. $4x^4 - x^2y^2 = x^2(4x^2 - y^2) = x^2[(2x)^2 - y^2] = x^2(2x + y)(2x - y)$

63. $x^4 - 81 = (x^2 + 9)(x^2 - 9) = (x^2 + 9)(x + 3)(x - 3)$

65. $16y^8 - 81z^4 = (4y^4 + 9z^2)(4y^4 - 9z^2) = (4y^4 + 9z^2)(2y^2 + 3z)(2y^2 - 3z)$

67. $a^2 + b^2$: sum of squares $\Rightarrow$ prime

69. $x^8y^8 - 1 = (x^4y^4 + 1)(x^4y^4 - 1) = (x^4y^4 + 1)(x^2y^2 + 1)(x^2y^2 - 1)$
$$= (x^4y^4 + 1)(x^2y^2 + 1)(xy + 1)(xy - 1)$$

71. $2a^3b - 242ab^3 = 2ab[a^2 - 121b^2] = 2ab[a^2 - (11b)^2] = 2ab(a + 11b)(a - 11b)$

73. $3a^{10} - 3a^2b^4 = 3a^2(a^8 - b^4) = 3a^2(a^4 + b^2)(a^4 - b^2) = 3a^2(a^4 + b^2)(a^2 + b)(a^2 - b)$

75. $3a^8 - 243a^4b^8 = 3a^4(a^4 - 81b^8) = 3a^4(a^2 + 9b^4)(a^2 - 9b^4) = 3a^4(a^2 + 9b^4)(a + 3b^2)(a - 3b^2)$

77. $a^2b^7 - 625a^2b^3 = a^2b^3(b^4 - 625) = a^2b^3(b^2 + 25)(b^2 - 25) = a^2b^3(b^2 + 25)(b + 5)(b - 5)$

79. $3a^5y + 6ay^5 = 3ay(a^4 + 2y^4)$

81. $144a^4 + 169b^4$: sum of squares $\Rightarrow$ prime

83. $2x^9y + 2xy^9 = 2xy(x^8 + y^8)$

85. $49(y + 1)^2 - x^2 = [7(y + 1)]^2 - x^2 = [7(y + 1) + x][7(y + 1) - x] = (7y + 7 + x)(7y + 7 - x)$

87. $y^3 - 16y - 3y^2 + 48 = y(y^2 - 16) - 3(y^2 - 16) = (y^2 - 16)(y - 3) = (y + 4)(y - 4)(y - 3)$

89. $50c^4d^2 - 8c^2d^4 = 2c^2d^2[25c^2 - 4d^2] = 2c^2d^2[(5c)^2 - (2d)^2] = 2c^2d^2(5c + 2d)(5c - 2d)$

91. **Answers may vary.**

93. $498 \cdot 502 = (500 - 2)(500 + 2) = 500^2 - 2^2 = 250,000 - 4 = 249,996$

Exercise 5.3 (page 327)

1. $1, 4$ **3.** $-2, 3$ **5.** $1, 4$ **7.** $-2, 9$

9. $x - 3 > 5$
$x > 8$

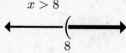

SECTION 5.3

11. $-3x - 5 \geq 4$
$-3x \geq 9$
$\dfrac{-3x}{-3} \leq \dfrac{9}{-3}$
$x \leq -3$

-3

13. $\dfrac{3(x-1)}{4} < 12$
$4 \cdot \dfrac{3x-3}{4} < 4(12)$
$3x - 3 < 48$
$3x < 51$
$x < 17$

17

15. $-2 < x \leq 4$
$-2 \quad 4$

17. $x^2 + 2xy + y^2 = (x+y)^2$

19. $x^2 + 5x + 6 = (x+2)(x+\boxed{3})$

21. $x^2 + x - 6 = (x\boxed{-}2)(x+\boxed{3})$

23. $x^2 + 5x - 6 = (x+\boxed{6})(x-\boxed{1})$

25. $y^2 + 6y + 8 = (y+\boxed{4})(y+\boxed{2})$

27. $x^2 - xy - 2y^2 = (x+\boxed{y})(x-\boxed{2y})$

29. $x^2 + 5x + 4 = (x+4)(x+1)$

31. $z^2 + 6z + 8 = (z+4)(z+2)$

33. $x^2 + 8x + 15 = (x+3)(x+5)$

35. $x^2 + 12x + 20 = (x+2)(x+10)$

37. $t^2 - 9t + 14 = (t-2)(t-7)$

39. $x^2 - 8x + 12 = (x-2)(x-6)$

41. $x^2 - 9x + 20 = (x-4)(x-5)$

43. $r^2 - 8r + 7 = (r-7)(r-1)$

45. $q^2 + 8q - 9 = (q+9)(q-1)$

47. $s^2 + 11s - 26 = (s+13)(s-2)$

49. $c^2 + 4c - 5 = (c+5)(c-1)$

51. $y^2 + 4y - 12 = (y+6)(y-2)$

53. $b^2 - 5b - 6 = (b-6)(b+1)$

55. $a^2 - 10a - 39 = (a-13)(a+3)$

57. $m^2 - 3m - 10 = (m-5)(m+2)$

59. $x^2 - 3x - 40 = (x-8)(x+5)$

61. $m^2 + 5mn - 14n^2 = (m+7n)(m-2n)$

63. $a^2 - 4ab - 12b^2 = (a-6b)(a+2b)$

65. $a^2 + 10ab + 9b^2 = (a+9b)(a+b)$

67. $m^2 - 11mn + 10n^2 = (m-10n)(m-n)$

69. $-x^2 - 7x - 10 = -(x^2 + 7x + 10)$
$= -(x+5)(x+2)$

71. $-y^2 - 2y + 15 = -(y^2 + 2y - 15)$
$= -(y+5)(y-3)$

73. $-t^2 - 4t + 32 = -(t^2 + 4t - 32)$
$= -(t+8)(t-4)$

75. $-r^2 + 14r - 40 = -(r^2 - 14r + 40) = -(r-4)(r-10)$

77. $a^2 + 3a + 10$: prime

79. $r^2 - 9r - 12$: prime

81. $2x^2 + 20x + 42 = 2\left(x^2 + 10x + 21\right)$
$= 2(x + 3)(x + 7)$

83. $3y^3 - 21y^2 + 18y = 3y\left(y^2 - 7y + 6\right)$
$= 3y(y - 6)(y - 1)$

85. $3z^2 - 15tz + 12t^2 = 3\left(z^2 - 5tz + 4t^2\right) = 3(z - 4t)(z - t)$

87. $-4x^2y - 4x^3 + 24xy^2 = -4x\left(xy + x^2 - 6y^2\right) = -4x\left(x^2 + xy - 6y^2\right) = -4x(x + 3y)(x - 2y)$

89. $x^2 + 4x + 4 - y^2 = (x + 2)(x + 2) - y^2 = (x + 2)^2 - y^2 = (x + 2 + y)(x + 2 - y)$

91. $b^2 - 6b + 9 - c^2 = (b - 3)(b - 3) - c^2 = (b - 3)^2 - c^2 = (b - 3 + c)(b - 3 - c)$

93. $x^2 + 6x + 5$: $a = 1, b = 6, c = 5 \Rightarrow$ key # $= ac = 1(5) = 5$.
Find two factors of 5 whose sum is $b = 6$: 1 and 5.
Rewrite and factor: $x^2 + 6x + 5 = x^2 + 1x + 5x + 5$
$= x(x + 1) + 5(x + 1) = (x + 1)(x + 5)$

95. $t^2 - 9t + 14$: $a = 1, b = -9, c = 14 \Rightarrow$ key # $= ac = 1(14) = 14$.
Find two factors of 14 whose sum is $b = -9$: -7 and -2.
Rewrite and factor: $t^2 - 9t + 14 = t^2 - 7t - 2t + 14$
$= t(t - 7) - 2(t - 7) = (t - 7)(t - 2)$

97. $a^2 + 6a - 16$: $a = 1, b = 6, c = -16 \Rightarrow$ key # $= ac = 1(-16) = -16$.
Find two factors of -16 whose sum is $b = 6$: 8 and -2.
Rewrite and factor: $a^2 + 6a - 16 = a^2 + 8a - 2a - 16$
$= a(a + 8) - 2(a + 8) = (a + 8)(a - 2)$

99. $y^2 - y - 30$: $a = 1, b = -1, c = -30 \Rightarrow$ key # $= ac = 1(-30) = -30$.
Find two factors of -30 whose sum is $b = -1$: 5 and -6.
Rewrite and factor: $y^2 - y - 30 = y^2 + 5y - 6y - 30$
$= y(y + 5) - 6(y + 5) = (y + 5)(y - 6)$

101. $x^2 + 6x + 9 = (x + 3)(x + 3)$
$= (x + 3)^2$

103. $y^2 - 8y + 16 = (y - 4)(y - 4)$
$= (y - 4)^2$

105. $u^2 - 18u + 81 = (u - 9)(u - 9)$
$= (u - 9)^2$

107. $x^2 + 4xy + 4y^2 = (x + 2y)(x + 2y)$
$= (x + 2y)^2$

109. $4 - 5x + x^2 = x^2 - 5x + 4$
$= (x - 4)(x - 1)$

111. $10y + 9 + y^2 = y^2 + 10y + 9$
$= (y + 9)(y + 1)$

113. $-r^2 + 2s^2 + rs = -r^2 + rs + 2s^2$
$$= -(r^2 - rs - 2s^2)$$
$$= -(r - 2s)(r + s)$$

115. $x^2 - 9x - 27$: prime

117. $-a^2 - 6ab - 5b^2 = -(a^2 + 6ab + 5b^2)$
$$= -(a + 5b)(a + b)$$

119. $12xy + 4x^2y - 72y = 4y(3x + x^2 - 18)$
$$= 4y(x^2 + 3x - 18)$$
$$= 4y(x + 6)(x - 3)$$

121. $r^2 - 2rs + 4s^2$: prime

123. $r^2 - 10rs + 25s^2 = (r - 5s)(r - 5s)$
$$= (r - 5s)^2$$

125. $a^2 + 6a + 9 - b^2 = (a + 3)(a + 3) - b^2 = (a + 3)^2 - b^2 = (a + 3 + b)(a + 3 - b)$

127. $t^2 + 18t + 81 = (t + 9)(t + 9) = (t + 9)^2$

129. Answers may vary.

131. Both answers check. Both may be factored more completely:
$(2x + 6)(x + 7) = 2(x + 3)(x + 7)$; $(x + 3)(2x + 14) = (x + 3) \cdot 2(x + 7)$

Exercise 5.4 (page 336)

1. $6x^2 + 7x + 2 = (2x + 1)(3x + \boxed{2})$

3. $6x^2 + x - 2 = (3x + \boxed{2})(2x - \boxed{1})$

5. $15x^2 - 7x - 4 = (5x - \boxed{4})(3x + 1)$

7. $l = f + (n - 1)d$
$l = f + nd - d$
$l - f + d = nd$
$\dfrac{l - f + d}{d} = \dfrac{nd}{d}$
$\dfrac{l - f + d}{d} = n$

9. the same as

11. ac method

13. $6x^2 + 5x - 1 = (x\boxed{+}1)(6x\boxed{-}1)$

15. $4x^2 + 4x - 3 = (2x + \boxed{3})(2x - \boxed{1})$

17. $12x^2 - 7xy + y^2 = (3x - \boxed{y})(4x - \boxed{y})$

19. $3a^2 + 7a + 2 = (3a + 1)(a + 2)$

21. $3a^2 + 10a + 3 = (3a + 1)(a + 3)$

23. $5t^2 + 13t + 6 = (5t + 3)(t + 2)$

25. $16x^2 + 16x + 3 = (4x + 3)(4x + 1)$

27. $5y^2 - 23y + 12 = (5y - 3)(y - 4)$

29. $2y^2 - 7y + 3 = (2y - 1)(y - 3)$

31. $16m^2 - 14m + 3 = (8m - 3)(2m - 1)$

33. $6x^2 - 7x + 2 = (2x - 1)(3x - 2)$

35. $3a^2 - 4a - 4 = (3a + 2)(a - 2)$

37. $12y^2 - y - 1 = (3y - 1)(4y + 1)$

39. $12y^2 - 5y - 2 = (4y + 1)(3y - 2)$

41. $10y^2 - 3y - 1 = (5y + 1)(2y - 1)$

43. $8q^2 + 10q - 3 = (4q - 1)(2q + 3)$

45. $10x^2 + 21x - 10 = (2x + 5)(5x - 2)$

47. $30x^2 - 23x - 14 = (5x + 2)(6x - 7)$

49. $8x^2 - 14x - 15 = (4x + 3)(2x - 5)$

51. $2x^2 + 3xy + y^2 = (2x + y)(x + y)$

53. $3x^2 - 4xy + y^2 = (3x - y)(x - y)$

55. $10p^2 - 11pq - 6q^2 = (2p - 3q)(5p + 2q)$

57. $6p^2 - pq - 2q^2 = (2p + q)(3p - 2q)$

59. $15 + 8a^2 - 26a = 8a^2 - 26a + 15$
$\qquad = (4a - 3)(2a - 5)$

61. $12x^2 + 10y^2 - 23xy = 12x^2 - 23xy + 10y^2$
$\qquad = (3x - 2y)(4x - 5y)$

63. $6x^2 + 7x + 2 = (2x + 1)(3x + 2)$

65. $-26x + 6x^2 - 20 = 6x^2 - 26x - 20 = 2(3x^2 - 13x - 10) = 2(3x + 2)(x - 5)$

67. $9x^2 - 12x + 4 = (3x - 2)(3x - 2)$
$\qquad = (3x - 2)^2$

69. $25x^2 + 30x + 9 = (5x + 3)(5x + 3)$
$\qquad = (5x + 3)^2$

71. $9a^2 + 24a + 16 = (3a + 4)(3a + 4)$
$\qquad = (3a + 4)^2$

73. $16x^2 - 40x + 25 = (4x - 5)(4x - 5)$
$\qquad = (4x - 5)^2$

75. $16x^2 + 8xy + y^2 - 9 = (16x^2 + 8xy + y^2) - 9 = (4x + y)(4x + y) - 9$
$\qquad = (4x + y)^2 - 3^2 = (4x + y + 3)(4x + y - 3)$

77. $9 - a^2 - 4ab - 4b^2 = 9 - (a^2 + 4ab + 4b^2) = 9 - (a + 2b)(a + 2b)$
$\qquad = 3^2 - (a + 2b)^2$
$\qquad = (3 + (a + 2b))(3 - (a + 2b))$
$\qquad = (3 + a + 2b)(3 - a - 2b)$

79. $-12y^2 - 12 + 25y = -(12y^2 - 25y + 12)$
$\qquad = -(3y - 4)(4y - 3)$

81. $5x^2 + 2 + x = 5x^2 + x + 2 \Rightarrow$ prime

83. $4x^2 + 10x - 6 = 2(2x^2 + 5x - 3)$
$\qquad = 2(2x - 1)(x + 3)$

85. $y^3 + 13y^2 + 12y = y(y^2 + 13y + 12)$
$\qquad = y(y + 12)(y + 1)$

87. $9y^3 + 3y^2 - 6y = 3y(3y^2 + y - 2)$
$\qquad = 3y(3y - 2)(y + 1)$

89. $6s^5 - 26s^4 - 20s^3 = 2s^3(3s^2 - 13s - 10)$
$\qquad = 2s^3(3s + 2)(s - 5)$

91. $12x^2 + 5xy - 3y^2 = (4x + 3y)(3x - y)$

93. $3b^2 + 3a^2 - ab = 3a^2 - ab + 3b^2 \Rightarrow$ prime

95. $25x^2 + 20xy + 4y^2 = (5x + 2y)(5x + 2y) = (5x + 2y)^2$

97. $-16x^4y^3 + 30x^3y^4 + 4x^2y^5 = -2x^2y^3(8x^2 - 15xy - 2y^2) = -2x^2y^3(8x + y)(x - 2y)$

99. $24a^2 + 14ab + 2b^2 = 2(12a^2 + 7ab + b^2) = 2(4a + b)(3a + b)$

101. **Answers may vary.**

103. $(6x + 1)(x + 6) = 6x^2 + 37x + 6;\ (6x - 1)(x - 6) = 6x^2 - 37x + 6 \Rightarrow b = \pm 37$
$(6x + 2)(x + 3) = 6x^2 + 20x + 6;\ (6x - 2)(x - 3) = 6x^2 - 20x + 6 \Rightarrow b = \pm 20$
$(6x + 3)(x + 2) = 6x^2 + 15x + 6;\ (6x - 3)(x - 2) = 6x^2 - 15x + 6 \Rightarrow b = \pm 15$
$(6x + 6)(x + 1) = 6x^2 + 12x + 6;\ (6x - 6)(x - 1) = 6x^2 - 12x + 6 \Rightarrow b = \pm 12$
$(2x + 1)(3x + 6) = 6x^2 + 15x + 6;\ (2x - 1)(3x - 6) = 6x^2 - 15x + 6 \Rightarrow b = \pm 15$
$(2x + 2)(3x + 3) = 6x^2 + 12x + 6;\ (2x - 2)(3x - 3) = 6x^2 - 12x + 6 \Rightarrow b = \pm 12$
$(2x + 3)(3x + 2) = 6x^2 + 13x + 6;\ (2x - 3)(3x - 2) = 6x^2 - 13x + 6 \Rightarrow b = \pm 13$
$(2x + 6)(3x + 1) = 6x^2 + 20x + 6;\ (2x - 6)(3x - 1) = 6x^2 - 20x + 6 \Rightarrow b = \pm 20$
$b = \pm 12,\ \pm 13,\ \pm 15,\ \pm 20,\ \pm 37$

Exercise 5.5 (page 341)

1. $8 = 2^3$ **3.** $-27 = (-3)^3$ **5.** $y^{12} = (y^4)^3$ **7.** $-y^9 = (-y^3)^3$

9. $3^2 = 9$ **11.** $(4y^2)^2 = 16y^4$ **13.** $(-2x)^2 = 4x^2$ **15.** $(-x^5)^2 = x^{10}$

17. $1 \times 10^{-13} = 0.0000000000001$ cm   **19.** sum of two cubes

21. $x^3 + y^3 = (x + y)\boxed{(x^2 - xy + y^2)}$

23. $a^3 + 8 = a^3 + 2^3$
$\qquad = (a + 2)(a^2 - 2a + 2^2)$
$\qquad = (a + 2)(a^2 - 2a + 4)$

25. $125x^3 + 8 = (5x)^3 + 2^3$
$\qquad = (5x + 2)((5x)^2 - 2(5x) + 2^2)$
$\qquad = (5x + 2)(25x^2 - 10x + 4)$

27. $y^3 + 1 = y^3 + 1^3$
$\qquad = (y + 1)(y^2 - 1y + 1^2)$
$\qquad = (y + 1)(y^2 - y + 1)$

29. $125 + a^3 = 5^3 + a^3$
$\qquad = (5 + a)(5^2 - 5a + a^2)$
$\qquad = (5 + a)(25 - 5a + a^2)$

31. $m^3 + n^3 = (m + n)(m^2 - mn + n^2)$ **33.** $x^3 + y^3 = (x + y)(x^2 - xy + y^2)$

35. $8u^3 + w^3 = (2u)^3 + w^3$
$\qquad = (2u + w)((2u)^2 - 2uw + w^2)$
$\qquad = (2u + w)(4u^2 - 2uw + w^2)$

37. $x^3 - y^3 = (x - y)(x^2 + xy + y^2)$

152

39. $x^3 - 8 = x^3 - 2^3$
$= (x - 2)(x^2 + 2x + 2^2)$
$= (x - 2)(x^2 + 2x + 4)$

41. $s^3 - t^3 = (s - t)(s^2 + st + t^2)$

43. $125p^3 - q^3 = (5p)^3 - q^3$
$= (5p - q)((5p)^2 + 5pq + q^2)$
$= (5p - q)(25p^2 + 5pq + q^2)$

45. $27a^3 - b^3 = (3a)^3 - b^3$
$= (3a - b)((3a)^2 + 3ab + b^2)$
$= (3a - b)(9a^2 + 3ab + b^2)$

47. $2x^3 + 54 = 2(x^3 + 27)$
$= 2(x^3 + 3^3)$
$= 2(x + 3)(x^2 - 3x + 3^2)$
$= 2(x + 3)(x^2 - 3x + 9)$

49. $-x^3 + 216 = -(x^3 - 216)$
$= -(x^3 - 6^3)$
$= -(x - 6)(x^2 + 6x + 6^2)$
$= -(x - 6)(x^2 + 6x + 36)$

51. $64m^3x - 8n^3x = 8x(8m^3 - n^3) = 8x((2m)^3 - n^3) = 8x(2m - n)((2m)^2 + 2mn + n^2)$
$= 8x(2m - n)(4m^2 + 2mn + n^2)$

53. $x^4y + 216xy^4 = xy(x^3 + 216y^3) = xy(x^3 + (6y)^3) = xy(x + 6y)(x^2 - 6xy + (6y)^2)$
$= xy(x + 6y)(x^2 - 6xy + 36y^2)$

55. $x^6 - 1 = (x^3)^2 - 1^2 = (x^3 + 1)(x^3 - 1) = (x^3 + 1^3)(x^3 - 1^3)$
$= (x + 1)(x^2 - 1x + 1^2)(x - 1)(x^2 + 1x + 1^2)$
$= (x + 1)(x^2 - x + 1)(x - 1)(x^2 + x + 1)$

57. $x^{12} - y^6 = (x^6)^2 - (y^3)^2 = (x^6 + y^3)(x^6 - y^3)$
$= ((x^2)^3 + y^3)((x^2)^3 - y^3)$
$= (x^2 + y)((x^2)^2 - x^2y + y^2)(x^2 - y)((x^2)^2 + x^2y + y^2)$
$= (x^2 + y)(x^4 - x^2y + y^2)(x^2 - y)(x^4 + x^2y + y^2)$

59. $y^3 + 8 = y^3 + 2^3 = (y + 2)(y^2 - 2y + 2^2) = (y + 2)(y^2 - 2y + 4)$

61. $27x^3 + 125 = (3x)^3 + 5^3 = (3x + 5)((3x)^2 - (3x)5 + 5^2) = (3x + 5)(9x^2 - 15x + 25)$

63. $64 - z^3 = 4^3 - z^3 = (4 - z)(4^2 + 4z + z^2) = (4 - z)(16 + 4z + z^2)$

65. $64x^3 + 27y^3 = (4x)^3 + (3y)^3 = (4x + 3y)((4x)^2 - (4x)(3y) + (3y)^2)$
$= (4x + 3y)(16x^2 - 12xy + 9y^2)$

67. $3(x^3 + y^3) - z(x^3 + y^3) = (x^3 + y^3)(3 - z) = (x + y)(x^2 - xy + y^2)(3 - z)$

69. $x(27y^3 - z^3) + 5(27y^3 - z^3) = (27y^3 - z^3)(x + 5)$
$$= ((3y)^3 - z^3)(x + 5)$$
$$= (3y - z)((3y)^2 + 3yz + z^2)(x + 5)$$
$$= (3y - z)(9y^2 + 3yz + z^2)(x + 5)$$

71. $(a^3x + b^3x) - (a^3y + b^3y) = x(a^3 + b^3) - y(a^3 + b^3) = (a^3 + b^3)(x - y)$
$$= (a + b)(a^2 - ab + b^2)(x - y)$$

73. $y^3(y^2 - 1) - 27(y^2 - 1) = (y^2 - 1)(y^3 - 27) = (y^2 - 1^2)(y^3 - 3^3)$
$$= (y + 1)(y - 1)(y - 3)(y^2 + 3y + 9)$$

75. **Answers may vary.**

77. $a^3 - b^3 = 11^3 - 7^3 = 1331 - 343 = \boxed{988}$
$(a - b)(a^2 + ab + b^2) = (11 - 7)(11^2 + 11 \cdot 7 + 7^2) = 4(121 + 77 + 49) = 4(247) = \boxed{988}$

Exercise 5.6 (page 344)

1. $3x^2 - 9x$: common factor $(3x)$

3. $125 + r^3s^3 = 5^3 + (rs)^3$: sum of 2 cubes

5. $x^2 + 36$: none (prime)

7. $25r^2 - s^4 = (5r)^2 - (s^2)^2$
difference of 2 squares

9. $2(t + 5) + t = 3(t + 2) + 4$
$2t + 10 + t = 3t + 6 + 4$
$3t + 10 = 3t + 10$
$10 = 10$
solution set: $\mathbb{R}$

11. $6 + 2(t + 3) = t + 3$
$6 + 2t + 6 = t + 3$
$2t + 12 = t + 3$
$t + 12 = 3$
$t = -9$

13. factors

15. binomials

17. $6x + 3 = 3(2x + 1)$

19. $x^2 + 10x + 9 = (x + 9)(x + 1)$

21. $8t^2 - 6t - 9 = (4t + 3)(2t - 3)$

23. $t^2 - 2t + 1 = (t - 1)(t - 1) = (t - 1)^2$

25. $2x^2 - 50 = 2(x^2 - 25) = 2(x^2 - 5^2) = 2(x + 5)(x - 5)$

27. $x^2 + 7x + 1 \Rightarrow$ prime

29. $-2x^5 + 128x^2 = -2x^2(x^3 - 64) = -2x^2(x^3 - 4^3) = -2x^2(x - 4)(x^2 + 4x + 16)$

31. $14t^3 - 40t^2 + 6t^4 = 2t^2(7t - 20 + 3t^2) = 2t^2(3t^2 + 7t - 20) = 2t^2(3t - 5)(t + 4)$

33. $6x^2 - x - 16 \Rightarrow$ prime

35. $6a^3 + 35a^2 - 6a = a(6a^2 + 35a - 6)$
$$= a(6a - 1)(a + 6)$$

37. $16x^2 - 40x^3 + 25x^4 = x^2(16 - 40x + 25x^2) = x^2(4 - 5x)(4 - 5x) = x^2(4 - 5x)^2$

39. $-84x^2 - 147x - 12x^3 = -12x^3 - 84x^2 - 147x = -3x(4x^2 + 28x + 49) = -3x(2x + 7)(2x + 7)$
$$= -3x(2x + 7)^2$$

41. $8x^6 - 8 = 8(x^6 - 1) = 8\left((x^3)^2 - 1^2\right) = 8(x^3 + 1)(x^3 - 1)$
$$= 8(x + 1)(x^2 - x + 1)(x - 1)(x^2 + x + 1)$$

43. $5x^3 - 5x^5 + 25x^2 = -5x^5 + 5x^3 + 25x^2$
$$= -5x^2(x^3 - x - 5)$$

45. $9x^2 + 12x + 16$: prime

47. $2ab^2 + 8ab - 24a = 2a(b^2 + 4b - 12)$
$$= 2a(b + 6)(b - 2)$$

49. $-8p^3q^7 - 4p^2q^3 = -4p^2q^3(2pq^4 + 1)$

51. $4a^2 - 4ab + b^2 - 9 = (4a^2 - 4ab + b^2) - 9 = (2a - b)^2 - 3^2 = (2a - b + 3)(2a - b - 3)$

53. $8a^3 - b^3 = (2a)^3 - b^3 = (2a - b)\left((2a)^2 + 2ab + b^2\right) = (2a - b)(4a^2 + 2ab + b^2)$

55. $x^2y^2 - 2x^2 - y^2 + 2 = x^2(y^2 - 2) - 1(y^2 - 2) = (y^2 - 2)(x^2 - 1) = (y^2 - 2)(x^2 - 1^2)$
$$= (y^2 - 2)(x + 1)(x - 1)$$

57. $a^2 + 2ab + b^2 - y^2 = (a + b)(a + b) - y^2 = (a + b)^2 - y^2 = (a + b + y)(a + b - y)$

59. $a^2(x - 3) - b^2(x - 3) = (x - 3)(a^2 - b^2) = (x - 3)(a + b)(a - b)$

61. $8p^6 - 27q^6 = (2p^2)^3 - (3q^2)^3 = (2p^2 - 3q^2)\left((2p^2)^2 + (2p^2)(3q^2) + (3q^2)^2\right)$
$$= (2p^2 - 3q^2)(4p^4 + 6p^2q^2 + 9q^4)$$

63. $125p^3 - 64y^3 = (5p)^3 - (4y)^3 = (5p - 4y)\left((5p)^2 + (5p)(4y) + (4y)^2\right)$
$$= (5p - 4y)(25p^2 + 20py + 16y^2)$$

65. $-16x^4y^2z + 24x^5y^3z^4 - 15x^2y^3z^7 = -x^2y^2z(16x^2 - 24x^3yz^3 + 15yz^6)$

67. $81p^4 - 16q^4 = (9p^2)^2 - (4q^2)^2 = (9p^2 + 4q^2)(9p^2 - 4q^2) = (9p^2 + 4q^2)(3p + 2q)(3p - 2q)$

69. $54x^3 + 250y^6 = 2(27x^3 + 125y^6) = 2\left((3x)^3 + (5y^2)^3\right)$
$$= 2(3x + 5y^2)\left((3x)^2 - (3x)(5y^2) + (5y^2)^2\right)$$
$$= 2(3x + 5y^2)(9x^2 - 15xy^2 + 25y^4)$$

71. $x^5 - x^3y^2 + x^2y^3 - y^5 = x^3(x^2 - y^2) + y^3(x^2 - y^2) = (x^2 - y^2)(x^3 + y^3)$
$$= (x + y)(x - y)(x + y)(x^2 - xy + y^2)$$

73. $2a^2c - 2b^2c + 4a^2d - 4b^2d = 2(a^2c - b^2c + 2a^2d - 2b^2d)$
$$= 2(c(a^2 - b^2) + 2d(a^2 - b^2))$$
$$= 2(a^2 - b^2)(c + 2d) = 2(a + b)(a - b)(c + 2d)$$

75. Answers may vary. ⟶ **77. Answers may vary.**

Exercise 5.7 (page 351)

1.
$$(x - 8)(x - 7) = 0$$
$$x - 8 = 0 \quad \textbf{or} \quad x - 7 = 0$$
$$x = 8 \qquad\qquad x = 7$$

3.
$$(x - 2)(x + 3) = 0$$
$$x - 2 = 0 \quad \textbf{or} \quad x + 3 = 0$$
$$x = 2 \qquad\qquad x = -3$$

5.
$$(x - 4)(x + 1) = 0$$
$$x - 4 = 0 \quad \textbf{or} \quad x + 1 = 0$$
$$x = 4 \qquad\qquad x = -1$$

7.
$$(2x - 5)(3x + 6) = 0$$
$$2x - 5 = 0 \quad \textbf{or} \quad 3x + 6 = 0$$
$$2x = 5 \qquad\qquad 3x = -6$$
$$x = \tfrac{5}{2} \qquad\qquad x = -2$$

9. $u^3u^2u^4 = u^{3+2+4} = u^9$

11. $\dfrac{a^3b^4}{a^2b^5} = a^1b^{-1} = \dfrac{a}{b}$

13. quadratic

15. second

17.
$$x^2 + 7x = 0$$
$$x(x + 7) = 0$$
$$x = 0 \quad \textbf{or} \quad x + 7 = 0$$
$$x = 0 \qquad\qquad x = -7$$

19.
$$x^2 - 2x + 1 = 0$$
$$(x - 1)(x - 1) = 0$$
$$x - 1 = 0 \quad \textbf{or} \quad x - 1 = 0$$
$$x = 1 \qquad\qquad x = 1$$

21.
$$x^2 - 3x = 0$$
$$x(x - 3) = 0$$
$$x = 0 \quad \textbf{or} \quad x - 3 = 0$$
$$x = 0 \qquad\qquad x = 3$$

23.
$$5x^2 + 7x = 0$$
$$x(5x + 7) = 0$$
$$x = 0 \quad \textbf{or} \quad 5x + 7 = 0$$
$$x = 0 \qquad\qquad 5x = -7$$
$$x = 0 \qquad\qquad x = -\tfrac{7}{5}$$

25.
$$x^2 - 7x = 0$$
$$x(x - 7) = 0$$
$$x = 0 \quad \textbf{or} \quad x - 7 = 0$$
$$x = 0 \qquad\qquad x = 7$$

27.
$$3x^2 + 8x = 0$$
$$x(3x + 8) = 0$$
$$x = 0 \quad \textbf{or} \quad 3x + 8 = 0$$
$$x = 0 \qquad\qquad 3x = -8$$
$$x = 0 \qquad\qquad x = -\tfrac{8}{3}$$

29.
$$x^2 - 25 = 0$$
$$(x + 5)(x - 5) = 0$$
$$x + 5 = 0 \quad \textbf{or} \quad x - 5 = 0$$
$$x = -5 \qquad\qquad x = 5$$

31.
$$9y^2 - 4 = 0$$
$$(3y + 2)(3y - 2) = 0$$
$$3y + 2 = 0 \quad \textbf{or} \quad 3y - 2 = 0$$
$$3y = -2 \qquad\qquad 3y = 2$$
$$y = -\tfrac{2}{3} \qquad\qquad y = \tfrac{2}{3}$$

33.
$$x^2 - 13x + 12 = 0$$
$$(x - 1)(x - 12) = 0$$
$$x - 1 = 0 \quad \textbf{or} \quad x - 12 = 0$$
$$x = 1 \qquad\qquad x = 12$$

35.
$$x^2 - 2x - 15 = 0$$
$$(x + 3)(x - 5) = 0$$
$$x + 3 = 0 \quad \textbf{or} \quad x - 5 = 0$$
$$x = -3 \qquad\qquad x = 5$$

37.
$$x^2 - 3x - 18 = 0$$
$$(x + 3)(x - 6) = 0$$
$$x + 3 = 0 \quad \textbf{or} \quad x - 6 = 0$$
$$x = -3 \qquad\qquad x = 6$$

39.
$$x^2 - x - 20 = 0$$
$$(x + 4)(x - 5) = 0$$
$$x + 4 = 0 \quad \textbf{or} \quad x - 5 = 0$$
$$x = -4 \qquad\qquad x = 5$$

41.
$$6x^2 + x = 2$$
$$6x^2 + x - 2 = 0$$
$$(2x - 1)(3x + 2) = 0$$
$$2x - 1 = 0 \quad \textbf{or} \quad 3x + 2 = 0$$
$$2x = 1 \qquad\qquad 3x = -2$$
$$x = \tfrac{1}{2} \qquad\qquad x = -\tfrac{2}{3}$$

43.
$$2x^2 - 5x = -2$$
$$2x^2 - 5x + 2 = 0$$
$$(2x - 1)(x - 2) = 0$$
$$2x - 1 = 0 \quad \textbf{or} \quad x - 2 = 0$$
$$2x = 1 \qquad\qquad x = 2$$
$$x = \tfrac{1}{2} \qquad\qquad x = 2$$

45.
$$x^2 = 49$$
$$x^2 - 49 = 0$$
$$(x + 7)(x - 7) = 0$$
$$x + 7 = 0 \quad \textbf{or} \quad x - 7 = 0$$
$$x = -7 \qquad\qquad x = 7$$

47.
$$4x^2 = 81$$
$$4x^2 - 81 = 0$$
$$(2x + 9)(2x - 9) = 0$$
$$2x + 9 = 0 \quad \textbf{or} \quad 2x - 9 = 0$$
$$2x = -9 \qquad\qquad 2x = 9$$
$$x = -\tfrac{9}{2} \qquad\qquad x = \tfrac{9}{2}$$

49.
$$x(6x + 5) = 6$$
$$6x^2 + 5x = 6$$
$$6x^2 + 5x - 6 = 0$$
$$(3x - 2)(2x + 3) = 0$$
$$3x - 2 = 0 \quad \textbf{or} \quad 2x + 3 = 0$$
$$3x = 2 \qquad\qquad 2x = -3$$
$$x = \tfrac{2}{3} \qquad\qquad x = -\tfrac{3}{2}$$

51.
$$(x + 1)(8x + 1) = 18x$$
$$8x^2 + 9x + 1 = 18x$$
$$8x^2 - 9x + 1 = 0$$
$$(x - 1)(8x - 1) = 0$$
$$x - 1 = 0 \quad \textbf{or} \quad 8x - 1 = 0$$
$$x = 1 \qquad\qquad 8x = 1$$
$$x = 1 \qquad\qquad x = \tfrac{1}{8}$$

53. $(x + 4)(x - 5)(x - 7) = 0$
$$x + 4 = 0 \quad \textbf{or} \quad x - 5 = 0 \quad \textbf{or} \quad x - 7 = 0$$
$$x = -4 \qquad\qquad x = 5 \qquad\qquad x = 7$$

55. $(x-1)(x^2+5x+6)=0$
$(x-1)(x+2)(x+3)=0$
$x-1=0$ **or** $x+2=0$ **or** $x+3=0$
$x=1$ $\qquad x=-2$ $\qquad x=-3$

57. $x^3+3x^2+2x=0$
$x(x^2+3x+2)=0$
$x(x+2)(x+1)=0$
$x=0$ **or** $x+2=0$ **or** $x+1=0$
$x=0$ $\qquad x=-2$ $\qquad x=-1$

59. $x^3-27x-6x^2=0$
$x(x^2-27-6x)=0$
$x(x^2-6x-27)=0$
$x(x-9)(x+3)=0$
$x=0$ **or** $x-9=0$ **or** $x+3=0$
$x=0$ $\qquad x=9$ $\qquad x=-3$

61. $\qquad\qquad 6x^3+20x^2=-6x$
$6x^3+20x^2+6x=0$
$2x(3x^2+10x+3)=0$
$2x(3x+1)(x+3)=0$
$2x=0$ **or** $3x+1=0$ **or** $x+3=0$
$x=0$ $\qquad 3x=-1$ $\qquad x=-3$
$x=0$ $\qquad x=-\frac{1}{3}$ $\qquad x=-3$

63. $\qquad\qquad x^3+7x^2=x^2-9x$
$x^3+6x^2+9x=0$
$x(x^2+6x+9)=0$
$x(x+3)(x+3)=0$
$x=0$ **or** $x+3=0$ **or** $x+3=0$
$x=0$ $\qquad x=-3$ $\qquad x=-3$

65. $\qquad x^2-4x=0$
$x(x-4)=0$
$x=0$ **or** $x-4=0$
$x=0$ $\qquad x=4$

67. $\qquad 9x^2+5x=0$
$x(9x+5)=0$
$x=0$ **or** $9x+5=0$
$x=0$ $\qquad 9x=-5$
$x=0$ $\qquad x=-\frac{5}{9}$

69. $(x+3)(x^2+2x-15)=0$
$(x+3)(x+5)(x-3)=0$
$x+3=0$ **or** $x+5=0$ **or** $x-3=0$
$x=-3$ $\qquad x=-5$ $\qquad x=3$

71. $\qquad x^2-4x-21=0$
$(x+3)(x-7)=0$
$x+3=0$ **or** $x-7=0$
$x=-3$ $\qquad x=7$

73. $\qquad 2y-8=-y^2$
$y^2+2y-8=0$
$(y+4)(y-2)=0$
$y+4=0$ **or** $y-2=0$
$y=-4$ $\qquad y=2$

75. $\qquad (p^2-81)(p+2)=0$
$(p+9)(p-9)(p+2)=0$
$p+9=0$ **or** $p-9=0$ **or** $p+2=0$
$p=-9$ $\qquad p=9$ $\qquad p=-2$

77.
$$15x^2 - 2 = 7x$$
$$15x^2 - 7x - 2 = 0$$
$$(3x - 2)(5x + 1) = 0$$
$$3x - 2 = 0 \quad \text{or} \quad 5x + 1 = 0$$
$$3x = 2 \qquad\qquad 5x = -1$$
$$x = \tfrac{2}{3} \qquad\qquad x = -\tfrac{1}{5}$$

79.
$$x^2 + 8 - 9x = 0$$
$$x^2 - 9x + 8 = 0$$
$$(x - 8)(x - 1) = 0$$
$$x - 8 = 0 \quad \text{or} \quad x - 1 = 0$$
$$x = 8 \qquad\qquad x = 1$$

81.
$$a^2 + 8a = -15$$
$$a^2 + 8a + 15 = 0$$
$$(a + 3)(a + 5) = 0$$
$$a + 3 = 0 \quad \text{or} \quad a + 5 = 0$$
$$a = -3 \qquad\qquad a = -5$$

83.
$$3x^2 - 8x = 3$$
$$3x^2 - 8x - 3 = 0$$
$$(3x + 1)(x - 3) = 0$$
$$3x + 1 = 0 \quad \text{or} \quad x - 3 = 0$$
$$3x = -1 \qquad\qquad x = 3$$
$$x = -\tfrac{1}{3} \qquad\qquad x = 3$$

85.
$$2x^2 + x - 3 = 0$$
$$(2x + 3)(x - 1) = 0$$
$$2x + 3 = 0 \quad \text{or} \quad x - 1 = 0$$
$$2x = -3 \qquad\qquad x = 1$$
$$x = -\tfrac{3}{2} \qquad\qquad x = 1$$

87.
$$14m^2 + 23m + 3 = 0$$
$$(7m + 1)(2m + 3) = 0$$
$$7m + 1 = 0 \quad \text{or} \quad 2m + 3 = 0$$
$$7m = -1 \qquad\qquad 2m = -3$$
$$m = -\tfrac{1}{7} \qquad\qquad m = -\tfrac{3}{2}$$

89.
$$(x + 2)(x^2 + x - 20) = 0$$
$$(x + 2)(x + 5)(x - 4) = 0$$
$$x + 2 = 0 \quad \text{or} \quad x + 5 = 0 \quad \text{or} \quad x - 4 = 0$$
$$x = -2 \qquad\qquad x = -5 \qquad\qquad x = 4$$

91.
$$z^2 - 81 = 0$$
$$(z + 9)(z - 9) = 0$$
$$z + 9 = 0 \quad \text{or} \quad z - 9 = 0$$
$$z = -9 \qquad\qquad z = 9$$

93.
$$4x^2 - 1 = 0$$
$$(2x + 1)(2x - 1) = 0$$
$$2x + 1 = 0 \quad \text{or} \quad 2x - 1 = 0$$
$$2x = -1 \qquad\qquad 2x = 1$$
$$x = -\tfrac{1}{2} \qquad\qquad x = \tfrac{1}{2}$$

95.
$$x^3 + 1.3x^2 - 0.3x = 0$$
$$10x^3 + 13x^2 - 3x = 0$$
$$x(10x^2 + 13x - 3) = 0$$
$$x(5x - 1)(2x + 3) = 0$$
$$x = 0 \quad \text{or} \quad 5x - 1 = 0 \quad \text{or} \quad 2x + 3 = 0$$
$$x = 0 \qquad\quad 5x = 1 \qquad\qquad 2x = -3$$
$$x = 0 \qquad\quad x = \tfrac{1}{5} \qquad\qquad x = -\tfrac{3}{2}$$

97. Answers may vary.

159

99.

$$3a^2 + 9a - 2a - 6 = 0$$
$$3a(a + 3) - 2(a + 3) = 0$$
$$(a + 3)(3a - 2) = 0$$

$a + 3 = 0 \quad$ **or** $\quad 3a - 2 = 0$
$\qquad a = -3 \qquad\qquad 3a = 2$
$\qquad\qquad\qquad\qquad a = \frac{2}{3}$

$$3a^2 + 9a - 2a - 6 = 0$$
$$3a^2 + 7a - 6 = 0$$
$$(a + 3)(3a - 2) = 0$$

$a + 3 = 0 \quad$ **or** $\quad 3a - 2 = 0$
$\qquad a = -3 \qquad\qquad 3a = 2$
$\qquad\qquad\qquad\qquad a = \frac{2}{3}$

Exercise 5.8 (page 356)

1. $A = lw$

3. $A = s^2$

5. $P = 2l + 2w$

7.
$$-2(5x + 2) = 3(2 - 3x)$$
$$-10x - 4 = 6 - 9x$$
$$-4 = 6 + x$$
$$-10 = x$$

9. Let w = the width and
$5w$ = the length.
Perimeter = 132
$$2(w) + 2(5w) = 132$$
$$2w + 10w = 132$$
$$12w = 132$$
$$w = 11$$
The area = $11(55) = 605$ ft^2.

11. analyze

13. Let x = the first integer. Then
$x + 4$ = the other integer.
$$x(x + 4) = 32$$
$$x^2 + 4x = 32$$
$$x^2 + 4x - 32 = 0$$
$$(x + 8)(x - 4) = 0$$
$x + 8 = 0 \quad$ **or** $\quad x - 4 = 0$
$\quad x = -8 \qquad\qquad x = 4$
The integers are -8 and -4, or 4 and 8.

15. Let x = the integer.
$$x^2 + 4 = 10x - 5$$
$$x^2 - 10x + 9 = 0$$
$$(x - 9)(x - 1) = 0$$
$x - 9 = 0 \quad$ **or** $\quad x - 1 = 0$
$\quad x = 9 \qquad\qquad x = 1$
The integer is 1 or 9.

17. Let $v = 144$ and $h = 0$:
$$h = vt - 16t^2$$
$$0 = 144t - 16t^2$$
$$0 = 16t(9 - t)$$
$16t = 0 \quad$ **or** $\quad 9 - t = 0$
$\quad t = 0 \qquad\qquad 9 = t$
Since $t = 0$ is when the object was first thrown, it will hit the ground in 9 seconds.

19. Let $v = 224$ and $h = 640$:
$$h = vt - 16t^2$$
$$640 = 224t - 16t^2$$
$$16t^2 - 224t + 640 = 0$$
$$16(t^2 - 14t + 40) = 0$$
$$16(t - 4)(t - 10) = 0$$
$t - 4 = 0 \quad$ **or** $\quad t - 10 = 0$
$\quad t = 4 \qquad\qquad t = 10$
The cannonball will be at a height of 640 feet after 4 seconds and after 10 seconds.

SECTION 5.8

21.
$$h = -16t^2 + 64$$
$$0 = -16t^2 + 64$$
$$16t^2 - 64 = 0$$
$$16(t^2 - 4) = 0$$
$$16(t + 2)(t - 2) = 0$$
$$t + 2 = 0 \quad \textbf{or} \quad t - 2 = 0$$
$$t = -2 \qquad\qquad t = 2$$

The value of $t = -2$ does not make sense, so the dive lasts 2 seconds.

23.
$$75^2 = h^2 + 72^2$$
$$5625 = h^2 + 5184$$
$$0 = h^2 - 441$$
$$0 = (h + 21)(h - 21)$$
$$h + 21 = 0 \quad \textbf{or} \quad h - 21 = 0$$
$$h = -21 \qquad\qquad h = 21$$

The value of $h = -21$ does not make sense, so the camper started at a height of 21 feet.

25. $\text{Area} = \text{Length} \cdot \text{Width}$
$$36 = (2w + 1)w$$
$$36 = 2w^2 + w$$
$$0 = 2w^2 + w - 36$$
$$0 = (2w + 9)(w - 4)$$
$$2w + 9 = 0 \quad \textbf{or} \quad w - 4 = 0$$
$$2w = -9 \qquad\qquad w = 4$$
$$w = -\tfrac{9}{2} \qquad\qquad w = 4$$

Since the answer $w = -\frac{9}{2}$ does not make sense, the dimensions are 4 m by 9 m.

27. Let $w = $ the width and $w + 2 = $ the length.
$\text{Area} = \text{Length} \cdot \text{Width}$
$$143 = (w + 2)w$$
$$143 = w^2 + 2w$$
$$0 = w^2 + 2w - 143$$
$$0 = (w + 13)(w - 11)$$
$$w + 13 = 0 \quad \textbf{or} \quad w - 11 = 0$$
$$w = -13 \qquad\qquad w = 11$$

Since the answer $w = -13$ does not make sense, the dimensions are 11 ft by 13 ft and the perimeter is 48 ft.

29. Let $b = $ the base and $5b - 2 = $ the height.
$$A = \frac{1}{2}bh$$
$$36 = \frac{1}{2}b(5b - 2)$$
$$72 = b(5b - 2)$$
$$72 = 5b^2 - 2b$$
$$0 = 5b^2 - 2b - 72$$
$$0 = (5b + 18)(b - 4)$$
$$5b + 18 = 0 \quad \textbf{or} \quad b - 4 = 0$$
$$5b = -18 \qquad\qquad b = 4$$
$$h = -\tfrac{18}{5} \qquad\qquad b = 4$$

Since the answer $b = -\frac{18}{5}$ does not make sense, the base is 4 in. and the height is 18 in.

31. Let $x = $ the base and the height.
$$\boxed{\text{Base}} + \boxed{\text{Height}} = \boxed{\text{Area}} - 6$$
$$x + x = \tfrac{1}{2}(x)(x) - 6$$
$$2x = \tfrac{1}{2}x^2 - 6$$
$$0 = \tfrac{1}{2}x^2 - 2x - 6$$
$$0 = x^2 - 4x - 12$$
$$0 = (x - 6)(x + 2)$$
$$x - 6 = 0 \quad \textbf{or} \quad x + 2 = 0$$
$$x = 6 \qquad\qquad x = -2$$

The answer $x = -2$ does not make sense, so the area is $\frac{1}{2}(6)(6) = 18$ square units.

SECTION 5.8

33.

$$\boxed{\text{Large area}} - \boxed{\text{Small area}} = \boxed{\text{Border area}}$$

$$(10 + 2w)(25 + 2w) - (10)(25) = 74$$
$$250 + 70w + 4w^2 - 250 = 74$$
$$4w^2 + 70w - 74 = 0$$
$$2(2w^2 + 35w - 37) = 0$$
$$2(2w + 37)(w - 1) = 0$$

$$2w + 37 = 0 \quad \textbf{or} \quad w - 1 = 0$$
$$2w = -37 \qquad\qquad w = 1$$
$$w = -\tfrac{37}{2} \qquad\qquad w = 1$$

Since the answer $w = -\tfrac{37}{2}$ does not make sense, the width should be 1 meter.

35. Let $w =$ the width and $2w + 1 =$ the height.

$$V = lwh$$
$$210 = 10w(2w + 1)$$
$$210 = 20w^2 + 10w$$
$$0 = 20w^2 + 10w - 210$$
$$0 = 10(2w^2 + w - 21)$$
$$0 = 10(2w + 7)(w - 3)$$

$$2w + 7 = 0 \quad \textbf{or} \quad w - 3 = 0$$
$$2w = -7 \qquad\qquad w = 3$$
$$w = -\tfrac{7}{2} \qquad\qquad w = 3$$

Since the answer $w = -\tfrac{7}{2}$ does not make sense, the width is 3 cm.

37. Let $x =$ one edge and $x - 3 =$ the other.

$$V = \frac{Bh}{3}$$
$$84 = \frac{x(x-3)(9)}{3}$$
$$252 = 9x^2 - 27x$$
$$0 = 9x^2 - 27x - 252$$
$$0 = 9(x + 4)(x - 7)$$

$$x + 4 = 0 \quad \textbf{or} \quad x - 7 = 0$$
$$x = -4 \qquad\qquad x = 7$$

Since the answer $x = -4$ does not make sense, the base is 7 cm by 4 cm.

39.

$$A = A_1 + A_2$$
$$= \pi r_1^2 + \pi r_2^2$$
$$= \pi\left(\frac{38}{2}\right)^2 + \pi\left(\frac{44}{2}\right)^2$$
$$= \pi(19)^2 + \pi(22)^2$$
$$= 361\pi + 484\pi = 845\pi \text{ m}^2$$

41.

$$a^2 + b^2 = c^2$$
$$x^2 + (x + 2)^2 = (x + 4)^2$$
$$x^2 + x^2 + 4x + 4 = x^2 + 8x + 16$$
$$2x^2 + 4x + 4 = x^2 + 8x + 16$$
$$x^2 - 4x - 12 = 0$$
$$(x - 6)(x + 2) = 0$$

$$(x - 6)(x + 2) = 0$$
$$x - 6 = 0 \quad \textbf{or} \quad x + 2 = 0$$
$$x = 6 \qquad\qquad x = -2$$

The value of $x = -2$ does not make sense, so $x = 6$. Then $x + 4 = 6 + 4 = 10$ and the height was 16 ft.

43. **Answers may vary.**

45. Let $w =$ the width and $w + 2 =$ the length.

$$\text{Area} = 18$$
$$w(w + 2) = 18$$
$$w^2 + 2w = 18$$
$$w^2 + 2w - 18 = 0$$

This is a prime trinomial, so it cannot be factored in order to solve.

Chapter 5 Review (page 361)

1. $24 = 8 \cdot 3 = 2^3 \cdot 3$

2. $45 = 9 \cdot 5 = 3^2 \cdot 5$

3. $96 = 12 \cdot 8 = 4 \cdot 3 \cdot 2^3 = 2^2 \cdot 3 \cdot 2^3 = 2^5 \cdot 3$

4. $102 = 2 \cdot 51 = 2 \cdot 3 \cdot 17$

5. $87 = 3 \cdot 29$

6. $99 = 9 \cdot 11 = 3^2 \cdot 11$

7. $2{,}050 = 50 \cdot 41 = 25 \cdot 2 \cdot 41 = 2 \cdot 5^2 \cdot 41$

8. $4{,}096 = 64 \cdot 64 = 2^6 \cdot 2^6 = 2^{12}$

9. $4x + 12y = 4(x + 3y)$

10. $5ax^2 + 15a = 5a(x^2 + 3)$

11. $7x^2 + 14x = 7x(x + 2)$

12. $9x^2 - 3x = 3x(3x - 1)$

13. $2x^3 + 4x^2 - 8x = 2x(x^2 + 2x - 4)$

14. $-ax - ay + az = -a(x + y - z)$

15. $ax + ay - a = a(x + y - 1)$

16. $x^2yz + xy^2z = xyz(x + y)$

17. $(a - b)x + (a - b)y = (a - b)(x + y)$

18. $(x + y)^2 + (x + y) = (x + y)(x + y) + 1(x + y) = (x + y)(x + y + 1)$

19. $2x^2(x + 2) + 6x(x + 2) = (x + 2)(2x^2 + 6x) = (x + 2)(2x)(x + 3) = 2x(x + 2)(x + 3)$

20. $5x(a + b)^2 - 10x(a + b) = 5x[(a + b)(a + b) - 2(a + b)] = 5x(a + b)(a + b - 2)$

21. $3p + 9q + ap + 3aq = 3(p + 3q) + a(p + 3q) = (p + 3q)(3 + a)$

22. $ar - 2as + 7r - 14s = a(r - 2s) + 7(r - 2s) = (r - 2s)(a + 7)$

23. $\begin{aligned}x^2 + ax + bx + ab &= x(x + a) + b(x + a) \\ &= (x + a)(x + b)\end{aligned}$

24. $\begin{aligned}xy + 2x - 2y - 4 &= x(y + 2) - 2(y + 2) \\ &= (y + 2)(x - 2)\end{aligned}$

25. $xa + yb + ya + xb = xa + ya + xb + yb = a(x + y) + b(x + y) = (x + y)(a + b)$

26. $x^3 - 4x^2 + 3x - 12 = x^2(x - 4) + 3(x - 4) = (x - 4)(x^2 + 3)$

27. $x^2 - 25 = x^2 - 5^2 = (x + 5)(x - 5)$

28. $x^2y^2 - 16 = (xy + 4)(xy - 4)$

29. $(x + 2)^2 - y^2 = (x + 2 + y)(x + 2 - y)$

30. $z^2 - (x + y)^2 = [z + (x + y)][z - (x + y)] = (z + x + y)(z - x - y)$

31. $2x^2y + 18y^3 = 2y(x^2 + 9y^2)$

32. $(x + y)^2 - z^2 = [(x + y) + z][(x + y) - z] = (x + y + z)(x + y - z)$

33. $x^2 + 7x + 10 = (x + 2)(x + 5)$

34. $x^2 - 8x + 15 = (x - 5)(x - 3)$

35. $x^2 + 2x - 24 = (x + 6)(x - 4)$

36. $x^2 - 4x - 12 = (x - 6)(x + 2)$

37. $2x^2 - 5x - 3 = (2x + 1)(x - 3)$

38. $3x^2 - 14x - 5 = (3x + 1)(x - 5)$

39. $15x^2 + x - 2 = (5x + 2)(3x - 1)$

40. $6x^2 + 3x - 3 = 3(2x^2 + x - 1)$
$= 3(2x - 1)(x + 1)$

41. $6x^3 + 17x^2 - 3x = x(6x^2 + 17x - 3)$
$= x(x + 3)(6x - 1)$

42. $4x^3 - 5x^2 - 6x = x(4x^2 - 5x - 6)$
$= x(x - 2)(4x + 3)$

43. $12x - 4x^3 - 2x^2 = -4x^3 - 2x^2 + 12x = -2x(2x^2 + x - 6) = -2x(2x - 3)(x + 2)$

44. $-4a^3 + 4a^2 b + 24ab^2 = -4a(a^2 - ab - 6b^2) = -4a(a - 3b)(a + 2b)$

45. $c^3 - 125 = c^3 - 5^3 = (c - 5)(c^2 + 5c + 25)$

46. $d^3 + 8 = d^3 + 2^3 = (d + 2)(d^2 - 2d + 4)$

47. $2x^3 + 54 = 2(x^3 + 27)$
$= 2(x^3 + 3^3)$
$= 2(x + 3)(x^2 - 3x + 9)$

48. $2ab^4 - 2ab = 2ab(b^3 - 1)$
$= 2ab(b^3 - 1^3)$
$= 2ab(b - 1)(b^2 + b + 1)$

49. $3x^2 y - xy^2 - 6xy + 2y^2 = y(3x^2 - xy - 6x + 2y) = y[x(3x - y) - 2(3x - y)]$
$= y(3x - y)(x - 2)$

50. $5x^2 - 5x - 30 = 5(x^2 - x - 6) = 5(x - 3)(x + 2)$

51. $2a^2 x + 2abx + a^3 + a^2 b = a(2ax + 2bx + a^2 + ab) = a[2x(a + b) + a(a + b)]$
$= a(a + b)(2x + a)$

52. $2x^2 - 8x - 11 \Rightarrow$ prime

53. $x^2 - 9 + ax + 3a = x^2 - 3^2 + ax + 3a = (x + 3)(x - 3) + a(x + 3) = (x + 3)(x - 3 + a)$

54. $10x^3 - 80y^3 = 10(x^3 - 8y^3) = 10[x^3 - (2y)^3] = 10(x - 2y)(x^2 + 2xy + 4y^2)$

55. $x^2 + 5x = 0$
$x(x + 5) = 0$
$x = 0 \quad \textbf{or} \quad x + 5 = 0$
$x = 0 \qquad\qquad x = -5$

56. $2x^2 - 6x = 0$
$2x(x - 3) = 0$
$2x = 0 \quad \textbf{or} \quad x - 3 = 0$
$x = 0 \qquad\qquad x = 3$

57.
$$3x^2 = 2x$$
$$3x^2 - 2x = 0$$
$$x(3x - 2) = 0$$
$$x = 0 \quad \textbf{or} \quad 3x - 2 = 0$$
$$3x = 2$$
$$x = \tfrac{2}{3}$$

58.
$$5x^2 + 25x = 0$$
$$5x(x + 5) = 0$$
$$5x = 0 \quad \textbf{or} \quad x + 5 = 0$$
$$x = \tfrac{0}{5} \qquad\qquad x = -5$$
$$x = 0$$

59.
$$y^2 - 49 = 0$$
$$(y + 7)(y - 7) = 0$$
$$y + 7 = 0 \quad \textbf{or} \quad y - 7 = 0$$
$$y = -7 \qquad\qquad y = 7$$

60.
$$x^2 - 25 = 0$$
$$(x + 5)(x - 5) = 0$$
$$x + 5 = 0 \quad \textbf{or} \quad x - 5 = 0$$
$$x = -5 \qquad\qquad x = 5$$

61.
$$a^2 - 9a + 20 = 0$$
$$(a - 5)(a - 4) = 0$$
$$a - 5 = 0 \quad \textbf{or} \quad a - 4 = 0$$
$$a = 5 \qquad\qquad a = 4$$

62.
$$(x - 1)(x + 4) = 6$$
$$x^2 + 3x - 4 = 6$$
$$x^2 + 3x - 10 = 0$$
$$(x + 5)(x - 2) = 0$$
$$x + 5 = 0 \quad \textbf{or} \quad x - 2 = 0$$
$$x = -5 \qquad\qquad x = 2$$

63.
$$2x - x^2 + 24 = 0$$
$$x^2 - 2x - 24 = 0$$
$$(x + 4)(x - 6) = 0$$
$$x + 4 = 0 \quad \textbf{or} \quad x - 6 = 0$$
$$x = -4 \qquad\qquad x = 6$$

64.
$$16 + x^2 - 10x = 0$$
$$x^2 - 10x + 16 = 0$$
$$(x - 2)(x - 8) = 0$$
$$x - 2 = 0 \quad \textbf{or} \quad x - 8 = 0$$
$$x = 2 \qquad\qquad x = 8$$

65.
$$2x^2 - 5x - 3 = 0$$
$$(2x + 1)(x - 3) = 0$$
$$2x + 1 = 0 \quad \textbf{or} \quad x - 3 = 0$$
$$2x = -1 \qquad\qquad x = 3$$
$$x = -\tfrac{1}{2} \qquad\qquad x = 3$$

66.
$$2x^2 + x - 3 = 0$$
$$(2x + 3)(x - 1) = 0$$
$$2x + 3 = 0 \quad \textbf{or} \quad x - 1 = 0$$
$$2x = -3 \qquad\qquad x = 1$$
$$x = -\tfrac{3}{2} \qquad\qquad x = 1$$

67.
$$16x^2 = 9$$
$$16x^2 - 9 = 0$$
$$(4x + 3)(4x - 3) = 0$$
$$4x + 3 = 0 \quad \textbf{or} \quad 4x - 3 = 0$$
$$4x = -3 \qquad\qquad 4x = 3$$
$$x = -\tfrac{3}{4} \qquad\qquad x = \tfrac{3}{4}$$

68.
$$9x^2 = 4$$
$$9x^2 - 4 = 0$$
$$(3x + 2)(3x - 2) = 0$$
$$3x + 2 = 0 \quad \textbf{or} \quad 3x - 2 = 0$$
$$3x = -2 \qquad\qquad 3x = 2$$
$$x = -\tfrac{2}{3} \qquad\qquad x = \tfrac{2}{3}$$

69.
$$x^3 - 7x^2 + 12x = 0$$
$$x(x^2 - 7x + 12) = 0$$
$$x(x-3)(x-4) = 0$$
$$x = 0 \ \text{ or } \ x - 3 = 0 \ \text{ or } \ x - 4 = 0$$
$$x = 0 \qquad\qquad x = 3 \qquad\qquad x = 4$$

70.
$$x^3 + 5x^2 + 6x = 0$$
$$x(x^2 + 5x + 6) = 0$$
$$x(x+2)(x+3) = 0$$
$$x = 0 \ \text{ or } \ x + 2 = 0 \ \text{ or } \ x + 3 = 0$$
$$x = 0 \qquad\qquad x = -2 \qquad\qquad x = -3$$

71.
$$2x^3 + 5x^2 = 3x$$
$$2x^3 + 5x^2 - 3x = 0$$
$$x(2x^2 + 5x - 3) = 0$$
$$x(2x-1)(x+3) = 0$$
$$x = 0 \ \text{ or } \ 2x - 1 = 0 \ \text{ or } \ x + 3 = 0$$
$$x = 0 \qquad\quad 2x = 1 \qquad\quad x = -3$$
$$x = 0 \qquad\quad x = \tfrac{1}{2} \qquad\quad x = -3$$

72.
$$3x^3 - 2x = x^2$$
$$3x^3 - x^2 - 2x = 0$$
$$x(3x^2 - x - 2) = 0$$
$$x(3x+2)(x-1) = 0$$
$$x = 0 \ \text{ or } \ 3x + 2 = 0 \ \text{ or } \ x - 1 = 0$$
$$x = 0 \qquad\quad 3x = -2 \qquad\quad x = 1$$
$$x = 0 \qquad\quad x = -\tfrac{2}{3} \qquad\quad x = 1$$

73. Let $x =$ one number. Then
$12 - x =$ the other number.
$$x(12 - x) = 35$$
$$12x - x^2 = 35$$
$$0 = x^2 - 12x + 35$$
$$0 = (x-5)(x-7)$$
$$x - 5 = 0 \ \text{ or } \ x - 7 = 0$$
$$x = 5 \qquad\qquad x = 7$$
The numbers are 5 and 7.

74. Let $x =$ the positive number.
$$3x^2 + 5x = 2$$
$$3x^2 + 5x - 2 = 0$$
$$(3x - 1)(x + 2) = 0$$
$$3x - 1 = 0 \ \text{ or } \ x + 2 = 0$$
$$3x = 1 \qquad\qquad x = -2$$
$$x = \tfrac{1}{3} \qquad\qquad x = -2$$
The answer of -2 is not positive,
so the number is $\tfrac{1}{3}$.

75. Let $w =$ the width and $w + 2 =$ the length.
Area = Length $\cdot$ Width
$$48 = (w + 2)w$$
$$0 = w^2 + 2w - 48$$
$$0 = (w + 8)(w - 6)$$
$$w + 8 = 0 \quad \text{ or } \quad w - 6 = 0$$
$$w = -8 \qquad\qquad w = 6$$
Since the answer $w = -8$ does not make
sense, the dimensions are 6 ft by 8 ft.

76. Let $w =$ the width and $2w + 3 =$ the length.
Area = Length $\cdot$ Width
$$27 = (2w + 3)w$$
$$0 = 2w^2 + 3w - 27$$
$$0 = (2w + 9)(w - 3)$$
$$2w + 9 = 0 \quad \text{ or } \quad w - 3 = 0$$
$$2w = -9 \qquad\qquad w = 3$$
$$w = -\tfrac{9}{2} \qquad\qquad w = 3$$
Since the answer $w = -\tfrac{9}{2}$ does not make
sense, the dimensions are 3 ft by 9 ft.

77. Let w = the width and $w + 3$ = the length.

$$\text{Area} = \text{Perimeter}$$
$$(w + 3)w = 2w + 2(w + 3)$$
$$w^2 + 3w = 2w + 2w + 6$$
$$w^2 + 3w = 4w + 6$$
$$w^2 - w - 6 = 0$$
$$(w + 2)(w - 3) = 0$$
$$w + 2 = 0 \quad \textbf{or} \quad w - 3 = 0$$
$$w = -2 \qquad\qquad w = 3$$

Since the answer $w = -2$ does not make sense, the dimensions are 3 ft by 6 ft.

78. Let x = the base and $x + 1$ = the height.

$$\text{Area} = \tfrac{1}{2}bh$$
$$36 = \tfrac{1}{2}x(x + 1)$$
$$72 = x(x + 1)$$
$$72 = x^2 + x$$
$$0 = x^2 + x - 72$$
$$0 = (x + 9)(x - 8)$$
$$x + 9 = 0 \quad \textbf{or} \quad x - 8 = 0$$
$$x = -9 \qquad\qquad x = 8$$

Since the answer $x = -9$ does not make sense, the base is 8 ft and the height is 9 ft.

Chapter 5 Test (page 366)

1. $120 = 8 \cdot 15 = 2^3 \cdot 3 \cdot 5$

2. $108 = 12 \cdot 9 = 4 \cdot 3 \cdot 3^2 = 2^2 \cdot 3^3$

3. $60ab^2c^3 + 30a^3b^2c - 25a = 5a(12b^2c^3 + 6a^2b^2c - 5)$

4. $3x^2(a + b) - 6xy(a + b) = 3x[x(a + b) - 2y(a + b)] = 3x(a + b)(x - 2y)$

5. $ax + ay + bx + by = a(x + y) + b(x + y)$
$$= (x + y)(a + b)$$

6. $x^2 - 64 = x^2 - 8^2 = (x + 8)(x - 8)$

7. $2a^2 - 32b^2 = 2(a^2 - 16b^2) = 2\left(a^2 - (4b)^2\right) = 2(a + 4b)(a - 4b)$

8. $16x^4 - 81y^4 = \left(4x^2\right)^2 - \left(9y^2\right)^2 = \left(4x^2 + 9y^2\right)\left(4x^2 - 9y^2\right)$
$$= \left(4x^2 + 9y^2\right)\left((2x)^2 - (3y)^2\right)$$
$$= \left(4x^2 + 9y^2\right)(2x + 3y)(2x - 3y)$$

9. $x^2 + 5x - 6 = (x + 6)(x - 1)$

10. $x^2 - 9x - 22 = (x - 11)(x + 2)$

11. $-x^2 - 10xy - 9y^2 = -\left(x^2 + 10xy + 9y^2\right)$
$$= -(x + y)(x + 9y)$$

12. $6x^2 - 30xy + 24y^2 = 6\left(x^2 - 5xy + 4y^2\right)$
$$= 6(x - 4y)(x - y)$$

13. $3x^2 + 13x + 4 = (3x + 1)(x + 4)$

14. $2a^2 + 5a - 12 = (2a - 3)(a + 4)$

15. $2x^2 + 3x - 1 \Rightarrow$ prime

16. $12 - 25x + 12x^2 = 12x^2 - 25x + 12$
$$= (4x - 3)(3x - 4)$$

17. $12a^2 + 6ab - 36b^2 = 6\left(2a^2 + ab - 6b^2\right)$
$$= 6(2a - 3b)(a + 2b)$$

18. $x^3 - 8y^3 = x^3 - (2y)^3$
$$= (x - 2y)\left(x^2 + 2xy + 4y^2\right)$$

19. $216 + 8a^3 = 8(27 + a^3) = 8(3^3 + a^3) = 8(3 + a)(9 - 3a + a^2)$

20. $x^9z^3 - y^3z^6 = z^3(x^9 - y^3z^3) = z^3\left((x^3)^3 - (yz)^3\right) = z^3(x^3 - yz)\left((x^3)^2 + x^3yz + (yz)^2\right)$
$$= z^3(x^3 - yz)(x^6 + x^3yz + y^2z^2)$$

21.
$$x^2 = -10x$$
$$x^2 + 10x = 0$$
$$x(x + 10) = 0$$
$$x = 0 \quad \textbf{or} \quad x + 10 = 0$$
$$x = 0 \qquad\qquad x = -10$$

22.
$$2x^2 + 5x + 3 = 0$$
$$(2x + 3)(x + 1) = 0$$
$$2x + 3 = 0 \quad \textbf{or} \quad x + 1 = 0$$
$$2x = -3 \qquad\qquad x = -1$$
$$x = -\tfrac{3}{2} \qquad\qquad x = -1$$

23.
$$16y^2 - 64 = 0$$
$$16(y^2 - 4) = 0$$
$$16(y + 2)(y - 2) = 0$$
$$y + 2 = 0 \quad \textbf{or} \quad y - 2 = 0$$
$$y = -2 \qquad\qquad y = 2$$

24.
$$-3(y - 6) + 2 = y^2 + 2$$
$$-3y + 18 + 2 = y^2 + 2$$
$$-3y + 20 = y^2 + 2$$
$$0 = y^2 + 3y - 18$$
$$0 = (y + 6)(y - 3)$$
$$y + 6 = 0 \quad \textbf{or} \quad y - 3 = 0$$
$$y = -6 \qquad\qquad y = 3$$

25.
$$10x^2 - 13x = 9$$
$$10x^2 - 13x - 9 = 0$$
$$(2x + 1)(5x - 9) = 0$$
$$2x + 1 = 0 \quad \textbf{or} \quad 5x - 9 = 0$$
$$2x = -1 \qquad\qquad 5x = 9$$
$$x = -\tfrac{1}{2} \qquad\qquad x = \tfrac{9}{5}$$

26.
$$10x^2 - x = 9$$
$$10x^2 - x - 9 = 0$$
$$(10x + 9)(x - 1) = 0$$
$$10x + 9 = 0 \quad \textbf{or} \quad x - 1 = 0$$
$$10x = -9 \qquad\qquad x = 1$$
$$x = -\tfrac{9}{10} \qquad\qquad x = 1$$

27.
$$10x^2 + 43x = 9$$
$$10x^2 + 43x - 9 = 0$$
$$(2x + 9)(5x - 1) = 0$$
$$2x + 9 = 0 \quad \textbf{or} \quad 5x - 1 = 0$$
$$2x = -9 \qquad\qquad 5x = 1$$
$$x = -\tfrac{9}{2} \qquad\qquad x = \tfrac{1}{5}$$

28.
$$10x^2 - 89x = 9$$
$$10x^2 - 89x - 9 = 0$$
$$(10x + 1)(x - 9) = 0$$
$$10x + 1 = 0 \quad \textbf{or} \quad x - 9 = 0$$
$$10x = -1 \qquad\qquad x = 9$$
$$x = -\tfrac{1}{10} \qquad\qquad x = 9$$

29. Let $v = 192$ and $h = 0$:
$$h = vt - 16t^2$$
$$0 = 192t - 16t^2$$
$$0 = 16t(12 - t)$$
$$16t = 0 \quad \textbf{or} \quad 12 - t = 0$$
$$t = 0 \qquad\qquad 12 = t$$
Since $t = 0$ is when the cannonball was fired, it will hit the ground in 12 seconds.

30. Let $h =$ the height and $h + 2 =$ the base.
$$A = \tfrac{1}{2}bh$$
$$40 = \tfrac{1}{2}(h + 2)h$$
$$80 = (h + 2)h$$
$$0 = h^2 + 2h - 80$$
$$0 = (h + 10)(h - 8)$$
$$h + 10 = 0 \quad \textbf{or} \quad h - 8 = 0$$
$$h = -10 \qquad\qquad h = 8$$
Since the answer $h = -10$ does not make sense, the height is 8 m and the base is 10 m.

Exercises 6.1 (page 376)

1. $\dfrac{14}{21} = \dfrac{2 \cdot 7}{3 \cdot 7} = \dfrac{2}{3}$

3. $\dfrac{12}{16} = \dfrac{3 \cdot 4}{4 \cdot 4} = \dfrac{3}{4}$

5. $\dfrac{15}{35} = \dfrac{3 \cdot 5}{7 \cdot 5} = \dfrac{3}{7}$

7. $\dfrac{-18}{54} = -\dfrac{1 \cdot 18}{3 \cdot 18} = -\dfrac{1}{3}$

9. If a, b and c are real numbers, then $(a + b) + c = a + (b + c)$.

11. 0 is the additive identity.

13. $\frac{7}{5}$ is the additive inverse of $-\frac{7}{5}$.

15. numerator

17. 0

19. negatives (or opposites)

21. $\dfrac{a}{b}$

23. factor; common

25. $y + 4 = 0$
$y = -4$

27. $x^2 - x - 2 = 0$
$(x + 1)(x - 2) = 0$
$x + 1 = 0$ **or** $x - 2 = 0$
$x = -1 \qquad x = 2$

29. $(x + 7)(2x - 1) = 0$
$x + 7 = 0$ **or** $2x - 1 = 0$
$x = -7 \qquad 2x = 1$
$\qquad\qquad x = \frac{1}{2}$

31. $3x^2 + x = 0$
$x(3x + 1) = 0$
$x = 0$ **or** $3x + 1 = 0$
$x = 0 \qquad 3x = -1$
$x = 0 \qquad x = -\frac{1}{3}$

33. $5x - 2 = 0$
$5x = 2$
$x = \frac{2}{5}$ D: $\left\{x | x \in \mathbb{R}, x \neq \frac{2}{5}\right\}$
or D: $\left(-\infty, \frac{2}{5}\right) \cup \left(\frac{2}{5}, \infty\right)$

35. $2m^2 - m - 3 = 0$
$(2m - 3)(m + 1) = 0$
$2m - 3 = 0$ **or** $m + 1 = 0$
$2m = 3 \qquad m = -1$
$m = \frac{3}{2} \qquad m = -1$
D: $\left\{m | m \in \mathbb{R}, m \neq \frac{3}{2}, -1\right\}$
or D: $\left(-\infty, -1\right) \cup \left(-1, \frac{3}{2}\right) \cup \left(\frac{3}{2}, \infty\right)$

37. $\dfrac{4x}{2} = \dfrac{2x \cdot 2}{1 \cdot 2} = 2x$

39. $\dfrac{-25y^2}{5y} = \dfrac{-5y \cdot 5y}{1 \cdot 5y} = -5y$

41. $\dfrac{6x^2 y}{6xy^2} = \dfrac{x \cdot 6xy}{y \cdot 6xy} = \dfrac{x}{y}$

43. $\dfrac{2x^2}{3y} \Rightarrow$ in simplest form

45. $\dfrac{x^2 + 7x}{2x + 14} = \dfrac{x(x + 7)}{2(x + 7)} = \dfrac{x}{2}$

47. $\dfrac{3x + 15}{x^2 - 25} = \dfrac{3(x + 5)}{(x + 5)(x - 5)} = \dfrac{3}{x - 5}$

49. $\dfrac{3x + 6}{2x + 1} = \dfrac{3(x + 2)}{2x + 1}$: in simplest form

51. $\dfrac{10x - 5}{18x - 9} = \dfrac{5(2x - 1)}{9(2x - 1)} = \dfrac{5}{9}$

53. $\dfrac{x+3}{3(x+3)} = \dfrac{1(x+3)}{3(x+3)} = \dfrac{1}{3}$

55. $\dfrac{5x+35}{x+7} = \dfrac{5(x+7)}{1(x+7)} = 5$

57. $\dfrac{5x^2+10x}{x-2} = \dfrac{5x(x+2)}{x-2}$: in simplest form

59. $\dfrac{3y^2+12y}{4+y} = \dfrac{3y(y+4)}{y+4} = 3y$

61. $\dfrac{x^2+3x+2}{x^2+x-2} = \dfrac{(x+2)(x+1)}{(x+2)(x-1)} = \dfrac{x+1}{x-1}$

63. $\dfrac{x^2-8x+15}{x^2-x-6} = \dfrac{(x-5)(x-3)}{(x+2)(x-3)} = \dfrac{x-5}{x+2}$

65. $\dfrac{2x^2-8x}{x^2-6x+8} = \dfrac{2x(x-4)}{(x-2)(x-4)} = \dfrac{2x}{x-2}$

67. $\dfrac{2a^3-16}{2a^2+4a+8} = \dfrac{2(a^3-8)}{2(a^2+2a+4)} = \dfrac{a^3-2^3}{a^2+2a+4} = \dfrac{(a-2)(a^2+2a+2^2)}{a^2+2a+4}$

$\qquad\qquad = \dfrac{(a-2)(a^2+2a+4)}{a^2+2a+4} = a-2$

69. $\dfrac{x^2-4x+4}{x^2-4} = \dfrac{(x-2)(x-2)}{(x+2)(x-2)} = \dfrac{x-2}{x+2}$

71. $\dfrac{30x^2-14x-8}{3x^2+4x+1} = \dfrac{2(15x^2-7x-4)}{3x^2+4x+1} = \dfrac{2(5x-4)(3x+1)}{(3x+1)(x+1)} = \dfrac{2(5x-4)}{x+1}$

73. $\dfrac{4(x+3)+4}{3(x+2)+6} = \dfrac{4x+12+4}{3x+6+6} = \dfrac{4x+16}{3x+12} = \dfrac{4(x+4)}{3(x+4)} = \dfrac{4}{3}$

75. $\dfrac{x^2+5x+4}{2(x+3)-(x+2)} = \dfrac{x^2+5x+4}{2x+6-x-2} = \dfrac{x^2+5x+4}{x+4} = \dfrac{(x+1)(x+4)}{x+4} = x+1$

77. $\dfrac{x^3+1}{ax+a+x+1} = \dfrac{x^3+1^3}{a(x+1)+1(x+1)} = \dfrac{(x+1)(x^2-1x+1^2)}{(x+1)(a+1)} = \dfrac{x^2-x+1}{a+1}$

79. $\dfrac{ab+b+2a+2}{ab+a+b+1} = \dfrac{b(a+1)+2(a+1)}{a(b+1)+1(b+1)} = \dfrac{(a+1)(b+2)}{(b+1)(a+1)} = \dfrac{b+2}{b+1}$

81. $\dfrac{x-y}{y-x} = \dfrac{-(y-x)}{y-x} = -1$

83. $\dfrac{6x-3y}{3y-6x} = \dfrac{-(3y-6x)}{3y-6x} = -1$

85. $\dfrac{45}{9a} = \dfrac{9\cdot5}{9\cdot a} = \dfrac{5}{a}$

87. $\dfrac{15x^2y}{5xy^2} = \dfrac{5xy\cdot3x}{5xy\cdot y} = \dfrac{3x}{y}$

89. $\dfrac{x^2+3x+2}{x^3+x^2} = \dfrac{(x+2)(x+1)}{x^2(x+1)} = \dfrac{x+2}{x^2}$

91. $\dfrac{3x+3y}{x^2+xy} = \dfrac{3(x+y)}{x(x+y)} = \dfrac{3}{x}$

93. $\dfrac{3y+xy}{3x+xy} = \dfrac{y(3+x)}{x(3+y)}$: in simplest form

95. $\dfrac{xz-2x}{yz-2y} = \dfrac{x(z-2)}{y(z-2)} = \dfrac{x}{y}$

97. $\dfrac{15x - 3x^2}{25y - 5xy} = \dfrac{3x(5 - x)}{5y(5 - x)} = \dfrac{3x}{5y}$

99. $\dfrac{4 + 2(x - 5)}{3x - 5(x - 2)} = \dfrac{4 + 2x - 10}{3x - 5x + 10} = \dfrac{2x - 6}{-2x + 10} = \dfrac{2(x - 3)}{-2(x - 5)} = \dfrac{x - 3}{5 - x}$ or $-\dfrac{x - 3}{x - 5}$

101. $\dfrac{x^2 + 4x - 77}{x^2 - 4x - 21} = \dfrac{(x + 11)(x - 7)}{(x + 3)(x - 7)} = \dfrac{x + 11}{x + 3}$

103. $\dfrac{xy + 3y + 3x + 9}{x^2 - 9} = \dfrac{y(x + 3) + 3(x + 3)}{(x + 3)(x - 3)} = \dfrac{(x + 3)(y + 3)}{(x + 3)(x - 3)} = \dfrac{y + 3}{x - 3}$

105. $\dfrac{2x^2 - 8}{x^2 - 3x + 2} = \dfrac{2(x^2 - 4)}{(x - 1)(x - 2)} = \dfrac{2(x + 2)(x - 2)}{(x - 1)(x - 2)} = \dfrac{2(x + 2)}{x - 1}$

107. $\dfrac{a + b - c}{c - a - b} = \dfrac{-(c - a - b)}{c - a - b} = -1$ **109.** $\dfrac{6a - 6b + 6c}{9a - 9b + 9c} = \dfrac{6(a - b + c)}{9(a - b + c)} = \dfrac{6}{9} = \dfrac{2}{3}$

111. $\dfrac{x - 7}{7 - x} = \dfrac{x - 7}{-(-7 + x)} = -\dfrac{x - 7}{x - 7} = -1$

113. $\dfrac{y^2 + 5(2y + 5)}{25 - y^2} = \dfrac{y^2 + 10y + 25}{25 - y^2} = \dfrac{(y + 5)(y + 5)}{(5 + y)(5 - y)} = \dfrac{y + 5}{5 - y}$

$\dfrac{y^2 + 5(2y + 5)}{25 - y^2} = \dfrac{y^2 + 10y + 25}{25 - y^2} = -\dfrac{y^2 + 10y + 25}{y^2 - 25} = -\dfrac{(y + 5)(y + 5)}{(y + 5)(y - 5)} = -\dfrac{y + 5}{y - 5}$

Exercises 6.2 (page 384)

1. $\dfrac{5}{2} \cdot \dfrac{3}{\boxed{5}} = \dfrac{3}{2}$ **3.** $\dfrac{x}{2} \cdot \dfrac{3}{\boxed{x}} = \dfrac{x \cdot 3}{2 \cdot x} = \dfrac{3}{2}$

5. $\dfrac{9}{\boxed{4}} \div \dfrac{3}{2} = \dfrac{3}{2}$ **7.** $\dfrac{\boxed{x^2}}{2} \div \dfrac{x}{3} = \dfrac{3x}{2}$

9. $2x^3 y^2 (-3x^2 y^4 z) = -6x^5 y^6 z$ **11.** $(5y)^{-3} = \dfrac{1}{(5y)^3} = \dfrac{1}{125y^3}$

13. $\dfrac{x^{3m}}{x^{4m}} = x^{-m} = \dfrac{1}{x^m}$

15. $-4\left(y^3 - 4y^2 + 3y - 2\right) + 6\left(-2y^2 + 4\right) - 4\left(-2y^3 - y\right)$
$= -4y^3 + 16y^2 - 12y + 8 - 12y^2 + 24 + 8y^3 + 4y$
$= 4y^3 + 4y^2 - 8y + 32$

SECTION 6.2

17. numerators; denominators

19. $\dfrac{ac}{bd}$

21. $\dfrac{d}{c}$

23. $\dfrac{5}{7} \cdot \dfrac{9}{13} = \dfrac{5 \cdot 9}{7 \cdot 13} = \dfrac{45}{91}$

25. $\dfrac{5x}{y} \cdot \dfrac{4x}{3y^2} = \dfrac{5x \cdot 4x}{y \cdot 3y^2} = \dfrac{20x^2}{3y^3}$

27. $\dfrac{z+7}{7} \cdot \dfrac{z+2}{z} = \dfrac{(z+7)(z+2)}{7z}$

29. $\dfrac{-3a}{a+2} \cdot \dfrac{a-1}{5} = \dfrac{-3a(a-1)}{5(a+2)}$

31. $\dfrac{2y}{z} \cdot \dfrac{z}{3} = \dfrac{2 \cdot y \cdot z}{z \cdot 3} = \dfrac{2y}{3}$

33. $\dfrac{5y}{7} \cdot \dfrac{7x}{5z} = \dfrac{5 \cdot y \cdot 7 \cdot x}{7 \cdot 5 \cdot z} = \dfrac{yx}{z}$

35. $\dfrac{x}{2x^2} \cdot \dfrac{-28xy}{7} = -\dfrac{28x^2y}{14x^2} = -2y$

37. $\dfrac{ab^2}{a^2b} \cdot \dfrac{b^2c^2}{abc} \cdot \dfrac{abc^2}{a^3c^2} = \dfrac{a^2b^5c^4}{a^6b^2c^3} = \dfrac{b^3c}{a^4}$

39. $\dfrac{x-2}{2} \cdot \dfrac{2x}{x-2} = \dfrac{2x(x-2)}{2(x-2)} = x$

41. $\dfrac{5y-5}{y-1} \cdot \dfrac{y}{10y^2} = \dfrac{5(y-1)}{y-1} \cdot \dfrac{y}{10y^2} = \dfrac{5y(y-1)}{10y^2(y-1)} = \dfrac{1}{2y}$

43. $\dfrac{3y-12}{y+8} \cdot \dfrac{y^2+8y}{y-4} = \dfrac{3(y-4)}{y+8} \cdot \dfrac{y(y+8)}{y-4} = \dfrac{3y(y-4)(y+8)}{(y+8)(y-4)} = 3y$

45. $\dfrac{5z-10}{z+2} \cdot \dfrac{3}{3z-6} = \dfrac{5(z-2)}{z+2} \cdot \dfrac{3}{3(z-2)} = \dfrac{15(z-2)}{3(z+2)(z-2)} = \dfrac{5}{z+2}$

47. $\dfrac{z^2+4z-5}{5z-5} \cdot \dfrac{5z}{z+5} = \dfrac{(z+5)(z-1)}{5(z-1)} \cdot \dfrac{5z}{z+5} = \dfrac{5z(z+5)(z-1)}{5(z-1)(z+5)} = z$

49. $\dfrac{(x+1)^2}{x+1} \cdot \dfrac{x+2}{x+1} = \dfrac{(x+2)(x+1)^2}{(x+1)^2} = x+2$

51. $\dfrac{m^2-2m-3}{2m+4} \cdot \dfrac{m^2-4}{m^2+3m+2} = \dfrac{(m-3)(m+1)}{2(m+2)} \cdot \dfrac{(m+2)(m-2)}{(m+2)(m+1)}$

$= \dfrac{(m-3)(m+1)(m+2)(m-2)}{2(m+2)(m+2)(m+1)} = \dfrac{(m-3)(m-2)}{2(m+2)}$

53. $\dfrac{abc^2}{a+1} \cdot \dfrac{c}{a^2b^2} \cdot \dfrac{a^2+a}{ac} = \dfrac{abc^2}{a+1} \cdot \dfrac{c}{a^2b^2} \cdot \dfrac{a(a+1)}{ac} = \dfrac{a^2bc^3(a+1)}{a^3b^2c(a+1)} = \dfrac{c^2}{ab}$

55. $\dfrac{x-5}{2x-8} \cdot (x-4) = \dfrac{x-5}{2(x-4)} \cdot \dfrac{x-4}{1} = \dfrac{(x-5)(x-4)}{2(x-4)} = \dfrac{x-5}{2}$

57. $(5x-10) \cdot \dfrac{x^2+2x}{x^2-4} = \dfrac{5(x-2)}{1} \cdot \dfrac{x(x+2)}{(x+2)(x-2)} = 5x$

172

59. $\dfrac{2}{3} \div \dfrac{1}{2} = \dfrac{2}{3} \cdot \dfrac{2}{1} = \dfrac{4}{3}$

61. $\dfrac{21}{14} \div \dfrac{5}{2} = \dfrac{21}{14} \cdot \dfrac{2}{5} = \dfrac{3 \cdot 7 \cdot 2}{2 \cdot 7 \cdot 5} = \dfrac{3}{5}$

63. $\dfrac{3x}{2} \div \dfrac{x}{2} = \dfrac{3x}{2} \cdot \dfrac{2}{x} = \dfrac{6x}{2x} = 3$

65. $\dfrac{x^2 y}{3xy} \div \dfrac{xy^2}{6y} = \dfrac{x^2 y}{3xy} \cdot \dfrac{6y}{xy^2} = \dfrac{6x^2 y^2}{3x^2 y^3} = \dfrac{2}{y}$

67. $\dfrac{x+2}{3x} \div \dfrac{x+2}{2} = \dfrac{x+2}{3x} \cdot \dfrac{2}{x+2} = \dfrac{2(x+2)}{3x(x+2)} = \dfrac{2}{3x}$

69. $\dfrac{x^2-4}{3x+6} \div \dfrac{x-2}{x+2} = \dfrac{x^2-4}{3x+6} \cdot \dfrac{x+2}{x-2} = \dfrac{(x+2)(x-2)(x+2)}{3(x+2)(x-2)} = \dfrac{x+2}{3}$

71. $\dfrac{y(y+2)}{y^2(y-3)} \div \dfrac{y^2(y+2)}{(y-3)^2} = \dfrac{y(y+2)}{y^2(y-3)} \cdot \dfrac{(y-3)^2}{y^2(y+2)} = \dfrac{y(y+2)(y-3)^2}{y^4(y-3)(y+2)} = \dfrac{y-3}{y^3}$

73. $\dfrac{x^2-x-6}{2x^2+9x+10} \div \dfrac{x^2-25}{2x^2+15x+25} = \dfrac{x^2-x-6}{2x^2+9x+10} \cdot \dfrac{2x^2+15x+25}{x^2-25}$

$\qquad\qquad = \dfrac{(x-3)(x+2)(2x+5)(x+5)}{(2x+5)(x+2)(x+5)(x-5)} = \dfrac{x-3}{x-5}$

75. $\dfrac{5x^2+13x-6}{x+3} \div \dfrac{5x^2-17x+6}{x-2} = \dfrac{5x^2+13x-6}{x+3} \cdot \dfrac{x-2}{5x^2-17x+6}$

$\qquad\qquad = \dfrac{(5x-2)(x+3)(x-2)}{(x+3)(5x-2)(x-3)} = \dfrac{x-2}{x-3}$

77. $\dfrac{x^2+7xy+12y^2}{x^2+2xy-8y^2} \cdot \dfrac{x^2-xy-2y^2}{x^2+4xy+3y^2} = \dfrac{(x+4y)(x+3y)}{(x+4y)(x-2y)} \cdot \dfrac{(x-2y)(x+y)}{(x+3y)(x+y)}$

$\qquad\qquad = \dfrac{(x+4y)(x+3y)(x-2y)(x+y)}{(x+4y)(x-2y)(x+3y)(x+y)} = 1$

79. $\dfrac{3x+9}{x+1} \div (x+3) = \dfrac{3(x+3)}{x+1} \div \dfrac{x+3}{1} = \dfrac{3(x+3)}{x+1} \cdot \dfrac{1}{x+3} = \dfrac{3(x+3)}{(x+1)(x+3)} = \dfrac{3}{x+1}$

81. $(3x+9) \div \dfrac{x^2-9}{6x} = \dfrac{3x+9}{1} \cdot \dfrac{6x}{x^2-9} = \dfrac{3(x+3)}{1} \cdot \dfrac{6x}{(x+3)(x-3)} = \dfrac{18x(x+3)}{(x+3)(x-3)} = \dfrac{18x}{x-3}$

83. $\dfrac{x}{3} \cdot \dfrac{9}{4} \div \dfrac{x^2}{6} = \dfrac{x}{3} \cdot \dfrac{9}{4} \cdot \dfrac{6}{x^2}$

$\qquad = \dfrac{x \cdot 3 \cdot 3 \cdot 2 \cdot 3}{3 \cdot 2 \cdot 2 \cdot x \cdot x} = \dfrac{9}{2x}$

85. $\dfrac{y^2}{2} \div \dfrac{4}{y} \cdot \dfrac{y^2}{8} = \dfrac{y^2}{2} \cdot \dfrac{y}{4} \cdot \dfrac{y^2}{8} = \dfrac{y^5}{64}$

87. $\dfrac{x^2-4}{2x+6} \div \dfrac{x+2}{4} \cdot \dfrac{x+3}{x-2} = \dfrac{x^2-4}{2x+6} \cdot \dfrac{4}{x+2} \cdot \dfrac{x+3}{x-2} = \dfrac{4(x+2)(x-2)(x+3)}{2(x+3)(x+2)(x-2)} = 2$

SECTION 6.2

89. $\dfrac{x^2 + x - 6}{x^2 - 4} \cdot \dfrac{x^2 + 2x}{x - 2} \div \dfrac{x^2 + 3x}{x + 2} = \dfrac{x^2 + x - 6}{x^2 - 4} \cdot \dfrac{x^2 + 2x}{x - 2} \cdot \dfrac{x + 2}{x^2 + 3x}$

$$= \dfrac{(x+3)(x-2)(x)(x+2)(x+2)}{(x+2)(x-2)(x-2)(x)(x+3)} = \dfrac{x+2}{x-2}$$

91. $\dfrac{x^2 - 1}{x^2 - 9} \left(\dfrac{x + 3}{x + 2} \div \dfrac{5}{x + 2} \right) = \dfrac{x^2 - 1}{x^2 - 9} \left(\dfrac{x + 3}{x + 2} \cdot \dfrac{x + 2}{5} \right) = \dfrac{(x+1)(x-1)(x+3)(x+2)}{5(x+3)(x-3)(x+2)}$

$$= \dfrac{(x+1)(x-1)}{5(x-3)}$$

93. $\dfrac{x - x^2}{x^2 - 4} \left(\dfrac{2x + 4}{x + 2} \div \dfrac{5}{x + 2} \right) = \dfrac{x - x^2}{x^2 - 4} \left(\dfrac{2x + 4}{x + 2} \cdot \dfrac{x + 2}{5} \right) = \dfrac{x - x^2}{x^2 - 4} \left(\dfrac{2(x+2)(x+2)}{5(x+2)} \right)$

$$= \dfrac{x(1 - x)}{(x+2)(x-2)} \cdot \dfrac{2(x+2)}{5}$$

$$= \dfrac{2x(1 - x)}{5(x - 2)}$$

95. $\dfrac{8z}{2x} \cdot \dfrac{16x}{3x} = \dfrac{4 \cdot 2 \cdot z \cdot 16 \cdot x}{2 \cdot x \cdot 3 \cdot x} = \dfrac{64z}{3x}$ **97.** $\dfrac{2x^2 z}{z} \cdot \dfrac{5x}{z} = \dfrac{10x^3 z}{z^2} = \dfrac{10x^3}{z}$

99. $\dfrac{3a^3 b}{25cd^3} \cdot \dfrac{-5cd^2}{6ab} \cdot \dfrac{10abc^2}{2bc^2 d} = -\dfrac{150a^4 b^2 c^3 d^2}{300ab^2 c^3 d^4} = -\dfrac{a^3}{2d^2}$

101. $\dfrac{3x}{y} \div \dfrac{2x}{4} = \dfrac{3x}{y} \cdot \dfrac{4}{2x} = \dfrac{3 \cdot x \cdot 2 \cdot 2}{y \cdot 2 \cdot x} = \dfrac{6}{y}$ **103.** $\dfrac{14}{7y} \div \dfrac{10}{5z} = \dfrac{14}{7y} \cdot \dfrac{5z}{10} = \dfrac{2 \cdot 7 \cdot 5 \cdot z}{7 \cdot y \cdot 2 \cdot 5} = \dfrac{z}{y}$

105. $\dfrac{(x + 7)^2}{x + 7} \div \dfrac{(x - 3)^2}{x + 7} = \dfrac{(x + 7)^2}{x + 7} \cdot \dfrac{x + 7}{(x - 3)^2} = \dfrac{(x + 7)^3}{(x + 7)(x - 3)^2} = \dfrac{(x + 7)^2}{(x - 3)^2}$

107. $\dfrac{x^2 - 16}{x - 4} \div \dfrac{3x + 12}{x} = \dfrac{x^2 - 16}{x - 4} \cdot \dfrac{x}{3x + 12} = \dfrac{(x+4)(x-4)x}{(x-4)(3)(x+4)} = \dfrac{x}{3}$

109. $\dfrac{a^2 - ab + b^2}{a^3 + b^3} \cdot \dfrac{ac + ad + bc + bd}{c^2 - d^2} = \dfrac{a^2 - ab + b^2}{(a + b)(a^2 - ab + b^2)} \cdot \dfrac{a(c + d) + b(c + d)}{(c + d)(c - d)}$

$$= \dfrac{a^2 - ab + b^2}{(a + b)(a^2 - ab + b^2)} \cdot \dfrac{(c + d)(a + b)}{(c + d)(c - d)}$$

$$= \dfrac{(a^2 - ab + b^2)(c + d)(a + b)}{(a + b)(a^2 - ab + b^2)(c + d)(c - d)} = \dfrac{1}{c - d}$$

111. $\dfrac{xw - xz + wy - yz}{x^2 + 2xy + y^2} \cdot \dfrac{x^3 - y^3}{z^2 - w^2} = \dfrac{x(w-z) + y(w-z)}{(x+y)(x+y)} \cdot \dfrac{(x-y)(x^2+xy+y^2)}{(z+w)(z-w)}$

$$= \dfrac{(w-z)(x+y)}{(x+y)(x+y)} \cdot \dfrac{(x-y)(x^2+xy+y^2)}{(z+w)(z-w)}$$

$$= \dfrac{(w-z)(x+y)(x-y)(x^2+xy+y^2)}{(x+y)(x+y)(z+w)(z-w)}$$

$$= -\dfrac{(x-y)(x^2+xy+y^2)}{(x+y)(z+w)}$$

113. $\dfrac{p^3 - p^2q + pq^2}{mp - mq + np - nq} \div \dfrac{q^3 + p^3}{q^2 - p^2} = \dfrac{p^3 - p^2q + pq^2}{mp - mq + np - nq} \cdot \dfrac{q^2 - p^2}{q^3 + p^3}$

$$= \dfrac{p(p^2 - pq + q^2)}{m(p-q) + n(p-q)} \cdot \dfrac{(q+p)(q-p)}{(q+p)(q^2 - pq + p^2)}$$

$$= \dfrac{p(p^2 - pq + q^2)}{(p-q)(m+n)} \cdot \dfrac{(q+p)(q-p)}{(q+p)(q^2 - pq + p^2)}$$

$$= \dfrac{p(p^2 - pq + q^2)(q+p)(q-p)}{(p-q)(m+n)(q+p)(q^2 - pq + p^2)} = -\dfrac{p}{m+n}$$

115. Answers may vary.   **117. Answers may vary.**

119. You always get the original value of x after simplifying.

Exercises 6.3 (page 395)

1. $\dfrac{12}{18} = \dfrac{2 \cdot 6}{3 \cdot 6} = \dfrac{2}{3}$; equal

3. $\dfrac{24}{42} = \dfrac{4 \cdot 6}{7 \cdot 6} = \dfrac{4}{7}$; not equal

5. $\dfrac{3x}{9} = \dfrac{3 \cdot x}{3 \cdot 3} = \dfrac{x}{3}$; equal

7. $\dfrac{5x}{3x} = \dfrac{x \cdot 5}{x \cdot 3} = \dfrac{5}{3}$; equal

9. $81 = 9 \cdot 9 = 3^2 \cdot 3^2 = 3^4$

11. $136 = 4 \cdot 34 = 2 \cdot 2 \cdot 2 \cdot 17 = 2^3 \cdot 17$

13. $102 = 2 \cdot 51 = 2 \cdot 3 \cdot 17$

15. $144 = 16 \cdot 9 = 2 \cdot 2 \cdot 2 \cdot 2 \cdot 3 \cdot 3 = 2^4 \cdot 3^2$

17. LCD

19. numerators; common denominator

21. $\dfrac{1}{8a} + \dfrac{1}{8a} = \dfrac{1+1}{8a} = \dfrac{2}{8a} = \dfrac{1}{4a}$

23. $\dfrac{2x}{y} + \dfrac{2x}{y} = \dfrac{2x+2x}{y} = \dfrac{4x}{y}$

25. $\dfrac{4y-1}{y-4} + \dfrac{5y+3}{y-4} = \dfrac{4y-1+5y+3}{y-4} = \dfrac{9y+2}{y-4}$

27. $\dfrac{3x-5}{x-2} + \dfrac{6x-13}{x-2} = \dfrac{3x-5+6x-13}{x-2} = \dfrac{9x-18}{x-2} = \dfrac{9(x-2)}{x-2} = 9$

29. $\dfrac{35}{72} - \dfrac{44}{72} = \dfrac{35 - 44}{72} = \dfrac{-9}{72} = -\dfrac{1}{8}$

31. $\dfrac{9y}{3x} - \dfrac{6y}{3x} = \dfrac{9y - 6y}{3x} = \dfrac{3y}{3x} = \dfrac{y}{x}$

33. $\dfrac{4y - 5}{2y} - \dfrac{3}{2y} = \dfrac{4y - 5 - 3}{2y} = \dfrac{4y - 8}{2y} = \dfrac{4(y - 2)}{2y} = \dfrac{2(y - 2)}{y}$

35. $\dfrac{6x - 5}{3xy} - \dfrac{3x - 5}{3xy} = \dfrac{6x - 5 - (3x - 5)}{3xy} = \dfrac{6x - 5 - 3x + 5}{3xy} = \dfrac{3x}{3xy} = \dfrac{1}{y}$

37. $\dfrac{3y - 2}{y + 3} - \dfrac{2y - 5}{y + 3} = \dfrac{3y - 2 - (2y - 5)}{y + 3} = \dfrac{3y - 2 - 2y + 5}{y + 3} = \dfrac{y + 3}{y + 3} = 1$

39. $\dfrac{5y + 3}{y - 4} - \dfrac{4y - 1}{y - 4} = \dfrac{5y + 3 - (4y - 1)}{y - 4} = \dfrac{5y + 3 - 4y + 1}{y - 4} = \dfrac{y + 4}{y - 4}$

41. $\dfrac{11x}{12} + \dfrac{7x}{12} - \dfrac{2x}{12} = \dfrac{11x + 7x - 2x}{12} = \dfrac{16x}{12} = \dfrac{4x}{3}$

43. $\dfrac{3x + 1}{x - 2} + \dfrac{5x + 2}{x - 2} - \dfrac{2x + 1}{x - 2} = \dfrac{3x + 1 + 5x + 2 - (2x + 1)}{x - 2} = \dfrac{3x + 1 + 5x + 2 - 2x - 1}{x - 2}$

$\qquad = \dfrac{6x + 2}{x - 2}$

45. $\dfrac{4b + 5}{b + 1} - \dfrac{6b - 2}{b + 1} + \dfrac{b - 7}{b + 1} = \dfrac{4b + 5 - (6b - 2) + b - 7}{b + 1} = \dfrac{4b + 5 - 6b + 2 + b - 7}{b + 1}$

$\qquad = \dfrac{-b}{b + 1} = -\dfrac{b}{b + 1}$

47. $\dfrac{x + 1}{x - 2} - \dfrac{2(x - 3)}{x - 2} + \dfrac{3(x + 1)}{x - 2} = \dfrac{x + 1 - 2(x - 3) + 3(x + 1)}{x - 2} = \dfrac{x + 1 - 2x + 6 + 3x + 3}{x - 2}$

$\qquad = \dfrac{2x + 10}{x - 2} = \dfrac{2(x + 5)}{x - 2}$

49. $\dfrac{21}{8} = \dfrac{21 \cdot 4}{8 \cdot 4} = \dfrac{84}{32}$

51. $\dfrac{8}{x} = \dfrac{8 \cdot xy}{x \cdot xy} = \dfrac{8xy}{x^2 y}$

53. $\dfrac{4x}{x + 3} = \dfrac{4x(x + 3)}{(x + 3)(x + 3)} = \dfrac{4x(x + 3)}{(x + 3)^2}$

55. $\dfrac{2y}{x} = \dfrac{2y(x + 1)}{x(x + 1)} = \dfrac{2y(x + 1)}{x^2 + x}$

57. $\dfrac{z}{z - 1} = \dfrac{z(z + 1)}{(z - 1)(z + 1)} = \dfrac{z(z + 1)}{z^2 - 1}$

59. $\dfrac{2}{x + 1} = \dfrac{2(x + 2)}{(x + 1)(x + 2)} = \dfrac{2(x + 2)}{x^2 + 3x + 2}$

61. $2x = 2 \cdot x$
$\quad 6x = 2 \cdot 3 \cdot x$
$\quad \text{LCD} = 2 \cdot 3 \cdot x = 6x$

63. $3x = 3 \cdot x$
$\quad 6y = 2 \cdot 3 \cdot y$
$\quad 9xy = 3^2 \cdot x \cdot y$
$\quad \text{LCD} = 2 \cdot 3^2 \cdot x \cdot y = 18xy$

65. $x^2 - 4 = (x+2)(x-2)$
$x + 2 = x + 2$
$\text{LCD} = (x+2)(x-2)$

67. $x^2 + 6x = x(x+6)$
$x + 6 = x + 6$
$x = x$
$\text{LCD} = x(x+6)$

69. $\dfrac{4x}{3y} + \dfrac{2x}{y} = \dfrac{4x}{3y} + \dfrac{2x \cdot 3}{y \cdot 3} = \dfrac{4x}{3y} + \dfrac{6x}{3y} = \dfrac{4x+6x}{3y} = \dfrac{10x}{3y}$

71. $\dfrac{x+2}{2x} + \dfrac{x-1}{3x} = \dfrac{(x+2)3}{2x \cdot 3} + \dfrac{(x-1)2}{3x \cdot 2} = \dfrac{3x+6}{6x} + \dfrac{2x-2}{6x} = \dfrac{3x+6+2x-2}{6x} = \dfrac{5x+4}{6x}$

73. $\dfrac{x+1}{x-1} + \dfrac{x-1}{x+1} = \dfrac{(x+1)(x+1)}{(x-1)(x+1)} + \dfrac{(x-1)(x-1)}{(x+1)(x-1)} = \dfrac{x^2+2x+1}{(x+1)(x-1)} + \dfrac{x^2-2x+1}{(x+1)(x-1)}$
$= \dfrac{2x^2+2}{(x+1)(x-1)}$

75. $\dfrac{x}{5x+2} + \dfrac{x-1}{x+2} = \dfrac{x(x+2)}{(5x+2)(x+2)} + \dfrac{(x-1)(5x+2)}{(x+2)(5x+2)} = \dfrac{x^2+2x}{(5x+2)(x+2)} + \dfrac{5x^2-3x-2}{(5x+2)(x+2)}$
$= \dfrac{6x^2-x-2}{(5x+2)(x+2)}$

77. $\dfrac{x-2}{x} + \dfrac{y+2}{y} = \dfrac{(x-2)y}{x \cdot y} + \dfrac{(y+2)x}{y \cdot x} = \dfrac{xy-2y}{xy} + \dfrac{xy+2x}{xy} = \dfrac{2xy+2x-2y}{xy}$

79. $\dfrac{3y}{x} + \dfrac{x+1}{y-1} = \dfrac{3y(y-1)}{x(y-1)} + \dfrac{(x+1)x}{(y-1)x} = \dfrac{3y^2-3y}{x(y-1)} + \dfrac{x^2+x}{x(y-1)} = \dfrac{3y^2-3y+x^2+x}{x(y-1)}$

81. $\dfrac{5}{x} - \dfrac{x+2}{x+1} = \dfrac{5(x+1)}{x(x+1)} - \dfrac{(x+2)x}{(x+1)x} = \dfrac{5(x+1)-x(x+2)}{x(x+1)} = \dfrac{5x+5-x^2-2x}{x(x+1)}$
$= \dfrac{-x^2+3x+5}{x(x+1)}$

83. $\dfrac{2x+3}{x+5} - \dfrac{x-1}{x+2} = \dfrac{(2x+3)(x+2)}{(x+5)(x+2)} - \dfrac{(x-1)(x+5)}{(x+2)(x+5)} = \dfrac{(2x+3)(x+2)-(x-1)(x+5)}{(x+5)(x+2)}$
$= \dfrac{2x^2+7x+6-(x^2+4x-5)}{(x+5)(x+2)}$
$= \dfrac{2x^2+7x+6-x^2-4x+5}{(x+5)(x+2)}$
$= \dfrac{x^2+3x+11}{(x+5)(x+2)}$

85. $\dfrac{x}{x-2} + \dfrac{4+2x}{x^2-4} = \dfrac{x}{x-2} + \dfrac{2(x+2)}{(x+2)(x-2)} = \dfrac{x}{x-2} + \dfrac{2}{x-2} = \dfrac{x+2}{x-2}$

87. $\dfrac{x+1}{x+2} - \dfrac{x^2+1}{x^2-x-6} = \dfrac{x+1}{x+2} - \dfrac{x^2+1}{(x+2)(x-3)} = \dfrac{(x+1)(x-3)}{(x+2)(x-3)} - \dfrac{x^2+1}{(x+2)(x-3)}$

$$= \dfrac{x^2-3x+x-3-x^2-1}{(x+2)(x-3)}$$

$$= \dfrac{-2x-4}{(x+2)(x-3)}$$

$$= \dfrac{-2(x+2)}{(x+2)(x-3)} = -\dfrac{2}{x-3}$$

89. $\dfrac{y+3}{y-1} - \dfrac{y+4}{1-y} = \dfrac{y+3}{y-1} + \dfrac{y+4}{y-1} = \dfrac{2y+7}{y-1}$ **91.** $\dfrac{x+5}{2x-y} - \dfrac{x-1}{y-2x} = \dfrac{x+5}{2x-y} + \dfrac{x-1}{2x-y}$

$$= \dfrac{2x+4}{2x-y}$$

93. $\dfrac{2x}{x^2-3x+2} + \dfrac{2x}{x-1} - \dfrac{x}{x-2} = \dfrac{2x}{(x-2)(x-1)} + \dfrac{2x}{x-1} - \dfrac{x}{x-2}$

$$= \dfrac{2x}{(x-2)(x-1)} + \dfrac{2x(x-2)}{(x-1)(x-2)} - \dfrac{x(x-1)}{(x-2)(x-1)}$$

$$= \dfrac{2x+2x(x-2)-x(x-1)}{(x-2)(x-1)}$$

$$= \dfrac{2x+2x^2-4x-x^2+x}{(x-2)(x-1)}$$

$$= \dfrac{x^2-x}{(x-2)(x-1)} = \dfrac{x(x-1)}{(x-2)(x-1)} = \dfrac{x}{x-2}$$

95. $\dfrac{a}{a-1} - \dfrac{2}{a+2} + \dfrac{3(a-2)}{a^2+a-2} = \dfrac{a}{a-1} - \dfrac{2}{a+2} + \dfrac{3(a-2)}{(a+2)(a-1)}$

$$= \dfrac{a(a+2)}{(a-1)(a+2)} - \dfrac{2(a-1)}{(a+2)(a-1)} + \dfrac{3(a-2)}{(a+2)(a-1)}$$

$$= \dfrac{a(a+2)-2(a-1)+3(a-2)}{(a+2)(a-1)}$$

$$= \dfrac{a^2+2a-2a+2+3a-6}{(a+2)(a-1)}$$

$$= \dfrac{a^2+3a-4}{(a+2)(a-1)} = \dfrac{(a+4)(a-1)}{(a+2)(a-1)} = \dfrac{a+4}{a+2}$$

97. $x^2-x-6 = (x+2)(x-3)$ **99.** $6y = 2\cdot3\cdot y, 8 = 2^3, \text{LCD} = 2^3\cdot3\cdot y = 24y$

$\qquad x^2-9 = (x+3)(x-3)$

$\text{LCD} = (x+2)(x-3)(x+3)$

101. $\dfrac{1}{2} + \dfrac{2}{3} = \dfrac{1\cdot3}{2\cdot3} + \dfrac{2\cdot2}{3\cdot2} = \dfrac{3}{6} + \dfrac{4}{6} = \dfrac{7}{6}$ **103.** $\dfrac{2}{3} - \dfrac{5}{6} = \dfrac{2\cdot2}{3\cdot2} - \dfrac{5}{6} = \dfrac{4}{6} - \dfrac{5}{6} = -\dfrac{1}{6}$

105. $\dfrac{2y}{5x} - \dfrac{y}{2} = \dfrac{2y \cdot 2}{5x \cdot 2} - \dfrac{y \cdot 5x}{2 \cdot 5x} = \dfrac{4y}{10x} - \dfrac{5xy}{10x} = \dfrac{4y - 5xy}{10x}$

107. $\dfrac{x+2}{x} + \dfrac{x-5}{x+2} = \dfrac{(x+2)(x+2)}{x(x+2)} + \dfrac{(x-5)x}{(x+2)x} = \dfrac{x^2 + 4x + 4}{x(x+2)} + \dfrac{x^2 - 5x}{x(x+2)} = \dfrac{2x^2 - x + 4}{x(x+2)}$

109. $\dfrac{y-7}{y^2} - \dfrac{y+7}{2y} = \dfrac{(y-7)2}{y^2 \cdot 2} - \dfrac{(y+7)y}{2y \cdot y} = \dfrac{2y - 14}{2y^2} - \dfrac{y^2 + 7y}{2y^2} = \dfrac{-y^2 - 5y - 14}{2y^2}$

$\qquad\qquad\qquad\qquad\qquad\qquad\qquad\qquad = -\dfrac{y^2 + 5y + 14}{2y^2}$

111. $\dfrac{5y}{8x} + \dfrac{4y}{8x} - \dfrac{y}{8x} = \dfrac{5y + 4y - y}{8x} = \dfrac{8y}{8x} = \dfrac{y}{x}$

113. $\dfrac{3x}{y+2} - \dfrac{3y}{y+2} + \dfrac{x+y}{y+2} = \dfrac{3x - 3y + x + y}{y+2} = \dfrac{4x - 2y}{y+2}$

115. $\dfrac{-a}{3a^2 - 27} + \dfrac{1}{3a+9} = \dfrac{-a}{3(a^2 - 9)} + \dfrac{1}{3(a+3)} = \dfrac{-a}{3(a+3)(a-3)} + \dfrac{1}{3(a+3)}$

$\qquad\qquad\qquad\qquad\qquad\qquad\qquad = \dfrac{-a}{3(a+3)(a-3)} + \dfrac{1(a-3)}{3(a+3)(a-3)}$

$\qquad\qquad\qquad\qquad\qquad\qquad\qquad = \dfrac{-a + 1(a-3)}{3(a+3)(a-3)}$

$\qquad\qquad\qquad\qquad\qquad\qquad\qquad = \dfrac{-a + a - 3}{3(a+3)(a-3)}$

$\qquad\qquad\qquad\qquad\qquad\qquad\qquad = \dfrac{-3}{3(a+3)(a-3)} = \dfrac{-1}{(a+3)(a-3)}$

117. $14 + \dfrac{10}{y^2} = \dfrac{14}{1} + \dfrac{10}{y^2} = \dfrac{14 \cdot y^2}{1 \cdot y^2} + \dfrac{10}{y^2} = \dfrac{14y^2}{y^2} + \dfrac{10}{y^2} = \dfrac{14y^2 + 10}{y^2}$

119. Answers may vary.

121. Answers may vary.

123. The subtraction needs to be distributed in the numerator of the 2nd fraction.

125. $\dfrac{a}{b} + \dfrac{c}{d} = \dfrac{a(d)}{b(d)} + \dfrac{c(b)}{d(b)}$

$\qquad = \dfrac{ad}{bd} + \dfrac{bc}{bd} = \dfrac{ad + bc}{bd}$

Exercises 6.4 (page 403)

1. $\dfrac{\frac{2}{3}}{\frac{1}{2}} = \dfrac{2}{3} \div \dfrac{1}{2} = \dfrac{2}{3} \cdot \dfrac{2}{1} = \dfrac{4}{3}$

3. $\dfrac{\frac{6}{7}}{\frac{8}{21}} = \dfrac{6}{7} \div \dfrac{8}{21} = \dfrac{6}{7} \cdot \dfrac{21}{8} = \dfrac{9}{4}$

5. $\dfrac{\frac{7}{8}}{\frac{49}{4}} = \dfrac{7}{8} \div \dfrac{49}{4} = \dfrac{7}{8} \cdot \dfrac{4}{49} = \dfrac{1}{14}$

7. $\dfrac{\frac{1}{2}}{2} = \dfrac{1}{2} \div 2 = \dfrac{1}{2} \div \dfrac{2}{1} = \dfrac{1}{2} \cdot \dfrac{1}{2} = \dfrac{1}{4}$

9. $\dfrac{\frac{2}{3}+1}{\frac{1}{3}+1} = \dfrac{\left(\frac{2}{3}+1\right)3}{\left(\frac{1}{3}+1\right)3} = \dfrac{\frac{2}{3}\cdot 3 + 1\cdot 3}{\frac{1}{3}\cdot 3 + 1\cdot 3}$
$= \dfrac{2+3}{1+3} = \dfrac{5}{4}$

11. $\dfrac{\frac{1}{2}+\frac{3}{4}}{\frac{3}{2}+\frac{1}{4}} = \dfrac{\left(\frac{1}{2}+\frac{3}{4}\right)4}{\left(\frac{3}{2}+\frac{1}{4}\right)4} = \dfrac{\frac{1}{2}\cdot 4 + \frac{3}{4}\cdot 4}{\frac{3}{2}\cdot 4 + \frac{1}{4}\cdot 4}$
$= \dfrac{2+3}{6+1} = \dfrac{5}{7}$

13. $t^5 t^2 t = t^{5+2+1} = t^8$

15. $-2r(r^3)^2 = -2rr^6 = -2r^7$

17. $\left(\dfrac{3r}{4r^3}\right)^{-4} = \left(\dfrac{4r^3}{3r}\right)^4 = \left(\dfrac{4r^2}{3}\right)^4 = \dfrac{(4r^2)^4}{3^4}$
$= \dfrac{256r^8}{81}$

19. $\left(\dfrac{6r^{-2}}{2r^3}\right)^{-2} = (3r^{-5})^{-2} = \dfrac{r^{10}}{9}$

21. complex fraction

23. single; divide

25. $\dfrac{\frac{2x}{y}}{\frac{4}{xy}} = \dfrac{2x}{y} \div \dfrac{4}{xy} = \dfrac{2x}{y}\cdot\dfrac{xy}{4} = \dfrac{2x^2 y}{4y} = \dfrac{x^2}{2}$

27. $\dfrac{\frac{5t^2}{9x^2}}{\frac{3t}{x^2 t}} = \dfrac{5t^2}{9x^2} \div \dfrac{3t}{x^2 t} = \dfrac{5t^2}{9x^2}\cdot\dfrac{x^2 t}{3t}$
$= \dfrac{5x^2 t^3}{27x^2 t} = \dfrac{5t^2}{27}$

29. $\dfrac{\frac{a}{b}}{\frac{a}{a+1}} = \dfrac{a}{b} \div \dfrac{a}{a+1} = \dfrac{a}{b}\cdot\dfrac{a+1}{a} = \dfrac{a+1}{b}$

31. $\dfrac{\frac{x}{y-1}}{\frac{x}{y+1}} = \dfrac{x}{y-1} \div \dfrac{x}{y+1} = \dfrac{x}{y-1}\cdot\dfrac{y+1}{x}$
$= \dfrac{y+1}{y-1}$

33. $\dfrac{\frac{1}{y}+3}{\frac{3}{y}-2} = \dfrac{\left(\frac{1}{y}+3\right)y}{\left(\frac{3}{y}-2\right)y} = \dfrac{\frac{1}{y}\cdot y + 3\cdot y}{\frac{3}{y}\cdot y - 2\cdot y}$
$= \dfrac{1+3y}{3-2y}$

35. $\dfrac{5+\frac{3}{x}}{3+\frac{2}{x}} = \dfrac{\left(5+\frac{3}{x}\right)x}{\left(3+\frac{2}{x}\right)x} = \dfrac{5\cdot x + \frac{3}{x}\cdot x}{3\cdot x + \frac{2}{x}\cdot x} = \dfrac{5x+3}{3x+2}$

37. $\dfrac{\frac{2}{a+2}+1}{\frac{3}{a+2}} = \dfrac{\left(\frac{2}{a+2}+1\right)(a+2)}{\left(\frac{3}{a+2}\right)(a+2)} = \dfrac{\frac{2}{a+2}(a+2)+1(a+2)}{3} = \dfrac{2+a+2}{3} = \dfrac{a+4}{3}$

39. $\dfrac{\frac{3}{x}+\frac{4}{x+1}}{\frac{2}{x+1}-\frac{3}{x}} = \dfrac{\left(\frac{3}{x}+\frac{4}{x+1}\right)\cdot x(x+1)}{\left(\frac{2}{x+1}-\frac{3}{x}\right)\cdot x(x+1)} = \dfrac{\frac{3}{x}\cdot x(x+1)+\frac{4}{x+1}\cdot x(x+1)}{\frac{2}{x+1}\cdot x(x+1)-\frac{3}{x}\cdot x(x+1)} = \dfrac{3(x+1)+4x}{2x-3(x+1)}$
$= \dfrac{3x+3+4x}{2x-3x-3} = \dfrac{7x+3}{-x-3}$

41. $\dfrac{\frac{3y}{x}-y}{y-\frac{y}{x}} = \dfrac{\left(\frac{3y}{x}-y\right)x}{\left(y-\frac{y}{x}\right)x} = \dfrac{\frac{3y}{x}\cdot x - y\cdot x}{y\cdot x - \frac{y}{x}\cdot x} = \dfrac{3y-xy}{xy-y} = \dfrac{y(3-x)}{y(x-1)} = \dfrac{3-x}{x-1}$

43. $\dfrac{1}{\frac{1}{x}+\frac{1}{y}} = \dfrac{1(xy)}{\left(\frac{1}{x}+\frac{1}{y}\right)xy} = \dfrac{xy}{\frac{1}{x}\cdot xy + \frac{1}{y}\cdot xy} = \dfrac{xy}{y+x}$

180

45. $\dfrac{\frac{2}{x}}{\frac{2}{y} - \frac{4}{x}} = \dfrac{\frac{2}{x}(xy)}{\left(\frac{2}{y} - \frac{4}{x}\right)(xy)} = \dfrac{\frac{2}{x}(xy)}{\frac{2}{y}(xy) - \frac{4}{x}(xy)} = \dfrac{2y}{2x - 4y} = \dfrac{2y}{2(x - 2y)} = \dfrac{y}{x - 2y}$

47. $\dfrac{\frac{3}{x} + \frac{2x}{y}}{\frac{4}{x}} = \dfrac{\left(\frac{3}{x} + \frac{2x}{y}\right)(xy)}{\frac{4}{x}(xy)} = \dfrac{\frac{3}{x}(xy) + \frac{2x}{y}(xy)}{4y} = \dfrac{3y + 2x^2}{4y}$

49. $\dfrac{\frac{1}{x+1}}{1 + \frac{1}{x+1}} = \dfrac{\left(\frac{1}{x+1}\right)(x+1)}{\left(1 + \frac{1}{x+1}\right)(x+1)} = \dfrac{\frac{1}{x+1}(x+1)}{1(x+1) + \frac{1}{x+1}(x+1)} = \dfrac{1}{x + 1 + 1} = \dfrac{1}{x+2}$

51. $\dfrac{\frac{x}{x+2}}{\frac{x}{x+2} + x} = \dfrac{\left(\frac{x}{x+2}\right)(x+2)}{\left(\frac{x}{x+2} + x\right)(x+2)} = \dfrac{\frac{x}{x+2}(x+2)}{\frac{x}{x+2}(x+2) + x(x+2)} = \dfrac{x}{x + x^2 + 2x} = \dfrac{x}{x^2 + 3x}$

$$= \dfrac{x}{x(x+3)} = \dfrac{1}{x+3}$$

53. $\dfrac{\frac{2}{x} - \frac{3}{x+1}}{\frac{2}{x+1} - \frac{3}{x}} = \dfrac{\left(\frac{2}{x} - \frac{3}{x+1}\right)(x)(x+1)}{\left(\frac{2}{x+1} - \frac{3}{x}\right)(x)(x+1)} = \dfrac{2(x+1) - 3x}{2x - 3(x+1)}$

$$= \dfrac{2x + 2 - 3x}{2x - 3x - 3} = \dfrac{-x + 2}{-x - 3} = \dfrac{-(x-2)}{-(x+3)} = \dfrac{x-2}{x+3}$$

55. $\dfrac{\frac{m}{m+2} - \frac{2}{m-1}}{\frac{3}{m+2} + \frac{m}{m-1}} = \dfrac{\left(\frac{m}{m+2} - \frac{2}{m-1}\right)(m+2)(m-1)}{\left(\frac{3}{m+2} + \frac{m}{m-1}\right)(m+2)(m-1)} = \dfrac{\frac{m}{m+2}(m+2)(m-1) - \frac{2}{m-1}(m+2)(m-1)}{\frac{3}{m+2}(m+2)(m-1) + \frac{m}{m-1}(m+2)(m-1)}$

$$= \dfrac{m(m-1) - 2(m+2)}{3(m-1) + m(m+2)}$$

$$= \dfrac{m^2 - m - 2m - 4}{3m - 3 + m^2 + 2m} = \dfrac{m^2 - 3m - 4}{m^2 + 5m - 3}$$

57. $\dfrac{x^{-2} + 1}{x^{-1} + 1} = \dfrac{\frac{1}{x^2} + 1}{\frac{1}{x} + 1} = \dfrac{\left(\frac{1}{x^2} + 1\right)x^2}{\left(\frac{1}{x} + 1\right)x^2} = \dfrac{\frac{1}{x^2} \cdot x^2 + 1x^2}{\frac{1}{x} \cdot x^2 + 1x^2} = \dfrac{1 + x^2}{x + x^2}$

59. $\dfrac{y^{-2} + 1}{y^{-2} - 1} = \dfrac{\frac{1}{y^2} + 1}{\frac{1}{y^2} - 1} = \dfrac{\left(\frac{1}{y^2} + 1\right)y^2}{\left(\frac{1}{y^2} - 1\right)y^2} = \dfrac{\frac{1}{y^2} \cdot y^2 + 1y^2}{\frac{1}{y^2} \cdot y^2 - 1y^2} = \dfrac{1 + y^2}{1 - y^2}$

61. $\dfrac{a^{-2} + a}{a + 1} = \dfrac{\frac{1}{a^2} + a}{a + 1} = \dfrac{\left(\frac{1}{a^2} + a\right)a^2}{(a + 1)a^2} = \dfrac{1 + a^3}{a^3 + a^2} = \dfrac{(1+a)(1 - a + a^2)}{a^2(a+1)} = \dfrac{a^2 - a + 1}{a^2}$

63. $\dfrac{2x^{-1} + 4x^{-2}}{2x^{-2} + x^{-1}} = \dfrac{\frac{2}{x} + \frac{4}{x^2}}{\frac{2}{x^2} + \frac{1}{x}} = \dfrac{\left(\frac{2}{x} + \frac{4}{x^2}\right)x^2}{\left(\frac{2}{x^2} + \frac{1}{x}\right)x^2} = \dfrac{2x + 4}{2 + x} = \dfrac{2(x + 2)}{x + 2} = 2$

65. $\dfrac{\frac{y}{x-1}}{\frac{y}{x}} = \dfrac{y}{x-1} \div \dfrac{y}{x} = \dfrac{y}{x-1} \cdot \dfrac{x}{y} = \dfrac{x}{x-1}$

67. $\dfrac{3 + \frac{3}{x-1}}{3 - \frac{3}{x}} = \dfrac{\left(3 + \frac{3}{x-1}\right)(x)(x-1)}{\left(3 - \frac{3}{x}\right)(x)(x-1)} = \dfrac{3x(x-1) + \frac{3}{x-1}(x)(x-1)}{3x(x-1) - \frac{3}{x}(x)(x-1)} = \dfrac{3x^2 - 3x + 3x}{3x^2 - 3x - 3(x-1)}$

$\qquad\qquad = \dfrac{3x^2}{3x^2 - 6x + 3}$

$\qquad\qquad = \dfrac{3x^2}{3(x^2 - 2x + 1)}$

$\qquad\qquad = \dfrac{x^2}{x^2 - 2x + 1} = \dfrac{x^2}{(x-1)^2}$

69. $\dfrac{\frac{2}{x+2}}{\frac{3}{x-3} + \frac{1}{x}} = \dfrac{\left(\frac{2}{x+2}\right)(x)(x+2)(x-3)}{\left(\frac{3}{x-3} + \frac{1}{x}\right)(x)(x+2)(x-3)} = \dfrac{2x(x-3)}{\frac{3}{x-3}(x)(x+2)(x-3) + \frac{1}{x}(x)(x+2)(x-3)}$

$\qquad\qquad = \dfrac{2x^2 - 6x}{3x(x+2) + (x+2)(x-3)}$

$\qquad\qquad = \dfrac{2x^2 - 6x}{3x^2 + 6x + x^2 - 3x + 2x - 6}$

$\qquad\qquad = \dfrac{2x^2 - 6x}{4x^2 + 5x - 6} = \dfrac{2x(x-3)}{(4x-3)(x+2)}$

71. $\dfrac{\frac{1}{x} + \frac{2}{x+1}}{\frac{2}{x-1} - \frac{1}{x}} = \dfrac{\left(\frac{1}{x} + \frac{2}{x+1}\right)(x)(x+1)(x-1)}{\left(\frac{2}{x-1} - \frac{1}{x}\right)(x)(x+1)(x-1)} = \dfrac{\frac{1}{x}(x)(x+1)(x-1) + \frac{2}{x+1}(x)(x+1)(x-1)}{\frac{2}{x-1}(x)(x+1)(x-1) - \frac{1}{x}(x)(x+1)(x-1)}$

$\qquad\qquad = \dfrac{(x+1)(x-1) + 2x(x-1)}{2x(x+1) - (x+1)(x-1)}$

$\qquad\qquad = \dfrac{x^2 - 1 + 2x^2 - 2x}{2x^2 + 2x - (x^2 - 1)}$

$\qquad\qquad = \dfrac{3x^2 - 2x - 1}{2x^2 + 2x - x^2 + 1}$

$\qquad\qquad = \dfrac{3x^2 - 2x - 1}{x^2 + 2x + 1} = \dfrac{(3x+1)(x-1)}{(x+1)^2}$

73. $\dfrac{\frac{1}{y^2+y} - \frac{1}{xy+x}}{\frac{1}{xy+x} - \frac{1}{y^2+y}} = \dfrac{\frac{1}{y(y+1)} - \frac{1}{x(y+1)}}{\frac{1}{x(y+1)} - \frac{1}{y(y+1)}} = \dfrac{\left(\frac{1}{y(y+1)} - \frac{1}{x(y+1)}\right)(x)(y)(y+1)}{\left(\frac{1}{x(y+1)} - \frac{1}{y(y+1)}\right)(x)(y)(y+1)}$

$\qquad\qquad = \dfrac{1x - 1y}{1y - 1x} = \dfrac{x-y}{-(x-y)} = -1$

75. $\dfrac{1 - 25y^{-2}}{1 + 10y^{-1} + 25y^{-2}} = \dfrac{1 - \frac{25}{y^2}}{1 + \frac{10}{y} + \frac{25}{y^2}} = \dfrac{\left(1 - \frac{25}{y^2}\right)y^2}{\left(1 + \frac{10}{y} + \frac{25}{y^2}\right)y^2} = \dfrac{y^2 - 25}{y^2 + 10y + 25}$

$\qquad\qquad = \dfrac{(y+5)(y-5)}{(y+5)(y+5)} = \dfrac{y-5}{y+5}$

77. Answers may vary.

79. $\dfrac{1}{1+1} = \dfrac{1}{2}$ $\quad \dfrac{1}{1+\frac{1}{2}} = \dfrac{1}{\frac{3}{2}} = \dfrac{2}{3}$ $\quad \dfrac{1}{1+\frac{1}{1+\frac{1}{2}}} = \dfrac{1}{1+\frac{2}{3}} = \dfrac{1}{\frac{5}{3}} = \dfrac{3}{5}$ $\quad \dfrac{1}{1+\frac{1}{1+\frac{1}{1+\frac{1}{2}}}} = \dfrac{1}{1+\frac{3}{5}} = \dfrac{1}{\frac{8}{5}} = \dfrac{5}{8}$

Exercises 6.5 (page 410)

1.
$$\frac{y}{3} + 6 = \frac{4y}{3}$$
$$3\left(\frac{y}{3} + 6\right) = 3\left(\frac{4y}{3}\right)$$
$$y + 18 = 4y$$
$$18 = 3y$$
$$6 = y$$

3.
$$\frac{z-3}{2} = z + 2$$
$$2\left(\frac{z-3}{2}\right) = 2(z+2)$$
$$z - 3 = 2z + 4$$
$$-3 = z + 4$$
$$-7 = z$$

5.
$$\frac{5(x+1)}{8} = x + 1$$
$$8\left(\frac{5(x+1)}{8}\right) = 8(x+1)$$
$$5(x+1) = 8(x+1)$$
$$5x + 5 = 8x + 8$$
$$5 = 3x + 8$$
$$-3 = 3x$$
$$-1 = x$$

7. $\text{LCD} = 2(x+5)$

9. $\text{LCD} = (y-1)(y+2)$

11. $x^2 - 9 = (x+3)(x-3)$
$\text{LCD} = (x+3)(x-3)$

13. $x^2 + 8x = x(x+8)$

15. $2x^2 + x - 3 = (2x+3)(x-1)$

17. $4x^2 + 10x - 6 = 2(2x^2 + 5x - 3) = 2(2x-1)(x+3)$

19. extraneous

21. LCD

23.
$$\frac{3}{x} + 2 = 3$$
$$x\left(\frac{3}{x} + 2\right) = x(3)$$
$$3 + 2x = 3x$$
$$3 = x$$

25.
$$\frac{x}{x+2} + 3 = \frac{2x}{x+2}$$
$$(x+2)\left(\frac{x}{x+2} + 3\right) = (x+2)\left(\frac{2x}{x+2}\right)$$
$$x + 3(x+2) = 2x$$
$$x + 3x + 6 = 2x$$
$$4x + 6 = 2x$$
$$6 = -2x$$
$$-3 = x$$

27.
$$\frac{2}{y+1} + 5 = \frac{12}{y+1}$$
$$(y+1)\left(\frac{2}{y+1} + 5\right) = (y+1)\left(\frac{12}{y+1}\right)$$
$$2 + 5(y+1) = 12$$
$$2 + 5y + 5 = 12$$
$$5y = 5$$
$$y = 1$$

29.
$$\frac{1}{x-1} + \frac{3}{x-1} = 1$$
$$(x-1)\left(\frac{1}{x-1} + \frac{3}{x-1}\right) = (x-1)(1)$$
$$1 + 3 = x - 1$$
$$5 = x$$

31.
$$\frac{a^2}{a+2} - \frac{4}{a+2} = a$$
$$(a+2)\left(\frac{a^2}{a+2} - \frac{4}{a+2}\right) = (a+2)a$$
$$a^2 - 4 = a^2 + 2a$$
$$-4 = 2a$$
$$-2 = a$$
-2 is extraneous $\Rightarrow \emptyset$

33.
$$\frac{x}{x-5} - \frac{5}{x-5} = 3$$
$$(x-5)\left(\frac{x}{x-5} - \frac{5}{x-5}\right) = (x-5)3$$
$$x - 5 = 3x - 15$$
$$-2x = -10$$
$$x = 5$$
5 is extraneous $\Rightarrow \emptyset$

35.
$$\frac{2x+1}{x+5} - 1 = \frac{3x-2}{x+5}$$
$$(x+5)\left(\frac{2x+1}{x+5} - 1\right) = (x+5) \cdot \frac{3x-2}{x+5}$$
$$2x + 1 - 1(x+5) = 3x - 2$$
$$2x + 1 - x - 5 = 3x - 2$$
$$x - 4 = 3x - 2$$
$$-4 = 2x - 2$$
$$-2 = 2x$$
$$-1 = x$$

37.
$$\frac{x-4}{x-3} + \frac{x-2}{x-3} = x - 3$$
$$(x-3)\left(\frac{x-4}{x-3} + \frac{x-2}{x-3}\right) = (x-3)(x-3)$$
$$x - 4 + x - 2 = x^2 - 6x + 9$$
$$0 = x^2 - 8x + 15$$
$$0 = (x-5)(x-3)$$
$$x - 5 = 0 \quad \textbf{or} \quad x - 3 = 0$$
$$x = 5 \qquad\qquad x = 3$$
$x = 3$ is extraneous, so $x = 5$ is the only solution.

39.
$$\frac{v}{v+2} + \frac{1}{v-1} = 1$$
$$(v+2)(v-1)\left(\frac{v}{v+2} + \frac{1}{v-1}\right) = (v+2)(v-1)1$$
$$v(v-1) + v + 2 = (v+2)(v-1)$$
$$v^2 - v + v + 2 = v^2 - v + 2v - 2$$
$$v^2 + 2 = v^2 + v - 2$$
$$2 = v - 2$$
$$4 = v$$

41.
$$\frac{u}{u-1} + \frac{1}{u} = \frac{u^2+1}{u^2-u}$$
$$\frac{u}{u-1} + \frac{1}{u} = \frac{u^2+1}{u(u-1)}$$
$$u(u-1)\left(\frac{u}{u-1} + \frac{1}{u}\right) = u(u-1)\left(\frac{u^2+1}{u(u-1)}\right)$$
$$u^2 + u - 1 = u^2 + 1$$
$$u - 1 = 1$$
$$u = 2$$

43.
$$\frac{5}{x} + \frac{3}{x+2} = \frac{-6}{x(x+2)}$$
$$x(x+2)\left(\frac{5}{x} + \frac{3}{x+2}\right) = x(x+2)\left(\frac{-6}{x(x+2)}\right)$$
$$5(x+2) + 3x = -6$$
$$5x + 10 + 3x = -6$$
$$8x = -16$$
$$x = -2: -2 \text{ is extraneous} \Rightarrow \emptyset$$

45.
$$\frac{-5}{s^2+s-2} + \frac{3}{s+2} = \frac{1}{s-1}$$
$$\frac{-5}{(s+2)(s-1)} + \frac{3}{s+2} = \frac{1}{s-1}$$
$$(s+2)(s-1)\left(\frac{-5}{(s+2)(s-1)} + \frac{3}{s+2}\right) = (s+2)(s-1)\left(\frac{1}{s-1}\right)$$
$$-5 + 3(s-1) = s+2$$
$$-5 + 3s - 3 = s+2$$
$$2s = 10$$
$$s = 5$$

47.
$$y + \frac{3}{4} = \frac{3y-50}{4y-24}$$
$$y + \frac{3}{4} = \frac{3y-50}{4(y-6)}$$
$$4(y-6)\left(y + \frac{3}{4}\right) = 4(y-6)\left(\frac{3y-50}{4(y-6)}\right)$$
$$4y(y-6) + 3(y-6) = 3y-50$$
$$4y^2 - 24y + 3y - 18 = 3y - 50$$
$$4y^2 - 24y + 32 = 0$$
$$4(y^2 - 6y + 8) = 0$$
$$4(y-2)(y-4) = 0$$
$$y - 2 = 0 \quad \textbf{or} \quad y - 4 = 0$$
$$y = 2 \qquad\qquad y = 4$$

49.
$$\frac{3}{5x-20} + \frac{4}{5} = \frac{3}{5x-20} - \frac{x}{5}$$
$$\frac{3}{5(x-4)} + \frac{4}{5} = \frac{3}{5(x-4)} - \frac{x}{5}$$
$$5(x-4)\left(\frac{3}{5(x-4)} + \frac{4}{5}\right) = 5(x-4)\left(\frac{3}{5(x-4)} - \frac{x}{5}\right)$$
$$3 + 4(x-4) = 3 - x(x-4)$$
$$3 + 4x - 16 = 3 - x^2 + 4x$$
$$x^2 - 16 = 0$$
$$(x+4)(x-4) = 0$$

$x + 4 = 0 \quad$ **or** $\quad x - 4 = 0 \; \Rightarrow x = 4$ is extraneous, so the only solution is $x = -4$.
$\qquad x = -4 \qquad\qquad x = 4$

51.
$$\frac{7}{q^2 - q - 2} + \frac{1}{q+1} = \frac{3}{q-2}$$
$$\frac{7}{(q-2)(q+1)} + \frac{1}{q+1} = \frac{3}{q-2}$$
$$(q-2)(q+1)\left(\frac{7}{(q-2)(q+1)} + \frac{1}{q+1}\right) = (q-2)(q+1)\left(\frac{3}{q-2}\right)$$
$$7 + q - 2 = 3(q+1)$$
$$q + 5 = 3q + 3$$
$$-2q = -2$$
$$q = 1$$

53.
$$\frac{3y}{3y-6} + \frac{8}{y^2-4} = \frac{2y}{2y+4}$$
$$\frac{3y}{3(y-2)} + \frac{8}{(y+2)(y-2)} = \frac{2y}{2(y+2)}$$
$$\frac{y}{y-2} + \frac{8}{(y+2)(y-2)} = \frac{y}{y+2}$$
$$(y+2)(y-2)\left(\frac{y}{y-2} + \frac{8}{(y+2)(y-2)}\right) = (y+2)(y-2)\left(\frac{y}{y+2}\right)$$
$$y(y+2) + 8 = y(y-2)$$
$$y^2 + 2y + 8 = y^2 - 2y$$
$$8 = -4y$$
$$-2 = y: \; -2 \text{ is extraneous} \Rightarrow \emptyset$$

55.
$$\frac{c-4}{4} = \frac{c+4}{8}$$
$$8\left(\frac{c-4}{4}\right) = 8\left(\frac{c+4}{8}\right)$$
$$2(c-4) = c+4$$
$$2c - 8 = c + 4$$
$$c = 12$$

57.
$$\frac{x}{5} - \frac{x}{3} = -8$$
$$15\left(\frac{x}{5} - \frac{x}{3}\right) = 15(-8)$$
$$3x - 5x = -120$$
$$-2x = -120$$
$$x = 60$$

59.
$$\frac{x+2}{2} - 3x = x + 8$$
$$2\left(\frac{x+2}{2} - 3x\right) = 2(x+8)$$
$$x + 2 - 6x = 2x + 16$$
$$-7x = 14$$
$$x = -2$$

61.
$$\frac{3r}{2} - \frac{3}{r} = \frac{3r}{2} + 3$$
$$2r\left(\frac{3r}{2} - \frac{3}{r}\right) = 2r\left(\frac{3r}{2} + 3\right)$$
$$3r^2 - 6 = 3r^2 + 6r$$
$$-6 = 6r$$
$$-1 = r$$

63.
$$\frac{1}{3} + \frac{2}{x-3} = 1$$
$$3(x-3)\left(\frac{1}{3} + \frac{2}{x-3}\right) = 3(x-3)(1)$$
$$x - 3 + 3(2) = 3x - 9$$
$$x - 3 + 6 = 3x - 9$$
$$12 = 2x$$
$$6 = x$$

65.
$$\frac{z-4}{z-3} = \frac{z+2}{z+1}$$
$$(z-3)(z+1)\left(\frac{z-4}{z-3}\right) = (z-3)(z+1)\left(\frac{z+2}{z+1}\right)$$
$$(z+1)(z-4) = (z-3)(z+2)$$
$$z^2 - 3z - 4 = z^2 - z - 6$$
$$-2z = -2$$
$$z = 1$$

67.
$$\frac{x-2}{x-3} + \frac{x-1}{x^2-8x+15} = 1$$
$$\frac{x-2}{x-3} + \frac{x-1}{(x-3)(x-5)} = 1$$
$$(x-3)(x-5)\left(\frac{x-2}{x-3} + \frac{x-1}{(x-3)(x-5)}\right) = (x-3)(x-5)(1)$$
$$(x-5)(x-2) + x - 1 = (x-3)(x-5)$$
$$x^2 - 7x + 10 + x - 1 = x^2 - 8x + 15$$
$$x^2 - 6x + 9 = x^2 - 8x + 15$$
$$-6x + 9 = -8x + 15$$
$$2x + 9 = 15$$
$$2x = 6$$
$$x = 3; \ 3 \text{ is extraneous} \Rightarrow \emptyset$$

69.
$$\frac{1}{a} + \frac{1}{b} = 1$$
$$ab\left(\frac{1}{a} + \frac{1}{b}\right) = ab(1)$$
$$b + a = ab$$
$$b = ab - a$$
$$b = a(b - 1)$$
$$\frac{b}{b-1} = a$$

71.
$$\frac{a}{b} + \frac{c}{d} = 1$$
$$bd\left(\frac{a}{b} + \frac{c}{d}\right) = bd(1)$$
$$ad + bc = bd$$
$$bc - bd = -ad$$
$$b(c - d) = -ad$$
$$b = \frac{-ad}{c-d} = \frac{ad}{d-c}$$

73.
$$\frac{1}{r} = \frac{1}{r_1} + \frac{1}{r_2}$$
$$rr_1r_2 \cdot \frac{1}{r} = rr_1r_2\left(\frac{1}{r_1} + \frac{1}{r_2}\right)$$
$$r_1r_2 = rr_2 + rr_1$$
$$r_1r_2 - rr_1 = rr_2$$
$$r_1(r_2 - r) = rr_2$$
$$r_1 = \frac{rr_2}{r_2 - r}$$

75.
$$\frac{1}{f} = \frac{1}{d_1} + \frac{1}{d_2}$$
$$fd_1d_2 \cdot \frac{1}{f} = fd_1d_2\left(\frac{1}{d_1} + \frac{1}{d_2}\right)$$
$$d_1d_2 = fd_2 + fd_1$$
$$d_1d_2 = f(d_2 + d_1)$$
$$\frac{d_1d_2}{d_2 + d_1} = f$$

77. **Answers may vary.**

79.
$$x = \frac{1}{x}$$
$$x(x) = x\left(\frac{1}{x}\right)$$
$$x^2 = 1$$
$$x^2 - 1 = 0$$
$$(x + 1)(x - 1) = 0$$
$$x + 1 = 0 \quad \textbf{or} \quad x - 1 = 0$$
$$x = -1 \qquad\qquad x = 1$$

The numbers 1 and -1 are equal to their own reciprocals.

Exercises 6.6 (page 416)

1. $i = pr$

3. $C = qd$

5.
$$x^2 - 5x - 6 = 0$$
$$(x - 6)(x + 1) = 0$$
$$x - 6 = 0 \quad \textbf{or} \quad x + 1 = 0$$
$$x = 6 \qquad\qquad x = -1$$

7.
$$(y-3)(y^2+5y+4)=0$$
$$(y-3)(y+4)(y+1)=0$$
$y-3=0$ **or** $y+4=0$ **or** $y+1=0$
$\quad y=3 \qquad\quad y=-4 \qquad\quad y=-1$

9.
$$y^3-y^2=0$$
$$y^2(y-1)=0$$
$y=0$ **or** $y=0$ **or** $y-1=0$
$\quad y=0 \qquad\quad y=0 \qquad\quad y=1$

11.
$$4(y-3)=-y^2$$
$$4y-12=-y^2$$
$$y^2+4y-12=0$$
$$(y-2)(y+6)=0$$
$y-2=0$ **or** $y+6=0$
$\quad y=2 \qquad\quad y=-6$

13. **Answers may vary.**

15. Let $x=$ the number.
$$\frac{2(3)}{4+x}=1$$
$$(4+x)\left(\frac{6}{4+x}\right)=(4+x)(1)$$
$$6=4+x$$
$$2=x \Rightarrow \text{The number is 2.}$$

17. Let $x=$ the number.
$$\frac{3+x}{4+2x}=\frac{4}{7}$$
$$7(4+2x)\left(\frac{3+x}{4+2x}\right)=7(4+2x)\left(\frac{4}{7}\right)$$
$$7(x+3)=4(4+2x)$$
$$7x+21=16+8x$$
$$5=x$$

The number is 5.

19. Let $x=$ minutes for both to grade the set.

$$\boxed{\begin{array}{c}\text{Teacher}\\\text{in 1 minute}\end{array}}+\boxed{\begin{array}{c}\text{Aide in}\\\text{1 minute}\end{array}}=\boxed{\begin{array}{c}\text{Total in}\\\text{1 minute}\end{array}}$$
$$\frac{1}{60}+\frac{1}{120}=\frac{1}{x}$$
$$120x\left(\frac{1}{60}+\frac{1}{120}\right)=120x\left(\frac{1}{x}\right)$$
$$2x+x=120$$
$$3x=120$$
$$x=40$$

It will take them 40 minutes to grade the set of quizzes.

21. Let $x=$ hours for both pipes to fill the pool.

$$\boxed{\begin{array}{c}\text{1st pipe}\\\text{in 1 hour}\end{array}}+\boxed{\begin{array}{c}\text{2nd pipe}\\\text{in 1 hour}\end{array}}=\boxed{\begin{array}{c}\text{Total in}\\\text{1 hour}\end{array}}$$
$$\frac{1}{5}+\frac{1}{4}=\frac{1}{x}$$
$$20x\left(\frac{1}{5}+\frac{1}{4}\right)=20x\left(\frac{1}{x}\right)$$
$$4x+5x=20$$
$$9x=20$$
$$x=\frac{20}{9}$$

The pool can be filled in $2\frac{2}{9}$ hours.

23. Let r = the rate for the heron.

	d	r	t
Goose	180	$r+10$	$\dfrac{180}{r+10}$
Heron	120	r	$\dfrac{120}{r}$

$$\boxed{\begin{array}{c}\text{Time for}\\\text{goose}\end{array}} = \boxed{\begin{array}{c}\text{Time for}\\\text{heron}\end{array}}$$

$$\frac{180}{r+10} = \frac{120}{r}$$

$$r(r+10)\left(\frac{180}{r+10}\right) = r(r+10)\left(\frac{120}{r}\right)$$

$$180r = 120(r+10)$$
$$180r = 120r + 1200$$
$$60r = 1200$$
$$20 = r$$

The heron's speed is 20 mph, while the goose's speed is 30 mph.

25. Let r = the rate of the plane.

	d	r	t
Plane	300	r	$\dfrac{300}{r}$
Car	120	$r-90$	$\dfrac{120}{r-90}$

$$\boxed{\begin{array}{c}\text{Time for}\\\text{plane}\end{array}} = \boxed{\begin{array}{c}\text{Time for}\\\text{car}\end{array}}$$

$$\frac{300}{r} = \frac{120}{r-90}$$

$$r(r-90)\left(\frac{300}{r}\right) = r(r-90)\left(\frac{120}{r-90}\right)$$

$$300(r-90) = 120r$$
$$300r - 27000 = 120r$$
$$180r = 27000$$
$$r = 150 \Rightarrow \text{The plane travels at 150 miles per hour.}$$

27. Let r = the lower interest rate.

	I	P	r
Lower rate CD	175	$\dfrac{175}{r}$	r
Higher rate CD	200	$\dfrac{200}{r+.01}$	$r+.01$

$$\boxed{\begin{array}{c}\text{Lower rate}\\\text{principal}\end{array}} = \boxed{\begin{array}{c}\text{Higher rate}\\\text{principal}\end{array}}$$

$$\frac{175}{r} = \frac{200}{r+.01}$$

$$r(r+.01)\left(\frac{175}{r}\right) = r(r+.01)\left(\frac{200}{r+.01}\right)$$

$$175(r+.01) = 200r$$
$$175r + 1.75 = 200r$$
$$1.75 = 25r$$
$$0.07 = r \Rightarrow \text{The rates are 7\% and 8\%.}$$

29. Let r = the lower interest rate.

	I	P	r
First fund	300	$\dfrac{300}{r}$	r
Second fund	60	$\dfrac{60}{r - .04}$	$r - .04$

$$\boxed{\begin{array}{c}\text{First}\\\text{principal}\end{array}} = \boxed{\begin{array}{c}\text{Second}\\\text{principal}\end{array}}$$

$$\frac{300}{r} = \frac{60}{r - .04}$$

$$r(r - .04)\left(\frac{300}{r}\right) = r(r - .04)\left(\frac{60}{r - .04}\right)$$

$$300(r - .04) = 60r$$

$$300r - 12 = 60r$$

$$-12 = -240r$$

$$0.05 = r \Rightarrow \text{The higher rate is 5\%.}$$

31. Let x = hours for pool to fill with drain open.

$$\boxed{\begin{array}{c}\text{Pipe in}\\\text{1 hour}\end{array}} - \boxed{\begin{array}{c}\text{Drain in}\\\text{1 hour}\end{array}} = \boxed{\begin{array}{c}\text{Total in}\\\text{1 hour}\end{array}}$$

$$\frac{1}{4} - \frac{1}{8} = \frac{1}{x}$$

$$8x\left(\frac{1}{4} - \frac{1}{8}\right) = 8x\left(\frac{1}{x}\right)$$

$$2x - x = 8$$

$$x = 8; \text{ The pool can be filled in 8 hours.}$$

33. Let x = the number bought.
Then each cost $\frac{120}{x}$.

$$\boxed{\begin{array}{c}\text{Number}\\\text{bought}\end{array}} \cdot \boxed{\begin{array}{c}\text{Amount charged}\\\text{for each}\end{array}} = 120$$

$$(x + 10) \cdot \left(\frac{120}{x} - 1\right) = 120$$

$$120 - x + \frac{1200}{x} - 10 = 120$$

$$\frac{1200}{x} - x - 10 = 0$$

$$x\left(\frac{1200}{x} - x - 10\right) = x(0)$$

$$-x^2 - 10x + 1200 = 0$$

$$-\left(x^2 + 10x - 1200\right) = 0$$

$$-(x + 40)(x - 30) = 0$$

$$x + 40 = 0 \qquad \textbf{or} \quad x - 30 = 0$$

$$x = -40 \qquad\qquad x = 30$$

The store can buy 30 at the regular price.

35. Let r = the speed of the current.

$$\boxed{\begin{array}{c}\text{Time}\\\text{downstream}\end{array}} = \boxed{\begin{array}{c}\text{Time}\\\text{upstream}\end{array}}$$

$$\frac{22}{18 + r} = \frac{14}{18 - r}$$

$$(18 + r)(18 - r)\left(\frac{22}{18 + r}\right) = (18 + r)(18 - r)\left(\frac{14}{18 - r}\right)$$

$$22(18 - r) = 14(18 + r)$$

$$396 - 22r = 252 + 14r$$

$$144 = 36r$$

$$4 = r$$

	d	r	t
Downstream	22	$18 + r$	$\dfrac{22}{18 + r}$
Upstream	14	$18 - r$	$\dfrac{14}{18 - r}$

The current has a speed of 4 miles per hour.

SECTION 6.6

37. Let x = the number. $\Rightarrow x + \dfrac{1}{x} = \dfrac{13}{6}$

$$6x\left(x + \dfrac{1}{x}\right) = 6x\left(\dfrac{13}{6}\right)$$
$$6x^2 + 6 = 13x$$
$$6x^2 - 13x + 6 = 0$$
$$(2x - 3)(3x - 2) = 0$$

$2x - 3 = 0$ **or** $3x - 2 = 0$
$2x = 3 \qquad\qquad 3x = 2$
$x = \dfrac{3}{2} \qquad\qquad x = \dfrac{2}{3}$ The numbers are $\dfrac{3}{2}$ and $\dfrac{2}{3}$.

39. Let r = the slower speed.

	d	r	t
G $\Rightarrow$ P	512	$r+20$	$\dfrac{512}{r+20}$
P $\Rightarrow$ U	528	r	$\dfrac{528}{r}$

$$\boxed{\text{Time to Poland}} + 4 = \boxed{\text{Time to Ukraine}}$$
$$\dfrac{512}{r+20} + 4 = \dfrac{528}{r}$$
$$r(r+20)\left(\dfrac{512}{r+20} + 4\right) = r(r+20)\left(\dfrac{528}{r}\right)$$

$$r(512) + 4r(r+20) = 528(r+20)$$
$$512r + 4r^2 + 80r = 528r + 10{,}560$$
$$4r^2 + 64r - 10{,}560 = 0$$
$$4\left(r^2 + 16r - 2640\right) = 0$$
$$(r+60)(r-44) = 0$$

$r + 60 = 0 \qquad$ **or** $\quad r - 44 = 0$
$\quad r = -60 \qquad\qquad\qquad r = 44$

The speeds are 44 and 64 miles per hour.

41. Let x = the number who contributed.

$$\boxed{\text{Original share}} = \boxed{\text{Share with more workers}} + 2$$
$$\dfrac{60}{x} = \dfrac{60}{x+5} + 2$$
$$x(x+5)\left(\dfrac{60}{x}\right) = x(x+5)\left(\dfrac{60}{x+5} + 2\right)$$
$$60(x+5) = 60x + 2x(x+5)$$
$$60x + 300 = 60x + 2x^2 + 10x$$
$$0 = 2x^2 + 10x - 300$$
$$0 = 2\left(x^2 + 5x - 150\right)$$
$$0 = 2(x+15)(x-10)$$

$x + 15 = 0 \qquad$ **or** $\quad x - 10 = 0$
$\quad x = -15 \qquad\qquad\qquad x = 10$

Since the answer cannot be negative, 10 workers must have contributed.

43. Let r = the still-water speed.

	d	r	t
Upstream	60	$r - 5$	$\dfrac{60}{r - 5}$
Downstream	60	$r + 5$	$\dfrac{60}{r + 5}$

$$\boxed{\text{Time upstream}} + \boxed{\text{Time downstream+}} = 5$$

$$\frac{60}{r - 5} + \frac{60}{r + 5} = 5$$

$$(r + 5)(r - 5)\left(\frac{60}{r - 5} + \frac{60}{r + 5}\right) = 5(r + 5)(r - 5)$$

$$60(r + 5) + 60(r - 5) = 5(r^2 - 25)$$

$$60r + 300 + 60r - 300 = 5r^2 - 125$$

$$0 = 5r^2 - 120r - 125$$

$$0 = 5(r^2 - 24r - 25)$$

$$0 = 5(r + 1)(r - 25)$$

$r + 1 = 0$ **or** $r - 25 = 0$ The still-water speed
$r = -1$ $r = 25$ should be 25 mph.

45. Answers may vary.   **47.** Answers may vary.

Exercises 6.7 (page 427)

1.
$$\frac{3}{5} \stackrel{?}{=} \frac{6}{10}$$
$$3(10) \stackrel{?}{=} 5(6)$$
$$30 = 30$$
proportion

3.
$$\frac{1}{2} \stackrel{?}{=} \frac{1}{4}$$
$$1(4) \stackrel{?}{=} 2(1)$$
$$4 \neq 2$$
not a proportion

5.
$$t = \frac{kl}{w}$$
$$20 = \frac{k(4)}{40}$$
$$20 = \frac{4k}{40}$$
$$800 = 4k \Rightarrow \boxed{200 = k}$$

7. $\left(x^4 x^{-5}\right)^2 = \left(x^{-1}\right)^2 = x^{-2} = \dfrac{1}{x^2}$

9. $\dfrac{3y^0 - 5y^0}{y^0} = \dfrac{3 - 5}{1} = \dfrac{-2}{1} = -2$

11. $470{,}000 = 4.7 \times 10^5$

13. $2.5 \times 10^{-3} = 0.0025$

15. unit costs; rates

17. extremes; means

19. direct

21. rational

23. joint; constant of proportionality

25. direct

27. neither

29.
$$\frac{x}{8} = \frac{28}{32}$$
$$32x = 224$$
$$x = 7$$

31.
$$\frac{r - 2}{3} = \frac{r}{5}$$
$$5(r - 2) = 3r$$
$$5r - 10 = 3r$$
$$2r = 10$$
$$r = 5$$

33.
$$\frac{x + 2}{x - 2} = \frac{7}{3}$$
$$3(x + 2) = 7(x - 2)$$
$$3x + 6 = 7x - 14$$
$$20 = 4x$$
$$5 = x$$

35.
$$\frac{5}{5z+3} = \frac{2z}{2z^2+6}$$
$$5(2z^2+6) = 2z(5z+3)$$
$$10z^2+30 = 10z^2+6z$$
$$30 = 6z$$
$$5 = z$$

37.
$$\frac{1}{x+3} = \frac{-2x}{x+5}$$
$$x+5 = -2x^2-6x$$
$$2x^2+7x+5 = 0$$
$$(2x+5)(x+1)$$
$$2x+5 = 0 \quad \textbf{or} \quad x+1 = 0$$
$$x = -\tfrac{5}{2} \qquad\qquad x = -1$$

39.
$$\frac{a-4}{a+2} = \frac{a-5}{a+1}$$
$$(a-4)(a+1) = (a+2)(a-5)$$
$$a^2-3a-4 = a^2-3a-10$$
$$-3a-4 = -3a-10$$
$$-4 \neq -10$$
$$\text{no solution} \Rightarrow \emptyset$$

41. $A = kp^2$

43. $a = \dfrac{k}{b^2}$

45. $B = kmn$

47. $X = \dfrac{kw}{q}$

49.
$$\frac{2}{c} = \frac{c-3}{2}$$
$$4 = c^2-3c$$
$$0 = c^2-3c-4$$
$$0 = (c+1)(c-4)$$
$$c+1 = 0 \quad \textbf{or} \quad c-4 = 0$$
$$c = -1 \qquad\qquad c = 4$$

51.
$$\frac{9}{5x} = \frac{3x}{15}$$
$$135 = 15x^2$$
$$0 = 15x^2-135$$
$$0 = 15(x+3)(x-3)$$
$$x+3 = 0 \quad \textbf{or} \quad x-3 = 0$$
$$x = -3 \qquad\qquad x = 3$$

53.
$$\frac{a+6}{a+4} = \frac{a-3}{a-4}$$
$$(a+6)(a-4) = (a+4)(a-3)$$
$$a^2+2a-24 = a^2+a-12$$
$$a = 12$$

55. $P = \dfrac{ka^2}{j^3}$

57. L varies jointly with m and n.

59. E varies jointly with a and the square of b.

61. X varies directly with x^2 and inversely with y^2.

63. R varies directly with L and inversely with d^2.

65. Let $c =$ the cost of 5 shirts.
$$\frac{2}{25} = \frac{5}{c}$$
$$2c = 125$$
$$c = 62.5$$
5 shirts will cost $62.50.

67. Let $g =$ gallons of gas for 323 miles.
$$\frac{38}{1} = \frac{323}{g}$$
$$38g = 323$$
$$g = \frac{323}{38} = 8.5$$
8.5 gallons of gas are needed.

69. Let $w =$ the width if it were a real house.

$$\frac{1 \text{ in.}}{1 \text{ ft}} = \frac{32 \text{ in.}}{w \text{ ft}}$$

$$w = 32$$

The width would be 32 feet.

71. Let $h =$ the actual height of the building.

$$\frac{7 \text{ in.}}{280 \text{ ft}} = \frac{2 \text{ in.}}{h \text{ ft}}$$

$$7h = 560$$

$$h = 80$$

The building is 80 feet tall.

73. Let $d =$ the dosage required.

$$\frac{0.006}{1} = \frac{d}{30}$$

$$0.18 = d$$

The dosage is 0.18 grams.

75. Let $h =$ the height of the tree.

$$\frac{6}{4} = \frac{h}{28}$$

$$168 = 4h$$

$$42 = h$$

The tree is 42 feet tall.

77.
$$\frac{20}{32} = \frac{w}{75}$$

$$1500 = 32w$$

$$w = \frac{1500}{32} = 46\frac{7}{8}$$

The width of the river is $46\frac{7}{8}$ feet.

79.
$$\frac{1350}{1} = \frac{x}{5}$$

$$6750 = x$$

The plane will descend 6750 feet.

81. $A = kr^2$

$A = \pi r^2$

$A = \pi (8 \text{ in.})^2$

$\boxed{A = 64\pi \text{ in.}^2}$

83.
$d = kg$ $d = 24g$

$288 = k(12)$ $d = 24(18)$

$24 = k$ $\boxed{d = 432 \text{ mi}}$

85.
$t = \dfrac{k}{x}$ $t = \dfrac{250}{x}$

$10 = \dfrac{k}{25}$ $t = \dfrac{250}{10}$

$250 = k$ $\boxed{t = 25 \text{ days}}$

87.
$V = \dfrac{k}{P}$ $V = \dfrac{120}{P}$

$20 = \dfrac{k}{6}$ $V = \dfrac{120}{10}$

$120 = k$ $\boxed{V = 12 \text{ in.}^3}$

89. $A_1 = klw$

$A_2 = k(3l)(3w) = 9klw = 9A_1$

The area is multiplied by 9.

91.
$c = kth$ $c = 75th$

$1800 = k(4)(6)$ $c = 75(10)(12)$

$1800 = 24k$ $\boxed{c = \$9000}$

$75 = k$

93.
$D = \dfrac{k}{wd^3}$ $D = \dfrac{281.6}{wd^3}$

$1.1 = \dfrac{k}{4(4)^3}$ $D = \dfrac{281.6}{2(8)^3}$

$1.1 = \dfrac{k}{256}$ $D = \dfrac{281.6}{1024}$

$281.6 = k$ $\boxed{D = 0.275 \text{ in.}}$

95.
$P = \dfrac{kT}{V}$ $P = \dfrac{\frac{1}{273}T}{V}$

$1 = \dfrac{k(273)}{1}$ $1 = \dfrac{\frac{1}{273}T}{2}$

$\dfrac{1}{273} = k$ $2 = \dfrac{1}{273}T$

            $\boxed{546 \text{ K} = T}$

97.
$$f = \frac{k}{l} \qquad f = \frac{512}{l}$$
$$256 = \frac{k}{2} \qquad f = \frac{512}{6}$$
$$512 = k \qquad \boxed{f = 85\frac{1}{3}}$$

99. $V = kC$
$6 = k(2)$
$3 = k \Rightarrow$ The resistance is 3 ohms.

101-103. Answers may vary.

105. This is not direct variation. For this to be direct variation, one temperature would have to be a constant multiple of the other.

Chapter 6 Review (page 433)

1.
$$(x+4)(x-2) = 0$$
$$x+4 = 0 \quad \text{or} \quad x-2 = 0$$
$$x = -4 \qquad\qquad x = 2$$

2.
$$x^2 + x - 6 = 0$$
$$(x+3)(x-2) = 0$$
$$x+3 = 0 \quad \text{or} \quad x-2 = 0$$
$$x = -3 \qquad\qquad x = 2$$

3. $x - 7 = 0$
$$x = 7 \ \text{ D: } \{x | x \in \mathbb{R}, x \neq 7\}$$
$$\text{or D: } (-\infty, 7) \cup (7, \infty)$$

4. $x^2 - 5x = 0$
$$x(x - 5) = 0$$
$$x = 0 \quad \text{or} \quad x - 5 = 0$$
$$x = 5$$
$$\text{D: } \{x | x \in \mathbb{R}, x \neq 0, 5\}$$
$$\text{or D: } (-\infty, 0) \cup (0, 5) \cup (5, \infty)$$

5. $-\dfrac{51}{153} = -\dfrac{51 \cdot 1}{51 \cdot 3} = -\dfrac{1}{3}$

6. $\dfrac{105}{45} = \dfrac{15 \cdot 7}{15 \cdot 3} = \dfrac{7}{3}$

7. $\dfrac{3x^2}{6x^3} = \dfrac{1}{2x}$

8. $\dfrac{5xy^2}{2x^2y^2} = \dfrac{5}{2x}$

9. $\dfrac{x^2}{x^2 + x} = \dfrac{x^2}{x(x+1)} = \dfrac{x}{x+1}$

10. $\dfrac{x+2}{x^2 + 2x} = \dfrac{x+2}{x(x+2)} = \dfrac{1}{x}$

11. $\dfrac{12x^2y}{4x^2y^2} = \dfrac{3}{y}$

12. $\dfrac{8x^2y}{2x(4xy)} = \dfrac{8x^2y}{8x^2y} = 1$

13. $\dfrac{7a - 5}{5 - 7a} = -1$

14. $\dfrac{x^2 - x - 56}{x^2 - 5x - 24} = \dfrac{(x-8)(x+7)}{(x-8)(x+3)} = \dfrac{x+7}{x+3}$

15. $\dfrac{2x^2 - 16x}{2x^2 - 18x + 16} = \dfrac{2x(x-8)}{2(x^2 - 9x + 8)} = \dfrac{2x(x-8)}{2(x-8)(x-1)} = \dfrac{x}{x-1}$

16. $\dfrac{a^2 + 2a + ab + 2b}{a^2 + 2ab + b^2} = \dfrac{a(a+2) + b(a+2)}{(a+b)(a+b)} = \dfrac{(a+2)(a+b)}{(a+b)(a+b)} = \dfrac{a+2}{a+b}$

17. $\dfrac{5x^2y}{3x} \cdot \dfrac{6x}{15y^2} = \dfrac{30x^3y}{45xy^2} = \dfrac{2x^2}{3y}$

18. $\dfrac{3x}{x^2-x} \cdot \dfrac{2x-2}{x^2} = \dfrac{3x(2x-2)}{(x^2-x)x^2} = \dfrac{6x(x-1)}{x^3(x-1)}$
$= \dfrac{6}{x^2}$

19. $\dfrac{x^2+3x+2}{x^2+2x} \cdot \dfrac{x}{x+1} = \dfrac{(x^2+3x+2)x}{(x^2+2x)(x+1)} = \dfrac{x(x+2)(x+1)}{x(x+2)(x+1)} = 1$

20. $\dfrac{x^2+x}{3x-15} \cdot \dfrac{6x-30}{x^2+2x+1} = \dfrac{(x^2+x)(6x-30)}{(3x-15)(x^2+2x+1)} = \dfrac{x(x+1)\cdot 6(x-5)}{3(x-5)\cdot(x+1)(x+1)} = \dfrac{2x}{x+1}$

21. $\dfrac{9xy}{14x^2} \div \dfrac{18y^2}{21xy} = \dfrac{9xy}{14x^2} \cdot \dfrac{21xy}{18y^2} = \dfrac{189x^2y^2}{252x^2y^2} = \dfrac{3}{4}$

22. $\dfrac{x^2+5x}{x^2+4x-5} \div \dfrac{x^2}{x-1} = \dfrac{x^2+5x}{x^2+4x-5} \cdot \dfrac{x-1}{x^2} = \dfrac{x(x+5)(x-1)}{x^2(x+5)(x-1)} = \dfrac{1}{x}$

23. $\dfrac{x^2-x-6}{2x-1} \div \dfrac{x^2-2x-3}{2x^2+x-1} = \dfrac{x^2-x-6}{2x-1} \cdot \dfrac{2x^2+x-1}{x^2-2x-3} = \dfrac{(x-3)(x+2)(2x-1)(x+1)}{(2x-1)(x-3)(x+1)}$
$= x+2$

24. $\dfrac{x^2-3x}{x^2-x-6} \div \dfrac{x^2-x}{x^2+x-2} = \dfrac{x^2-3x}{x^2-x-6} \cdot \dfrac{x^2+x-2}{x^2-x} = \dfrac{x(x-3)\cdot(x+2)(x-1)}{(x+2)(x-3)\cdot x(x-1)} = 1$

25. $\dfrac{x^2+4x+4}{x^2+x-6}\left(\dfrac{x-2}{x-1} \div \dfrac{x+2}{x^2+2x-3}\right) = \dfrac{x^2+4x+4}{x^2+x-6}\left(\dfrac{x-2}{x-1} \cdot \dfrac{x^2+2x-3}{x+2}\right)$
$= \dfrac{(x+2)(x+2)}{(x+3)(x-2)}\left(\dfrac{(x-2)(x+3)(x-1)}{(x-1)(x+2)}\right)$
$= \dfrac{(x+2)(x+2)(x-2)(x+3)(x-1)}{(x+3)(x-2)(x-1)(x+2)} = x+2$

26. $\dfrac{5x}{x+3} + \dfrac{x-8}{x+3} = \dfrac{5x+x-8}{x+3} = \dfrac{6x-8}{x+3} = \dfrac{2(3x-4)}{x+3}$

27. $\dfrac{y}{x-y} - \dfrac{x}{x-y} = \dfrac{y-x}{x-y} = -1$

28. $\dfrac{x}{x-1} + \dfrac{1}{x} = \dfrac{(x)(x)}{(x-1)(x)} + \dfrac{1(x-1)}{x(x-1)} = \dfrac{x^2}{x(x-1)} + \dfrac{x-1}{x(x-1)} = \dfrac{x^2+x-1}{x(x-1)}$

29. $\dfrac{1}{7} - \dfrac{1}{x} = \dfrac{1(x)}{7(x)} - \dfrac{1(7)}{x(7)} = \dfrac{x}{7x} - \dfrac{7}{7x} = \dfrac{x-7}{7x}$

30. $\dfrac{3}{x+1} - \dfrac{2}{x} = \dfrac{3(x)}{(x+1)(x)} - \dfrac{2(x+1)}{x(x+1)} = \dfrac{3x}{x(x+1)} - \dfrac{2x+2}{x(x+1)} = \dfrac{3x-2x-2}{x(x+1)} = \dfrac{x-2}{x(x+1)}$

31. $\dfrac{x+2}{2x} - \dfrac{2-x}{x^2} = \dfrac{(x+2)x}{2x(x)} - \dfrac{(2-x)2}{x^2(2)} = \dfrac{x^2+2x}{2x^2} - \dfrac{4-2x}{2x^2} = \dfrac{x^2+2x-4+2x}{2x^2}$

$\qquad\qquad\qquad\qquad\qquad\qquad\qquad\qquad\qquad = \dfrac{x^2+4x-4}{2x^2}$

32. $\dfrac{x}{x+2} + \dfrac{3}{x} - \dfrac{4}{x^2+2x} = \dfrac{x}{x+2} + \dfrac{3}{x} - \dfrac{4}{x(x+2)} = \dfrac{x(x)}{x(x+2)} + \dfrac{3(x+2)}{x(x+2)} - \dfrac{4}{x(x+2)}$

$\qquad\qquad\qquad\qquad\qquad\qquad\qquad = \dfrac{x^2}{x(x+2)} + \dfrac{3x+6}{x(x+2)} - \dfrac{4}{x(x+2)}$

$\qquad\qquad\qquad\qquad\qquad\qquad\qquad = \dfrac{x^2+3x+2}{x(x+2)} = \dfrac{(x+2)(x+1)}{x(x+2)} = \dfrac{x+1}{x}$

33. $\dfrac{2}{x-1} - \dfrac{3}{x+1} + \dfrac{x-5}{x^2-1} = \dfrac{2}{x-1} - \dfrac{3}{x+1} + \dfrac{x-5}{(x+1)(x-1)}$

$\qquad\qquad\qquad\qquad\qquad\qquad = \dfrac{2(x+1)}{(x-1)(x+1)} - \dfrac{3(x-1)}{(x+1)(x-1)} + \dfrac{x-5}{(x+1)(x-1)}$

$\qquad\qquad\qquad\qquad\qquad\qquad = \dfrac{2(x+1)-3(x-1)+x-5}{(x+1)(x-1)}$

$\qquad\qquad\qquad\qquad\qquad\qquad = \dfrac{2x+2-3x+3+x-5}{(x+1)(x-1)} = \dfrac{0}{(x+1)(x-1)} = 0$

34. $\dfrac{\frac{9}{4}}{\frac{4}{9}} = \dfrac{9}{4} \div \dfrac{4}{9} = \dfrac{9}{4} \cdot \dfrac{9}{4} = \dfrac{81}{16}$

35. $\dfrac{\frac{3}{2}+1}{\frac{2}{3}+1} = \dfrac{(\frac{3}{2}+1)6}{(\frac{2}{3}+1)6} = \dfrac{\frac{3}{2}\cdot 6 + 1\cdot 6}{\frac{2}{3}\cdot 6 + 1\cdot 6} = \dfrac{9+6}{4+6}$

$\qquad\qquad\qquad\qquad\qquad\qquad = \dfrac{15}{10} = \dfrac{3}{2}$

36. $\dfrac{\frac{1}{x}+1}{\frac{1}{x}-1} = \dfrac{(\frac{1}{x}+1)x}{(\frac{1}{x}-1)x} = \dfrac{\frac{1}{x}\cdot x + 1\cdot x}{\frac{1}{x}\cdot x - 1\cdot x}$

$\qquad\qquad\qquad = \dfrac{1+x}{1-x}$

37. $\dfrac{2+\frac{7}{x}}{3-\frac{1}{x^2}} = \dfrac{(2+\frac{7}{x})x^2}{(3-\frac{1}{x^2})x^2} = \dfrac{2x^2+7x}{3x^2-1}$

$\qquad\qquad\qquad = \dfrac{x(2x+7)}{3x^2-1}$

38. $\dfrac{\frac{2}{x-1}+\frac{x-1}{x+1}}{\frac{1}{x^2-1}} = \dfrac{\frac{2}{x-1}+\frac{x-1}{x+1}}{\frac{1}{(x+1)(x-1)}} = \dfrac{\left(\frac{2}{x-1}+\frac{x-1}{x+1}\right)(x+1)(x-1)}{\left(\frac{1}{(x+1)(x-1)}\right)(x+1)(x-1)} = \dfrac{2(x+1)+(x-1)(x-1)}{1}$

$\qquad\qquad\qquad\qquad\qquad\qquad\qquad\qquad = 2x+2+x^2-2x+1 = x^2+3$

39. $\dfrac{\frac{a}{b}+c}{\frac{b}{a}+c} = \dfrac{(\frac{a}{b}+c)ab}{(\frac{b}{a}+c)ab} = \dfrac{a^2+abc}{b^2+abc} = \dfrac{a(a+bc)}{b(b+ac)}$

40.
$$\frac{4}{x} = \frac{6}{x-3}$$
$$x(x-3)\left(\frac{4}{x}\right) = x(x-3)\left(\frac{6}{x-3}\right)$$
$$4(x-3) = 6x$$
$$4x - 12 = 6x$$
$$-12 = 2x$$
$$-6 = x \Rightarrow \text{The answer checks.}$$

41.
$$\frac{7}{x+3} = \frac{5}{x+1}$$
$$(x+3)(x+1)\left(\frac{7}{x+3}\right) = (x+3)(x+1)\left(\frac{5}{x+1}\right)$$
$$7(x+1) = 5(x+3)$$
$$7x + 7 = 5x + 15$$
$$2x = 8$$
$$x = 4 \Rightarrow \text{The answer checks.}$$

42.
$$\frac{2}{3x} + \frac{1}{x} = \frac{5}{9}$$
$$9x\left(\frac{2}{3x} + \frac{1}{x}\right) = 9x\left(\frac{5}{9}\right)$$
$$6 + 9 = 5x$$
$$15 = 5x$$
$$3 = x \Rightarrow \text{The answer checks.}$$

43.
$$\frac{2x}{x+4} = \frac{3}{x-1}$$
$$(x+4)(x-1)\left(\frac{2x}{x+4}\right) = (x+4)(x-1)\left(\frac{3}{x-1}\right)$$
$$2x(x-1) = 3(x+4)$$
$$2x^2 - 2x = 3x + 12$$
$$2x^2 - 5x - 12 = 0$$
$$(2x+3)(x-4) = 0$$

$$2x + 3 = 0 \quad \textbf{or} \quad x - 4 = 0 \quad \Rightarrow \text{Both answers check.}$$
$$2x = -3 \qquad\qquad x = 4$$
$$x = -\tfrac{3}{2} \qquad\qquad x = 4$$

44.

$$\frac{2}{x-1} + \frac{3}{x+4} = \frac{-5}{x^2+3x-4}$$

$$\frac{2}{x-1} + \frac{3}{x+4} = \frac{-5}{(x+4)(x-1)}$$

$$(x-1)(x+4)\left(\frac{2}{x-1} + \frac{3}{x+4}\right) = (x-1)(x+4)\left(\frac{-5}{(x+4)(x-1)}\right)$$

$$2(x+4) + 3(x-1) = -5$$

$$2x + 8 + 3x - 3 = -5$$

$$5x = -10$$

$$x = -2 \Rightarrow \text{The answer checks.}$$

45.

$$\frac{4}{x+2} - \frac{3}{x+3} = \frac{6}{x^2+5x+6}$$

$$\frac{4}{x+2} - \frac{3}{x+3} = \frac{6}{(x+2)(x+3)}$$

$$(x+2)(x+3)\left(\frac{4}{x+2} - \frac{3}{x+3}\right) = (x+2)(x+3)\left(\frac{6}{(x+2)(x+3)}\right)$$

$$4(x+3) - 3(x+2) = 6$$

$$4x + 12 - 3x - 6 = 6$$

$$x = 0 \Rightarrow \text{The answer checks.}$$

46.

$$\frac{1}{r} = \frac{1}{r_1} + \frac{1}{r_2}$$

$$rr_1r_2 \cdot \frac{1}{r} = rr_1r_2\left(\frac{1}{r_1} + \frac{1}{r_2}\right)$$

$$r_1r_2 = rr_2 + rr_1$$

$$r_1r_2 - rr_1 = rr_2$$

$$r_1(r_2 - r) = rr_2$$

$$r_1 = \frac{rr_2}{r_2 - r}$$

47.

$$E = 1 - \frac{T_2}{T_1}$$

$$T_1E = T_1\left(1 - \frac{T_2}{T_1}\right)$$

$$T_1E = T_1 - T_2$$

$$T_1E - T_1 = -T_2$$

$$T_1(E - 1) = -T_2$$

$$T_1 = \frac{-T_2}{E-1} = \frac{T_2}{1-E}$$

48.

$$H = \frac{RB}{R+B}$$

$$(R+B)H = (R+B)\left(\frac{RB}{R+B}\right)$$

$$RH + BH = RB$$

$$BH = RB - RH$$

$$BH = R(B - H)$$

$$\frac{BH}{B-H} = R, \text{ or } R = \frac{-BH}{H-B}$$

49. Let $x =$ hours for both pipes to empty.

1st pipe in 1 hour	+	2nd pipe in 1 hour	=	Total in 1 hour

$$\frac{1}{18} + \frac{1}{20} = \frac{1}{x}$$

$$180x\left(\frac{1}{18} + \frac{1}{20}\right) = 180x\left(\frac{1}{x}\right)$$

$$10x + 9x = 180$$

$$19x = 180$$

$$x = \frac{180}{19}$$

It can be emptied in $9\frac{9}{19}$ hours.

50. Let x = days for both working together.

| Painter in 1 day | + | Owner in 1 day | = | Total in 1 day |

$$\frac{1}{8} + \frac{1}{12} = \frac{1}{x}$$

$$24x\left(\frac{1}{8} + \frac{1}{12}\right) = 24x\left(\frac{1}{x}\right)$$

$$3x + 2x = 24$$

$$5x = 24$$

$$x = \frac{24}{5}$$

It can be painted in $4\frac{4}{5}$ days.

51. Let r = the rate at which he jogs.

	d	r	t
Jogs	10	r	$\frac{10}{r}$
Rides	30	$r + 10$	$\frac{30}{r+10}$

| Time he jogs | = | Time he rides |

$$\frac{10}{r} = \frac{30}{r+10}$$

$$r(r+10)\left(\frac{10}{r}\right) = r(r+10)\left(\frac{30}{r+10}\right)$$

$$10(r+10) = 30r$$

$$10r + 100 = 30r$$

$$100 = 20r$$

$$5 = r$$

He jogs 5 miles per hour.

52. Let r = the speed of the wind.

	d	r	t
Downwind	400	$360 + r$	$\frac{400}{360+r}$
Upwind	320	$360 - r$	$\frac{320}{360-r}$

| Time downwind | = | Time upwind |

$$\frac{400}{360+r} = \frac{320}{360-r}$$

$$(360+r)(360-r)\left(\frac{400}{360+r}\right) = (360+r)(360-r)\left(\frac{320}{360-r}\right)$$

$$400(360-r) = 320(360+r)$$

$$144000 - 400r = 115200 + 320r$$

$$28800 = 720r$$

$$40 = r \Rightarrow \text{The wind has a speed of 40 miles per hour.}$$

53.
$$\frac{x+1}{8} = \frac{4x-2}{24}$$

$$24(x+1) = 8(4x-2)$$

$$24x + 24 = 32x - 16$$

$$-8x = -40$$

$$x = 5$$

54.
$$\frac{1}{x+6} = \frac{x+10}{12}$$

$$12 = (x+6)(x+10)$$

$$12 = x^2 + 16x + 60$$

$$0 = x^2 + 16x + 48$$

$$0 = (x+12)(x+4)$$

$$x + 12 = 0 \quad \textbf{or} \quad x + 4 = 0$$

$$x = -12 \qquad\qquad x = -4$$

55. Let $h =$ the height of the tree.

$$\frac{44}{2.5} = \frac{h}{4}$$

$$176 = 2.5h$$

$70.4 = h \Rightarrow$ The tree is 70.4 feet tall.

56.
$$x = ky \qquad x = 6y$$
$$18 = k(3) \qquad x = 6(9)$$
$$6 = k \qquad \boxed{x = 54}$$

57.
$$x = \frac{k}{y} \qquad x = \frac{72}{y}$$
$$24 = \frac{k}{3} \qquad 12 = \frac{72}{y}$$
$$72 = k \qquad 12y = 72$$
$$\boxed{y = 6}$$

58.
$$x = kyz$$
$$24 = k(3)(4)$$
$$24 = 12k$$
$$\boxed{2 = k}$$

59.
$$x = \frac{kt}{y}$$
$$2 = \frac{k(8)}{64}$$
$$128 = 8k$$
$$\boxed{16 = k}$$

60.
$$T = kv \qquad T = kv$$
$$1575 = k(90{,}000) \qquad T = 0.0175(312{,}000)$$
$$\frac{1575}{90{,}000} = k \qquad \boxed{T = \$5460}$$
$$0.0175 = k$$

Chapter 6 Test (page 439)

1. $\dfrac{27x^3y^2}{45xy^3} = \dfrac{9 \cdot 3x^3y^2}{9 \cdot 5xy^3} = \dfrac{3x^2}{5y}$

2. $\dfrac{2x^2 - x - 3}{4x^2 - 9} = \dfrac{(2x-3)(x+1)}{(2x+3)(2x-3)} = \dfrac{x+1}{2x+3}$

3. $\dfrac{3(x+2) - 3}{2x - 4 - (x-5)} = \dfrac{3x + 6 - 3}{2x - 4 - x + 5}$

$= \dfrac{3x + 3}{x+1} = \dfrac{3(x+1)}{x+1} = 3$

4. $\dfrac{12x^2y}{15xyz} \cdot \dfrac{25y^2z}{16xt} = \dfrac{3 \cdot 4 \cdot 5 \cdot 5x^2y^3z}{3 \cdot 5 \cdot 4 \cdot 4x^2ytz} = \dfrac{5y^2}{4t}$

5. $\dfrac{x^2 + 3x + 2}{3x + 9} \cdot \dfrac{x+3}{x^2 - 4} = \dfrac{(x+2)(x+1)}{3(x+3)} \cdot \dfrac{x+3}{(x+2)(x-2)} = \dfrac{(x+2)(x+1)(x+3)}{3(x+3)(x+2)(x-2)} = \dfrac{x+1}{3(x-2)}$

6. $\dfrac{7ab^2}{24ac} \div \dfrac{21a^2b^3}{40abc^2} = \dfrac{7ab^2}{24ac} \cdot \dfrac{40abc^2}{21a^2b^3} = \dfrac{280a^2b^3c^2}{504a^3b^3c} = \dfrac{5c}{9a}$

7. $\dfrac{x^2 - x}{3x^2 + 6x} \div \dfrac{3x - 3}{3x^3 + 6x^2} = \dfrac{x^2 - x}{3x^2 + 6x} \cdot \dfrac{3x^3 + 6x^2}{3x - 3} = \dfrac{x(x-1)}{3x(x+2)} \cdot \dfrac{3x^2(x+2)}{3(x-1)}$

$= \dfrac{3x^3(x-1)(x+2)}{9x(x-1)(x+2)} = \dfrac{x^2}{3}$

8. $\dfrac{x^2+xy}{x-y} \cdot \dfrac{x^2-y^2}{x^2-2x} \div \dfrac{x^2+2xy+y^2}{x^2-4} = \dfrac{x^2+xy}{x-y} \cdot \dfrac{x^2-y^2}{x^2-2x} \cdot \dfrac{x^2-4}{x^2+2xy+y^2}$

$$= \dfrac{x(x+y)}{x-y} \cdot \dfrac{(x+y)(x-y)}{x(x-2)} \cdot \dfrac{(x+2)(x-2)}{(x+y)(x+y)}$$

$$= \dfrac{x(x+y)(x+y)(x-y)(x+2)(x-2)}{x(x-y)(x-2)(x+y)(x+y)} = x+2$$

9. $\dfrac{6x+5}{x-2} + \dfrac{3x-7}{x-2} = \dfrac{6x+5+3x-7}{x-2} = \dfrac{9x-2}{x-2}$

10. $\dfrac{3y+7}{2y+3} - \dfrac{3(y-2)}{2y+3} = \dfrac{3y+7-3(y-2)}{2y+3} = \dfrac{3y+7-3y+6}{2y+3} = \dfrac{13}{2y+3}$

11. $\dfrac{x+1}{x} + \dfrac{x-1}{x+1} = \dfrac{(x+1)(x+1)}{x(x+1)} + \dfrac{x(x-1)}{x(x+1)} = \dfrac{(x+1)(x+1)+x(x-1)}{x(x+1)}$

$$= \dfrac{x^2+2x+1+x^2-x}{x(x+1)} = \dfrac{2x^2+x+1}{x(x+1)}$$

12. $\dfrac{5x}{x-2} - 3 = \dfrac{5x}{x-2} - \dfrac{3}{1} = \dfrac{5x}{x-2} - \dfrac{3(x-2)}{x-2} = \dfrac{5x-3(x-2)}{x-2} = \dfrac{5x-3x+6}{x-2} = \dfrac{2x+6}{x-2}$

13. $\dfrac{\frac{9x^3}{xy^2}}{\frac{6y^2}{x^3y^4}} = \dfrac{9x^3}{xy^2} \div \dfrac{6y^2}{x^3y^4} = \dfrac{9x^3}{xy^2} \cdot \dfrac{x^3y^4}{6y^2} = \dfrac{9x^6y^4}{6xy^4} = \dfrac{3x^5}{2}$

14. $\dfrac{1+\frac{y}{x}}{\frac{y}{x}-1} = \dfrac{\left(1+\frac{y}{x}\right)x}{\left(\frac{y}{x}-1\right)x} = \dfrac{1 \cdot x + \frac{y}{x} \cdot x}{\frac{y}{x} \cdot x - 1 \cdot x}$

$$= \dfrac{x+y}{y-x}$$

15. $\dfrac{x}{10} - \dfrac{1}{2} = \dfrac{x}{5}$

$$10\left(\dfrac{x}{10} - \dfrac{1}{2}\right) = 10\left(\dfrac{x}{5}\right)$$

$$x - 5 = 2x$$

$$-5 = x$$

16.
$$\dfrac{1}{x+2} + \dfrac{1}{x-2} = \dfrac{4}{x^2-4}$$

$$\dfrac{1}{x+2} + \dfrac{1}{x-2} = \dfrac{4}{(x+2)(x-2)}$$

$$(x+2)(x-2)\left(\dfrac{1}{x+2} + \dfrac{1}{x-2}\right) = (x+2)(x-2)\left(\dfrac{4}{(x+2)(x-2)}\right)$$

$$x-2+x+2 = 4$$

$$2x = 4$$

$$x = 2; \text{ 2 is extraneous} \Rightarrow \emptyset$$

17.
$$\frac{7}{x+4} - \frac{1}{2} = \frac{3}{x+4}$$
$$2(x+4)\left(\frac{7}{x+4} - \frac{1}{2}\right) = 2(x+4)\left(\frac{3}{x+4}\right)$$
$$2(7) - (x+4) = 2(3)$$
$$14 - x - 4 = 6$$
$$-x = -4$$
$$x = 4$$

The answer checks.

18.
$$H = \frac{RB}{R+B}$$
$$(R+B)H = (R+B)\left(\frac{RB}{R+B}\right)$$
$$RH + BH = RB$$
$$RH = RB - BH$$
$$RH = B(R - H)$$
$$\frac{RH}{R-H} = B$$

19. Let x = hours working together.

$$\boxed{\begin{array}{c}\text{1st worker}\\\text{in 1 hour}\end{array}} + \boxed{\begin{array}{c}\text{2nd worker}\\\text{in 1 hour}\end{array}} = \boxed{\begin{array}{c}\text{Total in}\\\text{1 hour}\end{array}}$$
$$\frac{1}{7} + \frac{1}{9} = \frac{1}{x}$$
$$63x\left(\frac{1}{7} + \frac{1}{9}\right) = 63x\left(\frac{1}{x}\right)$$
$$9x + 7x = 63$$
$$16x = 63$$
$$x = \frac{63}{16}$$

They can finish in $3\frac{15}{16}$ hours.

20. Let r = the speed of the current.

$$\boxed{\begin{array}{c}\text{Time}\\\text{downstream}\end{array}} = \boxed{\begin{array}{c}\text{Time}\\\text{upstream}\end{array}}$$
$$\frac{28}{23+r} = \frac{18}{23-r}$$
$$(23+r)(23-r)\left(\frac{28}{23+r}\right) = (23+r)(23-r)\left(\frac{18}{23-r}\right)$$
$$28(23-r) = 18(23+r)$$
$$644 - 28r = 414 + 18r$$
$$230 = 46r$$
$$5 = r$$

The current has a speed of 5 miles per hour.

	d	r	t
Downstream	28	$23+r$	$\dfrac{28}{23+r}$
Upstream	18	$23-r$	$\dfrac{18}{23-r}$

21.
$$\frac{575}{\frac{1}{2}} = \frac{x}{7}$$
$$7(575) = \frac{1}{2}x$$
$$4025 = \frac{1}{2}x$$
$$2(4025) = 2\left(\frac{1}{2}x\right)$$
$$8050 = x \Rightarrow \text{The plane will lose 8,050 feet of altitude.}$$

22. $\dfrac{y}{y-1} = \dfrac{y-2}{y}$

$$y(y) = (y-1)(y-2)$$
$$y^2 = y^2 - 3y + 2$$
$$3y = 2$$
$$y = \dfrac{2}{3}$$

23. Let h = the height of the tree.

$$\dfrac{12}{2} = \dfrac{h}{3}$$
$$36 = 2h$$
$$18 = h$$

The tree is 18 feet tall.

24. $\dfrac{3}{x-2} = \dfrac{x+3}{2x}$

$$6x = (x+3)(x-2)$$
$$6x = x^2 + x - 6$$
$$0 = x^2 - 5x - 6$$
$$0 = (x-6)(x+1)$$
$$x - 6 = 0 \quad \textbf{or} \quad x + 1 = 0$$
$$x = 6 \qquad\qquad x = -1$$

25. $\quad V = \dfrac{k}{t} \qquad V = \dfrac{1100}{t}$

$$55 = \dfrac{k}{20} \qquad 75 = \dfrac{1100}{t}$$
$$1100 = k \qquad 75t = 1100$$
$$t = \dfrac{1100}{75}$$
$$\boxed{t = \dfrac{44}{3}}$$

26. **a.** $x - 3 = 0$
$$x = 3$$

b. D: $\{x \mid x \in \mathbb{R},\ x \neq 3\}$ or $(-\infty, 3) \cup (3, \infty)$

Cumulative Review Exercises (page 440)

1. $x^3 x^6 = x^{3+6} = x^9$

2. $\left(x^2\right)^5 = x^{2 \cdot 5} = x^{10}$

3. $\dfrac{x^5}{x^2} = x^{5-2} = x^3$

4. $\left(6x^4\right)^0 = 1$

5. $(3x^2 - 2x) + (6x^3 - 3x^2 - 1) = 3x^2 - 2x + 6x^3 - 3x^2 - 1 = 6x^3 - 2x - 1$

6. $(4x^3 - 2x) - (2x^3 - 2x^2 - 3x + 1) = 4x^3 - 2x - 2x^3 + 2x^2 + 3x - 1 = 2x^3 + 2x^2 + x - 1$

7. $3(5x^2 - 4x + 3) + 2(-x^2 + 2x - 4) = 15x^2 - 12x + 9 - 2x^2 + 4x - 8 = 13x^2 - 8x + 1$

8. $4(3x^2 - 4x - 1) - 2(-2x^2 + 4x - 3) = 12x^2 - 16x - 4 + 4x^2 - 8x + 6 = 16x^2 - 24x + 2$

9. $(2x^4 y^3)(-7x^5 y) = 2(-7)x^4 x^5 y^3 y = -14x^9 y^4$

10. $-5x^2(7x^3 - 2x^2 - 2) = -35x^5 + 10x^4 + 10x^2$

11. $(5x + 2)(4x + 1) = 20x^2 + 5x + 8x + 2 = 20x^2 + 13x + 2$

12. $(5x - 4y)(3x + 2y) = 15x^2 + 10xy - 12xy - 8y^2 = 15x^2 - 2xy - 8y^2$

13.
$$\begin{array}{r} x +\ \ 4 \\ x+3\ \overline{\smash{\big)}\ x^2+7x+12} \\ \underline{x^2+3x} \\ 4x+12 \\ \underline{4x+12} \\ 0 \end{array}$$

14.
$$\begin{array}{r} x^2 +\ \ x +\ 1 \\ 2x-3\ \overline{\smash{\big)}\ 2x^3-\ x^2-\ x-3} \\ \underline{2x^3-3x^2} \\ 2x^2-\ x \\ \underline{2x^2-3x} \\ 2x-3 \\ \underline{2x-3} \\ 0 \end{array}$$

15. $4xy^2 - 12x^2y^3 = 4xy^2(1 - 3xy)$

16. $3(a+b) + x(a+b) = (a+b)(3+x)$

17. $2a + 2b + ab + b^2 = 2(a+b) + b(a+b)$
$$= (a+b)(2+b)$$

18. $25p^4 - 16q^2 = \left(5p^2\right)^2 - \left(4q\right)^2$
$$= \left(5p^2 + 4q\right)\left(5p^2 - 4q\right)$$

19. $x^2 - 5x - 14 = (x-7)(x+2)$

20. $x^2 - xy - 6y^2 = (x-3y)(x+2y)$

21. $6a^2 - 7a - 20 = (2a-5)(3a+4)$

22. $8m^2 - 10mn - 3n^2 = (4m+n)(2m-3n)$

23. $p^3 - 27q^3 = p^3 - (3q)^3 = (p-3q)\left(p^2 + p(3q) + (3q)^2\right) = (p-3q)(p^2 + 3pq + 9q^2)$

24. $8r^3 + 64s^3 = 8\left(r^3 + 8s^3\right) = 8\left[r^3 + (2s)^3\right] = 8(r+2s)\left(r^2 - r \cdot 2s + (2s)^2\right)$
$$= 8(r+2s)\left(r^2 - 2rs + 4s^2\right)$$

25.
$$\frac{4}{5}x + 6 = 18$$
$$\frac{4}{5}x = 12$$
$$5 \cdot \frac{4}{5}x = 5(12)$$
$$4x = 60$$
$$x = 15$$

26.
$$5 - \frac{x+2}{3} = 7 - x$$
$$3\left(5 - \frac{x+2}{3}\right) = 3(7-x)$$
$$15 - (x+2) = 21 - 3x$$
$$15 - x - 2 = 21 - 3x$$
$$2x = 8$$
$$x = 4$$

27.
$$6x^2 - x - 2 = 0$$
$$(2x+1)(3x-2) = 0$$
$$2x+1 = 0 \quad \textbf{or} \quad 3x-2 = 0$$
$$2x = -1 \qquad\qquad 3x = 2$$
$$x = -\tfrac{1}{2} \qquad\qquad x = \tfrac{2}{3}$$

28.
$$5x^2 = 10x$$
$$5x^2 - 10x = 0$$
$$5x(x-2) = 0$$
$$5x = 0 \quad \textbf{or} \quad x-2 = 0$$
$$x = 0 \qquad\qquad x = 2$$

29.
$$x^2 + 6x + 5 = 0$$
$$(x+1)(x+5) = 0$$
$$x+1 = 0 \quad \textbf{or} \quad x+5 = 0$$
$$x = -1 \qquad\qquad x = -5$$

30.
$$2y^2 + 5y - 12 = 0$$
$$(2y-3)(y+4) = 0$$
$$2y-3 = 0 \quad \textbf{or} \quad y+4 = 0$$
$$2y = 3 \qquad\qquad y = -4$$
$$y = \tfrac{3}{2} \qquad\qquad y = -4$$

CUMULATIVE REVIEW EXERCISES

31. $4x - 7 > 1$
$4x > 8$
$x > 2$

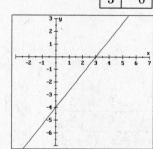

32. $7x - 9 < 5$
$7x < 14$
$x < 2$

33. $-2 < -x + 3 < 5$
$-5 < \quad -x \quad < 2$
$\dfrac{-5}{-1} > \quad \dfrac{-x}{-1} \quad > \dfrac{2}{-1}$
$5 > \quad x \quad > -2$
$-2 < \quad x \quad < 5$

34. $0 \leq \dfrac{4-x}{3} \leq 2$
$3(0) \leq 3 \cdot \dfrac{4-x}{3} \leq 3(2)$
$0 \leq \quad 4 - x \quad \leq 6$
$-4 \leq \quad -x \quad \leq 2$
$\dfrac{-4}{-1} \geq \quad \dfrac{-x}{-1} \quad \geq \dfrac{2}{-1}$
$4 \geq \quad x \quad \geq -2$
$-2 \leq \quad x \quad \leq 4$

35. $4x - 3y = 12$

x	y
0	-4
3	0

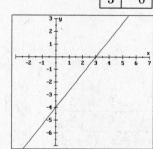

36. $3x + 4y = 4y + 12$
$3x = 12$
$x = 4$

x	y
4	0
4	2

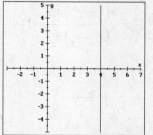

37. $f(-1) = 2(-1)^2 - 3 = 2(1) - 3$
$\qquad = 2 - 3 = -1$

38. $f(3) = 2(3)^2 - 3 = 2(9) - 3 = 18 - 3 = 15$

39. $f(-2) = 2(-2)^2 - 3 = 2(4) - 3$
$\qquad = 8 - 3 = 5$

40. $f(2x) = 2(2x)^2 - 3 = 2\left(4x^2\right) - 3$
$\qquad = 8x^2 - 3$

41. $\dfrac{x^2 + 3x + 2}{x^2 - 4} = \dfrac{(x+1)(x+2)}{(x+2)(x-2)} = \dfrac{x+1}{x-2}$

42. $\dfrac{x^2 + 2x - 15}{x^2 + 3x - 10} = \dfrac{(x+5)(x-3)}{(x+5)(x-2)} = \dfrac{x-3}{x-2}$

43. $\dfrac{x^2 + x - 6}{5x - 5} \cdot \dfrac{5x - 10}{x + 3} = \dfrac{(x+3)(x-2) \cdot 5(x-2)}{5(x-1) \cdot (x+3)} = \dfrac{(x-2)^2}{x-1}$

44. $\dfrac{p^2 - p - 6}{3p - 9} \div \dfrac{p^2 + 6p + 9}{p^2 - 9} = \dfrac{p^2 - p - 6}{3p - 9} \cdot \dfrac{p^2 - 9}{p^2 + 6p + 9} = \dfrac{(p-3)(p+2) \cdot (p+3)(p-3)}{3(p-3) \cdot (p+3)(p+3)}$

$\qquad\qquad\qquad\qquad\qquad\qquad\qquad\qquad\qquad\qquad\quad = \dfrac{(p+2)(p-3)}{3(p+3)}$

45. $\dfrac{3x}{x+2} + \dfrac{5x}{x+2} - \dfrac{7x - 2}{x+2} = \dfrac{3x + 5x - (7x - 2)}{x + 2} = \dfrac{3x + 5x - 7x + 2}{x + 2} = \dfrac{x+2}{x+2} = 1$

46. $\dfrac{x-1}{x+1} + \dfrac{x+1}{x-1} = \dfrac{(x-1)(x-1)}{(x+1)(x-1)} + \dfrac{(x+1)(x+1)}{(x-1)(x+1)} = \dfrac{x^2 - 2x + 1}{(x+1)(x-1)} + \dfrac{x^2 + 2x + 1}{(x+1)(x-1)}$

$\qquad\qquad\qquad\qquad\qquad\qquad\qquad\qquad\qquad\quad = \dfrac{2x^2 + 2}{(x+1)(x-1)} = \dfrac{2(x^2 + 1)}{(x+1)(x-1)}$

47. $\dfrac{a+1}{2a+4} - \dfrac{a^2}{2a^2 - 8} = \dfrac{a+1}{2(a+2)} - \dfrac{a^2}{2(a+2)(a-2)} = \dfrac{(a+1)(a-2)}{2(a+2)(a-2)} - \dfrac{a^2}{2(a+2)(a-2)}$

$\qquad\qquad\qquad\qquad\qquad\qquad\qquad\qquad\quad = \dfrac{a^2 - 2a + a - 2 - a^2}{2(a+2)(a-2)}$

$\qquad\qquad\qquad\qquad\qquad\qquad\qquad\qquad\quad = \dfrac{-a - 2}{2(a+2)(a-2)}$

$\qquad\qquad\qquad\qquad\qquad\qquad\qquad\qquad\quad = \dfrac{-(a+2)}{2(a+2)(a-2)} = -\dfrac{1}{2(a-2)}$

48. $\dfrac{\frac{1}{x} + \frac{1}{y}}{\frac{1}{x} - \frac{1}{y}} = \dfrac{\left(\frac{1}{x} + \frac{1}{y}\right)xy}{\left(\frac{1}{x} - \frac{1}{y}\right)xy} = \dfrac{\frac{1}{x} \cdot xy + \frac{1}{y} \cdot xy}{\frac{1}{x} \cdot xy - \frac{1}{y} \cdot xy} = \dfrac{y + x}{y - x}$

49.

$$
\begin{array}{r|rrr}
-5 & 1 & 9 & 20 \\
   &   & -5 & -20 \\
\hline
   & 1 & 4 & 0
\end{array}
$$

$\Rightarrow \boxed{x + 4}$

50.

$$
\begin{array}{r|rrrr}
1 & -1 & 2 & 4 & 3 \\
  &   & -1 & 1 & 5 \\
\hline
  & -1 & 1 & 5 & 8
\end{array}
$$

$\Rightarrow \boxed{-x^2 + x + 5 + \dfrac{8}{x-1}}$

Exercises 7.1 (page 452)

1. $2x + 4 = 6$
$\quad\ 2x = 2$
$\qquad x = 1$

3. $2x < 4$
$\quad\ x < 2$

5. $3x < -12$
$\quad \dfrac{3x}{3} < \dfrac{-12}{3}$
$\qquad x < -4$

7. $\left(\dfrac{t^3 t^5 t^{-6}}{t^2 t^{-4}}\right)^{-3} = \left(\dfrac{t^2 t^{-4}}{t^3 t^5 t^{-6}}\right)^3 = \left(\dfrac{t^{-2}}{t^2}\right)^3 = \left(\dfrac{1}{t^4}\right)^3 = \dfrac{1}{t^{12}}$

9. Let x = the number of pies made.

$$\boxed{\text{Expenses}} = \boxed{\text{Income}}$$

$$1068 + 3.50x = 9.50x$$
$$1068 = 6x$$
$$178 = x \Rightarrow \text{He must sell at least 179 pies to make a profit.}$$

11. equation

13. multiplied; divided

15. identity

17. contradiction

19. half-open

21. negative

23.
$$2x + 1 = 13$$
$$2x + 1 - 1 = 13 - 1$$
$$2x = 12$$
$$\frac{2x}{2} = \frac{12}{2}$$
$$x = 6$$

25.
$$3(x + 1) = 15$$
$$3x + 3 = 15$$
$$3x + 3 - 3 = 15 - 3$$
$$3x = 12$$
$$\frac{3x}{3} = \frac{12}{3}$$
$$x = 4$$

27.
$$2r - 5 = 1 - r$$
$$2r + r - 5 = 1 - r + r$$
$$3r - 5 + 5 = 1 + 5$$
$$3r = 6$$
$$\frac{3r}{3} = \frac{6}{3}$$
$$r = 2$$

29.
$$3(2y - 4) - 6 = 3y$$
$$6y - 12 - 6 = 3y$$
$$6y - 18 = 3y$$
$$6y - 3y - 18 = 3y - 3y$$
$$3y - 18 + 18 = 0 + 18$$
$$3y = 18$$
$$\frac{3y}{3} = \frac{18}{3}$$
$$y = 6$$

31.
$$5(5 - a) = 37 - 2a$$
$$25 - 5a = 37 - 2a$$
$$25 - 5a + 2a = 37 - 2a + 2a$$
$$25 - 25 - 3a = 37 - 25$$
$$-3a = 12$$
$$\frac{-3a}{-3} = \frac{12}{-3}$$
$$a = -4$$

33.
$$4(y + 1) = -2(4 - y)$$
$$4y + 4 = -8 + 2y$$
$$4y - 2y + 4 = -8 + 2y - 2y$$
$$2y + 4 - 4 = -8 - 4$$
$$2y = -12$$
$$\frac{2y}{2} = \frac{-12}{2}$$
$$y = -6$$

35.
$$\frac{x}{2} - \frac{x}{3} = 4$$
$$6\left(\frac{x}{2} - \frac{x}{3}\right) = 6(4)$$
$$6\left(\frac{x}{2}\right) - 6\left(\frac{x}{3}\right) = 24$$
$$3x - 2x = 24$$
$$x = 24$$

37.
$$\frac{x}{6} + 1 = \frac{x}{3}$$
$$6\left(\frac{x}{6} + 1\right) = 6\left(\frac{x}{3}\right)$$
$$6\left(\frac{x}{6}\right) + 6(1) = 2x$$
$$x + 6 = 2x$$
$$x - 2x + 6 = 2x - 2x$$
$$-x + 6 - 6 = 0 - 6$$
$$-x = -6$$
$$x = 6$$

39.
$$\frac{a+1}{3} + \frac{a-1}{5} = \frac{2}{15}$$
$$15\left(\frac{a+1}{3} + \frac{a-1}{5}\right) = 15\left(\frac{2}{15}\right)$$
$$15\left(\frac{a+1}{3}\right) + 15\left(\frac{a-1}{5}\right) = 2$$
$$5(a+1) + 3(a-1) = 2$$
$$5a + 5 + 3a - 3 = 2$$
$$8a + 2 - 2 = 2 - 2$$
$$8a = 0$$
$$\frac{8a}{8} = \frac{0}{8}$$
$$a = 0$$

41.
$$\frac{2z+3}{3} + \frac{3z-4}{6} = \frac{z-2}{2}$$
$$6\left(\frac{2z+3}{3} + \frac{3z-4}{6}\right) = 6 \cdot \frac{z-2}{2}$$
$$6\left(\frac{2z+3}{3}\right) + 6\left(\frac{3z-4}{6}\right) = 3(z-2)$$
$$2(2z+3) + 3z - 4 = 3z - 6$$
$$4z + 6 + 3z - 4 = 3z - 6$$
$$7z + 2 = 3z - 6$$
$$7z - 3z + 2 = 3z - 3z - 6$$
$$4z + 2 - 2 = -6 - 2$$
$$4z = -8$$
$$\frac{4z}{4} = \frac{-8}{4}$$
$$z = -2$$

43.
$$4(2 - 3t) + 6t = -6t + 8$$
$$8 - 12t + 6t = -6t + 8$$
$$8 - 6t = -6t + 8$$
$$-6t + 8 = -6t + 8$$
Identity; $\mathbb{R}, (-\infty, \infty)$

45.
$$3(x - 4) + 6 = -2(x + 4) + 5x$$
$$3x - 12 + 6 = -2x - 8 + 5x$$
$$3x - 6 = 3x - 8$$
$$3x - 3x - 6 = 3x - 3x - 8$$
$$-6 \neq -8$$
Contradiction; $\emptyset$

47.
$$V = \frac{1}{3}Bh$$
$$3V = 3 \cdot \frac{1}{3}Bh$$
$$3V = Bh$$
$$\frac{3V}{h} = \frac{Bh}{h}$$
$$\frac{3V}{h} = B, \text{ or } B = \frac{3V}{h}$$

49.
$$P = 2l + 2w$$
$$P - 2l = 2l - 2l + 2w$$
$$P - 2l = 2w$$
$$\frac{P - 2l}{2} = \frac{2w}{2}$$
$$\frac{P - 2l}{2} = w, \text{ or } w = \frac{P - 2l}{2}$$

51.
$$z = \frac{x - \mu}{\sigma}$$
$$\sigma z = \sigma \cdot \frac{x - \mu}{\sigma}$$
$$\sigma z = x - \mu$$
$$\sigma z + \mu = x, \text{ or } x = z\sigma + \mu$$

53.
$$y = mx + b$$
$$y - b = mx + b - b$$
$$y - b = mx$$
$$\frac{y - b}{m} = \frac{mx}{m}$$
$$\frac{y - b}{m} = x, \text{ or } x = \frac{y - b}{m}$$

55.
$$5x - 3 > 7$$
$$5x - 3 + 3 > 7 + 3$$
$$5x > 10$$
$$\frac{5x}{5} > \frac{10}{5}$$
$$x > 2$$
solution set: $(2, \infty)$

57.
$$6x + 5 \leq -13$$
$$6x + 5 - 5 \leq -13 - 5$$
$$6x \leq -18$$
$$\frac{6x}{6} \leq -\frac{18}{6}$$
$$x \leq -3$$
solution set: $(-\infty, -3]$

59.
$$-3x - 1 \leq 5$$
$$-3x - 1 + 1 \leq 5 + 1$$
$$-3x \leq 6$$
$$\frac{-3x}{-3} \geq \frac{6}{-3}$$
$$x \geq -2$$
solution set: $[-2, \infty)$

61.
$$-3(a + 2) > 2(a + 1)$$
$$-3a - 6 > 2a + 2$$
$$-3a - 2a - 6 > 2a - 2a + 2$$
$$-5a - 6 + 6 > 2 + 6$$
$$-5a > 8$$
$$\frac{-5a}{-5} < \frac{8}{-5}$$
$$a < -\frac{8}{5}$$
solution set: $\left(-\infty, -\frac{8}{5}\right)$

63.
$$\frac{1}{2}y + 2 \geq \frac{1}{3}y - 4$$
$$6\left(\frac{1}{2}y + 2\right) \geq 6\left(\frac{1}{3}y - 4\right)$$
$$3y + 12 \geq 2y - 24$$
$$3y - 2y + 12 - 12 \geq 2y - 2y - 24 - 12$$
$$y \geq -36$$
solution set: $[-36, \infty)$

65.
$$\frac{1}{5}y - \frac{8}{5} + 2 \leq \frac{2}{5} - \frac{1}{3}y$$
$$15\left(\frac{1}{5}y - \frac{8}{5} + 2\right) \leq 15\left(\frac{2}{5} - \frac{1}{3}y\right)$$
$$3y - 24 + 30 \leq 6 - 5y$$
$$3y + 6 \leq 6 - 5y$$
$$3y + 5y + 6 - 6 \leq 6 - 5y + 5y - 6$$
$$8y \leq 0$$
$$\frac{8y}{8} \leq \frac{0}{8}$$
$$y \leq 0$$
solution set: $(-\infty, 0]$

67.
$$\frac{x+1}{4} - \frac{2x-4}{2} > \frac{3-2x}{4}$$
$$4\left(\frac{x+1}{4} - \frac{2x-4}{2}\right) > 4\left(\frac{3-2x}{4}\right)$$
$$x+1 - 2(2x-4) > 3 - 2x$$
$$x + 1 - 4x + 8 > 3 - 2x$$
$$-3x + 9 > 3 - 2x$$
$$-3x + 2x + 9 - 9 > 3 - 2x + 2x - 9$$
$$-x > -6$$
$$\frac{-x}{-1} < \frac{-6}{-1}$$
$$x < 6$$
solution set: $(-\infty, 6)$

69.
$$\frac{3}{2}(x+4) > 10 - \frac{x}{2}$$
$$2 \cdot \frac{3}{2}(x+4) > 2\left(10 - \frac{x}{2}\right)$$
$$3(x+4) > 20 - x$$
$$3x + 12 > 20 - x$$
$$3x + x + 12 - 12 > 20 - x + x - 12$$
$$4x > 8$$
$$\frac{4x}{4} > \frac{8}{4}$$
$$x > 2$$
solution set: $(2, \infty)$

71.
$$-2 < -b + 3 < 5$$
$$-2 - 3 < -b + 3 - 3 < 5 - 3$$
$$-5 < -b < 2$$
$$\frac{-5}{-1} > \frac{-b}{-1} > \frac{2}{-1}$$
$$5 > b > -2, \text{ or } -2 < b < 5$$
solution set: $(-2, 5)$

73.
$$15 > 2x - 7 > 9$$
$$15 + 7 > 2x - 7 + 7 > 9 + 7$$
$$22 > 2x > 16$$
$$\frac{22}{2} > \frac{2x}{2} > \frac{16}{2}$$
$$11 > x > 8, \text{ or } 8 < x < 11$$
solution set: $(8, 11)$

75.
$$-6 < -3(x-4) \le 24$$
$$\frac{-6}{-3} > \frac{-3(x-4)}{-3} \ge \frac{24}{-3}$$
$$2 > x - 4 \ge -8$$
$$2 + 4 > x - 4 + 4 \ge -8 + 4$$
$$6 > x \ge -4, \text{ or } -4 \le x < 6$$
solution set: $[-4, 6)$

77. $0 \ge \dfrac{1}{2}x - 4 > 6$

This inequality indicates that $0 \ge 6$ (by the transitive property). Since this is not possible, there is no solution to the inequality. $\emptyset$

79.
$$x - 1 \le 2x + 5 < 3x$$
$$x - 1 \le 2x + 5 \quad \textbf{and} \quad 2x + 5 < 3x$$
$$-x \le 6 \qquad\qquad -x < -5$$
$$x \ge -6 \qquad\qquad x > 5$$

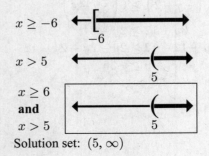

$x \ge -6$

$x > 5$

$x \ge 6$
and
$x > 5$

Solution set: $(5, \infty)$

81.
$$x + 5 \le 2x - 3 \le 5x - 1$$
$$x + 5 \le 2x - 3 \quad \textbf{and} \quad 2x - 3 \le 5x - 1$$
$$-x \le -8 \qquad\qquad -3x \le 2$$
$$x \ge 8 \qquad\qquad\quad x \ge -\tfrac{2}{3}$$

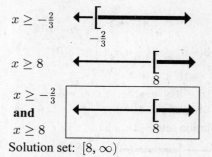

$x \ge -\tfrac{2}{3}$

$x \ge 8$

$x \ge -\tfrac{2}{3}$
and
$x \ge 8$

Solution set: $[8, \infty)$

83.
$$3x + 2 < 8 \qquad \text{or} \qquad 2x - 3 > 11$$
$$3x + 2 - 2 < 8 - 2 \qquad 2x - 3 + 3 > 11 + 3$$
$$3x < 6 \qquad\qquad\qquad 2x > 14$$
$$x < 2 \qquad\qquad\qquad x > 7$$
If $x < 2$ **or** $x > 7$, then the solution set is $(-\infty, 2) \cup (7, \infty)$.

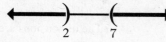

85.
$$-4(x + 2) \ge 12 \qquad \text{or} \qquad 3x + 8 < 11$$
$$-4x - 8 + 8 \ge 12 + 8 \qquad 3x + 8 - 8 < 11 - 8$$
$$-4x \ge 20 \qquad\qquad\qquad 3x < 3$$
$$\tfrac{-4x}{-4} \le \tfrac{20}{-4} \qquad\qquad\qquad x < 1$$
$$x \le -5$$
If $x \le -5$ **or** $x < 1$, then $x < 1$. The solution set is $(-\infty, 1)$.

87. $x < -3 \quad$ **and** $\quad x > 3$
It is impossible for a number to satisfy
both of these inequalities at the same time.
There is no solution $\Rightarrow \emptyset$.

89.
$$y(y + 2) = (y + 1)^2 - 1$$
$$y^2 + 2y = (y + 1)(y + 1) - 1$$
$$y^2 + 2y = y^2 + 2y + 1 - 1$$
$$y^2 + 2y = y^2 + 2y$$
Identity; $\mathbb{R}, \ (-\infty, \infty)$

91.
$$8 - 9y \ge -y$$
$$8 - 9y + y \ge -y + y$$
$$8 - 8 - 8y \ge 0 - 8$$
$$-8y \ge -8$$
$$\tfrac{-8y}{-8} \le \tfrac{-8}{-8}$$
$$y \le 1$$
solution set: $(-\infty, 1]$

93.
$$5(x - 2) \ge 0 \qquad \text{and} \quad -3x < 9$$
$$5x - 10 + 10 \ge 0 + 10 \qquad \tfrac{-3x}{-3} > \tfrac{9}{-3}$$
$$5x \ge 10 \qquad\qquad\qquad x > -3$$
$$x \ge 2$$
If $x \ge 2$ **and** $x > -3$, then $x \ge 2$.
The solution set is $[2, \infty)$.

213

95.
$$-6 \le \frac{1}{3}a + 1 < 0$$
$$3(-6) \le 3\left(\frac{1}{3}a + 1\right) < 3(0)$$
$$-18 \le a + 3 < 0$$
$$-18 - 3 \le a + 3 - 3 < 0 - 3$$
$$-21 \le a < -3$$
solution set: $[-21, -3)$

97.
$$P = L + \frac{s}{f}i$$
$$P - L = \frac{s}{f}i$$
$$f(P - L) = f \cdot \frac{s}{f}i$$
$$f(P - L) = si$$
$$\frac{f(P - L)}{i} = \frac{si}{i}$$
$$\frac{f(P - L)}{i} = s, \text{ or } s = \frac{f(P - L)}{i}$$

99. Let w = the width. Then $2w$ = the length.
$$\boxed{\text{Perimeter of garden}} = 72$$
$$w + 2w + w + 2w = 72$$
$$6w = 72$$
$$w = 12$$
The dimension are 12 m by 24 m.

101.
$$\boxed{\text{Total amount of fence}} = 150$$
$$x + x + x + x + x + 5 + x + 5 + x = 150$$
$$7x + 10 = 150$$
$$7x = 140$$
$$x = 20$$
The dimensions are 20 feet by 45 feet.

103. Let x = the length of the shorter piece.
Then $2x + 1$ = the length of the other piece.
$$\boxed{\begin{array}{c}\text{shorter}\\\text{length}\end{array}} + \boxed{\begin{array}{c}\text{other}\\\text{length}\end{array}} = \boxed{\text{total length}}$$
$$x + 2x + 1 = 22$$
$$3x + 1 = 22$$
$$3x = 21$$
$$x = 7$$
The pieces have lengths of 7 feet and 15 feet.

105. Let x = the amount invested at 9%.
$$0.08(10000) + 0.09x > 1250$$
$$800 + 0.09x > 1250$$
$$0.09x > 450$$
$$x > 5000$$
She must invest more than $5000 at 9%.

107. Let x = the # of songs downloaded.
$$210 + 1.10x \le 275$$
$$1.10x \le 65$$
$$x \le 59.09$$
The student can download at most 59 songs.

109. Answers may vary.

111. The 4 was not distributed correctly in the 2nd line. It should be $4x + 12 = 16$.

Exercises 7.2 (page 469)

1.
$$2x - 4y = 2$$
$$2(5) - 4(-2) \stackrel{?}{=} 2$$
$$10 + 8 \stackrel{?}{=} 2$$
$$18 \ne 2$$
not a solution

3.
$$3x + 5y = 5$$
$$3(5) + 5(-2) \stackrel{?}{=} 5$$
$$15 + (-10) \stackrel{?}{=} 5$$
$$5 = 5$$
solution

5. negative

SECTION 7.2

7. undefined

9. general

11. slope-intercept

13. $(x^3y^2)^3 = x^9y^6$

15. $(x^{-3}y^2)^{-4} = x^{12}y^{-8} = \dfrac{x^{12}}{y^8}$

17.
$$3(x+2) + x = 5x$$
$$3x + 6 + x = 5x$$
$$4x + 6 = 5x$$
$$6 = x$$

19.
$$\frac{5(2-x)}{3} - 1 = x + 5$$
$$\frac{10 - 5x}{3} = x + 6$$
$$3 \cdot \frac{10 - 5x}{3} = 3(x+6)$$
$$10 - 5x = 3x + 18$$
$$10 = 8x + 18$$
$$-8 = 8x$$
$$-1 = x$$

21. y-axis

23. vertical

25. $\left(\dfrac{a+c}{2}, \dfrac{b+d}{2}\right)$

27. change

29. rise

31. horizontal

33. perpendicular; reciprocals

35. $x + y = 4$

x	y
-1	5
0	4
2	2

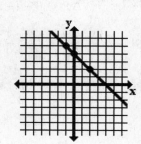

37. $2x - y = 3$

x	y
-1	-5
0	-3
3	3

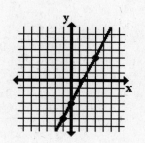

39.
$$3x + 4y = 12 \qquad 3x + 4y = 12$$
$$3x + 4(0) = 12 \qquad 3(0) + 4y = 12$$
$$3x = 12 \qquad\qquad 4y = 12$$
$$x = 4 \qquad\qquad\quad y = 3$$
x-intercept: $(4, 0)$ y-intercept: $(0, 3)$

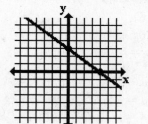

41.
$$y = -3x + 2 \qquad y = -3x + 2$$
$$0 = -3x + 2 \qquad y = -3(0) + 2$$
$$3x = 2 \qquad\qquad y = 2$$
$$x = \tfrac{2}{3} \qquad\qquad y\text{-intercept: } (0, 2)$$
x-intercept: $\left(\tfrac{2}{3}, 0\right)$

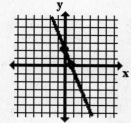

215

43. $x = 3$

vertical line with
x-coordinate of 3

$x = 3$

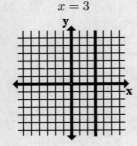

45. $-3y + 2 = 5$

$-3y = 3$

$y = -1$

horizontal line
with y-coordinate
of -1

$y = -1$

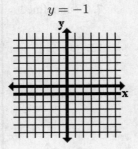

47. $x = \frac{x_1 + x_2}{2} = \frac{0 + 6}{2} = \frac{6}{2} = 3$

$y = \frac{y_1 + y_2}{2} = \frac{0 + 8}{2} = \frac{8}{2} = 4$

midpoint: $(3, 4)$

49. $x = \frac{x_1 + x_2}{2} = \frac{-2 + 3}{2} = \frac{1}{2}$

$y = \frac{y_1 + y_2}{2} = \frac{-8 + 4}{2} = \frac{-4}{2} = -2$

midpoint: $\left(\frac{1}{2}, -2\right)$

51. Points: $(0, -3)$ and $(4, 0)$

$m = \frac{\Delta y}{\Delta x} = \frac{0 - (-3)}{4 - 0} = \frac{3}{4}$

53. Points: $(-3, 4)$ and $(2, -3)$

$m = \frac{\Delta y}{\Delta x} = \frac{-3 - 4}{2 - (-3)} = \frac{-7}{5} = -\frac{7}{5}$

55. $m = \frac{\Delta y}{\Delta x} = \frac{9 - 0}{3 - 0} = \frac{9}{3} = 3$

57. $m = \frac{\Delta y}{\Delta x} = \frac{1 - 8}{6 - (-1)} = \frac{-7}{7} = -1$

59. $m = \frac{\Delta y}{\Delta x} = \frac{5 - 5}{-9 - 7} = \frac{0}{-16} = 0$

61. $m = \frac{\Delta y}{\Delta x} = \frac{-2 - (-5)}{-7 - (-7)} = \frac{3}{0}$: undefined

63. $3x + 2y = 12$

$2y = -3x + 12$

$\frac{2y}{2} = \frac{-3x + 12}{2}$

$y = -\frac{3}{2}x + 6$

$m = -\frac{3}{2}$

65. $3x = 4y - 2$

$-4y = -3x - 2$

$\frac{-4y}{-4} = \frac{-3x - 2}{-4}$

$y = \frac{3}{4}x + \frac{1}{2}$

$m = \frac{3}{4}$

67. $y = \frac{x - 4}{2}$

$y = \frac{1}{2}x - 2$

$m = \frac{1}{2}$

69. $4y = 3(y + 2)$

$4y = 3y + 6$

$y = 6$

$y = 0x + 6$

$m = 0$

71. $y = 3x + 4 \quad y = 3x - 7$

$m = 3 \qquad m = 3$

parallel

73. $x + y = 2 \qquad y = x + 5$

$y = -x + 2 \qquad m = 1$

$m = -1$

perpendicular

75. $y = 3x + 7$ $2y = 6x - 9$
$\quad m = 3$ $y = 3x - \dfrac{9}{2}$
$\qquad\qquad\qquad m = 3$
$\qquad\qquad$ parallel

77. $\quad x = 3y + 4$ $y = -3x + 7$
$\quad -3y = -x + 4$ $m = -3$
$\quad y = \frac{1}{3}x - \frac{4}{3}$
$\quad m = \frac{1}{3}$
$\qquad\qquad$ perpendicular

79. $y - y_1 = m(x - x_1)$
$y - 7 = 5(x - 0)$
$y - 7 = 5x$

81. $y - y_1 = m(x - x_1)$
$y - 0 = -3(x - 2)$
$y = -3(x - 2)$

83. $y = mx + b$
$y = 3x + 17$

85. $y = mx + b$
$5 = -7(7) + b$
$5 = -49 + b$
$54 = b$
$y = -7x + 54$

87. Find the slope of the line:
$m = \dfrac{\Delta y}{\Delta x} = \dfrac{4 - 0}{4 - 0} = \dfrac{4}{4} = 1$
$y = mx + b$
$0 = 1(0) + b$
$0 = 0 + b$
$0 = b$
$y = 1x + 0$, or $y = x$

89. Find the slope of the line:
$m = \dfrac{\Delta y}{\Delta x} = \dfrac{-3 - 4}{0 - 3} = \dfrac{-7}{-3} = \dfrac{7}{3}$
$y = mx + b$
$-3 = \dfrac{7}{3}(0) + b$
$-3 = 0 + b$
$-3 = b$
$y = \frac{7}{3}x - 3$

91. Find the slope of the given line:
$y = 4x - 7 \Rightarrow m = 4$
Use the parallel slope.
$y - y_1 = m(x - x_1)$
$y - 0 = 4(x - 0)$
$y = 4x$

93. Find the slope of the given line:
$4x - y = 7$
$-y = -4x + 7$
$y = 4x - 7 \Rightarrow m = 4$
Use the parallel slope.
$y - y_1 = m(x - x_1)$
$y - 5 = 4(x - 2)$
$y = 4x - 3$

95. Find the slope of the given line:
$y = 4x - 7 \Rightarrow m = 4$
Use the perpendicular slope.
$y - y_1 = m(x - x_1)$
$y - 0 = -\frac{1}{4}(x - 0)$
$y = -\frac{1}{4}x$

97. Find the slope of the given line: Use the perpendicular slope.

$$4x - y = 7$$
$$-y = -4x + 7$$
$$y = 4x - 7 \Rightarrow m = 4$$

$$y - y_1 = m(x - x_1)$$
$$y - 5 = -\frac{1}{4}(x - 2)$$
$$y - 5 = -\frac{1}{4}x + \frac{1}{2}$$
$$y = -\frac{1}{4}x + \frac{11}{2}$$

99. Let x represent the number of guests, and let y represent the total charge. Then we know two points on the line: $(100, 6775)$ and $(50, 3525)$. Find the slope and then use point-slope form to find the equation.

$$m = \frac{\Delta y}{\Delta x} = \frac{3525 - 6775}{50 - 100} = \frac{-3250}{-50} = 65 \qquad y - y_1 = m(x - x_1)$$
$$y - 6775 = 65(x - 100)$$
$$y = 65x + 275$$

101. Let x represent the number of feet used, and let y represent the total charge. Then we know two points on the line: $(340, 1359.80)$ and $(200, 839)$. Find the slope and then use point-slope form to find the equation.

$$m = \frac{\Delta y}{\Delta x} = \frac{839 - 1359.80}{200 - 340} = \frac{-520.80}{-140} = 3.72 \qquad y - y_1 = m(x - x_1)$$
$$y - 839 = 3.72(x - 200)$$
$$y = 3.72x + 95$$

103.
$$3y = 6x - 9$$
$$y = 2x - 3$$
$$m = 2; b = -3 \Rightarrow (0, -3)$$

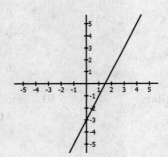

105.
$$3x - 2y = 8$$
$$-2y = -3x + 8$$
$$y = \frac{3}{2}x - 4$$
$$m = \frac{3}{2}; b = -4 \Rightarrow (0, -4)$$

107.
$$-2x + 4y = 12$$
$$4y = 2x + 12$$
$$y = \frac{1}{2}x + 3$$
$$m = \frac{1}{2}; b = 3 \Rightarrow (0, 3)$$

109. $y + 1 = x$

$y = x - 1$

$m = 1$

$b = -1; \; (0, -1)$

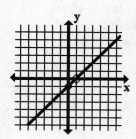

111. $x = \frac{3}{2}y - 3$

$2x = 3y - 6$

$-3y = -2x - 6$

$y = \frac{2}{3}x + 2$

$m = \frac{2}{3}$

$b = 2; \; (0, 2)$

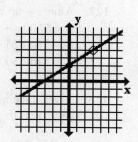

113. Notice that the line goes through $P(2, 5)$ and the point $(-1, 3)$. Find the slope:

$m = \dfrac{\Delta y}{\Delta x} = \dfrac{5 - 3}{2 - (-1)} = \dfrac{2}{3}$

Use point-slope form to find the equation:

$y - y_1 = m(x - x_1)$

$y - 5 = \dfrac{2}{3}(x - 2)$

$y - 5 = \dfrac{2}{3}x - \dfrac{4}{3}$

$y = \dfrac{2}{3}x - \dfrac{4}{3} + \dfrac{15}{3}$

$y = \dfrac{2}{3}x + \dfrac{11}{3}$

115. Notice that the line goes through $(2, -3)$ and the point $(-2, 4)$. Find the slope:

$m = \dfrac{\Delta y}{\Delta x} = \dfrac{4 - (-3)}{-2 - 2} = \dfrac{7}{-4} = -\dfrac{7}{4}$

Use point-slope form to find the equation:

$y - y_1 = m(x - x_1)$

$y - 4 = -\dfrac{7}{4}(x - (-2))$

$y - 4 = -\dfrac{7}{4}(x + 2)$

$y - 4 = -\dfrac{7}{4}x - \dfrac{7}{2}$

$y = -\dfrac{7}{4}x - \dfrac{7}{2} + 4$

$y = -\dfrac{7}{4}x + \dfrac{1}{2}$

117. Find the slope of the given line:

$x = \dfrac{5}{4}y - 2$

$4x = 5y - 8$

$-5y = -4x - 8$

$y = \dfrac{4}{5}x + \dfrac{8}{5} \Rightarrow m = \dfrac{4}{5}$

Use the parallel slope.

$y - y_1 = m(x - x_1)$

$y - (-2) = \dfrac{4}{5}(x - 4)$

$y + 2 = \dfrac{4}{5}x - \dfrac{16}{5}$

$y = \dfrac{4}{5}x - \dfrac{16}{5} - 2$

$y = \dfrac{4}{5}x - \dfrac{26}{5}$

119. Find the slope of the given line:

$x = \dfrac{5}{4}y - 2$

$4x = 5y - 8$

$-5y = -4x - 8$

$y = \dfrac{4}{5}x + \dfrac{8}{5} \Rightarrow m = \dfrac{4}{5}$

Use the perpendicular slope.

$y - y_1 = m(x - x_1)$

$y - (-2) = -\dfrac{5}{4}(x - 4)$

$y + 2 = -\dfrac{5}{4}x + 5$

$y = -\dfrac{5}{4}x + 3$

121. $x = \dfrac{x_1 + x_2}{2} = \dfrac{(a - b) + (a + b)}{2} = \dfrac{2a}{2} = a$

$y = \dfrac{y_1 + y_2}{2} = \dfrac{b + 3b}{2} = \dfrac{4b}{2} = 2b$

midpoint: $(a, 2b)$

219

123. After 5 years: Let $x = 5$.   After 10 years: Let $x = 10$.

$y = 7500x + 125000$ $y = 7500x + 125000$

$\qquad = 7500(5) + 125000$ $\qquad = 7500(10) + 125000$

$\qquad = \$162{,}500$ $\qquad = \$200{,}000$

125. To find the number of TVs produced, let $p = 150$:

$$p = \frac{1}{10}q + 130$$

$$150 = \frac{1}{10}q + 130$$

$$20 = \frac{1}{10}q$$

$$200 = q$$

200 TVs will be produced at a price of \$150.

127. $\text{slope} = \dfrac{\text{rise}}{\text{run}} = \dfrac{32 \text{ ft}}{1 \text{ mi}} = \dfrac{32 \text{ ft}}{5280 \text{ ft}} = \dfrac{1}{165}$

129. $m = \dfrac{\Delta y}{\Delta x} = \dfrac{2}{50} = \dfrac{1}{25}$; $\quad m = \dfrac{\Delta y}{\Delta x} = \dfrac{5}{50} = \dfrac{1}{10}$; $\quad m = \dfrac{\Delta y}{\Delta x} = \dfrac{8}{50} = \dfrac{4}{25}$

131. $m = \dfrac{\Delta \text{ temperature}}{\Delta \text{ time}} = \dfrac{1.4 - 0}{2040 - 2010} = \dfrac{1.4^\circ}{30 \text{ years}} = \dfrac{14^\circ}{300 \text{ years}} = \dfrac{7^\circ}{150 \text{ years}}$. or $\dfrac{7}{150}{}^\circ$ per year

133. a. $m = \dfrac{\Delta y}{\Delta x} = \dfrac{2}{25}$ **b.** $m = \dfrac{\Delta y}{\Delta x} = \dfrac{1}{20}$ (each part)

c. Design #1 has just one level, but the slope is steep. Design #2 has slopes which are less steep, but there are two levels.

135. Let x represent the number of years since the computer was purchased, and let y represent the value of the computer. Then we know two points on the depreciation line: $(0, 2350)$ and $(5, 200)$. Find the slope and then use point-slope form to find the depreciation equation.

$m = \dfrac{\Delta y}{\Delta x} = \dfrac{2350 - 200}{0 - 5} = \dfrac{2150}{-5} = -430 \qquad y - y_1 = m(x - x_1)$

$\qquad\qquad\qquad\qquad\qquad\qquad\qquad\qquad\qquad\quad y - 2350 = -430(x - 0)$

$\qquad\qquad\qquad\qquad\qquad\qquad\qquad\qquad\qquad\qquad\quad y = -430x + 2350$

137. Let x represent the number of years after the present, and let y represent the value of the house. Since the house is expected to appreciate \$4000 per year, the slope of the appreciation line is $m = 4000$. We also know one point on the appreciation line: $(0, 122000)$. Use slope-intercept form to find the equation of the appreciation line.

$y = mx + b$

$y = 4000x + 122{,}000$

139. Answers may vary.   **141. Answers may vary.**

143. If the line $y = ax + b$ passes through only quadrants I and II, then $a = 0$ and $b > 0$.

145. $Ax + By = C$
$$By = -Ax + C$$
$$y = -\frac{A}{B}x + \frac{C}{B} \Rightarrow m = -\frac{A}{B}$$

Exercises 7.3 (page 481)

1. 0

3. y-intercept

5. y

7. $\dfrac{y+2}{2} = 4(y+2)$
$$2 \cdot \frac{y+2}{2} = 2 \cdot 4(y+2)$$
$$y + 2 = 8(y+2)$$
$$y + 2 = 8y + 16$$
$$-7y = 14$$
$$y = -2$$

9. $\dfrac{2a}{3} + \dfrac{1}{2} = \dfrac{6a-1}{6}$
$$6\left(\frac{2a}{3} + \frac{1}{2}\right) = 6 \cdot \frac{6a-1}{6}$$
$$2(2a) + 3 = 6a - 1$$
$$4a + 3 = 6a - 1$$
$$-2a = -4$$
$$a = 2$$

11. relation

13. domain

15. vertical line test

17. cannot

19. first

21. domain $= \{3, 5, -4, 0\}$
range $= \{-2, 0, -5\}$; function

23. domain $= \{-2, 6, 5\}$
range $= \{3, 8, 5, 4\}$; not a function

25. The domain is the set of all x-coordinates on the graph. The domain is $(-\infty, 1]$.
The range is the set of all y-coordinates on the graph. The range is $(-\infty, \infty)$.
Since a vertical line can pass through the graph more than once, it is not the graph of a function.

27. The domain is the set of all x-coordinates on the graph. The domain is $(-\infty, \infty)$.
The range is the set of all y-coordinates on the graph. The range is $(-\infty, \infty)$.
Since each vertical line passes through the graph at most once, it is the graph of a function.

29. $f(x) = 3x$ $f(x) = 0$
$f(3) = 3(3) = 9$ $3x = 0$
$f(-1) = 3(-1) = -3$ $x = 0$

31. $f(x) = 2x - 3$ $f(x) = 0$
$f(3) = 2(3) - 3 = 3$ $2x - 3 = 0$
$f(-1) = 2(-1) - 3 = -5$ $2x = 3$
$$x = \tfrac{3}{2}$$

33. $f(x) = x^2$ $f(-2) = (-2)^2 = 4$ $f(2) = 2^2 = 4$ $f(3) = 3^2 = 9$

35. $f(x) = x^3 - 1$ $f(-2) = (-2)^3 - 1$ $f(-2) = (-2)^3 - 1$ $f(3) = 3^3 - 1 = 27 - 1$
$= -8 - 1 = -9$ $= -8 - 1 = -9$ $= 26$

37. $f(x) = |x| + 2$ $f(-2) = |-2| + 2$ $f(2) = |2| + 2$ $f(3) = |3| + 2$
$= 2 + 2 = 4$ $= 2 + 2 = 4$ $= 3 + 2 = 5$

39. $f(x) = x^2 - 2$ $f(-2) = (-2)^2 - 2$ $f(2) = 2^2 - 2 = 4 - 2$ $f(3) = 3^2 - 2 = 9 - 2$
$= 4 - 2 = 2$ $= 2$ $= 7$

41. $f(3) + f(2) = [2(3) + 1] + [2(2) + 1] = [7] + [5] = 12$

43. $f(b) - f(a) = [2(b) + 1] - [2(a) + 1] = [2b + 1] - [2a + 1] = 2b - 2a$

45. $x - 4 \neq 0 \Rightarrow x \neq 4 \Rightarrow$ Domain: $(-\infty, 4) \cup (4, \infty)$.

47. $x + 3 \neq 0 \Rightarrow x \neq -3 \Rightarrow$ Domain: $(-\infty, -3) \cup (-3, \infty)$.

49. $x^2 + 2 \neq 0 \Rightarrow$ This is always true $\Rightarrow$ Domain: $(-\infty, \infty)$.

51. $2x - 1 \neq 0 \Rightarrow x \neq \frac{1}{2} \Rightarrow$ Domain: $(-\infty, \frac{1}{2}) \cup (\frac{1}{2}, \infty)$.

53. $2x - 3y = 6$
$D = (-\infty, \infty)$
$R = (-\infty, \infty)$

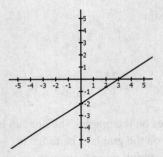

55. $f(x) = x^2 - 2x - 3$
$D = (-\infty, \infty)$
$R = [-4, \infty)$

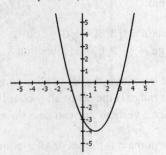

57. $f(x) = (x + 1)^2$
$f(2) = (2 + 1)^2 = 3^2 = 9$
$f(3) = (3 + 1)^2 = 4^2 = 16$
$f(-1) = (-1 + 1)^2$
$= 0^2 = 0$

59. $f(x) = 2x^2 - x$
$f(2) = 2(2)^2 - 2$
$= 2(4) - 2$
$= 8 - 2 = 6$
$f(3) = 2(3)^2 - 3$
$= 2(9) - 3$
$= 18 - 3 = 15$
$f(-1) = 2(-1)^2 - (-1)$
$= 2(1) + 1$
$= 2 + 1 = 3$

61. $f(x) = 7 + 5x$
$f(2) = 7 + 5(2) = 17$
$f(3) = 7 + 5(3) = 22$
$f(-1) = 7 + 5(-1) = 2$

63. $f(x) = 9 - 2x$ $f(2) = 9 - 2(2)$ $f(3) = 9 - 2(3)$ $f(-1) = 9 - 2(-1)$
$= 5$ $= 3$ $= 11$

65. $f(b) - f(0) = [2b + 1] - [2(0) + 1] = [2b + 1] - [1] = 2b$

67. $f(0) + f\left(-\frac{1}{2}\right) = [2(0) + 1] + \left[2\left(-\frac{1}{2}\right) + 1\right] = 0 + 1 - 1 + 1 = 1$

69. $g(x) = 2x$
$g(w) = 2w$
$g(w + 1) = 2(w + 1) = 2w + 2$

71. $g(x) = 3x - 5$
$g(w) = 3w - 5$
$g(w + 1) = 3(w + 1) - 5 = 3w + 3 - 5$
$= 3w - 2$

73. $y = 3x^2 + 2 \Rightarrow$ This is not a linear function because the exponent on x is 2.

75. $x = 3y - 4 \Rightarrow y = \frac{1}{3}x + \frac{4}{3} \Rightarrow$ This is a linear function $\left(m = \frac{1}{3}, b = \frac{4}{3} \right)$.

77. Let x represent the number of years after 2002, and let y represent the tuition. Then we know two points on the line: $(0, 4790)$ and $(10, 8240)$. Find the slope and then use point-slope form to find the function.
$$m = \frac{\Delta y}{\Delta x} = \frac{8240 - 4790}{10 - 0} = \frac{3450}{10} = 345$$
$$y - y_1 = m(x - x_1) \qquad f(16) = 345(16) + 4790$$
$$y - 4790 = 345(x - 0) \qquad = 5520 + 4790 = \$10{,}310$$
$$f(x) = 345x + 4790$$

79. Let x represent the number of years after 2000, and let y represent the number of violent crimes per 100,000 people. Then we know two points on the line: $(0, 507)$ and $(9, 429)$. Find the slope and then use point-slope form to find the function.
$$m = \frac{\Delta y}{\Delta x} = \frac{429 - 507}{9 - 0} = \frac{-78}{9} = -\frac{26}{3}$$
$$y - y_1 = m(x - x_1) \qquad f(21) = -\frac{26}{3}(21) + 507$$
$$y - 507 = -\frac{26}{3}(x - 0) \qquad = -182 + 507 = 325 \text{ violent crimes per } 100{,}000 \text{ people}$$
$$f(x) = -\frac{26}{3}x + 507 \qquad \text{To get the number of crimes for a city of } 400{,}000, \text{ multiply by 4:}$$
$$4(325) = 1300 \text{ violent crimes}$$

81. Let $t = 3$ and find s.
$$s = f(t) = -16t^2 + 256t$$
$$= -16(3)^2 + 256(3)$$
$$= -16(9) + 768$$
$$= -144 + 768$$
$$= 624$$
The bullet will have a height of 624 ft.

83. Let $t = 1.5$ and find h.
$$h = -16t^2 + 32t$$
$$h = -16(1.5)^2 + 32(1.5)$$
$$h = -16(2.25) + 48$$
$$h = -36 + 48 = 12$$
The dolphin will be 12 feet above the water.

85. Let $C = 25$ and find F.

$$F(C) = \frac{9}{5}C + 32$$
$$= \frac{9}{5}(25) + 32$$
$$= 45 + 32$$
$$= 77$$

The temperature is $77°F$.

87. Set $R(x)$ equal to $C(x)$ and solve for x.

$$R(x) = C(x)$$
$$120x = 57.50x + 12000$$
$$62.50x = 12000$$
$$x = \frac{12000}{62.50}$$
$$x = 192$$

The company must sell 192 players.

89. Answers may vary.

91. $f(x) + g(x) = 2x + 1 + x^2 = x^2 + 2x + 1 = g(x) + f(x)$. Yes, they are equal.

Exercises 7.4 (page 495)

1. $8x^3 - 12x^5 = 4x^2\underline{(2 - 3x^2)}$

3. $x^2 - 1 = \underline{(x + 1)(x - 1)}$

5. $x^2 + 5x - 6 = (x - 1)\underline{(x + 6)}$

7. 1.1×10^3 ft/sec $= 1,100$ ft/sec

9. $\frac{2}{3}(5t - 3) = 38$

$3 \cdot \frac{2}{3}(5t - 3) = 3(38)$

$2(5t - 3) = 114$

$10t - 6 = 114$

$10t = 120$

$t = 12$

11. $ab + ac$

13. prime

15. trial and error

17. $\left(x^2 + xy + y^2\right)$

19. $2x + 8 = \mathbf{2} \cdot x + \mathbf{2} \cdot 4 = \mathbf{2}(x + 4)$

21. $2x^2 - 6x = \mathbf{2x} \cdot x - \mathbf{2x} \cdot 3 = \mathbf{2x}(x - 3)$

23. $-3a - 6 = \mathbf{(-3)}(a) + \mathbf{(-3)}(2)$
$= -3(a + 2)$

25. $-6x^2 - 3xy = \mathbf{(-3x)}(2x) + \mathbf{(-3x)}(y)$
$= -3x(2x + y)$

27. $x^{n+2} + x^{n+3} = x^2(x^{n+2-2} + x^{n+3-2})$
$= x^2(x^n + x^{n+1})$

29. $2y^{n+2} - 3y^{n+3} = y^n(2y^{n+2-n} - 3y^{n+3-n})$
$= y^n(2y^2 - 3y^3)$

31. $ax + bx + ay + by = x(a + b) + y(a + b)$
$= (a + b)(x + y)$

33. $x^2 + yx + 2x + 2y = x(x + y) + 2(x + y)$
$= (x + y)(x + 2)$

35. $8a^3 - 2a^2 + 12a - 3 = 2a^2(4a - 1) + 3(4a - 1) = (4a - 1)(2a^2 + 3)$

37. $5x^3 + x^2 + 5x + 1 = x^2(5x + 1) + 1(5x + 1) = (5x + 1)(x^2 + 1)$

39. $x^2 - 4 = x^2 - 2^2 = (x + 2)(x - 2)$

41. $9y^2 - 64 = (3y)^2 - 8^2 = (3y + 8)(3y - 8)$

43. $x^4 - y^4 = (x^2 + y^2)(x^2 - y^2) = (x^2 + y^2)(x + y)(x - y)$

45. $2x^2 - 288 = 2(x^2 - 144)$
$\qquad = 2(x + 12)(x - 12)$

47. $x^2 + 5x + 6 = (x + 2)(x + 3)$

49. $a^2 + 5a - 52 \Rightarrow$ prime

51. $2x^2 - 11x + 5 = (2x - 1)(x - 5)$

53. $6y^2 + 7y + 2 = (2y + 1)(3y + 2)$

55. $3x^4 - 10x^3 + 3x^2 = x^2(3x^2 - 10x + 3)$
$\qquad = x^2(3x - 1)(x - 3)$

57. $-8x^3 - 18x + 24x^2 = -2x(4x^2 - 12x + 9) = -2x(2x - 3)(2x - 3) = -2x(2x - 3)^2$

59. $x^{2n} + 2x^n + 1 = (x^n + 1)(x^n + 1)$
$\qquad = (x^n + 1)^2$

61. $6x^{2n} + 7x^n - 3 = (3x^n - 1)(2x^n + 3)$

63. $x^{4n} + 2x^{2n}y^{2n} + y^{4n} = (x^{2n} + y^{2n})(x^{2n} + y^{2n}) = (x^{2n} + y^{2n})^2$

65. $y^{6n} + 2y^{3n}z + z^2 = (y^{3n} + z)(y^{3n} + z) = (y^{3n} + z)^2$

67. $x^2 + 4x + 4 - y^2 = (x^2 + 4x + 4) - y^2 = (x + 2)(x + 2) - y^2 = (x + 2)^2 - y^2$
$\qquad = [(x + 2) + y][(x + 2) - y]$
$\qquad = (x + 2 + y)(x + 2 - y)$

69. $x^2 + 2x + 1 - 9z^2 = (x^2 + 2x + 1) - 9z^2 = (x + 1)(x + 1) - 9z^2$
$\qquad = (x + 1)^2 - (3z)^2$
$\qquad = [(x + 1) + 3z][(x + 1) - 3z]$
$\qquad = (x + 1 + 3z)(x + 1 - 3z)$

71. $y^3 + 1 = y^3 + 1^3$
$\qquad = (y + 1)(y^2 - 1y + 1^2)$
$\qquad = (y + 1)(y^2 - y + 1)$

73. $8 + x^3 = 2^3 + x^3$
$\qquad = (2 + x)(2^2 - 2x + x^2)$
$\qquad = (2 + x)(4 - 2x + x^2)$

75. $a^3 - 64 = a^3 - 4^3$
$\qquad = (a - 4)(a^2 + 4a + 4^2)$
$\qquad = (a - 4)(a^2 + 4a + 16)$

77. $27 - y^3 = 3^3 - y^3$
$\qquad = (3 - y)(3^2 + 3y + y^2)$
$\qquad = (3 - y)(9 + 3y + y^2)$

79. $x^6 - 1 = (x^3)^2 - 1^2 = (x^3 + 1)(x^3 - 1) = (x^3 + 1^3)(x^3 - 1^3)$
$\qquad = (x + 1)(x^2 - 1x + 1^2)(x - 1)(x^2 + 1x + 1^2)$
$\qquad = (x + 1)(x^2 - x + 1)(x - 1)(x^2 + x + 1)$

81. $x^{12} - y^6 = \left(x^6\right)^2 - \left(y^3\right)^2 = \left(x^6 + y^3\right)\left(x^6 - y^3\right)$
$$= \left(\left(x^2\right)^3 + y^3\right)\left(\left(x^2\right)^3 - y^3\right)$$
$$= \left(x^2 + y\right)\left(\left(x^2\right)^2 - x^2 y + y^2\right)\left(x^2 - y\right)\left(\left(x^2\right)^2 + x^2 y + y^2\right)$$
$$= \left(x^2 + y\right)\left(x^4 - x^2 y + y^2\right)\left(x^2 - y\right)\left(x^4 + x^2 y + y^2\right)$$

83.
$$\frac{3a^2}{2} = \frac{1}{2} - a$$
$$2 \cdot \frac{3a^2}{2} = 2\left(\frac{1}{2} - a\right)$$
$$3a^2 = 1 - 2a$$
$$3a^2 + 2a - 1 = 0$$
$$(3a - 1)(a + 1) = 0$$
$$3a - 1 = 0 \quad \text{or} \quad a + 1 = 0$$
$$3a = 1 \qquad\qquad a = -1$$
$$a = \tfrac{1}{3}$$

85.
$$\frac{1}{2}x^2 - \frac{5}{4}x = -\frac{1}{2}$$
$$4\left(\frac{1}{2}x^2 - \frac{5}{4}x\right) = 4\left(-\frac{1}{2}\right)$$
$$2x^2 - 5x = -2$$
$$2x^2 - 5x + 2 = 0$$
$$(2x - 1)(x - 2) = 0$$
$$2x - 1 = 0 \quad \text{or} \quad x - 2 = 0$$
$$2x = 1 \qquad\qquad x = 2$$
$$x = \tfrac{1}{2}$$

87. $27z^3 + 12z^2 + 3z = \mathbf{3z} \cdot 9z^2 + \mathbf{3z} \cdot 4z + \mathbf{3z} \cdot 1 = \mathbf{3z}(9z^2 + 4z + 1)$

89. $15x^2 y - 10x^2 y^2 = \mathbf{5x^2 y} \cdot 3 - \mathbf{5x^2 y} \cdot 2y = \mathbf{5x^2 y}(3 - 2y)$

91. $13ab^2 c^3 - 26a^3 b^2 c = \mathbf{13ab^2 c} \cdot c^2 - \mathbf{13ab^2 c} \cdot 2a^2 = \mathbf{13ab^2 c}(c^2 - 2a^2)$

93. $24s^3 - 12s^2 t + 6st^2 = \mathbf{6s}(4s^2) - \mathbf{6s}(2st) + \mathbf{6s}(t^2) = \mathbf{6s}(4s^2 - 2st + t^2)$

95. $81a^4 - 49b^2 = (9a^2)^2 - (7b)^2$
$$= (9a^2 + 7b)(9a^2 - 7b)$$

97. $(x - y)^2 - z^2 = [(x - y) + z][(x - y) - z]$
$$= (x - y + z)(x - y - z)$$

99. $-a^2 + 4a + 32 = -(a^2 - 4a - 32)$
$$= -(a - 8)(a + 4)$$

101. $2x^2 + 8x - 42 = 2(x^2 + 4x - 21)$
$$= 2(x + 7)(x - 3)$$

103. $-3x^2 + 15x - 18 = -3(x^2 - 5x + 6)$
$$= -3(x - 2)(x - 3)$$

105. $a^2 - 3ab - 4b^2 = (a - 4b)(a + b)$

107. $8a^2 + 6a - 9 = (4a - 3)(2a + 3)$

109. $5x^2 + 4x + 1 \Rightarrow$ prime

111. $8x^2 - 10x + 3 = (4x - 3)(2x - 1)$

113. $2y^2 + yt - 6t^2 = (2y - 3t)(y + 2t)$

115. $-3a^2 + ab + 2b^2 = -(3a^2 - ab - 2b^2)$
$$= -(3a + 2b)(a - b)$$

117. $a^2 b^2 - 13ab^2 + 22b^2 = b^2(a^2 - 13a + 22)$
$$= b^2(a - 11)(a - 2)$$

119. $8a^3 + 27 = (2a)^3 + 3^3$
$= (2a + 3)\big((2a)^2 - 2a(3) + 3^2\big)$
$= (2a + 3)\big(4a^2 - 6a + 9\big)$

121. $a^3 + 8b^3 = a^3 + (2b)^3$
$= (a + 2b)\big(a^2 - 2ab + (2b)^2\big)$
$= (a + 2b)\big(a^2 - 2ab + 4b^2\big)$

123. $s^3 - t^3 = (s - t)(s^2 + st + t^2)$

125. $27a^3 - b^3 = (3a)^3 - b^3$
$= (3a - b)\big((3a)^2 + 3ab + b^2\big)$
$= (3a - b)\big(9a^2 + 3ab + b^2\big)$

127. $2x^3 - 32x = 2x(x^2 - 16)$
$= 2x(x + 4)(x - 4)$

129. $x^4 + 8x^2 + 15 = (x^2 + 5)(x^2 + 3)$

131. $y^4 - 13y^2 + 30 = (y^2 - 10)(y^2 - 3)$

133. $a^4 - 13a^2 + 36 = (a^2 - 4)(a^2 - 9) = (a + 2)(a - 2)(a + 3)(a - 3)$

135. $-x^3 + 216 = -\big(x^3 - 216\big)$
$= -\big(x^3 - 6^3\big)$
$= -(x - 6)\big(x^2 + 6x + 6^2\big)$
$= -(x - 6)\big(x^2 + 6x + 36\big)$

137. $a^2 - b^2 + a + b = (a^2 - b^2) + (a + b)$
$= (a + b)(a - b) + 1(a + b)$
$= (a + b)(a - b + 1)$

139. $24m^5n - 3m^2n^4 = 3m^2n\big(8m^3 - n^3\big) = 3m^2n\big((2m)^3 - n^3\big)$
$= 3m^2n(2m - n)\big((2m)^2 + 2mn + n^2\big)$
$= 3m^2n(2m - n)\big(4m^2 + 2mn + n^2\big)$

141. $4c^2 - a^2 - 6ab - 9b^2 = 4c^2 - (a^2 + 6ab + 9b^2) = 4c^2 - (a + 3b)(a + 3b)$
$= (2c)^2 - (a + 3b)^2$
$= [2c + (a + 3b)][2c - (a + 3b)]$
$= (2c + a + 3b)(2c - a - 3b)$

143. $x^4y + 216xy^4 = xy\big(x^3 + 216y^3\big) = xy\big(x^3 + (6y)^3\big) = xy(x + 6y)\big(x^2 - 6xy + (6y)^2\big)$
$= xy(x + 6y)\big(x^2 - 6xy + 36y^2\big)$

145.
$$x\left(3x + \frac{22}{5}\right) = 1$$
$$3x^2 + \frac{22}{5}x = 1$$
$$5\left(3x^2 + \frac{22}{5}x\right) = 5 \cdot 1$$
$$15x^2 + 22x = 5$$
$$15x^2 + 22x - 5 = 0$$
$$(3x + 5)(5x - 1) = 0$$
$3x + 5 = 0 \quad$ **or** $\quad 5x - 1 = 0$
$3x = -5 \qquad\qquad 5x = 1$
$x = -\frac{5}{3} \qquad\qquad x = \frac{1}{5}$

147-149. Answers may vary.

151. $x^2 - q^2 = x^2 + 0x - q^2 : a = 1, b = 0, c = -q^2 \Rightarrow b^2 - 4ac = 0^2 - 4(1)(-q^2) = 4q^2 = (2q)^2$.
Since $4q^2$ is a perfect square, the binomial is factorable, and the test works.

153. $x^{32} - y^{32} = (x^{16} + y^{16})(x^{16} - y^{16}) = (x^{16} + y^{16})(x^8 + y^8)(x^8 - y^8)$
$$= (x^{16} + y^{16})(x^8 + y^8)(x^4 + y^4)(x^4 - y^4)$$
$$= (x^{16} + y^{16})(x^8 + y^8)(x^4 + y^4)(x^2 + y^2)(x^2 - y^2)$$
$$= (x^{16} + y^{16})(x^8 + y^8)(x^4 + y^4)(x^2 + y^2)(x + y)(x - y)$$

Exercises 7.5 (page 507)

1. $\dfrac{6(x + 9)}{3(x + 9)} = \dfrac{6}{3} = 2$

3. $\dfrac{x + 5}{5 + x} = 1$

5. $\dfrac{x - 5}{5 - x} = \dfrac{x - 5}{-(x - 5)} = -1$

7. $\dfrac{x + 2}{(x - 1)(x + 2)} = \dfrac{1}{x - 1}$

9. ⟵—)——[—⟶
      -4 5

11. $x^2 - 5x - 6 = 0$
$(x - 6)(x + 1) = 0$
$x - 6 = 0$ **or** $x + 1 = 0$
$x = 6$ $x = -1$

13. $2x^2 - 5x - 3 = 0$
$(2x + 1)(x - 3) = 0$
$2x + 1 = 0$ **or** $x - 3 = 0$
$2x = -1$ $x = 3$
$x = -\frac{1}{2}$

15. $a^4 - 13a^2 + 36 = 0$
$(a^2 - 4)(a^2 - 9) = 0$
$(a + 2)(a - 2)(a + 3)(a - 3) = 0$
$a + 2 = 0$ **or** $a - 2 = 0$ **or** $a + 3 = 0$ **or** $a - 3 = 0$
$a = -2$ $a = 2$ $a = -3$ $a = 3$

17. $P = 2l + 2w$
$P - 2l = 2w$
$\dfrac{P - 2l}{2} = \dfrac{2w}{2}$
$\dfrac{P - 2l}{2} = w$, or $w = \dfrac{P - 2l}{2}$

19. $\dfrac{a}{b}$

21. $\dfrac{ad}{bc}$

23. $2x = 0$
$x = 0$

25. $x - 2 = 0$
$x = 2$

27. $3x + 2 = 0$
$3x = -2$
$x = -\frac{2}{3}$

29. $2x^2 - 5x - 3 = 0$
$(2x + 1)(x - 3) = 0$
$2x + 1 = 0$ **or** $x - 3 = 0$
$2x = -1$ $x = 3$
$x = -\frac{1}{2}$

31. $\dfrac{3x^2 - 12}{x^2 - x - 2} = \dfrac{3(x^2 - 4)}{(x - 2)(x + 1)} = \dfrac{3(x + 2)(x - 2)}{(x - 2)(x + 1)} = \dfrac{3(x + 2)}{x + 1}$

33. $\dfrac{x^2 + 2x + 1}{x^2 + 4x + 3} = \dfrac{(x + 1)(x + 1)}{(x + 1)(x + 3)} = \dfrac{x + 1}{x + 3}$

35. $\dfrac{4x^2 + 24x + 32}{16x^2 + 8x - 48} = \dfrac{4(x^2 + 6x + 8)}{8(2x^2 + x - 6)} = \dfrac{4(x + 4)(x + 2)}{8(2x - 3)(x + 2)} = \dfrac{x + 4}{2(2x - 3)}$

37. $\dfrac{x^3 + 8}{x^2 - 2x + 4} = \dfrac{(x + 2)(x^2 - 2x + 4)}{x^2 - 2x + 4}$
$= x + 2$

39. $\dfrac{y - x}{x^2 - y^2} = \dfrac{y - x}{(x + y)(x - y)} = \dfrac{-(x - y)}{(x + y)(x - y)}$
$= -\dfrac{1}{x + y}$

41. $\dfrac{x - y}{x^2 - y^2} = \dfrac{x - y}{(x + y)(x - y)} = \dfrac{1}{x + y}$

43. $\dfrac{3x^2 - 3y^2}{x^2 + 2xy + y^2} = \dfrac{3(x^2 - y^2)}{(x + y)(x + y)} = \dfrac{3(x + y)(x - y)}{(x + y)(x + y)} = \dfrac{3(x - y)}{x + y}$

45. $\dfrac{ax + by - ay - bx}{b^2 - a^2} = \dfrac{ax - ay - bx + by}{b^2 - a^2} = \dfrac{a(x - y) - b(x - y)}{b^2 - a^2} = \dfrac{(x - y)(a - b)}{(b + a)(b - a)} = -\dfrac{x - y}{a + b}$

47. $\dfrac{x^2 + 2x + 1}{x} \cdot \dfrac{x^2 - x}{x^2 - 1} = \dfrac{(x + 1)(x + 1)}{x} \cdot \dfrac{x(x - 1)}{(x + 1)(x - 1)} = x + 1$

49. $\dfrac{a + 6}{a^2 - 16} \cdot \dfrac{3a - 12}{3a + 18} = \dfrac{a + 6}{(a + 4)(a - 4)} \cdot \dfrac{3(a - 4)}{3(a + 6)} = \dfrac{1}{a + 4}$

51. $\dfrac{9x^2 + 3x - 20}{3x^2 - 7x + 4} \cdot \dfrac{3x^2 - 5x + 2}{9x^2 + 18x + 5} = \dfrac{(3x + 5)(3x - 4)}{(3x - 4)(x - 1)} \cdot \dfrac{(3x - 2)(x - 1)}{(3x + 5)(3x + 1)} = \dfrac{3x - 2}{3x + 1}$

53. $\dfrac{2p^2 - 5p - 3}{p^2 - 9} \cdot \dfrac{2p^2 + 5p - 3}{2p^2 + 5p + 2} = \dfrac{(2p + 1)(p - 3)}{(p + 3)(p - 3)} \cdot \dfrac{(2p - 1)(p + 3)}{(2p + 1)(p + 2)} = \dfrac{2p - 1}{p + 2}$

55. $\dfrac{x^2 - 16}{x^2 - 25} \div \dfrac{x + 4}{x - 5} = \dfrac{x^2 - 16}{x^2 - 25} \cdot \dfrac{x - 5}{x + 4} = \dfrac{(x + 4)(x - 4)}{(x + 5)(x - 5)} \cdot \dfrac{x - 5}{x + 4} = \dfrac{x - 4}{x + 5}$

57. $\dfrac{a^2 + 2a - 35}{12x} \div \dfrac{ax - 3x}{a^2 + 4a - 21} = \dfrac{a^2 + 2a - 35}{12x} \cdot \dfrac{a^2 + 4a - 21}{ax - 3x}$
$= \dfrac{(a + 7)(a - 5)}{12x} \cdot \dfrac{(a + 7)(a - 3)}{x(a - 3)} = \dfrac{(a + 7)^2(a - 5)}{12x^2}$

229

59. $\dfrac{x^2 - 6x + 9}{4 - x^2} \div \dfrac{x^2 - 9}{x^2 - 8x + 12} = \dfrac{x^2 - 6x + 9}{4 - x^2} \cdot \dfrac{x^2 - 8x + 12}{x^2 - 9} = \dfrac{(x-3)(x-3)}{(2+x)(2-x)} \cdot \dfrac{(x-6)(x-2)}{(x+3)(x-3)}$

$$= -\dfrac{(x-3)(x-6)}{(x+2)(x+3)}$$

61. $(2x^2 - 15x + 25) \div \dfrac{2x^2 - 3x - 5}{x + 1} = \dfrac{2x^2 - 15x + 25}{1} \cdot \dfrac{x+1}{2x^2 - 3x - 5}$

$$= \dfrac{(2x-5)(x-5)}{1} \cdot \dfrac{x+1}{(2x-5)(x+1)} = x - 5$$

63. $\dfrac{6a^2 - 7a - 3}{a^2 - 1} \div \dfrac{4a^2 - 12a + 9}{a^2 - 1} \cdot \dfrac{2a^2 - a - 3}{3a^2 - 2a - 1} = \dfrac{6a^2 - 7a - 3}{a^2 - 1} \cdot \dfrac{a^2 - 1}{4a^2 - 12a + 9} \cdot \dfrac{2a^2 - a - 3}{3a^2 - 2a - 1}$

$$= \dfrac{(3a+1)(2a-3)}{a^2 - 1} \cdot \dfrac{a^2 - 1}{(2a-3)(2a-3)} \cdot \dfrac{(2a-3)(a+1)}{(3a+1)(a-1)} = \dfrac{a+1}{a-1}$$

65. $\dfrac{4x^2 - 10x + 6}{x^4 - 3x^3} \div \dfrac{2x - 3}{2x^3} \cdot \dfrac{x-3}{2x-2} = \dfrac{2(2x^2 - 5x + 3)}{x^3(x-3)} \cdot \dfrac{2x^3}{2x-3} \cdot \dfrac{x-3}{2x-2}$

$$= \dfrac{2(2x-3)(x-1)}{x^3(x-3)} \cdot \dfrac{2x^3}{2x-3} \cdot \dfrac{x-3}{2(x-1)} = 2$$

67. $(x+3) \cdot \dfrac{x^2}{2x - 6} \div \dfrac{2}{x-3} = \dfrac{x+3}{1} \cdot \dfrac{x^2}{2x-6} \cdot \dfrac{x-3}{2} = \dfrac{x+3}{1} \cdot \dfrac{x^2}{2(x-3)} \cdot \dfrac{x-3}{2} = \dfrac{x^2(x+3)}{4}$

69. $\dfrac{2x^2 + 5x - 3}{x^2 + 2x - 3} \div \left(\dfrac{x^2 + 2x - 35}{x^2 - 6x + 5} \div \dfrac{x^2 - 9x + 14}{2x^2 - 5x + 2} \right)$

$$= \dfrac{2x^2 + 5x - 3}{x^2 + 2x - 3} \div \left(\dfrac{x^2 + 2x - 35}{x^2 - 6x + 5} \cdot \dfrac{2x^2 - 5x + 2}{x^2 - 9x + 14} \right)$$

$$= \dfrac{2x^2 + 5x - 3}{x^2 + 2x - 3} \div \left(\dfrac{(x+7)(x-5)}{(x-5)(x-1)} \cdot \dfrac{(2x-1)(x-2)}{(x-7)(x-2)} \right)$$

$$= \dfrac{2x^2 + 5x - 3}{x^2 + 2x - 3} \div \dfrac{(x+7)(2x-1)}{(x-1)(x-7)} = \dfrac{(2x-1)(x+3)}{(x+3)(x-1)} \cdot \dfrac{(x-1)(x-7)}{(x+7)(2x-1)} = \dfrac{x-7}{x+7}$$

71. $\dfrac{x}{x+4} + \dfrac{5}{x+4} = \dfrac{x+5}{x+4}$

73. $\dfrac{5x}{x+1} + \dfrac{3}{x+1} - \dfrac{2x}{x+1} = \dfrac{3x+3}{x+1}$

$$= \dfrac{3(x+1)}{x+1} = 3$$

75. $\dfrac{3x}{2x+2} + \dfrac{x+4}{2x+2} = \dfrac{4x+4}{2x+2} = \dfrac{4(x+1)}{2(x+1)} = 2$

77. $\dfrac{3(x^2 + x)}{x^2 - 5x + 6} + \dfrac{-3(x^2 - x)}{x^2 - 5x + 6} = \dfrac{3x^2 + 3x}{x^2 - 5x + 6} + \dfrac{-3x^2 + 3x}{x^2 - 5x + 6} = \dfrac{6x}{(x-3)(x-2)}$

79. $\dfrac{a+b}{3} + \dfrac{a-b}{7} = \dfrac{(a+b)7}{3(7)} + \dfrac{(a-b)3}{7(3)} = \dfrac{7a+7b}{21} + \dfrac{3a-3b}{21} = \dfrac{10a+4b}{21}$

81. $\dfrac{a}{2} + \dfrac{2a}{5} = \dfrac{a \cdot 5}{2 \cdot 5} + \dfrac{2a \cdot 2}{5 \cdot 2} = \dfrac{5a}{10} + \dfrac{4a}{10}$
$\qquad\qquad = \dfrac{9a}{10}$

83. $\dfrac{3}{4x} + \dfrac{2}{3x} = \dfrac{3 \cdot 3}{4x \cdot 3} + \dfrac{2 \cdot 4}{3x \cdot 4} = \dfrac{9}{12x} + \dfrac{8}{12x}$
$\qquad\qquad = \dfrac{17}{12x}$

85. $x + \dfrac{1}{x} = \dfrac{x}{1} + \dfrac{1}{x} = \dfrac{x(x)}{1(x)} + \dfrac{1}{x} = \dfrac{x^2}{x} + \dfrac{1}{x} = \dfrac{x^2+1}{x}$

87. $\dfrac{3}{x+2} + \dfrac{5}{x-4} = \dfrac{3(x-4)}{(x+2)(x-4)} + \dfrac{5(x+2)}{(x-4)(x+2)} = \dfrac{3x-12}{(x+2)(x-4)} + \dfrac{5x+10}{(x+2)(x-4)}$
$\qquad\qquad\qquad = \dfrac{8x-2}{(x+2)(x-4)} = \dfrac{2(4x-1)}{(x+2)(x-4)}$

89. $\dfrac{7}{x+3} + \dfrac{4x}{x+6} = \dfrac{7(x+6)}{(x+3)(x+6)} + \dfrac{4x(x+3)}{(x+6)(x+3)} = \dfrac{7x+42}{(x+3)(x+6)} + \dfrac{4x^2+12x}{(x+3)(x+6)}$
$\qquad\qquad\qquad = \dfrac{4x^2+19x+42}{(x+3)(x+6)}$

91. $1 + x - \dfrac{x}{x-5} = \dfrac{x+1}{1} - \dfrac{x}{x-5} = \dfrac{(x+1)(x-5)}{1(x-5)} - \dfrac{x}{x-5}$
$\qquad\qquad\qquad = \dfrac{x^2-4x-5}{x-5} - \dfrac{x}{x-5} = \dfrac{x^2-5x-5}{x-5}$

93. $\dfrac{8}{x^2-9} + \dfrac{2}{x-3} - \dfrac{6}{x} = \dfrac{8}{(x+3)(x-3)} + \dfrac{2}{x-3} - \dfrac{6}{x}$
$\qquad\qquad\qquad = \dfrac{8x}{x(x+3)(x-3)} + \dfrac{2x(x+3)}{x(x+3)(x-3)} - \dfrac{6(x+3)(x-3)}{x(x+3)(x-3)}$
$\qquad\qquad\qquad = \dfrac{8x}{x(x+3)(x-3)} + \dfrac{2x^2+6x}{x(x+3)(x-3)} - \dfrac{6(x^2-9)}{x(x+3)(x-3)}$
$\qquad\qquad\qquad = \dfrac{8x}{x(x+3)(x-3)} + \dfrac{2x^2+6x}{x(x+3)(x-3)} - \dfrac{6x^2-54}{x(x+3)(x-3)}$
$\qquad\qquad\qquad = \dfrac{-4x^2+14x+54}{x(x+3)(x-3)}$

95. $\dfrac{3}{x+1} - \dfrac{2}{x-1} + \dfrac{x+3}{x^2-1} = \dfrac{3}{x+1} - \dfrac{2}{x-1} + \dfrac{x+3}{(x+1)(x-1)}$

$$= \dfrac{3(x-1)}{(x+1)(x-1)} - \dfrac{2(x+1)}{(x+1)(x-1)} + \dfrac{x+3}{(x+1)(x-1)}$$

$$= \dfrac{3x-3}{(x+1)(x-1)} - \dfrac{2x+2}{(x+1)(x-1)} + \dfrac{x+3}{(x+1)(x-1)}$$

$$= \dfrac{2x-2}{(x+1)(x-1)} = \dfrac{2(x-1)}{(x+1)(x-1)} = \dfrac{2}{x+1}$$

97. $\dfrac{x}{x^2+5x+6} + \dfrac{x}{x^2-4} = \dfrac{x}{(x+2)(x+3)} + \dfrac{x}{(x+2)(x-2)}$

$$= \dfrac{x(x-2)}{(x+2)(x+3)(x-2)} + \dfrac{x(x+3)}{(x+2)(x-2)(x+3)}$$

$$= \dfrac{x^2-2x}{(x+2)(x+3)(x-2)} + \dfrac{x^2+3x}{(x+2)(x+3)(x-2)}$$

$$= \dfrac{2x^2+x}{(x+2)(x+3)(x-2)}$$

99. $\dfrac{\frac{1}{a} + \frac{1}{b}}{\frac{1}{a}} = \dfrac{\left(\frac{1}{a} + \frac{1}{b}\right) \cdot ab}{\frac{1}{a} \cdot ab} = \dfrac{\frac{1}{a} \cdot ab + \frac{1}{b} \cdot ab}{b} = \dfrac{b+a}{b}$

101. $\dfrac{\frac{y}{x} - \frac{x}{y}}{\frac{1}{x} + \frac{1}{y}} = \dfrac{\left(\frac{y}{x} - \frac{x}{y}\right) \cdot xy}{\left(\frac{1}{x} + \frac{1}{y}\right) \cdot xy} = \dfrac{\frac{y}{x} \cdot xy - \frac{x}{y} \cdot xy}{\frac{1}{x} \cdot xy + \frac{1}{y} \cdot xy} = \dfrac{y^2 - x^2}{y+x} = \dfrac{(y+x)(y-x)}{y+x} = y - x$

103. $\dfrac{\frac{1}{a} - \frac{1}{b}}{\frac{a}{b} - \frac{b}{a}} = \dfrac{\left(\frac{1}{a} - \frac{1}{b}\right) \cdot ab}{\left(\frac{a}{b} - \frac{b}{a}\right) \cdot ab} = \dfrac{\frac{1}{a} \cdot ab - \frac{1}{b} \cdot ab}{\frac{a}{b} \cdot ab - \frac{b}{a} \cdot ab} = \dfrac{b-a}{a^2-b^2} = \dfrac{b-a}{(a+b)(a-b)} = -\dfrac{1}{a+b}$

105. $\dfrac{\frac{1}{a+1} + 1}{\frac{3}{a-1} + 1} = \dfrac{\frac{1}{a+1} + \frac{a+1}{a+1}}{\frac{3}{a-1} + \frac{a-1}{a-1}} = \dfrac{\frac{a+2}{a+1}}{\frac{a+2}{a-1}} = \dfrac{a+2}{a+1} \div \dfrac{a+2}{a-1} = \dfrac{a+2}{a+1} \cdot \dfrac{a-1}{a+2} = \dfrac{a-1}{a+1}$

107. $\dfrac{x^{-1} + y^{-1}}{x^{-1} - y^{-1}} = \dfrac{\frac{1}{x} + \frac{1}{y}}{\frac{1}{x} - \frac{1}{y}} = \dfrac{\left(\frac{1}{x} + \frac{1}{y}\right)(xy)}{\left(\frac{1}{x} - \frac{1}{y}\right)(xy)} = \dfrac{y+x}{y-x}$

109. $\dfrac{x - y^{-2}}{y - x^{-2}} = \dfrac{x - \frac{1}{y^2}}{y - \frac{1}{x^2}} = \dfrac{\left(x - \frac{1}{y^2}\right)(x^2y^2)}{\left(y - \frac{1}{x^2}\right)(x^2y^2)} = \dfrac{x^3y^2 - x^2}{x^2y^3 - y^2} = \dfrac{x^2(xy^2 - 1)}{y^2(x^2y - 1)}$

111. $\dfrac{1+\frac{a}{b}}{1-\frac{a}{1-\frac{a}{b}}} = \dfrac{1+\frac{a}{b}}{1-\frac{a(b)}{\left(1-\frac{a}{b}\right)(b)}} = \dfrac{1+\frac{a}{b}}{1-\frac{ab}{b-a}} = \dfrac{\left(1+\frac{a}{b}\right)(b)(b-a)}{\left(1-\frac{ab}{b-a}\right)(b)(b-a)} = \dfrac{b(b-a)+a(b-a)}{b(b-a)-ab(b)}$

$\qquad\qquad = \dfrac{b^2-ab+ab-a^2}{b^2-ab-ab^2}$

$\qquad\qquad = \dfrac{b^2-a^2}{b(b-a-ab)}$

$\qquad\qquad = \dfrac{(b+a)(b-a)}{b(b-a-ab)}$

113. $a+\dfrac{a}{1+\frac{a}{a+1}} = a+\dfrac{a(a+1)}{\left(1+\frac{a}{a+1}\right)(a+1)} = a+\dfrac{a(a+1)}{a+1+a} = a+\dfrac{a^2+a}{2a+1} = \dfrac{a(2a+1)}{2a+1}+\dfrac{a^2+a}{2a+1}$

$\qquad\qquad = \dfrac{2a^2+a+a^2+a}{2a+1}$

$\qquad\qquad = \dfrac{3a^2+2a}{2a+1}$

115. $\qquad\dfrac{3}{y}-\dfrac{1}{12} = \dfrac{7}{4y}$

$\qquad\left(\dfrac{3}{y}-\dfrac{1}{12}\right)12y = \dfrac{7}{4y}\cdot 12y$

$\qquad\qquad 36-y = 21$

$\qquad\qquad\quad 15 = y$

The answer checks.

117. $\qquad\dfrac{y^2}{y+2}+4 = \dfrac{y+6}{y+2}+3$

$\qquad\left(\dfrac{y^2}{y+2}+4\right)(y+2) = \left(\dfrac{y+6}{y+2}+3\right)(y+2)$

$\qquad\quad y^2+4(y+2) = y+6+3(y+2)$

$\qquad\quad y^2+4y+8 = y+6+3y+6$

$\qquad\quad y^2+4y+8 = 4y+12$

$\qquad\qquad\quad y^2-4 = 0$

$\qquad\quad (y+2)(y-2) = 0$

$\qquad y+2=0 \quad\textbf{or}\quad y-2=0$

$\qquad\quad y=-2 \qquad\qquad y=2$

$y=2$ checks, but $y=-2$ does not and is extraneous.

119. $\dfrac{a^{-2}b^2}{x^{-1}y}\cdot\dfrac{a^4b^4}{x^2y^3} = \dfrac{a^2b^6}{xy^4}$

121. $\dfrac{m^2-n^2}{2x^2+3x-2}\cdot\dfrac{2x^2+5x-3}{n^2-m^2} = \dfrac{(m+n)(m-n)}{(2x-1)(x+2)}\cdot\dfrac{(2x-1)(x+3)}{(n+m)(n-m)} = -\dfrac{x+3}{x+2}$

123. $\dfrac{x^3+y^3}{x^3-y^3}\div\dfrac{x^2-xy+y^2}{x^2+xy+y^2} = \dfrac{x^3+y^3}{x^3-y^3}\cdot\dfrac{x^2+xy+y^2}{x^2-xy+y^2} = \dfrac{(x+y)(x^2-xy+y^2)}{(x-y)(x^2+xy+y^2)}\cdot\dfrac{x^2+xy+y^2}{x^2-xy+y^2}$

$\qquad\qquad = \dfrac{x+y}{x-y}$

125. $\dfrac{x+8}{x-3} - \dfrac{x-14}{3-x} = \dfrac{x+8}{x-3} - \dfrac{-x+14}{x-3} = \dfrac{2x-6}{x-3} = \dfrac{2(x-3)}{x-3} = 2$

127. $\dfrac{x-2}{x^2-3x} + \dfrac{2x-1}{x^2+3x} - \dfrac{2}{x^2-9} = \dfrac{x-2}{x(x-3)} + \dfrac{2x-1}{x(x+3)} - \dfrac{2}{(x+3)(x-3)}$

$\qquad\qquad = \dfrac{(x-2)(x+3)}{x(x-3)(x+3)} + \dfrac{(2x-1)(x-3)}{x(x-3)(x+3)} - \dfrac{2x}{x(x-3)(x+3)}$

$\qquad\qquad = \dfrac{x^2+x-6}{x(x-3)(x+3)} + \dfrac{2x^2-7x+3}{x(x-3)(x+3)} - \dfrac{2x}{x(x-3)(x+3)}$

$\qquad\qquad = \dfrac{3x^2-8x-3}{x(x-3)(x+3)} = \dfrac{(3x+1)(x-3)}{x(x-3)(x+3)} = \dfrac{3x+1}{x(x+3)}$

129. $\dfrac{x+y}{x^{-1}+y^{-1}} = \dfrac{x+y}{\frac{1}{x}+\frac{1}{y}} = \dfrac{(x+y)(xy)}{\left(\frac{1}{x}+\frac{1}{y}\right)(xy)} = \dfrac{(x+y)xy}{y+x} = xy$

131. $\qquad\qquad \dfrac{3}{x} + \dfrac{4}{x+1} = 2$

$\qquad\quad \left(\dfrac{3}{x} + \dfrac{4}{x+1}\right)(x)(x+1) = (2)(x)(x+1)$

$\qquad\qquad\quad 3(x+1) + 4x = 2x(x+1)$

$\qquad\qquad\quad 3x+3+4x = 2x^2 + 2x$

$\qquad\qquad\qquad\quad 0 = 2x^2 - 5x - 3$

$\qquad\qquad\qquad\quad 0 = (2x+1)(x-3)$

$\quad 2x + 1 = 0 \quad\textbf{or}\quad x - 3 = 0$

$\qquad 2x = -1 \qquad\qquad x = 3 \qquad$ Both answers check.

$\qquad\quad x = -\frac{1}{2}$

133. Answers may vary.　　　　　　**135. Answers may vary.**

137. $\dfrac{a-3b}{2b-a} = \dfrac{-(3b-a)}{-(a-2b)} = \dfrac{3b-a}{a-2b}.$

The two answers are the same.

139. You can divide out the 4's in parts a and d.

Exercises 7.6 (page 515)

1. $|8| = +8 = 8$　　　　　**3.** $|-5| = +5 = 5$　　　　　**5.** $|-6| = +6 = 6$

7. $-|5| = -(+5) = -5$　　**9.** $-|-20| = -(+20) = -20$　**11.** $2\pi - 4 > 0$, so
$\qquad\qquad\qquad\qquad\qquad\qquad\qquad\qquad\qquad\qquad\qquad |2\pi - 4| = 2\pi - 4$

13. $|2| = 2$; $|5| = 5$;　　　**15.** $|5| = 5$; $|-8| = 8$;　　　**17.** $|-2| = 2$; $|10| = 10$;
$\quad |2|$ is smaller.　　　　　$\quad |5|$ is smaller.　　　　　$\quad |-2|$ is smaller.

19. $|-3| = 3$; $-|-4| = -4$; $-|-4|$ is smaller.

21. $3(2a - 1) = 2a$
$6a - 3 = 2a$
$4a = 3$
$a = \dfrac{3}{4}$

23. $\dfrac{5x}{2} - 1 = \dfrac{x}{3} + 12$
$6\left(\dfrac{5x}{2} - 1\right) = 6\left(\dfrac{x}{3} + 12\right)$
$15x - 6 = 2x + 72$
$13x = 78$
$x = 6$

25. x

27. 0

29. $a = -b$

31. $|x| = 8$
$x = 8$ **or** $x = -8$

33. $|x| = 12$
$x = 12$ **or** $x = -12$

35. $|x - 3| = 6$
$x - 3 = 6$ **or** $x - 3 = -6$
$x = 9 \qquad\qquad x = -3$

37. $|x + 5| = 9$
$x + 5 = 9$ **or** $x + 5 = -9$
$x = 4 \qquad\qquad x = -14$

39. $|2x - 3| = 5$
$2x - 3 = 5$ **or** $2x - 3 = -5$
$2x = 8 \qquad\qquad 2x = -2$
$x = 4 \qquad\qquad x = -1$

41. $|3x + 2| = 16$
$3x + 2 = 16$ **or** $3x + 2 = -16$
$3x = 14 \qquad\qquad 3x = -18$
$x = \frac{14}{3} \qquad\qquad x = -6$

43. $|x + 3| + 7 = 10$
$|x + 3| = 3$
$x + 3 = 3$ **or** $x + 3 = -3$
$x = 0 \qquad\qquad x = -6$

45. $|0.3x - 3| - 2 = 7$
$|0.3x - 3| = 9$
$0.3x - 3 = 9$ **or** $0.3x - 3 = -9$
$0.3x = 12 \qquad\qquad 0.3x = -6$
$x = 40 \qquad\qquad x = -20$

47. $|x - 21| = -8$
Since an absolute value cannot be negative, there is no solution. $\emptyset$

49. $|6x + 2| + 8 = 5$
$|6x + 2| = -3$
Since an absolute value cannot be negative, there is no solution. $\emptyset$

51. $\left|\dfrac{3}{5}x - 4\right| - 2 = -2$
$\left|\dfrac{3}{5}x - 4\right| = 0$
$\dfrac{3}{5}x - 4 = 0$ **or** $\dfrac{3}{5}x - 4 = -0$
$\dfrac{3}{5}x = 4 \qquad\qquad \dfrac{3}{5}x = 4$
$x = \frac{20}{3} \qquad\qquad x = \frac{20}{3}$

53. $\left|\dfrac{5}{4}x + 1\right| - 2 = 5$
$\left|\dfrac{5}{4}x + 1\right| = 7$
$\dfrac{5}{4}x + 1 = 7$ **or** $\dfrac{5}{4}x + 1 = -7$
$\dfrac{5}{4}x = 6 \qquad\qquad \dfrac{5}{4}x = -8$
$x = \frac{24}{5} \qquad\qquad x = -\frac{32}{5}$

55.
$$|2x + 1| = |3x + 3|$$
$$2x + 1 = 3x + 3 \textbf{ or } 2x + 1 = -(3x + 3)$$
$$-x = 2 \qquad\qquad 2x + 1 = -3x - 3$$
$$x = -2 \qquad\qquad 5x = -4$$
$$x = -\tfrac{4}{5}$$

57.
$$|3x - 1| = |x + 5|$$
$$3x - 1 = x + 5 \textbf{ or } 3x - 1 = -(x + 5)$$
$$2x = 6 \qquad\qquad 3x - 1 = -x - 5$$
$$x = 3 \qquad\qquad 4x = -4$$
$$x = -1$$

59.
$$|4x + 3| = |9 - 2x|$$
$$4x + 3 = 9 - 2x \textbf{ or } 4x + 3 = -(9 - 2x)$$
$$6x = 6 \qquad\qquad 4x + 3 = -9 + 2x$$
$$x = 1 \qquad\qquad 2x = -12$$
$$x = -6$$

61.
$$|5x - 7| = |4x + 1|$$
$$5x - 7 = 4x + 1 \textbf{ or } 5x - 7 = -(4x + 1)$$
$$x = 8 \qquad\qquad 5x - 7 = -4x - 1$$
$$9x = 6$$
$$x = \tfrac{2}{3}$$

63.
$$|3x + 24| = 0$$
$$3x + 24 = 0 \qquad \textbf{or} \qquad 3x + 24 = -0$$
$$3x = -24 \qquad\qquad 3x = -24$$
$$x = -8 \qquad\qquad x = -8$$

65.
$$\left|\tfrac{x}{2} - 1\right| = 3$$
$$\tfrac{x}{2} - 1 = 3 \quad \textbf{or} \quad \tfrac{x}{2} - 1 = -3$$
$$\tfrac{x}{2} = 4 \qquad\qquad \tfrac{x}{2} = -2$$
$$x = 8 \qquad\qquad x = -4$$

67.
$$|3 - 4x| = 5$$
$$3 - 4x = 5 \qquad \textbf{or} \quad 3 - 4x = -5$$
$$-4x = 2 \qquad\qquad -4x = -8$$
$$x = -\tfrac{1}{2} \qquad\qquad x = 2$$

69.
$$\left|x + \tfrac{1}{3}\right| = |x - 3|$$
$$x + \tfrac{1}{3} = x - 3 \qquad \textbf{or} \qquad x + \tfrac{1}{3} = -(x - 3)$$
$$3x + 1 = 3x - 9 \qquad\qquad x + \tfrac{1}{3} = -x + 3$$
$$0 = -10 \qquad\qquad 3x + 1 = -3x + 9$$
(no solution from this part) $\qquad 6x = 8$
$$x = \tfrac{4}{3}$$

71.
$$|3x + 7| = -|8x - 2|$$
Since an absolute value cannot be negative, there is no solution. $\emptyset$

73.
$$\left|\frac{3x + 48}{3}\right| = 12$$
$$\frac{3x + 48}{3} = 12 \quad \textbf{or} \quad \frac{3x + 48}{3} = -12$$
$$3x + 48 = 36 \qquad\qquad 3x + 48 = -36$$
$$3x = -12 \qquad\qquad 3x = -84$$
$$x = -4 \qquad\qquad x = -28$$

75.
$$\left|\frac{8x+1}{7}\right| + 6 = 1$$
$$\left|\frac{8x+1}{7}\right| = -5$$
Since an absolute value cannot be negative, there is no solution. $\emptyset$

77. Graph $y = |0.75x + 0.12|$ and $y = 12.3$ and find the x-coordinate(s) of any point(s) of intersection:

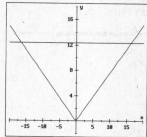

solutions: $-16.6, 16.2$

79. Answers may vary.

81.
$$|x| + k = 0$$
$$|x| = -k$$
If this equation has exactly two solutions, then $-k > 0$, or $k < 0$.

83. Answers may vary.

85. Answers may vary.

87.
$$x = |1 - x|$$
$x = 1 - x$ **or** $x = -(1 - x)$
$2x = 1$ $x = -1 + x$
$x = \frac{1}{2}$ $0 = -1$ no solution
                   from this part

Exercises 7.7 (page 521)

1. $x > -8$ and $x < 8$
$(-8, 8)$

3. $x < -1$ or $x > 1$
$(-\infty, -1) \cup (1, \infty)$

5. $x + 1 \le 2$ and $x + 1 \ge -2$
     $x \le 1$ $x \ge -3$
             $[-3, 1]$

7.
$$-5 < 2x + 3 < 5$$
$$-5 - 3 < 2x + 3 - 3 < 5 - 3$$
$$-8 < 2x < 2$$
$$-4 < x < 1$$
$$(-4, 1)$$

9.
$$A = p + prt$$
$$A - p = prt$$
$$\frac{A - p}{pr} = t, \text{ or } t = \frac{A - p}{pr}$$

11.
$$S = 2lw + 4wh$$
$$S - 4wh = 2lw$$
$$\frac{S - 4wh}{2w} = l, \text{ or } l = \frac{S - 4wh}{2w}$$

237

13. $-k < x < k$

15. $x < -k$ or $x > k$

17. $|x| < 8$
$-8 < x < 8$
solution set: $(-8, 8)$

19. $|2x| < 8$
$-8 < 2x < 8$
$\frac{-8}{2} < \frac{2x}{2} < \frac{8}{2}$
$-4 < x < 4$
solution set: $(-4, 4)$

21. $|x + 1| < 2$
$-2 < x + 1 < 2$
$-2 - 1 < x + 1 - 1 < 2 - 1$
$-3 < x < 1$
solution set: $(-3, 1)$

23. $|3x - 2| < 10$
$-10 < 3x - 2 < 10$
$-10 + 2 < 3x - 2 + 2 < 10 + 2$
$-8 < 3x < 12$
$-\frac{8}{3} < x < 4$
solution set: $\left(-\frac{8}{3}, 4\right)$

25. $|x + 9| \leq 12$
$-12 \leq x + 9 \leq 12$
$-12 - 9 \leq x + 9 - 9 \leq 12 - 9$
$-21 \leq x \leq 3$
solution set: $[-21, 3]$

27. $|4x - 1| \leq 7$
$-7 \leq 4x - 1 \leq 7$
$-7 + 1 \leq 4x - 1 + 1 \leq 7 + 1$
$-6 \leq 4x \leq 8$
$\frac{-6}{4} \leq \frac{4x}{4} \leq \frac{8}{4}$
$-\frac{3}{2} \leq x \leq 2$
solution set: $\left[-\frac{3}{2}, 2\right]$

29. $|5x| > 5$
$5x < -5$ **or** $5x > 5$
$x < -1$ $\qquad$ $x > 1$
solution set: $(-\infty, -1) \cup (1, \infty)$

31. $|x - 12| > 24$
$x - 12 < -24$ **or** $x - 12 > 24$
$x < -12$ $\qquad$ $x > 36$
solution set: $(-\infty, -12) \cup (36, \infty)$

33.
$$|x + 5| \geq 7$$
$$x + 5 \leq -7 \quad \textbf{or} \quad x + 5 \geq 7$$
$$x \leq -12 \qquad\qquad x \geq 2$$
solution set: $(-\infty, -12] \cup [2, \infty)$

$-12 \qquad 2$

35.
$$|2 - 3x| \geq 8$$
$$2 - 3x \leq -8 \quad \textbf{or} \quad 2 - 3x \geq 8$$
$$-3x \leq -10 \qquad\qquad -3x \geq 6$$
$$x \geq \tfrac{10}{3} \qquad\qquad x \leq -2$$
solution set: $(-\infty, -2] \cup \left[\tfrac{10}{3}, \infty\right)$

$-2 \qquad \tfrac{10}{3}$

37.
$$\left|\tfrac{1}{3}x + 7\right| + 5 > 6$$
$$\left|\tfrac{1}{3}x + 7\right| > 1$$
$$\tfrac{1}{3}x + 7 < -1 \quad \textbf{or} \quad \tfrac{1}{3}x + 7 > 1$$
$$x + 21 < -3 \qquad\qquad x + 21 > 3$$
$$x < -24 \qquad\qquad x > -18$$
solution set: $(-\infty, -24) \cup (-18, \infty)$

$-24 \qquad -18$

39.
$$|3x + 1| + 2 < 6$$
$$|3x + 1| < 4$$
$$-4 < 3x + 1 < 4$$
$$-5 < 3x < 3$$
$$-\tfrac{5}{3} < x < 1$$
solution set: $\left(-\tfrac{5}{3}, 1\right)$

$-\tfrac{5}{3} \qquad 1$

41.
$$-2|3x - 4| < 16$$
$$|3x - 4| > -8$$
Since an absolute value is always at least 0, this inequality is true for all real numbers.
solution set: $(-\infty, \infty)$

0

43.
$$|5x - 1| + 4 \leq 0$$
$$|5x - 1| \leq -4$$
Since an absolute value can never be negative, this inequality has no solution. $\emptyset$

45.
$$|2x + 1| + 2 \leq 2$$
$$|2x + 1| \leq 0$$
Since an absolute value can never be less than zero, the only solution for this inequality is when the absolute value is equal to 0.
$$|2x + 1| = 0$$
$$2x + 1 = 0 \quad \textbf{or} \quad 2x + 1 = -0$$
$$2x = -1 \qquad\qquad 2x = -1$$
$$x = -\tfrac{1}{2} \qquad\qquad x = -\tfrac{1}{2}$$
solution set: $\left[-\tfrac{1}{2}, -\tfrac{1}{2}\right]$

$-\tfrac{1}{2}$

47.
$$|4x + 3| > 0$$
$$4x + 3 < -0 \quad \textbf{or} \quad 4x + 3 > 0$$
$$4x < -3 \qquad\qquad 4x > -3$$
$$x < -\tfrac{3}{4} \qquad\qquad x > -\tfrac{3}{4}$$
solution set: $\left(-\infty, -\tfrac{3}{4}\right) \cup \left(-\tfrac{3}{4}, \infty\right)$

$-\tfrac{3}{4}$

SECTION 7.7

49.
$$3|2x + 5| \geq 9$$
$$|2x + 5| \geq 3$$
$$2x + 5 \leq -3 \quad \text{or} \quad 2x + 5 \geq 3$$
$$2x \leq -8 \qquad\qquad 2x \geq -2$$
$$x \leq -4 \qquad\qquad x \geq -1$$
solution set: $(-\infty, -4] \cup [-1, \infty)$

51.
$$\left|\tfrac{3}{5}x + \tfrac{7}{3}\right| < 2$$
$$-2 < \tfrac{3}{5}x + \tfrac{7}{3} < 2$$
$$15(-2) < 15\left(\tfrac{3}{5}x + \tfrac{7}{3}\right) < 15(2)$$
$$-30 < 9x + 35 < 30$$
$$-65 < 9x < -5$$
$$-\tfrac{65}{9} < x < -\tfrac{5}{9}$$
solution set: $\left(-\tfrac{65}{9}, -\tfrac{5}{9}\right)$

53.
$$-|3x + 1| < -8$$
$$|3x + 1| > 8$$
$$3x + 1 < -8 \quad \text{or} \quad 3x + 1 > 8$$
$$3x < -9 \qquad\qquad 3x > 7$$
$$x < -3 \qquad\qquad x > \tfrac{7}{3}$$
solution set: $(-\infty, -3) \cup \left(\tfrac{7}{3}, \infty\right)$

55.
$$\left|\tfrac{x-2}{3}\right| > 4$$
$$\tfrac{x-2}{3} < -4 \quad \text{or} \quad \tfrac{x-2}{3} > 4$$
$$x - 2 < -12 \qquad\qquad x - 2 > 12$$
$$x < -10 \qquad\qquad x > 14$$
solution set: $(-\infty, -10) \cup (14, \infty)$

57.
$$\left|\tfrac{1}{6}x + 6\right| + 2 < 2$$
$$\left|\tfrac{1}{6}x + 6\right| < 0$$
Since an absolute value can never be negative, this inequality has no solution. $\emptyset$

59.
$$\left|\tfrac{1}{7}x + 1\right| \leq 0$$
Since an absolute value can never be less than zero, the only solution for this inequality is when the absolute value is equal to 0.
$$\left|\tfrac{1}{7}x + 1\right| = 0$$
$$\tfrac{1}{7}x + 1 = 0 \quad \text{or} \quad \tfrac{1}{7}x + 1 = -0$$
$$\tfrac{1}{7}x = -1 \qquad\qquad \tfrac{1}{7}x = -1$$
$$x = -7 \qquad\qquad x = -7$$
solution set: $[-7, -7]$

61.
$$\left|\tfrac{1}{5}x - 5\right| + 4 > 4$$
$$\left|\tfrac{1}{5}x - 5\right| > 0$$
$$\tfrac{1}{5}x - 5 < -0 \quad \text{or} \quad \tfrac{1}{5}x - 5 > 0$$
$$\tfrac{1}{5}x < 5 \qquad\qquad \tfrac{1}{5}x > 5$$
$$x < 25 \qquad\qquad x > 25$$
solution set: $(-\infty, 25) \cup (25, \infty)$

63.
$$|3x - 2| + 2 \geq 0$$
$$|3x - 2| \geq -2$$
Since an absolute value is always at least 0, this inequality is true for all real numbers.
solution set: $(-\infty, \infty)$

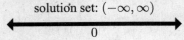

240

65. Find the x-coordinates of all points on the graph of $y = |0.5x + 0.7|$ that are below points on the graph of $y = 2.6$:

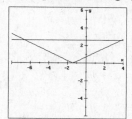

solution set: $(-6.6, 3.8)$

69. **Answers may vary.**

67. Find the x-coordinates of all points on the graph of $y = |2.15x - 3.05|$ that are above points on the graph of $y = 3.8$:

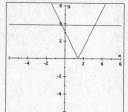

solution set: $(-\infty, -0.3) \cup (3.2, \infty)$

71. $|x| + |y| > |x + y|$ when x and y have different signs.

Chapter 7 Review (page 528)

1. $\begin{aligned} 3(y - 1) &= 24 \\ 3y - 3 &= 24 \\ 3y &= 27 \\ y &= 9 \end{aligned}$

2. $\begin{aligned} 2(x + 7) &= 42 \\ 2x + 14 &= 42 \\ 2x &= 28 \\ x &= 14 \end{aligned}$

3. $\begin{aligned} 13(x - 9) - 2 &= 7x - 5 \\ 13x - 117 - 2 &= 7x - 5 \\ 13x - 119 &= 7x - 5 \\ 13x - 7x - 119 &= 7x - 7x - 5 \\ 6x - 119 + 119 &= -5 + 119 \\ 6x &= 114 \\ \frac{6x}{6} &= \frac{114}{6} \\ x &= 19 \end{aligned}$

4. $\begin{aligned} \frac{8(x - 5)}{3} &= 2(x - 4) \\ 3 \cdot \frac{8(x - 5)}{3} &= 3 \cdot 2(x - 4) \\ 8(x - 5) &= 6(x - 4) \\ 8x - 40 &= 6x - 24 \\ 8x - 6x - 40 &= 6x - 6x - 24 \\ 2x - 40 + 40 &= -24 + 40 \\ 2x &= 16 \\ \frac{2x}{2} &= \frac{16}{2} \\ x &= 8 \end{aligned}$

5. $\begin{aligned} 2(x - 3) - x &= 5x - 4(x + 2) \\ 2x - 6 - x &= 5x - 4x - 8 \\ x - 6 &= x - 8 \\ x - x - 6 &= x - x - 8 \\ -6 &\neq -8 \\ \text{contradiction; } \emptyset \end{aligned}$

6. $\begin{aligned} 2(x + 7) - 8 &= 4x - 2(x - 3) \\ 2x + 14 - 8 &= 4x - 2x + 6 \\ 2x + 6 &= 2x + 6 \\ \text{identity; } \mathbb{R} \end{aligned}$

241

7.
$$V = \frac{1}{3}\pi r^2 h$$
$$3V = 3\left(\frac{1}{3}\pi r^2 h\right)$$
$$3V = \pi r^2 h$$
$$\frac{3V}{\pi r^2} = \frac{\pi r^2 h}{\pi r^2}$$
$$\frac{3V}{\pi r^2} = h, \text{ or } h = \frac{3V}{\pi r^2}$$

8.
$$V = \frac{1}{6}ab(x+y)$$
$$6V = 6 \cdot \frac{1}{6}ab(x+y)$$
$$6V = ab(x+y)$$
$$\frac{6V}{ab} = \frac{ab(x+y)}{ab}$$
$$\frac{6V}{ab} = x+y$$
$$\frac{6V}{ab} - y = x+y-y$$
$$\frac{6V}{ab} - y = x, \text{ or } x = \frac{6V}{ab} - y$$

9. Let x = the length of the first piece.
Then $4x$ = the length of the other piece.

| Sum of lengths | = | Total length |

$$x + 4x = 15$$
$$5x = 15$$
$$x = 3$$

He should cut the board 3 feet from one end.

10. Let w = the width of the rectangle. Then $2w - 5$ = the length of the rectangle.

| Perimeter | = 110 |

$$w + w + 2w - 5 + 2w - 5 = 110$$
$$6w - 10 = 110$$
$$6w = 120$$
$$w = 20$$

The dimensions are 20 feet by 35 feet, for an area of 700 ft^2.

11.
$$\frac{1}{3}y - 2 \geq \frac{1}{2}y + 2$$
$$6\left(\frac{1}{3}y - 2\right) \geq 6\left(\frac{1}{2}y + 2\right)$$
$$2y - 12 + 12 \geq 3y + 12 + 12$$
$$2y - 3y \geq 3y - 3y + 24$$
$$-y \geq 24$$
$$y \leq -24$$
solution set: $(-\infty, -24]$

$$-24$$

12.
$$\frac{7}{4}(x+3) < \frac{3}{8}(x-3)$$
$$8 \cdot \frac{7}{4}(x+3) < 8 \cdot \frac{3}{8}(x-3)$$
$$14(x+3) < 3(x-3)$$
$$14x + 42 < 3x - 9$$
$$14x < 3x - 51$$
$$11x < -51$$
$$x < -\frac{51}{11}$$
solution set: $\left(-\infty, -\frac{51}{11}\right)$

$$-\frac{51}{11}$$

13.
$$3 < 3x + 4 < 10$$
$$3 - 4 < 3x + 4 - 4 < 10 - 4$$
$$-1 < 3x < 6$$
$$-\frac{1}{3} < \frac{3x}{3} < \frac{6}{3}$$
$$-\frac{1}{3} < x < 2 \Rightarrow \text{solution set: } \left(-\frac{1}{3}, 2\right) \Rightarrow$$

$$-\frac{1}{3} \qquad 2$$

242

14.
$$4x > 3x + 2 > x - 3$$

| $4x > 3x + 2$ | and | $3x + 2 > x - 3$ |

$$4x > 3x + 2 \qquad\qquad 3x + 2 > x - 3$$
$$4x - 3x > 3x - 3x + 2 \qquad 3x + 2 - 2 > x - 3 - 2$$
$$x > 2 \qquad\qquad 3x - x > x - x - 5$$
$$x > 2 \qquad\qquad 2x > -5$$
$$\qquad\qquad\qquad x > -\tfrac{5}{2}$$

If $x > 2$ **and** $x > -\frac{5}{2}$, $x > 2$. The solution set is $(2, \infty)$.

15.
$$x + y = 4 \qquad x + y = 4$$
$$0 + y = 4 \qquad x + 0 = 4$$
$$y = 4 \qquad\qquad x = 4$$
$$(0, 4) \qquad\qquad (4, 0)$$

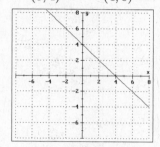

16.
$$2x - y = 8 \qquad 2x - y = 8$$
$$2(0) - y = 8 \qquad 2x - 0 = 8$$
$$-y = 8 \qquad\qquad 2x = 8$$
$$y = -8 \qquad\qquad x = 4$$
$$(0, -8) \qquad\qquad (4, 0)$$

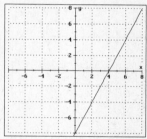

17.
$$y = 3x + 4 \qquad y = 3x + 4$$
$$y = 3(0) + 4 \qquad y = 3(1) + 4$$
$$y = 0 + 4 \qquad y = 3 + 4$$
$$y = 4 \qquad\qquad y = 7$$
$$(0, 4) \qquad\qquad (1, 7)$$

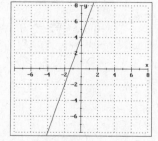

18.
$$x = 4 - 2y \qquad x = 4 - 2y$$
$$0 = 4 - 2y \qquad x = 4 - 2(0)$$
$$2y = 4 \qquad\qquad x = 4$$
$$y = 2 \qquad\qquad (4, 0)$$
$$(0, 2)$$

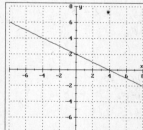

243

19. $y = 4$ (horizontal)

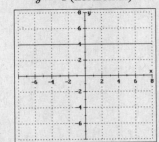

20. $x = -2$ (vertical)

21. $2(x + 3) = x + 2$
$2x + 6 = x + 2$
$x = -4$ (vertical)

22. $3y = 2(y - 1)$
$3y = 2y - 2$
$y = -2$ (horiozntal)

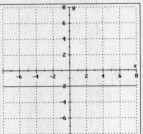

23. $x = \dfrac{x_1 + x_2}{2} = \dfrac{-3 + 6}{2} = \dfrac{3}{2}; \ y = \dfrac{y_1 + y_2}{2} = \dfrac{5 + 11}{2} = \dfrac{16}{2} = 8;$ midpoint: $\left(\dfrac{3}{2}, 8\right)$

24. $m = \dfrac{\Delta y}{\Delta x} = \dfrac{7 - 5}{5 - 3} = \dfrac{2}{2} = 1$

25. $m = \dfrac{\Delta y}{\Delta x} = \dfrac{12 - (-2)}{6 - (-3)} = \dfrac{14}{9}$

26. $m = \dfrac{\Delta y}{\Delta x} = \dfrac{-6 - 4}{-5 - (-3)} = \dfrac{-10}{-2} = 5$

27. $m = \dfrac{\Delta y}{\Delta x} = \dfrac{-9 - 4}{-6 - 5} = \dfrac{-13}{-11} = \dfrac{13}{11}$

28. $m = \dfrac{\Delta y}{\Delta x} = \dfrac{4 - 4}{1 - (-3)} = \dfrac{0}{4} = 0$

29. $m = \dfrac{\Delta y}{\Delta x} = \dfrac{8 - (-4)}{-5 - (-5)} = \dfrac{12}{0}$
no defined slope

30. $3x - 2y = 18$
$-2y = -3x + 18$
$\dfrac{-2y}{-2} = \dfrac{-3x}{-2} + \dfrac{18}{-2}$
$y = \dfrac{3}{2}x - 9 \Rightarrow m = \dfrac{3}{2}$

31. $x + 2y = 8$
$2y = -x + 8$
$\dfrac{2y}{2} = \dfrac{-x}{2} + \dfrac{8}{2}$
$y = -\dfrac{1}{2}x + 4 \Rightarrow m = -\dfrac{1}{2}$

32. $2(x+3) = 10$
$2x + 6 = 10$
$2x = 4$
$x = 2 \text{ (vertical)}$
The slope is undefined.

33. $3y + 2 = 9$
$3y = 7$
$y = \frac{7}{3} \text{ (horizontal)}$
$m = 0$

34. $m_1 \neq m_2 \Rightarrow$ not parallel
$m_1 \cdot m_2 = 4(-0.25) = -1 \Rightarrow$ perpendicular

35. $m_1 = m_2 \Rightarrow$ parallel

36. $m_1 \neq m_2 \Rightarrow$ not parallel
$m_1 \cdot m_2 = 0.5\left(-\frac{1}{2}\right) = -0.25$
$ \Rightarrow$ not perpendicular $\Rightarrow$ Neither

37. $m_1 \neq m_2 \Rightarrow$ not parallel
$m_1 \cdot m_2 = 5\left(\frac{-1}{5}\right) = -1$
$ \Rightarrow$ perpendicular

38. Let $y =$ sales and let $x =$ year of business.
$$m = \frac{\Delta y}{\Delta x} = \frac{y_2 - y_1}{x_1 - x_1} = \frac{120{,}000 - 66{,}000}{4 - 1} = \frac{54{,}000}{3} = \$18{,}000 \text{ per year}$$

39. $y - y_1 = m(x - x_1)$
$y - 5 = 3(x - 8)$
$y - 5 = 3x - 24$
$y = 3x - 19$

40. $m = \dfrac{\Delta y}{\Delta x} = \dfrac{-7 - 5}{6 - (-2)} = \dfrac{-12}{8} = -\dfrac{3}{2}$
$y - y_1 = m(x - x_1)$
$y - 5 = -\frac{3}{2}(x + 2)$
$y - 5 = -\frac{3}{2}x - 3$
$y = -\frac{3}{2}x + 2$

41. Find the slope of the given line:
$3x - 2y = 7$
$-2y = -3x + 7$
$y = \frac{3}{2}x - \frac{7}{2} \Rightarrow m = \frac{3}{2}$
Use the parallel slope:
$y - y_1 = m(x - x_1)$
$y - 5 = \frac{3}{2}(x - 3)$
$y - 5 = \frac{3}{2}x - \frac{9}{2}$
$y = \frac{3}{2}x - \frac{9}{2} + 5$
$y = \frac{3}{2}x + \frac{1}{2}$

42. Find the slope of the given line:
$y = -\frac{1}{4}x + 3 \Rightarrow m = -\frac{1}{4}$
Use the perpendicular slope:
$y - y_1 = m(x - x_1)$
$y + 5 = 4(x + 3)$
$y + 5 = 4x + 12$
$y = 4x + 7$

43. $y = \frac{2}{3}x + 1$

$m = \frac{2}{3}, b = 1 \Rightarrow (0, 1)$

44. $2x + 3y = 8$

$3y = -2x + 8$

$y = -\frac{2}{3}x + \frac{8}{3}; m_1 = -\frac{2}{3}$

$3x - 2y = 10$

$-2y = -3x + 10$

$y = \frac{3}{2}x - 5; m_2 = \frac{3}{2}$

$m_1 m_2 = -\frac{2}{3}\left(\frac{3}{2}\right) = -1 \Rightarrow$ perpendicular

45. Let x represent the number of years since the copy machine was purchased, and let y represent its value. We know two points on the line: $(0, 8700)$ and $(5, 100)$.

$$m = \frac{\Delta y}{\Delta x} = \frac{8700 - 100}{0 - 5} = \frac{8600}{-5} = -1720 \qquad y - y_1 = m(x - x_1)$$

$$y - 8700 = -1720(x - 0)$$

$$y = -1720x + 8700$$

46. $y = 6x - 4$ **is a function**, since each value of x corresponds to exactly one value of y.

47. $y = 4 - x$ **is a function**, since each value of x corresponds to exactly one value of y.

48. $y^2 = x$ **is not a function**, since $x = 9$ corresponds to both $y = 3$ and $y = -3$.

49. $|y| = x^2$ **is not a function**, since $x = 2$ corresponds to both $y = 4$ and $y = -4$.

50. $f(x) = 3x + 4$

$f(-3) = 3(-3) + 4 = -9 + 4 = -5$

51. $g(x) = x^2 - 3$

$g(8) = 8^2 - 3 = 64 - 3 = 61$

52. $g(x) = x^2 - 3$

$g(-2) = (-2)^2 - 3 = 4 - 3 = 1$

53. $f(x) = 3x + 4$

$f(5) = 3(5) + 4 = 15 + 4 = 19$

54. $2 - x \neq 0$

$2 \neq x$

Domain: $(-\infty, 2) \cup (2, \infty)$

55. $x - 3 \neq 0$

$x \neq 3$

Domain: $(-\infty, 3) \cup (3, \infty)$

56. Since the function is defined for all values of x, the domain is $(-\infty, \infty)$, or $\mathbb{R}$.

57. $f(x) = 6x + 2$
$D = (-\infty, \infty)$
$R = (-\infty, \infty)$

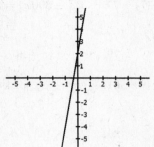

58. $f(x) = 3x - 10$
$D = (-\infty, \infty)$
$R = (-\infty, \infty)$

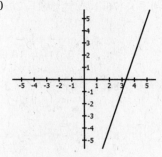

59. $f(x) = x^2 - 1$
$D = (-\infty, \infty)$
$R = [-1, \infty)$

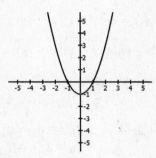

60. Every vertical line passes through the graph at most once, so it is a function.

61. Since a vertical line can pass through the graph more than once, it is not a function.

62. Since a vertical line can pass through the graph more than once, it is not a function.

63. Every vertical line passes through the graph at most once, so it is a function.

64. $3x + 6 = 3(x + 2)$

65. $5x^2y^3 - 10xy^2 = 5xy^2(xy - 2)$

66. $-8x^2y^3z^4 - 12x^4y^3z^2 = -4x^2y^3z^2(2z^2 + 3x^2)$

67. $12a^6b^4c^2 + 15a^2b^4c^6 = 3a^2b^4c^2(4a^4 + 5c^4)$

68. $x^3 + 2x^2 + 3x + 6 = x^2(x + 2) + 3(x + 2)$
$\qquad = (x + 2)(x^2 + 3)$

69. $ac + bc + 3a + 3b = c(a + b) + 3(a + b)$
$\qquad = (a + b)(c + 3)$

70. $x^{2n} + x^n = x^n(x^{2n-n} + 1) = x^n(x^n + 1)$

71. $y^{2n} - y^{3n} = y^{2n}(1 - y^{3n-2n}) = y^{2n}(1 - y^n)$

72. $x^4 + 4y + 4x^2 + x^2y = x^4 + x^2y + 4x^2 + 4y = x^2(x^2 + y) + 4(x^2 + y) = (x^2 + y)(x^2 + 4)$

73. $a^5 + b^2c + a^2c + a^3b^2 = a^5 + a^3b^2 + a^2c + b^2c = a^3(a^2 + b^2) + c(a^2 + b^2) = (a^2 + b^2)(a^3 + c)$

74. $z^2 + 25 = z^2 + 5^2$; sum of squares; prime

75. $y^2 - 121 = y^2 - 11^2 = (y + 11)(y - 11)$

76. $2x^4 - 98 = 2(x^4 - 49) = 2(x^2 + 7)(x^2 - 7)$

77. $3x^6 - 300x^2 = 3x^2(x^4 - 100)$
$$= 3x^2(x^2 + 10)(x^2 - 10)$$

78. $y^2 + 11y + 10 = (y + 10)(y + 1)$

79. $z^2 - 11z + 30 = (z - 5)(z - 6)$

80. $-2x^2 + 5x - 2 = -(2x^2 - 5x + 2)$
$$= -(2x - 1)(x - 2)$$

81. $-y^2 + 5y + 24 = -(y^2 - 5y - 24)$
$$= -(y - 8)(y + 3)$$

82. $y^3 + y^2 - 2y = y(y^2 + y - 2)$
$$= y(y + 2)(y - 1)$$

83. $2a^4 + 4a^3 - 6a^2 = 2a^2(a^2 + 2a - 3)$
$$= 2a^2(a + 3)(a - 1)$$

84. $15x^2 - 57xy - 12y^2 = 3(5x^2 - 19xy - 4y^2) = 3(5x + y)(x - 4y)$

85. $30x^2 + 65xy + 10y^2 = 5(6x^2 + 13xy + 2y^2) = 5(6x + y)(x + 2y)$

86. $x^2 + 4x + 4 - 4p^4 = (x^2 + 4x + 4) - 4p^4 = (x + 2)^2 - (2p^2)^2$
$$= (x + 2 + 2p^2)(x + 2 - 2p^2)$$

87. $y^2 + 3y + 2 + 2x + xy = (y^2 + 3y + 2) + 2x + xy = (y + 1)(y + 2) + x(2 + y)$
$$= (y + 2)(y + 1 + x)$$

88. $x^3 + 64 = x^3 + 4^3$
$$= (x + 4)(x^2 - 4x + 16)$$

89. $8y^3 - 512 = 8(y^3 - 64)$
$$= 8(y^3 - 4^3)$$
$$= 8(y - 4)(y^2 + 4y + 16)$$

90.
$$2x^2 - 18 = 0$$
$$2(x^2 - 9) = 0$$
$$2(x + 3)(x - 3) = 0$$
$$x + 3 = 0 \quad \textbf{or} \quad x - 3 = 0$$
$$x = -3 \qquad\qquad x = 3$$

91.
$$2x^2 + 9x = 5$$
$$2x^2 + 9x - 5 = 0$$
$$(2x - 1)(x + 5) = 0$$
$$2x - 1 = 0 \quad \textbf{or} \quad x + 5 = 0$$
$$2x = 1 \qquad\qquad x = -5$$
$$x = \tfrac{1}{2}$$

92.
$$5x^2 + 14x - 3 = 0$$
$$(5x - 1)(x + 3) = 0$$
$$5x - 1 = 0 \quad \textbf{or} \quad x + 3 = 0$$
$$5x = 1 \qquad\qquad x = -3$$
$$x = \tfrac{1}{5}$$

93.
$$2x^2 - 3x - 2 = 0$$
$$(2x + 1)(x - 2) = 0$$
$$2x + 1 = 0 \quad \textbf{or} \quad x - 2 = 0$$
$$2x = -1 \qquad\qquad x = 2$$
$$x = -\tfrac{1}{2}$$

94. $\dfrac{x^2 + 4x + 4}{x^2 - x - 6} \cdot \dfrac{x^2 - 9}{x^2 + 5x + 6} = \dfrac{(x + 2)(x + 2)}{(x + 2)(x - 3)} \cdot \dfrac{(x + 3)(x - 3)}{(x + 2)(x + 3)} = 1$

95. $\dfrac{x^3 - 64}{x^2 + 4x + 16} \div \dfrac{x^2 - 16}{x + 4} = \dfrac{x^3 - 64}{x^2 + 4x + 16} \cdot \dfrac{x + 4}{x^2 - 16} = \dfrac{(x - 4)(x^2 + 4x + 16)}{x^2 + 4x + 16} \cdot \dfrac{x + 4}{(x + 4)(x - 4)}$
$$= 1$$

96. $\dfrac{5y}{x-y} - \dfrac{3y+1}{x-y} = \dfrac{5y-3y-1}{x-y}$

$\qquad\qquad\qquad\;\; = \dfrac{2y-1}{x-y}$

97. $\dfrac{3x-1}{x^2+2} + \dfrac{3(x-2)}{x^2+2} = \dfrac{3x-1+3x-6}{x^2+2}$

$\qquad\qquad\qquad\qquad\;\; = \dfrac{6x-7}{x^2+2}$

98. $\dfrac{3}{x+2} + \dfrac{2}{x+3} = \dfrac{3(x+3)}{(x+2)(x+3)} + \dfrac{2(x+2)}{(x+2)(x+3)} = \dfrac{3x+9+2x+4}{(x+2)(x+3)} = \dfrac{5x+13}{(x+2)(x+3)}$

99. $\dfrac{4x}{x-4} - \dfrac{3}{x+3} = \dfrac{4x(x+3)}{(x-4)(x+3)} - \dfrac{3(x-4)}{(x-4)(x+3)} = \dfrac{4x^2+12x-3x+12}{(x-4)(x+3)} = \dfrac{4x^2+9x+12}{(x-4)(x+3)}$

100. $\dfrac{x^2+3x+2}{x^2-x-6} \cdot \dfrac{3x^2-3x}{x^2-3x-4} \div \dfrac{x^2+3x+2}{x^2-2x-8} = \dfrac{x^2+3x+2}{x^2-x-6} \cdot \dfrac{3x^2-3x}{x^2-3x-4} \cdot \dfrac{x^2-2x-8}{x^2+3x+2}$

$\qquad\qquad\qquad\qquad\qquad\qquad\qquad\qquad = \dfrac{(x+2)(x+1)}{(x-3)(x+2)} \cdot \dfrac{3x(x-1)}{(x-4)(x+1)} \cdot \dfrac{(x-4)(x+2)}{(x+2)(x+1)}$

$\qquad\qquad\qquad\qquad\qquad\qquad\qquad\qquad = \dfrac{3x(x-1)}{(x-3)(x+1)}$

101. $\dfrac{x^2-x-6}{x^2-3x-10} \div \dfrac{x^2-x}{x^2-5x} \cdot \dfrac{x^2-4x+3}{x^2-6x+9} = \dfrac{x^2-x-6}{x^2-3x-10} \cdot \dfrac{x^2-5x}{x^2-x} \cdot \dfrac{x^2-4x+3}{x^2-6x+9}$

$\qquad\qquad\qquad\qquad\qquad\qquad\qquad\qquad\qquad = \dfrac{(x-3)(x+2)}{(x-5)(x+2)} \cdot \dfrac{x(x-5)}{x(x-1)} \cdot \dfrac{(x-3)(x-1)}{(x-3)(x-3)} = 1$

102. $\dfrac{2x}{x+1} + \dfrac{3x}{x+2} + \dfrac{4x}{x^2+3x+2} = \dfrac{2x}{x+1} + \dfrac{3x}{x+2} + \dfrac{4x}{(x+2)(x+1)}$

$\qquad\qquad\qquad\qquad\qquad\qquad\;\; = \dfrac{2x(x+2)}{(x+1)(x+2)} + \dfrac{3x(x+1)}{(x+1)(x+2)} + \dfrac{4x}{(x+2)(x+1)}$

$\qquad\qquad\qquad\qquad\qquad\qquad\;\; = \dfrac{2x^2+4x+3x^2+3x+4x}{(x+1)(x+2)} = \dfrac{5x^2+11x}{(x+1)(x+2)}$

103. $\dfrac{5x}{x-3} + \dfrac{5}{x^2-5x+6} + \dfrac{x+3}{x-2} = \dfrac{5x}{x-3} + \dfrac{5}{(x-3)(x-2)} + \dfrac{x+3}{x-2}$

$\qquad\qquad\qquad\qquad\qquad\qquad\qquad = \dfrac{5x(x-2)}{(x-3)(x-2)} + \dfrac{5}{(x-3)(x-2)} + \dfrac{(x+3)(x-3)}{(x-3)(x-2)}$

$\qquad\qquad\qquad\qquad\qquad\qquad\qquad = \dfrac{5x^2-10x+5+x^2-9}{(x-3)(x-2)}$

$\qquad\qquad\qquad\qquad\qquad\qquad\qquad = \dfrac{6x^2-10x-4}{(x-3)(x-2)} = \dfrac{2(3x+1)(x-2)}{(x-3)(x-2)} = \dfrac{2(3x+1)}{x-3}$

104. $\dfrac{3(x+2)}{x^2-1} - \dfrac{2}{x+1} + \dfrac{4(x+3)}{x^2-2x+1} = \dfrac{3(x+2)}{(x+1)(x-1)} - \dfrac{2}{(x+1)} + \dfrac{4(x+3)}{(x-1)(x-1)}$

$\qquad = \dfrac{3(x+2)(x-1)}{(x+1)(x-1)(x-1)} - \dfrac{2(x-1)(x-1)}{(x+1)(x-1)(x-1)} + \dfrac{4(x+3)(x+1)}{(x+1)(x-1)(x-1)}$

$\qquad = \dfrac{3(x^2+x-2) - 2(x^2-2x+1) + 4(x^2+4x+3)}{(x+1)(x-1)(x-1)}$

$\qquad = \dfrac{3x^2+3x-6 - 2x^2+4x-2 + 4x^2+16x+12}{(x+1)(x-1)(x-1)} = \dfrac{5x^2+23x+4}{(x+1)(x-1)(x-1)}$

105. $\dfrac{x}{x^2+4x+4} + \dfrac{2x}{x^2-4} - \dfrac{x^2-4}{x-2}$

$\qquad = \dfrac{x}{(x+2)(x+2)} + \dfrac{2x}{(x+2)(x-2)} - \dfrac{x^2-4}{x-2}$

$\qquad = \dfrac{x(x-2)}{(x+2)(x+2)(x-2)} + \dfrac{2x(x+2)}{(x+2)(x-2)(x+2)} - \dfrac{(x^2-4)(x+2)(x+2)}{(x-2)(x+2)(x+2)}$

$\qquad = \dfrac{x^2-2x + 2x^2+4x - (x^2-4)(x^2+4x+4)}{(x+2)(x+2)(x-2)}$

$\qquad = \dfrac{3x^2+2x - (x^4+4x^3-16x-16)}{(x+2)(x+2)(x-2)}$

$\qquad = \dfrac{3x^2+2x - x^4-4x^3+16x+16}{(x+2)(x+2)(x-2)} = \dfrac{-x^4-4x^3+3x^2+18x+16}{(x+2)(x+2)(x-2)}$

106. $\dfrac{\frac{1}{x}+\frac{2}{y}}{\frac{2}{x}-\frac{1}{y}} = \dfrac{\left(\frac{1}{x}+\frac{2}{y}\right)xy}{\left(\frac{2}{x}-\frac{1}{y}\right)xy} = \dfrac{y+2x}{2y-x}$

107. $\dfrac{x^{-1}-y^{-1}}{x^{-1}+y^{-1}} = \dfrac{\frac{1}{x}-\frac{1}{y}}{\frac{1}{x}+\frac{1}{y}} = \dfrac{\left(\frac{1}{x}-\frac{1}{y}\right)xy}{\left(\frac{1}{x}+\frac{1}{y}\right)xy}$

$\qquad\qquad = \dfrac{y-x}{y+x}$

108.

$$\dfrac{2}{x+3} - \dfrac{x}{3-x} = -\dfrac{10}{x^2-9}$$

$$\dfrac{2}{x+3} - \dfrac{-x}{x-3} = -\dfrac{10}{(x+3)(x-3)}$$

$$\left(\dfrac{2}{x+3} + \dfrac{x}{x-3}\right)(x+3)(x-3) = \left(-\dfrac{10}{(x+3)(x-3)}\right)(x+3)(x-3)$$

$$2(x-3) + x(x+3) = -10$$

$$2x-6 + x^2+3x = -10$$

$$x^2+5x+4 = 0$$

$$(x+4)(x+1) = 0$$

$x+4=0 \quad$ **or** $\quad x+1=0$

$\qquad x=-4 \qquad\qquad x=-1 \qquad$ Both answers check.

109.
$$\frac{x}{x-2} + \frac{7}{x+5} = \frac{14}{x^2+3x-10}$$
$$\frac{x}{x-2} + \frac{7}{x+5} = \frac{14}{(x+5)(x-2)}$$
$$\left(\frac{x}{x-2} + \frac{7}{x+5}\right)(x+5)(x-2) = \left(\frac{14}{(x+5)(x-2)}\right)(x+5)(x-2)$$
$$x(x+5) + 7(x-2) = 14$$
$$x^2 + 5x + 7x - 14 = 14$$
$$x^2 + 12x - 28 = 0$$
$$(x+14)(x-2) = 0$$

$x + 14 = 0$ **or** $x - 2 = 0$ $x = -14$ checks, but $x = 2$ does not check
     $x = -14$ $x = 2$ and is extraneous.

110. $|4x + 2| = 10$
$4x + 2 = 10$ **or** $4x + 2 = -10$
   $4x = 8$ $4x = -12$
    $x = 2$ $x = -3$

111. $\left|\frac{3}{2}x - 4\right| = 0$
$\frac{3}{2}x - 4 = 0$ **or** $\frac{3}{2}x - 4 = -0$
   $\frac{3}{2}x = 4$ $\frac{3}{2}x = 4$
    $x = \frac{8}{3}$ $x = \frac{8}{3}$

112. $|3x + 2| + 7 = 2$
      $|3x + 2| = -5$
Since an absolute value cannot equal a
negative number, there is no solution. $\emptyset$

113. $|3x - 4| = |4x - 3|$
$3x - 4 = 4x - 3$ **or** $3x - 4 = -(4x - 3)$
   $-1 = x$ $3x - 4 = -4x + 3$
                    $7x = 7$
                     $x = 1$

114. $|2x + 7| < 3$
    $-3 < 2x + 7 < 3$
$-3 - 7 < 2x + 7 - 7 < 3 - 7$
     $-10 < 2x < -4$
      $-5 < x < -2$
solution set: $(-5, -2)$

115. $|3x - 8| \geq 4$
$3x - 8 \leq -4$ **or** $3x - 8 \geq 4$
   $3x \leq 4$ $3x \geq 12$
    $x \leq \frac{4}{3}$ $x \geq 4$
solution set: $\left(-\infty, \frac{4}{3}\right] \cup [4, \infty)$

116. $\left|\frac{3}{2}x - 14\right| \geq 0$
Since an absolute value is always at
least 0, this inequality is true for all
real #s. Solution set: $(-\infty, \infty)$

117. $\left|\frac{3}{2}x - 14\right| < 0$
Since an absolute value can never be
negative, there is no solution. $\emptyset$

Chapter 7 Test (page 538)

1.
$$9(x + 4) + 4 = 4(x - 5)$$
$$9x + 36 + 4 = 4x - 20$$
$$9x + 40 = 4x - 20$$
$$9x - 4x + 40 - 40 = 4x - 4x - 20 - 40$$
$$5x = -60$$
$$\frac{5x}{5} = \frac{-60}{5}$$
$$x = -12$$

2.
$$\frac{y - 1}{5} + 2 = \frac{2y - 3}{3}$$
$$15\left(\frac{y - 1}{5} + 2\right) = 15\left(\frac{2y - 3}{3}\right)$$
$$3(y - 1) + 30 = 10y - 15$$
$$3y - 3 + 30 = 10y - 15$$
$$3y + 27 = 10y - 15$$
$$3y - 10y + 27 - 27 = 10y - 10y - 15 - 27$$
$$-7y = -42$$
$$y = 6$$

3.
$$7 - 2(x - 3) = 8 - 2x + 5$$
$$7 - 2x + 6 = -2x + 13$$
$$-2x + 13 = -2x + 13$$
Identity; $(-\infty, \infty)$ or $\mathbb{R}$

4.
$$P = L + \frac{s}{f}i$$
$$P - L = \frac{s}{f}i$$
$$f(P - L) = f \cdot \frac{s}{f}i$$
$$f(P - L) = si$$
$$\frac{f(P - L)}{s} = \frac{si}{s}$$
$$\frac{f(P - L)}{s} = i, \text{ or } i = \frac{f(P - L)}{s}$$

5. Let $x =$ the length of the shortest piece.
Then $2x$ and $6x$ are the other lengths.
$$\boxed{\text{Total length}} = 20$$
$$x + 2x + 6x = 20$$
$$9x = 20$$
$$x = \frac{20}{9} \text{ feet}$$
$$\text{Longest} = 6\left(\frac{20}{9}\right) = \frac{\cancel{3} \cdot 2}{1} \cdot \frac{20}{\cancel{3} \cdot 3}$$
$$= \frac{40}{3} = 13\frac{1}{3} \text{ ft}$$
The longest piece is $13\frac{1}{3}$ ft. long.

6. Let $w =$ the width of the rectangle. Then
$w + 5 =$ the length of the rectangle.
$$\boxed{\text{Perimeter}} = 26$$
$$w + w + w + 5 + w + 5 = 26$$
$$4w + 10 = 26$$
$$4w = 16$$
$$w = 4$$
The dimensions are 4 cm by 9 cm, for an
area of 36 cm^2.

7.
$$-2(2x + 3) \geq 14$$
$$-4x - 6 + 6 \geq 14 + 6$$
$$-4x \geq 20$$
$$\frac{-4x}{-4} \leq \frac{20}{-4}$$
$$x \leq -5$$
solution set: $(-\infty, -5]$

8.
$$-2 < \frac{x - 4}{3} < 4$$
$$-6 < x - 4 < 12$$
$$-6 + 4 < x - 4 + 4 \leq 12 + 4$$
$$-2 < x < 16$$
solution set: $(-2, 16)$

9.
$$|2x + 3| - 1 = 10$$
$$|2x + 3| = 11$$
$$2x + 3 = 11 \quad \textbf{or} \quad 2x + 3 = -11$$
$$2x = 8 \qquad\qquad 2x = -14$$
$$x = 4 \qquad\qquad x = -7$$

10.
$$|3x + 4| = |x + 12|$$
$$3x + 4 = x + 12 \quad \textbf{or} \quad 3x + 4 = -(x + 12)$$
$$2x = 8 \qquad\qquad 3x + 4 = -x - 12$$
$$x = 4 \qquad\qquad 4x = -16$$
$$\qquad\qquad\qquad x = -4$$

11.
$$|x + 3| \leq 4$$
$$-4 \leq x + 3 \leq 4$$
$$-4 - 3 \leq x + 3 - 3 \leq 4 - 3$$
$$-7 \leq x \leq 1$$
solution set: $[-7, 1]$

12.
$$|2x - 4| > 22$$
$$2x - 4 < -22 \quad \textbf{or} \quad 2x - 4 > 22$$
$$2x < -18 \qquad\qquad 2x > 26$$
$$x < -9 \qquad\qquad x > 13$$
solution set: $(-\infty, -9) \cup (13, \infty)$

13.
$$2x - 5y = 10 \qquad 2x - 5y = 10$$
$$2(0) - 5y = 10 \qquad 2x - 5(0) = 10$$
$$-5y = 10 \qquad\qquad 2x = 10$$
$$y = -2 \qquad\qquad x = 5$$
$$(0, -2) \qquad\qquad (5, 0)$$

14.
$$x = \frac{x_1 + x_2}{2} = \frac{-3 + 4}{2} = \frac{1}{2}$$
$$y = \frac{y_1 + y_2}{2} = \frac{3 + (-2)}{2} = \frac{1}{2}$$
midpoint: $\left(\frac{1}{2}, \frac{1}{2} \right)$

15.
$$y = \frac{x-3}{5} \qquad y = \frac{x-3}{5}$$
$$0 = \frac{x-3}{5} \qquad y = \frac{0-3}{5}$$
$$0 = x-3 \qquad\quad y = -\frac{3}{5}$$
$$3 = x$$
x-intercept: $(3,0)$ y-intercept: $\left(0, -\frac{3}{5}\right)$

16.
$$x - 7 = 0$$
$$x = 7; \text{ vertical}$$

17. $m = \dfrac{\Delta y}{\Delta x} = \dfrac{4-8}{-2-6} = \dfrac{-4}{-8} = \dfrac{1}{2}$

18.
$$2x - 3y = 8$$
$$-3y = -2x + 8$$
$$y = \frac{2}{3}x - \frac{8}{3} \Rightarrow m = \frac{2}{3}$$

19. The graph of $x = 12$ is a vertical line, so the slope is undefined.

20. The graph of $y = 12$ is a horizontal line, so the slope is 0.

21.
$$y - y_1 = m(x - x_1)$$
$$y + 5 = \frac{2}{3}(x - 4)$$
$$y + 5 = \frac{2}{3}x - \frac{8}{3}$$
$$y = \frac{2}{3}x - \frac{8}{3} - 5$$
$$y = \frac{2}{3}x - \frac{23}{3}$$

22. $m = \dfrac{\Delta y}{\Delta x} = \dfrac{6 - (-10)}{-2 - (-4)} = \dfrac{16}{2} = 8$
$$y - y_1 = m(x - x_1)$$
$$y - 6 = 8(x + 2)$$
$$y - 6 = 8x + 16$$
$$y = 8x + 22$$

23.
$$-2(x - 3) = 3(2y + 5)$$
$$-2x + 6 = 6y + 15$$
$$6y + 15 = -2x + 6$$
$$6y = -2x - 9$$
$$y = \frac{-2}{6}x - \frac{9}{6}$$
$$y = -\frac{1}{3}x - \frac{3}{2}$$
$$m = -\frac{1}{3},\ b = -\frac{3}{2} \Rightarrow \left(0, -\frac{3}{2}\right)$$

24.
$$4x - y = 12$$
$$-y = -4x + 12$$
$$y = 4x - 12 \Rightarrow m = 4$$
$$y = \frac{1}{4}x + 3 \Rightarrow m = \frac{1}{4}$$
neither

25. $y = -\frac{2}{3}x + 4 \Rightarrow m = -\frac{2}{3}$
$$2y = 3x - 3$$
$$y = \frac{3}{2}x - \frac{3}{2} \Rightarrow m = \frac{3}{2}$$
perpendicular

26. $y = \frac{3}{2}x - 7 \Rightarrow m = \frac{3}{2}$
Use the parallel slope:
$$y - y_1 = m(x - x_1)$$
$$y - 0 = \frac{3}{2}(x - 0)$$
$$y = \frac{3}{2}x$$

27. $f(x) = x^3$
$D = (-\infty, \infty)$
$R = (-\infty, \infty)$

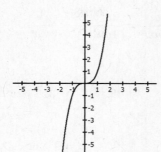

28. $f(x) = |x|$
$D = (-\infty, \infty)$
$R = [0, \infty)$

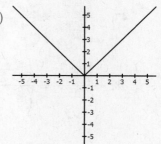

29. $f(x) = 3x + 1$
$f(3) = 3(3) + 1 = 9 + 1 = 10$

30. $g(x) = x^2 - 2$
$g(0) = (0)^2 - 2 = 0 - 2 = -2$

31. $f(x) = 3x + 1$
$f(a) = 3a + 1$

32. $g(x) = x^2 - 2$
$g(-x) = (-x)^2 - 2 = x^2 - 2$

33. function

34. not a function

35. $3xy^2 + 6x^2y = 3xy(y + 2x)$

36. $ax - xy + ay - y^2 = x(a - y) + y(a - y) = (a - y)(x + y)$

37. $x^2 - 49 = x^2 - 7^2 = (x + 7)(x - 7)$

38. $b^3 + 125 = (b + 5)(b^2 - 5b + 25)$

39. $b^2 - 3b + 5$; prime

40. $x^2 + 8x + 15 = (x + 3)(x + 5)$

41. $6b^2 + b - 2 = (3b + 2)(2b - 1)$

42. $6u^2 + 9u - 6 = 3(2u^2 + 3u - 2)$
$\qquad\qquad\qquad = 3(2u - 1)(u + 2)$

43. $x^2 + 6x + 9 - y^2 = (x^2 + 6x + 9) - y^2 = (x + 3)^2 - y^2 = (x + 3 + y)(x + 3 - y)$

44. $\dfrac{u^2 + 5u + 6}{u^2 - 4} \cdot \dfrac{u^2 - 5u + 6}{u^2 - 9} = \dfrac{(u + 2)(u + 3)}{(u + 2)(u - 2)} \cdot \dfrac{(u - 2)(u - 3)}{(u + 3)(u - 3)} = 1$

45. $\dfrac{x^3 + y^3}{4} \div \dfrac{x^2 - xy + y^2}{2x + 2y} = \dfrac{x^3 + y^3}{4} \cdot \dfrac{2x + 2y}{x^2 - xy + y^2} = \dfrac{(x + y)(x^2 - xy + y^2)}{4} \cdot \dfrac{2(x + y)}{x^2 - xy + y^2}$

$\qquad\qquad\qquad\qquad\qquad\qquad = \dfrac{(x + y)^2}{2}$

46. $\dfrac{x + 2}{x + 1} - \dfrac{x + 1}{x + 2} = \dfrac{(x + 2)(x + 2)}{(x + 1)(x + 2)} - \dfrac{(x + 1)(x + 1)}{(x + 1)(x + 2)} = \dfrac{x^2 + 4x + 4 - (x^2 + 2x + 1)}{(x + 1)(x + 2)}$

$\qquad\qquad\qquad\qquad\qquad = \dfrac{x^2 + 4x + 4 - x^2 - 2x - 1}{(x + 1)(x + 2)}$

$\qquad\qquad\qquad\qquad\qquad = \dfrac{2x + 3}{(x + 1)(x + 2)}$

47. $\dfrac{\frac{2u^2w^3}{v^2}}{\frac{4uw^4}{uv}} = \dfrac{2u^2w^3}{v^2} \div \dfrac{4uw^4}{uv} = \dfrac{2u^2w^3}{v^2} \cdot \dfrac{uv}{4uw^4} = \dfrac{2u^3w^3v}{4uv^2w^4} = \dfrac{u^2}{2vw}$

48. $\dfrac{\frac{x}{y} + \frac{1}{2}}{\frac{x}{2} - \frac{1}{y}} = \dfrac{\left(\frac{x}{y} + \frac{1}{2}\right)2y}{\left(\frac{x}{2} - \frac{1}{y}\right)2y} = \dfrac{2x + y}{xy - 2}$

49.
$$3x^2 = x$$
$$3x^2 - x = 0$$
$$x(3x - 1) = 0$$
$$x = 0 \quad \textbf{or} \quad 3x - 1 = 0$$
$$3x = 1$$
$$x = \tfrac{1}{3}$$

50.
$$\frac{a-1}{a+3} - \frac{1-2a}{3-a} = \frac{2-a}{a-3}$$
$$\frac{a-1}{a+3} - \frac{2a-1}{a-3} = \frac{2-a}{a-3}$$
$$\left(\frac{a-1}{a+3} - \frac{2a-1}{a-3}\right)(a+3)(a-3) = \left(\frac{2-a}{a-3}\right)(a+3)(a-3)$$
$$(a-1)(a-3) - (2a-1)(a+3) = (2-a)(a+3)$$
$$a^2 - 4a + 3 - (2a^2 + 5a - 3) = -a^2 - a + 6$$
$$a^2 - 4a + 3 - 2a^2 - 5a + 3 = -a^2 - a + 6$$
$$-a^2 - 9a + 6 = -a^2 - a + 6$$
$$-8a = 0$$
$$a = 0 \qquad \text{The answer checks.}$$

Exercises 8.1 (page 550)

1.
$$y = 3x \qquad y = \tfrac{1}{3}x + \tfrac{8}{3}$$
$$3 \overset{?}{=} 3(1) \qquad 3 \overset{?}{=} \tfrac{1}{3}(1) + \tfrac{8}{3}$$
$$3 = 3 \qquad 3 \overset{?}{=} \tfrac{1}{3} + \tfrac{8}{3}$$
$$3 = \tfrac{9}{3}$$
$(1, 3)$ is a solution.

3.
$$2x - 4y = 8 \qquad 4x + y = 9$$
$$2(2) - 4(-1) \overset{?}{=} 8 \qquad 4(2) + (-1) \overset{?}{=} 9$$
$$4 + 4 \overset{?}{=} 8 \qquad 8 - 1 \overset{?}{=} 9$$
$$8 = 8 \qquad 7 \neq 9$$
$(2, -1)$ is not a solution.

5. $850{,}000{,}000 = 8.5 \times 10^8$

7. $239 \times 10^3 = 2.39 \times 10^2 \times 10^3$
$$= 2.39 \times 10^5$$

9. system

11. inconsistent

13. dependent

15. $\begin{cases} x + y = 6 \\ x - y = 2 \end{cases}$

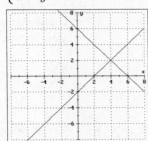

$(4, 2)$ is the solution.

17. $\begin{cases} 2x + y = 1 \\ x - 2y = -7 \end{cases}$

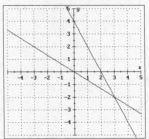

$(-1, 3)$ is the solution.

19. $\begin{cases} 2x + 3y = 0 \\ 2x + y = 4 \end{cases}$

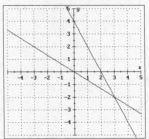

$(3, -2)$ is the solution.

21. $\begin{cases} y = 3 \\ x = 2 \end{cases}$

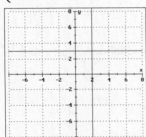

$(2, 3)$ is the solution.

23. $\begin{cases} x = 2 \\ y = \dfrac{4 - x}{2} \end{cases}$

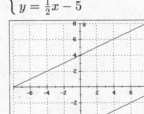

$(2, 1)$ is the solution.

25. $\begin{cases} \dfrac{5}{2}x + y = \dfrac{1}{2} \\ 2x - \dfrac{3}{2}y = 5 \end{cases}$

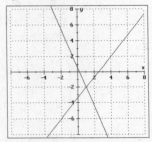

$(1, -2)$ is the solution.

27. $\begin{cases} 3x = 5 - 2y \\ 3x + 2y = 7 \end{cases}$

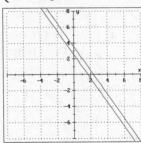

inconsistent system; $\emptyset$

29. $\begin{cases} x = 2y - 8 \\ y = \frac{1}{2}x - 5 \end{cases}$

inconsistent system; $\emptyset$

31. $\begin{cases} x = 3 - 2y \\ 2x + 4y = 6 \end{cases}$

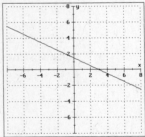

dependent system
$2x + 4y = 6$
$\qquad y = -\frac{1}{2}x + \frac{3}{2}$
$\left(x, -\frac{1}{2}x + \frac{3}{2}\right)$

33. $\begin{cases} 6x + 3y = 9 \\ y + 2x = 3 \end{cases}$

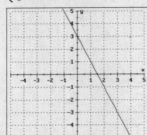

dependent system
$y + 2x = 3$
$y = -2x + 3$
$(x, -2x + 3)$

35. Graph $y = \frac{3}{2}x + 4$ and $y = -5$ and find the x-coordinate of any point of intersection.

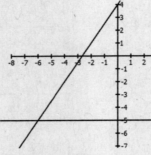

solution: $a = -6$

37. Graph $y = x + 2$ and $y = -x - 1$ and find the x-coordinate of any point of intersection.

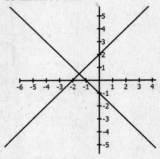

solution: $a = -\frac{3}{2}$

39. $\begin{cases} y = 5x \\ y = 5x + 3 \end{cases}$

Parallel lines: no solution

41. $\begin{cases} y = 6x + 1 \\ y = -6x + 1 \end{cases}$

Not parallel lines: one solution

43. $\begin{cases} x = \dfrac{11 - 2y}{3} \\ y = \dfrac{11 - 6x}{4} \end{cases}$

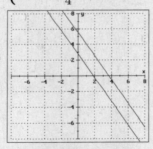

inconsistent system; $\emptyset$

45. $\begin{cases} \dfrac{5}{2}x + 3y = 6 \\ y = \dfrac{24 - 10x}{12} \end{cases}$

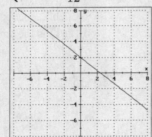

dependent system
$y = \dfrac{24 - 10x}{12}$
$y = 2 - \frac{5}{6}x$
$\left(x, 2 - \frac{5}{6}x\right)$

47. $\begin{cases} x = 13 - 4y \\ 3x = 4 + 2y \end{cases}$

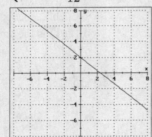

$\left(3, \frac{5}{2}\right)$ is the solution.

258

49. $\begin{cases} x = -\dfrac{3}{2}y \\ x = \dfrac{3}{2}y - 2 \end{cases}$

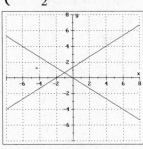

$\left(-1, \frac{2}{3}\right)$ is the solution.

51. $\begin{cases} y = 3.2x - 1.5 \\ y = -2.7x - 3.7 \end{cases}$

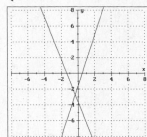

solution: $(-0.37, -2.69)$

53. $\begin{cases} 1.7x + 2.3y = 3.2 \\ y = 0.25x + 8.95 \end{cases}$

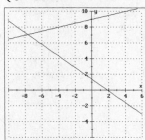

solution: $(-7.64, 7.04)$

55.
a. The point $(15, 2.0)$ is on the graph of the cost function, so it costs $2 million to manufacture 15,000 cameras.
b. The point $(20, 3.0)$ is on the graph of the revenue function, so there is a revenue of $3 million when 20,000 cameras are sold.
c. The graphs of the cost function and the revenue function meet at the point $(10, 1.5)$, so the revenue and cost functions are equal for 10,000 cameras.

57. $\begin{cases} 2x + 3y = 6 \\ 2x - 3y = 9 \end{cases}$

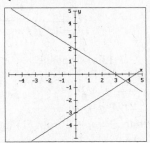

a. There is a possibility of a collision.
b. The danger point is at $(3.75, -0.5)$.
c. The collision is not certain, since the ships could be there at different times.

59. Let $x =$ the number of hours spent and let $y =$ the total cost.
$\begin{cases} y = 50x \\ y = 40x + 30 \end{cases}$

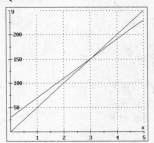

The lines meet at $(3, 150)$. Thus, the repair takes 3 hours.

61. $\begin{cases} y = 200x + 400 \\ y = 280x \end{cases}$

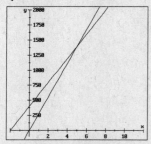

The lines meet at $(5, 1400)$.

63. **Answers may vary.**

65. One possible answer:
$\begin{cases} x + y = -3 \\ x - y = -7 \end{cases}$

Exercises 8.2 (page 563)

1.
$$3x - 4y = 7$$
$$3x - 4(x - 2) = 7$$
$$3x - 4x + 8 = 7$$
$$-x = -1$$
$$x = 1$$

3.
$$4x - 2y = 6$$
$$\underline{3x + 2y = 8}$$
$$7x \quad\quad = 14$$

5. $\left(a^3 b^4\right)^2 \left(ab^2\right)^3 = \left(a^3\right)^2 \left(b^4\right)^2 a^3 \left(b^2\right)^3$
$$= a^6 b^8 a^3 b^6 = a^9 b^{14}$$

7. $\left(\dfrac{-3x^3 y^4}{x^{-5} y^3}\right)^{-4} = \left(\dfrac{x^{-5} y^3}{-3x^3 y^4}\right)^4$
$$= \left(\dfrac{1}{-3x^8 y}\right)^4 = \dfrac{1}{81 x^{32} y^4}$$

9. setup; unit

11. parallelogram

13. Opposite

15. $\begin{cases} (1) \quad y = 3x \\ (2) \quad x + y = 8 \end{cases}$
Substitute $y = 3x$ from (1) into (2):
$$x + \boldsymbol{y} = 8$$
$$x + \boldsymbol{3x} = 8$$
$$4x = 8$$
$$x = 2$$
Substitute this and solve for y:
$$y = 3x = 3(2) = 6$$
Solution: $(2, 6)$

17. $\begin{cases} (1) \quad x - y = 2 \\ (2) \quad 2x + y = 13 \end{cases}$
Substitute $x = y + 2$ from (1) into (2):
$$2\boldsymbol{x} + y = 13$$
$$2(\boldsymbol{y + 2}) + y = 13$$
$$2y + 4 + y = 13$$
$$3y = 9$$
$$y = 3$$
Substitute this and solve for x:
$$x = y + 2 = 3 + 2 = 5$$
Solution: $(5, 3)$

19. $\dfrac{x}{2} + \dfrac{y}{2} = 6 \Rightarrow \times 2 \quad x + y = 12$

    $\dfrac{x}{2} - \dfrac{y}{2} = -2 \Rightarrow \times 2 \quad x - y = -4$

    $\begin{cases} (1) \quad x + y = 12 \\ (2) \quad x - y = -4 \end{cases}$

    Substitute $x = -y + 12$ from (1) into (2):

    $x - y = -4$

    $-y + 12 - y = -4$

    $-2y + 12 = -4$

    $-2y = -16$

    $y = 8$

    Substitute this and solve for x:

    $x = -y + 12 = -8 + 12 = 4$

    Solution: $(4, 8)$

21. $x = \dfrac{2}{3}y \qquad \Rightarrow \times 3 \qquad 3x = 2y$

    $y = 4x + 5$

    $\begin{cases} (1) \quad 3x = 2y \\ (2) \quad y = 4x + 5 \end{cases}$

    Substitute $y = 4x + 5$ from (2) into (1):

    $3x = 2y$

    $3x = 2(4x + 5)$

    $3x = 8x + 10$

    $-5x = 10$

    $x = -2$

    Substitute this and solve for y:

    $y = 4x + 5 = 4(-2) + 5 = -3$

    Solution: $(-2, -3)$

23. $\begin{array}{l} 2x - y = -5 \\ \underline{2x + y = -3} \\ 4x \quad\;\; = -8 \\ x \quad\;\;\; = -2 \end{array}$ Substitute and solve for y:

    $2x + y = -3$

    $2(-2) + y = -3$

    $-4 + y = -3$

    $y = 1$

    The solution is $(-2, 1)$.

25. $\begin{array}{l} 3x + 5y = 2 \Rightarrow \times (3) \\ \underline{4x - 3y = 22} \Rightarrow \times (5) \end{array}$ $\begin{array}{l} 9x + 15y = 6 \\ \underline{20x - 15y = 110} \\ 29x \quad\quad = 116 \\ x \quad\quad\;\; = 4 \end{array}$ $\begin{array}{l} 3x + 5y = 2 \\ 3(4) + 5y = 2 \\ 5y = -10 \\ y = -2 \end{array}$ Solution: $\boxed{(4, -2)}$

27. $\begin{array}{l} 5x - 2y = 19 \Rightarrow \times (2) \\ \underline{3x + 4y = 1} \end{array}$ $\begin{array}{l} 10x - 4y = 38 \\ \underline{3x + 4y = 1} \\ 13x \quad\quad = 39 \\ x \quad\quad\;\; = 3 \end{array}$ $\begin{array}{l} 3x + 4y = 1 \\ 3(3) + 4y = 1 \\ 4y = -8 \\ y = -2 \end{array}$ Solution: $\boxed{(3, -2)}$

29. $\begin{array}{l} 4x + 6y = 5 \Rightarrow \times (-2) \\ \underline{8x - 9y = 3} \end{array}$ $\begin{array}{l} -8x - 12y = -10 \\ \underline{8x - 9y = 3} \\ -21y = -7 \\ y = \dfrac{-7}{-21} = \dfrac{1}{3} \end{array}$ $\begin{array}{l} 4x + 6y = 5 \\ 4x + 6\left(\dfrac{1}{3}\right) = 5 \\ 4x + 2 = 5 \\ 4x = 3 \\ x = \dfrac{3}{4} \end{array}$ Solution: $\boxed{\left(\dfrac{3}{4}, \dfrac{1}{3}\right)}$

31. $\dfrac{5}{6}x + \dfrac{2}{3}y = \dfrac{7}{6} \Rightarrow \times 6 \quad 5x + 4y = 7 \Rightarrow \times 7 \quad 35x + 28y = 49$

$\dfrac{10}{7}x - \dfrac{4}{9}y = \dfrac{17}{21} \Rightarrow \times 63 \quad 90x - 28y = 51 \Rightarrow \quad \underline{90x - 28y = 51}$

$$125x \qquad = 100$$
$$x \qquad = \tfrac{100}{125} = \tfrac{4}{5}$$

$$5x + 4y = 7$$
$$5 \cdot \dfrac{4}{5} + 4y = 7$$
$$4 + 4y = 7$$
$$4y = 3$$
$$y = \tfrac{3}{4} \Rightarrow \text{Solution: } \boxed{\left(\tfrac{4}{5}, \tfrac{3}{4}\right)}$$

33. $\dfrac{3}{4}x + \dfrac{2}{3}y = 7 \Rightarrow \times 12 \quad 9x + 8y = 84 \Rightarrow \times 2 \qquad 18x + 16y = 168 \qquad 6x - 5y = 180$

$\dfrac{3}{5}x - \dfrac{1}{2}y = 18 \Rightarrow \times 10 \quad 6x - 5y = 180 \Rightarrow \times (-3) \quad \underline{-18x + 15y = -540} \qquad 6x - 5(-12) = 180$

$$31y = -372 \qquad 6x + 60 = 180$$
$$y = -12 \qquad 6x = 120$$
$$x = 20$$

Solution: $\boxed{(20, -12)}$

35. $3x = 2y - 4 \Rightarrow 3x - 2y = -4 \Rightarrow \times (-2) \quad -6x + 4y = 8 \qquad \boxed{\text{No solution, } \emptyset}$

$\underline{6x - 4y = -4} \Rightarrow \underline{6x - 4y = -4} \Rightarrow \qquad \underline{6x - 4y = -4}$

$$0 \neq 4$$

37. $x = 5y + 2 \Rightarrow x - 5y = 2 \Rightarrow \times (-3) \quad -3x + 15y = -6 \qquad \boxed{\text{No solution, } \emptyset}$

$\underline{3x = 15y + 10} \Rightarrow \underline{3x - 15y = 10} \Rightarrow \qquad \underline{3x - 15y = 10}$

$$0 \neq 4$$

39. $12x = 4y + 6 \Rightarrow 12x - 4y = 6 \Rightarrow \qquad 12x - 4y = 6 \qquad 6x - 2y = 3 \Rightarrow y = 3x - \tfrac{3}{2}$

$\underline{6x - 2y = 3} \Rightarrow \underline{6x - 2y = 3} \Rightarrow \times (-2) \quad \underline{-12x + 4y = -6} \qquad \boxed{\begin{array}{l}\text{Dependent equations} \\ \left(x, 3x - \tfrac{3}{2}\right)\end{array}}$

$$0 = 0$$

41. $y - 2x = 6 \Rightarrow -2x + y = 6 \Rightarrow \times (2) \quad -4x + 2y = 12 \qquad y - 2x = 6 \Rightarrow y = 2x + 6$

$\underline{4x + 12 = 2y} \Rightarrow \underline{4x - 2y = -12} \Rightarrow \qquad \underline{4x - 2y = -12} \qquad \boxed{\begin{array}{l}\text{Dependent equations} \\ (x, 2x + 6)\end{array}}$

$$0 = 0$$

43. Let $x = 0.6666\overline{6}$. Then $10x = 6.6666\overline{6}$.

$$10x = 6.6666\overline{6}$$
$$\underline{x = 0.6666\overline{6}}$$
$$9x = 6$$
$$\dfrac{9x}{9} = \dfrac{6}{9}$$
$$x = \tfrac{2}{3} \quad \text{The fraction is } \tfrac{2}{3}.$$

45. Let $x = -0.34\overline{898989}$.

$$\begin{aligned} &\text{Then } 100x = -34.89\overline{898989}. \\ &100x = -34.89898989 \\ &\underline{x = -0.34898989} \\ &99x = -34.55 \end{aligned}$$

$$\begin{aligned} 99x &= -34.55 \\ \frac{99x}{99} &= -\frac{34.55}{99} \\ x &= -\frac{34.55(100)}{99(100)} \\ x &= -\frac{3455}{9900} = -\frac{691}{1980} \end{aligned}$$

The fraction is $-\frac{691}{1980}$.

47. $\begin{cases} (1) & y = 2x \\ (2) & x + y = 6 \end{cases}$

Substitute $y = 2x$ from (1) into (2):
$$\begin{aligned} x + \boldsymbol{y} &= 6 \\ x + \boldsymbol{2x} &= 6 \\ 3x &= 6 \\ x &= 2 \end{aligned}$$
Substitute this and solve for y:
$$y = 2x = 2(2) = 4$$
Solution: $(2, 4)$

49. $\begin{cases} (1) & 3x - 4y = 9 \\ (2) & x + 2y = 8 \end{cases}$

Substitute $x = -2y + 8$ from (2) into (1):
$$\begin{aligned} 3\boldsymbol{x} - 4y &= 9 \\ 3(\boldsymbol{-2y + 8}) - 4y &= 9 \\ -6y + 24 - 4y &= 9 \\ -10y &= -15 \\ y &= \tfrac{3}{2} \end{aligned}$$
Substitute this and solve for x:
$$x = -2y + 8 = -2\left(\frac{3}{2}\right) + 8 = 5$$
Solution: $\left(5, \tfrac{3}{2}\right)$

51. $\begin{aligned} x - y &= 6 \\ \underline{x + y} &= 2 \\ 2x &= 8 \\ x &= 4 \end{aligned}$ Substitute and solve for y:
$$\begin{aligned} x + y &= 2 \\ 4 + y &= 2 \\ y &= -2 \end{aligned}$$
Solution: $\boxed{(4, -2)}$

53. $\begin{aligned} 2x + 2y &= -1 \Rightarrow \times(-2) \\ \underline{3x + 4y} &= 0 \end{aligned}$ $\begin{aligned} -4x - 4y &= 2 \\ \underline{3x + 4y} &= 0 \\ -x &= 2 \\ x &= -2 \end{aligned}$ $\begin{aligned} 2x + 2y &= -1 \\ 2(-2) + 2y &= -1 \\ -4 + 2y &= -1 \\ 2y &= 3 \\ y &= \tfrac{3}{2} \end{aligned}$ Solution: $\boxed{\left(-2, \tfrac{3}{2}\right)}$

55. $\begin{aligned} \frac{2}{5}x - \frac{1}{6}y &= \frac{7}{10} \Rightarrow \times 30 \\ \frac{3}{4}x - \frac{2}{3}y &= \frac{19}{8} \Rightarrow \times 24 \end{aligned}$ $\begin{aligned} 12x - 5y &= 21 \Rightarrow \times 3 \\ \underline{18x - 16y = 57 \Rightarrow \times(-2)} \end{aligned}$ $\begin{aligned} 36x - 15y &= 63 \\ \underline{-36x + 32y = -114} \\ 17y &= -51 \\ y &= -3 \end{aligned}$

$$\begin{aligned} 12x - 5y &= 21 \\ 12x - 5(-3) &= 21 \\ 12x + 15 &= 21 \\ 12x &= 6 \\ x = \tfrac{6}{12} = \tfrac{1}{2} &\Rightarrow \text{Solution: } \boxed{\left(\tfrac{1}{2}, -3\right)} \end{aligned}$$

57. $\quad \begin{aligned} 3x + 5y &= -14 \Rightarrow \times (3) \\ 2x - 3y &= 16 \Rightarrow \times (5) \end{aligned}$ $\quad \begin{aligned} 9x + 15y &= -42 \\ 10x - 15y &= 80 \\ \hline 19x &= 38 \\ x &= 2 \end{aligned}$ $\quad \begin{aligned} 3x + 5y &= -14 \\ 3(2) + 5y &= -14 \\ 5y &= -20 \\ y &= -4 \end{aligned}$ $\quad$ Solution: $\boxed{(2, -4)}$

For problems #59-61, begin each problem by letting $m = \frac{1}{x}$ and $n = \frac{1}{y}$. Solve for m and n, and then solve for x and y.

59. $\quad \begin{aligned} \frac{1}{x} + \frac{1}{y} &= \frac{5}{6} \Rightarrow \\ \frac{1}{x} - \frac{1}{y} &= \frac{1}{6} \Rightarrow \end{aligned}$ $\quad \begin{aligned} m + n &= \frac{5}{6} \\ m - n &= \frac{1}{6} \\ \hline 2m &= \frac{6}{6} \\ 2m &= 1 \\ m &= \frac{1}{2} \end{aligned}$

Solve for n:
$$m + n = \frac{5}{6}$$
$$\frac{1}{2} + n = \frac{5}{6}$$
$$n = \frac{5}{6} - \frac{1}{2}$$
$$n = \frac{1}{3}$$

Solve for x:
$$m = \frac{1}{x}$$
$$\frac{1}{2} = \frac{1}{x}$$
$$2 = x$$

Solve for y:
$$n = \frac{1}{y}$$
$$\frac{1}{3} = \frac{1}{y}$$
$$3 = y$$

Solution: $\boxed{(2, 3)}$

61. $\quad \begin{aligned} \frac{1}{x} + \frac{2}{y} &= -1 \Rightarrow \\ \frac{2}{x} - \frac{1}{y} &= -7 \Rightarrow \end{aligned}$ $\quad \begin{aligned} m + 2n &= -1 \Rightarrow \\ 2m - n &= -7 \Rightarrow \times 2 \end{aligned}$ $\quad \begin{aligned} m + 2n &= -1 \\ 4m - 2n &= -14 \\ \hline 5m &= -15 \\ m &= -3 \end{aligned}$

Solve for n:
$$m + 2n = -1$$
$$-3 + 2n = -1$$
$$2n = 2$$
$$n = 1$$

Solve for x:
$$m = \frac{1}{x}$$
$$-3 = \frac{1}{x}$$
$$-\frac{1}{3} = x$$

Solve for y:
$$n = \frac{1}{y}$$
$$1 = \frac{1}{y}$$
$$1 = y$$

Solution: $\boxed{\left(-\frac{1}{3}, 1\right)}$

63. Let $x =$ the cost of the pair of shoes and $y =$ the cost of the sweater.

$\quad \begin{aligned} (1) \quad & x + y = 110 \\ (2) \quad & y = x + 20 \end{aligned}$ $\quad \begin{aligned} x + \mathbf{y} &= 110 \\ x + \mathbf{x + 20} &= 110 \\ 2x &= 90 \\ x &= 45 \end{aligned}$ $\quad \begin{aligned} y &= x + 20 \\ y &= 45 + 20 \\ y &= 65 \end{aligned}$ $\quad$ The sweater cost \$65.

65. $\quad \begin{aligned} (1) \quad & R_1 + R_2 = 1375 \\ (2) \quad & R_1 = R_2 + 125 \end{aligned}$ $\quad \begin{aligned} \mathbf{R_1} + R_2 &= 1375 \\ \mathbf{R_2 + 125} + R_2 &= 1375 \\ 2R_2 &= 1250 \\ R_2 &= 625 \end{aligned}$ $\quad \begin{aligned} R_1 &= R_2 + 125 \\ R_1 &= 625 + 125 \\ R_1 &= 750 \end{aligned}$ $\quad$ The resistances are $R_1 = 750$ ohms and $R_2 = 625$ ohms.

67. Let l = the length of the field and w = the width of the field.

$$2w + 2l = 72 \Rightarrow \times(-1) \quad -2w - 2l = -72 \qquad 2w + 2l = 72$$
$$\underline{3w + 2l = 88} \Rightarrow \qquad \underline{3w + 2l = 88} \qquad 2(16) + 2l = 72$$
$$\qquad\qquad\qquad\qquad\qquad w = 16 \qquad 32 + 2l = 72$$
$$2l = 40$$
$$l = 20$$

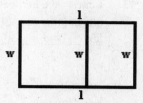

The dimensions of the field are 20 meters by 16 meters.

69. Let x = the amount invested at 10% and y = the amount invested at 12%.

$$x + y = 8000 \Rightarrow \qquad x + y = 8000 \Rightarrow \times(-10) \quad -10x - 10y = -80000$$
$$\underline{0.10x + 0.12y = 900} \Rightarrow \times100 \quad \underline{10x + 12y = 90000} \Rightarrow \qquad \underline{10x + 12y = 90000}$$
$$2y = 10000$$
$$y = 5000$$

$$x + y = 8000$$
$$x + 5000 = 8000$$
$$x = 3000 \Rightarrow \$3000 \text{ was invested at 10\%, while } \$5000 \text{ was invested at 12\%.}$$

71. Let x = the # of ounces of the 8% solution and y = the # of ounces of the 15% solution.

$$x + y = 100 \Rightarrow \qquad x + y = 100 \Rightarrow \times(-8) \quad -8x - 8y = -800$$
$$\underline{0.08x + 0.15y = 12.2} \Rightarrow \times100 \quad \underline{8x + 15y = 1220} \Rightarrow \qquad \underline{8x + 15y = 1220}$$
$$7y = 420$$
$$y = 60$$

$$x + y = 100$$
$$x + 60 = 100$$
$$x = 40 \Rightarrow 40 \text{ oz of the 8\% and 60 oz of the 15\% solution should be used.}$$

73. Let c = the speed of the car and p = the speed of the plane. Remember: distance = rate · time, so time = $\dfrac{\text{distance}}{\text{rate}}$. Form one equation from the fact that the car travels 50 miles in the same time that the plane travels 180 miles:

Car time = Plane time $\Rightarrow \dfrac{50}{c} = \dfrac{180}{p} \Rightarrow 50p = 180c \Rightarrow 50p - 180c = 0$.

Form a second equation from the relationship given between the rates: $p = c + 143$.

(1) $\quad 50p - 180c = 0 \qquad\qquad\qquad\qquad 50\boldsymbol{p} - 180c = 0 \qquad$ The car's speed is 55 mph.

(2) $\qquad\qquad p = c + 143 \quad\quad 50\boldsymbol{(c + 143)} - 180c = 0$

$$50c + 7150 - 180c = 0$$
$$-130c = -7150$$
$$c = 55$$

75. Let r = the number of racing bikes and m = the number of mountain bikes.

$$
\begin{array}{rll}
60r + 90m = 15900 & \Rightarrow \times(-7) & -420r - 630m = -111300 \\
55r + 70m = 13075 & \Rightarrow \times 9 & \underline{495r + 630m = 117675} \\
& & 75r = 6375 \\
& & r = 85
\end{array}
$$

$$
\begin{aligned}
60r + 90m &= 15900 \\
60(85) + 90m &= 15900 \\
5100 + 90m &= 15900 \\
90m &= 10800 \\
m &= 120
\end{aligned}
$$

85 racing bikes and 120 mountain bikes can be built.

77. Let x = the number of plates produced at the break-even point. Let C_1 = the cost of the first machine: $C_1 = 300 + 2x$. Let C_2 = the cost of the second machine: $C_2 = 500 + x$. To find the break point, find x such that $C_1 = C_2$:

$$
\begin{aligned}
C_1 &= C_2 \\
300 + 2x &= 500 + x \\
x &= 200 \Rightarrow \text{The break point is 200 plates.}
\end{aligned}
$$

79. Let x = the number of computers sold at the break-even point. Let C = the costs of the store: $C = 8925 + 850x$. Let R = the revenue of the store: $R = 1275x$. To find the break-even point, find x such that $C = R$:

$$
\begin{aligned}
C &= R \\
8925 + 850x &= 1275x \\
8925 &= 425x \\
21 &= x \Rightarrow \text{The break-even point is 21 computers.}
\end{aligned}
$$

81. Let x = the number of pieces of software sold at the break-even point. Let C = the costs of the business: $C = 18375 + 5.45x$. Let R = the revenue of the business: $R = 29.95x$. To find the break-even point, find x such that $C = R$:

$$
\begin{aligned}
C &= R \\
18375 + 5.45x &= 29.95x \\
18375 &= 24.50x \\
750 &= x \Rightarrow \text{The break-even point is 750 pieces of software.}
\end{aligned}
$$

83. Let x = the number of gallons of paint A sold at the break-even point. Let C = the costs of the company: $C = 32500 + 13x$. Let R = the revenue of the business: $R = 18x$. To find the break-even point, find x such that $C = R$:

$$
\begin{aligned}
C &= R \\
32500 + 13x &= 18x \\
32500 &= 5x \\
6500 &= x \Rightarrow \text{The break-even point is 6500 gallons of paint A (per month).}
\end{aligned}
$$

85. Calculate the profit made using each process (profit = revenue − cost)
A: revenue − cost = $18(6000) - (32500 + 13 \cdot 6000) = 108000 - 110500 = -2500$
B: revenue − cost = $18(6000) - (80600 + 5 \cdot 6000) = 108000 - 110600 = -2600$
Since the loss is less with process A, process A should be used.

87. Let $x =$ the number of pumps using process A sold at the break-even point. Let $C =$ the costs of the company: $C = 12390 + 29x$. Let $R =$ the revenue of the business: $R = 50x$. To find the break-even point, find x such that $C = R$:

$$C = R$$
$$12390 + 29x = 50x$$
$$12390 = 21x$$
$$590 = x \Rightarrow \text{The break-even point is 590 pumps using process A (per month).}$$

89. Calculate the profit made using each process (profit = revenue − cost)
A: revenue − cost $= 550(50) - (12390 + 550 \cdot 29) = 27500 - 28340 = -840$
B: revenue − cost $= 550(50) - (20460 + 550 \cdot 17) = 27500 - 29810 = -2310$
Since the loss is less with process A, process A should be used.

91. Calculate the profit made using each process (profit = revenue − cost)
A: revenue − cost $= 650(50) - (12390 + 650 \cdot 29) = 32500 - 31240 = 1260$
B: revenue − cost $= 650(50) - (20460 + 650 \cdot 17) = 32500 - 31510 = 990$
Since the profit is more with process A, process A should be used.

93. Let $x =$ the measure of the first angle and $y =$ the measure of the second angle.

$$
\begin{array}{ll}
x + y = 180 & x + y = 180 \quad \text{The angles have measures of } 35° \text{ and } 145°. \\
\underline{x - y = 110} & 145 + y = 180 \\
2x \quad\;\; = 290 & \quad\; y = 35 \\
x \quad\;\; = 145 &
\end{array}
$$

95. $\angle A$ and $\angle B$ are supplementary $\Rightarrow 2x + y + 3x = 180 \Rightarrow 5x + y = 180$.
$\angle A$ and $\angle D$ are supplementary $\Rightarrow 2x + y + y = 180 \Rightarrow 2x + 2y = 180$.

$$
\begin{array}{lll}
5x + \;\; y = 180 \Rightarrow \times(-2) & -10x - 2y = -360 & \text{Solve for } y: \\
\underline{2x + 2y = 180} \Rightarrow & \underline{2x + 2y = \quad 180} & 2x + 2y = 180 \\
& -8x \quad\quad = -180 & 2(22.5) + 2y = 180 \\
& x \quad\quad = \;\; 22.5 & 45 + 2y = 180 \\
& & 2y = 135 \\
& & y = 67.5
\end{array}
$$

97. Let $x =$ the measure of the angle of the range of motion and $y =$ the measure of the second angle.

$$
\begin{array}{llll}
(1) \quad x + y = 90 & \boldsymbol{x} + y = 90 & x = 4y & \text{The range of motion is } 72°. \\
(2) \quad\quad x = 4y & \boldsymbol{4y} + y = 90 & x = 4(18) & \\
& 5y = 90 & x = 72 & \\
& y = 18 & &
\end{array}
$$

99. Set $X_L = X_C$:

$$X_L = X_C$$

$$2\pi f L = \frac{1}{2\pi f C}$$

$$2\pi f L \cdot 2\pi f C = \frac{1}{2\pi f C} \cdot 2\pi f C$$

$$4\pi^2 f^2 L C = 1$$

$$\frac{4\pi^2 f^2 L C}{4\pi^2 L C} = \frac{1}{4\pi^2 L C}$$

$$f^2 = \frac{1}{4\pi^2 L C}$$

101-103. Answers may vary.

Exercises 8.3 (page 576)

1. Let p represent the number of pizzas sold.
Let c represent the number of calzones sold.

3. Let d represent the amount of money Danielle invested.
Let k represent the amount of money Kinley invested.

5. Let n represent the number of milliliters of the 9% saline solution.
Let t represent the number of milliliters of the 20% saline solution.

7. Let a represent the cost of one adult ticket.
Let s represent the cost of one student ticket.

9. $p + c = 52$; The number of pizzas sold plus the number of calzones sold equals the total number of items sold (52).

11. $d + k = 15{,}000$; The amount of money Danielle invested plus the amount of money Kinley invested equals the total amount of money invested (15,000).

13. $n + t = 25$; The number of milliliters of 9% saline solution plus the number of milliliters of 20% saline solution equals the total number of milliliters needed (25).

15. $a + s = 273$; The number of adult tickets sold plus the number of student tickets sold equals the total number of tickets sold (273).

17. $x < 4$

19. $-1 < x \le 2$

21. $9 \cdot 9 \cdot 9 \cdot 9 \cdot a = 9^4 a$

23. $x \cdot x \cdot x \cdot y \cdot y \cdot y \cdot y = x^3 y^4$

25. variable

27. system

29. Let x = President's salary.   Substitute $x = y + 207400$ into (2):   Substitute; solve for x:

Let y = Vice President's salary. $x + y = 592600$ $x = y + 207400$

$\begin{cases} (1) & x = y + 207400 \\ (2) & x + y = 592600 \end{cases}$ $y + 207400 + y = 592600$ $x = 192600 + 207400$

$2y = 385200$ $x = 400000$

$y = 192600$

The president makes \$400,000 and the vice president makes \$192,600.

31. Let x = older amount.   Substitute $x = 2y + 5000$ into (2):   The younger son will get

Let y = younger amount. $x + y = 742250$ \$245,750.

$\begin{cases} (1) & x = 2y + 5000 \\ (2) & x + y = 742250 \end{cases}$ $2y + 5000 + y = 742250$

$3y = 737250$

$y = 245750$

33. Let x = the length of one piece.   Substitute $x = y + 5$ into (2):   Substitute; solve for x:

Let y = the other length. $x + y = 25$ $x = y + 5$

$\begin{cases} (1) & x = y + 5 \\ (2) & x + y = 25 \end{cases}$ $y + 5 + y = 25$ $x = 10 + 5 = 15$

$2y = 20$ The lengths are 10 ft and 15 ft.

$y = 10$

35. Let w = the width.   Substitute $l = w + 5$ into (2):   Substitute; solve for l:

Let l = the length. $2l + 2w = 110$ $l = w + 5$

$\begin{cases} (1) & l = w + 5 \\ (2) & 2l + 2w = 110 \end{cases}$ $2(w + 5) + 2w = 110$ $l = 25 + 5 = 30$

$2w + 10 + 2w = 110$ The dimensions are 25 ft by 30 ft.

$4w = 100$

$w = 25$

37. Let x = the length.   Substitute $x = 2y - 3$ into (2):   Substitute; solve for x:

Let y = the width. $2x + 2y = 48$ $x = 2y - 3$

$\begin{cases} (1) & x = 2y - 3 \\ (2) & 2x + 2y = 48 \end{cases}$ $2(2y - 3) + 2y = 48$ $x = 2(9) - 3 = 15$

$4y - 6 + 2y = 48$ The area is $(9)(15) = 135$ ft^2.

$6y = 54$

$y = 9$

39. Let x = the width.   Substitute $x = y - 26$ into (2):   Substitute; solve for x:

Let y = the length. $2x + 2y = 332$ $x = y - 26$

$\begin{cases} (1) & x = y - 26 \\ (2) & 2x + 2y = 332 \end{cases}$ $2(y - 26) + 2y = 332$ $x = 96 - 26 = 70$

$2y - 52 + 2y = 332$ The area is $(96)(70) = 6{,}720$ ft^2.

$4y = 384$

$y = 96$

41. Let $x =$ Bill's principal and $y =$ Janette's principal.

$$x + y = 5000 \qquad x + y = 5000 \Rightarrow \times(-5) \quad -5x - 5y = -25000$$
$$\underline{0.05x + 0.07y = 310} \Rightarrow \times(100) \quad \underline{5x + 7y = 31000} \qquad \qquad \underline{5x + 7y = 31000}$$
$$ 2y = 6000$$
$$ y = 3000$$

$$x + y = 5000 \quad \text{Bill invested \$2000.}$$
$$x + 3000 = 5000$$
$$x = 2000$$

43. Let $x =$ number of student tickets and $y =$ number of nonstudent tickets.

$$x + y = 350 \Rightarrow \times(-1) \quad -x - y = -350 \qquad x + y = 350 \quad \text{There were 250 student}$$
$$\underline{x + 2y = 450} \qquad\qquad \underline{x + 2y = 450} \quad x + 100 = 350 \quad \text{tickets sold.}$$
$$ y = 100 \qquad x = 250$$

45. Let $b =$ speed of boat in still water and $c =$ speed of current.

	d	r	t	Equation $(d = r \cdot t)$
Downstream	24	$b+c$	2	$24 = (b+c) \cdot 2 \Rightarrow 2b + 2c = 24$
Upstream	24	$b-c$	3	$24 = (b-c) \cdot 3 \Rightarrow 3b - 3c = 24$

$$2b + 2c = 24 \Rightarrow \times(3) \quad 6b + 6c = 72 \quad \text{The speed of the boat is 10 mph in still water.}$$
$$\underline{3b - 3c = 24} \Rightarrow \times(2) \quad \underline{6b - 6c = 48}$$
$$ 12b = 120$$
$$ b = 10$$

47. Let $p =$ airspeed of plane and $w =$ speed of wind.

	d	r	t	Equation $(d = r \cdot t)$
With wind	600	$p+w$	2	$600 = (p+w) \cdot 2 \Rightarrow 2p + 2w = 600$
Against wind	600	$p-w$	3	$600 = (p-w) \cdot 3 \Rightarrow 3p - 3w = 600$

$$2p + 2w = 600 \Rightarrow \times(3) \quad 6p + 6w = 1800 \qquad 2p + 2w = 600 \quad \text{The speed of the wind is}$$
$$\underline{3p - 3w = 600} \Rightarrow \times(2) \quad \underline{6p - 6w = 1200} \quad 2(250) + 2w = 600 \quad \text{50 mph.}$$
$$ 12p = 3000 \qquad 500 + 2w = 600$$
$$ p = 250 \qquad\qquad 2w = 100$$
$$ w = 50$$

49. Let $x =$ liters of first solution and $y =$ liters of second solution.

	Fractional part that is alcohol	Number of liters of solution	Number of liters of alcohol
First solution	0.40	x	$0.40x$
Second solution	0.55	y	$0.55y$
Final solution	0.50	15	$0.50(15) = 7.5$

$$x + y = 15 \qquad\qquad x + y = 15 \Rightarrow \times(-40) \quad -40x - 40y = -600$$
$$\underline{0.40x + 0.55y = 7.5} \Rightarrow \times(100) \quad \underline{40x + 55y = 750} \qquad\qquad \underline{40x + 55y = 750}$$
$$ 15y = 150$$
$$ y = 10$$

continued on next page...

49. continued

$x + y = 15$ The chemist should use 5 L of the 40% solution and 10 L of the 55% solution.
$x + 10 = 15$
$x = 5$

51. Let x = pounds of peanuts and y = pounds of cashews.

	Cost per pound	Number of pounds	Total value
Peanuts	3	x	$3x$
Cashews	6	y	$6y$
Mixture	4	48	$4(48) = 192$

$\begin{aligned} x + y &= 48 \Rightarrow \times (-3) \\ 3x + 6y &= 192 \end{aligned}$ $\begin{aligned} -3x - 3y &= -144 \\ 3x + 6y &= 192 \\ \hline 3y &= 48 \\ y &= 16 \end{aligned}$ $\begin{aligned} x + y &= 48 \\ x + 16 &= 48 \\ x &= 32 \end{aligned}$ The merchant should use 32 pounds of peanuts and 16 pounds of cashews.

53. Let c = the cost.   Substitute: The break point is about 9.9 years.
Let n = number of years. $2250 + 412n = 1715 + 466n$
$\begin{cases} (1) & c = 2250 + 412n \\ (2) & c = 1715 + 466n \end{cases}$ $\begin{aligned} -54n &= -535 \\ n &\approx 9.9 \end{aligned}$

55. Let c = the cost.   Substitute: The break point is 250 tires.
Let n = number of tires. $600 + 15n = 1100 + 13n$
$\begin{cases} (1) & c = 600 + 15n \\ (2) & c = 1100 + 13n \end{cases}$ $\begin{aligned} 2n &= 500 \\ n &= 250 \end{aligned}$

57. Let x = cost of contact cleaner and y = cost of soaking solution.

$\begin{aligned} 2x + 3y &= 29.40 \Rightarrow \times (-2) \\ 3x + 2y &= 28.60 \Rightarrow \times (3) \end{aligned}$ $\begin{aligned} -4x - 6y &= -58.80 \\ 9x + 6y &= 85.80 \\ \hline 5x &= 27 \\ x &= 5.40 \end{aligned}$ $\begin{aligned} 2x + 3y &= 29.40 \\ 2(5.40) + 3y &= 29.40 \\ 10.80 + 3y &= 29.40 \\ 3y &= 18.60 \\ y &= 6.20 \end{aligned}$ The contact cleaner costs \$5.40, while the soaking solution costs \$6.20.

59. Let x = number of horses and y = number of cows.

$\begin{cases} (1) & y = 5x \\ (2) & x + y = 168 \end{cases}$ Substitute $y = 5x$ from (1) into (2):   Substitute and solve for y:
$\begin{aligned} x + y &= 168 \\ x + 5x &= 168 \\ 6x &= 168 \\ x &= 28 \end{aligned}$ $\begin{aligned} y &= 5x \\ y &= 5(28) = 140 \end{aligned}$
He has 140 cows.

61. Let x = time for causes.   Substitute $x = 4y$ into (2):   Substitute; solve for x:
Let y = time for outcome. $\begin{aligned} x + y &= 30 \\ 4y + y &= 30 \\ 5y &= 30 \\ y &= 6 \end{aligned}$ $\begin{aligned} x &= 4y \\ x &= 4(6) = 24 \end{aligned}$
$\begin{cases} (1) & x = 4y \\ (2) & x + y = 30 \end{cases}$ 24 minutes will be devoted to the causes, and 6 minutes to the outcome.

SECTION 8.3

63. Let x and y represent the integers.   Substitute $x = 3y$ into (2):   Substitute; solve for x:

$$\begin{cases} (1) & x = 3y \\ (2) & x + y = 112 \end{cases}$$

$$x + y = 112$$
$$3y + y = 112$$
$$4y = 112$$
$$y = 28$$

$$x = 3y$$
$$x = 3(28) = 84$$

The integers are 28 and 84.

65. Let x and y represent the integers.

$$\begin{array}{l} 3x + y = 29 \Rightarrow \times (-2) \\ x + 2y = 18 \end{array}$$

$$\begin{array}{r} -6x - 2y = -58 \\ x + 2y = 18 \\ \hline -5x = -40 \\ x = 8 \end{array}$$

$$\begin{array}{l} x + 2y = 18 \\ 8 + 2y = 18 \\ 2y = 10 \\ y = 5 \end{array}$$

The integers are 5 and 8.

67. Let $r =$ the lower rate. Then $r + 0.02 =$ the higher rate.

$$\boxed{\text{Interest on \$950}} + \boxed{\text{Interest on \$1,200}} = \boxed{\text{Total interest}}$$

$$950r + 1200(r + 0.02) = 88.50$$
$$950r + 1200r + 24 = 88.50$$
$$2150r = 64.50$$
$$r = 0.03 \Rightarrow \text{The lower rate is 3\%.}$$

69. Let $x =$ number of inexpensive radios and $y =$ number of expensive radios.

$$\begin{array}{l} x + y = 25 \Rightarrow \times (-87) \\ 87x + 119y = 2495 \end{array}$$

$$\begin{array}{r} -87x - 87y = -2175 \\ 87x + 119y = 2495 \\ \hline 32y = 320 \\ y = 10 \end{array}$$

$$\begin{array}{l} x + y = 25 \\ x + 10 = 25 \\ x = 15 \end{array}$$

There were 15 inexpensive radios sold.

71. Let $x =$ cost of a gallon of paint and $y =$ cost of a brush.

$$\begin{array}{l} 8x + 3y = 135 \Rightarrow \times (2) \\ 6x + 2y = 100 \Rightarrow \times (-3) \end{array}$$

$$\begin{array}{r} 16x + 6y = 270 \\ -18x - 6y = -300 \\ \hline -2x = -30 \\ x = 15 \end{array}$$

$$\begin{array}{l} 6x + 2y = 100 \\ 6(15) + 2y = 100 \\ 90 + 2y = 100 \\ 2y = 10 \\ y = 5 \end{array}$$

The paint costs \$15 per gallon, while each brush costs \$5.

73. **Answers may vary.**

75. Let $x =$ weight of one nail, $y =$ weight of one bolt and $z =$ weight of one nut.

$$\begin{array}{l} 3x + y = 3z \\ y + z = 5x \end{array}$$

Solve the second equation for y:

$$y = 5x - z$$

Substitute for y in the first equation and solve for z.

$$3x + y = 3z$$
$$3x + 5x - z = 3z$$
$$8x = 4z$$
$$2x = z$$

It will take 2 nails to balance 1 nut.

272

Exercises 8.4 (page 586)

1.
$$2x + y - 3z = 0 \qquad 3x - 2y + 4z = 5 \qquad 4x + 2y - 6z = 0$$
$$2(1) + 1 - 3(1) \overset{?}{=} 0 \quad 3(1) - 2(1) + 4(1) \overset{?}{=} 5 \quad 4(1) + 2(1) - 6(1) \overset{?}{=} 0$$
$$2 + 1 - 3 \overset{?}{=} 0 \qquad 3 - 2 + 4 \overset{?}{=} 5 \qquad 4 + 2 - 6 \overset{?}{=} 0$$
$$0 = 0 \qquad\qquad 5 = 5 \qquad\qquad 0 = 0$$

$(1, 1, 1)$ is a solution to the system.

3. $m = \dfrac{\Delta y}{\Delta x} = \dfrac{3 - (-5)}{2 - (-1)} = \dfrac{8}{3}$
5. $f(0) = 2(0)^2 + 1 = 2(0) + 1 = 0 + 1 = 1$

7. $f(3s) = 2(3s)^2 + 1 = 2(9s^2) + 1$
 $= 18s^2 + 1$
9. plane
11. infinitely

13.
$$x + y - z = 2 \qquad 2x + y + 3z = 7 \qquad x - \tfrac{1}{3}y + \tfrac{2}{3}z = 1$$
$$1 + 2 - 1 \overset{?}{=} 2 \quad 2(1) + 2 + 3(1) \overset{?}{=} 7 \quad 1 - \tfrac{1}{3}(2) + \tfrac{2}{3}(1) \overset{?}{=} 1$$
$$2 = 2 \qquad\qquad 2 + 2 + 3 \overset{?}{=} 7 \qquad\qquad 1 - \tfrac{2}{3} + \tfrac{2}{3} \overset{?}{=} 1$$
$$7 = 7 \qquad\qquad 1 = 1$$

$(1, 2, 1)$ is a solution to the system.

15.
$$\begin{array}{ll} (1) & x + y + z = 4 \\ (2) & 2x + y - z = 1 \\ (3) & 2x - 3y + z = 1 \end{array} \quad \begin{array}{l} (1) \quad x + y + z = 4 \\ (2) \quad \underline{2x + y - z = 1} \\ (4) \quad 3x + 2y = 5 \end{array} \quad \begin{array}{l} (2) \quad 2x + y - z = 1 \\ (3) \quad \underline{2x - 3y + z = 1} \\ (5) \quad 4x - 2y = 2 \end{array}$$

$$\begin{array}{ll} (4) & 3x + 2y = 5 \\ (5) & \underline{4x - 2y = 2} \\ & 7x = 7 \\ & x = 1 \end{array} \qquad \begin{array}{l} 3x + 2y = 5 \\ 3(1) + 2y = 5 \\ 3 + 2y = 5 \\ 2y = 2 \\ y = 1 \end{array} \qquad \begin{array}{l} x + y + z = 4 \\ 1 + 1 + z = 4 \\ 2 + z = 4 \\ z = 2 \end{array} \quad \boxed{\text{The solution is } (1, 1, 2).}$$

17.
$$\begin{array}{ll} (1) & 4x + 3z = 4 \\ (2) & 2y - 6z = -1 \\ (3) & 8x + 4y + 3z = 9 \end{array} \quad \begin{array}{ll} (2) & 2y - 6z = -1 \\ 2 \cdot (1) & \underline{8x + 6z = 8} \\ (4) & 8x + 2y = 7 \end{array} \quad \begin{array}{ll} (2) & 2y - 6z = -1 \\ 2 \cdot (3) & \underline{16x + 8y + 6z = 18} \\ (5) & 16x + 10y = 17 \end{array}$$

$$\begin{array}{l} 8x + 2y = 7 \Rightarrow \times(-2) \\ \underline{16x + 10y = 17} \Rightarrow \end{array} \quad \begin{array}{l} -16x - 4y = -14 \\ \underline{16x + 10y = 17} \\ 6y = 3 \\ y = \tfrac{1}{2} \end{array} \quad \begin{array}{l} 8x + 2y = 7 \\ 8x + 2\left(\tfrac{1}{2}\right) = 7 \\ 8x + 1 = 7 \\ 8x = 6 \\ x = \tfrac{3}{4} \end{array} \quad \begin{array}{l} 4x + 3z = 4 \\ 4\left(\tfrac{3}{4}\right) + 3z = 4 \\ 3 + 3z = 4 \\ 3z = 1 \\ z = \tfrac{1}{3} \end{array}$$

Solution: $\boxed{\left(\tfrac{3}{4}, \tfrac{1}{2}, \tfrac{1}{3}\right)}$

19.
$$\begin{array}{ll} (1) & 3a - 2b - c = 4 \\ (2) & 6a - 4b - 2c = 10 \\ (3) & a + 3b + c = 2 \end{array}$$

$$\begin{array}{ll} -2 \cdot (1) & -6a + 4b + 2c = -8 \\ (2) & \underline{6a - 4b - 2c = 10} \\ (4) & 0 = 2 \end{array}$$

Since equation (4) is always false, there is no solution. The system is inconsistent. $\boxed{\varnothing}$

21.
$$\begin{array}{ll} (1) & 7x - 2y - z = 1 \\ (2) & 9x - 6y + z = 7 \\ (3) & x - 2y + z = 3 \end{array}$$

$$\begin{array}{ll} -3 \cdot (1) & -21x + 6y + 3z = -3 \\ (2) & \underline{9x - 6y + z = 7} \\ (4) & -12x \quad + 4z = 4 \end{array}$$

$$\begin{array}{ll} (2) & 9x - 6y + z = 7 \\ -3 \cdot (3) & \underline{-3x + 6y - 3z = -9} \\ (5) & 6x \quad - 2z = -2 \end{array}$$

$$\begin{array}{llll} -12x + 4z = 4 & \Rightarrow & -12x + 4z = 4 \\ \underline{6x - 2z = -2} & \Rightarrow \times 2 & \underline{12x - 4z = -4} \\ & & 0 = 0 & \text{Dependent equations} \end{array}$$

Dependent equations $x - 2y + z = 3$
If $x = x$, then $x - 2y + 3x + 1 = 3$
$6x - 2z = -2$ $-2y = -4x + 2$
$\quad -2z = -6x - 2$ $y = 2x - 1$
$\quad\quad z = 3x + 1$ $\boxed{(x, 2x - 1, 3x + 1)}$

23.
$$\begin{array}{ll} (1) & x + 2y + z = 1 \\ (2) & 3x + y + 3z = 3 \\ (3) & -2x + y - 2z = -2 \end{array}$$

$$\begin{array}{ll} 2 \cdot (1) & 2x + 4y + 2z = 2 \\ (3) & \underline{-2x + y - 2z = -2} \\ (4) & 5y = 0 \\ & y = 0 \end{array}$$

$$\begin{array}{ll} (1) & x + 2y + z = 1 \\ (2) & 3x + y + 3z = 3 \\ (3) & -2x + y - 2z = -2 \end{array}$$

$$\begin{array}{ll} (1) & x + 2(0) + z = 1 \\ (2) & 3x + 0 + 3z = 3 \\ (3) & -2x + 0 - 2z = -2 \end{array}$$

$$\begin{array}{ll} (1) & x + z = 1 \\ (2) & 3x + 3z = 3 \\ (3) & -2x - 2z = -2 \end{array}$$

$$\begin{array}{ll} 2 \cdot (1) & 2x + 2z = 2 \\ (3) & \underline{-2x - 2z = -2} \\ (4) & 0 = 0 \end{array}$$

Dependent equations: $x + z = 1 \Rightarrow z = 1 - x$

Solution: $\boxed{(x, 0, 1 - x)}$

25.
$$\begin{array}{ll} (1) & a + 3b + c = 10 \\ (2) & 3a + b + c = 12 \\ (3) & a + b + 3c = 8 \end{array}$$

$$\begin{array}{ll} -3 \cdot (1) & -3a - 9b - 3c = -30 \\ (3) & \underline{a + b + 3c = 8} \\ (4) & -2a - 8b = -22 \end{array}$$

$$\begin{array}{ll} -3 \cdot (2) & -9a - 3b - 3c = -36 \\ (3) & \underline{a + b + 3c = 8} \\ (5) & -8a - 2b = -28 \end{array}$$

$$\begin{array}{llll} -2a - 8b = -22 & \Rightarrow \times(-4) & 8a + 32b = 88 \\ -8a - 2b = -28 & \Rightarrow & \underline{-8a - 2b = -28} \\ & & 30b = 60 \\ & & b = 2 \end{array}$$

$$\begin{array}{ll} -2a - 8b = -22 & a + 3b + c = 10 \\ -2a - 8(2) = -22 & 3 + 3(2) + c = 10 \\ -2a - 16 = -22 & 3 + 6 + c = 10 \\ -2a = -6 & c = 1 \\ a = 3 \end{array}$$

Solution: $\boxed{(3, 2, 1)}$

27.
$$x + \tfrac{1}{3}y + z = 13 \quad \Rightarrow \times 3 \quad (1) \; 3x + y + 3z = 39$$
$$\tfrac{1}{2}x - y + \tfrac{1}{3}z = -2 \quad \Rightarrow \times 6 \quad (2) \; 3x - 6y + 2z = -12$$
$$x + \tfrac{1}{2}y - \tfrac{1}{3}z = 2 \quad \Rightarrow \times 6 \quad (3) \; 6x + 3y - 2z = 12$$

$$
\begin{array}{ll}
(2) & 3x - 6y + 2z = -12 \\
6 \cdot (1) & 18x + 6y + 18z = 234 \\
\hline
(4) & 21x \qquad + 20z = 222
\end{array}
\qquad
\begin{array}{ll}
(2) & 3x - 6y + 2z = -12 \\
2 \cdot (3) & 12x + 6y - 4z = 24 \\
\hline
(5) & 15x \qquad - 2z = 12
\end{array}
$$

$$
\begin{array}{l}
21x + 20z = 222 \Rightarrow \\
\underline{15x - 2z = 12} \Rightarrow \times 10
\end{array}
\quad
\begin{array}{l}
21x + 20z = 222 \\
\underline{150x - 20z = 120} \\
171x \qquad = 342 \\
x \qquad = 2
\end{array}
\quad
\begin{array}{l}
15x - 2z = 12 \\
15(2) - 2z = 12 \\
30 - 2z = 12 \\
-2z = -18 \\
z = 9
\end{array}
\quad
\begin{array}{l}
3x + y + 3z = 39 \\
3(2) + y + 3(9) = 39 \\
6 + y + 27 = 39 \\
y = 6
\end{array}
$$

Solution: $\boxed{(2, 6, 9)}$

29.
$$
\begin{array}{ll}
(1) & x - 3y + z = 1 \\
(2) & 2x - y - 2z = 2 \\
(3) & x + 2y - 3z = -1
\end{array}
\qquad
\begin{array}{ll}
-2 \cdot (1) & -2x + 6y - 2z = -2 \\
(2) & 2x - y - 2z = 2 \\
\hline
(4) & 5y - 4z = 0
\end{array}
\qquad
\begin{array}{ll}
-2 \cdot (3) & -2x - 4y + 6z = 2 \\
(2) & 2x - y - 2z = 2 \\
\hline
(5) & -5y + 4z = 4
\end{array}
$$

$$
\begin{array}{l}
5y - 4z = 0 \\
\underline{-5y + 4z = 4} \\
0 \neq 4
\end{array}
\quad
\begin{array}{l}
\text{Since this equation is always false, there is no} \\
\text{solution.} \; \boxed{\text{The system is inconsistent.}}
\end{array}
$$

31.
$$
\begin{array}{ll}
(1) & 2x + 3y + 4z = 6 \\
(2) & 2x - 3y - 4z = -4 \\
(3) & 4x + 6y + 8z = 12
\end{array}
\qquad
\begin{array}{ll}
(1) & 2x + 3y + 4z = 6 \\
(2) & 2x - 3y - 4z = -4 \\
\hline
(4) & 4x \qquad = 2 \\
& x \qquad = \tfrac{1}{2}
\end{array}
\qquad
\begin{array}{l}
\text{Substitute } x = \tfrac{1}{2} \text{ into (1):} \\
2\left(\tfrac{1}{2}\right) + 3y + 4z = 6 \\
1 + 3y + 4z = 6 \\
y = \tfrac{5}{3} - \tfrac{4}{3}z
\end{array}
$$

Since (3) is a multiple of (1), the equations are dependent: $\boxed{\left(\tfrac{1}{2}, \tfrac{5}{3} - \tfrac{4}{3}z, z\right)}$

33.
$$
\begin{array}{ll}
(1) & 2x + 2y + 3z = 10 \\
(2) & 3x + y - z = 0 \\
(3) & x + y + 2z = 6
\end{array}
\qquad
\begin{array}{ll}
(1) & 2x + 2y + 3z = 10 \\
3 \cdot (2) & 9x + 3y - 3z = 0 \\
\hline
(4) & 11x + 5y = 10
\end{array}
\qquad
\begin{array}{ll}
(3) & x + y + 2z = 6 \\
2 \cdot (2) & 6x + 2y - 2z = 0 \\
\hline
(5) & 7x + 3y = 6
\end{array}
$$

$$
\begin{array}{l}
11x + 5y = 10 \Rightarrow \times 3 \\
\underline{7x + 3y = 6} \Rightarrow \times (-5)
\end{array}
\quad
\begin{array}{l}
33x + 15y = 30 \\
\underline{-35x - 15y = -30} \\
-2x \qquad = 0 \\
x \qquad = 0
\end{array}
\quad
\begin{array}{l}
11x + 5y = 10 \\
11(0) + 5y = 10 \\
0 + 5y = 10 \\
5y = 10 \\
y = 2
\end{array}
\quad
\begin{array}{l}
x + y + 2z = 6 \\
0 + 2 + 2z = 6 \\
2 + 2z = 6 \\
2z = 4 \\
z = 2
\end{array}
$$

Solution: $\boxed{(0, 2, 2)}$

35. Let $x =$ the number of \$5 statues, $y =$ the number of \$4 statues and $z =$ the number of \$3 statues.

(1) $\qquad x + y + z = 180 \qquad$ (total number made) $\qquad -3 \cdot (1) \quad -3x - 3y - 3z = -540$

(2) $\qquad 5x + 4y + 3z = 650 \qquad$ (total cost) $\qquad\qquad\qquad$ (2) $\quad \dfrac{5x + 4y + 3z = \quad 650}{2x + \; y \qquad\quad = \quad 110}$

(3) $\quad 20x + 12y + 9z = 2100 \qquad$ (total revenue) $\qquad\qquad$ (4)

$-9 \cdot (1) \quad -9x - 9y - 9z = -1620 \qquad 2x + \; y = 110 \Rightarrow \times(-3) \quad -6x - 3y = -330$

$\qquad$ (3) $\quad 20x + 12y + 9z = \quad 2100 \qquad 11x + 3y = 480 \Rightarrow \qquad\qquad \dfrac{11x + 3y = \quad 480}{5x \qquad\quad = \quad 150}$

$\qquad$ (5) $\quad \dfrac{11x + \; 3y \qquad\quad = \quad 480}{} \qquad\qquad\qquad\qquad\qquad\qquad\qquad\qquad x \qquad = \quad 30$

$2x + y = 110 \qquad\quad x + y + z = 180 \qquad$ 30 of the \$5, 50 of the \$4 and 100 of the \$3 statues

$2(30) + y = 110 \qquad 30 + 50 + z = 180 \qquad$ should be made.

$60 + y = 110 \qquad\qquad\qquad z = 100$

$y = 50$

37. Substitute the coordinates of each point for x and y in the equation $y = ax^2 + bx + c$.

$y = ax^2 + bx + c \qquad\qquad y = ax^2 + bx + c \qquad\qquad y = ax^2 + bx + c$

$0 = a(0)^2 + b(0) + c \qquad -4 = a(2)^2 + b(2) + c \qquad 0 = a(4)^2 + b(4) + c$

$0 = c \qquad\qquad\qquad\qquad -4 = 4a + 2b + c \qquad\qquad 0 = 16a + 4b + c$

Solve the system of equations formed from the three equations:

(1) $\qquad\qquad c = 0 \qquad 4a + 2b = -4 \Rightarrow \times(-2) \quad -8a - 4b = 8 \qquad 16a + 4b = 0$

(2) $\quad 4a + 2b + c = -4 \qquad \dfrac{16a + 4b = \quad 0}{} \Rightarrow \qquad \dfrac{16a + 4b = 0}{8a \qquad\quad = 8} \qquad 16(1) + 4b = 0$

(3) $\quad 16a + 4b + c = 0 \qquad\qquad\qquad\qquad\qquad\qquad\qquad a \qquad = 1 \qquad 16 + 4b = 0$

$\qquad\qquad\qquad\qquad\qquad\qquad\qquad\qquad\qquad\qquad\qquad\qquad\qquad\qquad\qquad\qquad 4b = -16$

$\qquad\qquad\qquad\qquad\qquad\qquad\qquad\qquad\qquad\qquad\qquad\qquad\qquad\qquad\qquad\qquad b = -4$

The equation is $y = x^2 - 4x$.

39. Let $x =$ the first integer, $y =$ the second integer and $z =$ the third integer.

$x + y + z = 19 \qquad \Rightarrow \quad$ (1) $x + y + z = 19 \qquad$ (1) $\qquad x + \; y + z = 19$

$z = 2y \qquad\qquad\quad \Rightarrow \quad$ (2) $-2y + z = 0 \qquad$ (3) $\quad \dfrac{-x + \; y \qquad\quad = \quad 5}{2y + z = 24}$

$y = x + 5 \qquad\qquad \Rightarrow \quad$ (3) $-x + y = 5 \qquad$ (4)

$2y + \; z = 24 \qquad 2y + z = 24 \qquad x + y + z = 19 \qquad$ The integers

$\dfrac{-2y + \; z = \; 0}{2z = 24} \qquad 2y + 12 = 24 \qquad x + 6 + 12 = 19 \qquad$ are 1, 6 and 12.

$\qquad z = 12 \qquad\qquad 2y = 12 \qquad\qquad x = 1$

$\qquad\qquad\qquad\qquad\qquad y = 6$

41. Let A, B and C represent the measures of the three angles.

$A + B + C = 180 \qquad\qquad \Rightarrow \quad$ (1) $A + B + C = 180 \qquad$ (1) $\quad A + B + C = \; 180$

$A = B + C - 80 \qquad\quad \Rightarrow \quad$ (2) $A - B - C = -80 \qquad$ (2) $\quad \dfrac{A - B - C = -80}{2A \qquad\qquad\quad = \quad 100}$

$C = 2B - 50 \qquad\qquad\quad \Rightarrow \quad$ (3) $-2B + C = -50 \qquad$ (4)

$\qquad\qquad\qquad\qquad\qquad\qquad\qquad\qquad\qquad\qquad\qquad\qquad\qquad\qquad\qquad A \qquad\qquad = \quad 50$

$A + B + C = 180 \qquad -1 \cdot (3) \quad 2B - C = \; 50 \qquad B + C = 130 \qquad$ The angles have measures

$50 + B + C = 180 \qquad$ (5) $\qquad \dfrac{B + C = 130}{3B \qquad\quad = \; 180} \qquad 60 + C = 130 \qquad$ of $A = 50°$, $B = 60°$ and

(5) $\quad B + C = 130 \qquad\qquad\qquad\qquad B \qquad = \; 60 \qquad\qquad C = 70 \qquad\qquad C = 70°$.

43. Let A = the units of food A, B = the units of food B and C = the units of food C.

(1) $\quad A + 2B + 2C = 11$ (fat) $\qquad$ (1) $\qquad A + 2B + 2C = 11$

(2) $\qquad A + B + C = 6$ (carbohydrate) $\quad -2 \cdot (2) \quad -2A - 2B - 2C = -12$

(3) $\quad 2A + B + 2C = 10$ (protein) $\qquad$ (4) $\quad \overline{-A = -1}$

$$A = 1$$

(1) $\qquad A + 2B + 2C = 11$ $\qquad -3A - 2C = -9$ $\qquad A + B + C = 6$

$-2 \cdot (3) \quad \underline{-4A - 2B - 4C = -20}$ $\qquad -3(1) - 2C = -9$ $\qquad 1 + B + 3 = 6$

(5) $\quad \overline{-3A - 2C = -9}$ $\qquad -3 - 2C = -9$ $\qquad B + 4 = 6$

$$-2C = -6 \qquad\qquad B = 2$$
$$C = 3$$

1 unit of food A, 2 units of food B and 3 units of food C should be used.

45. Let x = the number of \$5 tickets, y = the number of \$3 tickets and z = the number of \$2 tickets.

(1) $\qquad x + y + z = 750$ (total sold) $\qquad 2 \cdot (1) \quad 2x + 2y + 2z = 1500$

(2) $\qquad x = 2z \Rightarrow x - 2z = 0$ (twice as many) $\quad$ (2) $\quad x - 2z = 0$

(3) $\quad 5x + 3y + 2z = 2625$ (total revenue) $\qquad$ (4) $\quad \overline{3x + 2y = 1500}$

(2) $\quad x - 2z = 0$ $\qquad 3x + 2y = 1500 \Rightarrow \times(-2) \quad -6x - 4y = -3000$

(3) $\quad \underline{5x + 3y + 2z = 2625}$ $\qquad 6x + 3y = 2625 \Rightarrow \qquad\qquad \underline{6x + 3y = 2625}$

(5) $\quad \overline{6x + 3y = 2625}$ $\qquad\qquad\qquad\qquad\qquad\qquad\qquad -y = -375$
$$y = 375$$

$$3x + 2y = 1500 \qquad\qquad x + y + z = 750 \qquad \text{250 of the \$5, 375 of the \$3 and 125 of the \$2}$$
$$3x + 2(375) = 1500 \qquad 250 + 375 + z = 750 \qquad \text{tickets were sold.}$$
$$3x + 750 = 1500 \qquad\qquad 625 + z = 750$$
$$3x = 750 \qquad\qquad\qquad z = 125$$
$$x = 250$$

47. Let x = the number of totem poles, y = the number of bears and z = the number of deer.

(1) $\quad 2x + 2y + z = 14$ (carving) $\qquad -2 \cdot (1) \quad -4x - 4y - 2z = -28$

(2) $\quad x + 2y + 2z = 15$ (sanding) $\qquad$ (2) $\quad \underline{x + 2y + 2z = 15}$

(3) $\quad 3x + 2y + 2z = 21$ (painting) $\qquad$ (4) $\quad \overline{-3x - 2y = -13}$

$-2 \cdot (1) \quad -4x - 4y - 2z = -28$ $\qquad -3x - 2y = -13 \Rightarrow \times(-1) \quad 3x + 2y = 13$

(3) $\quad \underline{3x + 2y + 2z = 21}$ $\qquad\qquad -x - 2y = -7 \Rightarrow \qquad \underline{-x - 2y = -7}$

(5) $\quad \overline{-x - 2y = -7}$ $\qquad\qquad\qquad\qquad\qquad\qquad\qquad 2x = 6$
$$x = 3$$

$$3x + 2y = 13 \qquad\qquad 2x + 2y + z = 14$$
$$3(3) + 2y = 13 \qquad 2(3) + 2(2) + z = 14$$
$$9 + 2y = 13 \qquad\qquad 6 + 4 + z = 14$$
$$2y = 4 \qquad\qquad\qquad z = 4 \qquad \boxed{\text{3 totem poles, 2 bears and 4 deer should be made.}}$$
$$y = 2$$

SECTION 8.4

49. Let $x =$ the % of nitrogen, $y =$ the % of oxygen and $z =$ the % of other gases.

$$
\begin{array}{ll}
(1) & x + y + z = 100 \\
(2) & x = 3(y+z) + 12 \Rightarrow \quad x - 3y - 3z = 12 \\
(3) & z = y - 20 \Rightarrow \quad\quad\quad -y + z = -20
\end{array}
$$

$$
\begin{array}{ll}
(1) & x + y + z = 100 \\
(3) & \quad -y + z = -20 \\
(4) & x \quad\quad + 2z = 80
\end{array}
$$

$$
\begin{array}{ll}
3 \cdot (1) & 3x + 3y + 3z = 300 \\
(2) & x - 3y - 3z = 12 \\
(5) & \overline{4x \quad\quad\quad = 312} \\
& x \quad\quad\quad = 78
\end{array}
$$

$$
\begin{aligned}
x + 2z &= 80 \\
78 + 2z &= 80 \\
2z &= 2 \\
z &= 1
\end{aligned}
$$

$$
\begin{aligned}
x + y + z &= 100 \\
78 + y + 1 &= 100 \\
y + 79 &= 100 \\
y &= 21
\end{aligned}
$$

The composition is 78% nitrogen, 21% oxygen, and 1% other gases.

51. Substitute the coordinates of each point for x and y in the equation $x^2 + y^2 + cx + dy + e = 0$.

$$
\begin{aligned}
x^2 + y^2 + cx + dy + e &= 0 \\
(1)^2 + (3)^2 + c(1) + d(3) + e &= 0 \\
1 + 9 + c + 3d + e &= 0 \\
c + 3d + e &= -10
\end{aligned}
$$

$$
\begin{aligned}
x^2 + y^2 + cx + dy + e &= 0 \\
(3)^2 + (1)^2 + c(3) + d(1) + e &= 0 \\
9 + 1 + 3c + d + e &= 0 \\
3c + d + e &= -10
\end{aligned}
$$

$$
\begin{aligned}
x^2 + y^2 + cx + dy + e &= 0 \\
(1)^2 + (-1)^2 + c(1) + d(-1) + e &= 0 \\
1 + 1 + c - d + e &= 0 \\
c - d + e &= -2
\end{aligned}
$$

$$
\begin{array}{ll}
(1) & c + 3d + e = -10 \\
(2) & 3c + d + e = -10 \\
(3) & c - d + e = -2
\end{array}
$$

$$
\begin{array}{ll}
(1) & c + 3d + e = -10 \\
-1 \cdot (2) & -3c - d - e = 10 \\
(4) & \overline{-2c + 2d \quad = 0}
\end{array}
$$

$$
\begin{array}{ll}
(1) & c + 3d + e = -10 \\
-1 \cdot (3) & -c + d - e = 2 \\
(5) & \overline{4d \quad = -8} \\
& d \quad = -2
\end{array}
$$

$$
\begin{aligned}
-2c + 2d &= 0 \\
-2c + 2(-2) &= 0 \\
-2c - 4 &= 0 \\
-2c = 4 &\Rightarrow c = -2
\end{aligned}
$$

$$
\begin{aligned}
c + 3d + e &= -10 \\
-2 + 3(-2) + e &= -10 \\
-2 - 6 + e &= -10 \\
-8 + e &= -10 \\
e &= -2
\end{aligned}
$$

The equation is $x^2 + y^2 - 2x - 2y - 2 = 0$.

53. Answers may vary.

55.

$$
\begin{array}{ll}
(1) & x + y + z + w = 3 \\
(2) & x - y - z - w = -1 \\
(3) & x + y - z - w = 1 \\
(4) & x + y - z + w = 3
\end{array}
$$

$$
\begin{array}{ll}
(1) & x + y + z + w = 3 \\
(2) & \underline{x - y - z - w = -1} \\
& 2x \quad\quad\quad = 2 \\
& x \quad\quad\quad = 1
\end{array}
$$

$$
\begin{array}{ll}
(1) & x + y + z + w = 3 \\
(3) & \underline{x + y - z - w = 1} \\
& 2x + 2y \quad\quad = 4
\end{array}
$$

$$
\begin{array}{ll}
(1) & x + y + z + w = 3 \\
(4) & \underline{x + y - z + w = 3} \\
& 2x + 2y \quad + 2w = 6
\end{array}
$$

$$
\begin{aligned}
2x + 2y &= 4 \\
2(1) + 2y &= 4 \\
2y &= 2 \\
y &= 1
\end{aligned}
$$

$$
\begin{aligned}
2x + 2y + 2w &= 6 \\
2(1) + 2(1) + 2w &= 6 \\
4 + 2w &= 6 \\
2w &= 2 \\
w &= 1
\end{aligned}
$$

$$
\begin{aligned}
x + y + z + w &= 3 \\
1 + 1 + z + 1 &= 3 \\
3 + z &= 3 \\
z &= 0
\end{aligned}
$$

The solution is $(1, 1, 0, 1)$.

278

Exercises 8.5 (page 595)

Note: The notation $3R_1 + R_3 \Rightarrow R_2$ means to multiply Row #1 of the previous matrix by 3, add that result to Row #3 of the previous matrix, and write the final result in Row #2 of the current matrix.

1. $\begin{cases} 2x - y = 3 \\ x + 5y = 7 \end{cases}$

3. $\begin{cases} 3x - 2y + 4z = 1 \\ 5x + 2y - 3z = -7 \\ -x + 9y + 8z = 0 \end{cases}$

5. $470{,}000{,}000 = 4.7 \times 10^8$

7. $75 \times 10^4 = 7.5 \times 10^1 \times 10^4 = 7.5 \times 10^5$

9. matrix

11. 3; columns

13. augmented; coefficient

15. $\begin{bmatrix} 2 & -3 \\ 4 & 2 \end{bmatrix}$

17. yes

19. $\begin{bmatrix} 3 & 1 & 2 \\ 4 & 5 & 7 \end{bmatrix} \overset{-R_1 + R_2 \Rightarrow R_2}{\Rightarrow} \begin{bmatrix} 3 & 1 & 2 \\ 1 & 4 & \boxed{5} \end{bmatrix}$

21. $\begin{bmatrix} -5 & -1 & 3 \\ -2 & 3 & 1 \end{bmatrix} \overset{2R_2 \Rightarrow R_2}{\Rightarrow} \begin{bmatrix} -5 & -1 & 3 \\ -4 & 6 & \boxed{2} \end{bmatrix}$

23. $\begin{bmatrix} 2 & -3 & | & 3 \\ 2 & -1 & | & 5 \end{bmatrix} \overset{R_1 + (-R_2) \Rightarrow R_2}{\Rightarrow} \begin{bmatrix} 2 & -3 & | & 3 \\ 0 & -2 & | & -2 \end{bmatrix} \overset{-\frac{1}{2}R_2 \Rightarrow R_2}{\Rightarrow} \begin{bmatrix} 2 & -3 & | & 3 \\ 0 & 1 & | & 1 \end{bmatrix}$

From R_2, $y = 1$. From R_1:

$$2x - 3y = 3$$
$$2x - 3(1) = 3$$
$$2x - 3 = 3$$
$$2x = 6 \Rightarrow x = 3$$

The solution is $(3, 1)$.

25. $\begin{bmatrix} 1 & 2 & | & -2 \\ 3 & -1 & | & 8 \end{bmatrix} \overset{-3R_1 + R_2 \Rightarrow R_2}{\Rightarrow} \begin{bmatrix} 1 & 2 & | & -2 \\ 0 & -7 & | & 14 \end{bmatrix} \overset{-\frac{1}{7}R_2 \Rightarrow R_2}{\Rightarrow} \begin{bmatrix} 1 & 2 & | & -2 \\ 0 & 1 & | & -2 \end{bmatrix}$

From R_2, $y = -2$. From R_1:

$$x + 2y = -2$$
$$x + 2(-2) = -2$$
$$x - 4 = -2 \Rightarrow x = 2$$

The solution is $(2, -2)$.

27.
$$-R_1 + R_2 \Rightarrow R_2$$
$$-R_1 + R_3 \Rightarrow R_3$$

$$\begin{bmatrix} 1 & 1 & 1 & | & 6 \\ 1 & 2 & 1 & | & 8 \\ 1 & 1 & 2 & | & 9 \end{bmatrix} \Rightarrow \begin{bmatrix} 1 & 1 & 1 & | & 6 \\ 0 & 1 & 0 & | & 2 \\ 0 & 0 & 1 & | & 3 \end{bmatrix}$$

From R_3, $z = 3$. From R_2, $y = 2$. From R_1: The solution is $(1, 2, 3)$.

$$x + y + z = 6$$
$$x + 2 + 3 = 6$$
$$x + 5 = 6$$
$$x = 1$$

29.
$$-R_1 + R_3 \Rightarrow R_3 \qquad -R_2 + R_3 \Rightarrow R_3$$

$$\begin{bmatrix} 1 & -1 & 0 & | & 1 \\ 0 & 1 & 1 & | & 1 \\ 1 & 0 & 1 & | & 2 \end{bmatrix} \Rightarrow \begin{bmatrix} 1 & -1 & 0 & | & 1 \\ 0 & 1 & 1 & | & 1 \\ 0 & 1 & 1 & | & 1 \end{bmatrix} \Rightarrow \begin{bmatrix} 1 & -1 & 0 & | & 1 \\ 0 & 1 & 1 & | & 1 \\ 0 & 0 & 0 & | & 0 \end{bmatrix}$$

From R_2:   From R_1:   The solution is

$$y + z = 1 \qquad\qquad x - y = 1 \qquad (2 - z, 1 - z, z).$$
$$\quad y = 1 - z \quad x - (1 - z) = 1$$
$$x - 1 + z = 1$$
$$x + z = 2$$
$$x = 2 - z$$

31.
$$-3R_1 + R_2 \Rightarrow R_2 \qquad -\tfrac{1}{4} R_2 \Rightarrow R_2$$
$$-2R_1 + R_3 \Rightarrow R_3 \quad -4R_3 + R_2 \Rightarrow R_3$$

$$\begin{bmatrix} 1 & 1 & | & 3 \\ 3 & -1 & | & 1 \\ 2 & 1 & | & 4 \end{bmatrix} \Rightarrow \begin{bmatrix} 1 & 1 & | & 3 \\ 0 & -4 & | & -8 \\ 0 & -1 & | & -2 \end{bmatrix} \Rightarrow \begin{bmatrix} 1 & 1 & | & 3 \\ 0 & 1 & | & 2 \\ 0 & 0 & | & 0 \end{bmatrix}$$

From R_2, $y = 2$. From R_1: The solution is $(1, 2)$.

$$x + y = 3$$
$$x + 2 = 3$$
$$x = 1$$

33.
$$-2R_2 + R_1 \Rightarrow R_2 \qquad \tfrac{1}{3} R_2 \Rightarrow R_2$$
$$2R_3 + R_1 \Rightarrow R_3 \quad -\tfrac{7}{3} R_2 + R_3 \Rightarrow R_3$$

$$\begin{bmatrix} 2 & 1 & | & 7 \\ 1 & -1 & | & 2 \\ -1 & 3 & | & -2 \end{bmatrix} \Rightarrow \begin{bmatrix} 2 & 1 & | & 7 \\ 0 & 3 & | & 3 \\ 0 & 7 & | & 3 \end{bmatrix} \Rightarrow \begin{bmatrix} 2 & 1 & | & 7 \\ 0 & 1 & | & 1 \\ 0 & 0 & | & -4 \end{bmatrix}$$

R_3: $0x + 0y = -4$, or $0 = -4$, which is an impossible equation. $\emptyset$

35. $\begin{bmatrix} 1 & 3 & 2 & | & 4 \\ -1 & -2 & -1 & | & -2 \end{bmatrix} \Rightarrow \overset{R_1 + R_2 \Rightarrow R_2}{\begin{bmatrix} 1 & 3 & 2 & | & 4 \\ 0 & 1 & 1 & | & 2 \end{bmatrix}}$

From R_2:   From R_1:   The solution is

$y + z = 2$ $x + 3y + 2z = 4$ $(z - 2, 2 - z, z)$.

    $y = 2 - z$ $x + 3(2 - z) + 2z = 4$

                $x + 6 - 3z + 2z = 4$

                     $x - z = -2$

                       $x = z - 2$

37. $\begin{bmatrix} -3 & -2 & 2 & | & 0 \\ -1 & 1 & -1 & | & -5 \end{bmatrix} \Rightarrow \overset{-3R_2 + R_1 \Rightarrow R_2}{\begin{bmatrix} -3 & -2 & 2 & | & 0 \\ 0 & -5 & 5 & | & 15 \end{bmatrix}} \Rightarrow \overset{-\frac{1}{5}R_2 \Rightarrow R_2}{\begin{bmatrix} -3 & -2 & 2 & | & 0 \\ 0 & 1 & -1 & | & -3 \end{bmatrix}}$

From R_2:   From R_1:   The solution is

$y - z = -3$ $-3x - 2y + 2z = 0$ $(2, z - 3, z)$.

    $y = z - 3$ $-3x - 2(z - 3) + 2z = 0$

               $-3x - 2z + 6 + 2z = 0$

                    $-3x = -6$

                        $x = 2$

39. $\begin{bmatrix} 5 & -4 & | & 8 \\ 10 & 3 & | & -6 \end{bmatrix} \Rightarrow \overset{-2R_1 + R_2 \Rightarrow R_2}{\begin{bmatrix} 5 & -4 & | & 8 \\ 0 & 11 & | & -22 \end{bmatrix}} \Rightarrow \overset{\frac{1}{11}R_2 \Rightarrow R_2}{\begin{bmatrix} 5 & -4 & | & 8 \\ 0 & 1 & | & -2 \end{bmatrix}}$

From R_2, $y = -2$.   From R_1:   The solution is $(0, -2)$.

                 $5x - 4y = 8$

              $5x - 4(-2) = 8$

                $5x + 8 = 8$

                 $5x = 0 \Rightarrow x = 0$

41. $\begin{cases} 5a = 24 + 2b \\ 5b = 3a + 16 \end{cases} \Rightarrow \begin{cases} 5a - 2b = 24 \\ -3a + 5b = 16 \end{cases}$

$\begin{bmatrix} 5 & -2 & | & 24 \\ -3 & 5 & | & 16 \end{bmatrix} \Rightarrow \overset{2R_2 + R_1 \Rightarrow R_1}{\begin{bmatrix} -1 & 8 & | & 56 \\ -3 & 5 & | & 16 \end{bmatrix}} \Rightarrow \overset{-R_1 \Rightarrow R_1}{\begin{bmatrix} 1 & -8 & | & -56 \\ -3 & 5 & | & 16 \end{bmatrix}} \Rightarrow \overset{3R_1 + R_2 \Rightarrow R_2}{\begin{bmatrix} 1 & -8 & | & -56 \\ 0 & -19 & | & -152 \end{bmatrix}}$

$\overset{-\frac{1}{19}R_2 \Rightarrow R_2}{\Rightarrow} \begin{bmatrix} 1 & -8 & | & -56 \\ 0 & 1 & | & 8 \end{bmatrix}$ From R_2, $b = 8$.   From R_1:   The solution is $(8, 8)$.

                                     $a - 8b = -56$

                                    $a - 8(8) = -56$

                                  $a - 64 = -56 \Rightarrow a = 8$

43.

$$-3R_2 + R_1 \Rightarrow R_2$$
$$-3R_3 + R_1 \Rightarrow R_3 \qquad\qquad \tfrac{2}{7}R_2 + R_3 \Rightarrow R_3$$

$$\begin{bmatrix} 3 & 1 & -3 & 5 \\ 1 & -2 & 4 & 10 \\ 1 & 1 & 1 & 13 \end{bmatrix} \Rightarrow \begin{bmatrix} 3 & 1 & -3 & 5 \\ 0 & 7 & -15 & -25 \\ 0 & -2 & -6 & -34 \end{bmatrix} \Rightarrow \begin{bmatrix} 3 & 1 & -3 & 5 \\ 0 & 7 & -15 & -25 \\ 0 & 0 & -\frac{72}{7} & -\frac{288}{7} \end{bmatrix}$$

From R_3:   From R_2:   From R_1:

$$-\frac{72}{7}c = -\frac{288}{7} \qquad 7b - 15c = -25 \qquad 3a + b - 3c = 5$$

$$-\frac{7}{72}\left(-\frac{72}{7}c\right) = -\frac{7}{72}\left(-\frac{288}{7}\right) \qquad 7b - 15(4) = -25 \qquad 3a + 5 - 3(4) = 5$$

$$\qquad\qquad\qquad\qquad\qquad 7b = 35 \qquad\qquad 3a - 7 = 5$$

$$c = 4 \qquad\qquad\qquad b = 5 \qquad\qquad\qquad 3a = 12$$

$$a = 4$$

The solution is $(4, 5, 4)$.

45.

$$-2R_1 + R_2 \Rightarrow R_2$$
$$-4R_1 + R_3 \Rightarrow R_3 \qquad\qquad -R_2 + R_3 \Rightarrow R_3$$

$$\begin{bmatrix} 1 & 2 & 2 & 2 \\ 2 & 1 & -1 & 1 \\ 4 & 5 & 3 & 3 \end{bmatrix} \Rightarrow \begin{bmatrix} 1 & 2 & 2 & 2 \\ 0 & -3 & -5 & -3 \\ 0 & -3 & -5 & -5 \end{bmatrix} \Rightarrow \begin{bmatrix} 1 & 2 & 2 & 2 \\ 0 & -3 & -5 & -3 \\ 0 & 0 & 0 & -2 \end{bmatrix}$$

The last row indicates that $0x + 0y + 0z = -2$. This is false, so there is no solution. $\emptyset$

47.

$$-2R_2 + R_1 \Rightarrow R_2 \qquad -\tfrac{1}{7}R_2 \Rightarrow R_2$$
$$2R_3 + R_1 \Rightarrow R_3 \qquad -\tfrac{9}{7}R_2 + R_3 \Rightarrow R_3$$

$$\begin{bmatrix} 2 & -1 & 4 \\ 1 & 3 & 2 \\ -1 & -4 & -2 \end{bmatrix} \Rightarrow \begin{bmatrix} 2 & -1 & 4 \\ 0 & -7 & 0 \\ 0 & -9 & 0 \end{bmatrix} \Rightarrow \begin{bmatrix} 2 & -1 & 4 \\ 0 & 1 & 0 \\ 0 & 0 & 0 \end{bmatrix}$$

From R_2, $y = 0$.   From R_1:   The solution is $(2, 0)$.

$$2x - y = 4$$
$$2x - 0 = 4$$
$$2x = 4$$
$$x = 2$$

49.

$$-R_2 + R_1 \Rightarrow R_2 \qquad\quad \tfrac{1}{2}R_2 \Rightarrow R_2$$
$$-3R_1 + R_3 \Rightarrow R_3 \qquad 4R_2 + R_3 \Rightarrow R_3$$

$$\begin{bmatrix} 1 & 3 & 7 \\ 1 & 1 & 3 \\ 3 & 1 & 5 \end{bmatrix} \Rightarrow \begin{bmatrix} 1 & 3 & 7 \\ 0 & 2 & 4 \\ 0 & -8 & -16 \end{bmatrix} \Rightarrow \begin{bmatrix} 1 & 3 & 7 \\ 0 & 1 & 2 \\ 0 & 0 & 0 \end{bmatrix}$$

From R_2, $y = 2$.   From R_1:   The solution is $(1, 2)$.

$$x + 3y = 7$$
$$x + 3(2) = 7$$
$$x + 6 = 7 \Rightarrow x = 1$$

51. $\begin{bmatrix} 5 & -2 & | & 4 \\ 2 & -4 & | & -8 \end{bmatrix} \Rightarrow \overset{\frac{1}{2}R_2 \Rightarrow R_2}{\begin{bmatrix} 5 & -2 & | & 4 \\ 1 & -2 & | & -4 \end{bmatrix}} \Rightarrow \overset{R_1 + (-5R_2) \Rightarrow R_2}{\begin{bmatrix} 5 & -2 & | & 4 \\ 0 & 8 & | & 24 \end{bmatrix}} \Rightarrow \overset{\frac{1}{8}R_2 \Rightarrow R_2}{\begin{bmatrix} 5 & -2 & | & 4 \\ 0 & 1 & | & 3 \end{bmatrix}}$

From R_2, $y = 3$. From R_1: The solution is $(2, 3)$.

$$5x - 2y = 4$$
$$5x - 2(3) = 4$$
$$5x = 10 \Rightarrow x = 2$$

53. $\begin{bmatrix} 2 & 1 & | & -4 \\ 6 & 3 & | & 1 \end{bmatrix} \Rightarrow \overset{-3R_1 + R_2 \Rightarrow R_2}{\begin{bmatrix} 2 & 1 & | & -4 \\ 0 & 0 & | & 13 \end{bmatrix}}$ The last equation is impossible. $\emptyset$

55. $\begin{bmatrix} 3 & -2 & 4 & | & 4 \\ 1 & 1 & 1 & | & 3 \\ 6 & -2 & -3 & | & 10 \end{bmatrix} \Rightarrow \overset{\substack{-3R_2 + R_1 \Rightarrow R_2 \\ -2R_1 + R_3 \Rightarrow R_3}}{\begin{bmatrix} 3 & -2 & 4 & | & 4 \\ 0 & -5 & 1 & | & -5 \\ 0 & 2 & -11 & | & 2 \end{bmatrix}} \Rightarrow \overset{\frac{2}{5}R_2 + R_3 \Rightarrow R_3}{\begin{bmatrix} 3 & -2 & 4 & | & 4 \\ 0 & -5 & 1 & | & -5 \\ 0 & 0 & -\frac{53}{5} & | & 0 \end{bmatrix}}$

From R_3, $z = 0$. From R_2: From R_1: The solution is $(2, 1, 0)$.

$$\begin{aligned} -5y + z &= -5 & 3x - 2y + 4z &= 4 \\ -5y + 0 &= -5 & 3x - 2(1) + 4(0) &= 4 \\ -5y &= -5 & 3x - 2 &= 4 \\ y &= 1 & 3x &= 6 \\ & & x &= 2 \end{aligned}$$

57. $\begin{bmatrix} 3 & -1 & | & 9 \\ -6 & 2 & | & -18 \end{bmatrix} \Rightarrow \overset{2R_1 + R_2 \Rightarrow R_2}{\begin{bmatrix} 3 & -1 & | & 9 \\ 0 & 0 & | & 0 \end{bmatrix}}$

From R_2: The solution is

$$3x - y = 9 \qquad (x, 3x - 9).$$
$$y = 3x - 9$$

59. $\begin{bmatrix} 1 & 1 & 1 & | & 6 \\ 1 & -1 & 1 & | & 2 \end{bmatrix} \Rightarrow \overset{-R_1 + R_2 \Rightarrow R_2}{\begin{bmatrix} 1 & 1 & 1 & | & 6 \\ 0 & -2 & 0 & | & -4 \end{bmatrix}} \Rightarrow -2y = -4 \Rightarrow y = 2$

From R_1: The solution is

$$x + y + z = 6 \qquad (4 - z, 2, z).$$
$$x + 2 + z = 6$$
$$x = 4 - z$$

61.

$$-2R_1 + R_2 \Rightarrow R_2$$
$$-3R_1 + R_3 \Rightarrow R_3 \quad R_2 + (-R_3) \Rightarrow R_3$$

$$\begin{bmatrix} 1 & 2 & 1 & | & 1 \\ 2 & -1 & 2 & | & 2 \\ 3 & 1 & 3 & | & 3 \end{bmatrix} \Rightarrow \begin{bmatrix} 1 & 2 & 1 & | & 1 \\ 0 & -5 & 0 & | & 0 \\ 0 & -5 & 0 & | & 0 \end{bmatrix} \Rightarrow \begin{bmatrix} 1 & 2 & 1 & | & 1 \\ 0 & -5 & 0 & | & 0 \\ 0 & 0 & 0 & | & 0 \end{bmatrix}$$

From R_2:   From R_1:   The solution is
$$-5y = 0 \quad\quad x + 2y + z = 1 \quad\quad\quad (x, 0, 1-x).$$
$$y = 0 \quad\quad x + 2(0) + z = 1$$
$$z = 1 - x$$

63.

$$R_1 + R_2 \Rightarrow R_2 \quad\quad\quad\quad\quad\quad\quad\quad -\tfrac{1}{4}R_2 \Rightarrow R_2$$
$$-2R_1 + R_3 \Rightarrow R_3 \quad\quad R_2 \Leftrightarrow R_3 \quad\quad \tfrac{1}{4}R_3 \Rightarrow R_3$$

$$\begin{bmatrix} 2 & 1 & 3 & | & 3 \\ -2 & -1 & 1 & | & 5 \\ 4 & -2 & 2 & | & 2 \end{bmatrix} \Rightarrow \begin{bmatrix} 2 & 1 & 3 & | & 3 \\ 0 & 0 & 4 & | & 8 \\ 0 & -4 & -4 & | & -4 \end{bmatrix} \Rightarrow \begin{bmatrix} 2 & 1 & 3 & | & 3 \\ 0 & -4 & -4 & | & -4 \\ 0 & 0 & 4 & | & 8 \end{bmatrix} \Rightarrow \begin{bmatrix} 2 & 1 & 3 & | & 3 \\ 0 & 1 & 1 & | & 1 \\ 0 & 0 & 1 & | & 2 \end{bmatrix}$$

From R_3, $z = 2$.   From R_2:   From R_1:   The solution is $(-1, -1, 2)$.
$$y + z = 1 \quad\quad\quad 2x + y + 3z = 3$$
$$y + 2 = 1 \quad\quad\quad 2x + (-1) + 3(2) = 3$$
$$y = -1 \quad\quad\quad 2x - 1 + 6 = 3$$
$$2x + 5 = 3$$
$$2x = -2$$
$$x = -1$$

65. Let $x =$ the number of nickels, $y =$ the number of dimes, $\quad \begin{cases} x + y + z = 64 \\ 5x + 10y + 25z = 600 \\ 10x + 5y + 25z = 500 \end{cases}$

and $z =$ the number of quarters.

$$-5R_1 + R_2 \Rightarrow R_2 \quad\quad\quad\quad\quad\quad\quad\quad\quad \tfrac{1}{5}R_2 \Rightarrow R_2$$
$$-10R_1 + R_3 \Rightarrow R_3 \quad\quad R_2 + R_3 \Rightarrow R_3 \quad\quad \tfrac{1}{35}R_3 \Rightarrow R_3$$

$$\begin{bmatrix} 1 & 1 & 1 & | & 64 \\ 5 & 10 & 25 & | & 600 \\ 10 & 5 & 25 & | & 500 \end{bmatrix} \Rightarrow \begin{bmatrix} 1 & 1 & 1 & | & 64 \\ 0 & 5 & 20 & | & 280 \\ 0 & -5 & 15 & | & -140 \end{bmatrix} \Rightarrow \begin{bmatrix} 1 & 1 & 1 & | & 64 \\ 0 & 5 & 20 & | & 280 \\ 0 & 0 & 35 & | & 140 \end{bmatrix} \Rightarrow \begin{bmatrix} 1 & 1 & 1 & | & 64 \\ 0 & 1 & 4 & | & 56 \\ 0 & 0 & 1 & | & 4 \end{bmatrix}$$

From R_3, $z = 4$.   From R_2, $y + 4z = 56$ From R_3, $x + y + z = 64$ There are 20 nickels,
$$y + 4(4) = 56 \quad\quad\quad x + 40 + 4 = 64 \quad\quad\text{40 dimes, and 4 quarters.}$$
$$y + 16 = 56 \quad\quad\quad\quad x + 44 = 64$$
$$y = 40 \quad\quad\quad\quad\quad x = 20$$

67. Plug the coordinates of the points into the general equation to form and solve a system of equations.

$$y = ax^2 + bx + c \quad\quad\quad\quad y = ax^2 + bx + c \quad\quad\quad\quad y = ax^2 + bx + c$$
$$1 = a(0)^2 + b(0) + c \quad\quad\quad 2 = a(1)^2 + b(1) + c \quad\quad\quad 4 = a(-1)^2 + b(-1) + c$$
$$1 = c \quad\quad\quad\quad\quad\quad\quad\quad 2 = a + b + c \quad\quad\quad\quad\quad 4 = a - b + c$$

continued on next page...

284

67. continued

$$\begin{cases} a+b+c=2 \\ a-b+c=4 \\ \phantom{a+b+{}}c=1 \end{cases} \Rightarrow \begin{bmatrix} 1 & 1 & 1 & 2 \\ 1 & -1 & 1 & 4 \\ 0 & 0 & 1 & 1 \end{bmatrix} \xRightarrow{-R_2+R_1 \Rightarrow R_2} \begin{bmatrix} 1 & 1 & 1 & 2 \\ 0 & 2 & 0 & -2 \\ 0 & 0 & 1 & 1 \end{bmatrix} \xRightarrow{\frac{1}{2}R_2 \Rightarrow R_2} \begin{bmatrix} 1 & 1 & 1 & 2 \\ 0 & 1 & 0 & -1 \\ 0 & 0 & 1 & 1 \end{bmatrix}$$

From R_3, $c = 1$.   From R_2, $b = -1$. From R_1:

$$a + b + c = 2$$
$$a + (-1) + 1 = 2$$
$$a = 2$$

The equation is $y = 2x^2 - x + 1$.

69. Let $x =$ the measure of the first angle and $y =$ the measure of the second angle. Form and solve this

system of equations: $\begin{cases} x + y = 180 \\ y = x + 28 \end{cases} \Rightarrow \begin{cases} x + y = 180 \\ -x + y = 28 \end{cases}$

$$\begin{bmatrix} 1 & 1 & 180 \\ -1 & 1 & 28 \end{bmatrix} \xRightarrow{R_1+R_2 \Rightarrow R_2} \begin{bmatrix} 1 & 1 & 180 \\ 0 & 2 & 208 \end{bmatrix} \xRightarrow{\frac{1}{2}R_2 \Rightarrow R_2} \begin{bmatrix} 1 & 1 & 180 \\ 0 & 1 & 104 \end{bmatrix}$$

From R_2, $y = 104$.   From R_1:   The angles have measures

$$x + y = 180$$ of 76° and 104°.
$$x + 104 = 180$$
$$x = 76$$

71. Let $x =$ the measure of the first angle and $y =$ the measure of the second angle. Form and solve this

system of equations: $\begin{cases} x + y = 90 \\ y = x + 28 \end{cases} \Rightarrow \begin{cases} x + y = 90 \\ -x + y = 28 \end{cases}$

$$\begin{bmatrix} 1 & 1 & 90 \\ -1 & 1 & 28 \end{bmatrix} \xRightarrow{R_1+R_2 \Rightarrow R_2} \begin{bmatrix} 1 & 1 & 90 \\ 0 & 2 & 118 \end{bmatrix} \xRightarrow{\frac{1}{2}R_2 \Rightarrow R_2} \begin{bmatrix} 1 & 1 & 90 \\ 0 & 1 & 59 \end{bmatrix}$$

From R_2, $y = 59$.   From R_1:   The angles have measures

$$x + y = 90$$ of 31° and 59°.
$$x + 59 = 90$$
$$x = 31$$

73. Let A, B and C represent the measures of the three angles.

$$\begin{cases} A+B+C=180 \\ B = A + 25 \\ C = 2A - 5 \end{cases} \Rightarrow \begin{cases} A+B+C=180 \\ -A+B=25 \\ -2A+C=-5 \end{cases}$$

$$\begin{bmatrix} 1 & 1 & 1 & 180 \\ -1 & 1 & 0 & 25 \\ -2 & 0 & 1 & -5 \end{bmatrix} \xRightarrow[2R_1+R_3 \Rightarrow R_3]{R_1+R_2 \Rightarrow R_2} \begin{bmatrix} 1 & 1 & 1 & 180 \\ 0 & 2 & 1 & 205 \\ 0 & 2 & 3 & 355 \end{bmatrix} \xRightarrow{-R_2+R_3 \Rightarrow R_3} \begin{bmatrix} 1 & 1 & 1 & 180 \\ 0 & 2 & 1 & 205 \\ 0 & 0 & 2 & 150 \end{bmatrix} \xRightarrow{\frac{1}{2}R_3 \Rightarrow R_3} \begin{bmatrix} 1 & 1 & 1 & 180 \\ 0 & 2 & 1 & 205 \\ 0 & 0 & 1 & 75 \end{bmatrix}$$

continued on next page...

73. **continued**

From R_3, $C = 75$.

From R_2:

$$2B + C = 205$$
$$2B + 75 = 205$$
$$2B = 130$$
$$B = 65$$

From R_1:

$$A + B + C = 180$$
$$A + 65 + 75 = 180$$
$$A + 140 = 180$$
$$A = 40$$

The angles have measures of $40°$, $65°$ and $75°$.

75. **Answers may vary.**

77. The last equation represents the equation $0x + 0y + 0z = k$, or $0 = k$. If $k = 0$, then the system can be solved. However, if $k \neq 0$, the system will have no solution.

Exercises 8.6 (page 605)

1. $ad - bc = 4(-1) - 3(2)$
$= -4 - 6 = -10$

3. $ad - bc = 0(5) - 8(2)$
$= 0 - 16 = -16$

5. $\begin{vmatrix} -21 & -1 \\ 7 & 5 \end{vmatrix}$

7. $5(x - 2) - (3 - x) = x + 2$
$5x - 10 - 3 + x = x + 2$
$6x - 13 = x + 2$
$5x = 15$
$x = 3$

9. $\dfrac{5}{3}(5x + 6) - 10 = 0$

$3 \cdot \dfrac{5}{3}(5x + 6) - 3 \cdot 10 = 3 \cdot 0$

$5(5x + 6) - 30 = 0$
$25x + 30 - 30 = 0$
$25x = 0$
$x = 0$

11. number; square

13. $\begin{vmatrix} a_2 & c_2 \\ a_3 & c_3 \end{vmatrix}$

15. Cramer's rule

17. consistent; independent

19. $\begin{vmatrix} 4 & 2 \\ -3 & 2 \end{vmatrix} = 4(2) - 2(-3)$
$= 8 + 6 = 14$

21. $\begin{vmatrix} -2 & 5 \\ 1 & -3 \end{vmatrix} = -2(-3) - 5(1)$
$= 6 - 5 = 1$

23. $\begin{vmatrix} 1 & 1 & 1 \\ 1 & 0 & 2 \\ 0 & 2 & 0 \end{vmatrix} = 1\begin{vmatrix} 0 & 2 \\ 2 & 0 \end{vmatrix} - 1\begin{vmatrix} 1 & 2 \\ 0 & 0 \end{vmatrix} + 1\begin{vmatrix} 1 & 0 \\ 0 & 2 \end{vmatrix} = 1(-4) - 1(0) + 1(2) = -4 - 0 + 2 = -2$

25. $\begin{vmatrix} -1 & 2 & 1 \\ 2 & 1 & -3 \\ 1 & 1 & 1 \end{vmatrix} = -1\begin{vmatrix} 1 & -3 \\ 1 & 1 \end{vmatrix} - 2\begin{vmatrix} 2 & -3 \\ 1 & 1 \end{vmatrix} + 1\begin{vmatrix} 2 & 1 \\ 1 & 1 \end{vmatrix} = -1(4) - 2(5) + 1(1) = -4 - 10 + 1$

$$= -13$$

27. $\begin{vmatrix} 1 & -2 & 3 \\ -2 & 1 & 1 \\ -3 & -2 & 1 \end{vmatrix} = 1\begin{vmatrix} 1 & 1 \\ -2 & 1 \end{vmatrix} - (-2)\begin{vmatrix} -2 & 1 \\ -3 & 1 \end{vmatrix} + 3\begin{vmatrix} -2 & 1 \\ -3 & -2 \end{vmatrix} = 1(3) + 2(1) + 3(7) = 3 + 2 + 21$

$$= 26$$

29. $\begin{vmatrix} 2 & 4 & 6 \\ 1 & 3 & 5 \\ 9 & 8 & 7 \end{vmatrix} = 2\begin{vmatrix} 3 & 5 \\ 8 & 7 \end{vmatrix} - 4\begin{vmatrix} 1 & 5 \\ 9 & 7 \end{vmatrix} + 6\begin{vmatrix} 1 & 3 \\ 9 & 8 \end{vmatrix} = 2(-19) - 4(-38) + 6(-19)$

$$= -38 + 152 - 114 = 0$$

31. $x = \dfrac{\begin{vmatrix} 5 & 3 \\ -9 & -2 \end{vmatrix}}{\begin{vmatrix} 4 & 3 \\ 3 & -2 \end{vmatrix}} = \dfrac{-10 - (-27)}{-8 - 9} = \dfrac{17}{-17} = -1; \; y = \dfrac{\begin{vmatrix} 4 & 5 \\ 3 & -9 \end{vmatrix}}{\begin{vmatrix} 4 & 3 \\ 3 & -2 \end{vmatrix}} = \dfrac{-36 - 15}{-17} = \dfrac{-51}{-17} = 3$

solution: $(-1, 3)$

33. $x = \dfrac{\begin{vmatrix} 6 & 1 \\ 2 & -1 \end{vmatrix}}{\begin{vmatrix} 1 & 1 \\ 1 & -1 \end{vmatrix}} = \dfrac{-6 - 2}{-1 - 1} = \dfrac{-8}{-2} = 4; \; y = \dfrac{\begin{vmatrix} 1 & 6 \\ 1 & 2 \end{vmatrix}}{\begin{vmatrix} 1 & 1 \\ 1 & -1 \end{vmatrix}} = \dfrac{2 - 6}{-2} = \dfrac{-4}{-2} = 2; \; \text{solution: } (4, 2)$

35. $\begin{cases} 4x = 2y - 5 \\ y = \dfrac{4x - 5}{2} \end{cases} \Rightarrow \begin{cases} 4x = 2y - 5 \\ 2y = 4x - 5 \end{cases} \Rightarrow \begin{cases} 4x - 2y = -5 \\ -4x + 2y = -5 \end{cases}$

$x = \dfrac{\begin{vmatrix} -5 & -2 \\ -5 & 2 \end{vmatrix}}{\begin{vmatrix} 4 & -2 \\ -4 & 2 \end{vmatrix}} = \dfrac{-10 - 10}{8 - 8} = \dfrac{-20}{0} \Rightarrow \text{denominator} = 0, \text{numerator} \neq 0 \Rightarrow \emptyset$

37. $\begin{cases} 2x = \dfrac{y + 5}{2} \\ y = 4x - 6 \end{cases} \Rightarrow \begin{cases} 4x = y + 5 \\ y = 4x - 6 \end{cases} \Rightarrow \begin{cases} 4x - y = 5 \\ -4x + y = -6 \end{cases}$

$x = \dfrac{\begin{vmatrix} 5 & -1 \\ -6 & 1 \end{vmatrix}}{\begin{vmatrix} 4 & -1 \\ -4 & 1 \end{vmatrix}} = \dfrac{5 - 6}{4 - 4} = \dfrac{-1}{0} \Rightarrow \text{denominator} = 0, \text{numerator} \neq 0 \Rightarrow \emptyset$

39. $\begin{cases} 2x + 3y = 9 \\ y = -\dfrac{2}{3}x + 3 \end{cases} \Rightarrow \begin{cases} 2x + 3y = 9 \\ 3y = -2x + 9 \end{cases} \Rightarrow \begin{cases} 2x + 3y = 9 \\ 2x + 3y = 9 \end{cases}$

$x = \dfrac{\begin{vmatrix} 9 & 3 \\ 9 & 3 \end{vmatrix}}{\begin{vmatrix} 2 & 3 \\ 2 & 3 \end{vmatrix}} = \dfrac{27 - 27}{6 - 6} = \dfrac{0}{0} \Rightarrow$ denominator $= 0$, numerator $= 0 \Rightarrow$ dependent equations

solution: $\left(x, -\frac{2}{3}x + 3\right)$

41. $\begin{cases} 4x - 3y = 6 \\ y = \dfrac{4x - 6}{3} \end{cases} \Rightarrow \begin{cases} 4x - 3y = 6 \\ 3y = 4x - 6 \end{cases} \Rightarrow \begin{cases} 4x - 3y = 6 \\ -4x + 3y = -6 \end{cases}$

$x = \dfrac{\begin{vmatrix} 6 & -3 \\ -6 & 3 \end{vmatrix}}{\begin{vmatrix} 4 & -3 \\ -4 & 3 \end{vmatrix}} = \dfrac{18 - 18}{12 - 12} = \dfrac{0}{0} \Rightarrow$ denominator $= 0$, numerator $= 0 \Rightarrow$ dependent equations

$y = \frac{4x-6}{3} = \frac{4}{3}x - 2$; solution: $\left(x, \frac{4}{3}x - 2\right)$

Note: In the following problems, D stands for the denominator determinant, while N_x, N_y and N_z stand for the numerator determinants for x, y and z, respectively.

43. $D = \begin{vmatrix} 1 & 1 & 1 \\ 1 & 1 & -1 \\ 1 & -1 & 1 \end{vmatrix} = 1 \begin{vmatrix} 1 & -1 \\ -1 & 1 \end{vmatrix} - 1 \begin{vmatrix} 1 & -1 \\ 1 & 1 \end{vmatrix} + 1 \begin{vmatrix} 1 & 1 \\ 1 & -1 \end{vmatrix} = 1(0) - 1(2) + 1(-2) = -4$

$N_x = \begin{vmatrix} 4 & 1 & 1 \\ 0 & 1 & -1 \\ 2 & -1 & 1 \end{vmatrix} = 4 \begin{vmatrix} 1 & -1 \\ -1 & 1 \end{vmatrix} - 1 \begin{vmatrix} 0 & -1 \\ 2 & 1 \end{vmatrix} + 1 \begin{vmatrix} 0 & 1 \\ 2 & -1 \end{vmatrix} = 4(0) - 1(2) + 1(-2) = -4$

$N_y = \begin{vmatrix} 1 & 4 & 1 \\ 1 & 0 & -1 \\ 1 & 2 & 1 \end{vmatrix} = 1 \begin{vmatrix} 0 & -1 \\ 2 & 1 \end{vmatrix} - 4 \begin{vmatrix} 1 & -1 \\ 1 & 1 \end{vmatrix} + 1 \begin{vmatrix} 1 & 0 \\ 1 & 2 \end{vmatrix} = 1(2) - 4(2) + 1(2) = -4$

$N_z = \begin{vmatrix} 1 & 1 & 4 \\ 1 & 1 & 0 \\ 1 & -1 & 2 \end{vmatrix} = 1 \begin{vmatrix} 1 & 0 \\ -1 & 2 \end{vmatrix} - 1 \begin{vmatrix} 1 & 0 \\ 1 & 2 \end{vmatrix} + 4 \begin{vmatrix} 1 & 1 \\ 1 & -1 \end{vmatrix} = 1(2) - 1(2) + 4(-2) = -8$

$x = \dfrac{N_x}{D} = \dfrac{-4}{-4} = 1$; $y = \dfrac{N_y}{D} = \dfrac{-4}{-4} = 1$; $z = \dfrac{N_z}{D} = \dfrac{-8}{-4} = 2 \Rightarrow$ solution: $(1, 1, 2)$

45. $D = \begin{vmatrix} 1 & 1 & 2 \\ 1 & 2 & 1 \\ 2 & 1 & 1 \end{vmatrix} = 1 \begin{vmatrix} 2 & 1 \\ 1 & 1 \end{vmatrix} - 1 \begin{vmatrix} 1 & 1 \\ 2 & 1 \end{vmatrix} + 2 \begin{vmatrix} 1 & 2 \\ 2 & 1 \end{vmatrix} = 1(1) - 1(-1) + 2(-3) = -4$

$N_x = \begin{vmatrix} 7 & 1 & 2 \\ 8 & 2 & 1 \\ 9 & 1 & 1 \end{vmatrix} = 7 \begin{vmatrix} 2 & 1 \\ 1 & 1 \end{vmatrix} - 1 \begin{vmatrix} 8 & 1 \\ 9 & 1 \end{vmatrix} + 2 \begin{vmatrix} 8 & 2 \\ 9 & 1 \end{vmatrix} = 7(1) - 1(-1) + 2(-10) = -12$

continued on next page

45. **continued**

$$N_y = \begin{vmatrix} 1 & 7 & 2 \\ 1 & 8 & 1 \\ 2 & 9 & 1 \end{vmatrix} = 1\begin{vmatrix} 8 & 1 \\ 9 & 1 \end{vmatrix} - 7\begin{vmatrix} 1 & 1 \\ 2 & 1 \end{vmatrix} + 2\begin{vmatrix} 1 & 8 \\ 2 & 9 \end{vmatrix} = 1(-1) - 7(-1) + 2(-7) = -8$$

$$N_z = \begin{vmatrix} 1 & 1 & 7 \\ 1 & 2 & 8 \\ 2 & 1 & 9 \end{vmatrix} = 1\begin{vmatrix} 2 & 8 \\ 1 & 9 \end{vmatrix} - 1\begin{vmatrix} 1 & 8 \\ 2 & 9 \end{vmatrix} + 7\begin{vmatrix} 1 & 2 \\ 2 & 1 \end{vmatrix} = 1(10) - 1(-7) + 7(-3) = -4$$

$$x = \frac{N_x}{D} = \frac{-12}{-4} = 3; \ y = \frac{N_y}{D} = \frac{-8}{-4} = 2; \ z = \frac{N_z}{D} = \frac{-4}{-4} = 1 \Rightarrow \text{solution: } (3, 2, 1)$$

47. $D = \begin{vmatrix} 2 & 1 & -1 \\ 1 & 2 & 2 \\ 4 & 5 & 3 \end{vmatrix} = 2\begin{vmatrix} 2 & 2 \\ 5 & 3 \end{vmatrix} - 1\begin{vmatrix} 1 & 2 \\ 4 & 3 \end{vmatrix} + (-1)\begin{vmatrix} 1 & 2 \\ 4 & 5 \end{vmatrix} = 2(-4) - 1(-5) - 1(-3) = 0$

$$N_x = \begin{vmatrix} 1 & 1 & -1 \\ 2 & 2 & 2 \\ 3 & 5 & 3 \end{vmatrix} = 1\begin{vmatrix} 2 & 2 \\ 5 & 3 \end{vmatrix} - 1\begin{vmatrix} 2 & 2 \\ 3 & 3 \end{vmatrix} + (-1)\begin{vmatrix} 2 & 2 \\ 3 & 5 \end{vmatrix} = 1(-4) - 1(0) - 1(4) = -8$$

$$x = \frac{N_x}{D} = \frac{-8}{0} \Rightarrow \text{denominator} = 0, \text{ numerator} \neq 0 \Rightarrow \emptyset$$

49. $D = \begin{vmatrix} 2 & 3 & 4 \\ 2 & -3 & -4 \\ 4 & 6 & 8 \end{vmatrix} = 2\begin{vmatrix} -3 & -4 \\ 6 & 8 \end{vmatrix} - 3\begin{vmatrix} 2 & -4 \\ 4 & 8 \end{vmatrix} + 4\begin{vmatrix} 2 & -3 \\ 4 & 6 \end{vmatrix} = 2(0) - 3(32) + 4(24) = 0$

$$N_x = \begin{vmatrix} 6 & 3 & 4 \\ -4 & -3 & -4 \\ 12 & 6 & 8 \end{vmatrix} = 6\begin{vmatrix} -3 & -4 \\ 6 & 8 \end{vmatrix} - 3\begin{vmatrix} -4 & -4 \\ 12 & 8 \end{vmatrix} + 4\begin{vmatrix} -4 & -3 \\ 12 & 6 \end{vmatrix} = 6(0) - 3(16) + 4(12) = 0$$

$$x = \frac{N_x}{D} = \frac{0}{0} \Rightarrow \text{denominator} = 0, \text{ numerator} = 0 \Rightarrow \text{dependent equations.}$$

$$-R_1 + R_2 \Rightarrow R_2$$
$$-2R_1 + R_3 \Rightarrow R_3$$

$$\begin{bmatrix} 2 & 3 & 4 & | & 6 \\ 2 & -3 & -4 & | & -4 \\ 4 & 6 & 8 & | & 12 \end{bmatrix} \Rightarrow \begin{bmatrix} 2 & 3 & 4 & | & 6 \\ 0 & -6 & -8 & | & -10 \\ 0 & 0 & 0 & | & 0 \end{bmatrix}$$

From R_2: $\qquad$ From R_1: $\qquad$ The solution is

$-6y - 8z = -10$ $\qquad$ $2x + 3y + 4z = 6$ $\qquad$ $\left(\frac{1}{2}, \frac{5}{3} - \frac{4}{3}z, z\right)$.

$\qquad y = \frac{5}{3} - \frac{4}{3}z$ $\qquad$ $2x + 3\left(\frac{5}{3} - \frac{4}{3}z\right) + 4z = 6$

$\qquad\qquad\qquad\qquad 2x + 5 - 4z + 4z = 6$

$\qquad\qquad\qquad\qquad\qquad x = \frac{1}{2}$

51. $\begin{vmatrix} 2a & b \\ b & 2a \end{vmatrix} = 2a(2a) - b(b) = 4a^2 - b^2$

53. $\begin{vmatrix} a & 2a & -a \\ 2 & -1 & 3 \\ 1 & 2 & -3 \end{vmatrix} = a\begin{vmatrix} -1 & 3 \\ 2 & -3 \end{vmatrix} - 2a\begin{vmatrix} 2 & 3 \\ 1 & -3 \end{vmatrix} + (-a)\begin{vmatrix} 2 & -1 \\ 1 & 2 \end{vmatrix} = a(-3) - 2a(-9) - a(5)$

$$= -3a + 18a - 5a = 10a$$

55. $\begin{vmatrix} 1 & a & b \\ 1 & 2a & 2b \\ 1 & 3a & 3b \end{vmatrix} = 1\begin{vmatrix} 2a & 2b \\ 3a & 3b \end{vmatrix} - a\begin{vmatrix} 1 & 2b \\ 1 & 3b \end{vmatrix} + b\begin{vmatrix} 1 & 2a \\ 1 & 3a \end{vmatrix} = 1(0) - a(b) + b(a) = 0 - ab + ab = 0$

57. $\begin{vmatrix} 2 & -3 & 4 \\ -1 & 2 & 4 \\ 3 & -3 & 1 \end{vmatrix} = -23$

59. $\begin{vmatrix} 2 & 1 & -3 \\ -2 & 2 & 4 \\ 1 & -2 & 2 \end{vmatrix} = 26$

61. $x = \dfrac{\begin{vmatrix} 0 & 3 \\ -4 & -6 \end{vmatrix}}{\begin{vmatrix} 2 & 3 \\ 4 & -6 \end{vmatrix}} = \dfrac{0 - (-12)}{-12 - 12} = \dfrac{12}{-24} = -\dfrac{1}{2}$; $y = \dfrac{\begin{vmatrix} 2 & 0 \\ 4 & -4 \end{vmatrix}}{\begin{vmatrix} 2 & 3 \\ 4 & -6 \end{vmatrix}} = \dfrac{-8 - 0}{-24} = \dfrac{-8}{-24} = \dfrac{1}{3}$

solution: $\left(-\frac{1}{2}, \frac{1}{3}\right)$

63. $\begin{cases} y = \dfrac{-2x + 1}{3} \\ 3x - 2y = 8 \end{cases} \Rightarrow \begin{cases} 3y = -2x + 1 \\ 3x - 2y = 8 \end{cases} \Rightarrow \begin{cases} 2x + 3y = 1 \\ 3x - 2y = 8 \end{cases}$

$x = \dfrac{\begin{vmatrix} 1 & 3 \\ 8 & -2 \end{vmatrix}}{\begin{vmatrix} 2 & 3 \\ 3 & -2 \end{vmatrix}} = \dfrac{-2 - 24}{-4 - 9} = \dfrac{-26}{-13} = 2$; $y = \dfrac{\begin{vmatrix} 2 & 1 \\ 3 & 8 \end{vmatrix}}{\begin{vmatrix} 2 & 3 \\ 3 & -2 \end{vmatrix}} = \dfrac{16 - 3}{-13} = \dfrac{13}{-13} = -1$; solution: $(2, -1)$

65. $\begin{cases} x = \dfrac{5y - 4}{2} \\ y = \dfrac{3x - 1}{5} \end{cases} \Rightarrow \begin{cases} 2x = 5y - 4 \\ 5y = 3x - 1 \end{cases} \Rightarrow \begin{cases} 2x - 5y = -4 \\ -3x + 5y = -1 \end{cases}$

$x = \dfrac{\begin{vmatrix} -4 & -5 \\ -1 & 5 \end{vmatrix}}{\begin{vmatrix} 2 & -5 \\ -3 & 5 \end{vmatrix}} = \dfrac{-20 - 5}{10 - 15} = \dfrac{-25}{-5} = 5$; $y = \dfrac{\begin{vmatrix} 2 & -4 \\ -3 & -1 \end{vmatrix}}{\begin{vmatrix} 2 & -5 \\ -3 & 5 \end{vmatrix}} = \dfrac{-2 - 12}{-5} = \dfrac{-14}{-5} = \dfrac{14}{5}$

solution: $\left(5, \frac{14}{5}\right)$

67. $D = \begin{vmatrix} 2 & 1 & 1 \\ 1 & -2 & 3 \\ 1 & 1 & -4 \end{vmatrix} = 2\begin{vmatrix} -2 & 3 \\ 1 & -4 \end{vmatrix} - 1\begin{vmatrix} 1 & 3 \\ 1 & -4 \end{vmatrix} + 1\begin{vmatrix} 1 & -2 \\ 1 & 1 \end{vmatrix} = 2(5) - 1(-7) + 1(3) = 20$

$N_x = \begin{vmatrix} 5 & 1 & 1 \\ 10 & -2 & 3 \\ -3 & 1 & -4 \end{vmatrix} = 5\begin{vmatrix} -2 & 3 \\ 1 & -4 \end{vmatrix} - 1\begin{vmatrix} 10 & 3 \\ -3 & -4 \end{vmatrix} + 1\begin{vmatrix} 10 & -2 \\ -3 & 1 \end{vmatrix}$

$= 5(5) - 1(-31) + 1(4) = 60$

$N_y = \begin{vmatrix} 2 & 5 & 1 \\ 1 & 10 & 3 \\ 1 & -3 & -4 \end{vmatrix} = 2\begin{vmatrix} 10 & 3 \\ -3 & -4 \end{vmatrix} - 5\begin{vmatrix} 1 & 3 \\ 1 & -4 \end{vmatrix} + 1\begin{vmatrix} 1 & 10 \\ 1 & -3 \end{vmatrix}$

$= 2(-31) - 5(-7) + 1(-13) = -40$

continued on next page...

67. continued

$$N_z = \begin{vmatrix} 2 & 1 & 5 \\ 1 & -2 & 10 \\ 1 & 1 & -3 \end{vmatrix} = 2\begin{vmatrix} -2 & 10 \\ 1 & -3 \end{vmatrix} - 1\begin{vmatrix} 1 & 10 \\ 1 & -3 \end{vmatrix} + 5\begin{vmatrix} 1 & -2 \\ 1 & 1 \end{vmatrix} = 2(-4) - 1(-13) + 5(3) = 20$$

$$x = \frac{N_x}{D} = \frac{60}{20} = 3; \ y = \frac{N_y}{D} = \frac{-40}{20} = -2; \ z = \frac{N_z}{D} = \frac{20}{20} = 1 \Rightarrow \text{solution: } (3, -2, 1)$$

69. $$D = \begin{vmatrix} 4 & 0 & 3 \\ 0 & 2 & -6 \\ 8 & 4 & 3 \end{vmatrix} = 4\begin{vmatrix} 2 & -6 \\ 4 & 3 \end{vmatrix} - 0\begin{vmatrix} 0 & -6 \\ 8 & 3 \end{vmatrix} + 3\begin{vmatrix} 0 & 2 \\ 8 & 4 \end{vmatrix} = 4(30) - 0(48) + 3(-16) = 72$$

$$N_x = \begin{vmatrix} 4 & 0 & 3 \\ -1 & 2 & -6 \\ 9 & 4 & 3 \end{vmatrix} = 4\begin{vmatrix} 2 & -6 \\ 4 & 3 \end{vmatrix} - 0\begin{vmatrix} -1 & -6 \\ 9 & 3 \end{vmatrix} + 3\begin{vmatrix} -1 & 2 \\ 9 & 4 \end{vmatrix} = 4(30) - 0(51) + 3(-22) = 54$$

$$N_y = \begin{vmatrix} 4 & 4 & 3 \\ 0 & -1 & -6 \\ 8 & 9 & 3 \end{vmatrix} = 4\begin{vmatrix} -1 & -6 \\ 9 & 3 \end{vmatrix} - 4\begin{vmatrix} 0 & -6 \\ 8 & 3 \end{vmatrix} + 3\begin{vmatrix} 0 & -1 \\ 8 & 9 \end{vmatrix} = 4(51) - 4(48) + 3(8) = 36$$

$$N_z = \begin{vmatrix} 4 & 0 & 4 \\ 0 & 2 & -1 \\ 8 & 4 & 9 \end{vmatrix} = 4\begin{vmatrix} 2 & -1 \\ 4 & 9 \end{vmatrix} - 0\begin{vmatrix} 0 & -1 \\ 8 & 9 \end{vmatrix} + 4\begin{vmatrix} 0 & 2 \\ 8 & 4 \end{vmatrix} = 4(22) - 0(8) + 4(-16) = 24$$

$$x = \frac{N_x}{D} = \frac{54}{72} = \frac{3}{4}; \ y = \frac{N_y}{D} = \frac{36}{72} = \frac{1}{2}; \ z = \frac{N_z}{D} = \frac{24}{72} = \frac{1}{3} \Rightarrow \text{solution: } \left(\frac{3}{4}, \frac{1}{2}, \frac{1}{3}\right)$$

71. $$\begin{cases} x + y = 1 \Rightarrow & x + y = 1 \\ \frac{1}{2}y + z = \frac{5}{2} \Rightarrow & y + 2z = 5 \\ x - z = -3 \Rightarrow & x - z = -3 \end{cases}$$

$$D = \begin{vmatrix} 1 & 1 & 0 \\ 0 & 1 & 2 \\ 1 & 0 & -1 \end{vmatrix} = 1\begin{vmatrix} 1 & 2 \\ 0 & -1 \end{vmatrix} - 1\begin{vmatrix} 0 & 2 \\ 1 & -1 \end{vmatrix} + 0\begin{vmatrix} 0 & 1 \\ 1 & 0 \end{vmatrix} = 1(-1) - 1(-2) + 0(-1) = 1$$

$$N_x = \begin{vmatrix} 1 & 1 & 0 \\ 5 & 1 & 2 \\ -3 & 0 & -1 \end{vmatrix} = 1\begin{vmatrix} 1 & 2 \\ 0 & -1 \end{vmatrix} - 1\begin{vmatrix} 5 & 2 \\ -3 & -1 \end{vmatrix} + 0\begin{vmatrix} 5 & 1 \\ -3 & 0 \end{vmatrix} = 1(-1) - 1(1) + 0(3) = -2$$

$$N_y = \begin{vmatrix} 1 & 1 & 0 \\ 0 & 5 & 2 \\ 1 & -3 & -1 \end{vmatrix} = 1\begin{vmatrix} 5 & 2 \\ -3 & -1 \end{vmatrix} - 1\begin{vmatrix} 0 & 2 \\ 1 & -1 \end{vmatrix} + 0\begin{vmatrix} 0 & 5 \\ 1 & -3 \end{vmatrix} = 1(1) - 1(-2) + 0(-5) = 3$$

$$N_z = \begin{vmatrix} 1 & 1 & 1 \\ 0 & 1 & 5 \\ 1 & 0 & -3 \end{vmatrix} = 1\begin{vmatrix} 1 & 5 \\ 0 & -3 \end{vmatrix} - 1\begin{vmatrix} 0 & 5 \\ 1 & -3 \end{vmatrix} + 1\begin{vmatrix} 0 & 1 \\ 1 & 0 \end{vmatrix} = 1(-3) - 1(-5) + 1(-1) = 1$$

$$x = \frac{N_x}{D} = \frac{-2}{1} = -2; \ y = \frac{N_y}{D} = \frac{3}{1} = 3; \ z = \frac{N_z}{D} = \frac{1}{1} = 1 \Rightarrow \text{solution: } (-2, 3, 1)$$

73. $\begin{vmatrix} x & 2 \\ -3 & 1 \end{vmatrix} = 2$

$x + 6 = 2$

$x = -4$

75. $\begin{vmatrix} x & -2 \\ 3 & 1 \end{vmatrix} = \begin{vmatrix} 4 & 2 \\ x & 3 \end{vmatrix}$

$x - (-6) = 12 - 2x$

$x + 6 = 12 - 2x$

$3x = 6$

$x = 2$

77. $\begin{cases} 2x + y = 180 \\ y = x + 30 \end{cases} \Rightarrow \begin{cases} 2x + y = 180 \\ -x + y = 30 \end{cases}$

$x = \dfrac{\begin{vmatrix} 180 & 1 \\ 30 & 1 \end{vmatrix}}{\begin{vmatrix} 2 & 1 \\ -1 & 1 \end{vmatrix}} = \dfrac{180 - 30}{2 - (-1)} = \dfrac{150}{3} = 50°; \; y = \dfrac{\begin{vmatrix} 2 & 180 \\ -1 & 30 \end{vmatrix}}{\begin{vmatrix} 2 & 1 \\ -1 & 1 \end{vmatrix}} = \dfrac{60 - (-180)}{3} = \dfrac{240}{3} = 80°$

79. Let $x =$ the amount invested in HiTech, $y =$ the amount invested in SaveTel and $z =$ the amount invested in HiGas. Form and solve the following system of equations:

$\begin{cases} x + y + z = 20000 \Rightarrow \\ 0.10x + 0.05y + 0.06z = 0.066(20000) \Rightarrow \\ y + z = 3x \Rightarrow \end{cases}$ $\begin{aligned} & x + y + z = 20000 \\ & 10x + 5y + 6x = 132000 \\ & -3x + y + z = 0 \end{aligned}$

$D = \begin{vmatrix} 1 & 1 & 1 \\ 10 & 5 & 6 \\ -3 & 1 & 1 \end{vmatrix} = 1\begin{vmatrix} 5 & 6 \\ 1 & 1 \end{vmatrix} - 1\begin{vmatrix} 10 & 6 \\ -3 & 1 \end{vmatrix} + 1\begin{vmatrix} 10 & 5 \\ -3 & 1 \end{vmatrix} = 1(-1) - 1(28) + 1(25) = -4$

$N_x = \begin{vmatrix} 20000 & 1 & 1 \\ 132000 & 5 & 6 \\ 0 & 1 & 1 \end{vmatrix} = 20000\begin{vmatrix} 5 & 6 \\ 1 & 1 \end{vmatrix} - 1\begin{vmatrix} 132000 & 6 \\ 0 & 1 \end{vmatrix} + 1\begin{vmatrix} 132000 & 5 \\ 0 & 1 \end{vmatrix}$

$= 20000(-1) - 1(132000) + 1(132000) = -20000$

$N_y = \begin{vmatrix} 1 & 20000 & 1 \\ 10 & 132000 & 6 \\ -3 & 0 & 1 \end{vmatrix} = 1\begin{vmatrix} 132000 & 6 \\ 0 & 1 \end{vmatrix} - 20000\begin{vmatrix} 10 & 6 \\ -3 & 1 \end{vmatrix} + 1\begin{vmatrix} 10 & 132000 \\ -3 & 0 \end{vmatrix}$

$= 1(132000) - 20000(28) + 1(396000) = -32000$

$N_z = \begin{vmatrix} 1 & 1 & 20000 \\ 10 & 5 & 132000 \\ -3 & 1 & 0 \end{vmatrix} = 1\begin{vmatrix} 5 & 132000 \\ 1 & 0 \end{vmatrix} - 1\begin{vmatrix} 10 & 132000 \\ -3 & 0 \end{vmatrix} + 20000\begin{vmatrix} 10 & 5 \\ -3 & 1 \end{vmatrix}$

$= 1(-132000) - 1(396000) + 20000(25) = -28000$

$x = \dfrac{N_x}{D} = \dfrac{-20000}{-4} = 5000; \; y = \dfrac{N_y}{D} = \dfrac{-32000}{-4} = 8000; \; z = \dfrac{N_z}{D} = \dfrac{-28000}{-4} = 7000$

He should invest \$5000 in HiTech, \$8000 in SaveTel and \$7000 in HiGas.

81. **Answers may vary.**

83.
$$\begin{vmatrix} x & y & 1 \\ -2 & 3 & 1 \\ 3 & 5 & 1 \end{vmatrix} = 0$$

$$x\begin{vmatrix} 3 & 1 \\ 5 & 1 \end{vmatrix} - y\begin{vmatrix} -2 & 1 \\ 3 & 1 \end{vmatrix} + 1\begin{vmatrix} -2 & 3 \\ 3 & 5 \end{vmatrix} = 0$$

$$x(3-5) - y(-2-3) + 1(-10-9) = 0$$

$$-2x + 5y - 19 = 0$$

$$2x - 5y = -19 \qquad \text{Verify that both points satisfy this equation.}$$

85.
$$\begin{vmatrix} 1 & 0 & 2 & 1 \\ 2 & 1 & 1 & 3 \\ 1 & 1 & 1 & 1 \\ 2 & 1 & 1 & 1 \end{vmatrix} = 1\begin{vmatrix} 1 & 1 & 3 \\ 1 & 1 & 1 \\ 1 & 1 & 1 \end{vmatrix} - 0\begin{vmatrix} 2 & 1 & 3 \\ 1 & 1 & 1 \\ 2 & 1 & 1 \end{vmatrix} + 2\begin{vmatrix} 2 & 1 & 3 \\ 1 & 1 & 1 \\ 2 & 1 & 1 \end{vmatrix} - 1\begin{vmatrix} 2 & 1 & 1 \\ 1 & 1 & 1 \\ 2 & 1 & 1 \end{vmatrix}$$

$$= 1(0) - 0(???) + 2(-2) - 1(0) = -4$$

Exercises 8.7 (page 617)

1. $y > 3x + 2$

$0 \overset{?}{>} 3(0) + 2$

$0 \overset{?}{>} 0 + 2$

$0 \not> 2$

not a solution.

3. $y > 3x + 2$

$4 \overset{?}{>} 3(-2) + 2$

$4 \overset{?}{>} -6 + 2$

$4 > -4$

solution.

5. $y > x - 2$

$y > \Rightarrow$ ABOVE

7. $x - y > -1$

$-y > -x - 1$

$y < x + 1$

$y < \Rightarrow$ BELOW

9. $4x - 6 = 18$

$4x = 24$

$x = 6$

11.
$$A = P + Prt$$
$$A - P = Prt$$
$$\frac{A-P}{Pr} = \frac{Prt}{Pr}$$
$$\frac{A-P}{Pr} = t, \text{ or } t = \frac{A-P}{Pr}$$

13. $7x + 4(x-6) = 7x + 4x - 24 = 11x - 24$

15. $3(y-x) + 5y + 8x = 3y - 3x + 5y + 8x$
$$= 5x + 8y$$

17. inequality

19. boundary

21. inequalities

23. doubly shaded

25. a. does not include boundary

b. does include boundary

27. a.
$$5x - 3y \geq 0$$
$$5(1) - 3(1) \overset{?}{\geq} 0$$
$$5 - 3 \overset{?}{\geq} 0$$
$$2 \geq 0$$
$(1, 1)$ is a solution.

b.
$$5x - 3y \geq 0$$
$$5(-2) - 3(-3) \overset{?}{\geq} 0$$
$$-10 + 9 \overset{?}{\geq} 0$$
$$-1 \not\geq 0$$
$(-2, -3)$ is not a solution.

27. c.
$$5x - 3y \geq 0$$
$$5(0) - 3(0) \overset{?}{\geq} 0$$
$$0 - 0 \overset{?}{\geq} 0$$
$$0 \geq 0$$
$(0, 0)$ is a solution.

d.
$$5x - 3y \geq 0$$
$$5\left(\frac{1}{5}\right) - 3\left(\frac{4}{3}\right) \overset{?}{\geq} 0$$
$$1 - 4 \overset{?}{\geq} 0$$
$$-3 \not\geq 0$$
$\left(\frac{1}{5}, \frac{4}{3}\right)$ is not a solution.

29. a.
$$x + y > 4$$
$$0 + 4 \overset{?}{>} 4$$
$$4 \not> 4$$
$(0, 4)$ is not a solution.

b.
$$x + y > 4$$
$$1 + 5 \overset{?}{>} 4$$
$$6 > 4$$
$(1, 5)$ is a solution.

c.
$$x + y > 4$$
$$-1 + \frac{1}{2} \overset{?}{>} 4$$
$$-\frac{1}{2} \not> 4$$
$\left(-1, \frac{1}{2}\right)$ is not a solution.

d.
$$x + y > 4$$
$$-\frac{3}{4} + 7 \overset{?}{>} 4$$
$$\frac{25}{4} > 4$$
$\left(-\frac{3}{4}, 7\right)$ is a solution.

31. Boundary (solid)   Test point: $(0, 0)$

$y = x + 2$

x	y
0	2
2	4

$$y \leq x + 2$$
$$0 \overset{?}{\leq} 0 + 2$$
$$0 \leq 2$$
same half-plane

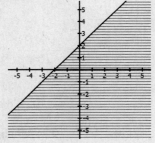

33. Boundary (solid)   Test point: $(1, 1)$

$y = 4x$

x	y
0	0
1	4

$$y \leq 4x$$
$$1 \overset{?}{\leq} 4(1)$$
$$1 \leq 4$$
same half-plane

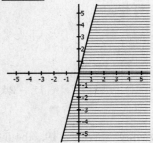

35. Boundary (dotted)

$y = x - 3$

x	y
0	-3
3	0

Test point: $(0, 0)$

$y > x - 3$

$0 \overset{?}{>} 0 - 3$

$0 > -3$

same half-plane

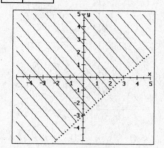

37. Boundary (dotted)

$y = 2x - 4$

x	y
0	-4
2	0

Test point: $(0, 0)$

$y > 2x - 4$

$0 \overset{?}{>} 2(0) - 4$

$0 \overset{?}{>} 0 - 4$

$0 > -4$

same half-plane

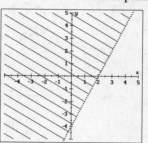

39. Boundary (solid)

$2x - y = 4$

x	y
0	-4
2	0

Test point: $(0, 0)$

$2x - y \leq 4$

$2(0) - 0 \overset{?}{\leq} 4$

$0 - 0 \overset{?}{\leq} 4$

$0 \leq 4$

same half-plane

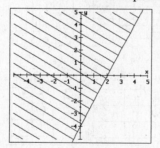

41. Boundary (solid)

$x - 2y = 4$

x	y
0	-2
4	0

Test point: $(0, 0)$

$x - 2y \leq 4$

$0 - 2(0) \overset{?}{\leq} 4$

$0 - 0 \overset{?}{\leq} 4$

$0 \leq 4$

same half-plane

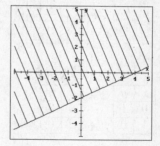

43. $x + 2y \leq 3$ $2x - y \geq 1$

x	y
1	1
3	0

x	y
0	-1
2	3

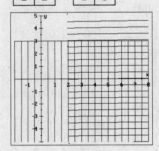

45. $x + y < -1$ $x - y > -1$

x	y
0	-1
-1	0

x	y
0	1
-1	0

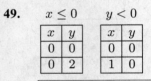

47. $x > 2$ $y \leq 3$

x	y
2	0
2	2

x	y
0	3
1	3

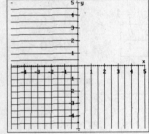

49. $x \leq 0$ $y < 0$

x	y
0	0
0	2

x	y
0	0
1	0

51. $x + y < 1$ $x + y > 3$

x	y
0	1
1	0

x	y
0	3
3	0

solution set: $\emptyset$

53. $y \leq -\frac{4}{3}x - 2$ $4x + 3y > 15$

x	y
0	-2
-3	2

x	y
0	5
3	1

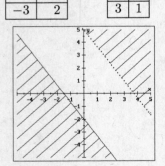

solution set: $\emptyset$

55. Boundary   Test point: $(1, 1)$
(dotted) $y < 3x$

$y = 3x$ $1 \overset{?}{<} 3(1)$

x	y
0	0
1	3

$1 < 3$

same half-plane

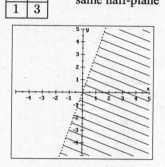

57. $2x - y < 4$ $x + y \geq -1$

x	y
0	-4
2	0

x	y
0	-1
-1	0

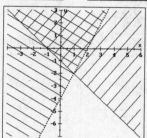

59. Boundary   Test point: $(0, 0)$
(dotted) $y < 2 - 3x$

$y = 2 - 3x$ $0 \overset{?}{<} 2 - 3(0)$

x	y
0	2
1	-1

$0 \overset{?}{<} 2 - 0$

$0 < 2$

same half-plane

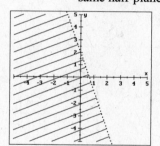

61. Boundary   Test point: $(0, 0)$
(dotted) $x < 2$

$x = 2$ $0 < 2$

x	y
2	0
2	4

same half-plane

63. Boundary (solid)

$y + 9x = 3$

x	y
0	3
1	−6

Test point: $(0, 0)$

$y + 9x \geq 3$

$0 + 9(0) \overset{?}{\geq} 3$

$0 + 0 \overset{?}{\geq} 3$

$0 \not\geq 3$

opposite half-plane

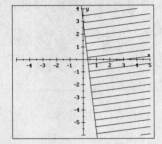

65. Boundary (solid)

$4x + 3y = 12$

x	y
0	4
3	0

Test point: $(0, 0)$

$4x + 3y \leq 12$

$4(0) + 3(0) \overset{?}{\leq} 12$

$0 + 0 \overset{?}{\leq} 12$

$0 \leq 12$

same half-plane

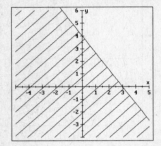

67. Boundary (solid)

$y = 1$

x	y
0	1
4	1

Test point: $(0, 0)$

$y \leq 1$

$0 \leq 1$

same half-plane

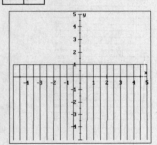

69. $3x + 4y > -7$ $2x - 3y \geq 1$

x	y
−1	−1
−5	2

x	y
2	1
−1	−1

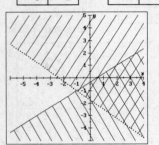

71. $2x - 4y > -6$ $3x + y \geq 5$

x	y
1	2
-3	0

x	y
0	5
1	2

73. $\dfrac{x}{2} + \dfrac{y}{3} \geq 2$ $\dfrac{x}{2} - \dfrac{y}{2} < -1$

$3x + 2y \geq 12$ $x - y < -2$

x	y
0	6
4	0

x	y
0	2
-2	0

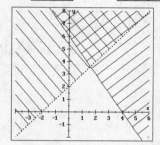

75. $\boxed{\text{Cake cost}} + \boxed{\text{Pie cost}} \leq 120$

$3x + 4y \leq 120$

Boundary   Test point: $(0,0)$

(solid) $3x + 4y \leq 120$

$3x + 4y = 120$ $3(0) + 4(0) \overset{?}{\leq} 120$

x	y
0	30
40	0

$0 + 0 \overset{?}{\leq} 120$

$0 \leq 120$

same half-plane

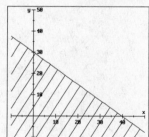

Solutions: $(10, 10), (20, 10), (10, 20)$

77. $\boxed{\text{Leather}} + \boxed{\text{Nylon}} \geq 4400$

$100x + 88y \geq 4400$

Boundary   Test point: $(0,0)$

(solid) $100x + 88y \geq 4400$

x	y
0	50
44	0

$100(0) + 88(0) \overset{?}{\geq} 4400$

$0 + 0 \overset{?}{\geq} 4400$

$0 \ngeq 4400$

opposite half-plane

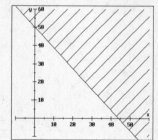

Solutions: $(50, 50), (30, 40), (40, 40)$

79. ☐R. stock☐ + ☐M. stock☐ ≤ 8000

$$40x + 50y \leq 8000$$

Boundary   Test point: $(0, 0)$

(solid) $40x + 50y \leq 8000$

$40x + 50y = 8000$ $40(0) + 50(0) \overset{?}{\leq} 8000$

x	y
0	160
200	0

$0 + 0 \overset{?}{\leq} 8000$

$0 \leq 8000$

same half-plane

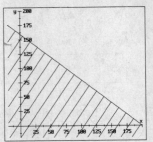

Solutions: $(80, 40), (80, 80), (120, 40)$

81. $10x + 15y \geq 30$ $10x + 15y \leq 60$

x	y
0	2
3	0

x	y
0	4
6	0

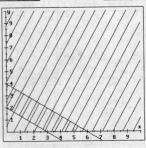

1 $10 CD and 2 $15 CDs

4 $10 CDs and 1 $15 CD

83. $150x + 100y \leq 900$ $y > x$

x	y
0	9
6	0

x	y
0	0
1	1

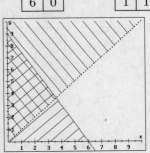

2 desk chairs and 4 side chairs

1 desk chair and 5 side chairs

85-91. Answers may vary.

Exercise 8.8   (page 629)

1. $P = 2x + 5y = 2(0) + 5(4) = 20$

3.
$$x + y = 5 \Rightarrow \times (-2) \quad -2x - 2y = -10$$
$$\underline{3x + 2y = 12} \Rightarrow \qquad \underline{3x + 2y = \;\; 12}$$
$$x \qquad = 2$$

Substitute and solve for y:

$x + y = 5$

$2 + y = 5$

$\quad y = 3$

The solution is $(2, 3)$.

5. $m = \dfrac{\Delta y}{\Delta x} = \dfrac{-4 - 5}{-1 - 3} = \dfrac{-9}{-4} = \dfrac{9}{4}$

7.
$y - y_1 = m(x - x_1)$

$y - 5 = \frac{9}{4}(x - 3)$

$y - 5 = \frac{9}{4}x - \frac{27}{4}$

$\quad y = \frac{9}{4}x - \frac{7}{4}$

9. constraints

11. objective

13.

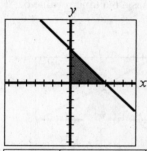

Vertex	$P = 2x + 3y$	Maximum?
$(0, 0)$	$= 0$	No
$(4, 0)$	$= 8$	No
$(0, 4)$	$= 12$	YES

P has a maximum value of 12 at $(0, 4)$.

15.

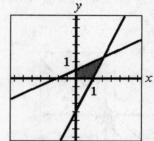

Vertex	$P = y + \frac{1}{2}x$	Maximum?
$(0, 0)$	$= 0$	No
$(1, 0)$	$= \frac{1}{2}$	No
$\left(\frac{5}{3}, \frac{4}{3}\right)$	$= \frac{13}{6}$	YES
$\left(0, \frac{1}{2}\right)$	$= \frac{1}{2}$	No

P has a maximum value of $\frac{13}{6}$ at $\left(\frac{5}{3}, \frac{4}{3}\right)$.

17.

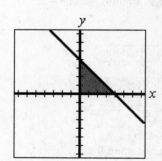

Vertex	$P = 5x + 12y$	Minimum?
$(4, 0)$	$= 20$	No
$(0, 4)$	$= 48$	No
$(0, 0)$	$= 0$	YES

P has a minimum value of 0 at $(0, 0)$.

19.

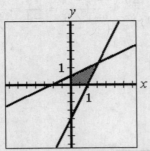

Vertex	$P = 3y + x$	Minimum?
$(0, 0)$	$= 0$	YES
$(1, 0)$	$= 1$	No
$\left(\frac{5}{3}, \frac{4}{3}\right)$	$= \frac{17}{3}$	No
$\left(0, \frac{1}{2}\right)$	$= \frac{3}{2}$	No

P has a minimum value of 0 at $(0, 0)$.

21.

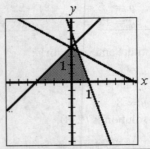

Vertex	$P = 2x + y$	Maximum?
$(1, 0)$	$= 2$	No
$\left(\frac{3}{7}, \frac{12}{7}\right)$	$= \frac{18}{7}$	YES
$(0, 2)$	$= 2$	No
$(-2, 0)$	$= -4$	No

P has a maximum value of $\frac{18}{7}$ at $\left(\frac{3}{7}, \frac{12}{7}\right)$.

23.

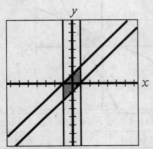

Vertex	$P = 3x - 2y$	Maximum?
$(1, 0)$	$= 3$	YES
$(1, 2)$	$= -1$	No
$(-1, 0)$	$= -3$	No
$(-1, -2)$	$= 1$	No

P has a maximum value of 3 at $(1, 0)$.

25.

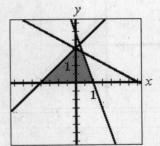

Vertex	$P = 6x + 2y$	Minimum?
$(1, 0)$	$= 6$	No
$\left(\frac{3}{7}, \frac{12}{7}\right)$	$= 6$	No
$(0, 2)$	$= 4$	No
$(-2, 0)$	$= -12$	YES

P has a minimum value of -12 at $(-2, 0)$.

27.

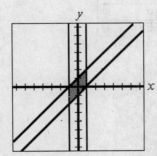

Vertex	$P = 2x - 2y$	Minimum?
$(1, 0)$	$= 2$	No
$(1, 2)$	$= -2$	YES
$(-1, 0)$	$= -2$	YES
$(-1, -2)$	$= 2$	No

P has a minimum value of -2 at $(1, 2)$ and at $(-1, 0)$ (and at all points along the boundary between them).

29. Let x = the number of tables made, and let y = the number of chairs made.

$$\begin{cases} x \geq 0 \qquad\qquad y \geq 0 \\ 2x + 3y \leq 42 \quad \text{(Tom's hours)} \\ 6x + 2y \leq 42 \quad \text{(Carlos' hours)} \end{cases}$$

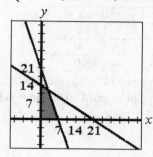

Profit = $P = 100x + 80y$.

Vertex	$P = 100x + 80y$
$(0, 0)$	$= 0$
$(7, 0)$	$= 700$
$(3, 12)$	$= 1260$
$(0, 14)$	$= 1120$

The maximum income of $1260 results when they make 3 tables and 12 chairs.

31. Let x = the number of IBMs stocked, and let y = the number of Macintoshes stocked.

$$\begin{cases} 20 \leq x \leq 30 \\ 30 \leq y \leq 50 \\ x + y \leq 60 \quad \text{(Total stock)} \end{cases}$$

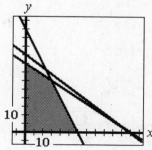

Let the objective function be the total commissions:
$P = 50x + 40y$.

Vertex	$P = 50x + 40y$
$(20, 30)$	$= 2200$
$(30, 30)$	$= 2700$
$(20, 40)$	$= 2600$

The maximum amount of commissions is $2700, which results from stocking 30 of each type.

33. Let x = the number of DVD players made, and let y = the number of TVs made.

$$\begin{cases} x \geq 0 \qquad\qquad y \geq 0 \\ 3x + 4y \leq 180 \quad \text{(electronics hours)} \\ 2x + 3y \leq 120 \quad \text{(assembly hours)} \\ 2x + y \leq 60 \quad \text{(finishing hours)} \end{cases}$$

Let the objective function be the profit:
$P = 40x + 32y$.

Vertex	$P = 40x + 32y$
$(0, 0)$	$= 0$
$(30, 0)$	$= 1200$
$(15, 30)$	$= 1560$
$(0, 40)$	$= 1280$

The maximum profit of $1560 results when they make 15 DVD players and 30 TVs.

35. Let $x =$ the amount in stocks, and let $y =$ the amount in bonds.

$$\begin{cases} x \geq 100000 \\ y \geq 50000 \\ x + y \leq 200000 \quad \text{(total amount)} \end{cases}$$

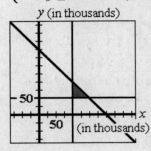

Let the objective function be the income:
$P = 0.09x + 0.07y.$

Vertex	$P = 0.09x + 0.07y$
$(100000, 50000)$	$= 12500$
$(150000, 50000)$	$= 17000$
$(100000, 100000)$	$= 16000$

The maximum income of $17,000 results when she invests $150,000 in stocks and $50,000 in bonds.

37. Answers may vary.

39. Answers may vary.

Chapter 8 Review   (page 634)

1. $\begin{cases} 2x + y = 11 \\ -x + 2y = 7 \end{cases}$

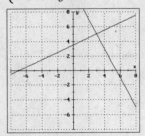

$(3, 5)$ is the solution.

2. $\begin{cases} 3x + 2y = 0 \\ 2x - 3y = -13 \end{cases}$

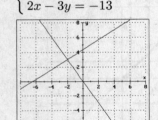

$(-2, 3)$ is the solution.

3. $\begin{cases} \frac{1}{2}x + \frac{1}{3}y = 2 \\ y = 6 - \frac{3}{2}x \end{cases}$

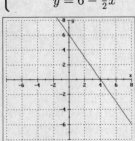

dependent equations; $\left(x, 6 - \frac{3}{2}x\right)$

4. $\begin{cases} \frac{1}{3}x - \frac{1}{2}y = 1 \\ 6x - 9y = 2 \end{cases}$

inconsistent system; $\emptyset$

5. $\begin{cases} (1) & y = 2x + 5 \\ (2) & 3x + 4y = 9 \end{cases}$

Substitute $y = 2x + 5$ from (1) into (2):

$$3x + 4\boldsymbol{y} = 9$$
$$3x + 4(\boldsymbol{2x + 5}) = 9$$
$$3x + 8x + 20 = 9$$
$$11x = -11$$
$$x = -1$$

Substitute this and solve for y:

$$y = 2x + 5 = 2(-1) + 5 = 3$$

Solution: $\boxed{(-1, 3)}$

6. $\begin{cases} (1) & y = 3x + 5 \\ (2) & 3x - y = -5 \end{cases}$

Substitute $y = 3x + 5$ from (1) into (2):

$$3x - \boldsymbol{y} = -5$$
$$3x - (\boldsymbol{3x + 5}) = -5$$
$$3x - 3x - 5 = -5$$
$$-5 = -5$$

Dependent equations

$\boxed{(x, 3x + 5)}$

7. $\begin{cases} (1) & x + 2y = 11 \\ (2) & 2x - y = 2 \end{cases}$

Substitute $x = -2y + 11$ from (1) into (2):

$$2\boldsymbol{x} - y = 2$$
$$2(\boldsymbol{-2y + 11}) - y = 2$$
$$-4y + 22 - y = 2$$
$$-5y = -20$$
$$y = 4$$

Substitute this and solve for x:

$$x = -2y + 11 = -2(4) + 11 = 3$$

Solution: $\boxed{(3, 4)}$

8. $\begin{cases} (1) & 2x + 3y = -2 \\ (2) & 3x + 5y = -2 \end{cases}$

Substitute $x = \frac{-3y - 2}{2}$ from (1) into (2):

$$3\boldsymbol{x} + 5y = -2$$
$$3 \cdot \frac{\boldsymbol{-3y - 2}}{2} + 5y = -2$$
$$3(-3y - 2) + 10y = -4$$
$$-9y - 6 + 10y = -4$$
$$y = 2$$

Substitute this and solve for x:

$$x = \frac{-3y - 2}{2} = \frac{-3(2) - 2}{2} = \frac{-8}{2} = -4$$

Solution: $\boxed{(-4, 2)}$

9.

$$\begin{array}{ll} x - y = 1 & \Rightarrow \times(2) \quad 2x - 2y = 2 \\ 5x + 2y = -16 & \Rightarrow \qquad\quad 5x + 2y = -16 \\ \hline & \qquad\qquad\quad 7x \quad\;\; = -14 \\ & \qquad\qquad\quad\; x \quad\;\; = -2 \end{array}$$

Substitute and solve for y:

$$x - y = 1$$
$$-2 - y = 1$$
$$-y = 3$$
$$y = -3 \quad \text{Solution: } \boxed{(-2, -3)}$$

10.

$$\begin{array}{ll} 3x + 2y = 1 & \Rightarrow \times 3 \quad 9x + 6y = 3 \\ 2x - 3y = 5 & \Rightarrow \times 2 \quad 4x - 6y = 10 \\ \hline & \qquad\quad 13x \quad\;\; = 13 \\ & \qquad\quad\; x \quad\;\; = 1 \end{array}$$

Substitute and solve for y:

$$3x + 2y = 1$$
$$3(1) + 2y = 1$$
$$3 + 2y = 1$$
$$2y = -2$$
$$y = -1 \quad \text{Solution: } \boxed{(1, -1)}$$

11.

$$\begin{array}{lll} 2x + 6 = -3y & \Rightarrow & 2x + 3y = -6 \Rightarrow \\ -2x = 3y + 6 & \Rightarrow & -2x - 3y = 6 \Rightarrow \\ & & \hline \qquad\quad 0 = 0 \end{array}$$

Dependent equations

$$2x + 3y = -6$$
$$y = \frac{-2x - 6}{3}$$
$$y = -\frac{2}{3}x - 2 \Rightarrow \boxed{\left(x, -\frac{2}{3}x - 2\right)}$$

12. $y = \dfrac{2x-3}{2} \Rightarrow \times 2 \quad 2y = 2x - 3 \Rightarrow -2x + 2y = -3$ Inconsistent system

$x = \dfrac{2y+7}{2} \Rightarrow \times 2 \quad 2x = 2y + 7 \Rightarrow \quad 2x - 2y = \quad 7$ No solutions: $\boxed{\emptyset}$

$$0 \neq 4$$

13. Let x and y represent the integers. Substitute $x = 3y$ into (2): Substitute; solve for x:

$\begin{cases} (1) \quad x = 3y \\ (2) \quad x + y = 84 \end{cases}$

$x + y = 84$
$3y + y = 84$
$4y = 84$
$y = 21$

$x = 3y$
$x = 3(21) = 63$
The integers are 21 and 63.

14. Let w = the width. Substitute $l = 3w$ into (2): Substitute; solve for l:

Let l = the length. $\quad 2l + 2w = 24$

$\begin{cases} (1) \quad l = 3w \\ (2) \quad 2l + 2w = 24 \end{cases}$

$2(3w) + 2w = 24$
$6w + 2w = 24$
$8w = 24$
$w = 3$

$l = 3w$
$l = 3(3) = 9$
The dimensions are 3 ft by 9 ft.

15. Let x = the cost of a grapefruit. Substitute $x = y + 15$ into (2): Substitute; solve for x:

Let y = the cost of an orange. $\quad x + y = 85$

$\begin{cases} (1) \quad x = y + 15 \\ (2) \quad x + y = 85 \end{cases}$

$y + 15 + y = 85$
$2y = 70$
$y = 35$

$x = y + 15$
$x = 35 + 15 = 50$
A grapefruit costs 50¢.

16. Let x = the electric bill. Substitute $x = y - 23$ into (2): The gas bill was $66.

Let y = the gas bill. $\quad x + y = 109$

$\begin{cases} (1) \quad x = y - 23 \\ (2) \quad x + y = 109 \end{cases}$

$y - 23 + y = 109$
$2y = 132$
$y = 66$

17. Let x = cost of a gallon of milk and y = cost of one dozen eggs.

$2x + 3y = 6.80 \Rightarrow \times (-2) \quad -4x - 6y = -13.60$ A gallon of milk costs $1.69.

$3x + 2y = 7.35 \Rightarrow \times (3) \quad \dfrac{9x + 6y = \quad 22.05}{}$

$$5x = 8.45$$
$$x = 1.69$$

18. Let x = principal at 4% and y = principal at 3%.

$x + \quad y = 4500 \qquad x + y = \quad 4500 \Rightarrow \times (-3) \quad -3x - 3y = -13500$

$\dfrac{0.04x + 0.03y = \quad 160}{} \Rightarrow \times (100) \quad \dfrac{4x + 3y = 16000}{}$

$$\dfrac{4x + 3y = \quad 16000}{x = \quad 2500}$$

$x + y = 4500$ He invested $2000 at the 3% rate.

$2500 + y = 4500$

$y = 2000$

19. Let b = speed of boat in still water and c = speed of current.

	d	r	t	Equation $(d = r \cdot t)$
Downstream	56	$b + c$	4	$56 = (b + c) \cdot 4 \Rightarrow 4b + 4c = 56$
Upstream	56	$b - c$	7	$56 = (b - c) \cdot 7 \Rightarrow 7b - 7c = 56$

$4b + 4c = 56 \Rightarrow \times (7) \quad 28b + 28c = 392$ $4b + 4c = 56$ The speed of the current

$\underline{7b - 7c = 56 \Rightarrow \times (4)} \quad \underline{28b - 28c = 224} \quad 4(11) + 4c = 56$ is 3 mph.

$ 56b = 616 \quad\quad 44 + 4c = 56$

$ b = 11 \quad\quad\quad 4c = 12$

$ c = 3$

20. Let x = milliliters of first solution and y = milliliters of second solution.

	Fractional part that is saline	Number of milliliters of solution	Number of milliliters of saline
First solution	0.1	x	$0.1x$
Second solution	0.6	y	$0.6y$
Final solution	0.3	50	$0.3(50) = 15$

$x + y = 50$ $x + y = 50 \Rightarrow \times (-1) \quad -x - y = -50$

$\underline{0.1x + 0.6y = 15} \Rightarrow \times (10) \quad \underline{x + 6y = 150} \quad \underline{x + 6y = 150}$

$ 5y = 100$

$ y = 20$

$x + y = 50$ The chemist should use 30 mL of the 10% solution and 20 mL of the 60% solution.

$x + 20 = 50$

$ x = 30$

21.

$(1) \quad x + y + z = 6$ $(1) \quad x + y + z = 6$ $(1) \quad x + y + z = 6$

$(2) \quad x - y - z = -4$ $(2) \quad \underline{x - y - z = -4}$ $(3) \quad \underline{-x + y - z = -2}$

$(3) \quad -x + y - z = -2$ $(4) \quad 2x = 2$ $(5) \quad 2y = 4$

$ x = 1$ $ y = 2$

$x + y + z = 6$

$1 + 2 + z = 6$

$ 3 + z = 6$

$ z = 3$ Solution: $\boxed{(1, 2, 3)}$

22.

$(1) \quad 2x + 3y + z = -5$ $(1) \quad 2x + 3y + z = -5$ $(3) \quad 3x + y + 2z = 4$

$(2) \quad -x + 2y - z = -6$ $(2) \quad \underline{-x + 2y - z = -6}$ $2 \cdot (2) \quad \underline{-2x + 4y - 2z = -12}$

$(3) \quad 3x + y + 2z = 4$ $(4) \quad x + 5y = -11$ $(5) \quad x + 5y = -8$

$x + 5y = -11 \Rightarrow \times (-1) \quad -x - 5y = 11$

$\underline{x + 5y = -8} \Rightarrow \quad \underline{x + 5y = -8}$

$ 0 = 3$

Since this is an impossible equation, there is no solution. It is an inconsistent system. $\boxed{\varnothing}$

23.

$$\begin{array}{c} R_1 + R_2 \Rightarrow R_2 \\ 3R_1 + R_3 \Rightarrow R_3 \end{array}$$

$$\begin{bmatrix} -1 & 5 & 2 & | & -6 \\ 1 & -5 & -1 & | & 6 \\ 3 & -15 & -6 & | & 18 \end{bmatrix} \Rightarrow \begin{bmatrix} -1 & 5 & 2 & | & -6 \\ 0 & 0 & 1 & | & 0 \\ 0 & 0 & 0 & | & 0 \end{bmatrix}$$

From R_2, $z = 0$.   From R_1:

$-x + 5y + z = -6$

$-x + 5y + 0 = -6$

$y = \frac{1}{5}x - \frac{6}{5}$

Solution: $\boxed{(x, \frac{1}{5}x - \frac{6}{5}, 0)}$

24.

$$\begin{array}{cc} -2R_1 + R_2 \Rightarrow R_2 & -R_2 \Rightarrow R_2 \\ -4R_1 + R_3 \Rightarrow R_3 & -R_3 + R_2 \Rightarrow R_3 \end{array} \qquad \frac{1}{4}R_3 \Rightarrow R_3$$

$$\begin{bmatrix} 1 & 1 & 1 & | & 6 \\ 2 & -1 & 1 & | & 1 \\ 4 & 1 & -1 & | & 5 \end{bmatrix} \Rightarrow \begin{bmatrix} 1 & 1 & 1 & | & 6 \\ 0 & -3 & -1 & | & -11 \\ 0 & -3 & -5 & | & -19 \end{bmatrix} \Rightarrow \begin{bmatrix} 1 & 1 & 1 & | & 6 \\ 0 & 3 & 1 & | & 11 \\ 0 & 0 & 4 & | & 8 \end{bmatrix} \Rightarrow \begin{bmatrix} 1 & 1 & 1 & | & 6 \\ 0 & 3 & 1 & | & 11 \\ 0 & 0 & 1 & | & 2 \end{bmatrix}$$

From R_3, $z = 2$.   From R_2:   From R_1:

$3y + z = 11$ $x + y + z = 6$

$3y + 2 = 11$ $x + 3 + 2 = 6$

$3y = 9$ $x = 1$

$y = 3$

Solution: $\boxed{(1, 3, 2)}$

25.

$$\begin{array}{cc} -R_1 + R_2 \Rightarrow R_2 & -\frac{1}{3}R_2 \Rightarrow R_2 \\ -2R_1 + R_3 \Rightarrow R_3 & -3R_3 + R_2 \Rightarrow R_3 \end{array}$$

$$\begin{bmatrix} 1 & 1 & | & 3 \\ 1 & -2 & | & -3 \\ 2 & 1 & | & 4 \end{bmatrix} \Rightarrow \begin{bmatrix} 1 & 1 & | & 3 \\ 0 & -3 & | & -6 \\ 0 & -1 & | & -2 \end{bmatrix} \Rightarrow \begin{bmatrix} 1 & 1 & | & 3 \\ 0 & 1 & | & 2 \\ 0 & 0 & | & 0 \end{bmatrix}$$

From R_2, $y = 2$.   From R_1:   Solution: $\boxed{(1, 2)}$

$x + y = 3$

$x + 2 = 3$

$x = 1$

26.

$$-2R_1 + R_2 \Rightarrow R_2$$

$$\begin{bmatrix} 1 & -3 & 1 & | & 4 \\ 2 & -5 & 3 & | & 6 \end{bmatrix} \Rightarrow \begin{bmatrix} 1 & -3 & 1 & | & 4 \\ 0 & 1 & 1 & | & -2 \end{bmatrix}$$

From R_2:   From R_1:

$y + z = -2$ $x - 3y + z = 4$

$y = -2 - z$ $x - 3(-2 - z) + z = 4$

$x + 6 + 3z + z = 4$

$x + 4z = -2$

$x = -2 - 4z$

Solution: $\boxed{(-2 - 4z, -2 - z, z)}$

27. $\begin{vmatrix} 3 & 1 \\ -2 & 4 \end{vmatrix} = 3(4) - 1(-2) = 12 + 2 = 14$ **28.** $\begin{vmatrix} -4 & 5 \\ 6 & -2 \end{vmatrix} = -4(-2) - 5(6) = 8 - 30$

$= -22$

29. $\begin{vmatrix} -1 & 2 & -1 \\ 2 & -1 & 3 \\ 1 & -2 & 2 \end{vmatrix} = -1 \begin{vmatrix} -1 & 3 \\ -2 & 2 \end{vmatrix} - 2 \begin{vmatrix} 2 & 3 \\ 1 & 2 \end{vmatrix} + (-1) \begin{vmatrix} 2 & -1 \\ 1 & -2 \end{vmatrix} = -1(4) - 2(1) - 1(-3) = -3$

30. $\begin{vmatrix} 3 & -2 & 2 \\ 1 & -2 & -2 \\ 2 & 1 & -1 \end{vmatrix} = 3 \begin{vmatrix} -2 & -2 \\ 1 & -1 \end{vmatrix} - (-2) \begin{vmatrix} 1 & -2 \\ 2 & -1 \end{vmatrix} + 2 \begin{vmatrix} 1 & -2 \\ 2 & 1 \end{vmatrix} = 3(4) + 2(3) + 2(5) = 28$

31. $x = \dfrac{\begin{vmatrix} 10 & 4 \\ 1 & -3 \end{vmatrix}}{\begin{vmatrix} 3 & 4 \\ 2 & -3 \end{vmatrix}} = \dfrac{-30-4}{-9-8} = \dfrac{-34}{-17} = 2; \; y = \dfrac{\begin{vmatrix} 3 & 10 \\ 2 & 1 \end{vmatrix}}{\begin{vmatrix} 3 & 4 \\ 2 & -3 \end{vmatrix}} = \dfrac{3-20}{-17} = \dfrac{-17}{-17} = 1$

Solution: $\boxed{(2,1)}$

32. $x = \dfrac{\begin{vmatrix} -17 & -5 \\ 3 & 2 \end{vmatrix}}{\begin{vmatrix} 2 & -5 \\ 3 & 2 \end{vmatrix}} = \dfrac{-34+15}{4+15} = \dfrac{-19}{19} = -1; \; y = \dfrac{\begin{vmatrix} 2 & -17 \\ 3 & 3 \end{vmatrix}}{\begin{vmatrix} 2 & -5 \\ 3 & 2 \end{vmatrix}} = \dfrac{6+51}{19} = \dfrac{57}{19} = 3$

Solution: $\boxed{(-1,3)}$

33. $D = \begin{vmatrix} 1 & 2 & 1 \\ 2 & 1 & 1 \\ 1 & 1 & 2 \end{vmatrix} = 1 \begin{vmatrix} 1 & 1 \\ 1 & 2 \end{vmatrix} - 2 \begin{vmatrix} 2 & 1 \\ 1 & 2 \end{vmatrix} + 1 \begin{vmatrix} 2 & 1 \\ 1 & 1 \end{vmatrix} = 1(1) - 2(3) + 1(1) = -4$

$N_x = \begin{vmatrix} 0 & 2 & 1 \\ 3 & 1 & 1 \\ 5 & 1 & 2 \end{vmatrix} = 0 \begin{vmatrix} 1 & 1 \\ 1 & 2 \end{vmatrix} - 2 \begin{vmatrix} 3 & 1 \\ 5 & 2 \end{vmatrix} + 1 \begin{vmatrix} 3 & 1 \\ 5 & 1 \end{vmatrix} = 0(1) - 2(1) + 1(-2) = -4$

$N_y = \begin{vmatrix} 1 & 0 & 1 \\ 2 & 3 & 1 \\ 1 & 5 & 2 \end{vmatrix} = 1 \begin{vmatrix} 3 & 1 \\ 5 & 2 \end{vmatrix} - 0 \begin{vmatrix} 2 & 1 \\ 1 & 2 \end{vmatrix} + 1 \begin{vmatrix} 2 & 3 \\ 1 & 5 \end{vmatrix} = 1(1) - 0(3) + 1(7) = 8$

$N_z = \begin{vmatrix} 1 & 2 & 0 \\ 2 & 1 & 3 \\ 1 & 1 & 5 \end{vmatrix} = 1 \begin{vmatrix} 1 & 3 \\ 1 & 5 \end{vmatrix} - 2 \begin{vmatrix} 2 & 3 \\ 1 & 5 \end{vmatrix} + 0 \begin{vmatrix} 2 & 1 \\ 1 & 1 \end{vmatrix} = 1(2) - 2(7) + 0(1) = -12$

$x = \dfrac{N_x}{D} = \dfrac{-4}{-4} = 1; \; y = \dfrac{N_y}{D} = \dfrac{8}{-4} = -2; \; z = \dfrac{N_z}{D} = \dfrac{-12}{-4} = 3 \Rightarrow$ solution: $(1, -2, 3)$

34. $D = \begin{vmatrix} 1 & 1 & 1 \\ -1 & 1 & -1 \\ -1 & -1 & 1 \end{vmatrix} = 1 \begin{vmatrix} 1 & -1 \\ -1 & 1 \end{vmatrix} - 1 \begin{vmatrix} -1 & -1 \\ -1 & 1 \end{vmatrix} + 1 \begin{vmatrix} -1 & 1 \\ -1 & -1 \end{vmatrix} = 1(0) - 1(-2) + 1(2) = 0$

$N_x = \begin{vmatrix} 1 & 1 & 1 \\ 3 & 1 & -1 \\ 5 & -1 & 1 \end{vmatrix} = 1 \begin{vmatrix} 1 & -1 \\ -1 & 1 \end{vmatrix} - 1 \begin{vmatrix} 3 & -1 \\ 5 & 1 \end{vmatrix} + 1 \begin{vmatrix} 3 & 1 \\ 5 & -1 \end{vmatrix} = 1(0) - 1(8) + 1(-8) = -16$

$x = \dfrac{N_x}{D} = \dfrac{-16}{0} \Rightarrow$ denominator $= 0$, numerator $\neq 0 \Rightarrow \emptyset$

CHAPTER 8 REVIEW

35. Boundary   Test point: $(0, 0)$
(solid)

$y = x + 2$

$y \geq x + 2$

$$0 \overset{?}{\geq} 0 + 2$$

$$0 \not\geq 2$$

opposite half-plane

x	y
0	2
-2	0

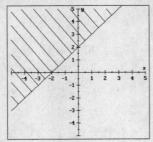

36. Boundary   Test point: $(0, 0)$
(dotted)

$x = 3$ $x < 3$

$0 < 3$

same half-plane

x	y
3	0
3	4

37. $5x + 3y < 15$ $3x - y > 3$

x	y
0	5
3	0

x	y
0	-3
1	0

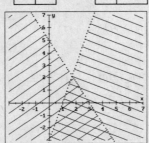

38. $5x - 3y \geq 5$ $3x + 2y \geq 3$

x	y
1	0
4	5

x	y
1	0
3	-3

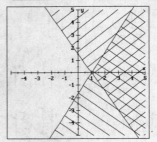

39. $x \geq 3y$ $y < 3x$

x	y
0	0
3	1

x	y
0	0
1	3

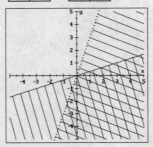

40. $x \geq 0$ $x \leq 3$

x	y
0	0
0	4

x	y
3	1
3	0

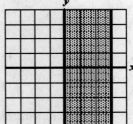

310

41. $10x + 20y \geq 40$ $10x + 20y \leq 60$

x	y
0	2
4	0

x	y
0	3
6	0

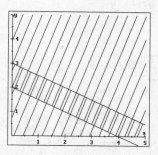

3 shirts and 1 pair of pants
1 shirt and 2 pairs of pants

42.

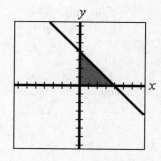

Vertex	$P = 3x + y$	Maximum?
$(4, 0)$	$= 12$	YES
$(0, 4)$	$= 4$	No
$(0, 0)$	$= 0$	No

P has a maximum value of 12 at $(4, 0)$.

43. Let $x =$ the number of bags of fertilizer X, and let $y =$ the number of bags of fertilizer Y.

$$\begin{cases} x \geq 0 \qquad\qquad y \geq 0 \\ 6x + 10y \leq 20000 \quad \text{(Nitrogen)} \\ 8x + 6y \leq 16400 \quad \text{(Phosphorus)} \\ 6x + 4y \leq 12000 \quad \text{(Potash)} \end{cases}$$

Let the objective function be the total profit:
$P = 6x + 5y$.

Vertex	$P = 6x + 5y$
$(2000, 0)$	$= 12000$
$(1600, 600)$	$= 12600$
$(1000, 1400)$	$= 13000$
$(0, 2000)$	$= 10000$

The maximum profit is $13,000,
which results from using 1000 bags of fertilizer X
and 1400 bags of fertilizer Y.

Chapter 8 Test (page 642)

1.
$$\begin{cases} 2x + y = 5 \\ y = 2x - 3 \end{cases}$$

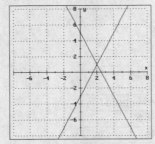

$(2, 1)$ is the solution.

2.
$$\begin{cases} (1) & 2x - 4y = 14 \\ (2) & x = -2y + 7 \end{cases}$$

Substitute $x = -2y + 7$ from (2) into (1):
$$2\boldsymbol{x} - 4y = 14$$
$$2(\boldsymbol{-2y + 7}) - 4y = 14$$
$$-4y + 14 - 4y = 14$$
$$-8y = 0$$
$$y = 0$$

Substitute this and solve for x:
$$x = -2y + 7$$
$$x = -2(0) + 7 = 7$$

Solution: $(7, 0)$

3.
$$\begin{array}{l} 2x + 3y = -5 \Rightarrow \times 2 \\ 3x - 2y = 12 \Rightarrow \times 3 \end{array} \quad \begin{array}{l} 4x + 6y = -10 \\ \underline{9x - 6y = 36} \\ 13x = 26 \\ x = 2 \end{array} \quad \begin{array}{l} 2x + 3y = -5 \\ 2(2) + 3y = -5 \\ 4 + 3y = -5 \\ 3y = -9 \\ y = -3 \end{array}$$

Solution: $\boxed{(2, -3)}$

4.
$$\begin{array}{l} \dfrac{x}{2} - \dfrac{y}{4} = -4 \Rightarrow \times 4 \\ x + y = -2 \Rightarrow \end{array} \quad \begin{array}{l} 2x - y = -16 \\ \underline{x + y = -2} \\ 3x = -18 \\ x = -6 \end{array} \quad \begin{array}{l} x + y = -2 \\ -6 + y = -2 \\ y = 4 \end{array}$$

Solution: $\boxed{(-6, 4)}$

5.
$$3(x + y) = x - 3 \Rightarrow \quad 2x + 3y = -3$$
$$-y = \frac{2x + 3}{3} \Rightarrow \quad 2x + 3y = -3$$

$$\begin{array}{l} 2x + 3y = -3 \Rightarrow \\ 2x + 3y = -3 \Rightarrow \times(-1) \end{array} \quad \begin{array}{l} 2x + 3y = -3 \\ \underline{-2x - 3y = 3} \\ 0 = 0 \end{array}$$

The equation is an identity, so the system has infinitely many solutions. $\Rightarrow$ dependent equations

6. See **#5**. Since the system has at least one solution, it is a consistent system.

7.
$$\begin{bmatrix} 1 & 3 & -2 \\ 4 & -2 & -1 \end{bmatrix} \Rightarrow \overset{-5R_1 + R_2 \Rightarrow R_2}{\begin{bmatrix} 1 & 3 & -2 \\ -1 & -17 & \boxed{9} \end{bmatrix}}$$

8.
$$\begin{bmatrix} -1 & 3 & 6 \\ 3 & -2 & 4 \end{bmatrix} \Rightarrow \overset{-2R_1 + R_2 \Rightarrow R_2}{\begin{bmatrix} -1 & 3 & 6 \\ 5 & -8 & \boxed{-8} \end{bmatrix}}$$

9.
$$\left[\begin{array}{ccc|c} 1 & 1 & 1 & 4 \\ 1 & 1 & -1 & 6 \\ 2 & -3 & 1 & -1 \end{array} \right]$$

For #10-12, find the denominator determinant:

$$D = \begin{vmatrix} 1 & 1 & 1 \\ 1 & 1 & -1 \\ 2 & -3 & 1 \end{vmatrix} = 1 \begin{vmatrix} 1 & -1 \\ -3 & 1 \end{vmatrix} - 1 \begin{vmatrix} 1 & -1 \\ 2 & 1 \end{vmatrix} + 1 \begin{vmatrix} 1 & 1 \\ 2 & -3 \end{vmatrix} = 1(-2) - 1(3) + 1(-5) = -10$$

10. $N_x = \begin{vmatrix} 4 & 1 & 1 \\ 6 & 1 & -1 \\ -1 & -3 & 1 \end{vmatrix} = 4 \begin{vmatrix} 1 & -1 \\ -3 & 1 \end{vmatrix} - 1 \begin{vmatrix} 6 & -1 \\ -1 & 1 \end{vmatrix} + 1 \begin{vmatrix} 6 & 1 \\ -1 & -3 \end{vmatrix} = 4(-2) - 1(5) + 1(-17)$

$$= -30$$

$$x = \frac{N_x}{D} = \frac{-30}{-10} = 3$$

11. $N_y = \begin{vmatrix} 1 & 4 & 1 \\ 1 & 6 & -1 \\ 2 & -1 & 1 \end{vmatrix} = 1 \begin{vmatrix} 6 & -1 \\ -1 & 1 \end{vmatrix} - 4 \begin{vmatrix} 1 & -1 \\ 2 & 1 \end{vmatrix} + 1 \begin{vmatrix} 1 & 6 \\ 2 & -1 \end{vmatrix} = 1(5) - 4(3) + 1(-13) = -20$

$$y = \frac{N_y}{D} = \frac{-20}{-10} = 2$$

12. $N_z = \begin{vmatrix} 1 & 1 & 4 \\ 1 & 1 & 6 \\ 2 & -3 & -1 \end{vmatrix} = 1 \begin{vmatrix} 1 & 6 \\ -3 & -1 \end{vmatrix} - 1 \begin{vmatrix} 1 & 6 \\ 2 & -1 \end{vmatrix} + 4 \begin{vmatrix} 1 & 1 \\ 2 & -3 \end{vmatrix} = 1(17) - 1(-13) + 4(-5) = 10$

$$z = \frac{N_z}{D} = \frac{10}{-10} = -1$$

13. $\begin{bmatrix} 1 & 1 & | & 4 \\ 2 & -1 & | & 2 \end{bmatrix} \overset{-2R_1 + R_2 \Rightarrow R_2}{\Rightarrow} \begin{bmatrix} 1 & 1 & | & 4 \\ 0 & -3 & | & -6 \end{bmatrix} \overset{-\frac{1}{3}R_2 \Rightarrow R_2}{\Rightarrow} \begin{bmatrix} 1 & 1 & | & 4 \\ 0 & 1 & | & 2 \end{bmatrix}$

From R_2, $y = 2$.   From R_1:   Solution: $\boxed{(2, 2)}$

$$x + y = 4$$
$$x + 2 = 4$$
$$x = 2$$

14. $\begin{bmatrix} 1 & 1 & | & 2 \\ 1 & -1 & | & -4 \\ 2 & 1 & | & 1 \end{bmatrix} \overset{\begin{array}{c} -R_2 + R_1 \Rightarrow R_2 \\ -2R_1 + R_3 \Rightarrow R_3 \end{array}}{\Rightarrow} \begin{bmatrix} 1 & 1 & | & 2 \\ 0 & 2 & | & 6 \\ 0 & -1 & | & -3 \end{bmatrix} \overset{\begin{array}{c} \frac{1}{2}R_2 \Rightarrow R_2 \\ 2R_3 + R_2 \Rightarrow R_3 \end{array}}{\Rightarrow} \begin{bmatrix} 1 & 1 & | & 2 \\ 0 & 1 & | & 3 \\ 0 & 0 & | & 0 \end{bmatrix}$

From R_2, $y = 3$.   From R_1:   Solution: $\boxed{(-1, 3)}$

$$x + y = 2$$
$$x + 3 = 2$$
$$x = -1$$

15. $\begin{vmatrix} 2 & -3 \\ 4 & 5 \end{vmatrix} = 2(5) - (-3)(4) = 10 - (-12)$

$= 22$

16. $\begin{vmatrix} -3 & -4 \\ -2 & 3 \end{vmatrix} = -3(3) - (-4)(-2)$

$= -9 - 8 = -17$

17. $\begin{vmatrix} 1 & 2 & 0 \\ 2 & 0 & 3 \\ 1 & -2 & 2 \end{vmatrix} = 1 \begin{vmatrix} 0 & 3 \\ -2 & 2 \end{vmatrix} - 2 \begin{vmatrix} 2 & 3 \\ 1 & 2 \end{vmatrix} + 0 \begin{vmatrix} 2 & 0 \\ 1 & -2 \end{vmatrix} = 1(6) - 2(1) + 0(-4) = 4$

18. $\begin{vmatrix} 1 & -2 & 3 \\ 3 & 1 & -2 \\ 2 & -4 & 6 \end{vmatrix} = 1 \begin{vmatrix} 1 & -2 \\ -4 & 6 \end{vmatrix} - (-2) \begin{vmatrix} 3 & -2 \\ 2 & 6 \end{vmatrix} + 3 \begin{vmatrix} 3 & 1 \\ 2 & -4 \end{vmatrix} = 1(-2) + 2(22) + 3(-14) = 0$

19. $\begin{vmatrix} -6 & -1 \\ -6 & 1 \end{vmatrix}$

20. $\begin{vmatrix} 1 & -1 \\ 3 & 1 \end{vmatrix}$

21. $x = \dfrac{\begin{vmatrix} -6 & -1 \\ -6 & 1 \end{vmatrix}}{\begin{vmatrix} 1 & -1 \\ 3 & 1 \end{vmatrix}} = \dfrac{-6 - 6}{1 - (-3)} = \dfrac{-12}{4} = -3$

22. $y = \dfrac{\begin{vmatrix} 1 & -6 \\ 3 & -6 \end{vmatrix}}{\begin{vmatrix} 1 & -1 \\ 3 & 1 \end{vmatrix}} = \dfrac{-6 - (-18)}{1 - (-3)} = \dfrac{12}{4} = 3$

23. Let x and y represent the numbers. Substitute $x = 3y + 2$. Substitute; solve for x:

$\begin{cases} (1) & x = 3y + 2 \\ (2) & x + y = -18 \end{cases}$

$\begin{aligned} x + y &= -18 \\ 3y + 2 + y &= -18 \\ 4y &= -20 \\ y &= -5 \end{aligned}$

$\begin{aligned} x &= 3y + 2 \\ x &= 3(-5) + 2 \\ x &= -15 + 2 = -13 \end{aligned}$

The numbers are -5 and -13.
The product is $(-5)(-13) = 65$.

24. Let $x =$ number of adult tickets and $y =$ number of child tickets.

$\begin{array}{r} x + y = 7 \Rightarrow \times (-14) \\ \underline{21x + 14y = 119} \end{array}$

$\begin{array}{r} -14x - 14y = -98 \\ \underline{21x + 14y = 119} \\ 7x \quad\;\; = 21 \\ x \quad\;\; = 3 \end{array}$
He bought 3 adult tickets.

25. Let $x =$ principal at 3% and $y =$ principal at 4%.

$\begin{array}{l} x + y = 10000 \\ \underline{0.03x + 0.04y = 340} \Rightarrow \times (100) \end{array}$

$\begin{array}{l} x + y = 10000 \Rightarrow \times (-4) \\ 3x + 4y = 34000 \end{array}$

$\begin{array}{r} -4x - 4y = -40000 \\ \underline{3x + 4y = 34000} \\ -x \;\; = -6000 \\ x \;\; = 6000 \end{array}$

$\begin{aligned} x + y &= 10000 \\ 6000 + y &= 10000 \\ y &= 4000 \end{aligned}$

She invested $4000 at the 4% rate.

314

26. Let b = speed of boat in still water and c = speed of current.

	d	r	t	Equation $(d = r \cdot t)$
Downstream	8	$b+c$	2	$8 = (b+c) \cdot 2 \Rightarrow 2b + 2c = 8$
Upstream	8	$b-c$	4	$8 = (b-c) \cdot 4 \Rightarrow 4b - 4c = 8$

$2b + 2c = 8 \Rightarrow \times (2) \quad 4b + 4c = 16 \qquad 2b + 2c = 8$ The speed of the current

$\underline{4b - 4c = 8} \Rightarrow \qquad \underline{4b - 4c = 8} \qquad 2(3) + 2c = 8$ is 1 mph.

$\qquad\qquad\qquad\qquad \begin{array}{rl} 8b &= 24 \\ b &= 3 \end{array} \qquad \begin{array}{l} 6 + 2c = 8 \\ 2c = 2 \\ c = 1 \end{array}$

27. $x + y < 3 \qquad x - y < 1$

x	y
0	3
3	0

x	y
0	-1
1	0

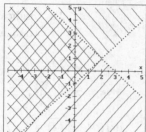

28. $2x + 3y \le 6 \qquad x \ge 2$

x	y
0	2
3	0

x	y
2	0
2	3

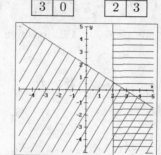

29. $\begin{cases} 2x - 3y \ge 6 \\ y \le -x + 1 \end{cases}$

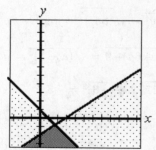

30.

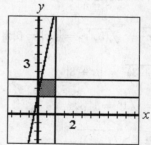

Vertex	$P = 3x - y$	Maximum?
$(0, 1)$	$= -1$	No
$(1, 1)$	$= 2$	YES
$(1, 2)$	$= 1$	No
$\left(\frac{1}{3}, 2\right)$	$= -1$	No

P has a maximum value of 2 at $(1, 1)$.

Exercises 9.1 (page 656)

1. $(5)^2 = 25, (-5)^2 = 25$

3. $(2x)^2 = 4x^2, (-2x)^2 = 4x^2$

5. $(2)^3 = 8$

7. $(3)^4 = 81, (-3)^4 = 81$

9. $\dfrac{x^2 + 7x + 10}{x^2 - 4} = \dfrac{(x+2)(x+5)}{(x+2)(x-2)} = \dfrac{x+5}{x-2}$

11. $\dfrac{x^2 - x - 6}{x^2 - 2x - 3} \cdot \dfrac{x^2 - 1}{x^2 + x - 2} = \dfrac{(x-3)(x+2)}{(x-3)(x+1)} \cdot \dfrac{(x+1)(x-1)}{(x+2)(x-1)} = 1$

13. $\dfrac{3}{m+1} + \dfrac{3m}{m-1} = \dfrac{3(m-1)}{(m+1)(m-1)} + \dfrac{3m(m+1)}{(m+1)(m-1)} = \dfrac{3m^2 + 6m - 3}{(m+1)(m-1)} = \dfrac{3(m^2 + 2m - 1)}{(m+1)(m-1)}$

15. $(5x^2)^2; 6^2 = 36$ **17.** positive   **19.** 5; left   **21.** radical; index; radicand

23. $|x|$ **25.** x **27.** even   **29.** $7y^2$

31. $a^2 + b^3$ **33.** $10^2 = 100; 10, -10$ **35.** $7^2 = 49; 7, -7$

37. $\sqrt{121} = \sqrt{11^2} = 11$ **39.** $-\sqrt{64} = -\sqrt{8^2} = -8$ **41.** $-\sqrt{\frac{25}{49}} = -\sqrt{\left(\frac{5}{7}\right)^2} = -\frac{5}{7}$

43. $\sqrt{-4}$: not a real number   **45.** $\sqrt{0.16} = \sqrt{(0.4)^2} = 0.4$

47. $\sqrt{25x^2} = \sqrt{(5x)^2} = |5x| = 5|x|$ **49.** $\sqrt{64y^4} = \sqrt{(8y^2)^2} = |8y^2| = 8y^2$

51. $\sqrt{(x+3)^2} = |x+3|$ **53.** $\sqrt{a^2 + 6a + 9} = \sqrt{(a+3)^2} = |a+3|$

55. $\sqrt[3]{1} = \sqrt[3]{1^3} = 1$ **57.** $\sqrt[3]{-27} = \sqrt[3]{(-3)^3} = -3$

59. $\sqrt[3]{-\frac{64}{27}} = \sqrt[3]{\left(-\frac{4}{3}\right)^3} = -\frac{4}{3}$ **61.** $\sqrt[3]{0.064} = \sqrt[3]{(0.4)^3} = 0.4$

63. $\sqrt[3]{125y^3} = \sqrt[3]{(5y)^3} = 5y$ **65.** $\sqrt[3]{-1000p^3q^3} = \sqrt[3]{(-10pq)^3} = -10pq$

67. $-\sqrt[5]{243} = -\sqrt[5]{3^5} = -3$ **69.** $-\sqrt[4]{16} = -\sqrt[4]{2^4} = -2$

71. $\sqrt[6]{64} = \sqrt[6]{2^6} = 2$ **73.** $\sqrt[4]{\frac{16}{625}} = \sqrt[4]{\left(\frac{2}{5}\right)^4} = \frac{2}{5}$

75. $\sqrt[4]{-256} \Rightarrow$ not a real number   **77.** $-\sqrt[5]{-\frac{1}{32}} = -\sqrt[5]{\left(-\frac{1}{2}\right)^5} = -\left(-\frac{1}{2}\right) = \frac{1}{2}$

79. $\sqrt[4]{81x^4} = \sqrt[4]{(3x)^4} = |3x| = 3|x|$

81. $\sqrt[3]{8a^3} = \sqrt[3]{(2a)^3} = 2a$

83. $\sqrt[4]{\frac{1}{16}x^4} = \sqrt[4]{\left(\frac{1}{2}x\right)^4} = \left|\frac{1}{2}x\right| = \frac{1}{2}|x|$

85. $\sqrt[4]{x^{12}} = \sqrt[4]{(x^3)^4} = |x^3|$

87. $\sqrt[5]{-x^5} = \sqrt[5]{(-x)^5} = -x$

89. $\sqrt[3]{-27a^6} = \sqrt[3]{(-3a^2)^3} = -3a^2$

91. $f(x) = \sqrt{x+4}$; Shift $y = \sqrt{x}$ left 4.

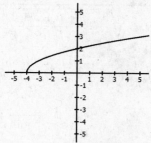

$D = [-4, \infty); R = [0, \infty)$

93. $f(x) = -\sqrt{x} - 3$; Reflect $y = \sqrt{x}$ about the x-axis and shift down 3.

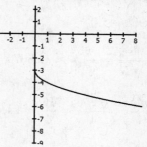

$D = [0, \infty); R = (-\infty, -3]$

95. $f(x) = \sqrt[3]{x} - 1$; Shift $y = \sqrt[3]{x}$ down 1.

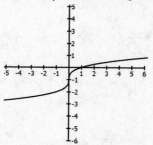

$D = (-\infty, \infty); R = (-\infty, \infty)$

97. $f(x) = -\sqrt[3]{x-1} - 2$; Reflect $y = \sqrt[3]{x}$ about the x-axis. Shift right 1 and down 2.

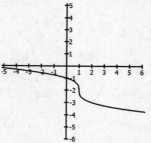

$D = (-\infty, \infty); R = (-\infty, \infty)$

99. $\sqrt{(-4)^2} = \sqrt{16} = 4$

101. $\sqrt{-36}$: not a real number

103. $\sqrt{(-5b)^2} = |-5b| = 5|b|$

105. $\sqrt{t^2 + 24t + 144} = \sqrt{(t+12)^2}$
$= |t+12|$

107. $\sqrt[3]{-\frac{1}{8}m^6n^3} = \sqrt[3]{\left(-\frac{1}{2}m^2n\right)^3} = -\frac{1}{2}m^2n$

109. $\sqrt[3]{0.008z^9} = \sqrt[3]{(0.2z^3)^3} = 0.2z^3$

111. $\sqrt[25]{(x+2)^{25}} = x+2$

113. $\sqrt[8]{0.00000001x^{16}y^8} = \sqrt[8]{(0.1x^2y)^8} = |0.1x^2y| = 0.1x^2|y|$

115. $\sqrt{12} \approx 3.4641$

117. $\sqrt{679.25} \approx 26.0624$

119. $t = 2\pi\sqrt{\dfrac{l}{32}} = 2\pi\sqrt{\dfrac{13}{32}} \approx 4.0048 \approx 4$ sec

121. mean $= \dfrac{2+5+5+6+7}{5} = \dfrac{25}{5} = 5$

Original term	Mean	Difference (term−mean)	Square of difference
2	5	−3	9
5	5	0	0
5	5	0	0
6	5	1	1
7	5	2	4

st. dev. $= \sqrt{\dfrac{9+0+0+1+4}{5}} \approx 1.67$

123. $s_{\bar{x}} = \dfrac{s}{\sqrt{N}} = \dfrac{65}{\sqrt{30}} \approx 11.8673$

125. $r = \sqrt{\dfrac{A}{\pi}} = \sqrt{\dfrac{9\pi}{\pi}} = \sqrt{9} = 3$ units

127. $t = \dfrac{\sqrt{s}}{4} = \dfrac{\sqrt{256}}{4} = \dfrac{16}{4} = 4$ seconds

129. $I = \sqrt{\dfrac{P}{18}} = \sqrt{\dfrac{980}{18}} \approx \sqrt{54.44} \approx 7.4$ amps

131. Answers may vary.

133. $\sqrt{x^2 - 4x + 4} = \sqrt{(x-2)^2} = |x-2|.$
$|x-2| = x - 2$ when $x - 2 \geq 0$, or $x \geq 2$.

Exercises 9.2 (page 663)

1. $\sqrt{625} = \sqrt{(25)^2} = 25$

3. $\sqrt{7^2 + 24^2} = \sqrt{49 + 576} = \sqrt{625} = 25$

5. $\sqrt{5^2 - 3^2} = \sqrt{25 - 9} = \sqrt{16} = 4$

7. $(2x + 5)(3x - 4) = 6x^2 + 7x - 20$

9. $(4a - 3b)(5a - 2b) = 20a^2 - 23ab + 6b^2$

11. hypotenuse

13. $a^2 + b^2 = c^2$

15. positive

17. $c^2 = a^2 + b^2$
$c^2 = 6^2 + 8^2$
$c^2 = 36 + 64$
$c^2 = 100$
$c = \sqrt{100}$
$c = 10$ ft

19. $c^2 = a^2 + b^2$
$82^2 = a^2 + 18^2$
$6724 = a^2 + 324$
$6400 = a^2$
$\sqrt{6400} = a$
80 m $= a$

21.
$$c^2 = a^2 + b^2$$
$$50^2 = 14^2 + b^2$$
$$2500 = 196 + b^2$$
$$2304 = b^2$$
$$\sqrt{2304} = b$$
$$48 \text{ in.} = b$$

23.
$$c^2 = a^2 + b^2$$
$$c^2 = \left(\sqrt{8}\right)^2 + \left(\sqrt{8}\right)^2$$
$$c^2 = 8 + 8$$
$$c^2 = 16$$
$$c = \sqrt{16}$$
$$c = 4 \text{ mi}$$

25. $d = \sqrt{(x_2 - x_1)^2 + (y_2 - y_1)^2} = \sqrt{(3 - 0)^2 + (-4 - 0)^2} = \sqrt{3^2 + (-4)^2} = \sqrt{9 + 16}$
$$= \sqrt{25} = 5$$

27. $d = \sqrt{(x_2 - x_1)^2 + (y_2 - y_1)^2} = \sqrt{(-2 - 10)^2 + (3 - 8)^2} = \sqrt{(-12)^2 + (-5)^2} = \sqrt{144 + 25}$
$$= \sqrt{169} = 13$$

29. $d = \sqrt{(x_2 - x_1)^2 + (y_2 - y_1)^2} = \sqrt{[-1 - (-7)]^2 + (4 - 12)^2} = \sqrt{6^2 + (-8)^2} = \sqrt{36 + 64}$
$$= \sqrt{100} = 10$$

31. $d = \sqrt{(x_2 - x_1)^2 + (y_2 - y_1)^2} = \sqrt{(-3 - 12)^2 + [2 - (-6)]^2} = \sqrt{(-15)^2 + 8^2} = \sqrt{225 + 64}$
$$= \sqrt{289} = 17$$

33. $d = \sqrt{(x_2 - x_1)^2 + (y_2 - y_1)^2} = \sqrt{[-5 - (-3)]^2 + (-5 - 5)^2} = \sqrt{(-2)^2 + (-10)^2}$
$$= \sqrt{4 + 100} = \sqrt{104} \approx 10.2$$

35. $d = \sqrt{(x_2 - x_1)^2 + (y_2 - y_1)^2} = \sqrt{[4 - (-9)]^2 + (7 - 3)^2} = \sqrt{13^2 + 4^2} = \sqrt{169 + 16}$
$$= \sqrt{185} \approx 13.6$$

37. Let x = distance to 2nd
$$x^2 = 90^2 + 90^2$$
$$x^2 = 8100 + 8100$$
$$x^2 = 16200$$
$$x \approx 127 \text{ ft}$$

39. Refer to the diagram provided. The 3rd baseman is at B, so $\overline{BC}$
has a length of 10 ft. Let $\overline{AC}$ and $\overline{AB}$ both have a length x.
$$x^2 + x^2 = 10^2$$
$$2x^2 = 100$$
$$x^2 = 50$$
$$x = \sqrt{50} \approx 7.1 \text{ ft}$$
continued on next page...

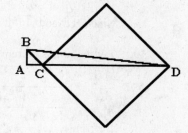

39. continued

From **#37**, the length of $\overline{CD} \approx 127.3$ ft, so $\overline{AD}$ has a length of about $127.3 + 7.1 = 134.4$ ft. Let $y =$ the length of $\overline{BD}$.

$$y^2 = (7.1)^2 + (134.4)^2$$
$$y^2 = 18113.77$$
$$y = \sqrt{18113.77} \approx 135 \text{ ft}$$

41. Let $x =$ the length of the diagonal.

$$7^2 + 7^2 = x^2$$
$$49 + 49 = x^2$$
$$98 = x^2$$
$$\sqrt{98} = x$$
$$9.9 \text{ cm} \approx x$$

43. Let $x =$ half the length of the stretched wire.

$$x^2 = 20^2 + 1^2$$
$$x^2 = 401$$
$$x = 20.025 \text{ ft}$$

The stretched wire has a length of 40.05 ft. It has been stretched by 0.05 ft.

45.
$$d^2 = 5^2 + 12^2$$
$$d^2 = 25 + 144$$
$$d^2 = 169$$
$$d = 13 \text{ ft}$$

47.
$$37^2 = 9^2 + h^2$$
$$1369 = 81 + h^2$$
$$1288 = h^2$$
$$35.9 = h$$

The ladder will reach.

49. Let $x =$ direct distance from A to D.

$$x^2 = 52^2 + (105 + 60)^2$$
$$x^2 = 52^2 + 165^2$$
$$x^2 = 29929$$
$$x = 173 \text{ yd}$$

51. $A = 6\sqrt[3]{V^2} = 6\sqrt[3]{8^2} = 6\sqrt[3]{64} = 6(4)$
$$= 24 \text{ cm}^2$$

53. $d = \sqrt{a^2 + b^2 + c^2} = \sqrt{12^2 + 24^2 + 17^2} = \sqrt{1009} = 31.76 \text{ in.} \Rightarrow$ The racket will not fit.

55. $d = \sqrt{a^2 + b^2 + c^2} = \sqrt{21^2 + 21^2 + 3^2} = \sqrt{891} = 29.8 \text{ in.} \Rightarrow$ The femur will not fit.

57. Let the points be represented by $A(5, 1)$, $B(7, 0)$ and $C(3, 0)$. Find the length of $\overline{AB}$ and $\overline{AC}$:

$\overline{AB}: \sqrt{(5-7)^2 + (1-0)^2} = \sqrt{(-2)^2 + 1^2} = \sqrt{5}$
$\overline{AC}: \sqrt{(5-3)^2 + (1-0)^2} = \sqrt{(2)^2 + 1^2} = \sqrt{5}$

Since $\overline{AB}$ and $\overline{AC}$ have the same length, $(5, 1)$ is equidistant from $(7, 0)$ and $(3, 0)$.

59. Let the points be represented by $A(-2, 4)$, $B(2, 8)$ and $C(6, 4)$. Find the length of each side:

$\overline{AB}: \sqrt{(-2-2)^2 + (4-8)^2} = \sqrt{(-4)^2 + (-4)^2} = \sqrt{32}$
$\overline{AC}: \sqrt{(-2-6)^2 + (4-4)^2} = \sqrt{(-8)^2 + 0^2} = \sqrt{64}$
$\overline{BC}: \sqrt{(2-6)^2 + (8-4)^2} = \sqrt{(-4)^2 + 4^2} = \sqrt{32}$

Since $\overline{AB}$ and $\overline{BC}$ have the same length, the triangle is isosceles.

61. Answers may vary.

63. $I = \dfrac{703w}{h^2} = \dfrac{703(104)}{(54.1)^2} = \dfrac{73112}{2926.81} \approx 25$

65. $v^2 = 64d$
$v^2 = 64(4)$
$v^2 = 256$
$v = \sqrt{256} = 16$ ft/sec

Exercises 9.3 (page 672)

1. $-7^2 = -1 \cdot 7 \cdot 7 = -49$ **3.** $5^3 = 5 \cdot 5 \cdot 5 = 125$ **5.** $4^{-2} = \dfrac{1}{4^2} = \dfrac{1}{16}$

7. $\sqrt{4^3} = \sqrt{64} = 8$ **9.** $\sqrt[3]{8^2} = \sqrt[3]{64} = 4$ **11.** $8x - 7 \leq 1$
$8x \leq 8$
$x \leq 1$

13. $\dfrac{4}{5}(r - 3) > \dfrac{2}{3}(r + 2)$
$15 \cdot \dfrac{4}{5}(r - 3) > 15 \cdot \dfrac{2}{3}(r + 2)$
$12(r - 3) > 10(r + 2)$
$2r > 56$
$r > 28$

15. Let x = pints of water added (0% alcohol).

$\boxed{\begin{array}{c}\text{Alcohol}\\\text{at start}\end{array}} + \boxed{\begin{array}{c}\text{Alcohol}\\\text{added}\end{array}} = \boxed{\begin{array}{c}\text{Alcohol}\\\text{at end}\end{array}}$

$0.20(5) + 0(x) = 0.15(5 + x)$
$1 + 0 = 0.75 + 0.15x$
$0.25 = 0.15x$
$x = \dfrac{0.25}{0.15} = \dfrac{5}{3}$

$1\frac{2}{3}$ pints of water should be added.

17. $a \cdot a \cdot a \cdot a$ **19.** a^{mn} **21.** $\dfrac{a^n}{b^n}$ **23.** $\dfrac{1}{a^n}; 0$

25. $\left(\dfrac{b}{a}\right)^n$ **27.** $|x|$ **29.** $5^{1/4} = \sqrt[4]{5}$ **31.** $8^{1/5} = \sqrt[5]{8}$

33. $(13a)^{1/7} = \sqrt[7]{13a}$ **35.** $\left(\frac{1}{2}x^3y\right)^{1/4} = \sqrt[4]{\frac{1}{2}x^3y}$ **37.** $(6a^3b)^{1/4} = \sqrt[4]{6a^3b}$

39. $(x^2 + y^2)^{1/2} = \sqrt{x^2 + y^2}$ **41.** $49^{1/2} = \sqrt{49} = 7$ **43.** $27^{1/3} = \sqrt[3]{27} = 3$

45. $\left(\frac{1}{9}\right)^{1/2} = \sqrt{\frac{1}{9}} = \frac{1}{3}$ **47.** $\left(\frac{1}{8}\right)^{1/3} = \sqrt[3]{\frac{1}{8}} = \frac{1}{2}$ **49.** $-81^{1/4} = -\sqrt[4]{81} = -3$

51. $(-25)^{1/2} = \sqrt{-25}$; not a real number

53. $\sqrt{7} = 7^{1/2}$ **55.** $\sqrt[4]{3a} = (3a)^{1/4}$ **57.** $5\sqrt[7]{b} = 5b^{1/7}$

321

59. $\sqrt[6]{\frac{1}{7}abc} = \left(\frac{1}{7}abc\right)^{1/6}$ **61.** $\sqrt[5]{\frac{1}{2}mn} = \left(\frac{1}{2}mn\right)^{1/5}$ **63.** $\sqrt[3]{x^2 + y^2} = (x^2 + y^2)^{1/3}$

65. $(25y^2)^{1/2} = \left[(5y)^2\right]^{1/2} = |5y| = 5|y|$ **67.** $(243x^5)^{1/5} = \left[(3x)^5\right]^{1/5} = 3x$

69. $\left[(x+1)^4\right]^{1/4} = |x+1|$ **71.** $(-81x^{12})^{1/4} \Rightarrow$ not a real number

73. $25^{3/2} = \left(25^{1/2}\right)^3 = 5^3 = 125$ **75.** $81^{3/4} = \left(81^{1/4}\right)^3 = 3^3 = 27$

77. $(-27x^3)^{4/3} = \left[(-27x^3)^{1/3}\right]^4 = [-3x]^4$ **79.** $\left(\frac{1}{8}\right)^{2/3} = \left[\left(\frac{1}{8}\right)^{1/3}\right]^2 = \left(\frac{1}{2}\right)^2 = \frac{1}{4}$
$\qquad\qquad = 81x^4$

81. $4^{-1/2} = \dfrac{1}{4^{1/2}} = \dfrac{1}{2}$ **83.** $4^{-3/2} = \dfrac{1}{4^{3/2}} = \dfrac{1}{\left(4^{1/2}\right)^3} = \dfrac{1}{2^3} = \dfrac{1}{8}$

85. $(16x^2)^{-3/2} = \dfrac{1}{(16x^2)^{3/2}} = \dfrac{1}{\left[(16x^2)^{1/2}\right]^3} = \dfrac{1}{(4x)^3} = \dfrac{1}{64x^3}$

87. $(-27y^3)^{-2/3} = \dfrac{1}{(-27y^3)^{2/3}} = \dfrac{1}{\left[(-27y^3)^{1/3}\right]^2} = \dfrac{1}{(-3y)^2} = \dfrac{1}{9y^2}$

89. $\left(\dfrac{1}{4}\right)^{-3/2} = \left(\dfrac{4}{1}\right)^{3/2} = 4^{3/2} = \left(4^{1/2}\right)^3 = 2^3 = 8$

91. $\left(\dfrac{27}{8}\right)^{-4/3} = \left(\dfrac{8}{27}\right)^{4/3} = \left[\left(\dfrac{8}{27}\right)^{1/3}\right]^4 = \left(\dfrac{2}{3}\right)^4 = \dfrac{16}{81}$

93. $5^{4/9}5^{4/9} = 5^{4/9+4/9} = 5^{8/9}$ **95.** $\left(4^{1/5}\right)^3 = 4^{(1/5)\cdot 3} = 4^{3/5}$ **97.** $6^{-2/3}6^{-4/3} = 6^{-6/3} = 6^{-2}$
$\qquad\qquad\qquad\qquad\qquad\qquad\qquad\qquad\qquad\qquad\qquad\qquad\qquad\qquad\qquad = \frac{1}{36}$

99. $\dfrac{9^{4/5}}{9^{3/5}} = 9^{4/5 - 3/5} = 9^{1/5}$ **101.** $\dfrac{7^{1/2}}{7^0} = 7^{1/2 - 0} = 7^{1/2}$ **103.** $\dfrac{2^{5/6}2^{1/3}}{2^{1/2}} = \dfrac{2^{7/6}}{2^{1/2}} = 2^{4/6}$
$\qquad\qquad\qquad\qquad\qquad\qquad\qquad\qquad\qquad\qquad\qquad\qquad\qquad\qquad\qquad\qquad = 2^{2/3}$

105. $\left(x^{1/4} \cdot x^{3/2}\right)^8 = x^{8/4} \cdot x^{24/2} = x^{14}$ **107.** $\left(a^{2/3}\right)^{1/3} = a^{(2/3)(1/3)} = a^{2/9}$

109. $y^{1/3}\left(y^{2/3} + y^{5/3}\right) = y^{3/3} + y^{6/3} = y + y^2$

111. $x^{3/5}\left(x^{7/5} - x^{2/5} + 1\right) = x^{10/5} - x^{5/5} + x^{3/5} = x^2 - x + x^{3/5}$

113. $\left(x^{1/2} + y^{1/2}\right)\left(x^{1/2} - y^{1/2}\right) = x^{2/2} - x^{1/2}y^{1/2} + x^{1/2}y^{1/2} - y^{2/2} = x - y$

115. $\left(x^{2/3} + y^{2/3}\right)^2 = \left(x^{2/3} + y^{2/3}\right)\left(x^{2/3} + y^{2/3}\right) = x^{4/3} + x^{2/3}y^{2/3} + x^{2/3}y^{2/3} + y^{4/3}$
$$= x^{4/3} + 2x^{2/3}y^{2/3} + y^{4/3}$$

117. $\sqrt[6]{p^3} = (p^3)^{1/6} = p^{3/6} = p^{1/2} = \sqrt{p}$

119. $\sqrt[4]{25b^2} = (5^2 b^2)^{1/4} = 5^{1/2}b^{1/2} = \sqrt{5b}$

121. $16^{1/4} = \sqrt[4]{16} = 2$

123. $32^{1/5} = \sqrt[5]{32} = 2$

125. $0^{1/3} = \sqrt[3]{0} = 0$

127. $(-27)^{1/3} = \sqrt[3]{-27} = -3$

129. $\left(25x^4\right)^{3/2} = \left[\left(25x^4\right)^{1/2}\right]^3 = \left(5x^2\right)^3$
$$= 125x^6$$

131. $\left(\dfrac{8x^3}{27}\right)^{2/3} = \left[\left(\dfrac{8x^3}{27}\right)^{1/3}\right]^2 = \left(\dfrac{2x}{3}\right)^2$
$$= \dfrac{4x^2}{9}$$

133. $\left(-32p^5\right)^{-2/5} = \dfrac{1}{\left(-32p^5\right)^{2/5}} = \dfrac{1}{\left[\left(-32p^5\right)^{1/5}\right]^2} = \dfrac{1}{\left(-2p\right)^2} = \dfrac{1}{4p^2}$

135. $\left(-\dfrac{8x^3}{27}\right)^{-1/3} = \left(-\dfrac{27}{8x^3}\right)^{1/3} = -\dfrac{3}{2x}$

137. $\left(a^{1/2}b^{1/3}\right)^{3/2} = a^{3/4}b^{1/2}$

139. $\left(mn^{-2/3}\right)^{-3/5} = m^{-3/5}n^{2/5} = \dfrac{n^{2/5}}{m^{3/5}}$

141. $\dfrac{(4x^3y)^{1/2}}{(9xy)^{1/2}} = \dfrac{2x^{3/2}y^{1/2}}{3x^{1/2}y^{1/2}} = \dfrac{2x}{3}$

143. $\left(27x^{-3}\right)^{-1/3} = (27)^{-1/3}x = \dfrac{1}{3}x$

145. $x^{4/3}\left(x^{2/3} + 3x^{5/3} - 4\right) = x^{6/3} + 3x^{9/3} - 4x^{4/3} = x^2 + 3x^3 - 4x^{4/3}$

147. $\left(x^{-1/2} - x^{1/2}\right)^2 = \left(x^{-1/2} - x^{1/2}\right)\left(x^{-1/2} - x^{1/2}\right) = x^{-2/2} - x^{1/2}x^{-1/2} - x^{1/2}x^{-1/2} + x^{2/2}$
$$= x^{-1} - 1 - 1 + x = \dfrac{1}{x} - 2 + x$$

Problems 149-155 are to be solved using a calculator. The keystrokes needed to solve each problem using a TI-83 graphing calculator appear in each solution. There may be other solutions. Keystrokes for other calculators may be slightly different.

149. $\boxed{1}\ \boxed{5}\ \boxed{\wedge}\ \boxed{(}\ \boxed{1}\ \boxed{\div}\ \boxed{3}\ \boxed{)}\ \boxed{\text{ENTER}}\ \{2.4662...\} \Rightarrow 2.47$

151. $\boxed{1}\ \boxed{.}\ \boxed{0}\ \boxed{4}\ \boxed{5}\ \boxed{\wedge}\ \boxed{(}\ \boxed{1}\ \boxed{\div}\ \boxed{5}\ \boxed{)}\ \boxed{\text{ENTER}}\ \{1.0088...\} \Rightarrow 1.01$

153. $\boxed{1}\ \boxed{7}\ \boxed{\wedge}\ \boxed{(}\ \boxed{(-)}\ \boxed{1}\ \boxed{\div}\ \boxed{2}\ \boxed{)}\ \boxed{\text{ENTER}}\ \{0.2425...\} \Rightarrow 0.24$

155. $\boxed{(}\ \boxed{(-)}\ \boxed{.}\ \boxed{2}\ \boxed{5}\ \boxed{)}\ \boxed{\wedge}\ \boxed{(}\ \boxed{(-)}\ \boxed{1}\ \boxed{\div}\ \boxed{5}\ \boxed{)}\ \boxed{\text{ENTER}}\ \{-1.3195...\} \Rightarrow -1.32$

157. Answers may vary.

159. $16^{2/4} = 2^2 = 4;\ 16^{1/2} = 4 \Rightarrow$ Yes

323

Exercises 9.4 (page 681)

1. $28 = 4 \cdot 7$ **3.** $18 = 9 \cdot 2$ **5.** $16 = 8 \cdot 2$ **7.** $54 = 27 \cdot 2$

9. $12x^2 + 12x^2 = 24x^2$

11. $7x^3y(-3x^4y^{-5}) = -21x^7y^{-4} = -\dfrac{21x^7}{y^4}$

13. $(3t + 2)^2 = (3t + 2)(3t + 2)$
$\qquad = 9t^2 + 12t + 4$

15.
$$
\begin{array}{r}
3p + \quad 4 + \frac{-5}{2p-5} \\
2p - 5 \,\overline{)\,6p^2 - 7p - 25} \\
\underline{6p^2 - 15p} \\
8p - 25 \\
\underline{8p - 20} \\
-5
\end{array}
$$

17. $\sqrt[n]{a}\,\sqrt[n]{b}$

19. like

21. $\sqrt{6}\,\sqrt{6} = \sqrt{36} = 6$

23. $\sqrt{t}\,\sqrt{t} = \sqrt{t^2} = t$

25. $\sqrt[3]{5x^2}\,\sqrt[3]{25x} = \sqrt[3]{125x^3} = 5x$

27. $\dfrac{\sqrt{500}}{\sqrt{5}} = \sqrt{\dfrac{500}{5}} = \sqrt{100} = 10$

29. $\dfrac{\sqrt{98x^3}}{\sqrt{2x}} = \sqrt{\dfrac{98x^3}{2x}} = \sqrt{49x^2} = 7x$

31. $\dfrac{\sqrt{180ab^4}}{\sqrt{5ab^2}} = \sqrt{\dfrac{180ab^4}{5ab^2}} = \sqrt{36b^2} = 6b$

33. $\dfrac{\sqrt[3]{48}}{\sqrt[3]{6}} = \sqrt[3]{\dfrac{48}{6}} = \sqrt[3]{8} = 2$

35. $\dfrac{\sqrt[3]{189a^4}}{\sqrt[3]{7a}} = \sqrt[3]{\dfrac{189a^4}{7a}} = \sqrt[3]{27a^3} = 3a$

37. $\sqrt{20} = \sqrt{4 \cdot 5} = \sqrt{4}\,\sqrt{5} = 2\sqrt{5}$

39. $-\sqrt{200} = -\sqrt{100 \cdot 2} = -\sqrt{100}\,\sqrt{2}$
$\qquad = -10\sqrt{2}$

41. $\sqrt[3]{80} = \sqrt[3]{8 \cdot 10} = \sqrt[3]{8}\,\sqrt[3]{10} = 2\sqrt[3]{10}$

43. $\sqrt[3]{-81} = \sqrt[3]{-27 \cdot 3} = \sqrt[3]{-27}\,\sqrt[3]{3} = -3\sqrt[3]{3}$

45. $\sqrt[4]{32} = \sqrt[4]{16 \cdot 2} = \sqrt[4]{16}\,\sqrt[4]{2} = 2\sqrt[4]{2}$

47. $\sqrt[5]{96} = \sqrt[5]{32 \cdot 3} = \sqrt[5]{32}\,\sqrt[5]{3} = 2\sqrt[5]{3}$

49. $\sqrt{\dfrac{5}{49a^2}} = \dfrac{\sqrt{5}}{\sqrt{49a^2}} = \dfrac{\sqrt{5}}{7a}$

51. $\sqrt[3]{\dfrac{7a^3}{64}} = \dfrac{\sqrt[3]{7a^3}}{\sqrt[3]{64}} = \dfrac{\sqrt[3]{a^3}\,\sqrt[3]{7}}{4} = \dfrac{a\sqrt[3]{7}}{4}$

53. $\sqrt[4]{\dfrac{3p^4}{10{,}000q^4}} = \dfrac{\sqrt[4]{3p^4}}{\sqrt[4]{10{,}000q^4}} = \dfrac{\sqrt[4]{p^4}\,\sqrt[4]{3}}{10q}$
$\qquad = \dfrac{p\sqrt[4]{3}}{10q}$

55. $\sqrt[5]{\dfrac{3m^{15}}{32n^{10}}} = \dfrac{\sqrt[5]{3m^{15}}}{\sqrt[5]{32n^{10}}} = \dfrac{\sqrt[5]{m^{15}}\,\sqrt[5]{3}}{2n^2}$
$\qquad = \dfrac{m^3\sqrt[5]{3}}{2n^2}$

57. $\sqrt{63y^3} = \sqrt{9y^2 \cdot 7y} = \sqrt{9y^2}\sqrt{7y}$
$= 3y\sqrt{7y}$

59. $\sqrt{48a^5} = \sqrt{16a^4 \cdot 3a} = \sqrt{16a^4}\sqrt{3a}$
$= 4a^2\sqrt{3a}$

61. $-\sqrt{112a^3} = -\sqrt{16a^2 \cdot 7a} = -\sqrt{16a^2}\sqrt{7a} = -4a\sqrt{7a}$

63. $\sqrt{175a^2b^3} = \sqrt{25a^2b^2 \cdot 7b} = \sqrt{25a^2b^2}\sqrt{7b} = 5ab\sqrt{7b}$

65. $\dfrac{-\sqrt{150ab^4}}{\sqrt{2a}} = -\sqrt{\dfrac{150ab^4}{2a}} = -\sqrt{75b^4} = -\sqrt{25b^4}\sqrt{3} = -5b^2\sqrt{3}$

67. $\dfrac{\sqrt[3]{-108x^6y}}{\sqrt[3]{2y}} = \sqrt[3]{\dfrac{-108x^6y}{2y}} = \sqrt[3]{-54x^6} = \sqrt[3]{-27x^6}\sqrt[3]{2} = -3x^2\sqrt[3]{2}$

69. $\sqrt[3]{16x^{12}y^3} = \sqrt[3]{8x^{12}y^3 \cdot 2} = \sqrt[3]{8x^{12}y^3}\sqrt[3]{2}$
$= 2x^4y\sqrt[3]{2}$

71. $\sqrt{\dfrac{z^2}{16x^2}} = \dfrac{\sqrt{z^2}}{\sqrt{16x^2}} = \dfrac{z}{4x}$

73. $\sqrt{3} + \sqrt{27} = \sqrt{3} + \sqrt{9}\sqrt{3}$
$= \sqrt{3} + 3\sqrt{3} = 4\sqrt{3}$

75. $\sqrt{2} - \sqrt{8} = \sqrt{2} - \sqrt{4}\sqrt{2}$
$= \sqrt{2} - 2\sqrt{2} = -\sqrt{2}$

77. $\sqrt{98} - \sqrt{50} = \sqrt{49}\sqrt{2} - \sqrt{25}\sqrt{2} = 7\sqrt{2} - 5\sqrt{2} = 2\sqrt{2}$

79. $3\sqrt{125} + 4\sqrt{45} = 3\sqrt{25}\sqrt{5} + 4\sqrt{9}\sqrt{5} = 3(5)\sqrt{5} + 4(3)\sqrt{5} = 15\sqrt{5} + 12\sqrt{5} = 27\sqrt{5}$

81. $3\sqrt{54} + 2\sqrt{28} - 4\sqrt{96} = 3\sqrt{9}\sqrt{6} + 2\sqrt{4}\sqrt{7} - 4\sqrt{16}\sqrt{6} = 3(3)\sqrt{6} + 2(2)\sqrt{7} - 4(4)\sqrt{6}$
$= 9\sqrt{6} + 4\sqrt{7} - 16\sqrt{6}$
$= 4\sqrt{7} - 7\sqrt{6}$

83. $\sqrt[3]{24} + \sqrt[3]{81} = \sqrt[3]{8}\sqrt[3]{3} + \sqrt[3]{27}\sqrt[3]{3} = 2\sqrt[3]{3} + 3\sqrt[3]{3} = 5\sqrt[3]{3}$

85. $\sqrt[3]{32} - \sqrt[3]{108} = \sqrt[3]{8}\sqrt[3]{4} - \sqrt[3]{27}\sqrt[3]{4} = 2\sqrt[3]{4} - 3\sqrt[3]{4} = -\sqrt[3]{4}$

87. $2\sqrt[3]{125} - 5\sqrt[3]{64} = 2(5) - 5(4) = 10 - 20 = -10$

89. $2\sqrt[3]{16} - \sqrt[3]{54} - 3\sqrt[3]{128} = 2\sqrt[3]{8}\sqrt[3]{2} - \sqrt[3]{27}\sqrt[3]{2} - 3\sqrt[3]{64}\sqrt[3]{2} = 2(2)\sqrt[3]{2} - 3\sqrt[3]{2} - 3(4)\sqrt[3]{2}$
$= 4\sqrt[3]{2} - 3\sqrt[3]{2} - 12\sqrt[3]{2} = -11\sqrt[3]{2}$

91. $\sqrt[3]{81x^5} - \sqrt[3]{24x^5} = \sqrt[3]{27x^3}\sqrt[3]{3x^2} - \sqrt[3]{8x^3}\sqrt{3x^2} = 3x\sqrt[3]{3x^2} - 2x\sqrt[3]{3x^2} = x\sqrt[3]{3x^2}$

93. $\sqrt{25yz^2} + \sqrt{9yz^2} = \sqrt{25z^2}\sqrt{y} + \sqrt{9z^2}\sqrt{y} = 5z\sqrt{y} + 3z\sqrt{y} = 8z\sqrt{y}$

95. $\sqrt{y^5} - \sqrt{9y^5} - \sqrt{25y^5} = \sqrt{y^4}\sqrt{y} - \sqrt{9y^4}\sqrt{y} - \sqrt{25y^4}\sqrt{y} = y^2\sqrt{y} - 3y^2\sqrt{y} - 5y^2\sqrt{y}$
$= -7y^2\sqrt{y}$

97. $3\sqrt[3]{-2x^4} - \sqrt[3]{54x^4} = 3\sqrt[3]{-x^3}\sqrt[3]{2x} - \sqrt[3]{27x^3}\sqrt[3]{2x} = -3x\sqrt[3]{2x} - 3x\sqrt[3]{2x} = -6x\sqrt[3]{2x}$

99. $a = b = \frac{2}{3}; c = \frac{2}{3}\sqrt{2}$

101. $b = a = 5\sqrt{2}; c = 5\sqrt{2}\sqrt{2} = 5(2) = 10$

103. $a = b = \frac{5\sqrt{2}}{\sqrt{2}} = 5$

105. $a = b = \frac{7\sqrt{2}}{\sqrt{2}} = 7$

107. $b = 5\sqrt{3}; c = 2(5) = 10$

109. $a = \frac{9\sqrt{3}}{\sqrt{3}} = 9; c = 2(9) = 18$

111. $a = \frac{24}{2} = 12; b = 12\sqrt{3}$

113. $a = \frac{15}{2}; b = \frac{15}{2}\sqrt{3}$

115. $4\sqrt{2x} + 6\sqrt{2x} = 10\sqrt{2x}$

117. $\sqrt[4]{32x^{12}y^4} = \sqrt[4]{16x^{12}y^4 \cdot 2}$
$= \sqrt[4]{16x^{12}y^4}\sqrt[4]{2} = 2x^3y\sqrt[4]{2}$

119. $\sqrt[4]{\frac{5x}{16z^4}} = \frac{\sqrt[4]{5x}}{\sqrt[4]{16z^4}} = \frac{\sqrt[4]{5x}}{2z}$

121. $\sqrt{98} - \sqrt{50} - \sqrt{72} = \sqrt{49}\sqrt{2} - \sqrt{25}\sqrt{2} - \sqrt{36}\sqrt{2} = 7\sqrt{2} - 5\sqrt{2} - 6\sqrt{2} = -4\sqrt{2}$

123. $3\sqrt[3]{27} + 12\sqrt[3]{216} = 3(3) + 12(6) = 9 + 72 = 81$

125. $23\sqrt[4]{768} + \sqrt[4]{48} = 23\sqrt[4]{256}\sqrt[4]{3} + \sqrt[4]{16}\sqrt[4]{3} = 23(4)\sqrt[4]{3} + 2\sqrt[4]{3} = 92\sqrt[4]{3} + 2\sqrt[4]{3} = 94\sqrt[4]{3}$

127. $4\sqrt[4]{243} - \sqrt[4]{48} = 4\sqrt[4]{81}\sqrt[4]{3} - \sqrt[4]{16}\sqrt[4]{3} = 4(3)\sqrt[4]{3} - 2\sqrt[4]{3} = 12\sqrt[4]{3} - 2\sqrt[4]{3} = 10\sqrt[4]{3}$

129. $6\sqrt[3]{5y} + 3\sqrt[3]{5y} = 9\sqrt[3]{5y}$

131. $10\sqrt[6]{12xyz} - \sqrt[6]{12xyz} = 9\sqrt[6]{12xyz}$

133. $\sqrt[5]{x^6y^2} + \sqrt[5]{32x^6y^2} + \sqrt[5]{x^6y^2} = \sqrt[5]{x^5}\sqrt[5]{xy^2} + \sqrt[5]{32x^5}\sqrt[5]{xy^2} + \sqrt[5]{x^5}\sqrt[5]{xy^2}$
$= x\sqrt[5]{xy^2} + 2x\sqrt[5]{xy^2} + x\sqrt[5]{xy^2} = 4x\sqrt[5]{xy^2}$

135. $\sqrt{x^2 + 2x + 1} + \sqrt{x^2 + 2x + 1} = \sqrt{(x+1)^2} + \sqrt{(x+1)^2} = x + 1 + x + 1 = 2x + 2$

137. $x = 2.00;$
$h = 2\sqrt{2} \approx 2.83$

139. $h = 2(5) = 10.00;$
$x = 5\sqrt{3} \approx 8.66$

141. $x = \frac{9.37}{2} \approx 4.69;$
$y \approx 4.69\sqrt{3} \approx 8.11$

143. $x = y = \frac{17.12}{\sqrt{2}} \approx 12.11$

145. $x = 5\sqrt{3} \approx 8.66$ mm

$h = 2x = 2\left(5\sqrt{3}\right) = 10\sqrt{3} \approx 17.32$ mm

147. **Answers may vary.**

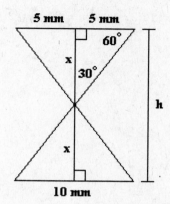

149. $\sqrt{3} + \sqrt{3^2} + \sqrt{3^3} + \sqrt{3^4} + \sqrt{3^5} = \sqrt{3} + 3 + \sqrt{3^2}\sqrt{3} + 3^2 + \sqrt{3^4}\sqrt{3}$

$\qquad\qquad\qquad\qquad\qquad\quad = \sqrt{3} + 3 + 3\sqrt{3} + 9 + 9\sqrt{3} = 12 + 13\sqrt{3}$

Exercises 9.5 (page 691)

1. $\sqrt{5}\sqrt{5} = \sqrt{25} = 5$

3. $\sqrt{xy}\sqrt{xy} = \sqrt{x^2y^2} = xy$

5. $\sqrt[3]{5}\sqrt[3]{5}\sqrt[3]{5} = \sqrt[3]{125} = 5$

7. $\sqrt[3]{a^2b}\sqrt[3]{ab^2} = \sqrt[3]{a^3b^3} = ab$

9. $\dfrac{2}{3-a} = 1$

$\quad 2 = 3 - a$

$\quad a = 1$

11. $\dfrac{8}{b-2} + \dfrac{3}{2-b} = -\dfrac{1}{b}$

$\quad \dfrac{8}{b-2} + \dfrac{-3}{b-2} = -\dfrac{1}{b}$

$\quad\quad \dfrac{5}{b-2} = \dfrac{-1}{b}$

$\quad\quad 5b = -b + 2$

$\quad\quad 6b = 2$

$\quad\quad b = \dfrac{2}{6} = \dfrac{1}{3}$

13. $2; \sqrt{7}; \sqrt{5}$

15. $\sqrt{x} - 1$

17. conjugate

19. $\sqrt{2}\sqrt{8} = \sqrt{16} = 4$

21. $\sqrt{5}\sqrt{10} = \sqrt{50} = \sqrt{25}\sqrt{2} = 5\sqrt{2}$

23. $2\sqrt{3}\sqrt{6} = 2\sqrt{18} = 2\sqrt{9}\sqrt{2} = 2(3)\sqrt{2}$

$\qquad\qquad\qquad = 6\sqrt{2}$

25. $\sqrt[3]{5}\sqrt[3]{25} = \sqrt[3]{125} = 5$

27. $\sqrt[3]{5r^2s}\sqrt[3]{2r} = \sqrt[3]{10r^3s} = \sqrt[3]{r^3}\sqrt[3]{10s}$
$$= r\sqrt[3]{10s}$$

29. $\sqrt{ab^3}\sqrt{ab} = \sqrt{a^2b^4} = ab^2$

31. $\left(2\sqrt{3x^3y}\right)\left(5\sqrt{6xy^2}\right) = 10\sqrt{18x^4y^3} = 10\sqrt{9x^4y^2}\sqrt{2y} = 10(3x^2y)\sqrt{2y} = 30x^2y\sqrt{2y}$

33. $\sqrt{x(x+3)}\sqrt{x^3(x+3)} = \sqrt{x^4(x+3)^2} = x^2(x+3)$

35. $3\sqrt{5}\left(4-\sqrt{5}\right) = 12\sqrt{5} - 3\sqrt{25} = 12\sqrt{5} - 3(5) = 12\sqrt{5} - 15$

37. $3\sqrt{2}\left(4\sqrt{3}+2\sqrt{7}\right) = 12\sqrt{6} + 6\sqrt{14}$

39. $\left(\sqrt{2}+1\right)\left(\sqrt{2}-3\right) = \sqrt{4} - 3\sqrt{2} + \sqrt{2} - 3 = 2 - 2\sqrt{2} - 3 = -1 - 2\sqrt{2}$

41. $\left(3\sqrt{2}+2\right)\left(\sqrt{2}+1\right) = 3\sqrt{4} + 3\sqrt{2} + 2\sqrt{2} + 2 = 6 + 5\sqrt{2} + 2 = 8 + 5\sqrt{2}$

43. $(4\sqrt{x}+3)(2\sqrt{x}-5) = 8\sqrt{x^2} - 20\sqrt{x} + 6\sqrt{x} - 15 = 8x - 14\sqrt{x} - 15$

45. $\left(\sqrt{3a}+\sqrt{6}\right)\left(\sqrt{2a}-\sqrt{3}\right) = \sqrt{6a^2} - \sqrt{9a} + \sqrt{12a} - \sqrt{18}$
$$= \sqrt{a^2}\sqrt{6} - \sqrt{9}\sqrt{a} + \sqrt{4}\sqrt{3a} - \sqrt{9}\sqrt{2}$$
$$= a\sqrt{6} - 3\sqrt{a} + 2\sqrt{3a} - 3\sqrt{2}$$

47. $\sqrt{\dfrac{1}{7}} = \dfrac{\sqrt{1}}{\sqrt{7}} = \dfrac{1\sqrt{7}}{\sqrt{7}\sqrt{7}} = \dfrac{\sqrt{7}}{7}$

49. $\sqrt{\dfrac{2}{3}} = \dfrac{\sqrt{2}}{\sqrt{3}} = \dfrac{\sqrt{2}\sqrt{3}}{\sqrt{3}\sqrt{3}} = \dfrac{\sqrt{6}}{3}$

51. $\dfrac{\sqrt{24}}{\sqrt{5}} = \dfrac{\sqrt{24}\sqrt{5}}{\sqrt{5}\sqrt{5}} = \dfrac{\sqrt{120}}{5} = \dfrac{\sqrt{4}\sqrt{30}}{5}$
$$= \dfrac{2\sqrt{30}}{5}$$

53. $\dfrac{\sqrt{7}}{\sqrt{12}} = \dfrac{\sqrt{7}\sqrt{3}}{\sqrt{12}\sqrt{3}} = \dfrac{\sqrt{21}}{\sqrt{36}} = \dfrac{\sqrt{21}}{6}$

55. $\dfrac{3}{\sqrt[3]{16}} = \dfrac{3\sqrt[3]{4}}{\sqrt[3]{16}\sqrt[3]{4}} = \dfrac{3\sqrt[3]{4}}{\sqrt[3]{64}} = \dfrac{3\sqrt[3]{4}}{4}$

57. $\dfrac{\sqrt[3]{7}}{\sqrt[3]{72}} = \dfrac{\sqrt[3]{7}\sqrt[3]{3}}{\sqrt[3]{72}\sqrt[3]{3}} = \dfrac{\sqrt[3]{21}}{\sqrt[3]{216}} = \dfrac{\sqrt[3]{21}}{6}$

59. $\dfrac{\sqrt{8x^2y}}{\sqrt{xy}} = \sqrt{\dfrac{8x^2y}{xy}} = \sqrt{8x} = 2\sqrt{2x}$

61. $\dfrac{\sqrt{10xy^2}}{\sqrt{2xy^3}} = \sqrt{\dfrac{10xy^2}{2xy^3}} = \sqrt{\dfrac{5}{y}} = \dfrac{\sqrt{5}\sqrt{y}}{\sqrt{y}\sqrt{y}}$
$$= \dfrac{\sqrt{5y}}{y}$$

63. $\dfrac{1}{\sqrt{2}-1} = \dfrac{1\left(\sqrt{2}+1\right)}{\left(\sqrt{2}-1\right)\left(\sqrt{2}+1\right)} = \dfrac{\sqrt{2}+1}{\sqrt{4}-1} = \dfrac{\sqrt{2}+1}{2-1} = \dfrac{\sqrt{2}+1}{1} = \sqrt{2}+1$

65. $\dfrac{\sqrt{2}}{\sqrt{5}+3} = \dfrac{\sqrt{2}\left(\sqrt{5}-3\right)}{\left(\sqrt{5}+3\right)\left(\sqrt{5}-3\right)} = \dfrac{\sqrt{2}\left(\sqrt{5}-3\right)}{\sqrt{25}-9} = \dfrac{\sqrt{2}\left(\sqrt{5}-3\right)}{5-9} = \dfrac{\sqrt{2}\left(\sqrt{5}-3\right)}{-4}$

$$= \dfrac{\sqrt{10}-3\sqrt{2}}{-4}$$

$$= \dfrac{3\sqrt{2}-\sqrt{10}}{4}$$

67. $\dfrac{\sqrt{3}+1}{\sqrt{3}-1} = \dfrac{\left(\sqrt{3}+1\right)\left(\sqrt{3}+1\right)}{\left(\sqrt{3}-1\right)\left(\sqrt{3}+1\right)} = \dfrac{\sqrt{9}+\sqrt{3}+\sqrt{3}+1}{\sqrt{9}-1} = \dfrac{4+2\sqrt{3}}{2} = \dfrac{2\left(2+\sqrt{3}\right)}{2}$

$$= 2+\sqrt{3}$$

69. $\dfrac{\sqrt{7}-\sqrt{2}}{\sqrt{2}+\sqrt{7}} = \dfrac{\left(\sqrt{7}-\sqrt{2}\right)\left(\sqrt{2}-\sqrt{7}\right)}{\left(\sqrt{2}+\sqrt{7}\right)\left(\sqrt{2}-\sqrt{7}\right)} = \dfrac{\sqrt{14}-7-2+\sqrt{14}}{\sqrt{4}-\sqrt{49}} = \dfrac{2\sqrt{14}-9}{-5} = \dfrac{9-2\sqrt{14}}{5}$

71. $\dfrac{2}{\sqrt{x}+1} = \dfrac{2\left(\sqrt{x}-1\right)}{\left(\sqrt{x}+1\right)\left(\sqrt{x}-1\right)} = \dfrac{2\left(\sqrt{x}-1\right)}{\sqrt{x^2}-1} = \dfrac{2\left(\sqrt{x}-1\right)}{x-1}$

73. $\dfrac{x}{\sqrt{x}-4} = \dfrac{x\left(\sqrt{x}+4\right)}{\left(\sqrt{x}-4\right)\left(\sqrt{x}+4\right)} = \dfrac{x\left(\sqrt{x}+4\right)}{\sqrt{x^2}-16} = \dfrac{x\left(\sqrt{x}+4\right)}{x-16}$

75. $\dfrac{\sqrt{x}-\sqrt{y}}{\sqrt{x}+\sqrt{y}} = \dfrac{\left(\sqrt{x}-\sqrt{y}\right)\left(\sqrt{x}-\sqrt{y}\right)}{\left(\sqrt{x}+\sqrt{y}\right)\left(\sqrt{x}-\sqrt{y}\right)} = \dfrac{\sqrt{x^2}-\sqrt{xy}-\sqrt{xy}+\sqrt{y^2}}{\sqrt{x^2}-\sqrt{y^2}} = \dfrac{x-2\sqrt{xy}+y}{x-y}$

77. $\dfrac{\sqrt{3}+1}{2} = \dfrac{\left(\sqrt{3}+1\right)\left(\sqrt{3}-1\right)}{2\left(\sqrt{3}-1\right)} = \dfrac{\sqrt{9}-1}{2\left(\sqrt{3}-1\right)} = \dfrac{2}{2\left(\sqrt{3}-1\right)} = \dfrac{1}{\sqrt{3}-1}$

79. $\dfrac{\sqrt{x}+3}{x} = \dfrac{\left(\sqrt{x}+3\right)\left(\sqrt{x}-3\right)}{x\left(\sqrt{x}-3\right)} = \dfrac{\sqrt{x^2}-9}{x\left(\sqrt{x}-3\right)} = \dfrac{x-9}{x\left(\sqrt{x}-3\right)}$

81. $\left(3\sqrt[3]{9}\right)\left(2\sqrt[3]{3}\right) = 6\sqrt[3]{27} = 6(3) = 18$

83. $\sqrt{5ab}\,\sqrt{5a} = \sqrt{25a^2 b} = \sqrt{25a^2}\,\sqrt{b}$
$= 5a\sqrt{b}$

85. $\sqrt[3]{a^5 b}\,\sqrt[3]{16ab^5} = \sqrt[3]{16a^6 b^6} = \sqrt[3]{8a^6 b^6}\,\sqrt[3]{2}$
$= 2a^2 b^2 \sqrt[3]{2}$

87. $\sqrt[3]{6x^2(y+z)^2}\,\sqrt[3]{18x(y+z)} = \sqrt[3]{108x^3(y+z)^3} = \sqrt[3]{27x^3(y+z)^3}\,\sqrt[3]{4} = 3x(y+z)\sqrt[3]{4}$

89. $-2\sqrt{5x}\left(4\sqrt{2x}-3\sqrt{3}\right) = -8\sqrt{10x^2}+6\sqrt{15x} = -8x\sqrt{10}+6\sqrt{15x}$

91. $\left(\sqrt{5z}+\sqrt{3}\right)\left(\sqrt{5z}+\sqrt{3}\right)=\sqrt{25z^2}+\sqrt{15z}+\sqrt{15z}+\sqrt{9}=5z+2\sqrt{15z}+3$

93. $\left(3\sqrt{2r}-2\right)^2=\left(3\sqrt{2r}-2\right)\left(3\sqrt{2r}-2\right)=9\sqrt{4r^2}-6\sqrt{2r}-6\sqrt{2r}+4=18r-12\sqrt{2r}+4$

95. $-2\left(\sqrt{3x}+\sqrt{3}\right)^2=-2\left(\sqrt{3x}+\sqrt{3}\right)\left(\sqrt{3x}+\sqrt{3}\right)$

$$=-2\left(\sqrt{9x^2}+\sqrt{9x}+\sqrt{9x}+\sqrt{9}\right)$$

$$=-2(3x+3\sqrt{x}+3\sqrt{x}+3)=-2(3x+6\sqrt{x}+3)=-6x-12\sqrt{x}-6$$

97. $\dfrac{3}{\sqrt[3]{9}}=\dfrac{3\sqrt[3]{3}}{\sqrt[3]{9}\sqrt[3]{3}}=\dfrac{3\sqrt[3]{3}}{\sqrt[3]{27}}=\dfrac{3\sqrt[3]{3}}{3}=\sqrt[3]{3}$

99. $\dfrac{1}{\sqrt[4]{4}}=\dfrac{1\sqrt[4]{4}}{\sqrt[4]{4}\sqrt[4]{4}}=\dfrac{\sqrt[4]{4}}{\sqrt[4]{16}}=\dfrac{\sqrt[4]{4}}{2}$

101. $\dfrac{1}{\sqrt[5]{16}}=\dfrac{1\sqrt[5]{2}}{\sqrt[5]{16}\sqrt[5]{2}}=\dfrac{\sqrt[5]{2}}{\sqrt[5]{32}}=\dfrac{\sqrt[5]{2}}{2}$

103. $\dfrac{\sqrt[3]{4a^2}}{\sqrt[3]{2ab}}=\sqrt[3]{\dfrac{4a^2}{2ab}}=\sqrt[3]{\dfrac{2a}{b}}=\dfrac{\sqrt[3]{2a}\sqrt[3]{b^2}}{\sqrt[3]{b}\sqrt[3]{b^2}}$

$$=\dfrac{\sqrt[3]{2ab^2}}{b}$$

105. $\dfrac{2z-1}{\sqrt{2z}-1}=\dfrac{(2z-1)\left(\sqrt{2z}+1\right)}{\left(\sqrt{2z}-1\right)\left(\sqrt{2z}+1\right)}=\dfrac{(2z-1)\left(\sqrt{2z}+1\right)}{\sqrt{4z^2}-1}=\dfrac{(2z-1)\left(\sqrt{2z}+1\right)}{2z-1}$

$$=\sqrt{2z}+1$$

107. $\dfrac{\sqrt{x}+\sqrt{y}}{\sqrt{x}}=\dfrac{\left(\sqrt{x}+\sqrt{y}\right)\left(\sqrt{x}-\sqrt{y}\right)}{\sqrt{x}\left(\sqrt{x}-\sqrt{y}\right)}=\dfrac{\sqrt{x^2}-\sqrt{y^2}}{\sqrt{x^2}-\sqrt{xy}}=\dfrac{x-y}{x-\sqrt{xy}}$

109. If the area of the aperture is again cut in half, the area will equal $9\pi/4$ cm^2.

$$A=\pi r^2 \qquad\qquad f\text{-number}=\dfrac{f}{d}$$

$$\dfrac{9\pi}{4}=\pi\left(\dfrac{d}{2}\right)^2 \qquad\qquad =\dfrac{12}{3}$$

$$\dfrac{9\pi}{4}=\dfrac{\pi d^2}{4} \qquad\qquad =4$$

$$36\pi=4\pi d^2 \qquad\qquad \boxed{f/4}$$

$$\dfrac{36\pi}{4\pi}=d^2$$

$$9=d^2$$

$$3=d$$

111. $r=\sqrt{\dfrac{A}{\pi}}=\dfrac{\sqrt{A}}{\sqrt{\pi}}=\dfrac{\sqrt{A}\sqrt{\pi}}{\sqrt{\pi}\sqrt{\pi}}=\dfrac{\sqrt{A\pi}}{\pi}$

113. $\dfrac{8}{\sqrt{2}}=\dfrac{8\sqrt{2}}{\sqrt{2}\sqrt{2}}=\dfrac{8\sqrt{2}}{2}=4\sqrt{2}$ cm

115. shorter leg: $\dfrac{6}{\sqrt{3}}=\dfrac{6\sqrt{3}}{\sqrt{3}\sqrt{3}}=\dfrac{6\sqrt{3}}{3}=2\sqrt{3}$ ft; hypotenuse: $2\left(2\sqrt{3}\right)=4\sqrt{3}$ ft

117. Answers may vary.

119. $\dfrac{\sqrt{x}-3}{4} = \dfrac{\left(\sqrt{x}-3\right)\left(\sqrt{x}+3\right)}{4\left(\sqrt{x}+3\right)} = \dfrac{x-9}{4\left(\sqrt{x}+3\right)}$

Exercises 9.6 (page 699)

1. $\left(x^{1/2}\right)^2 = x^{2/2} = x$

3. $\left[(x+3)^{1/2}\right]^2 = (x+3)^{2/2} = x+3$

5. $(x+7)^2 = (x+7)(x+7) = x^2 + 14x + 49$

7. $f(-2) = 3(-2)^2 - 4(-2) + 2 = 22$

9. $f(3) = 3(3)^2 - 4(3) + 2 = 17$

11. $x^n = y^n$; power rule

13. square

15. extraneous

17.
$$\sqrt{4x+5} = 3$$
$$\left(\sqrt{4x+5}\right)^2 = 3^2$$
$$4x + 5 = 9$$
$$4x = 4$$
$$x = 1$$
The answer checks.

19.
$$\sqrt{4x-8} + 4 = 10$$
$$\sqrt{4x-8} = 6$$
$$\left(\sqrt{4x-8}\right)^2 = 6^2$$
$$4x - 8 = 36$$
$$4x = 44$$
$$x = 11$$
The answer checks.

21.
$$\sqrt[3]{7n-1} = 3$$
$$\left(\sqrt[3]{7n-1}\right)^3 = 3^3$$
$$7n - 1 = 27$$
$$7n = 28$$
$$n = 4$$
The answer checks.

23.
$$x = \frac{\sqrt{12x-5}}{2}$$
$$2x = \sqrt{12x-5}$$
$$(2x)^2 = \left(\sqrt{12x-5}\right)^2$$
$$4x^2 = 12x - 5$$
$$4x^2 - 12x + 5 = 0$$
$$(2x-1)(2x-5) = 0$$
$$2x - 1 = 0 \quad \textbf{or} \quad 2x - 5 = 0$$
$$2x = 1 \qquad\qquad 2x = 5$$
$$x = \tfrac{1}{2} \qquad\qquad x = \tfrac{5}{2}$$
Both answers check.

25.
$$\sqrt{2y - 5} + 4 = y$$
$$\sqrt{2y - 5} = y - 4$$
$$\left(\sqrt{2y - 5}\right)^2 = (y - 4)^2$$
$$2y - 5 = y^2 - 8y + 16$$
$$0 = y^2 - 10y + 21$$
$$0 = (y - 7)(y - 3)$$
$$y - 7 = 0 \quad \textbf{or} \quad y - 3 = 0$$
$$y = 7 \qquad\qquad y = 3$$
$$\text{solution} \qquad\quad \text{extraneous}$$

27.
$$\sqrt{-5x + 24} = 6 - x$$
$$\left(\sqrt{-5x + 24}\right)^2 = (6 - x)^2$$
$$-5x + 24 = 36 - 12x + x^2$$
$$0 = x^2 - 7x + 12$$
$$0 = (x - 3)(x - 4)$$
$$x - 3 = 0 \quad \textbf{or} \quad x - 4 = 0$$
$$x = 3 \qquad\qquad x = 4$$
$$\text{solution} \qquad\quad \text{solution}$$

29.
$$\sqrt[3]{x^3 - 1} = x - 1$$
$$\left(\sqrt[3]{x^3 - 1}\right)^3 = (x - 1)^3$$
$$x^3 - 1 = x^3 - 3x^2 + 3x - 1$$
$$3x^2 - 3x = 0$$
$$3x(x - 1) = 0$$
$$3x = 0 \quad \textbf{or} \quad x - 1 = 0$$
$$x = 0 \qquad\qquad x = 1$$
$$\text{solution} \qquad\quad \text{solution}$$

31.
$$\sqrt[3]{x^3 + 56} - 2 = x$$
$$\sqrt[3]{x^3 + 56} = x + 2$$
$$\left(\sqrt[3]{x^3 + 56}\right)^3 = (x + 2)^3$$
$$x^3 + 56 = x^3 + 6x^2 + 12x + 8$$
$$0 = 6x^2 + 12x - 48$$
$$0 = 6(x - 2)(x + 4)$$
$$x - 2 = 0 \quad \textbf{or} \quad x + 4 = 0$$
$$x = 2 \qquad\qquad x = -4$$
$$\text{solution} \qquad\quad \text{solution}$$

33.
$$2\sqrt{4x + 1} = \sqrt{x + 4}$$
$$\left(2\sqrt{4x + 1}\right)^2 = \left(\sqrt{x + 4}\right)^2$$
$$4(4x + 1) = x + 4$$
$$16x + 4 = x + 4$$
$$15x = 0$$
$$x = 0$$
The answer checks.

35.
$$\sqrt{x + 5} = \sqrt{7 - x}$$
$$\left(\sqrt{x + 5}\right)^2 = \left(\sqrt{7 - x}\right)^2$$
$$x + 5 = 7 - x$$
$$2x = 2$$
$$x = 1$$
The answer checks.

37.
$$1 + \sqrt{z} = \sqrt{z + 3}$$
$$\left(1 + \sqrt{z}\right)^2 = \left(\sqrt{z + 3}\right)^2$$
$$1 + 2\sqrt{z} + z = z + 3$$
$$2\sqrt{z} = 2$$
$$\left(2\sqrt{z}\right)^2 = 2^2$$
$$4z = 4$$
$$z = 1$$
The answer checks.

39.
$$\sqrt{4s+1} - \sqrt{6s} = -1$$
$$\sqrt{4s+1} = \sqrt{6s} - 1$$
$$\left(\sqrt{4s+1}\right)^2 = \left(\sqrt{6s} - 1\right)^2$$
$$4s + 1 = 6s - 2\sqrt{6s} + 1$$
$$2\sqrt{6s} = 2s$$
$$\left(2\sqrt{6s}\right)^2 = (2s)^2$$
$$24s = 4s^2$$
$$0 = 4s^2 - 24s$$
$$0 = 4s(s - 6)$$
$$4s = 0 \quad \textbf{or} \quad s - 6 = 0$$
$$s = 0 \qquad\qquad s = 6$$
$$\text{extraneous} \qquad \text{solution}$$

41.
$$\sqrt{y} + \sqrt{5} = \sqrt{y + 15}$$
$$\left(\sqrt{y} + \sqrt{5}\right)^2 = \left(\sqrt{y + 15}\right)^2$$
$$y + 2\sqrt{5y} + 5 = y + 15$$
$$2\sqrt{5y} = 10$$
$$\left(2\sqrt{5y}\right)^2 = 10^2$$
$$20y = 100$$
$$y = 5$$
The answer checks.

43.
$$\sqrt{3x} - \sqrt{x+1} = \sqrt{x-2}$$
$$\left(\sqrt{3x} - \sqrt{x+1}\right)^2 = \left(\sqrt{x-2}\right)^2$$
$$3x - 2\sqrt{3x(x+1)} + x + 1 = x - 2$$
$$3x + 3 = 2\sqrt{3x^2 + 3x}$$
$$(3x + 3)^2 = \left(2\sqrt{3x^2 + 3x}\right)^2$$

$$(3x + 3)^2 = \left(2\sqrt{3x^2 + 3x}\right)^2$$
$$9x^2 + 18x + 9 = 12x^2 + 12x$$
$$0 = 3x^2 - 6x - 9$$
$$0 = 3(x - 3)(x + 1)$$
$$x - 3 = 0 \quad \textbf{or} \quad x + 1 = 0$$
$$x = 3 \qquad\qquad x = -1$$
$$\text{solution} \qquad\quad \text{extraneous}$$

45.
$$v = \sqrt{2gh}$$
$$v^2 = \left(\sqrt{2gh}\right)^2$$
$$v^2 = 2gh$$
$$\frac{v^2}{2h} = \frac{2gh}{2h}$$
$$\frac{v^2}{2h} = g, \text{ or } g = \frac{v^2}{2h}$$

47.
$$T = 2\pi\sqrt{\frac{l}{32}}$$
$$T^2 = \left(2\pi\sqrt{\frac{l}{32}}\right)^2$$
$$T^2 = 4\pi^2 \cdot \frac{l}{32}$$

$$32(T^2) = 32\left(4\pi^2 \cdot \frac{l}{32}\right)$$
$$32T^2 = 4\pi^2 l$$
$$\frac{32T^2}{4\pi^2} = \frac{4\pi^2 l}{4\pi^2}$$
$$\frac{8T^2}{\pi^2} = l, \text{ or } l = \frac{8T^2}{\pi^2}$$

49.
$$r = \sqrt[3]{\frac{A}{P}} - 1$$
$$r + 1 = \sqrt[3]{\frac{A}{P}}$$
$$(r + 1)^3 = \left(\sqrt[3]{\frac{A}{P}}\right)^3$$

$$(r + 1)^3 = \left(\sqrt[3]{\frac{A}{P}}\right)^3$$
$$(r + 1)^3 = \frac{A}{P}$$
$$P(r + 1)^3 = P\left(\frac{A}{P}\right)$$
$$P(r + 1)^3 = A, \text{ or } A = P(r + 1)^3$$

333

51.

$$L_A = L_B \sqrt{1 - \frac{v^2}{c^2}}$$

$$L_A^2 = \left(L_B \sqrt{1 - \frac{v^2}{c^2}} \right)^2$$

$$L_A^2 = L_B^2 \left(1 - \frac{v^2}{c^2} \right)$$

$$\frac{L_A^2}{L_B^2} = 1 - \frac{v^2}{c^2}$$

$$\frac{L_A^2}{L_B^2} = 1 - \frac{v^2}{c^2}$$

$$\frac{v^2}{c^2} = 1 - \frac{L_A^2}{L_B^2}$$

$$c^2 \left(\frac{v^2}{c^2} \right) = c^2 \left(1 - \frac{L_A^2}{L_B^2} \right)$$

$$v^2 = c^2 \left(1 - \frac{L_A^2}{L_B^2} \right)$$

53.

$$\sqrt{y+2} = 4 - y$$

$$\left(\sqrt{y+2} \right) = (4-y)^2$$

$$y + 2 = 16 - 8y + y^2$$

$$0 = y^2 - 9y + 14$$

$$0 = (y-2)(y-7)$$

$$y - 2 = 0 \quad \textbf{or} \quad y - 7 = 0$$

$$y = 2 \qquad\qquad y = 7$$

solution   extraneous

55.

$$2\sqrt{x} = \sqrt{5x - 16}$$

$$\left(2\sqrt{x} \right)^2 = \left(\sqrt{5x - 16} \right)^2$$

$$4x = 5x - 16$$

$$-x = -16$$

$$x = 16$$

The answer checks.

57.

$$\sqrt{2y+1} = 1 - 2\sqrt{y}$$

$$\left(\sqrt{2y+1} \right)^2 = \left(1 - 2\sqrt{y} \right)^2$$

$$2y + 1 = 1 - 4\sqrt{y} + 4y$$

$$4\sqrt{y} = 2y$$

$$\left(4\sqrt{y} \right)^2 = (2y)^2$$

$$16y = 4y^2$$

$$0 = 4y^2 - 16y$$

$$0 = 4y(y-4)$$

$$4y = 0 \quad \textbf{or} \quad y - 4 = 0$$

$$y = 0 \qquad\qquad y = 4$$

solution   extraneous

59.

$$\sqrt{2x+5} + \sqrt{x+2} = 5$$

$$\sqrt{2x+5} = 5 - \sqrt{x+2}$$

$$\left(\sqrt{2x+5} \right)^2 = \left(5 - \sqrt{x+2} \right)^2$$

$$2x + 5 = 25 - 10\sqrt{x+2} + x + 2$$

$$10\sqrt{x+2} = -x + 22$$

$$\left(10\sqrt{x+2} \right)^2 = (-x+22)^2$$

$$\left(10\sqrt{x+2} \right)^2 = (-x+22)^2$$

$$100(x+2) = x^2 - 44x + 484$$

$$0 = x^2 - 144x + 284$$

$$0 = (x-142)(x-2)$$

$$x - 142 = 0 \quad \textbf{or} \quad x - 2 = 0$$

$$x = 142 \qquad\qquad x = 2$$

extraneous   solution

334

61.
$$5r + 4 = \sqrt{5r + 20} + 4r$$
$$r + 4 = \sqrt{5r + 20}$$
$$(r + 4)^2 = \left(\sqrt{5r + 20}\right)^2$$
$$r^2 + 8r + 16 = 5r + 20$$
$$r^2 + 3r - 4 = 0$$
$$(r + 4)(r - 1) = 0$$
$$r + 4 = 0 \quad \textbf{or} \quad r - 1 = 0$$
$$r = -4 \qquad\qquad r = 1$$
$$\text{solution} \qquad\quad \text{solution}$$

63.
$$\sqrt{x}\sqrt{x + 6} = 4$$
$$\left(\sqrt{x}\sqrt{x + 6}\right)^2 = 4^2$$
$$x(x + 6) = 16$$
$$x^2 + 6x - 16 = 0$$
$$(x - 2)(x + 8) = 0$$
$$x - 2 = 0 \quad \textbf{or} \quad x + 8 = 0$$
$$x = 2 \qquad\qquad x = -8$$
$$\text{solution} \qquad\quad \text{extraneous}$$

65.
$$\sqrt[4]{x^4 + 4x^2 - 4} = -x$$
$$\left(\sqrt[4]{x^4 + 4x^2 - 4}\right)^4 = (-x)^4$$
$$x^4 + 4x^2 - 4 = x^4$$
$$4x^2 - 4 = 0$$
$$4(x + 1)(x - 1) = 0$$
$$x + 1 = 0 \quad \textbf{or} \quad x - 1 = 0$$
$$x = -1 \qquad\qquad x = 1$$
$$\text{solution} \qquad\quad \text{extraneous}$$

67.
$$2 + \sqrt{u} = \sqrt{2u + 7}$$
$$\left(2 + \sqrt{u}\right)^2 = \left(\sqrt{2u + 7}\right)^2$$
$$4 + 4\sqrt{u} + u = 2u + 7$$
$$4\sqrt{u} = u + 3$$
$$\left(4\sqrt{u}\right)^2 = (u + 3)^2$$
$$16u = u^2 + 6u + 9$$
$$0 = u^2 - 10u + 9$$
$$0 = (u - 9)(u - 1)$$
$$u - 9 = 0 \quad \textbf{or} \quad u - 1 = 0$$
$$u = 9 \qquad\qquad u = 1$$
$$\text{solution} \qquad\quad \text{solution}$$

69.
$$u = \sqrt[4]{u^4 - 6u^2 + 24}$$
$$u^4 = \left(\sqrt[4]{u^4 - 6u^2 + 24}\right)^4$$
$$u^4 = u^4 - 6u^2 + 24$$
$$6u^2 - 24 = 0$$
$$6(u + 2)(u - 2) = 0$$
$$u + 2 = 0 \quad \textbf{or} \quad u - 2 = 0$$
$$u = -2 \qquad\qquad u = 2$$
$$\text{extraneous} \qquad\quad \text{solution}$$

71.
$$\sqrt{x - 5} - \sqrt{x + 3} = 4$$
$$\sqrt{x - 5} = \sqrt{x + 3} + 4$$
$$\left(\sqrt{x - 5}\right)^2 = \left(\sqrt{x + 3} + 4\right)^2$$
$$x - 5 = x + 3 + 8\sqrt{x + 3} + 16$$
$$-24 = 8\sqrt{x + 3}$$
$$-3 = \sqrt{x + 3}$$
$$(-3)^2 = \left(\sqrt{x + 3}\right)^2$$
$$9 = x + 3$$
$$6 = x$$
$$\text{extraneous} \Rightarrow \emptyset$$

73. $\sqrt{x+8} - \sqrt{x-4} = -2$

$\sqrt{x+8} = \sqrt{x-4} - 2$

$\left(\sqrt{x+8}\right)^2 = \left(\sqrt{x-4} - 2\right)^2$

$x + 8 = x - 4 - 4\sqrt{x-4} + 4$

$4\sqrt{x-4} = -8$

$\sqrt{x-4} = -2$

$\left(\sqrt{x-4}\right)^2 = (-2)^2$

$x - 4 = 4$

$x = 8$

extraneous $\Rightarrow \emptyset$

75. $\sqrt{z-1} + \sqrt{z+2} = 3$

$\sqrt{z-1} = 3 - \sqrt{z+2}$

$\left(\sqrt{z-1}\right)^2 = \left(3 - \sqrt{z+2}\right)^2$

$z - 1 = 9 - 6\sqrt{z+2} + z + 2$

$6\sqrt{z+2} = 12$

$\sqrt{z+2} = 2$

$\left(\sqrt{z+2}\right)^2 = 2^2$

$z + 2 = 4$

$z = 2$: The answer checks.

77. $\sqrt{\sqrt{a} + \sqrt{a+8}} = 2$

$\left(\sqrt{\sqrt{a} + \sqrt{a+8}}\right)^2 = 2^2$

$\sqrt{a} + \sqrt{a+8} = 4$

$\left(\sqrt{a} + \sqrt{a+8}\right)^2 = 4^2$

$a + 2\sqrt{a(a+8)} + a + 8 = 16$

$2\sqrt{a^2 + 8a} = -2a + 8$

$\left(2\sqrt{a^2 + 8a}\right)^2 = (-2a + 8)^2$

$4a^2 + 32a = 4a^2 - 32a + 64$

$64a = 64$

$a = 1$: The answer checks.

79. $\dfrac{\sqrt{2x}}{\sqrt{x+2}} = \sqrt{x-1}$

$\left(\dfrac{\sqrt{2x}}{\sqrt{x+2}}\right)^2 = \left(\sqrt{x-1}\right)^2$

$\dfrac{2x}{x+2} = x - 1$

$2x = x^2 + x - 2$

$0 = x^2 - x - 2$

$0 = (x-2)(x+1)$

$x - 2 = 0$ **or** $x + 1 = 0$

$\quad x = 2 \qquad\qquad x = -1$

solution $\qquad\quad$ extraneous

81.
$$s = 1.45\sqrt{r}$$
$$65 = 1.45\sqrt{r}$$
$$65^2 = \left(1.45\sqrt{r}\right)^2$$
$$4225 = 2.1025r$$
$$2010 \text{ ft} \approx r$$

83.
$$v = \sqrt[3]{\dfrac{P}{0.02}}$$
$$v = \sqrt[3]{\dfrac{500}{0.02}}$$
$$v = \sqrt[3]{25000}$$
$$v \approx 29 \text{ mph}$$

85. $r = 1 - \sqrt[n]{\dfrac{S}{C}} = 1 - \sqrt[15]{\dfrac{900}{22000}} \approx 1 - \sqrt[15]{0.040909} \approx 1 - 0.808 \approx 0.192 \approx 19\%$

87.
$$\sqrt{5x} = \sqrt{100 - 3x^2}$$
$$\left(\sqrt{5x}\right)^2 = \left(\sqrt{100 - 3x^2}\right)^2$$
$$5x = 100 - 3x^2$$
$$3x^2 + 5x - 100 = 0$$
$$(x - 5)(3x + 20) = 0$$

$x - 5 = 0$ **or** $\quad 3x + 20 = 0$

$\qquad x = 5 \qquad\qquad\qquad 3x = -20$

solution $\qquad\qquad\qquad\qquad x = -\dfrac{20}{3}$

$\qquad\qquad\qquad\qquad$ does not make sense

The equilibrium price is \$5.

89.
$$r = \sqrt[4]{\dfrac{8kl}{\pi R}}$$
$$r^4 = \left(\sqrt[4]{\dfrac{8kl}{\pi R}}\right)^4$$
$$r^4 = \dfrac{8kl}{\pi R}$$
$$\pi R r^4 = 8kl$$
$$R = \dfrac{8kl}{\pi r^4}$$

91. **Answers may vary.**

93.
$$\sqrt[3]{2x} = \sqrt{x}$$
$$\left(\sqrt[3]{2x}\right)^2 = \left(\sqrt{x}\right)^2$$
$$\left[\left(\sqrt[3]{2x}\right)^2\right]^3 = x^3$$
$$4x^2 = x^3$$
$$0 = x^3 - 4x^2$$
$$0 = x^2(x - 4)$$
$$x = 0 \text{ or } x = 4$$

Exercises 9.7 (page 710)

1. $\sqrt{-9}$: not real

3. $\sqrt{a}$ if $a \geq 0$: real

5. $-\sqrt{-1}$: not real

7. $3 - \sqrt{5}$

9. $5 + 3x + 7 + 4x = 7x + 12$

11. $\dfrac{x^2 - x - 6}{9 - x^2} \cdot \dfrac{x^2 + x - 6}{x^2 - 4} = \dfrac{(x - 3)(x + 2)}{(3 + x)(3 - x)} \cdot \dfrac{(x + 3)(x - 2)}{(x + 2)(x - 2)} = -1$

SECTION 9.7

13. Let $w =$ the speed of the wind.

	Rate	Time	Dist.
Downwind	$200 + w$	$\frac{330}{200+w}$	330
Upwind	$200 - w$	$\frac{330}{200-w}$	330

$$\frac{330}{200+w} + \frac{330}{200-w} = \frac{10}{3}$$

$$\left(\frac{330}{200+w} + \frac{330}{200-w}\right)3(200+w)(200-w) = \frac{10}{3} \cdot 3(200+w)(200-w)$$

$$330(3)(200-w) + 330(3)(200+w) = 10(40,000 - w^2)$$

$$198,000 - 990w + 198,000 + 990w = 400,000 - 10w^2$$

$$10w^2 = 4000$$

$$w^2 = 400$$

$$w = 20 \text{ mph}$$

15. i

17. $-i$

19. imaginary

21. $\dfrac{\sqrt{a}}{\sqrt{b}}$

23. $5; 7$

25. conjugates

27. $\sqrt{-25} = \sqrt{25(-1)} = \sqrt{25}\sqrt{-1} = 5i$

29. $\sqrt{-121} = \sqrt{121(-1)} = \sqrt{121}\sqrt{-1} = 11i$

31. $\sqrt{-7} = \sqrt{7(-1)} = \sqrt{7}\sqrt{-1} = i\sqrt{7}$

33. $\sqrt{-8} = \sqrt{8(-1)} = \sqrt{8}\sqrt{-1} = \sqrt{4}\sqrt{2} \cdot i$
$= 2i\sqrt{2}$

35. $9 = 9 + 0i$

37. $9 + \sqrt{-9} = 9 + 3i$

39. $3 + 7i \overset{?}{=} \sqrt{9} + (5 + 2)i$
$3 + 7i \overset{?}{=} 3 + 7i$
They are equal.

41. $\sqrt{4} + \sqrt{-4} \overset{?}{=} 2 - 2i$
$2 + \sqrt{-1 \cdot 4} \overset{?}{=} 2 - 2i$
$2 + 2i = 2 - 2i$
They are not equal.

43. $(4 + 7i) + (3 - 5i) = 4 + 7i + 3 - 5i$
$= 7 + 2i$

45. $(7 - 3i) - (4 + 2i) = 7 - 3i - 4 - 2i$
$= 3 - 5i$

47. $(8 + 5i) + (7 + 2i) = 8 + 5i + 7 + 2i$
$= 15 + 7i$

49. $(1 + i) - 2i + (5 - 7i) = 1 + i - 2i + 5 - 7i$
$= 6 - 8i$

51. $6i(3 - 4i) = 18i - 24i^2 = 18i - 24(-1) = 18i + 24 = 24 + 18i$

53. $-5i(5 - 5i) = -25i + 25i^2 = -25i + 25(-1) = -25i - 25 = -25 - 25i$

55. $(2 + i)(3 - i) = 6 - 2i + 3i - i^2 = 6 + i - (-1) = 6 + i + 1 = 7 + i$

57. $(5 + 2i)(4 - 6i) = 20 - 30i + 8i - 12i^2 = 20 - 22i - 12(-1) = 20 - 22i + 12 = 32 - 22i$

59. $(4 + \sqrt{3}i)(3 - \sqrt{3}i) = 12 - 4\sqrt{3}i + 3\sqrt{3}i - 3i^2 = 12 - \sqrt{3}i - 3(-1) = 15 - \sqrt{3}i$

61. $(4 + 3i)^2 = (4 + 3i)(4 + 3i) = 16 + 12i + 12i + 9i^2 = 16 + 24i - 9 = 7 + 24i$

63. $(6 - 5i)(6 + 5i) = 36 + 30i - 30i - 25i^2 = 36 - 25(-1) = 36 + 25 = 61$

65. $(3 + 4i)(3 - 4i) = 9 - 12i + 12i - 16i^2 = 9 - 16(-1) = 9 + 16 = 25$

67. $\dfrac{5}{2 - i} = \dfrac{5(2 + i)}{(2 - i)(2 + i)} = \dfrac{5(2 + i)}{4 - i^2} = \dfrac{5(2 + i)}{5} = 2 + i$

69. $\dfrac{10}{5 - i} = \dfrac{10(5 + i)}{(5 - i)(5 + i)} = \dfrac{10(5 + i)}{25 - i^2} = \dfrac{10(5 + i)}{26} = \dfrac{5(5 + i)}{13} = \dfrac{25 + 5i}{13} = \dfrac{25}{13} + \dfrac{5}{13}i$

71. $\dfrac{3 - 2i}{3 + 2i} = \dfrac{(3 - 2i)(3 - 2i)}{(3 + 2i)(3 - 2i)} = \dfrac{9 - 6i - 6i + 4i^2}{9 - 4i^2} = \dfrac{5 - 12i}{13} = \dfrac{5}{13} - \dfrac{12}{13}i$

73. $\dfrac{3 + 2i}{3 + i} = \dfrac{(3 + 2i)(3 - i)}{(3 + i)(3 - i)} = \dfrac{9 - 3i + 6i - 2i^2}{9 - i^2} = \dfrac{11 + 3i}{10} = \dfrac{11}{10} + \dfrac{3}{10}i$

75. $\dfrac{-12}{7 - \sqrt{-1}} = \dfrac{-12}{7 - i} = \dfrac{-12(7 + i)}{(7 - i)(7 + i)} = \dfrac{-84 - 12i}{49 - i^2} = \dfrac{-84 - 12i}{50} = \dfrac{-84}{50} - \dfrac{12}{50}i = -\dfrac{42}{25} - \dfrac{6}{25}i$

77. $\dfrac{7}{5 - \sqrt{-9}} = \dfrac{7}{5 - 3i} = \dfrac{7(5 + 3i)}{(5 - 3i)(5 + 3i)} = \dfrac{7(5 + 3i)}{25 - 9i^2} = \dfrac{7(5 + 3i)}{34} = \dfrac{35 + 21i}{34} = \dfrac{35}{34} + \dfrac{21}{34}i$

79. $i^5 = i^4 i^1 = 1i = i$

81. $i^{10} = i^8 i^2 = (i^4)^2 i^2 = 1^2(-1) = -1$

83. $i^{37} = i^{36} i^1 = (i^4)^9 i = 1^9 i = i$

85. $i^{27} = i^{24} i^3 = (i^4)^6 i^3 = 1^6 i^3 = i^3 = -i$

87. $i^{64} = (i^4)^{16} = 1^{16} = 1$

89. $i^{97} = i^{96} i^1 = (i^4)^{24} i = 1^{24} i = i$

91. $5i^3 + 2i^2 = 5(-i) + 2(-1) = -2 - 5i$

93. $\dfrac{1}{i} = \dfrac{1i^3}{ii^3} = \dfrac{i^3}{i^4} = \dfrac{i^3}{1} = i^3 = -i = 0 - i$

95. $\dfrac{4}{5i^3} = \dfrac{4i}{5i^3 i} = \dfrac{4i}{5i^4} = \dfrac{4i}{5(1)} = \dfrac{4}{5}i = 0 + \dfrac{4}{5}i$

97. $\dfrac{3i}{8\sqrt{-9}} = \dfrac{3i}{8(3i)} = \dfrac{1}{8} = \dfrac{1}{8} + 0i$

99. $\dfrac{-3}{5i^5} = \dfrac{-3i^3}{5i^5 i^3} = \dfrac{-3(-i)}{5i^8} = \dfrac{3i}{5} = 0 + \dfrac{3}{5}i$

101. $|6 + 8i| = \sqrt{6^2 + 8^2} = \sqrt{36 + 64} = \sqrt{100} = 10$

103. $|12 - 5i| = \sqrt{12^2 + (-5)^2} = \sqrt{144 + 25} = \sqrt{169} = 13$

105. $|5 + 7i| = \sqrt{5^2 + 7^2} = \sqrt{25 + 49} = \sqrt{74}$

107. $\left|\dfrac{3}{5} - \dfrac{4}{5}i\right| = \sqrt{\left(\dfrac{3}{5}\right)^2 + \left(-\dfrac{4}{5}\right)^2} = \sqrt{\dfrac{9}{25} + \dfrac{16}{25}} = \sqrt{\dfrac{25}{25}} = \sqrt{1} = 1$

109. $8 + 5i \stackrel{?}{=} 2^3 + \sqrt{25}i^3$

$8 + 5i \stackrel{?}{=} 8 + 5(-i)$

$8 + 5i \stackrel{?}{=} 8 - 5i$; not equal

111. $(8 - \sqrt{-1})(-2 - \sqrt{-16}) = (8 - i)(-2 - 4i) = -16 - 32i + 2i + 4i^2 = -16 - 30i - 4$
$$= -20 - 30i$$

113. $(2 + 3i)^2 = (2 + 3i)(2 + 3i) = 4 + 6i + 6i + 9i^2 = 4 + 12i - 9 = -5 + 12i$

115. $\dfrac{5i}{6 + 2i} = \dfrac{5i(6 - 2i)}{(6 + 2i)(6 - 2i)} = \dfrac{30i - 10i^2}{36 - 4i^2} = \dfrac{10 + 30i}{40} = \dfrac{10}{40} + \dfrac{30}{40}i = \dfrac{1}{4} + \dfrac{3}{4}i$

117. $\dfrac{\sqrt{5} - \sqrt{3}i}{\sqrt{5} + \sqrt{3}i} = \dfrac{\left(\sqrt{5} - \sqrt{3}i\right)\left(\sqrt{5} - \sqrt{3}i\right)}{\left(\sqrt{5} + \sqrt{3}i\right)\left(\sqrt{5} - \sqrt{3}i\right)} = \dfrac{5 - \sqrt{15}i - \sqrt{15}i + 3i^2}{5 - 3i^2} = \dfrac{2 - 2\sqrt{15}i}{8}$

$$= \dfrac{2}{8} - \dfrac{2\sqrt{15}}{8}i$$

$$= \dfrac{1}{4} - \dfrac{\sqrt{15}}{4}i$$

119. $\left(\dfrac{i}{3 + 2i}\right)^2 = \dfrac{i^2}{(3 + 2i)^2} = \dfrac{-1}{(3 + 2i)(3 + 2i)} = \dfrac{-1}{9 + 12i + 4i^2} = \dfrac{-1}{5 + 12i} = \dfrac{-1(5 - 12i)}{(5 + 12i)(5 - 12i)}$

$$= \dfrac{-5 + 12i}{25 - 144i^2}$$

$$= -\tfrac{5}{169} + \tfrac{12}{169}i$$

121. $(5 + 3i) - (3 - 5i) + \sqrt{-1} = 5 + 3i - 3 + 5i + i = 2 + 9i$

123. $(-8 - \sqrt{3}i) - (7 - 3\sqrt{3}i) = -8 - \sqrt{3}i - 7 + 3\sqrt{3}i = -15 + 2\sqrt{3}i$

125. $(2 + i)(2 - i)(1 + i) = (4 - 2i + 2i - i^2)(1 + i) = 5(1 + i) = 5 + 5i$

127. $(3 + i)[(3 - 2i) + (2 + i)] = (3 + i)(5 - i) = 15 - 3i + 5i - i^2 = 16 + 2i$

129. $V = IR = (2 - 3i)(2 + i) = 4 + 2i - 6i - 3i^2 = 4 - 4i - 3(-1) = 4 - 4i + 3 = 7 - 4i$ volts

131. $Z = \dfrac{V}{I} = \dfrac{1.7 + 0.5i}{0.5i} = \dfrac{(1.7 + 0.5i)i}{(0.5i)i} = \dfrac{1.7i + 0.5i^2}{0.5i^2} = \dfrac{-0.5 + 1.7i}{-0.5} = 1 - 3.4i$

133. Answers may vary.

135.
$$x^2 - 2x + 26 = 0$$
$$(1 - 5i)^2 - 2(1 - 5i) + 26 = 0$$
$$(1 - 5i)(1 - 5i) - 2 + 10i + 26 = 0$$
$$1 - 10i + 25i^2 + 24 + 10i = 0$$
$$1 - 10i - 25 + 24 + 10i = 0$$
$$0 = 0$$

137.
$$x^4 - 3x^2 - 4 = 0$$
$$i^4 - 3i^2 - 4 = 0$$
$$1 - 3(-1) - 4 = 0$$
$$1 + 3 - 4 = 0$$
$$0 = 0$$

139. $\dfrac{3 - i}{2} = \dfrac{(3 - i)(3 + i)}{2(3 + i)} = \dfrac{9 - i^2}{2(3 + i)} = \dfrac{10}{2(3 + i)} = \dfrac{5}{3 + i}$

Chapter 9 Review (page 716)

1. $\sqrt{81} = \sqrt{9^2} = 9$ **2.** $-\sqrt{169} = -\sqrt{13^2} = -13$ **3.** $-\sqrt{36} = -\sqrt{6^2} = -6$

4. $\sqrt{225} = \sqrt{15^2} = 15$ **5.** $\sqrt[3]{-27} = \sqrt[3]{(-3)^3} = -3$ **6.** $-\sqrt[3]{216} = -\sqrt[3]{6^3} = -6$

7. $\sqrt[4]{625} = \sqrt[4]{5^4} = 5$ **8.** $\sqrt[5]{-32} = \sqrt[5]{(-2)^5} = -2$

9. $\sqrt{25x^2} = \sqrt{5^2 x^2} = |5x| = 5|x|$ **10.** $\sqrt{x^2 + 6x + 9} = \sqrt{(x + 3)^2} = |x + 3|$

11. $\sqrt[3]{27a^6 b^3} = 3a^2 b$ **12.** $\sqrt[4]{256x^8 y^4} = |4x^2 y| = 4x^2 |y|$

13. $y = f(x) = \sqrt{x + 2}$; Shift $y = \sqrt{x}$ left 2.
D: $[-2, \infty)$, R: $[0, \infty)$

14. $y = f(x) = -\sqrt{x - 1}$; Reflect $y = \sqrt{x}$
about the x-axis and shift right 1.
D: $[1, \infty)$, R: $(-\infty, 0]$

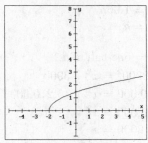

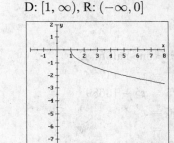

15. $y = f(x) = -\sqrt{x} + 2$; Reflect $y = \sqrt{x}$
about the x-axis and shift up 2.
D: $[0, \infty)$, R: $(-\infty, 2]$

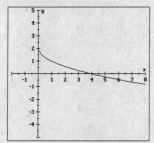

16. $y = f(x) = -\sqrt[3]{x} + 3$; Reflect $y = \sqrt[3]{x}$
about the x-axis and shift up 3.
D: $(-\infty, \infty)$, R: $(-\infty, \infty)$

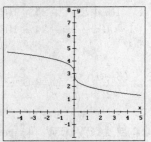

17. mean $= \frac{4+8+12+16+20}{5} = \frac{60}{5} = 12$

18.

Original term	Mean	Difference (term−mean)	Square of difference
4	12	−8	64
8	12	−4	16
12	12	0	0
16	12	4	16
20	12	8	64

$$s = \sqrt{\frac{64 + 16 + 0 + 16 + 64}{5}}$$
$$= \sqrt{\frac{160}{5}}$$
$$= \sqrt{32}$$
$$\approx 5.7$$

19. $d = 1.4\sqrt{h}$
$d = 1.4\sqrt{4.7} \approx 3$ miles

20. $d = 1.4\sqrt{h}$
$4 = 1.4\sqrt{h}$
$16 = 1.96h$
8.2 ft $= h$

21. Let $x =$ one-half of d.
$125^2 = x^2 + 117^2$
$15625 = x^2 + 13689$
$1936 = x^2$
$44 = x$
$d = 2x = 2(44) = 88$ yd

22. Let $x =$ one-half of d.
$8900^2 = x^2 + 3900^2$
$79,210,000 = x^2 + 15,210,000$
$64,000,000 = x^2$
$8000 = x$
$d = 2x = 2(8000) = 16,000$ yd (≈ 9 mi)

23. $d = \sqrt{(x_2 - x_1)^2 + (y_2 - y_1)^2} = \sqrt{(9 - 1)^2 + (-1 - 5)^2} = \sqrt{8^2 + (-6)^2} = \sqrt{100} = 10$

24. $d = \sqrt{(x_2 - x_1)^2 + (y_2 - y_1)^2} = \sqrt{(-2 - (-4))^2 + (8 - 6)^2} = \sqrt{(2)^2 + (2)^2} = \sqrt{8} \approx 2.83$

25. $81^{1/2} = 9$ **26.** $-49^{1/2} = -7$ **27.** $9^{3/2} = \left(9^{1/2}\right)^3 = 3^3 = 27$

28. $16^{3/2} = \left(16^{1/2}\right)^3 = 4^3$
$= 64$

29. $(-27)^{1/3} = -3$

30. $-8^{2/3} = -(8^{1/3})^2 = -2^2$
$= -4$

31. $8^{-2/3} = \dfrac{1}{8^{2/3}} = \dfrac{1}{(8^{1/3})^2} = \dfrac{1}{2^2} = \dfrac{1}{4}$

32. $64^{-1/3} = \dfrac{1}{64^{1/3}} = \dfrac{1}{4}$

33. $-49^{5/2} = -(49^{1/2})^5 = -7^5 = -16,807$

34. $\dfrac{1}{16^{5/2}} = \dfrac{1}{(16^{1/2})^5} = \dfrac{1}{4^5} = \dfrac{1}{1024}$

35. $\left(\dfrac{27}{125}\right)^{-2/3} = \left(\dfrac{125}{27}\right)^{2/3} = \left(\left(\dfrac{125}{27}\right)^{1/3}\right)^2 = \left(\dfrac{5}{3}\right)^2 = \dfrac{25}{9}$

36. $\left(\dfrac{4}{9}\right)^{-3/2} = \left(\dfrac{9}{4}\right)^{3/2} = \left[\left(\dfrac{9}{4}\right)^{1/2}\right]^3 = \left(\dfrac{3}{2}\right)^3 = \dfrac{27}{8}$

37. $(27x^3y)^{1/3} = 3xy^{1/3}$

38. $(16x^2y^4)^{1/4} = 2x^{1/2}y$

39. $(25x^3y^4)^{3/2} = 125x^{9/2}y^6$

40. $(8u^2v^3)^{-2/3} = \dfrac{1}{4u^{4/3}v^2}$

41. $5^{1/4}5^{1/2} = 5^{1/4+1/2} = 5^{3/4}$

42. $a^{5/9}a^{2/9} = a^{5/9+2/9} = a^{7/9}$

43. $u^{1/2}\left(u^{1/2} - u^{-1/2}\right) = u^{2/2} - u^0 = u - 1$

44. $v^{2/3}\left(v^{1/3} + v^{4/3}\right) = v^{3/3} + v^{6/3} = v + v^2$

45. $\left(x^{1/2} + y^{1/2}\right)^2 = \left(x^{1/2} + y^{1/2}\right)\left(x^{1/2} + y^{1/2}\right) = x^{2/2} + x^{1/2}y^{1/2} + x^{1/2}y^{1/2} + y^{2/2}$
$= x + 2x^{1/2}y^{1/2} + y$

46. $\left(a^{2/3} + b^{2/3}\right)\left(a^{2/3} - b^{2/3}\right) = a^{4/3} - a^{2/3}b^{2/3} + a^{2/3}b^{2/3} - b^{4/3} = a^{4/3} - b^{4/3}$

47. $\sqrt[6]{5^2} = 5^{2/6} = 5^{1/3} = \sqrt[3]{5}$

48. $\sqrt[8]{x^4} = x^{4/8} = x^{1/2} = \sqrt{x}$

49. $\sqrt[9]{27a^3b^6} = (3^3a^3b^6)^{1/9} = 3^{3/9}a^{3/9}b^{6/9} = 3^{1/3}a^{1/3}b^{2/3} = \sqrt[3]{3ab^2}$

50. $\sqrt[4]{25a^2b^2} = 5^{2/4}a^{2/4}b^{2/4} = 5^{1/2}a^{1/2}b^{1/2} = \sqrt{5ab}$

51. $\sqrt{176} = \sqrt{16 \cdot 11} = \sqrt{16}\sqrt{11} = 4\sqrt{11}$

52. $\sqrt[3]{250} = \sqrt[3]{125 \cdot 2} = \sqrt[3]{125}\sqrt[3]{2} = 5\sqrt[3]{2}$

53. $\sqrt[4]{32} = \sqrt[4]{16 \cdot 2} = \sqrt[4]{16}\sqrt[4]{2} = 2\sqrt[4]{2}$

54. $\sqrt[5]{96} = \sqrt[5]{32 \cdot 3} = \sqrt[5]{32}\sqrt[5]{3} = 2\sqrt[5]{3}$

55. $\sqrt{8x^3} = \sqrt{4x^2}\sqrt{2x} = 2x\sqrt{2x}$

56. $\sqrt{24x^6y^5} = \sqrt{4x^6y^4}\sqrt{6y} = 2x^3y^2\sqrt{6y}$

57. $\sqrt[3]{16x^5y^4} = \sqrt[3]{8x^3y^3}\sqrt[3]{2x^2y} = 2xy\sqrt[3]{2x^2y}$

58. $\sqrt[3]{54x^7y^3} = \sqrt[3]{27x^6y^3}\sqrt[3]{2x} = 3x^2y\sqrt[3]{2x}$

59. $\dfrac{\sqrt{32x^3}}{\sqrt{2x}} = \sqrt{\dfrac{32x^3}{2x}} = \sqrt{16x^2} = 4x$

60. $\dfrac{\sqrt[3]{16x^5}}{\sqrt[3]{2x^2}} = \sqrt[3]{\dfrac{16x^5}{2x^2}} = \sqrt[3]{8x^3} = 2x$

61. $\sqrt[3]{\dfrac{2a^2b}{27x^3}} = \dfrac{\sqrt[3]{2a^2b}}{3x}$

62. $\sqrt{\dfrac{17xy}{64a^4}} = \dfrac{\sqrt{17xy}}{8a^2}$

63. $\sqrt{12} + \sqrt{27} = \sqrt{4}\sqrt{3} + \sqrt{9}\sqrt{3}$
$\qquad = 2\sqrt{3} + 3\sqrt{3} = 5\sqrt{3}$

64. $\sqrt{18} - \sqrt{32} = \sqrt{9}\sqrt{2} - \sqrt{16}\sqrt{2}$
$\qquad = 3\sqrt{2} - 4\sqrt{2} = -\sqrt{2}$

65. $2\sqrt[3]{3} - \sqrt[3]{24} = 2\sqrt[3]{3} - \sqrt[3]{8}\sqrt[3]{3} = 2\sqrt[3]{3} - 2\sqrt[3]{3} = 0$

66. $\sqrt[4]{32} + 2\sqrt[4]{162} = \sqrt[4]{16}\sqrt[4]{2} + 2\sqrt[4]{81}\sqrt[4]{2} = 2\sqrt[4]{2} + 2(3)\sqrt[4]{2} = 2\sqrt[4]{2} + 6\sqrt[4]{2} = 8\sqrt[4]{2}$

67. $2x\sqrt{8} + 2\sqrt{200x^2} + \sqrt{50x^2} = 2x\sqrt{4}\sqrt{2} + 2\sqrt{100x^2}\sqrt{2} + \sqrt{25x^2}\sqrt{2}$
$\qquad = 2x(2)\sqrt{2} + 2(10x)\sqrt{2} + 5x\sqrt{2}$
$\qquad = 4x\sqrt{2} + 20x\sqrt{2} + 5x\sqrt{2} = 29x\sqrt{2}$

68. $3\sqrt{27a^3} - 2a\sqrt{3a} + 5\sqrt{75a^3} = 3\sqrt{9a^2}\sqrt{3a} - 2a\sqrt{3a} + 5\sqrt{25a^2}\sqrt{3a}$
$\qquad = 3(3a)\sqrt{3a} - 2a\sqrt{3a} + 5(5a)\sqrt{3a}$
$\qquad = 9a\sqrt{3a} - 2a\sqrt{3a} + 25a\sqrt{3a} = 32a\sqrt{3a}$

69. $\sqrt[3]{54} - 3\sqrt[3]{16} + 4\sqrt[3]{128} = \sqrt[3]{27}\sqrt[3]{2} - 3\sqrt[3]{8}\sqrt[3]{2} + 4\sqrt[3]{64}\sqrt[3]{2}$
$\qquad = 3\sqrt[3]{2} - 3(2)\sqrt[3]{2} + 4(4)\sqrt[3]{2} = 3\sqrt[3]{2} - 6\sqrt[3]{2} + 16\sqrt[3]{2} = 13\sqrt[3]{2}$

70. $2\sqrt[4]{32x^5} + 4\sqrt[4]{162x^5} - 5x\sqrt[4]{512x} = 2\sqrt[4]{16x^4}\sqrt[4]{2x} + 4\sqrt[4]{81x^4}\sqrt[4]{2x} - 5x\sqrt[4]{256}\sqrt[4]{2x}$
$\qquad = 2(2x)\sqrt[4]{2x} + 4(3x)\sqrt[4]{2x} - 5x(4)\sqrt[4]{2x}$
$\qquad = 4x\sqrt[4]{2x} + 12x\sqrt[4]{2x} - 20x\sqrt[4]{2x} = -4x\sqrt[4]{2x}$

71. $\text{hypotenuse} = 7\sqrt{2}$ m

72. $\text{shorter leg} = \dfrac{1}{2}\left(12\sqrt{3}\right) = 6\sqrt{3}$ cm
$\text{longer leg} = \sqrt{3}\left(6\sqrt{3}\right) = 6(3) = 18$ cm

73. $x = 5\sqrt{2} \approx 7.07$ in.

74. $x = \sqrt{3}\left(\dfrac{1}{2} \cdot 10\right) = 5\sqrt{3} \approx 8.66$ cm

75. $\left(5\sqrt{3}\right)\left(2\sqrt{5}\right) = 10\sqrt{15}$

76. $2\sqrt{6}\sqrt{216} = 2\sqrt{6}\sqrt{36}\sqrt{6} = 2(6)(6) = 72$

77. $\sqrt{5x}\sqrt{4x} = \sqrt{20x^2} = \sqrt{4x^2}\sqrt{5} = 2x\sqrt{5}$

78. $\sqrt[3]{3x^2}\sqrt[3]{9x^4} = \sqrt[3]{27x^6} = 3x^2$

79. $-\sqrt[3]{2x^2}\sqrt[3]{4x} = -\sqrt[3]{8x^3} = -2x$

80. $-\sqrt[4]{256x^5y^{11}}\sqrt[4]{625x^9y^3} = -\sqrt[4]{256x^4y^8}\sqrt[4]{xy^3}\sqrt[4]{625x^8}\sqrt[4]{xy^3} = -4xy^2(5x^2)\sqrt[4]{x^2y^6}$
$\qquad = -20x^3y^2\sqrt[4]{y^4}\sqrt[4]{x^2y^2}$
$\qquad = -20x^3y^3\sqrt[4]{x^2y^2} = -20x^3y^3\sqrt{xy}$

81. $\sqrt{3}\left(\sqrt{27}-4\right)=\sqrt{81}-4\sqrt{3}=9-4\sqrt{3}$ **82.** $\sqrt{5}\left(\sqrt{5}-6\right)=\sqrt{25}-6\sqrt{5}=5-6\sqrt{5}$

83. $\sqrt{5}\left(\sqrt{2}-1\right)=\sqrt{10}-\sqrt{5}$ **84.** $\sqrt{3}\left(\sqrt{3}+\sqrt{2}\right)=\sqrt{9}+\sqrt{6}=3+\sqrt{6}$

85. $\left(\sqrt{5}+1\right)\left(\sqrt{5}-1\right)=\sqrt{25}-\sqrt{5}+\sqrt{5}-1=4$

86. $\left(\sqrt{3}+\sqrt{2}\right)\left(\sqrt{3}+\sqrt{2}\right)=\sqrt{9}+\sqrt{6}+\sqrt{6}+\sqrt{4}=5+2\sqrt{6}$

87. $\left(\sqrt{x}+\sqrt{y}\right)\left(\sqrt{x}-\sqrt{y}\right)=\sqrt{x^2}-\sqrt{xy}+\sqrt{xy}-\sqrt{y^2}=x-y$

88. $\left(2\sqrt{u}+3\right)\left(3\sqrt{u}-4\right)=6\sqrt{u^2}-8\sqrt{u}+9\sqrt{u}-12=6u+\sqrt{u}-12$

89. $\dfrac{1}{\sqrt{3}}=\dfrac{1\sqrt{3}}{\sqrt{3}\sqrt{3}}=\dfrac{\sqrt{3}}{3}$ **90.** $\dfrac{\sqrt{3}}{\sqrt{5}}=\dfrac{\sqrt{3}\sqrt{5}}{\sqrt{5}\sqrt{5}}=\dfrac{\sqrt{15}}{5}$

91. $\dfrac{x}{\sqrt{xy}}=\dfrac{x\sqrt{xy}}{\sqrt{xy}\sqrt{xy}}=\dfrac{x\sqrt{xy}}{xy}=\dfrac{\sqrt{xy}}{y}$

92. $\dfrac{\sqrt[3]{uv}}{\sqrt[3]{u^5v^7}}=\dfrac{\sqrt[3]{uv}\sqrt[3]{uv^2}}{\sqrt[3]{u^5v^7}\sqrt[3]{uv^2}}=\dfrac{\sqrt[3]{u^2v^3}}{\sqrt[3]{u^6v^9}}=\dfrac{v\sqrt[3]{u^2}}{u^2v^3}=\dfrac{\sqrt[3]{u^2}}{u^2v^2}$

93. $\dfrac{2}{\sqrt{2}-1}=\dfrac{2\left(\sqrt{2}+1\right)}{\left(\sqrt{2}-1\right)\left(\sqrt{2}+1\right)}=\dfrac{2\left(\sqrt{2}+1\right)}{\sqrt{4}-1}=\dfrac{2\left(\sqrt{2}+1\right)}{2-1}=\dfrac{2\left(\sqrt{2}+1\right)}{1}=2\left(\sqrt{2}+1\right)$

94. $\dfrac{\sqrt{2}}{\sqrt{3}-1}=\dfrac{\sqrt{2}\left(\sqrt{3}+1\right)}{\left(\sqrt{3}-1\right)\left(\sqrt{3}+1\right)}=\dfrac{\sqrt{2}\left(\sqrt{3}+1\right)}{\sqrt{9}-1}=\dfrac{\sqrt{2}\left(\sqrt{3}+1\right)}{3-1}=\dfrac{\sqrt{2}\left(\sqrt{3}+1\right)}{2}$

$$=\dfrac{\sqrt{6}+\sqrt{2}}{2}$$

95. $\dfrac{2x-32}{\sqrt{x}+4}=\dfrac{(2x-32)(\sqrt{x}-4)}{(\sqrt{x}+4)(\sqrt{x}-4)}=\dfrac{2(x-16)(\sqrt{x}-4)}{\sqrt{x^2}-16}=\dfrac{2(x-16)(\sqrt{x}-4)}{x-16}=2(\sqrt{x}-4)$

96. $\dfrac{\sqrt{a}+1}{\sqrt{a}-1}=\dfrac{(\sqrt{a}+1)(\sqrt{a}+1)}{(\sqrt{a}-1)(\sqrt{a}+1)}=\dfrac{\sqrt{a^2}+\sqrt{a}+\sqrt{a}+1}{\sqrt{a^2}-1}=\dfrac{a+2\sqrt{a}+1}{a-1}$

97. $\dfrac{\sqrt{3}}{5}=\dfrac{\sqrt{3}\sqrt{3}}{5\sqrt{3}}=\dfrac{3}{5\sqrt{3}}$ **98.** $\dfrac{\sqrt[3]{9}}{3}=\dfrac{\sqrt[3]{9}\sqrt[3]{3}}{3\sqrt[3]{3}}=\dfrac{\sqrt[3]{27}}{3\sqrt[3]{3}}=\dfrac{3}{3\sqrt[3]{3}}=\dfrac{1}{\sqrt[3]{3}}$

99. $\dfrac{3-\sqrt{x}}{2} = \dfrac{\left(3-\sqrt{x}\right)\left(3+\sqrt{x}\right)}{2\left(3+\sqrt{x}\right)} = \dfrac{9-\sqrt{x^2}}{2\left(3+\sqrt{x}\right)} = \dfrac{9-x}{2\left(3+\sqrt{x}\right)}$

100. $\dfrac{\sqrt{a}-\sqrt{b}}{\sqrt{a}} = \dfrac{\left(\sqrt{a}-\sqrt{b}\right)\left(\sqrt{a}+\sqrt{b}\right)}{\sqrt{a}\left(\sqrt{a}+\sqrt{b}\right)} = \dfrac{\sqrt{a^2}-\sqrt{b^2}}{\sqrt{a^2}+\sqrt{ab}} = \dfrac{a-b}{a+\sqrt{ab}}$

101.
$$\sqrt{3y-11} = \sqrt{y+7}$$
$$\left(\sqrt{3y-11}\right)^2 = \left(\sqrt{y+7}\right)^2$$
$$3y-11 = y+7$$
$$2y = 18$$
$$y = 9$$
The answer checks.

102.
$$u = \sqrt{25u-144}$$
$$u^2 = \left(\sqrt{25u-144}\right)^2$$
$$u^2 = 25u-144$$
$$u^2 - 25u + 144 = 0$$
$$(u-9)(u-16) = 0$$
$$u = 9 \text{ or } u = 16; \text{ Both answers check.}$$

103.
$$\sqrt{2x+6}+1 = x$$
$$\sqrt{2x+6} = x-1$$
$$\left(\sqrt{2x+6}\right)^2 = (x-1)^2$$
$$2x+6 = x^2 - 2x + 1$$
$$0 = x^2 - 4x - 5$$
$$0 = (x-5)(x+1)$$
$$x-5 = 0 \quad \textbf{or} \quad x+1 = 0$$
$$x = 5 \qquad\qquad x = -1$$
solution $\qquad$ extraneous

104.
$$\sqrt{z+1}+\sqrt{z} = 2$$
$$\sqrt{z+1} = 2-\sqrt{z}$$
$$\left(\sqrt{z+1}\right)^2 = \left(2-\sqrt{z}\right)^2$$
$$z+1 = 4 - 4\sqrt{z} + z$$
$$4\sqrt{z} = 3$$
$$\left(4\sqrt{z}\right)^2 = 3^2$$
$$16z = 9$$
$$z = \tfrac{9}{16}; \text{ The answer checks.}$$

105.
$$\sqrt{2x+5}-\sqrt{2x} = 1$$
$$\sqrt{2x+5} = 1+\sqrt{2x}$$
$$\left(\sqrt{2x+5}\right)^2 = \left(1+\sqrt{2x}\right)^2$$
$$2x+5 = 1 + 2\sqrt{2x} + 2x$$
$$4 = 2\sqrt{2x}$$
$$4^2 = \left(2\sqrt{2x}\right)^2$$
$$16 = 8x$$
$$2 = x; \text{ The answer checks.}$$

106.
$$\sqrt[3]{x^3+8} = x+2$$
$$\left(\sqrt[3]{x^3+8}\right)^3 = (x+2)^3$$
$$x^3+8 = x^3 + 6x^2 + 12x + 8$$
$$0 = 6x^2 + 12x$$
$$0 = 6x(x+2)$$
$$x = 0 \text{ or } x = -2; \text{ Both answers check.}$$

107. $(6+3i)+(4-15i) = 6+3i+4-15i = 10-12i$

108. $(-5-22i)-(-7+28i) = -5-22i+7-28i = 2-50i$

109. $(-32+\sqrt{-144})-(64+\sqrt{-81}) = -32+\sqrt{144i^2}-64-\sqrt{81i^2} = -32+12i-64-9i$
$$= -96+3i$$

110. $(-8 + \sqrt{-8}) + (6 - \sqrt{-32}) = -8 + \sqrt{4i^2}\sqrt{2} + 6 - \sqrt{16i^2}\sqrt{2}$
$$= -8 + 2i\sqrt{2} + 6 - 4i\sqrt{2} = -2 - 2\sqrt{2}i$$

111. $(2 - 7i)(-3 + 4i) = -6 + 8i + 21i - 28i^2 = -6 + 29i + 28 = 22 + 29i$

112. $(-4 + 5i)(3 + 2i) = -12 - 8i + 15i + 10i^2 = -12 + 7i - 10 = -22 + 7i$

113. $(5 - \sqrt{-27})(-6 + \sqrt{-12}) = (5 - 3i\sqrt{3})(-6 + 2i\sqrt{3}) = -30 + 10i\sqrt{3} + 18i\sqrt{3} - 6i^2(3)$
$$= -30 + 28i\sqrt{3} + 18 = -12 + 28\sqrt{3}i$$

114. $(2 + \sqrt{-128})(3 - \sqrt{-98}) = (2 + 8i\sqrt{2})(3 - 7i\sqrt{2}) = 6 - 14i\sqrt{2} + 24i\sqrt{2} - 56i^2(2)$
$$= 6 + 10i\sqrt{2} + 112 = 118 + 10\sqrt{2}i$$

115. $\dfrac{3}{4i} = \dfrac{3}{4i} \cdot \dfrac{i}{i} = \dfrac{3i}{4i^2} = \dfrac{3i}{-4} = -\dfrac{3}{4}i = 0 - \dfrac{3}{4}i$ **116.** $\dfrac{-2}{5i^3} = \dfrac{-2}{5i^3} \cdot \dfrac{i}{i} = \dfrac{-2i}{5i^4} = \dfrac{-2i}{5} = 0 - \dfrac{2}{5}i$

117. $\dfrac{6}{2 + i} = \dfrac{6}{2 + i} \cdot \dfrac{2 - i}{2 - i} = \dfrac{6(2 - i)}{4 - i^2} = \dfrac{6(2 - i)}{5} = \dfrac{12 - 6i}{5} = \dfrac{12}{5} - \dfrac{6}{5}i$

118. $\dfrac{7}{3 - i} = \dfrac{7}{3 - i} \cdot \dfrac{3 + i}{3 + i} = \dfrac{7(3 + i)}{9 - i^2} = \dfrac{7(3 + i)}{10} = \dfrac{21 + 7i}{10} = \dfrac{21}{10} + \dfrac{7}{10}i$

119. $\dfrac{4 + i}{4 - i} = \dfrac{4 + i}{4 - i} \cdot \dfrac{4 + i}{4 + i} = \dfrac{16 + 8i + i^2}{16 - i^2} = \dfrac{15 + 8i}{17} = \dfrac{15}{17} + \dfrac{8}{17}i$

120. $\dfrac{3 - i}{3 + i} = \dfrac{3 - i}{3 + i} \cdot \dfrac{3 - i}{3 - i} = \dfrac{9 - 6i + i^2}{9 - i^2} = \dfrac{8 - 6i}{10} = \dfrac{8}{10} - \dfrac{6}{10}i = \dfrac{4}{5} - \dfrac{3}{5}i$

121. $\dfrac{3}{5 + \sqrt{-4}} = \dfrac{3}{5 + 2i} = \dfrac{3}{5 + 2i} \cdot \dfrac{5 - 2i}{5 - 2i} = \dfrac{3(5 - 2i)}{25 - 4i^2} = \dfrac{3(5 - 2i)}{29} = \dfrac{15 - 6i}{29} = \dfrac{15}{29} - \dfrac{6}{29}i$

122. $\dfrac{2}{3 - \sqrt{-9}} = \dfrac{2}{3 - 3i} = \dfrac{2}{3 - 3i} \cdot \dfrac{3 + 3i}{3 + 3i} = \dfrac{2(3 + 3i)}{9 - 9i^2} = \dfrac{2(3 + 3i)}{18} = \dfrac{3 + 3i}{9} = \dfrac{3}{9} + \dfrac{3}{9}i = \dfrac{1}{3} + \dfrac{1}{3}i$

123. $|9 + 12i| = \sqrt{9^2 + 12^2} = \sqrt{81 + 144} = \sqrt{225} = 15$

124. $|24 - 10i| = \sqrt{24^2 + (-10)^2} = \sqrt{576 + 100} = \sqrt{676} = 26$

125. $i^{39} = i^{36}i^3 = (i^4)^9 i^3 = 1^9(-i) = -i$ **126.** $i^{52} = (i^4)^{13} = 1^{13} = 1$

Chapter 9 Test   (page 722)

1. $\sqrt{36} = \sqrt{6^2} = 6$ **2.** $\sqrt{125} = \sqrt{25}\sqrt{5} = 5\sqrt{5}$

3. $\sqrt{4x^2} = \sqrt{(2x)^2} = |2x| = 2|x|$

4. $\sqrt[3]{8x^3} = \sqrt[3]{(2x)^3} = 2x$

5. $f(x) = \sqrt{x-2}$; Shift $y = \sqrt{x}$ right 2.

domain $= [2, \infty)$, range $= [0, \infty)$

6. $f(x) = \sqrt[3]{x} + 3$; Shift $y = \sqrt[3]{x}$ up 3.

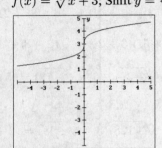

domain $= (-\infty, \infty)$, range $= (-\infty, \infty)$

7.
$$53^2 = 45^2 + h^2$$
$$2809 = 2025 + h^2$$
$$784 = h^2$$
$$28 \text{ in.} = h$$

8.
$$2^2 = \left(\frac{w}{2}\right)^2 + (1.9)^2$$
$$4 = \frac{w^2}{4} + 3.61$$
$$0.39 = \frac{w^2}{4}$$
$$1.56 = w^2$$
$$1.25 \text{ meters} = w$$

9. $d = \sqrt{(x_2 - x_1)^2 + (y_2 - y_1)^2} = \sqrt{(0-6)^2 + (0-8)^2} = \sqrt{(-6)^2 + (-8)^2} = \sqrt{100} = 10$

10. $d = \sqrt{(x_2 - x_1)^2 + (y_2 - y_1)^2} = \sqrt{(22 - (-2))^2 + (12 - 5)^2} = \sqrt{24^2 + 7^2} = \sqrt{625} = 25$

11. $81^{1/4} = 3$

12. $-64^{2/3} = -\left(64^{1/3}\right)^2 = -4^2 = -16$

13. $36^{-3/2} = \dfrac{1}{36^{3/2}} = \dfrac{1}{(36^{1/2})^3} = \dfrac{1}{6^3} = \dfrac{1}{216}$

14. $\left(-\dfrac{8}{27}\right)^{-2/3} = \left(-\dfrac{27}{8}\right)^{2/3} = \left[\left(-\dfrac{27}{8}\right)^{1/3}\right]^2 = \left(-\dfrac{3}{2}\right)^2 = \dfrac{9}{4}$

15. $\dfrac{2^{5/3}2^{1/6}}{2^{1/2}} = \dfrac{2^{10/6}2^{1/6}}{2^{3/6}} = 2^{10/6 + 1/6 - 3/6} = 2^{8/6} = 2^{4/3}$

16. $\dfrac{(8x^3y)^{1/2}(8xy^5)^{1/2}}{(x^3y^6)^{1/3}} = \dfrac{8^{1/2}x^{3/2}y^{1/2}8^{1/2}x^{1/2}y^{5/2}}{x^{3/3}y^{6/3}} = \dfrac{8^{2/2}x^{4/2}y^{6/2}}{xy^2} = \dfrac{8x^2y^3}{xy^2} = 8xy$

17. $\sqrt{108} = \sqrt{36}\sqrt{3} = 6\sqrt{3}$

18. $\sqrt{250x^3y^5} = \sqrt{25x^2y^4}\sqrt{10xy} = 5xy^2\sqrt{10xy}$

19. $\dfrac{\sqrt[3]{240x^{10}y^4}}{\sqrt[3]{6xy}} = \sqrt[3]{\dfrac{240x^{10}y^4}{6xy}} = \sqrt[3]{40x^9y^3} = \sqrt[3]{8x^9y^3}\,\sqrt[3]{5} = 2x^3y\sqrt[3]{5}$

20. $\sqrt{\dfrac{3a^5}{48a^7}} = \sqrt{\dfrac{1}{16a^2}} = \dfrac{1}{4a}$

21. $\sqrt{45x^2} = \sqrt{9x^2}\,\sqrt{5} = 3|x|\sqrt{5}$

22. $\sqrt{48x^6} = \sqrt{16x^6}\,\sqrt{3} = 4|x^3|\sqrt{3}$

23. $\sqrt[3]{16x^9} = \sqrt[3]{8x^9}\,\sqrt[3]{2} = 2x^3\sqrt[3]{2}$

24. $\sqrt{18x^4y^9} = \sqrt{9x^4y^8}\,\sqrt{2y} = |3x^2y^4|\sqrt{2y} = 3x^2y^4\sqrt{2y}$

25. $\sqrt{12} - \sqrt{27} = \sqrt{4}\,\sqrt{3} - \sqrt{9}\,\sqrt{3} = 2\sqrt{3} - 3\sqrt{3} = -\sqrt{3}$

26. $2\sqrt[3]{40} - \sqrt[3]{5000} + 4\sqrt[3]{625} = 2\sqrt[3]{8}\,\sqrt[3]{5} - \sqrt[3]{1000}\,\sqrt[3]{5} + 4\sqrt[3]{125}\,\sqrt[3]{5} = 2(2)\sqrt[3]{5} - 10\sqrt[3]{5} + 4(5)\sqrt[3]{5}$
$$= 4\sqrt[3]{5} - 10\sqrt[3]{5} + 20\sqrt[3]{5}$$
$$= 14\sqrt[3]{5}$$

27. $2\sqrt{48y^5} - 3y\sqrt{12y^3} = 2\sqrt{16y^4}\,\sqrt{3y} - 3y\sqrt{4y^2}\,\sqrt{3y} = 2(4y^2)\sqrt{3y} - 3y(2y)\sqrt{3y}$
$$= 8y^2\sqrt{3y} - 6y^2\sqrt{3y} = 2y^2\sqrt{3y}$$

28. $\sqrt[4]{768z^5} + z\sqrt[4]{48z} = \sqrt[4]{256z^4}\,\sqrt[4]{3z} + z\sqrt[4]{16}\,\sqrt[4]{3z} = 4z\sqrt[4]{3z} + 2z\sqrt[4]{3z} = 6z\sqrt[4]{3z}$

29. $-2\sqrt{xy}\left(3\sqrt{x} + \sqrt{xy^3}\right) = -6\sqrt{x^2y} - 2\sqrt{x^2y^4} = -6x\sqrt{y} - 2xy^2$

30. $\left(3\sqrt{2} + \sqrt{3}\right)\left(2\sqrt{2} - 3\sqrt{3}\right) = 6\sqrt{4} - 9\sqrt{6} + 2\sqrt{6} - 3\sqrt{9} = 12 - 7\sqrt{6} - 9 = 3 - 7\sqrt{6}$

31. $\dfrac{1}{\sqrt{5}} = \dfrac{1\sqrt{5}}{\sqrt{5}\sqrt{5}} = \dfrac{\sqrt{5}}{5}$

32. $\dfrac{3t-1}{\sqrt{3t}-1} = \dfrac{(3t-1)\left(\sqrt{3t}+1\right)}{\left(\sqrt{3t}-1\right)\left(\sqrt{3t}+1\right)} = \dfrac{(3t-1)\left(\sqrt{3t}+1\right)}{\sqrt{9t^2}-1} = \dfrac{(3t-1)\left(\sqrt{3t}+1\right)}{3t-1} = \sqrt{3t}+1$

33. $\dfrac{\sqrt{3}}{\sqrt{7}} = \dfrac{\sqrt{3}\sqrt{3}}{\sqrt{7}\sqrt{3}} = \dfrac{3}{\sqrt{21}}$

34. $\dfrac{\sqrt{a}+\sqrt{b}}{\sqrt{a}-\sqrt{b}} = \dfrac{\left(\sqrt{a}+\sqrt{b}\right)\left(\sqrt{a}-\sqrt{b}\right)}{\left(\sqrt{a}-\sqrt{b}\right)\left(\sqrt{a}-\sqrt{b}\right)} = \dfrac{\sqrt{a^2}-\sqrt{b^2}}{\sqrt{a^2}-\sqrt{ab}-\sqrt{ab}+\sqrt{b^2}} = \dfrac{a-b}{a-2\sqrt{ab}+b}$

35. $\sqrt[3]{6n+4} - 4 = 0$

$\sqrt[3]{6n+4} = 4$

$\left(\sqrt[3]{6n+4}\right)^3 = 4^3$

$6n + 4 = 64$

$6n = 60$

$n = 10$

The answer checks.

36. $1 - \sqrt{u} = \sqrt{u-3}$

$\left(1 - \sqrt{u}\right)^2 = \left(\sqrt{u-3}\right)^2$

$1 - 2\sqrt{u} + u = u - 3$

$4 = 2\sqrt{u}$

$4^2 = \left(2\sqrt{u}\right)^2$

$16 = 4u$

$4 = u$

extraneous $\Rightarrow \emptyset$

37. $(2+4i) + (-3+7i) = 2 + 4i - 3 + 7i = -1 + 11i$

38. $(3 - \sqrt{-9}) - (-1 + \sqrt{-16}) = 3 - 3i + 1 - 4i = 4 - 7i$

39. $2i(3 - 4i) = 6i - 8i^2 = 6i + 8 = 8 + 6i$

40. $(3 + 2i)(-4 - i) = -12 - 3i - 8i - 2i^2 = -12 - 11i + 2 = -10 - 11i$

41. $\dfrac{1}{\sqrt{2}i} = \dfrac{1}{\sqrt{2}i} \cdot \dfrac{i\sqrt{2}}{i\sqrt{2}} = \dfrac{i\sqrt{2}}{2i^2} = -\dfrac{\sqrt{2}i}{2} = 0 - \dfrac{\sqrt{2}}{2}i$

42. $\dfrac{2+i}{3-i} = \dfrac{2+i}{3-i} \cdot \dfrac{3+i}{3+i} = \dfrac{6 + 5i + i^2}{9 - i^2} = \dfrac{5 + 5i}{10} = \dfrac{1}{2} + \dfrac{1}{2}i$

Exercises 10.1 (page 736)

1. $(x+5)(x-5) = x^2 - 5x + 5x - 25 = x^2 - 25$

3. $(x+2)^2 = (x+2)(x+2) = x^2 + 2x + 2x + 4 = x^2 + 4x + 4$

5. $(x-7)^2 = (x-7)(x-7) = x^2 - 7x - 7x + 49 = x^2 - 14x + 49$

7. $\dfrac{t+9}{2} + \dfrac{t+2}{5} = \dfrac{8}{5} + 4t$

$5(t+9) + 2(t+2) = 2(8) + 40t$

$5t + 45 + 2t + 4 = 16 + 40t$

$33 = 33t$

$1 = t$

9. $3(t-3) + 3t \leq 2(t+1) + t + 1$

$3t - 9 + 3t \leq 2t + 2 + t + 1$

$3t \leq 12$

$t \leq 4$

11. $x = \sqrt{c}; x = -\sqrt{c}$

13. positive or negative

15.
$$x^2 - 81 = 0$$
$$(x + 9)(x - 9) = 0$$
$$x + 9 = 0 \quad \textbf{or} \quad x - 9 = 0$$
$$x = -9 \qquad\qquad x = 9$$

17.
$$2y^2 - 72 = 0$$
$$2(y^2 - 36) = 0$$
$$2(y + 6)(y - 6) = 0$$
$$y + 6 = 0 \quad \textbf{or} \quad y - 6 = 0$$
$$y = -6 \qquad\qquad y = 6$$

19.
$$y^2 + 7y + 12 = 0$$
$$(y + 3)(y + 4) = 0$$
$$y + 3 = 0 \quad \textbf{or} \quad y + 4 = 0$$
$$y = -3 \qquad\qquad y = -4$$

21.
$$6s^2 + 11s - 10 = 0$$
$$(2s + 5)(3s - 2) = 0$$
$$2s + 5 = 0 \quad \textbf{or} \quad 3s - 2 = 0$$
$$s = -\tfrac{5}{2} \qquad\qquad s = \tfrac{2}{3}$$

23.
$$x^2 = 36$$
$$x = \pm\sqrt{36} = \pm 6$$

25.
$$z^2 = 5$$
$$z = \pm\sqrt{5}$$

27.
$$(y + 3)^2 = 9$$
$$y + 3 = \pm\sqrt{9}$$
$$y + 3 = \pm 3$$
$$y = -3 \pm 3$$
$$y = 0 \text{ or } y = -6$$

29.
$$(x - 2)^2 - 5 = 0$$
$$(x - 2)^2 = 5$$
$$x - 2 = \pm\sqrt{5}$$
$$x = 2 \pm\sqrt{5}$$

31.
$$16p^2 + 49 = 0$$
$$16p^2 = -49$$
$$p^2 = -\frac{49}{16}$$
$$p = \pm\sqrt{-\frac{49}{16}}$$
$$p = \pm\frac{7}{4}i$$

33.
$$4m^2 + 81 = 0$$
$$4m^2 = -81$$
$$m^2 = -\frac{81}{4}$$
$$m = \pm\sqrt{-\frac{81}{4}}$$
$$m = \pm\frac{9}{2}i$$

35. $x^2 + 4x; \; \frac{1}{2} \cdot 4 = 2; \; 2^2 = \boxed{4}$

37. $x^2 - 3x; \; \frac{1}{2} \cdot 3 = \frac{3}{2}; \; \left(\frac{3}{2}\right)^2 = \boxed{\frac{9}{4}}$

39.
$$x^2 + 4x - 12 = 0$$
$$x^2 + 4x = 12$$
$$x^2 + 4x + 4 = 12 + 4$$
$$(x + 2)^2 = 16$$
$$x + 2 = \pm 4$$
$$x = -2 \pm 4$$
$$x = 2 \text{ or } x = -6$$

41.
$$x^2 - 6x + 8 = 0$$
$$x^2 - 6x = -8$$
$$x^2 - 6x + 9 = -8 + 9$$
$$(x - 3)^2 = 1$$
$$x - 3 = \pm 1$$
$$x = 3 \pm 1$$
$$x = 4 \text{ or } x = 2$$

43.
$$6x^2 + 11x + 3 = 0$$
$$6x^2 + 11x = -3$$
$$x^2 + \frac{11}{6}x = -\frac{1}{2}$$
$$x^2 + \frac{11}{6}x + \frac{121}{144} = -\frac{1}{2} + \frac{121}{144}$$
$$\left(x + \frac{11}{12}\right)^2 = \frac{49}{144}$$
$$x + \frac{11}{12} = \pm\frac{7}{12}$$
$$x = -\frac{11}{12} \pm \frac{7}{12}$$
$$x = -\frac{4}{12} = -\frac{1}{3} \text{ or } x = -\frac{18}{12} = -\frac{3}{2}$$

45.
$$6x^2 - 7x - 5 = 0$$
$$6x^2 - 7x = 5$$
$$x^2 - \frac{7}{6}x = \frac{5}{6}$$
$$x^2 - \frac{7}{6}x + \frac{49}{144} = \frac{5}{6} + \frac{49}{144}$$
$$\left(x - \frac{7}{12}\right)^2 = \frac{169}{144}$$
$$x - \frac{7}{12} = \pm\frac{13}{12}$$
$$x = \frac{7}{12} \pm \frac{13}{12}$$
$$x = \frac{20}{12} = \frac{5}{3} \text{ or } x = -\frac{6}{12} = -\frac{1}{2}$$

47.
$$9 - 6r = 8r^2$$
$$-8r^2 - 6r = -9$$
$$r^2 + \frac{3}{4}r = \frac{9}{8}$$
$$r^2 + \frac{3}{4}r + \frac{9}{64} = \frac{9}{8} + \frac{9}{64}$$
$$\left(r + \frac{3}{8}\right)^2 = \frac{81}{64}$$
$$r + \frac{3}{8} = \pm\frac{9}{8}$$
$$r = -\frac{3}{8} \pm \frac{9}{8}$$
$$r = \frac{6}{8} = \frac{3}{4} \text{ or } r = -\frac{12}{8} = -\frac{3}{2}$$

49.
$$x + 1 = 2x^2$$
$$-2x^2 + x = -1$$
$$x^2 - \frac{1}{2}x = \frac{1}{2}$$
$$x^2 - \frac{1}{2}x + \frac{1}{16} = \frac{1}{2} + \frac{1}{16}$$
$$\left(x - \frac{1}{4}\right)^2 = \frac{9}{16}$$
$$x - \frac{1}{4} = \pm\frac{3}{4}$$
$$x = \frac{1}{4} \pm \frac{3}{4}$$
$$x = \frac{4}{4} = 1 \text{ or } x = -\frac{2}{4} = -\frac{1}{2}$$

51.
$$\frac{7x + 1}{5} = -x^2$$
$$7x + 1 = -5x^2$$
$$5x^2 + 7x = -1$$
$$x^2 + \frac{7}{5}x = -\frac{1}{5}$$
$$x^2 + \frac{7}{5}x + \frac{49}{100} = -\frac{1}{5} + \frac{49}{100}$$
$$\left(x + \frac{7}{10}\right)^2 = \frac{29}{100}$$
$$x + \frac{7}{10} = \pm\frac{\sqrt{29}}{10}$$
$$x = -\frac{7}{10} \pm \frac{\sqrt{29}}{10}$$

53.
$$3x^2 - 6x + 1 = 0$$
$$3x^2 - 6x = -1$$
$$x^2 - 2x = -\frac{1}{3}$$
$$x^2 - 2x + 1 = -\frac{1}{3} + 1$$
$$(x - 1)^2 = \frac{2}{3}$$
$$x - 1 = \pm\sqrt{\frac{2}{3}}$$
$$x - 1 = \pm\frac{\sqrt{2}}{\sqrt{3}} \cdot \frac{\sqrt{3}}{\sqrt{3}}$$
$$x - 1 = \pm\frac{\sqrt{6}}{3}$$
$$x = 1 \pm \frac{\sqrt{6}}{3}$$

55. $p^2 + 2p + 2 = 0$
$$p^2 + 2p = -2$$
$$p^2 + 2p + 1 = -2 + 1$$
$$(p+1)^2 = -1$$
$$p + 1 = \pm\sqrt{-1}$$
$$p + 1 = \pm i$$
$$p = -1 \pm i$$

57. $y^2 + 8y + 18 = 0$
$$y^2 + 8y = -18$$
$$y^2 + 8y + 16 = -18 + 16$$
$$(y+4)^2 = -2$$
$$y + 4 = \pm\sqrt{-2}$$
$$y = -4 \pm i\sqrt{2}$$

59. $9x - 8 = x^2$
$$0 = x^2 - 9x + 8$$
$$0 = (x-8)(x-1)$$
$$x - 8 = 0 \quad \textbf{or} \quad x - 1 = 0$$
$$x = 8 \qquad\qquad x = 1$$

61. $3x^2 - 16 = 0$
$$3x^2 = 16$$
$$x^2 = \frac{16}{3}$$
$$x = \pm\sqrt{\frac{16}{3}} = \pm\frac{4}{\sqrt{3}} = \pm\frac{4\sqrt{3}}{3}$$

63. $(s-7)^2 - 9 = 0$
$$(s-7)^2 = 9$$
$$s - 7 = \pm\sqrt{9}$$
$$s - 7 = \pm 3$$
$$s = 7 \pm 3$$
$$s = 10 \text{ or } s = 4$$

65. $(x+5)^2 - 3 = 0$
$$(x+5)^2 = 3$$
$$x + 5 = \pm\sqrt{3}$$
$$x = -5 \pm\sqrt{3}$$

67. $2z^2 - 5z + 2 = 0$
$$(2z-1)(z-2) = 0$$
$$2z - 1 = 0 \quad \textbf{or} \quad z - 2 = 0$$
$$z = \tfrac{1}{2} \qquad\qquad z = 2$$

69. $3m^2 - 2m - 3 = 0$
$$m^2 - \frac{2}{3}m = 1$$
$$m^2 - \frac{2}{3}m + \frac{1}{9} = 1 + \frac{1}{9}$$
$$\left(m - \frac{1}{3}\right)^2 = \frac{10}{9}$$
$$m - \frac{1}{3} = \pm\sqrt{\frac{10}{9}}$$
$$m = \frac{1}{3} \pm \frac{\sqrt{10}}{3}$$

71. $5x^2 + 15x = 0$
$$5x(x+3) = 0$$
$$5x = 0 \quad \textbf{or} \quad x + 3 = 0$$
$$x = 0 \qquad\qquad x = -3$$

73. $x^2 + 5x + 4 = 0$
$$(x+4)(x+1) = 0$$
$$x + 4 = 0 \quad \textbf{or} \quad x + 1 = 0$$
$$x = -4 \qquad\qquad x = -1$$

75. $x^2 - 9x - 10 = 0$
$(x + 1)(x - 10) = 0$
$x + 1 = 0$ **or** $x - 10 = 0$
$x = -1$ $\qquad x = 10$

77. $2x^2 - x + 8 = 0$
$2x^2 - x = -8$
$x^2 - \frac{1}{2}x = -4$
$x^2 - \frac{1}{2}x + \frac{1}{16} = -4 + \frac{1}{16}$
$\left(x - \frac{1}{4}\right)^2 = \frac{-63}{16}$
$x - \frac{1}{4} = \pm \frac{\sqrt{-63}}{4}$
$x = \frac{1}{4} \pm \frac{i\sqrt{9}\sqrt{7}}{4}$
$x = \frac{1}{4} \pm \frac{3\sqrt{7}}{4}i$

79. $2d^2 = 3h$
$d^2 = \frac{3h}{2}$
$d = \sqrt{\frac{3h}{2}}$
$d = \frac{\sqrt{3h}}{\sqrt{2}} \cdot \frac{\sqrt{2}}{\sqrt{2}}$
$d = \frac{\sqrt{6h}}{2}$

81. $E = mc^2$
$\frac{E}{m} = c^2$
$\sqrt{\frac{E}{m}} = c$
$\frac{\sqrt{E}}{\sqrt{m}} \cdot \frac{\sqrt{m}}{\sqrt{m}} = c$
$\frac{\sqrt{Em}}{m} = c$

83. $f(x) = 0$
$2x^2 + x - 5 = 0$
$x^2 + \frac{1}{2}x = \frac{5}{2}$
$x^2 + \frac{1}{2}x + \frac{1}{16} = \frac{5}{2} + \frac{1}{16}$
$\left(x + \frac{1}{4}\right)^2 = \frac{41}{16}$
$x + \frac{1}{4} = \pm \sqrt{\frac{41}{16}}$
$x + \frac{1}{4} = \pm \frac{\sqrt{41}}{4}$
$x = -\frac{1}{4} \pm \frac{\sqrt{41}}{4}$

85. $f(x) = 0$
$x^2 + x - 3 = 0$
$x^2 + x = 3$
$x^2 + x + \frac{1}{4} = 3 + \frac{1}{4}$
$\left(x + \frac{1}{2}\right)^2 = \frac{13}{4}$
$x + \frac{1}{2} = \pm \sqrt{\frac{13}{4}}$
$x = -\frac{1}{2} \pm \frac{\sqrt{13}}{2}$

87.
$$s = 16t^2$$
$$256 = 16t^2$$
$$16 = t^2$$
$$\pm\sqrt{16} = t$$
$$\pm 4 = t$$
$t = 4$ is the only answer that makes sense, so it will take 4 seconds.

89.
$$s^2 = 10.5l$$
$$s^2 = 10.5(495)$$
$$s^2 = 5197.5$$
$$s = \pm\sqrt{5197.5}$$
$$s \approx \pm 72.09$$
s must be positive, so the speed was about 72 mph.

91.
$$A = P(1+r)^t$$
$$9193.60 = 8500(1+r)^2$$
$$\frac{9193.60}{8500} = (1+r)^2$$
$$1.0816 = (1+r)^2$$
$$\pm\sqrt{1.0816} = \sqrt{(1+r)^2}$$
$$\pm 1.04 = 1+r$$
$$-1 \pm 1.04 = r$$
$r = 0.04$ or $r = -2.04$; r must be positive, so $r = 0.04$, or 4%.

93.
$$A = 100$$
$$lw = 100$$
$$(1.9x)(x) = 100$$
$$1.9x^2 = 100$$
$$x^2 = \frac{100}{1.9}$$
$$x = \pm\sqrt{\frac{100}{1.9}} \approx \pm 7.25$$
x must be positive, so $x = 7\frac{1}{4}$ ft.
$1.9x \approx 1.9(7.25) \approx 13.775$
$13.775 \approx 13.75$, which is $13\frac{3}{4}$ ft

95. **Answers may vary.**

97. $\left(\frac{1}{2}\sqrt{3}\right)^2 = \left(\frac{\sqrt{3}}{2}\right)^2 = \frac{3}{4}$

Exercises 10.2 (page 743)

1. $5x^2 - 3x = 4 \Rightarrow 5x^2 - 3x - 4 = 0$
$a = 5, b = -3, c = -4$

3.
$$Ax + By = C$$
$$By = -Ax + C$$
$$y = \frac{-Ax + C}{B}$$

5. $\sqrt{45} = \sqrt{9}\sqrt{5} = 3\sqrt{5}$

7. $\frac{5}{\sqrt{5}} = \frac{5}{\sqrt{5}} \cdot \frac{\sqrt{5}}{\sqrt{5}} = \frac{5\sqrt{5}}{5} = \sqrt{5}$

9. $7; -4; -9$

11. $x^2 + 6x + 5 = 0$
$a = 1, b = 6, c = 5$
$$x = \frac{-b \pm \sqrt{b^2 - 4ac}}{2a}$$
$$= \frac{-6 \pm \sqrt{6^2 - 4(1)(5)}}{2(1)}$$

$$x = \frac{-6 \pm \sqrt{6^2 - 4(1)(5)}}{2(1)}$$
$$= \frac{-6 \pm \sqrt{36 - 20}}{2}$$
$$= \frac{-6 \pm \sqrt{16}}{2} = \frac{-6 \pm 4}{2}$$
$$x = -\frac{10}{2} = -5 \text{ or } x = -\frac{2}{2} = -1$$

355

13. $x^2 - 5x - 14 = 0$
$a = 1, b = -5, c = -14$
$$x = \frac{-b \pm \sqrt{b^2 - 4ac}}{2a}$$
$$= \frac{5 \pm \sqrt{(-5)^2 - 4(1)(-14)}}{2(1)}$$
$$= \frac{5 \pm \sqrt{25 + 56}}{2}$$
$$= \frac{5 \pm \sqrt{81}}{2} = \frac{5 \pm 9}{2}$$
$x = \frac{14}{2} = 7$ or $x = -\frac{4}{2} = -2$

15. $\qquad x^2 + 9x = -20$
$x^2 + 9x + 20 = 0$
$a = 1, b = 9, c = 20$
$$x = \frac{-b \pm \sqrt{b^2 - 4ac}}{2a}$$
$$= \frac{-9 \pm \sqrt{9^2 - 4(1)(20)}}{2(1)}$$
$$= \frac{-9 \pm \sqrt{81 - 80}}{2}$$
$$= \frac{-9 \pm \sqrt{1}}{2} = \frac{-9 \pm 1}{2}$$
$x = -\frac{8}{2} = -4$ or $x = -\frac{10}{2} = -5$

17. $2x^2 - x - 3 = 0$
$a = 2, b = -1, c = -3$
$$x = \frac{-b \pm \sqrt{b^2 - 4ac}}{2a}$$
$$= \frac{1 \pm \sqrt{(-1)^2 - 4(2)(-3)}}{2(2)}$$
$$= \frac{1 \pm \sqrt{1 + 24}}{4}$$
$$= \frac{1 \pm \sqrt{25}}{4} = \frac{1 \pm 5}{4}$$
$x = \frac{6}{4} = \frac{3}{2}$ or $x = -\frac{4}{4} = -1$

19. $\qquad 15x^2 - 14x = 8$
$15x^2 - 14x - 8 = 0$
$a = 15, b = -14, c = -8$
$$x = \frac{-b \pm \sqrt{b^2 - 4ac}}{2a}$$
$$= \frac{14 \pm \sqrt{(-14)^2 - 4(15)(-8)}}{2(15)}$$
$$= \frac{14 \pm \sqrt{196 + 480}}{30}$$
$$= \frac{14 \pm \sqrt{676}}{30} = \frac{14 \pm 26}{30}$$
$x = \frac{40}{30} = \frac{4}{3}$ or $x = -\frac{12}{30} = -\frac{2}{5}$

21. $\qquad 8u = -4u^2 - 3$
$4u^2 + 8u + 3 = 0$
$a = 4, b = 8, c = 3$
$$u = \frac{-b \pm \sqrt{b^2 - 4ac}}{2a}$$
$$= \frac{-8 \pm \sqrt{8^2 - 4(4)(3)}}{2(4)}$$
$$= \frac{-8 \pm \sqrt{64 - 48}}{8}$$
$$= \frac{-8 \pm \sqrt{16}}{8} = \frac{-8 \pm 4}{8}$$
$u = \frac{-4}{8} = -\frac{1}{2}$ or $u = \frac{-12}{8} = -\frac{3}{2}$

23. $5x^2 + 5x + 1 = 0$
$a = 5, b = 5, c = 1$
$$x = \frac{-b \pm \sqrt{b^2 - 4ac}}{2a}$$
$$= \frac{-5 \pm \sqrt{5^2 - 4(5)(1)}}{2(5)}$$
$$= \frac{-5 \pm \sqrt{25 - 20}}{10}$$
$$= \frac{-5 \pm \sqrt{5}}{10}$$
$$= \frac{-5}{10} \pm \frac{\sqrt{5}}{10} = -\frac{1}{2} \pm \frac{\sqrt{5}}{10}$$

25. $5x^2 + 2x - 1 = 0$
$a = 5, b = 2, c = -1$

$$x = \frac{-b \pm \sqrt{b^2 - 4ac}}{2a}$$

$$= \frac{-2 \pm \sqrt{2^2 - 4(5)(-1)}}{2(5)}$$

$$= \frac{-2 \pm \sqrt{4 + 20}}{10}$$

$$= \frac{-2 \pm \sqrt{24}}{10}$$

$$= \frac{-2}{10} \pm \frac{\sqrt{4}\sqrt{6}}{10}$$

$$= -\frac{1}{5} \pm \frac{2\sqrt{6}}{10} = -\frac{1}{5} \pm \frac{\sqrt{6}}{5}$$

27. $x^2 + 2x + 2 = 0$
$a = 1, b = 2, c = 2$

$$x = \frac{-b \pm \sqrt{b^2 - 4ac}}{2a}$$

$$= \frac{-2 \pm \sqrt{2^2 - 4(1)(2)}}{2(1)}$$

$$= \frac{-2 \pm \sqrt{4 - 8}}{2}$$

$$= \frac{-2 \pm \sqrt{-4}}{2}$$

$$= \frac{-2 \pm \sqrt{-1 \cdot 4}}{2}$$

$$= \frac{-2 \pm 2i}{2} = \frac{-2}{2} \pm \frac{2i}{2} = -1 \pm i$$

29. $x^2 + 5x + 7 = 0$
$a = 1, b = 5, c = 7$

$$x = \frac{-b \pm \sqrt{b^2 - 4ac}}{2a}$$

$$= \frac{-5 \pm \sqrt{5^2 - 4(1)(7)}}{2(1)}$$

$$= \frac{-5 \pm \sqrt{25 - 28}}{2}$$

$$= \frac{-5 \pm \sqrt{-3}}{2}$$

$$= \frac{-5 \pm \sqrt{-1 \cdot 3}}{2}$$

$$= \frac{-5 \pm i\sqrt{3}}{2} = -\frac{5}{2} \pm \frac{\sqrt{3}}{2}i$$

31. $3x^2 - 4x = -2$
$3x^2 - 4x + 2 = 0$
$a = 3, b = -4, c = 2$

$$x = \frac{-b \pm \sqrt{b^2 - 4ac}}{2a}$$

$$= \frac{4 \pm \sqrt{(-4)^2 - 4(3)(2)}}{2(3)}$$

$$= \frac{4 \pm \sqrt{16 - 24}}{6}$$

$$= \frac{4 \pm \sqrt{-8}}{6}$$

$$= \frac{4 \pm \sqrt{-1 \cdot 4 \cdot 2}}{6}$$

$$= \frac{4 \pm 2i\sqrt{2}}{6} = \frac{4}{6} \pm \frac{2\sqrt{2}}{6}i = \frac{2}{3} \pm \frac{\sqrt{2}}{3}i$$

33. $3x^2 - 2x = -3$
$3x^2 - 2x + 3 = 0$
$a = 3, b = -2, c = 3$

$$x = \frac{-b \pm \sqrt{b^2 - 4ac}}{2a}$$

$$= \frac{2 \pm \sqrt{(-2)^2 - 4(3)(3)}}{2(3)}$$

$$= \frac{2 \pm \sqrt{4 - 36}}{6}$$

$$x = \frac{2 \pm \sqrt{4 - 36}}{6}$$

$$= \frac{2 \pm \sqrt{-32}}{6}$$

$$= \frac{2 \pm \sqrt{-1 \cdot 16 \cdot 2}}{6}$$

$$= \frac{2 \pm 4i\sqrt{2}}{6}$$

$$= \frac{2}{6} \pm \frac{4\sqrt{2}}{6}i = \frac{1}{3} \pm \frac{2\sqrt{2}}{3}i$$

35.
$$C = \frac{N^2 - N}{2}$$
$$2C = N^2 - N$$
$$N^2 - N - 2C = 0$$
$$a = 1, b = -1, c = -2C$$
$$N = \frac{-b \pm \sqrt{b^2 - 4ac}}{2a}$$
$$= \frac{1 \pm \sqrt{(-1)^2 - 4(1)(-2C)}}{2(1)}$$
$$= \frac{1 \pm \sqrt{1 + 8C}}{2}$$
$$= \frac{1}{2} \pm \frac{\sqrt{1 + 8C}}{2}$$

37.
$$x^2 - kx = -ay$$
$$x^2 - kx + ay = 0$$
$$a = 1, b = -k, c = ay$$
$$x = \frac{-b \pm \sqrt{b^2 - 4ac}}{2a}$$
$$= \frac{k \pm \sqrt{(-k)^2 - 4(1)(ay)}}{2(1)}$$
$$= \frac{k \pm \sqrt{k^2 - 4ay}}{2} = \frac{k}{2} \pm \frac{\sqrt{k^2 - 4ay}}{2}$$

39.
$$6x^2 - 5x - 4 = 0$$
$$a = 6, b = -5, c = -4$$
$$x = \frac{-b \pm \sqrt{b^2 - 4ac}}{2a}$$
$$= \frac{5 \pm \sqrt{(-5)^2 - 4(6)(-4)}}{2(6)}$$
$$= \frac{5 \pm \sqrt{25 + 96}}{12}$$
$$= \frac{5 \pm \sqrt{121}}{12} = \frac{5 \pm 11}{12}$$
$$x = \frac{16}{12} = \frac{4}{3} \text{ or } x = -\frac{6}{12} = -\frac{1}{2}$$

41.
$$\frac{x^2}{2} + \frac{5}{2}x = -1$$
$$x^2 + 5x = -2$$
$$x^2 + 5x + 2 = 0$$
$$a = 1, b = 5, c = 2$$
$$x = \frac{-b \pm \sqrt{b^2 - 4ac}}{2a}$$
$$= \frac{-5 \pm \sqrt{5^2 - 4(1)(2)}}{2(1)}$$
$$= \frac{-5 \pm \sqrt{25 - 8}}{2}$$
$$= \frac{-5 \pm \sqrt{17}}{2} = -\frac{5}{2} \pm \frac{\sqrt{17}}{2}$$

43.
$$3x^2 - 2 = 2x$$
$$3x^2 - 2x - 2 = 0$$
$$a = 3, b = -2, c = -2$$
$$x = \frac{-b \pm \sqrt{b^2 - 4ac}}{2a}$$
$$= \frac{2 \pm \sqrt{(-2)^2 - 4(3)(-2)}}{2(3)}$$
$$= \frac{2 \pm \sqrt{4 + 24}}{6}$$

$$x = \frac{2 \pm \sqrt{4 + 24}}{6}$$
$$= \frac{2 \pm \sqrt{28}}{6}$$
$$= \frac{2 \pm \sqrt{4}\sqrt{7}}{6}$$
$$= \frac{2}{6} \pm \frac{2\sqrt{7}}{6} = \frac{1}{3} \pm \frac{\sqrt{7}}{3}$$

45. $16y^2 + 8y - 3 = 0$
$a = 16, b = 8, c = -3$

$$y = \frac{-b \pm \sqrt{b^2 - 4ac}}{2a}$$

$$= \frac{-8 \pm \sqrt{8^2 - 4(16)(-3)}}{2(16)}$$

$$= \frac{-8 \pm \sqrt{64 + 192}}{32}$$

$$= \frac{-8 \pm \sqrt{256}}{32} = \frac{-8 \pm 16}{32}$$

$$y = \frac{8}{32} = \frac{1}{4} \text{ or } y = \frac{-24}{32} = -\frac{3}{4}$$

47. $$f(x) = 0$$
$$4x^2 + 4x - 19 = 0$$
$$a = 4, b = 4, c = -19$$

$$x = \frac{-b \pm \sqrt{b^2 - 4ac}}{2a}$$

$$= \frac{-4 \pm \sqrt{4^2 - 4(4)(-19)}}{2(4)}$$

$$= \frac{-4 \pm \sqrt{16 + 304}}{8}$$

$$= \frac{-4 \pm \sqrt{320}}{8}$$

$$= -\frac{4}{8} \pm \frac{8\sqrt{5}}{8} = -\frac{1}{2} \pm \sqrt{5}$$

49. $$f(x) = 0$$
$$3x^2 + 2x + 2 = 0$$
$$a = 3, b = 2, c = 2$$

$$x = \frac{-b \pm \sqrt{b^2 - 4ac}}{2a}$$

$$= \frac{-2 \pm \sqrt{2^2 - 4(3)(2)}}{2(3)}$$

$$= \frac{-2 \pm \sqrt{4 - 24}}{6}$$

$$= \frac{-2 \pm \sqrt{-20}}{6}$$

$$= -\frac{2}{6} \pm \frac{2i\sqrt{5}}{6} = -\frac{1}{3} \pm \frac{\sqrt{5}}{3}i$$

51. $0.7x^2 - 3.5x - 25 = 0$
$a = 0.7, b = -3.5, c = -25$

$$x = \frac{-b \pm \sqrt{b^2 - 4ac}}{2a}$$

$$= \frac{3.5 \pm \sqrt{(-3.5)^2 - 4(0.7)(-25)}}{2(0.7)}$$

$$= \frac{3.5 \pm \sqrt{12.25 + 70}}{1.4}$$

$$= \frac{3.5 \pm \sqrt{82.25}}{1.4}$$

$$= \frac{3.5 \pm 9.069}{1.4}$$

$x = 8.98 \text{ or } x = -3.98$

53. $(x - 3)(x - 5) = 0$
$x^2 - 8x + 15 = 0$

55. $(x - 2)(x - 3)(x + 4) = 0$
$(x^2 - 5x + 6)(x + 4) = 0$
$x^3 - x^2 - 14x + 24 = 0$

57. $\text{length} \cdot \text{width} = \text{Area}$
$(x + 4)x = 96$
$x^2 + 4x - 96 = 0$
$(x + 12)(x - 8) = 0$
$x = -12 \text{ or } x = 8$
Since the width is positive, the dimensions are 8 ft by 12 ft.

59. Let s = the length of a side.
$\text{Area} = \text{perimeter}$
$s^2 = 4s$
$s^2 - 4s = 0$
$s(s - 4) = 0$
$s = 0 \text{ or } s = 4$
Since the length cannot be 0, the length of a side is 4 units.

61. Let b represent the base.
Then $3b + 5$ represents the height.

$\frac{1}{2}$base $\cdot$ height $=$ Area

$\frac{1}{2}b(3b + 5) = 6$

$b(3b + 5) = 12$

$3b^2 + 5b - 12 = 0$

$(3b - 4)(b + 3) = 0$

$b = \frac{4}{3}$ or $b = -3$

Since the base is positive, it must be $\frac{4}{3}$ cm.

65. Let x and $x + 1$ represent the integers.

$x^2 + (x + 1)^2 = 85$

$x^2 + x^2 + 2x + 1 = 85$

$2x^2 + 2x - 84 = 0$

$2(x + 7)(x - 6) = 0$

$x = -7$ or $x = 6$; The integers are 6 & 7, or -7 & -6.

63. Let x and $x + 2$ represent the integers.

$x(x + 2) = 168$

$x^2 + 2x - 168 = 0$

$(x + 14)(x - 12) = 0$

$x = -14$ or $x = 12$

The integers are 12 & 14, or -14 & -12.

67. Let $r =$ the slower rate. Then $r + 20 =$ the faster rate.

	Rate	Time	Dist.
Slower	r	$\frac{150}{r}$	150
Faster	$r + 20$	$\frac{150}{r+20}$	150

$$\boxed{\begin{matrix}\text{Faster}\\\text{time}\end{matrix}} + 2 = \boxed{\begin{matrix}\text{Slower}\\\text{time}\end{matrix}}$$

$$\frac{150}{r + 20} + 2 = \frac{150}{r}$$

$$\left(\frac{150}{r + 20} + 2\right)(r)(r + 20) = \frac{150}{r} \cdot r(r + 20)$$

$$150r + 2r(r + 20) = 150(r + 20)$$

$$2r^2 + 40r - 3000 = 0$$

$$2(r + 50)(r - 30) = 0$$

$r = -50$ or $r = 30$ $r = 30$ is the only answer that makes sense.

Her original speed was 30 mph.

69. Let $x =$ the number of 10¢ increases. Then the ticket price will be $4 + 0.10x$, while the projected attendance will be $300 - 5x$, for total receipts of $(4 + 0.10x)(300 - 5x)$.

$$\text{Total} = 1248$$
$$(4 + 0.10x)(300 - 5x) = 1248$$
$$1200 + 10x - 0.5x^2 = 1248$$
$$-0.5x^2 + 10x - 48 = 0$$
$$x^2 - 20x + 96 = 0$$
$$(x - 12)(x - 8) = 0$$
$$x = 12 \text{ or } x = 8 \Rightarrow 4 + 0.10(12) = 5.20; 4 + 0.10(8) = 4.80$$

The ticket price would be either $5.20 or $4.80.

71. Let $x =$ the number of additional subscribers. Then the profit per subscriber will be $20 + 0.01x$, for a total profit of $(20 + 0.01x)(3000 + x)$.

$$\text{Total profit} = 120000$$
$$(20 + 0.01x)(3000 + x) = 120000$$
$$60,000 + 50x + 0.01x^2 = 120000$$
$$0.01x^2 + 50x - 60,000 = 0$$
$$x^2 + 5000x - 6,000,000 = 0$$
$$(x + 6000)(x - 1000) = 0$$
$$x = -6000 \text{ (impossible) or } x = 1000 \Rightarrow \text{The total number of subscribers would be 4000.}$$

73. Let $w =$ the constant width.

$$\text{Frame area} = \text{Picture area}$$
$$(12 + 2w)(10 + 2w) - 12(10) = 12(10)$$
$$120 + 44w + 4w^2 - 240 = 0$$
$$4w^2 + 44w - 120 = 0$$
$$4(w^2 + 11w - 30) = 0$$
$$w^2 + 11w - 30 = 0 \Rightarrow a = 1, b = 11, c = -30$$
$$w = \frac{-11 \pm \sqrt{11^2 - 4(1)(-30)}}{2(1)}$$
$$= \frac{-11 \pm \sqrt{121 + 120}}{2} = \frac{-11 \pm \sqrt{241}}{2} = \frac{-11 \pm 15.52}{2} = \frac{4.52}{2} \text{ or } \frac{-26.52}{2} \text{ (impossible)}$$
$$w = 2.26 \text{ in.}$$

75. $P = -0.0072x^2 + 0.4904x + 58.2714$

$65 = -0.0072x^2 + 0.4904x + 58.2714$

$0 = -0.0072x^2 + 0.4904x - 6.7286$

$a = -0.0072, b = 0.4904, c = -6.7286$

$x = \dfrac{-b \pm \sqrt{b^2 - 4ac}}{2a}$

$= \dfrac{-0.4904 \pm \sqrt{(0.4904)^2 - 4(-0.0072)(-6.7286)}}{2(-0.0072)} \approx \dfrac{-0.4904 \pm \sqrt{0.0467}}{-0.0144} \approx 19.05 \text{ or } 49.06$

Since $x \le 42$, the only answer that works is 19.05. The model indicates the desired result happened in 1985.

77. Let $[\text{H}^+]$ (and then $[\text{A}^-]$) $= x$ and $[\text{HA}] = 0.1 - x$.

$$\frac{[\text{H}^+][\text{A}^-]}{[\text{HA}]} = 4 \times 10^{-4}$$

$$\frac{x^2}{0.1 - x} = 4 \times 10^{-4}$$

$$x^2 = 4 \times 10^{-5} - \left(4 \times 10^{-4}\right)x$$

$x^2 + \left(4 \times 10^{-4}\right)x - 4 \times 10^{-5} = 0$

$x = \dfrac{-b \pm \sqrt{b^2 - 4ac}}{2a} = \dfrac{-4 \times 10^{-4} \pm \sqrt{\left(4 \times 10^{-4}\right)^2 - 4(1)(-4 \times 10^{-5})}}{2(1)}$

$\approx \dfrac{-4 \times 10^{-4} \pm 0.012655}{2} \approx \dfrac{0.012255}{2} \text{ or } -\dfrac{0.013055}{2} \text{ (impossible)}$

The concentration is about $0.00613 \text{ M} = 6.13 \times 10^{-3} \text{ M}$.

79. Answers may vary.

81. $x^2 + 2\sqrt{2}x - 6 = 0$

$a = 1, b = 2\sqrt{2}, c = -6$

$x = \dfrac{-b \pm \sqrt{b^2 - 4ac}}{2a}$

$= \dfrac{-2\sqrt{2} \pm \sqrt{\left(2\sqrt{2}\right)^2 - 4(1)(-6)}}{2(1)}$

$= \dfrac{-2\sqrt{2} \pm \sqrt{8 + 24}}{2}$

$= \dfrac{-2\sqrt{2} \pm \sqrt{32}}{2} = \dfrac{-2\sqrt{2} \pm 4\sqrt{2}}{2}$

$x = \dfrac{2\sqrt{2}}{2} = \sqrt{2} \text{ or } x = \dfrac{-6\sqrt{2}}{2} = -3\sqrt{2}$

83. $x^2 - 3ix - 2 = 0$

$a = 1, b = -3i, c = -2$

$x = \dfrac{-b \pm \sqrt{b^2 - 4ac}}{2a}$

$= \dfrac{3i \pm \sqrt{(-3i)^2 - 4(1)(-2)}}{2(1)}$

$= \dfrac{3i \pm \sqrt{9i^2 + 8}}{2}$

$= \dfrac{3i \pm \sqrt{-1}}{2} = \dfrac{3i \pm i}{2}$

$x = \dfrac{4i}{2} = 2i \text{ or } x = \dfrac{2i}{2} = i$

Exercises 10.3 (page 752)

1. $b^2 - 4ac = (1)^2 - 4(1)(1)$
$$= 1 - 4 = -3$$

3. If $1 - i$ is a solution, then so is $1 + i$.
Sum: $(1 - i) + (1 + i) = 2 = -b$
Product: $(1 + i)(1 - i) = 1 - i^2 = 2 = c$
Yes, $1 - i$ is a solution.

5. If $-1 + i$ is a solution, then so is $-1 - i$.
Sum: $(-1 + i) + (-1 - i) = -2 \neq -b$
No, $-1 + i$ is not a solution.

7.
$$\frac{1}{4} + \frac{1}{t} = \frac{1}{2t}$$
$$\left(\frac{1}{4} + \frac{1}{t}\right)4t = \frac{1}{2t} \cdot 4t$$
$$t + 4 = 2$$
$$t = -2$$

9. $m = \dfrac{\Delta y}{\Delta x} = \dfrac{1 - (-2)}{4 - (-3)} = \dfrac{3}{7}$

11. $b^2 - 4ac$

13. rational; unequal

15. $9x^2 + 6x + 1 = 0; a = 9, b = 6, c = 1$
$b^2 - 4ac = (6)^2 - 4(9)(1)$
$$= 36 - 36 = 0$$
The solutions are rational and equal.

17. $10x^2 + 2x + 3 = 0; a = 10, b = 2, c = 3$
$b^2 - 4ac = 2^2 - 4(10)(3)$
$$= 4 - 120 = -116$$
The solutions are complex conjugates.

19.
$$6x^2 = 8x - 1$$
$6x^2 - 8x + 1 = 0; a = 6, b = -8, c = 1$
$b^2 - 4ac = (-8)^2 - 4(6)(1)$
$$= 64 - 24 = 40$$
The solutions are irrational and unequal.

21.
$$x(2x - 3) = 20$$
$2x^2 - 3x - 20 = 0; a = 2, b = -3, c = -20$
$b^2 - 4ac = (-3)^2 - 4(2)(-20)$
$$= 9 + 160 = 169$$
The solutions are rational and unequal.

23. $x^2 + kx + 16 = 0; a = 1, b = k, c = 16$
Set the discriminant equal to 0:
$$b^2 - 4ac = 0$$
$$k^2 - 4(1)(16) = 0$$
$$k^2 - 64 = 0$$
$$k^2 = 64$$
$$k = \pm 8$$

25.
$$4x^2 + 9 = kx$$
$4x^2 - kx + 9 = 0; a = 4, b = -k, c = 9$
Set the discriminant equal to 0:
$$b^2 - 4ac = 0$$
$$(-k)^2 - 4(4)(9) = 0$$
$$k^2 - 144 = 0$$
$$k^2 = 144$$
$$k = \pm 12$$

27.
$$x^4 - 17x^2 + 16 = 0$$
$$(x^2 - 16)(x^2 - 1) = 0$$
$x^2 - 16 = 0$ **or** $x^2 - 1 = 0$
$\quad x^2 = 16 \qquad\qquad x^2 = 1$
$\quad\quad x = \pm 4 \qquad\qquad x = \pm 1$

29.
$$x^4 - 3x^2 = -2$$
$$x^4 - 3x^2 + 2 = 0$$
$$(x^2 - 2)(x^2 - 1) = 0$$
$$x^2 - 2 = 0 \quad \textbf{or} \quad x^2 - 1 = 0$$
$$x^2 = 2 \qquad\qquad x^2 = 1$$
$$x = \pm\sqrt{2} \qquad\quad x = \pm 1$$

31.
$$x^4 = 6x^2 - 5$$
$$x^4 - 6x^2 + 5 = 0$$
$$(x^2 - 5)(x^2 - 1) = 0$$
$$x^2 - 5 = 0 \quad \textbf{or} \quad x^2 - 1 = 0$$
$$x^2 = 5 \qquad\qquad x^2 = 1$$
$$x = \pm\sqrt{5} \qquad\quad x = \pm 1$$

33.
$$2x^4 - 10x^2 = -8$$
$$2x^4 - 10x^2 + 8 = 0$$
$$2(x^2 - 4)(x^2 - 1) = 0$$
$$x^2 - 4 = 0 \quad \textbf{or} \quad x^2 - 1 = 0$$
$$x^2 = 4 \qquad\qquad x^2 = 1$$
$$x = \pm 2 \qquad\quad x = \pm 1$$

35.
$$x - 7\sqrt{x} + 10 = 0$$
$$(\sqrt{x} - 2)(\sqrt{x} - 5) = 0$$
$$\sqrt{x} - 2 = 0 \quad \textbf{or} \quad \sqrt{x} - 5 = 0$$
$$\sqrt{x} = 2 \qquad\qquad \sqrt{x} = 5$$
$$(\sqrt{x})^2 = 2^2 \qquad (\sqrt{x})^2 = 5^2$$
$$x = 4 \qquad\qquad x = 25$$
$$\text{Solution} \qquad\qquad \text{Solution}$$

37.
$$2x - \sqrt{x} = 3$$
$$2x - \sqrt{x} - 3 = 0$$
$$(2\sqrt{x} - 3)(\sqrt{x} + 1) = 0$$
$$2\sqrt{x} - 3 = 0 \quad \textbf{or} \quad \sqrt{x} + 1 = 0$$
$$2\sqrt{x} = 3 \qquad\qquad \sqrt{x} = -1$$
$$(2\sqrt{x})^2 = 3^2 \qquad (\sqrt{x})^2 = (-1)^2$$
$$4x = 9 \qquad\qquad x = 1$$
$$x = \frac{9}{4} \qquad\qquad \text{Extraneous}$$
$$\text{Solution}$$

39.
$$2x + x^{1/2} - 3 = 0$$
$$(2x^{1/2} + 3)(x^{1/2} - 1) = 0$$
$$2x^{1/2} + 3 = 0 \quad \textbf{or} \quad x^{1/2} - 1 = 0$$
$$2x^{1/2} = -3 \qquad\qquad x^{1/2} = 1$$
$$x^{1/2} = -\frac{3}{2} \qquad\qquad (x^{1/2})^2 = 1^2$$
$$(x^{1/2})^2 = \left(-\frac{3}{2}\right)^2 \qquad\qquad x = 1$$
$$x = \frac{9}{4} \qquad\qquad \text{Solution}$$
$$\text{Extraneous}$$

41.
$$3x + 5x^{1/2} + 2 = 0$$
$$(3x^{1/2} + 2)(x^{1/2} + 1) = 0$$
$$3x^{1/2} + 2 = 0 \quad \textbf{or} \quad x^{1/2} + 1 = 0$$
$$3x^{1/2} = -2 \qquad\qquad x^{1/2} = -1$$
$$x^{1/2} = -\frac{2}{3} \qquad\qquad (x^{1/2})^2 = (-1)^2$$
$$(x^{1/2})^2 = \left(-\frac{2}{3}\right)^2 \qquad\qquad x = 1$$
$$x = \frac{4}{9} \qquad\qquad \text{Extraneous}$$
$$\text{Extraneous}$$
solution set: $\emptyset$

43.
$$x^{2/3} + 5x^{1/3} + 6 = 0$$
$$(x^{1/3} + 2)(x^{1/3} + 3) = 0$$
$$x^{1/3} + 2 = 0 \quad \textbf{or} \quad x^{1/3} + 3 = 0$$
$$x^{1/3} = -2 \qquad\qquad x^{1/3} = -3$$
$$(x^{1/3})^3 = (-2)^3 \qquad (x^{1/3})^3 = (-3)^3$$
$$x = -8 \qquad\qquad x = -27$$
$$\text{Solution} \qquad\qquad \text{Solution}$$

45.
$$3m^{2/3} - m^{1/3} - 2 = 0$$
$$\left(3m^{1/3} + 2\right)\left(m^{1/3} - 1\right) = 0$$

| $3m^{1/3} + 2 = 0$ | **or** | $m^{1/3} - 1 = 0$ |

$$3m^{1/3} = -2 \qquad\qquad m^{1/3} = 1$$
$$m^{1/3} = -\tfrac{2}{3} \qquad\qquad \left(m^{1/3}\right)^3 = 1^3$$
$$\left(m^{1/3}\right)^3 = \left(-\tfrac{2}{3}\right)^3 \qquad\qquad m = 1$$
$$m = -\tfrac{8}{27} \qquad\qquad \text{Solution}$$
$$\text{Solution}$$

47.
$$x + 10 + \frac{9}{x} = 0$$
$$x\left(x + 10 + \frac{9}{x}\right) = x(0)$$
$$x^2 + 10x + 9 = 0$$
$$(x + 9)(x + 1) = 0$$

| $x + 9 = 0$ | **or** | $x + 1 = 0$ |
| $x = -9$ | | $x = -1$ |

49.
$$x + 3 = \frac{28}{x}$$
$$x + 3 - \frac{28}{x} = 0$$
$$x\left(x + 3 - \frac{28}{x}\right) = x(0)$$
$$x^2 + 3x - 28 = 0$$
$$(x + 7)(x - 4) = 0$$

| $x + 7 = 0$ | **or** | $x - 4 = 0$ |
| $x = -7$ | | $x = 4$ |

51.
$$\frac{1}{x - 1} + \frac{3}{x + 1} = 2$$
$$\left(\frac{1}{x - 1} + \frac{3}{x + 1}\right)(x - 1)(x + 1) = 2(x + 1)(x - 1)$$
$$1(x + 1) + 3(x - 1) = 2(x^2 - 1)$$
$$x + 1 + 3x - 3 = 2x^2 - 2$$
$$0 = 2x^2 - 4x$$
$$0 = 2x(x - 2) \qquad 2x = 0 \quad \textbf{or} \quad x - 2 = 0$$
$$x = 0 \qquad\qquad x = 2$$

53.
$$\frac{1}{x + 2} + \frac{24}{x + 3} = 13$$
$$\left(\frac{1}{x + 2} + \frac{24}{x + 3}\right)(x + 2)(x + 3) = 13(x + 2)(x + 3)$$
$$1(x + 3) + 24(x + 2) = 13(x^2 + 5x + 6)$$
$$x + 3 + 24x + 48 = 13x^2 + 65x + 78$$
$$0 = 13x^2 + 40x + 27$$
$$0 = (13x + 27)(x + 1) \qquad 13x + 27 = 0 \quad \textbf{or} \quad x + 1 = 0$$
$$13x = -27 \qquad\qquad x = -1$$
$$x = -\tfrac{27}{13}$$

55.
$$x^{-4} - 2x^{-2} + 1 = 0$$
$$(x^{-2} - 1)(x^{-2} - 1) = 0$$
$$x^{-2} - 1 = 0 \quad \textbf{or} \quad x^{-2} - 1 = 0$$
$$x^{-2} = 1 \qquad\qquad x^{-2} = 1$$
$$\frac{1}{x^2} = 1 \qquad\qquad \frac{1}{x^2} = 1$$
$$1 = x^2 \qquad\qquad 1 = x^2$$
$$\pm 1 = x \qquad\qquad \pm 1 = x$$
(double roots)

57.
$$8a^{-2} - 10a^{-1} - 3 = 0$$
$$(2a^{-1} - 3)(4a^{-1} + 1) = 0$$
$$2a^{-1} - 3 = 0 \quad \textbf{or} \quad 4a^{-1} + 1 = 0$$
$$2a^{-1} = 3 \qquad\qquad 4a^{-1} = -1$$
$$\frac{2}{a} = 3 \qquad\qquad \frac{4}{a} = -1$$
$$2 = 3a \qquad\qquad 4 = -a$$
$$\tfrac{2}{3} = a \qquad\qquad -4 = a$$

59.
$$x^2 + y^2 = r^2$$
$$x^2 = r^2 - y^2$$
$$x = \pm\sqrt{r^2 - y^2}$$

61. $xy^2 + 5xy + 3 = 0; a = x, b = 5x, c = 3$
$$y = \frac{-b \pm \sqrt{b^2 - 4ac}}{2a}$$
$$= \frac{-5x \pm \sqrt{(5x)^2 - 4(x)(3)}}{2x}$$
$$= \frac{-5x \pm \sqrt{25x^2 - 12x}}{2x}$$

63. $12x^2 - 5x - 2 = 0; a = 12, b = -5, c = -2$
$$(4x + 1)(3x - 2) = 0$$

$$4x + 1 = 0 \quad \textbf{or} \quad 3x - 2 = 0 \qquad -\frac{b}{a} = -\frac{-5}{12} = \frac{5}{12} \qquad\qquad \frac{c}{a} = \frac{-2}{12} = -\frac{1}{6}$$
$$4x = -1 \qquad\qquad 3x = 2 \qquad -\frac{1}{4} + \frac{2}{3} = -\frac{3}{12} + \frac{8}{12} = \frac{5}{12} \qquad \left(-\frac{1}{4}\right)\left(\frac{2}{3}\right) = -\frac{1}{6}$$
$$x = -\frac{1}{4} \qquad\qquad x = \frac{2}{3}$$

65. $2x^2 + 5x + 1 = 0; a = 2, b = 5, c = 1; -\dfrac{b}{a} = -\dfrac{5}{2}; \dfrac{c}{a} = \dfrac{1}{2}$
$$x = \frac{-b \pm \sqrt{b^2 - 4ac}}{2a} = \frac{-5 \pm \sqrt{5^2 - 4(2)(1)}}{2(2)} = \frac{-5 \pm \sqrt{17}}{4} = -\frac{5}{4} \pm \frac{\sqrt{17}}{4}$$
$$\frac{-5 + \sqrt{17}}{4} + \frac{-5 - \sqrt{17}}{4} = \frac{-10}{4} = -\frac{5}{2}$$
$$\left(\frac{-5 + \sqrt{17}}{4}\right)\left(\frac{-5 - \sqrt{17}}{4}\right) = \frac{25 + 5\sqrt{17} - 5\sqrt{17} - 17}{16} = \frac{8}{16} = \frac{1}{2}$$

67. $3x^2 - 2x + 4 = 0; a = 3, b = -2, c = 4; -\dfrac{b}{a} = -\dfrac{-2}{3} = \dfrac{2}{3}; \dfrac{c}{a} = \dfrac{4}{3}$

$x = \dfrac{-b \pm \sqrt{b^2 - 4ac}}{2a} = \dfrac{-(-2) \pm \sqrt{(-2)^2 - 4(3)(4)}}{2(3)} = \dfrac{2 \pm \sqrt{-44}}{6} = \dfrac{1}{3} \pm \dfrac{i\sqrt{11}}{3}$

$\dfrac{1 + i\sqrt{11}}{3} + \dfrac{1 - i\sqrt{11}}{3} = \dfrac{2}{3}$

$\left(\dfrac{1 + i\sqrt{11}}{3}\right)\left(\dfrac{1 - i\sqrt{11}}{3}\right) = \dfrac{1 - i\sqrt{11} + i\sqrt{11} - 11i^2}{9} = \dfrac{1 + 11}{9} = \dfrac{12}{9} = \dfrac{4}{3}$

69. $x^2 + 2x + 5 = 0; a = 1, b = 2, c = 5; -\dfrac{b}{a} = -\dfrac{2}{1} = -2; \dfrac{c}{a} = \dfrac{5}{1} = 5$

$x = \dfrac{-b \pm \sqrt{b^2 - 4ac}}{2a} = \dfrac{-2 \pm \sqrt{2^2 - 4(1)(5)}}{2(1)} = \dfrac{-2 \pm \sqrt{-16}}{2} = \dfrac{-2 \pm 4i}{2} = -1 \pm 2i$

$(-1 + 2i) + (-1 - 2i) = -2; \ (-1 + 2i)(-1 - 2i) = 1 + 2i - 2i - 4i^2 = 1 + 4 = 5$

71. $1492x^2 + 1776x - 1984 = 0$

$a = 1492, b = 1776, c = -1984$

$b^2 - 4ac = (1776)^2 - 4(1492)(-1984)$

$\qquad = 3{,}154{,}176 + 11{,}840{,}512$

$\qquad = 14{,}994{,}688$

The solutions are real numbers.

73.
$$4x - 5\sqrt{x} - 9 = 0$$
$$(4\sqrt{x} - 9)(\sqrt{x} + 1) = 0$$

$4\sqrt{x} - 9 = 0$	**or**	$\sqrt{x} + 1 = 0$
$4\sqrt{x} = 9$		$\sqrt{x} = -1$
$\left(4\sqrt{x}\right)^2 = 9^2$		$\left(\sqrt{x}\right)^2 = (-1)^2$
$16x = 81$		$x = 1$
$x = \frac{81}{16}$		Extraneous
Solution		

75.
$$3x^{2/3} - x^{1/3} - 2 = 0$$
$$\left(x^{1/3} - 1\right)\left(3x^{1/3} + 2\right) = 0$$

$x^{1/3} - 1 = 0$	**or**	$3x^{1/3} + 2 = 0$
$x^{1/3} = 1$		$3x^{1/3} = -2$
$\left(x^{1/3}\right)^3 = (1)^3$		$\left(3x^{1/3}\right)^3 = (-2)^3$
$x = 1$		$27x = -8$
Solution		$x = -\frac{8}{27}$
		Solution

77.
$$2x^4 + 24 = 26x^2$$
$$2x^4 - 26x^2 + 24 = 0$$
$$2(x^2 - 12)(x^2 - 1) = 0$$

$x^2 - 12 = 0$	**or**	$x^2 - 1 = 0$
$x^2 = 12$		$x^2 = 1$
$x = \pm\sqrt{12}$		$x = \pm 1$
$= \pm 2\sqrt{3}$		

79.
$$x^4 - 24 = -2x^2$$
$$x^4 + 2x^2 - 24 = 0$$
$$(x^2 - 4)(x^2 + 6) = 0$$

$x^2 - 4 = 0$	**or**	$x^2 + 6 = 0$
$x^2 = 4$		$x^2 = -6$
$x = \pm\sqrt{4}$		$x = \pm\sqrt{-6}$
$x = \pm 2$		$x = \pm i\sqrt{6}$

81. Let $y = 2x - 1$.

$$4(2x - 1)^2 - 3(2x - 1) - 1 = 0$$
$$4y^2 - 3y - 1 = 0$$
$$(4y + 1)(y - 1) = 0$$

$4y + 1 = 0$ **or** $y - 1 = 0$
$4y = -1$ $\qquad\qquad y = 1$
$y = -\frac{1}{4}$ $\qquad\qquad y = 1$
$2x - 1 = -\frac{1}{4}$ $\qquad 2x - 1 = 1$
$2x = \frac{3}{4}$ $\qquad\qquad 2x = 2$
$x = \frac{3}{8}$ $\qquad\qquad x = 1$

83.
$$x^{-2/3} - 2x^{-1/3} - 3 = 0$$
$$(x^{-1/3} - 3)(x^{-1/3} + 1) = 0$$

$x^{-1/3} - 3 = 0$ **or** $x^{-1/3} + 1 = 0$
$x^{-1/3} = 3$ $\qquad\qquad x^{-1/3} = -1$
$\left(x^{-1/3}\right)^{-3} = 3^{-3}$ $\qquad \left(x^{-1/3}\right)^{-3} = (-1)^{-3}$
$x = \frac{1}{3^3}$ $\qquad\qquad x = \frac{1}{(-1)^3}$
$x = \frac{1}{27}$ $\qquad\qquad x = -1$

85.
$$x + \frac{2}{x - 2} = 0$$
$$\left(x + \frac{2}{x - 2}\right)(x - 2) = 0(x - 2)$$
$$x(x - 2) + 2 = 0$$
$$x^2 - 2x + 2 = 0$$
$$x^2 - 2x = -2$$
$$x^2 - 2x + 1 = -2 + 1$$
$$(x - 1)^2 = -1$$
$$x - 1 = \pm\sqrt{-1}$$
$$x = 1 \pm i$$

87. Let $y = m + 1$.

$$8(m + 1)^{-2} - 30(m + 1)^{-1} + 7 = 0$$
$$8y^{-2} - 30y^{-1} + 7 = 0$$
$$(4y^{-1} - 1)(2y^{-1} - 7) = 0$$

$4y^{-1} - 1 = 0$ **or** $2y^{-1} - 7 = 0$
$4y^{-1} = 1$ $\qquad\qquad 2y^{-1} = 7$
$\frac{4}{y} = 1$ $\qquad\qquad \frac{2}{y} = 7$
$4 = y$ $\qquad\qquad 2 = 7y$
$\qquad\qquad\qquad \frac{2}{7} = y$
$4 = m + 1$ $\qquad\qquad \frac{2}{7} = m + 1$
$3 = m$ $\qquad\qquad -\frac{5}{7} = m$

89.
$$I = \frac{k}{d^2}$$
$$Id^2 = k$$
$$d^2 = \frac{k}{I}$$
$$d = \pm\sqrt{\frac{k}{I}} = \pm\frac{\sqrt{kI}}{I}$$

91.
$$\sigma = \sqrt{\frac{\Sigma x^2}{N} - \mu^2}$$
$$\sigma^2 = \frac{\Sigma x^2}{N} - \mu^2$$
$$\mu^2 = \frac{\Sigma x^2}{N} - \sigma^2$$

93. $(k - 1)x^2 + (k - 1)x + 1 = 0$
$a = k - 1, b = k - 1, c = 1$

Set the discriminant equal to 0:
$$b^2 - 4ac = 0$$
$$(k - 1)^2 - 4(k - 1)(1) = 0$$
$$k^2 - 2k + 1 - 4k + 4 = 0$$
$$k^2 - 6k + 5 = 0$$
$$(k - 5)(k - 1) = 0$$

$k - 5 = 0$ **or** $k - 1 = 0$
$k = 5$ $\qquad\qquad k = 1$: does not check

95. $(k+4)x^2 + 2kx + 9 = 0$
$a = k+4, b = 2k, c = 9$
Set the discriminant equal to 0:
$$b^2 - 4ac = 0$$
$$(2k)^2 - 4(k+4)(9) = 0$$
$$4k^2 - 36(k+4) = 0$$
$$4k^2 - 36k - 144 = 0$$
$$4(k^2 - 9k - 36) = 0$$
$$4(k-12)(k+3) = 0$$
$$k - 12 = 0 \quad \textbf{or} \quad k + 3 = 0$$
$$k = 12 \qquad\qquad k = -3$$

97. $$3x^2 + 4x = k$$
$$3x^2 + 4x - k = 0$$
$$a = 3, b = 4, c = -k$$
Set the discriminant less than 0:
$$b^2 - 4ac < 0$$
$$4^2 - 4(3)(-k) < 0$$
$$16 + 12k < 0$$
$$12k < -16$$
$$k < -\frac{16}{12}, \text{ or } k < -\frac{4}{3}$$

99. Answers may vary.

101. No

Exercises 10.4 (page 766)

1. $x = 3$

3. $x = -1$

5. $\begin{aligned} f(x) &= 0 \\ x^2 - 9 &= 0 \\ x^2 &= 9 \\ x &= \pm 3 \end{aligned}$

7. $\begin{aligned} 3x + 5 &= 5x - 15 \\ -2x &= -20 \\ x &= 10 \end{aligned}$

9. Let t = the time of the second train.
Then $t + 3$ = the time of the first train.

	Rate	Time	Dist.
First	30	$t+3$	$30(t+3)$
Second	55	t	$55t$

1st distance = 2nd distance
$$30(t+3) = 55t$$
$$30t + 90 = 55t$$
$$-25t = -90$$
$$t = \frac{-90}{-25} = \frac{18}{5} = 3\frac{3}{5} \text{ hours}$$

11. $f(x) = ax^2 + bx + c;$
$a \neq 0$

13. maximum; minimum; vertex

15. upward

17. to the right

19. upward

21. $f(x) = x^2$
vertex: $(0, 0)$; opens U

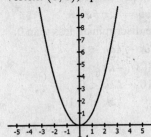

23. $f(x) = -2x^2$
vertex: $(0, 0)$; opens D

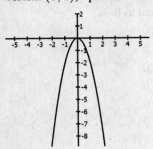

25. $f(x) = x^2 + 2$
Shift $f(x) = x^2$ U2.

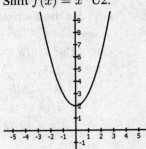

27. $f(x) = x^2 - 1$
Shift $f(x) = x^2$ D1.

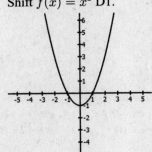

29. $f(x) = -(x - 2)^2$; opens D
Shift $f(x) = x^2$ R2.

31. $f(x) = -(x - 3)^2$; opens D
Shift $f(x) = x^2$ R3.

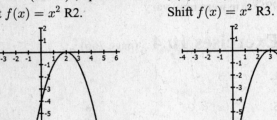

33. $f(x) = (x - 3)^2 + 2$
Shift $f(x) = x^2$ U2 and R3.

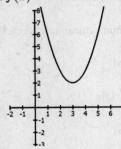

35. $y = (x - 4)^2 + 5$
Shift $y = x^2$ U5 and R4.

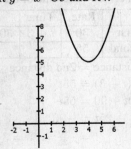

37. $y = (x - 1)^2 + 2$; $V(1, 2)$; axis: $x = 1$

39. $y = 2(x + 3)^2 - 4$; $V(-3, -4)$; axis: $x = -3$

41. $y = -3x^2 \Rightarrow y = -3(x - 0)^2 + 0$
$V(0, 0)$; axis: $x = 0$

43. $y = 2x^2 - 4x$
$y = 2(x^2 - 2x)$
$y = 2(x^2 - 2x + 1) - 2$
$y = 2(x - 1)^2 - 2$
$V(1, -2)$; axis: $x = 1$

45. $f(x) = 3x^2 + 6x + 1$

$x = -\dfrac{b}{2a} = -\dfrac{6}{2(3)} = -1$

$y = 3(-1)^2 + 6(-1) + 1 = -2$

vertex: $(-1, -2)$

47. $f(x) = -3x^2 + 12x + 4$

$x = -\dfrac{b}{2a} = -\dfrac{12}{2(-3)} = 2$

$y = -3(2)^2 + 12(2) + 4 = 16$

vertex: $(2, 16)$

49. $f(x) = -2x^2 + 4x + 1$

$x = -\dfrac{b}{2a} = -\dfrac{4}{2(-2)} = 1$

$y = -2(1)^2 + 4(1) + 1 = 3$

vertex: $(1, 3)$; opens D

axis: $x = 1$

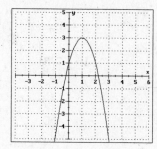

51. $f(x) = 3x^2 - 12x + 10$

$x = -\dfrac{b}{2a} = -\dfrac{-12}{2(3)} = 2$

$y = 3(2)^2 - 12(2) + 10 = -2$

vertex: $(2, -2)$; opens U

axis: $x = 2$

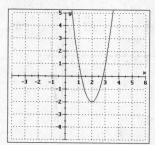

53. $y = -4x^2 + 16x + 5$

$y = -4(x^2 - 4x) + 5$

$y = -4(x^2 - 4x + 4) + 5 + 16$

$y = -4(x - 2)^2 + 21$

$V(2, 21)$; axis: $x = 2$

55. $y - 7 = 6x^2 - 5x$

$y = 6x^2 - 5x + 7$

$x = -\dfrac{b}{2a} = -\dfrac{-5}{2(6)} = \dfrac{5}{12}$

$y = 6x^2 - 5x + 7$

$\quad = 6\left(\dfrac{5}{12}\right)^2 - 5\left(\dfrac{5}{12}\right) + 7$

$\quad = \dfrac{25}{24} - \dfrac{25}{12} + 7 = \dfrac{143}{24}$

$V\left(\dfrac{5}{12}, \dfrac{143}{24}\right)$; axis: $x = \dfrac{5}{12}$

57. $f(x) = x^2 + x - 6$

$f(x) = \left(x + \frac{1}{2}\right)^2 - \frac{25}{4}$

vertex: $\left(-\frac{1}{2}, -\frac{25}{4}\right)$; opens U

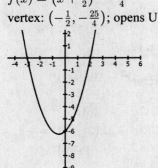

59. $y = 2x^2 - x + 1$

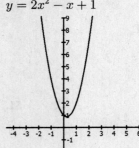

$V(0.25, 0.88)$

61. $y = 7 + x - x^2$

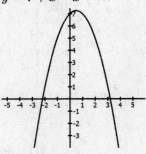

$V(0.5, 7.25)$

63. $y = x^2 + 9x - 10$

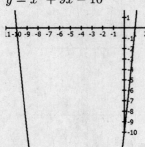

solution set: $\{-10, 1\}$

65. $y = 0.5x^2 - 0.7x - 3$

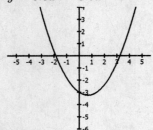

solution set: $\{-1.85, 3.25\}$

67. $y - 3 = (x + 7)^2$

$y = (x + 7)^2 + 3$

$V(-7, 3)$

69. Since the graph of the height equation is a parabola, the max. height occurs at the vertex.

$s = 48t - 16t^2$

$s = -16(t^2 - 3t)$

$s = -16\left(t^2 - 3t + \frac{9}{4}\right) + 36$

$s = -16\left(t - \frac{3}{2}\right)^2 + 36$

$V\left(\frac{3}{2}, 36\right) \Rightarrow$ max. height $= 36$ ft

The maximum height is 36 feet, which occurs after $\frac{3}{2}$ seconds (1.5 seconds).

71. Let $w =$ the width of the rectangle.

Then $200 - w =$ the length.

$A = w(200 - w)$

$A = -w^2 + 200w$

$A = -(w^2 - 200w + 10000) + 10000$

$A = -(w - 100)^2 + 10000$

dim: 100 ft by 100 ft; area $= 10{,}000$ ft^2

73. Replace p with x.

Graph $y = 50x(1 - x)$ and find the x-coordinate(s) when $y = 9.375$.

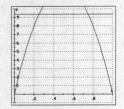

$p = 0.25$ or $p = 0.75$

75. Let $w =$ the width of the rectangle.
Then $150 - w =$ the length.
$A = w(150 - w)$
$A = -w^2 + 150w$
$A = -(w^2 - 150w + 5625) + 5625$
$A = -(w - 75)^2 + 5625$
dim: 75 ft by 75 ft; area $= 5625$ ft^2

77. Replace t with x.
Graph $y = H = 3.3x^2 - 59.4x + 281.3$
and find the y-coordinate of the vertex:

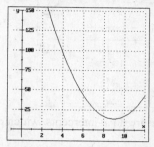

The minimum level was 14 feet.

79. Graph $y = R = -\frac{x^2}{1000} + 10x$ and find
the x-coordinate of the vertex:

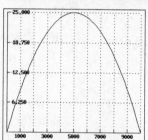

5000 stereos should be sold.

81. Graph $y = R = -\frac{x^2}{728} + 9x$ and find the
x- and y- coordinates of the vertex:

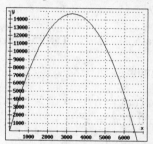

Max. revenue $= \$14,742$; # radios $= 3276$

83. Let $x =$ the number of \$1 increases to the
price. Then the sales will be $4000 - 100x$,
and the revenue will be
$(30 + x)(4000 - 100x)$. Find the vertex
of the parabola $y = (30 + x)(4000 - 100x)$.

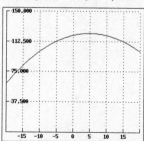

The price should increase \$5, to a total of \$35.

85. **Answers may vary.**

373

87. Graph $y = x^2 + x + 1$ to find x-intercept(s):

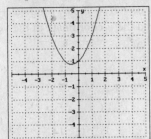

There are no x-intercepts, which means there is no real solution to the equation.

Exercises 10.5 (page 779)

1. the graph of $f(x) = x^2$ shifted up 3

3. the graph of $f(x) = x^2$ shifted left 3

5. 31, 37

7. $a \cdot b = b \cdot a$

9. 1

11. cubic

13. vertical

15. 5; upward

17. 5; to the right

19. x-axis

21. asymptote

23. $f(x) = x^3 - 3$
Shift $f(x) = x^3$ D 3.

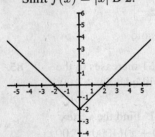

25. $f(x) = |x| - 2$
Shift $f(x) = |x|$ D 2.

27. $f(x) = (x - 1)^3$
Shift $f(x) = x^3$ R 1.

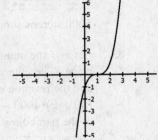

374

29. $f(x) = |x - 1|$
Shift $f(x) = |x|$ R 1.

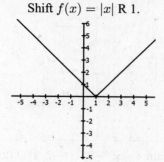

31. $f(x) = |x - 2| - 1$
Shift $f(x) = |x|$ R2 D1.

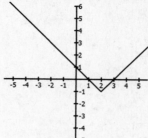

33. $f(x) = (x + 1)^3 - 2$
Shift $f(x) = x^3$ L1 D2.

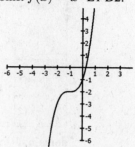

35. $f(x) = -x^2$
Reflect $f(x) = x^2$ about
x-axis.

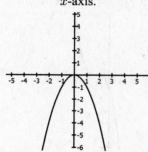

37. $f(x) = -|x + 1| - 2$
Reflect $f(x) = |x|$ about
x-axis, shift L1 D2.

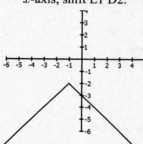

39. $f(x) = -x^3 + 2$
Reflect $f(x) = x^3$ about
x-axis, shift U2.

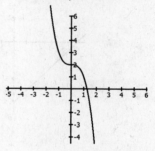

41. $f(x) = -(x + 1)^3 - 4$
Reflect $f(x) = x^3$ about
x-axis, shift L1 D4.

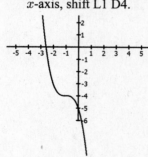

43. $t = f(r) = \dfrac{600}{r}$
$f(30) = \frac{600}{30}$
$\quad = 20$ hrs

45. $t = f(r) = \dfrac{600}{r}$
$f(50) = \frac{600}{50}$
$\quad = 12$ hrs

375

47. $f(x) = \frac{x}{x-2}$

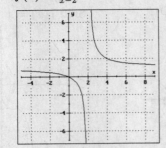

domain: $(-\infty, 2) \cup (2, \infty)$
range: $(-\infty, 1) \cup (1, \infty)$

49. $f(x) = \frac{x+1}{x^2-4}$

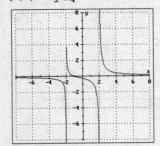

domain: $(-\infty, -2) \cup (-2, 2) \cup (2, \infty)$
range: $(-\infty, \infty)$

51. $\quad f(x) = |x| - 5$
Shift $f(x) = |x|$ D 5.

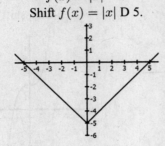

53. $\quad f(x) = (x-1)^3$
Shift $f(x) = x^3$ R 1.

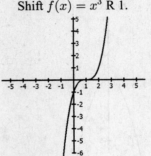

55. $f(x) = x^2 + 8$
$x: [-4, 4], \ y: [-4, 4]$

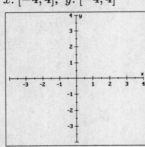

$x: [-7, 7], \ y: [-2, 12]$

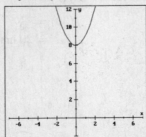

57. $f(x) = |x + 5|$
$x\colon [-4, 4], \; y\colon [-4, 4]$

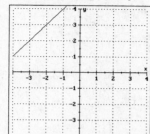

59. $f(x) = (x - 6)^2$
$x\colon [-4, 4], \; y\colon [-4, 4]$

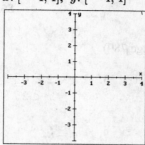

61. $f(x) = x^3 + 8$
$x\colon [-4, 4], \; y\colon [-4, 4]$

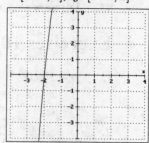

$x\colon [-12, 2], \; y\colon [-4, 10]$

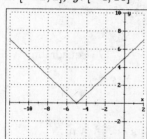

$x\colon [-2, 12], \; y\colon [-2, 12]$

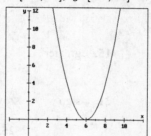

$x\colon [-10, 10], \; y\colon [-4, 16]$

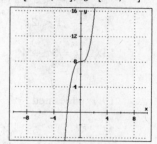

63. $C = f(p) = \dfrac{50{,}000p}{100 - p}$

$f(20) = \dfrac{50{,}000(20)}{100 - 20}$

$\quad = \dfrac{1{,}000{,}000}{80}$

$\quad = 12500$

$C = \$12{,}500$

65. $C = f(p) = \dfrac{50{,}000p}{100 - p}$

$f(50) = \dfrac{50{,}000(50)}{100 - 50}$

$\quad = \dfrac{2{,}500{,}000}{50}$

$\quad = 50{,}000.00$

$C = \$50{,}000$

67. $c = f(x) = 1.25x + 700$

69. Refer to **#67**. $c = f(250) = 1.25(250) + 700 = \$1012.50$

71. Refer to **#68**. $\bar{c} = f(300) = \dfrac{1.25(300) + 700}{300} = \dfrac{1075}{300} \approx \$3.58$

73. $c = f(n) = 0.114n + 15$

75. Refer to **#73**. $c = f(1850) = 0.114(1850) + 15 = 210.90 + 15 = \$225.90$

77. Refer to **#74**. $\bar{c} = f(2500) = \dfrac{0.114(2500) + 15}{2500} = \dfrac{300}{2500} = \$0.12 = 12¢$

79. $c = f(x) = 350x + 5000$

81. Refer to #79. $c = f(120) = 350(120) + 5000 = \$47,000$

83. **Answers may vary.**   **85.** No.

Exercises 10.6 (page 789)

1. $x - 2 = 0$
$ x = 2$

3. $x - 2 < 0$
$ x < 2$

5. $x + 3 > 0$
$ x > -3$

7. $\dfrac{1}{x} < 2$
$x \cdot \dfrac{1}{x} < x \cdot 2 \ (x > 0)$
$ 1 < 2x$

9. $y = kx$

11. $t = kxy$

13. $y = 3x - 4$
$ m = 3$

15. greater

17. quadratic

19. undefined

21. sign

23. $x^2 - 5x + 4 < 0$
$(x - 4)(x - 1) < 0$
$x - 4 \ \text{-------} 0 \ ++++$
$x - 1 \ \text{---} 0+++++++++$

solution set: $(1, 4)$

25. $x^2 - 8x + 15 > 0$
$(x - 5)(x - 3) > 0$
$x - 5 \ \text{----------} 0++++$
$x - 3 \ \text{---} \ 0++++++++++$

solution set: $(-\infty, 3) \cup (5, \infty)$

27. $x^2 + x - 12 \le 0$
$(x - 3)(x + 4) \le 0$
$x - 3 \ \text{---------} 0++++$
$x + 4 \ \text{---} \ 0++++++++++$

solution set: $[-4, 3]$

29. $x^2 + 2x \ge 15$
$x^2 + 2x - 15 \ge 0$
$(x - 3)(x + 5) \ge 0$
$x - 3 \ \text{---------} 0++++$
$x + 5 \ \text{---} \ 0++++++++++$

solution set: $(-\infty, -5] \cup [3, \infty)$

31. $x^2 + 8x < -16$
$x^2 + 8x + 16 < 0$
$(x + 4)(x + 4) < 0$
$x + 4 \ \text{----} \ 0+++++++$
$x + 4 \ \text{----} \ 0+++++++$

Since the product is never negative, there is no solution.

33. $x^2 \ge 9$
$x^2 - 9 \ge 0$
$(x - 3)(x + 3) \ge 0$
$x - 3 \ \text{---------} 0++++$
$x + 3 \ \text{---} \ 0++++++++++$

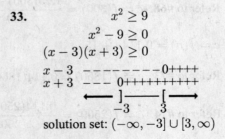

solution set: $(-\infty, -3] \cup [3, \infty)$

35.
$$\frac{1}{x} < 2$$

$$\frac{1}{x} - 2 < 0$$

$$\frac{1}{x} - \frac{2x}{x} < 0$$

$$\frac{1 - 2x}{x} < 0$$

$1 - 2x$ $\quad ++++++++\ 0---$
x $\quad\quad ---0++++++++++$

solution set: $(-\infty, 0) \cup \left(\frac{1}{2}, \infty\right)$

37.
$$\frac{4}{x} \geq 2$$

$$\frac{4}{x} - 2 \geq 0$$

$$\frac{4}{x} - \frac{2x}{x} \geq 0$$

$$\frac{4 - 2x}{x} \geq 0$$

$4 - 2x$ $\quad +++++++++++0---$
x $\quad\quad ---0+++++++++++$

solution set: $(0, 2]$

39.
$$\frac{x^2 - x - 12}{x - 1} < 0$$

$$\frac{(x-4)(x+3)}{x-1} < 0$$

$x - 4$ $\quad --------\ ---\ 0++++$
$x - 1$ $\quad --------\ 0+++\ +++++$
$x + 3$ $\quad ---\ 0++++++\ ++++\ +++++$

solution set: $(-\infty, -3) \cup (1, 4)$

41.
$$\frac{x^2 + x - 20}{x + 2} \geq 0$$

$$\frac{(x-4)(x+5)}{x+2} \geq 0$$

$x - 4$ $\quad --------\ ----\ 0++++$
$x + 2$ $\quad --------\ 0++++\ +++++$
$x + 5$ $\quad ---\ 0++++++++++\ +++++$

solution set: $[-5, -2) \cup [4, \infty)$

43.
$$\frac{x^2 - 4x + 4}{x + 4} < 0$$

$$\frac{(x-2)(x-2)}{x+4} < 0$$

$x - 2$ $\quad --------\ 0++++$
$x - 2$ $\quad --------\ 0++++$
$x + 4$ $\quad ---\ 0++++++\ +++++$

solution set: $(-\infty, -4)$

45.
$$\frac{6x^2 - 5x + 1}{2x + 1} > 0$$

$$\frac{(2x-1)(3x-1)}{2x+1} > 0$$

$2x - 1$ $\quad --------\ ----\ 0++++$
$3x - 1$ $\quad --------\ 0+++++++++$
$2x + 1$ $\quad ---\ 0++++++++++++++++$

solution set: $\left(-\frac{1}{2}, \frac{1}{3}\right) \cup \left(\frac{1}{2}, \infty\right)$

47.
$$\frac{3}{x - 2} < \frac{4}{x} \qquad \frac{-x + 8}{x(x-2)} < 0$$

$$\frac{3}{x-2} - \frac{4}{x} < 0$$

$-x + 8$ $\quad +++++++++++++++0---$
$x - 2$ $\quad\quad -------\ 0++++++++++$
$$\frac{3x}{x(x-2)} - \frac{4(x-2)}{x(x-2)} < 0$$
x $\quad\quad ---\ 0+++++\ +++++++++$

$$\frac{-x + 8}{x(x-2)} < 0$$

solution set: $(0, 2) \cup (8, \infty)$

49.
$$\frac{-5}{x+2} \geq \frac{4}{2-x}$$

$$\frac{-5}{x+2} - \frac{4}{2-x} \geq 0$$

$$\frac{-5(2-x)}{(x+2)(2-x)} - \frac{4(x+2)}{(x+2)(2-x)} \geq 0$$

$$\frac{x-18}{(x+2)(2-x)} \geq 0$$

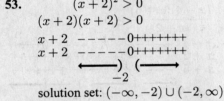

x − 18 − − − − − − − − − − − − 0++++
2 − x ++++++++++++ 0− − − − − −
x + 2 − − − 0+++++++++++++++++

solution set: $(-\infty, -2) \cup (2, 18]$

51.
$$\frac{7}{x-3} \geq \frac{2}{x+4}$$

$$\frac{7}{x-3} - \frac{2}{x+4} \geq 0$$

$$\frac{7(x+4)}{(x-3)(x+4)} - \frac{2(x-3)}{(x-3)(x+4)} \geq 0$$

$$\frac{5x+34}{(x-3)(x+4)} \geq 0$$

x − 3 − − − − − − − − − − − − 0++++
x + 4 − − − − − − − − 0++++ +++++
5x + 34 − − − 0++++++ +++++ +++++

solution set: $\left[-\frac{34}{5}, -4\right) \cup (3, \infty)$

53.
$$(x+2)^2 > 0$$
$$(x+2)(x+2) > 0$$

x + 2 − − − − − 0+++++++
x + 2 − − − − − 0+++++++

solution set: $(-\infty, -2) \cup (-2, \infty)$

55. $y < x^2 + 1$

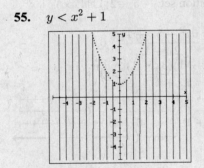

57. $y \leq x^2 + 5x + 6$

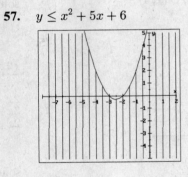

59. $y \geq (x-1)^2$

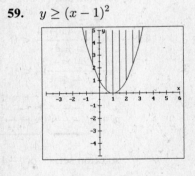

61. $-x^2 - y + 6 > -x$
$-x^2 + x + 6 > y$

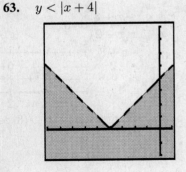

63. $y < |x+4|$

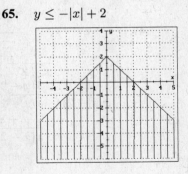

65. $y \leq -|x| + 2$

380

67.
$$2x^2 - 50 < 0$$
$$2(x-5)(x+5) < 0$$

$x - 5$ $------- 0{+}{+}{+}{+}$
$x + 5$ $--- 0{+}{+}{+}{+}{+}{+} {+}{+}{+}{+}$

$$\longleftarrow \;\underset{-5}{(}\!\!\rule[0.5ex]{2.5em}{0.4pt}\!\!\underset{5}{)}\longrightarrow$$

solution set: $(-5, 5)$

69.
$$-\frac{5}{x} < 3$$
$$-\frac{5}{x} - 3 < 0$$
$$\frac{-5}{x} - \frac{3x}{x} < 0$$
$$\frac{-5 - 3x}{x} < 0$$

$-5 - 3x$ $\;{+}{+}{+}\; 0 ---------$
x $\qquad ----------0{+}{+}{+}{+}$

$$\longleftarrow \;\underset{-\frac{5}{3}}{)}\!\!\rule[0.5ex]{2.5em}{0.4pt}\!\!\underset{0}{(}\longrightarrow$$

solution set: $\left(-\infty, -\frac{5}{3}\right) \cup (0, \infty)$

71.
$$\frac{x}{x+4} \le \frac{1}{x+1}$$
$$\frac{x}{x+4} - \frac{1}{x+1} \le 0$$
$$\frac{x(x+1)}{(x+4)(x+1)} - \frac{1(x+4)}{(x+4)(x+1)} \le 0$$
$$\frac{x^2 - 4}{(x+4)(x+1)} \le 0$$
$$\frac{(x+2)(x-2)}{(x+4)(x+1)} \le 0$$

$x - 2$ $\;------- \quad ----0{+}{+}{+}$
$x + 1$ $\;---------- \; 0{+}{+}{+}{+} \;{+}{+}{+}{+}$
$x + 2$ $\;------ \; 0{+}{+}{+}{+} \;{+}{+}{+}{+}{+} \;{+}{+}{+}{+}$
$x + 4$ $\;-- \; 0{+}{+}{+}{+}{+}{+}{+}{+}{+} \;{+}{+}{+}{+}{+} \;{+}{+}{+}{+}$

$$\longleftarrow \;\underset{-4}{(}\!\!\rule[0.5ex]{1.5em}{0.4pt}\!\!\underset{-2}{]}\!\!\rule[0.5ex]{1em}{0.4pt}\!\!\underset{-1}{(}\!\!\rule[0.5ex]{1.5em}{0.4pt}\!\!\underset{2}{]}\longrightarrow$$

solution set: $(-4, -2] \cup (-1, 2]$

73.
$$\frac{x}{x+16} > \frac{1}{x+1}$$
$$\frac{x}{x+16} - \frac{1}{x+1} > 0$$
$$\frac{x(x+1)}{(x+16)(x+1)} - \frac{1(x+16)}{(x+16)(x+1)} > 0$$
$$\frac{x^2 - 16}{(x+16)(x+1)} > 0$$
$$\frac{(x+4)(x-4)}{(x+16)(x+1)} > 0$$

$x - 4$ $\;------- \quad ---- \;---0{+}{+}{+}$
$x + 1$ $\;------- \quad ---- \; 0{+}{+}{+}{+} \;{+}{+}{+}{+}$
$x + 4$ $\;------- \; 0{+}{+}{+}{+} \;{+}{+}{+}{+}{+} \;{+}{+}{+}{+}$
$x + 16$ $\;--- \; 0{+}{+}{+}{+} \;{+}{+}{+}{+}{+} \;{+}{+}{+}{+}{+} \;{+}{+}{+}{+}$

$$\longleftarrow \;\underset{-16}{)}\!\!\rule[0.5ex]{1.5em}{0.4pt}\!\!\underset{-4}{(}\!\!\rule[0.5ex]{1.5em}{0.4pt}\!\!\underset{-1}{)}\!\!\rule[0.5ex]{1.5em}{0.4pt}\!\!\underset{4}{(}\longrightarrow$$

solution set: $(-\infty, -16) \cup (-4, -1) \cup (4, \infty)$

75. $x^2 - 2x - 3 < 0$

Graph $y = x^2 - 2x - 3$ and find the x-coordinates of points below the x-axis.

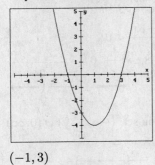

$(-1, 3)$

77. $\frac{x+3}{x-2} > 0$

Graph $y = (x + 3)/(x - 2)$ and find the x-coordinates of points above the x-axis.

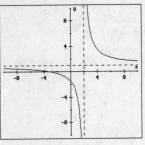

$(-\infty, -3) \cup (2, \infty)$

79. **Answers may vary.**

81. It will be positive if 4, 2 or 0 factors are negative.

Chapter 10 Review (page 795)

1.
$$12x^2 + x - 6 = 0$$
$$(4x + 3)(3x - 2) = 0$$
$$4x + 3 = 0 \quad \text{or} \quad 3x - 2 = 0$$
$$x = -\tfrac{3}{4} \qquad\qquad x = \tfrac{2}{3}$$

2.
$$6x^2 + 17x + 5 = 0$$
$$(2x + 5)(3x + 1) = 0$$
$$2x + 5 = 0 \quad \text{or} \quad 3x + 1 = 0$$
$$x = -\tfrac{5}{2} \qquad\qquad x = -\tfrac{1}{3}$$

3.
$$15x^2 + 2x - 8 = 0$$
$$(3x - 2)(5x + 4) = 0$$
$$3x - 2 = 0 \quad \text{or} \quad 5x + 4 = 0$$
$$x = \tfrac{2}{3} \qquad\qquad x = -\tfrac{4}{5}$$

4.
$$(x + 3)^2 = 16$$
$$x + 3 = \pm\sqrt{16}$$
$$x + 3 = \pm 4$$
$$x = -3 \pm 4$$
$$x = 1 \quad \text{or} \quad x = -7$$

5.
$$x^2 + 8x + 12 = 0$$
$$x^2 + 8x = -12$$
$$x^2 + 8x + 16 = -12 + 16$$
$$(x + 4)^2 = 4$$
$$x + 4 = \pm 2$$
$$x = -4 \pm 2$$
$$x = -2 \quad \text{or} \quad x = -6$$

6. $2x^2 - 9x + 7 = 0$

$$x^2 - \frac{9}{2}x + \frac{7}{2} = 0$$

$$x^2 - \frac{9}{2}x = -\frac{7}{2}$$

$$x^2 - \frac{9}{2}x + \frac{81}{16} = -\frac{56}{16} + \frac{81}{16}$$

$$\left(x - \frac{9}{4}\right)^2 = \frac{25}{16}$$

$$x - \frac{9}{4} = \pm\frac{5}{4}$$

$$x = \frac{9}{4} \pm \frac{5}{4}$$

$$x = \frac{7}{2} \quad \text{or} \quad x = 1$$

7. $2x^2 - x - 5 = 0$

$$x^2 - \frac{1}{2}x - \frac{5}{2} = 0$$

$$x^2 - \frac{1}{2}x = \frac{5}{2}$$

$$x^2 - \frac{1}{2}x + \frac{1}{16} = \frac{5}{2} + \frac{1}{16}$$

$$\left(x - \frac{1}{4}\right)^2 = \frac{41}{16}$$

$$x - \frac{1}{4} = \pm\frac{\sqrt{41}}{4}$$

$$x = \frac{1}{4} \pm \frac{\sqrt{41}}{4}$$

8. $x^2 - 5x - 6 = 0$

$a = 1, b = -5, c = -6$

$$x = \frac{-b \pm \sqrt{b^2 - 4ac}}{2a}$$

$$= \frac{-(-5) \pm \sqrt{(-5)^2 - 4(1)(-6)}}{2(1)}$$

$$= \frac{5 \pm \sqrt{25 + 24}}{2}$$

$$= \frac{5 \pm \sqrt{49}}{2} = \frac{5 \pm 7}{2}$$

$$x = \frac{12}{2} = 6 \text{ or } x = \frac{-2}{2} = -1$$

9. $x^2 = 7x$

$x^2 - 7x = 0$

$a = 1, b = -7, c = 0$

$$x = \frac{-b \pm \sqrt{b^2 - 4ac}}{2a}$$

$$= \frac{-(-7) \pm \sqrt{(-7)^2 - 4(1)(0)}}{2(1)}$$

$$= \frac{7 \pm \sqrt{49 + 0}}{2}$$

$$= \frac{7 \pm \sqrt{49}}{2} = \frac{7 \pm 7}{2}$$

$$x = \frac{14}{2} = 7 \text{ or } x = \frac{0}{2} = 0$$

10. $2x^2 + 13x - 7 = 0$

$a = 2, b = 13, c = -7$

$$x = \frac{-b \pm \sqrt{b^2 - 4ac}}{2a}$$

$$= \frac{-(13) \pm \sqrt{13^2 - 4(2)(-7)}}{2(2)}$$

$$= \frac{-13 \pm \sqrt{169 + 56}}{4}$$

$$= \frac{-13 \pm \sqrt{225}}{4} = \frac{-13 \pm 15}{4}$$

$$x = \frac{2}{4} = \frac{1}{2} \text{ or } x = \frac{-28}{4} = -7$$

11. $5x^2 - 2x - 3 = 0$

$a = 5, b = -2, c = -3$

$$x = \frac{-b \pm \sqrt{b^2 - 4ac}}{2a}$$

$$= \frac{-(-2) \pm \sqrt{(-2)^2 - 4(5)(-3)}}{2(5)}$$

$$= \frac{2 \pm \sqrt{4 + 60}}{10}$$

$$= \frac{2 \pm \sqrt{64}}{10} = \frac{2 \pm 8}{10}$$

$$x = \frac{10}{10} = 1 \text{ or } x = \frac{-6}{10} = -\frac{3}{5}$$

12.
$$2x^2 = x + 2$$
$$2x^2 - x - 2 = 0$$
$$a = 2, b = -1, c = -2$$
$$x = \frac{-b \pm \sqrt{b^2 - 4ac}}{2a}$$
$$= \frac{-(-1) \pm \sqrt{(-1)^2 - 4(2)(-2)}}{2(2)}$$
$$= \frac{1 \pm \sqrt{1 + 16}}{4}$$
$$= \frac{1 \pm \sqrt{17}}{4} = \frac{1}{4} \pm \frac{\sqrt{17}}{4}$$

13. $x^2 + x + 2 = 0$
$$a = 1, b = 1, c = 2$$
$$x = \frac{-b \pm \sqrt{b^2 - 4ac}}{2a}$$
$$= \frac{-1 \pm \sqrt{1^2 - 4(1)(2)}}{2(1)}$$
$$= \frac{-1 \pm \sqrt{1 - 8}}{2}$$
$$= \frac{-1 \pm \sqrt{-7}}{2} = -\frac{1}{2} \pm \frac{\sqrt{7}}{2}i$$

14. Let w represent the original width.
Then $w + 2$ represents the original length.
The new dimensions are then $2w$ and
$2(w + 2) = 2w + 4$.
 Old Area + 72 = New Area
$$w(w + 2) + 72 = 2w(2w + 4)$$
$$w^2 + 2w + 72 = 4w^2 + 8w$$
$$0 = 3w^2 + 6w - 72$$
$$0 = 3(w + 6)(w - 4)$$
$w = -6$ or $w = 4$
Since the width is positive, the
dimensions are 4 cm by 6 cm.

15. Let w represent the original width.
Then $w + 1$ represents the original length.
The new dimensions are then $2w$ and
$3(w + 1) = 3w + 3$.
 Old Area + 30 = New Area
$$w(w + 1) + 30 = 2w(3w + 3)$$
$$w^2 + w + 30 = 6w^2 + 6w$$
$$0 = 5w^2 + 5w - 30$$
$$0 = 5(w + 3)(w - 2)$$
$w = -3$ or $w = 2$
Since the width is positive, the
dimensions are 2 ft by 3 ft.

16. When the rocket hits the ground, $h = 0$:
$$h = 112t - 16t^2$$
$$0 = 112t - 16t^2$$
$$0 = 16t(7 - t)$$
$$t = 0 \quad \text{or} \quad t = 7$$
It hits the ground after 7 seconds.

17. The maximum height occurs at the vertex:
$$h = 112t - 16t^2$$
$$h = -16t^2 + 112t$$
$$h = -16(t^2 - 7t)$$
$$h = -16\left(t^2 - 7t + \frac{49}{4}\right) + 196$$

$$h = -16\left(t^2 - 7t + \frac{49}{4}\right) + 196$$
$$h = -16\left(t - \frac{7}{2}\right)^2 + 196$$
Vertex: $\left(\frac{7}{2}, 196\right) \Rightarrow$ max. height = 196 ft

18. $2x^2 + 5x - 4 = 0$
$a = 2, b = 5, c = -4$
$b^2 - 4ac = 5^2 - 4(2)(-4)$
$\qquad = 25 + 32 = 57$
two irrational unequal solutions

19. $4x^2 - 5x + 7 = 0$
$a = 4, b = -5, c = 7$
$b^2 - 4ac = (-5)^2 - 4(4)(7)$
$\qquad = 25 - 112 = -87$
two complex conjugate solutions

20. $(k-8)x^2 + (k+16)x = -49$
$(k-8)x^2 + (k+16)x + 49 = 0$
$a = k-8, b = k+16, c = 49$
Set the discriminant equal to 0:
$$b^2 - 4ac = 0$$
$$(k+16)^2 - 4(k-8)(49) = 0$$
$$k^2 + 32k + 256 - 196k + 1568 = 0$$
$$k^2 - 164k + 1824 = 0$$
$$(k-12)(k-152) = 0$$
$k - 12 = 0 \quad$ **or** $\quad k - 152 = 0$
$\quad k = 12 \qquad\qquad k = 152$

21. $3x^2 + 4x = k + 1$
$3x^2 + 4x - k - 1 = 0$
$a = 3, b = 4, c = -k - 1$
Set the discriminant ≥ 0:
$$b^2 - 4ac \geq 0$$
$$4^2 - 4(3)(-k-1) \geq 0$$
$$16 + 12k + 12 \geq 0$$
$$12k \geq -28$$
$$k \geq -\frac{28}{12}$$
$$k \geq -\frac{7}{3}$$

22. $\qquad x - 8x^{1/2} + 7 = 0$
$\qquad \left(x^{1/2} - 7\right)\left(x^{1/2} - 1\right) = 0$
$x^{1/2} - 7 = 0 \quad$ **or** $\quad x^{1/2} - 1 = 0$
$\quad x^{1/2} = 7 \qquad\qquad x^{1/2} = 1$
$\left(x^{1/2}\right)^2 = (7)^2 \qquad \left(x^{1/2}\right)^2 = 1^2$
$\qquad x = 49 \qquad\qquad\quad x = 1$
$\quad$ Solution. $\qquad\qquad$ Solution.

23. $\qquad a^{2/3} + a^{1/3} - 6 = 0$
$\qquad \left(a^{1/3} - 2\right)\left(a^{1/3} + 3\right) = 0$
$a^{1/3} - 2 = 0 \quad$ **or** $\quad a^{1/3} + 3 = 0$
$\quad a^{1/3} = 2 \qquad\qquad a^{1/3} = -3$
$\left(a^{1/3}\right)^3 = (2)^3 \qquad \left(a^{1/3}\right)^3 = (-3)^3$
$\qquad a = 8 \qquad\qquad\quad a = -27$
$\quad$ Solution. $\qquad\qquad$ Solution.

24. $\qquad \dfrac{1}{x+1} - \dfrac{1}{x} = -\dfrac{1}{x+1}$
$\left(\dfrac{1}{x+1} - \dfrac{1}{x}\right)(x)(x+1) = -\dfrac{1}{x+1}(x)(x+1)$
$\qquad 1(x) - 1(x+1) = -x$
$\qquad x - x - 1 = -x$
$\qquad\qquad -1 = -x$
$\qquad\qquad\quad 1 = x$

CHAPTER 10 REVIEW

25.
$$\frac{6}{x+2} + \frac{6}{x+1} = 5$$
$$\left(\frac{6}{x+2} + \frac{6}{x+1}\right)(x+2)(x+1) = 5(x+2)(x+1)$$
$$6(x+1) + 6(x+2) = 5(x^2 + 3x + 2)$$
$$6x + 6 + 6x + 12 = 5x^2 + 15x + 10$$
$$0 = 5x^2 + 3x - 8$$
$$0 = (5x+8)(x-1) \qquad 5x+8=0 \quad \textbf{or} \quad x-1=0$$
$$x = -\tfrac{8}{5} \qquad\qquad x = 1$$

26. $3x^2 - 14x + 3 = 0$

sum $= -\dfrac{b}{a} = -\dfrac{-14}{3} = \dfrac{14}{3}$

27. $3x^2 - 14x + 3 = 0$

product $= \dfrac{c}{a} = \dfrac{3}{3} = 1$

28. $y = 2x^2 - 3$
$y = 2(x-0)^2 - 3$

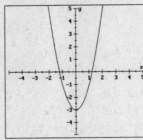

vertex: $(0, -3)$

29. $y = -2x^2 - 1$
$y = -2(x-0)^2 - 1$

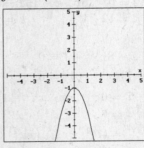

vertex: $(0, -1)$

30. $y = -4(x-2)^2 + 1$

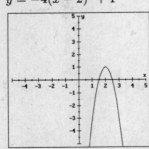

vertex: $(2, 1)$

31. $y = 5x^2 + 10x - 1$
$\quad= 5(x^2 + 2x) - 1$
$\quad= 5(x^2 + 2x + 1) - 1 - 5$
$\quad= 5(x+1)^2 - 6$

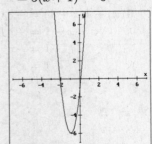

vertex: $(-1, -6)$

386

32. $f(x) = 4x^2 - 16x - 3;\ x = -\dfrac{b}{2a} = -\dfrac{-16}{2(4)} = 2;\ y = 4(2)^2 - 16(2) - 3 = -19$

vertex: $(2, -19)$; axis: $x = 2$

33. $f(x) = |x| - 3$: Shift
$f(x) = |x|$ D 3.

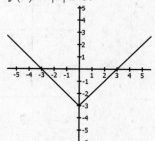

34. $f(x) = |x| - 4$: Shift
$f(x) = |x|$ D 4.

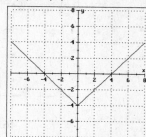

35. $f(x) = (x - 2)^3$: Shift
$f(x) = x^3$ R 2.

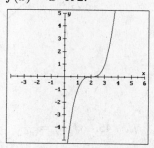

36. $f(x) = (x + 4)^2 - 3$: Shift
$f(x) = x^2$ D 3 and L 4.

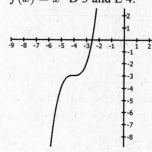

37. $f(x) = -x^3 - 2$: Reflect
$f(x) = x^3$ about the x-
axis and shift D 2.

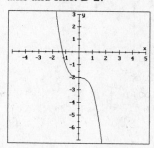

38. $f(x) = -|x - 1| + 2$
Reflect $f(x) = |x|$ about
the x-axis and shift U 2, R 1.

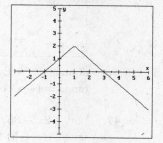

39. $f(x) = x^2 - 3$: Shift
$f(x) = x^2$ D 3.

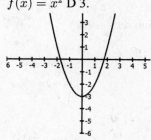

40. $f(x) = |x| - 4$: Shift
$f(x) = |x|$ D 4.

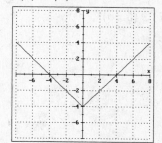

41. $f(x) = (x - 2)^3$: Shift
$f(x) = x^3$ R 2.

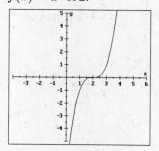

42. $f(x) = |x + 4| - 3$: Shift $f(x) = |x|$ D 3 and L 4.

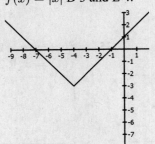

43. $f(x) = -x^3 - 2$: Reflect $f(x) = x^3$ about the x-axis and shift D 2.

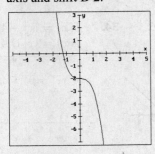

44. $f(x) = -|x - 1| + 2$ Reflect $f(x) = |x|$ about the x-axis and shift U 2.

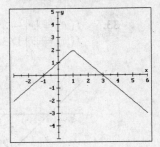

45. $f(x) = \frac{2}{x-2}$

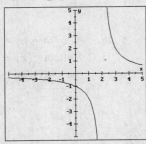

domain $= (-\infty, 2) \cup (2, \infty)$
range $= (-\infty, 0) \cup (0, \infty)$

46. $f(x) = \frac{x}{x+3}$

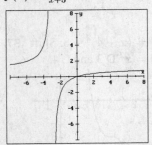

domain $= (-\infty, -3) \cup (-3, \infty)$
range $= (-\infty, 1) \cup (1, \infty)$

47. $x^2 + 2x - 35 > 0$
$(x + 7)(x - 5) > 0$

$x - 5 \quad ---------0\,++++$
$x + 7 \quad ---\;0++++++\;++++$

$\longleftarrow \quad)\!\!-\!\!-\!\!-\!\!-\!\!(\longrightarrow$
$\qquad\quad -7 \qquad 5$

solution set: $(-\infty, -7) \cup (5, \infty)$

48. $x^2 + 7x - 18 < 0$
$(x - 2)(x + 9) < 0$

$x - 2 \quad --------- 0\,++++$
$x + 9 \quad --- \; 0+++++++++$

$\longleftarrow \quad (\!\!-\!\!-\!\!-\!\!-\!\!) \longrightarrow$
$\qquad\quad -9 \qquad 2$

solution set: $(-9, 2)$

49.
$$\frac{3}{x} \le 5$$

$$\frac{3}{x} - 5 \le 0$$

$$\frac{3}{x} - \frac{5x}{x} \le 0$$

$$\frac{3 - 5x}{x} \le 0$$

$3 - 5x$ ++++++++ 0 ---
x ---0+++++++++

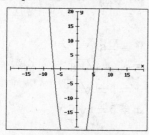

solution set: $(-\infty, 0) \cup [\frac{3}{5}, \infty)$

50.
$$\frac{2x^2 - x - 28}{x - 1} > 0$$

$$\frac{(2x + 7)(x - 4)}{x - 1} > 0$$

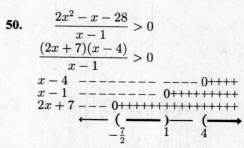

$x - 4$ --------- ----0++++
$x - 1$ -------- 0+++++++++
$2x + 7$ --- 0++++++++++++++++++

solution set: $\left(-\frac{7}{2}, 1\right) \cup (4, \infty)$

51. $x^2 + 2x - 35 > 0$
Graph $y = x^2 + 2x - 35$
and find the x-coordinates
of points above the x-axis.

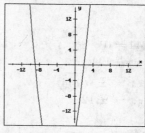

$(-\infty, -7) \cup (5, \infty)$

52. $x^2 + 7x - 18 < 0$
Graph $y = x^2 + 7x - 18$
and find the x-coordinates
of points below the x-axis.

$(-9, 2)$

53. $\frac{3}{x} \le 5 \Rightarrow \frac{3}{x} - 5 \le 0$
Graph $y = (3/x) - 5$
and find the x-coordinates
of points below or on the x-axis.

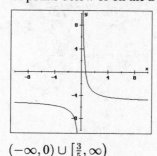

$(-\infty, 0) \cup [\frac{3}{5}, \infty)$

54. $\frac{2x^2 - x - 28}{x - 1} > 0$
Graph $y = (2x^2 - x - 28)/(x - 1)$
and find the x-coordinates
of points above the x-axis.

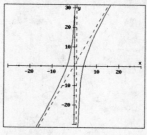

$\left(-\frac{7}{2}, 1\right) \cup (4, \infty)$

55. $y < \dfrac{1}{2}x^2 - 1$

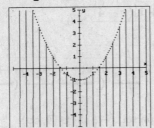

56. $y \geq -|x|$

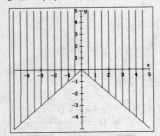

Chapter 10 Test (page 801)

1.
$$x^2 - 3x - 28 = 0$$
$$(x + 4)(x - 7) = 0$$
$$x + 4 = 0 \quad \textbf{or} \quad x - 7 = 0$$
$$x = -4 \qquad\qquad x = 7$$

2.
$$x(6x + 19) = -15$$
$$6x^2 + 19x + 15 = 0$$
$$(2x + 3)(3x + 5) = 0$$
$$2x + 3 = 0 \quad \textbf{or} \quad 3x + 5 = 0$$
$$x = -\tfrac{3}{2} \qquad\qquad x = -\tfrac{5}{3}$$

3.
$$x^2 - 144 = 0$$
$$x^2 = 144$$
$$x = \pm\sqrt{144}$$
$$x = \pm 12$$

4.
$$(x - 5)^2 = -6$$
$$x - 5 = \pm\sqrt{-6}$$
$$x = 5 \pm \sqrt{-6}$$
$$x = 5 \pm i\sqrt{6}$$

5.
$$x^2 + 6x + 7 = 0$$
$$x^2 + 6x = -7$$
$$x^2 + 6x + 9 = -7 + 9$$
$$(x + 3)^2 = 2$$
$$x + 3 = \pm\sqrt{2}$$
$$x = -3 \pm \sqrt{2}$$

6.
$$x^2 - 5x - 3 = 0$$
$$x^2 - 5x = 3$$
$$x^2 - 5x + \frac{25}{4} = 3 + \frac{25}{4}$$
$$\left(x - \frac{5}{2}\right)^2 = \frac{37}{4}$$
$$x - \frac{5}{2} = \pm\sqrt{\frac{37}{4}}$$
$$x = \frac{5}{2} \pm \frac{\sqrt{37}}{2}$$

7. $2x^2 + 5x + 1 = 0$
$a = 2, b = 5, c = 1$

$$x = \frac{-b \pm \sqrt{b^2 - 4ac}}{2a}$$

$$= \frac{-5 \pm \sqrt{5^2 - 4(2)(1)}}{2(2)}$$

$$= \frac{-5 \pm \sqrt{25 - 8}}{4}$$

$$= \frac{-5 \pm \sqrt{17}}{4} = -\frac{5}{4} \pm \frac{\sqrt{17}}{4}$$

8. $x^2 - 2x + 6 = 0$
$a = 1, b = -2, c = 6$

$$x = \frac{-b \pm \sqrt{b^2 - 4ac}}{2a}$$

$$= \frac{-(-2) \pm \sqrt{(-2)^2 - 4(1)(6)}}{2(1)}$$

$$= \frac{2 \pm \sqrt{4 - 24}}{2}$$

$$= \frac{2 \pm \sqrt{-20}}{2} = \frac{2 \pm 2i\sqrt{5}}{2} = 1 \pm i\sqrt{5}$$

9. $4x^2 + 3x + 10 = 0$
$a = 4, b = 3, c = 10$
$b^2 - 4ac = 3^2 - 4(4)(10)$
$\qquad\qquad = 9 - 160 = -151$
nonreal solutions

10. $4x^2 - 2kx + k - 1 = 0$
$a = 4, b = -2k, c = k - 1$
Set the discriminant equal to 0:

$$b^2 - 4ac = 0$$
$$(-2k)^2 - 4(4)(k - 1) = 0$$
$$4k^2 - 16k + 16 = 0$$
$$4(k - 2)(k - 2) = 0$$

$k - 2 = 0 \quad$ **or** $\quad k - 2 = 0$
$\quad k = 2 \qquad\qquad\quad k = 2$

11. Let $x =$ the length of the shorter leg.
Then $x + 14 =$ the other length.
$$x^2 + (x + 14)^2 = 26^2$$
$$x^2 + x^2 + 28x + 196 = 676$$
$$2x^2 + 28x - 480 = 0$$
$$2(x + 24)(x - 10) = 0$$
$x = -24 \quad$ or $\quad x = 10$
The shorter leg is 10 inches long.

12.
$$2y - 3y^{1/2} + 1 = 0$$
$$\left(2y^{1/2} - 1\right)\left(y^{1/2} - 1\right) = 0$$

$2y^{1/2} - 1 = 0 \quad$ **or** $\quad y^{1/2} - 1 = 0$
$2y^{1/2} = 1 \qquad\qquad\quad y^{1/2} = 1$
$\left(y^{1/2}\right)^2 = \left(\frac{1}{2}\right)^2 \qquad \left(y^{1/2}\right)^2 = 1^2$
$\quad y = \frac{1}{4} \qquad\qquad\qquad y = 1$
Solution $\qquad\qquad$ Solution

13. $y = \frac{1}{2}x^2 - 4 = \frac{1}{2}(x - 0)^2 - 4$

vertex: $(0, -4)$

14. $y = -3x^2 + 12x - 5$
$\quad = -3\left(x^2 - 4x\right) - 5$
$\quad = -3\left(x^2 - 4x + 4\right) - 5 + 12$
$\quad = -3(x - 2)^2 + 7$
Vertex: $(2, 7)$; axis: $x = 2$

15. $f(x) = (x-3)^2 + 1$
Shift $f(x) = x^2$ R3 U1.

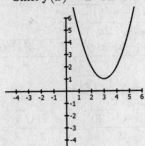

16. $f(x) = x^3$

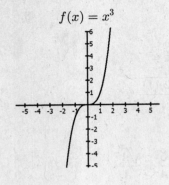

17. $f(x) = |x| - 2$
Shift $f(x) = |x|$ D2.

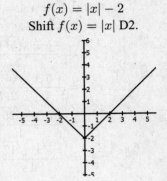

18. $f(x) = (x+1)^3 - 3$
Shift $f(x) = x^3$ L1 D3.

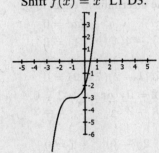

19. $f(x) = -|x-2| + 1$
Reflect $f(x) = |x|$ about
x-axis, shift R2 U1.

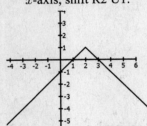

20. $y \le -x^2 + 3$

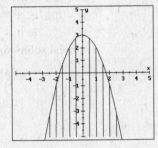

21. $x^2 - 2x - 8 > 0$
$(x+2)(x-4) > 0$

$x - 4 \quad -------- 0 ++++$
$x + 2 \quad --- 0++++++ \ ++++$

$\longleftarrow \quad)\underset{-2}{}\text{———}\underset{4}{(}\longrightarrow$

solution set: $(-\infty, -2) \cup (4, \infty)$

22. $x^2 + 5x - 6 \le 0$
$(x-1)(x+6) \le 0$

$x - 1 \quad --------- 0++++$
$x + 6 \quad --- 0++++++++++$

$\longleftarrow \quad [\underset{-6}{}\text{———}\underset{1}{]}\longrightarrow$

solution set: $[-6, 1]$

23. $\dfrac{x-2}{x+3} \le 0$

$x - 2 \quad ------- 0 \ ++++$
$x + 3 \quad --- 0++++++ \ ++++$

$\longleftarrow \quad (\underset{-3}{}\text{———}\underset{2}{]}\longrightarrow$

solution set: $(-3, 2]$

24. $|y| \le x$

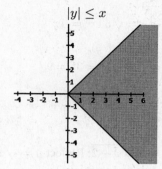

25. $f(x) < |x - 2| + 3$

Cumulative Review Exercises (page 802)

1. $y = f(x) = 5x^2 - 2$
domain $= (-\infty, \infty)$
range $= [-2, \infty)$

2. $y = f(x) = -|x - 4|$
domain $= (-\infty, \infty)$
range $= (-\infty, 0]$

3. $y - y_1 = m(x - x_1)$
$y + 5 = 4(x + 3)$
$y = 4x + 7$

4. $2x + 3y = 6$ $\qquad y - y_1 = m(x - x_1)$
$\quad 3y = -2x + 6$
$\qquad y = -\dfrac{2}{3}x + 2 \qquad y + 2 = -\dfrac{2}{3}(x - 0)$
$\qquad\qquad\qquad\qquad\qquad y = -\dfrac{2}{3}x - 2$

5. $(2a^2 + 4a - 7) - 2(3a^2 - 4a) = 2a^2 + 4a - 7 - 6a^2 + 8a = -4a^2 + 12a - 7$

6. $(4x - 3)(5x + 2) = 20x^2 + 8x - 15x - 6 = 20x^2 - 7x - 6$

7. $x^4 - 16y^4 = (x^2 + 4y^2)(x^2 - 4y^2) = (x^2 + 4y^2)(x + 2y)(x - 2y)$

8. $15x^2 - 2x - 8 = (5x - 4)(3x + 2)$

9. $\quad x^2 - 8x - 9 = 0$
$(x - 9)(x + 1) = 0$
$x - 9 = 0 \quad \textbf{or} \quad x + 1 = 0$
$\qquad x = 9 \qquad\qquad x = -1$

10. $\qquad 6a^3 - 2a = a^2$
$\quad 6a^3 - a^2 - 2a = 0$
$\quad a(6a^2 - a - 2) = 0$
$a(3a - 2)(2a + 1) = 0$
$a = 0 \quad \textbf{or} \quad 3a - 2 = 0 \quad \textbf{or} \quad 2a + 1 = 0$
$\qquad\qquad\qquad\qquad a = \dfrac{2}{3} \qquad\qquad a = -\dfrac{1}{2}$

11. $\sqrt{36a^2b^4} = 6ab^2$

12. $\sqrt{48t^3} = \sqrt{16t^2}\sqrt{3t} = 4t\sqrt{3t}$

13. $\sqrt[3]{-64y^6} = -4y^2$

14. $\sqrt[3]{\dfrac{128x^4}{2x}} = \sqrt[3]{64x^3} = 4x$

15. $8^{-1/3} = \dfrac{1}{8^{1/3}} = \dfrac{1}{2}$

16. $27^{2/3} = \left(27^{1/3}\right)^2 = 3^2 = 9$

17. $\dfrac{y^{2/3}y^{5/3}}{y^{1/3}} = \dfrac{y^{7/3}}{y^{1/3}} = y^{6/3} = y^2$

18. $\dfrac{x^{5/3}x^{1/2}}{x^{3/4}} = \dfrac{x^{13/6}}{x^{3/4}} = x^{17/12}$

19. $f(x) = \sqrt{x-2}$; Shift $y = \sqrt{x}$ right 2.

20. $f(x) = -\sqrt{x+2}$; Reflect $y = \sqrt{x}$ about the x-axis and shift left 2.

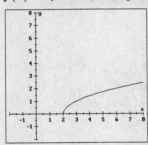

$D = [2, \infty),\ R = [0, \infty)$

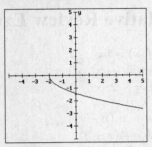

$D = [-2, \infty),\ R = (-\infty, 0]$

21. $\left(x^{2/3} - x^{1/3}\right)\left(x^{2/3} + x^{1/3}\right) = x^{4/3} + x^{3/3} - x^{3/3} - x^{2/3} = x^{4/3} - x^{2/3}$

22. $\left(x^{-1/2} + x^{1/2}\right)^2 = \left(x^{-1/2} + x^{1/2}\right)\left(x^{-1/2} + x^{1/2}\right) = x^{-2/2} + x^0 + x^0 + x^{2/2} = x + 2 + \dfrac{1}{x}$

23. $\sqrt{50} - \sqrt{8} + \sqrt{32} = \sqrt{25}\sqrt{2} - \sqrt{4}\sqrt{2} + \sqrt{16}\sqrt{2} = 5\sqrt{2} - 2\sqrt{2} + 4\sqrt{2} = 7\sqrt{2}$

24. $-3\sqrt[4]{32} - 2\sqrt[4]{162} + 5\sqrt[4]{48} = -3\sqrt[4]{16}\sqrt[4]{2} - 2\sqrt[4]{81}\sqrt[4]{2} + 5\sqrt[4]{16}\sqrt[4]{3}$

$$= -3(2)\sqrt[4]{2} - 2(3)\sqrt[4]{2} + 5(2)\sqrt[4]{3}$$

$$= -6\sqrt[4]{2} - 6\sqrt[4]{2} + 10\sqrt[4]{3} = -12\sqrt[4]{2} + 10\sqrt[4]{3}$$

25. $3\sqrt{2}(2\sqrt{3} - 4\sqrt{12}) = 6\sqrt{6} - 12\sqrt{24} = 6\sqrt{6} - 12\sqrt{4}\sqrt{6} = 6\sqrt{6} - 24\sqrt{6} = -18\sqrt{6}$

26. $\dfrac{5}{\sqrt[3]{x}} = \dfrac{5}{\sqrt[3]{x}} \cdot \dfrac{\sqrt[3]{x^2}}{\sqrt[3]{x^2}} = \dfrac{5\sqrt[3]{x^2}}{\sqrt[3]{x^3}} = \dfrac{5\sqrt[3]{x^2}}{x}$

27. $\dfrac{\sqrt{x}+2}{\sqrt{x}-1} = \dfrac{\sqrt{x}+2}{\sqrt{x}-1} \cdot \dfrac{\sqrt{x}+1}{\sqrt{x}+1} = \dfrac{x+3\sqrt{x}+2}{x-1}$

28. $\sqrt[6]{x^3y^3} = (x^3y^3)^{1/6} = x^{3/6}y^{3/6} = x^{1/2}y^{1/2} = \sqrt{xy}$

29.

$$5\sqrt{x+2} = x+8$$
$$\left(5\sqrt{x+2}\right)^2 = (x+8)^2$$
$$25(x+2) = x^2 + 16x + 64$$
$$25x + 50 = x^2 + 16x + 64$$
$$0 = x^2 - 9x + 14$$
$$0 = (x-7)(x-2)$$
$$x = 7 \quad \text{or} \quad x = 2 \quad \text{(Both check.)}$$

30.

$$\sqrt{x} + \sqrt{x+2} = 2$$
$$\sqrt{x} = 2 - \sqrt{x+2}$$
$$\left(\sqrt{x}\right)^2 = \left(2 - \sqrt{x+2}\right)^2$$
$$x = 4 - 4\sqrt{x+2} + x + 2$$
$$4\sqrt{x+2} = 6$$
$$\left(4\sqrt{x+2}\right)^2 = 6^2$$
$$16(x+2) = 36$$
$$16x + 32 = 36$$
$$16x = 4$$
$$x = \frac{4}{16} = \frac{1}{4}$$

31. hypotenuse $= 3\sqrt{2}$ in.

32. hypotenuse $= 2 \cdot \dfrac{3}{\sqrt{3}} = \dfrac{6\sqrt{3}}{3} = 2\sqrt{3}$ in.

33. $d = \sqrt{(-2-4)^2 + (6-14)^2} = \sqrt{(-6)^2 + (-8)^2} = \sqrt{36+64} = \sqrt{100} = 10$

34. $x^2 + 25 = 0$
$$x^2 = -25$$
$$x = \pm\sqrt{-25} = \pm 5i$$

35.

$$2x^2 + x - 3 = 0$$
$$x^2 + \frac{1}{2}x - \frac{3}{2} = 0$$
$$x^2 + \frac{1}{2}x = \frac{3}{2}$$
$$x^2 + \frac{1}{2}x + \frac{1}{16} = \frac{3}{2} + \frac{1}{16}$$

$$\left(x + \frac{1}{4}\right)^2 = \frac{25}{16}$$
$$x + \frac{1}{4} = \pm\frac{5}{4}$$
$$x = -\frac{1}{4} \pm \frac{5}{4}$$
$$x = \frac{4}{4} = 1 \quad \text{or} \quad x = -\frac{6}{4} = -\frac{3}{2}$$

36. $3x^2 + 4x - 1 = 0$
$a = 3, b = 4, c = -1$
$$x = \frac{-b \pm \sqrt{b^2 - 4ac}}{2a}$$
$$= \frac{-4 \pm \sqrt{4^2 - 4(3)(-1)}}{2(3)}$$

$$x = \frac{-4 \pm \sqrt{16+12}}{6}$$
$$= \frac{-4 \pm \sqrt{28}}{6}$$
$$= \frac{-4 \pm 2\sqrt{7}}{6} = -\frac{2}{3} \pm \frac{\sqrt{7}}{3}$$

CUMULATIVE REVIEW EXERCISES

37. $y = \frac{1}{2}x^2 + 5 = \frac{1}{2}(x-0)^2 + 5$

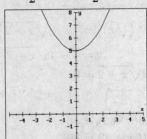

vertex: $(0,5)$; axis: $x=0$

38. $y \le -x^2 + 3$

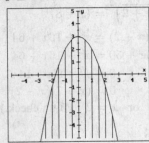

vertex: $(0,3)$

39. $(5+7i)+(8-10i) = 5+7i+8-10i = 13-3i$

40. $(7-4i)-(12+3i) = 7-4i-12-3i = -5-7i$

41. $(6+5i)(6-5i) = 36-30i+30i-25i^2 = 36+25 = 61$

42. $(3+i)(3-3i) = 9-9i+3i-3i^2 = 9-6i+3 = 12-6i$

43. $(3-2i)-(4+i)^2 = 3-2i-(16+8i+i^2) = 3-2i-(15+8i) = 3-2i-15-8i$
$$= -12-10i$$

44. $\dfrac{5}{3-i} = \dfrac{5}{3-i} \cdot \dfrac{3+i}{3+i} = \dfrac{5(3+i)}{9-i^2} = \dfrac{5(3+i)}{10} = \dfrac{3+i}{2} = \dfrac{3}{2} + \dfrac{1}{2}i$

45. $|3+2i| = \sqrt{3^2+2^2} = \sqrt{9+4} = \sqrt{13}$

46. $|5-6i| = \sqrt{5^2+(-6)^2} = \sqrt{25+36} = \sqrt{61}$

47.
$$2x^2 + 4x = k$$
$$2x^2 + 4x - k = 0$$
$$a = 2, b = 4, c = -k$$
Set the discriminant equal to 0:
$$b^2 - 4ac = 0$$
$$4^2 - 4(2)(-k) = 0$$
$$16 + 8k = 0$$
$$8k = -16$$
$$k = -2$$

48.
$$a - 7a^{1/2} + 12 = 0$$
$$\left(a^{1/2}-3\right)\left(a^{1/2}-4\right) = 0$$
$$a^{1/2} - 3 = 0 \quad \textbf{or} \quad a^{1/2} - 4 = 0$$
$$a^{1/2} = 3 \qquad\qquad a^{1/2} = 4$$
$$\left(a^{1/2}\right)^2 = 3^2 \qquad \left(a^{1/2}\right)^2 = 4^2$$
$$a = 9 \qquad\qquad a = 16$$
Solution $\qquad$ Solution

396

49.
$$x^2 - x - 6 > 0$$
$$(x + 2)(x - 3) > 0$$
$$x - 3 \quad --------0 ++++$$
$$x + 2 \quad --- 0++++++ ++++$$

$$\longleftarrow \overset{)}{\underset{-2}{}} \rule{1cm}{0.4pt} \overset{(}{\underset{3}{}} \longrightarrow$$

solution set: $(-\infty, -2) \cup (3, \infty)$

50.
$$\frac{x - 3}{x + 2} \leq 0$$
$$x - 3 \quad --------0 \quad ++++$$
$$x + 2 \quad --- 0++++++ \quad ++++$$

$$\longleftarrow \overset{(}{\underset{-2}{}} \rule{1cm}{0.4pt} \overset{]}{\underset{3}{}} \longrightarrow$$

solution set: $(-2, 3]$

51. $f(x) = -|x - 5| - 2$
Reflect $f(x) = |x|$ about
the x-axis, shift 5R D2.

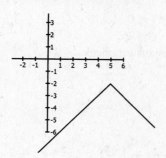

52. $f(x) = (x - 1)^3 + 4$
Shift $f(x) = x^3$ 1R U4.

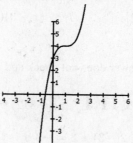

53.
$$\frac{5}{x} - 10 > 0$$
$$\frac{5}{x} - \frac{10x}{x} > 0$$
$$\frac{5 - 10x}{x} > 0$$
$$5 - 10x \quad +++++++++0---$$
$$x \qquad ---0+++++++++$$

$$\longleftarrow \overset{(}{\underset{0}{}} \rule{1cm}{0.4pt} \overset{)}{\underset{\frac{1}{2}}{}} \longrightarrow$$

solution set: $\left(0, \frac{1}{2}\right)$

54.
$$\frac{x - 3}{2} \geq \frac{x + 3}{x}$$
$$\frac{x - 3}{2} - \frac{x + 3}{x} \geq 0$$
$$\frac{x(x - 3)}{2x} - \frac{2(x + 3)}{2x} \geq 0$$
$$\frac{x^2 - 5x - 6}{2x} \geq 0$$
$$\frac{(x - 6)(x + 1)}{2x} \geq 0$$
$$x - 6 \quad -------- \quad ----0++++$$
$$2x \qquad -------- \quad 0+++++++++$$
$$x + 1 \quad --- 0++++++++++++++$$

$$\longleftarrow \overset{[}{\underset{-1}{}} \rule{1cm}{0.4pt} \overset{)}{\underset{0}{}} \rule{1cm}{0.4pt} \overset{[}{\underset{6}{}} \longrightarrow$$

solution set: $[-1, 0) \cup [6, \infty)$

55. $\dfrac{x^2 - 9}{x^2 - 3x} \div \dfrac{x^2 + 9x + 18}{x} = \dfrac{x^2 - 9}{x^2 - 3x} \cdot \dfrac{x}{x^2 + 9x + 18} = \dfrac{(x + 3)(x - 3)}{x(x - 3)} \cdot \dfrac{x}{(x + 3)(x + 6)} = \dfrac{1}{x + 6}$

56. $\dfrac{x + 5}{x + 7} - \dfrac{x - 3}{x - 4} = \dfrac{(x + 5)(x - 4)}{(x + 7)(x - 4)} - \dfrac{(x - 3)(x + 7)}{(x - 4)(x + 7)} = \dfrac{x^2 + x - 20}{(x - 4)(x + 7)} - \dfrac{x^2 + 4x - 21}{(x - 4)(x + 7)}$

$$= \dfrac{-3x + 1}{(x - 4)(x + 7)}$$

57. $\dfrac{\frac{1}{x}+4}{\frac{2}{x^2}+3} = \dfrac{\left(\frac{1}{x}+4\right)x^2}{\left(\frac{2}{x^2}+3\right)x^2} = \dfrac{x+4x^2}{2+3x^2} = \dfrac{4x^2+x}{3x^2+2}$

58.

$$\dfrac{x-2}{x-3} - \dfrac{1}{x} = \dfrac{1}{x-3}$$

$$\left(\dfrac{x-2}{x-3} - \dfrac{1}{x}\right) \cdot x(x-3) = \dfrac{1}{x-3} \cdot x(x-3)$$

$$x(x-2) - (x-3) = x$$

$$x^2 - 2x - x + 3 = x$$

$$x^2 - 4x + 3 = 0$$

$$(x-3)(x-1) = 0$$

$$x - 3 = 0 \quad \textbf{or} \quad x - 1 = 0$$

$$x = 3 \qquad\qquad x = 1$$

This answer does not check and is extraneous.　　　This answer checks

Exercises 11.1 　(page 812)

1. $f(-3) = 2(-3) + 3 = -3$

3. $g(x+1) = 4(x+1)^2 - 2 = 4(x^2 + 2x + 1) - 2 = 4x^2 + 8x + 4 - 2 = 4x^2 + 8x + 2$

5. $(2x+3) + \left(4x^2 - 2\right) = 2x + 3 + 4x^2 - 2$　　**7.**　$(2x+3) - \left(4x^2 - 2\right) = 2x + 3 - 4x^2 + 2$
　　　$= 4x^2 + 2x + 1$ 　　　　　　　　　　　　　　　　　　　$= -4x^2 + 2x + 5$

9. $\dfrac{5x^2 - 13x - 6}{9 - x^2} = \dfrac{(5x+2)(x-3)}{(3+x)(3-x)} = -\dfrac{5x+2}{x+3}$

11. $\dfrac{8 + 2x - x^2}{12 + x - 3x^2} \div \dfrac{3x^2 + 5x - 2}{3x - 1} = \dfrac{x^2 - 2x - 8}{3x^2 - x - 12} \cdot \dfrac{3x - 1}{3x^2 + 5x - 2}$

$$= \dfrac{(x-4)(x+2)}{3x^2 - x - 12} \cdot \dfrac{3x - 1}{(3x-1)(x+2)} = \dfrac{x-4}{3x^2 - x - 12}$$

13. $f(x) + g(x)$ 　　　　　**15.** $f(x)g(x)$ 　　　　　　　**17.** domain

19. $f(g(x))$ 　　　　　　**21.** given point

23. $(f+g)(x) = f(x) + g(x) = 3x + 4x = 7x$ 　　**25.** $(g - f)(x) = g(x) - f(x) = 4x - 3x = x$
　　domain $= (-\infty, \infty)$ 　　　　　　　　　　　　　domain $= (-\infty, \infty)$

27. $(f \cdot g)(x) = f(x) \cdot g(x) = 3x \cdot 4x = 12x^2$ 　　**29.** $(g/f)(x) = \dfrac{g(x)}{f(x)} = \dfrac{4x}{3x} = \dfrac{4}{3}$ (for $x \neq 0$)
　　domain $= (-\infty, \infty)$ 　　　　　　　　　　　　　domain $= (-\infty, 0) \cup (0, \infty)$

31. $(f+g)(x) = f(x) + g(x) = 2x + 1 + x - 3 = 3x - 2$; domain $= (-\infty, \infty)$

33. $(g - f)(x) = g(x) - f(x) = (x - 3) - (2x + 1) = x - 3 - 2x - 1 = -x - 4$; domain $= (-\infty, \infty)$

35. $(f \cdot g)(x) = f(x) \cdot g(x) = (2x + 1)(x - 3) = 2x^2 - 5x - 3$; domain $= (-\infty, \infty)$

37. $(g/f)(x) = \dfrac{g(x)}{f(x)} = \dfrac{x - 3}{2x + 1}$; domain $= \left(-\infty, -\dfrac{1}{2}\right) \cup \left(-\dfrac{1}{2}, \infty\right)$

39. $(f \circ g)(4) = f(g(4)) = f(4^2 - 1) = f(15) = 2(15) + 1 = 31$

41. $(g \circ f)(-3) = g(f(-3)) = g(2(-3) + 1) = g(-5) = (-5)^2 - 1 = 24$

43. $(f \circ g)(0) = f(g(0)) = f(0^2 - 1) = f(-1) = 2(-1) + 1 = -1$

45. $(f \circ g)\left(\dfrac{1}{4}\right) = f\left(g\left(\dfrac{1}{4}\right)\right) = f\left(\left(\dfrac{1}{4}\right)^2 - 1\right) = f\left(-\dfrac{15}{16}\right) = 2\left(-\dfrac{15}{16}\right) + 1 = -\dfrac{15}{8} + 1 = -\dfrac{7}{8}$

47. $(f \circ g)(x) = f(g(x)) = f(x^2 - 1) = 2(x^2 - 1) + 1 = 2x^2 - 2 + 1 = 2x^2 - 1$

49. $(g \circ f)(3x) = g(f(3x)) = g(2(3x) + 1) = g(6x + 1) = (6x + 1)^2 - 1 = 36x^2 + 12x + 1 - 1$
$$= 36x^2 + 12x$$

51. $\dfrac{f(x + h) - f(x)}{h} = \dfrac{4(x + h) + 5 - (4x + 5)}{h} = \dfrac{4x + 4h + 5 - 4x - 5}{h} = \dfrac{4h}{h} = 4$

53. $\dfrac{f(x + h) - f(x)}{h} = \dfrac{(x + h)^2 - x^2}{h} = \dfrac{x^2 + 2xh + h^2 - x^2}{h} = \dfrac{2xh + h^2}{h} = 2x + h$

55. $\dfrac{f(x + h) - f(x)}{h} = \dfrac{2(x + h)^2 - 1 - (2x^2 - 1)}{h} = \dfrac{2x^2 + 4xh + 2h^2 - 1 - 2x^2 + 1}{h}$
$$= \dfrac{4xh + 2h^2}{h} = 4x + 2h$$

57. $\dfrac{f(x + h) - f(x)}{h} = \dfrac{(x + h)^2 + (x + h) - (x^2 + x)}{h} = \dfrac{x^2 + 2xh + h^2 + x + h - x^2 - x}{h}$
$$= \dfrac{2xh + h^2 + h}{h} = 2x + h + 1$$

59. $\dfrac{f(x + h) - f(x)}{h} = \dfrac{(x + h)^2 + 3(x + h) - 4 - (x^2 + 3x - 4)}{h}$
$$= \dfrac{x^2 + 2xh + h^2 + 3x + 3h - 4 - x^2 - 3x + 4}{h}$$
$$= \dfrac{2xh + h^2 + 3h}{h} = 2x + h + 3$$

61. $\dfrac{f(x+h) - f(x)}{h} = \dfrac{2(x+h)^2 + 3(x+h) - 7 - (2x^2 + 3x - 7)}{h}$

$= \dfrac{2x^2 + 4xh + 2h^2 + 3x + 3h - 7 - 2x^2 - 3x + 7}{h}$

$= \dfrac{4xh + 2h^2 + 3h}{h} = 4x + 2h + 3$

63. $(f - g)(x) = f(x) - g(x) = (3x - 2) - (2x^2 + 1) = 3x - 2 - 2x^2 - 1 = -2x^2 + 3x - 3$
domain $= (-\infty, \infty)$

65. $(f/g)(x) = \dfrac{f(x)}{g(x)} = \dfrac{3x - 2}{2x^2 + 1}$; domain $= (-\infty, \infty)$

67. $(f - g)(x) = f(x) - g(x) = (x^2 - 1) - (x^2 - 4) = x^2 - 1 - x^2 + 4 = 3$; domain $= (-\infty, \infty)$

69. $(g/f)(x) = \dfrac{g(x)}{f(x)} = \dfrac{x^2 - 4}{x^2 - 1} = \dfrac{(x+2)(x-2)}{(x+1)(x-1)}$; domain $= (-\infty, -1) \cup (-1, 1) \cup (1, \infty)$

71. $(f \circ g)(4) = f(g(4)) = f(4^2 + 4) = f(20) = 3(20) - 2 = 58$

73. $(g \circ f)(-2) = g(f(-2)) = g(3(-2) - 2) = g(-8) = (-8)^2 + (-8) = 56$

75. $(g \circ f)(0) = g(f(0)) = g(3(0) - 2) = g(-2) = (-2)^2 + (-2) = 2$

77. $(g \circ f)(x) = g(f(x)) = g(3x - 2) = (3x - 2)^2 + 3x - 2 = 9x^2 - 12x + 4 + 3x - 2$
$\qquad = 9x^2 - 9x + 2$

79. $\dfrac{f(x) - f(a)}{x - a} = \dfrac{(4x - 7) - (4a - 7)}{x - a} = \dfrac{4x - 7 - 4a + 7}{x - a} = \dfrac{4x - 4a}{x - a} = \dfrac{4(x - a)}{x - a} = 4$

81. $\dfrac{f(x) - f(a)}{x - a} = \dfrac{x^2 - a^2}{x - a} = \dfrac{(x + a)(x - a)}{x - a} = x + a$

83. $\dfrac{f(x) - f(a)}{x - a} = \dfrac{(2x^2 - 1) - (2a^2 - 1)}{x - a} = \dfrac{2x^2 - 1 - 2a^2 + 1}{x - a} = \dfrac{2x^2 - 2a^2}{x - a}$

$= \dfrac{2(x + a)(x - a)}{x - a}$

$= 2(x + a) = 2x + 2a$

85. $\dfrac{f(x) - f(a)}{x - a} = \dfrac{(x^2 + x) - (a^2 + a)}{x - a} = \dfrac{x^2 + x - a^2 - a}{x - a} = \dfrac{x^2 - a^2 + x - a}{x - a}$

$= \dfrac{(x + a)(x - a) + 1(x - a)}{x - a}$

$= \dfrac{(x - a)(x + a + 1)}{x - a} = x + a + 1$

87. $\dfrac{f(x) - f(a)}{x - a} = \dfrac{(x^2 + 3x - 4) - (a^2 + 3a - 4)}{x - a} = \dfrac{x^2 + 3x - 4 - a^2 - 3a + 4}{x - a}$

$\qquad\qquad\qquad = \dfrac{x^2 - a^2 + 3x - 3a}{x - a}$

$\qquad\qquad\qquad = \dfrac{(x + a)(x - a) + 3(x - a)}{x - a}$

$\qquad\qquad\qquad = \dfrac{(x - a)(x + a + 3)}{x - a} = x + a + 3$

89. $\dfrac{f(x) - f(a)}{x - a} = \dfrac{(2x^2 + 3x - 7) - (2a^2 + 3a - 7)}{x - a} = \dfrac{2x^2 + 3x - 7 - 2a^2 - 3a + 7}{x - a}$

$\qquad\qquad\qquad = \dfrac{2x^2 - 2a^2 + 3x - 3a}{x - a}$

$\qquad\qquad\qquad = \dfrac{2(x + a)(x - a) + 3(x - a)}{x - a}$

$\qquad\qquad\qquad = \dfrac{(x - a)(2(x + a) + 3)}{x - a}$

$\qquad\qquad\qquad = 2(x + a) + 3 = 2x + 2a + 3$

91. $(f \circ g)(x) = f(g(x)) = f(2x - 5) = (2x - 5) + 1 = 2x - 4$

$\quad\ \ (g \circ f)(x) = g(f(x)) = g(x + 1) = 2(x + 1) - 5 = 2x + 2 - 5 = 2x - 3$

93. $f(a) = a^2 + 2a - 3;\ f(h) = h^2 + 2h - 3 \Rightarrow f(a) + f(h) = a^2 + h^2 + 2a + 2h - 6$

$\quad\ \ f(a + h) = (a + h)^2 + 2(a + h) - 3 = a^2 + 2ah + h^2 + 2a + 2h - 3$

$\qquad\qquad\qquad\qquad = a^2 + h^2 + 2ah + 2a + 2h - 3$

95. $\dfrac{f(x + h) - f(x)}{h} = \dfrac{(x + h)^3 - 1 - (x^3 - 1)}{h} = \dfrac{x^3 + 3x^2 h + 3xh^2 + h^3 - 1 - x^3 + 1}{h}$

$\qquad\qquad\qquad = \dfrac{3x^2 h + 3xh^2 + h^3}{h}$

$\qquad\qquad\qquad = \dfrac{h(3x^2 + 3xh + h^2)}{h} = 3x^2 + 3xh + h^2$

97. $F(t) = 2700 - 200t;\ C(F) = \frac{5}{9}(F - 32)$

$\quad\ \ C(F(t)) = C(2700 - 200t) = \frac{5}{9}(2700 - 200t - 32) = \frac{5}{9}(2668 - 200t)$

99. Answers may vary.

101. It is associative. Examples will vary.

Exercises 11.2   (page 821)

1. $f(0) = 2(0) + 1 = 1;\ (0, 1)$

3. $f(3) = 2(3) + 1 = 7;\ (3, 7)$

5. $f(-1) = 2(-1) + 1 = -1;\ (-1, -1)$

7. $7 - \sqrt{-49} = 7 - \sqrt{49i^2} = 7 - 7i$

9. $(3+4i)(2-3i) = 6 - 9i + 8i - 12i^2 = 6 - i - 12(-1) = 6 - i + 12 = 18 - i$

11. $|6 - 8i| = \sqrt{6^2 + (-8)^2} = \sqrt{36 + 64} = \sqrt{100} = 10$

13. one-to-one   **15.** 2   **17.** x

19. Each input has a different output.
one-to-one

21. The inputs $x = 2$ and $x = -2$ have the same output. not one-to-one

23. one-to-one   **25.** one-to-one   **27.** not one-to-one   **29.** one-to-one

31. inverse $= \{(3, 4), (2, 3), (1, 2)\}$. Since each x-coordinate is paired with only one y-coordinate, the inverse relation **is a function**.

33. inverse $= \{(2, 1), (3, 2), (3, 1), (5, 1)\}$. Since $x = 3$ is paired with more than one y-coordinate, the inverse relation **is not a function**.

35.

$$f(x) = 4x + 1$$
$$y = 4x + 1$$
$$x = 4y + 1$$
$$x - 1 = 4y$$
$$\frac{x - 1}{4} = y$$
$$\frac{1}{4}x - \frac{1}{4} = y$$
$$f^{-1}(x) = \tfrac{1}{4}x - \tfrac{1}{4}$$

$f \circ f^{-1}$
$$f[f^{-1}(x)] = f\left(\tfrac{1}{4}x - \tfrac{1}{4}\right)$$
$$= 4 \cdot \left(\tfrac{1}{4}x - \tfrac{1}{4}\right) + 1$$
$$= x - 1 + 1$$
$$= x$$

$f^{-1} \circ f$
$$f^{-1}[f(x)] = f^{-1}(4x + 1)$$
$$= \tfrac{1}{4}(4x + 1) - \tfrac{1}{4}$$
$$= x + \tfrac{1}{4} - \tfrac{1}{4}$$
$$= x$$

37.

$$x + 4 = 5y$$
$$y = f(x) = \frac{x + 4}{5}$$
$$x = \frac{y + 4}{5}$$
$$5x = y + 4$$
$$5x - 4 = y$$
$$f^{-1}(x) = 5x - 4$$

$f \circ f^{-1}$
$$f[f^{-1}(x)] = f(5x - 4)$$
$$= \frac{(5x - 4) + 4}{5}$$
$$= \frac{5x}{5}$$
$$= x$$

$f^{-1} \circ f$
$$f^{-1}[f(x)] = f^{-1}\left(\frac{x + 4}{5}\right)$$
$$= 5 \cdot \left(\frac{x + 4}{5}\right) - 4$$
$$= x + 4 - 4$$
$$= x$$

39.
$$y = 4x + 3$$
$$x = 4y + 3$$
$$x - 3 = 4y$$
$$\frac{x-3}{4} = y;\ f^{-1}(x) = \frac{x-3}{4}$$

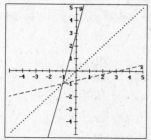

41.
$$x = \frac{y-2}{3}$$
$$y = \frac{x-2}{3};\ f^{-1}(x) = \frac{x-2}{3}$$

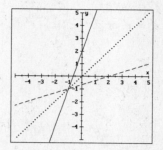

43.
$$y = x^2 + 9$$
$$x = y^2 + 9$$
$$x - 9 = y^2$$
$$\pm\sqrt{x-9} = y$$
The relation **is not** a function.

45.
$$y = x^3$$
$$x = y^3$$
$$\sqrt[3]{x} = y;\ f^{-1}(x) = \sqrt[3]{x}$$
The relation **is** a function.

47.
$$y = x^2 + 1;\ x \geq 0$$
inverse: $x = y^2 + 1;\ y \geq 0$
$$x - 1 = y^2;\ y \geq 0$$
$$\pm\sqrt{x-1} = y;\ y \geq 0$$
$$\sqrt{x-1} = y$$

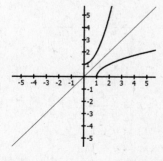

49.
$$y = \sqrt{x},\ x \geq 0$$
inverse: $x = \sqrt{y},\ y \geq 0$
$$x^2 = y,\ y \geq 0$$

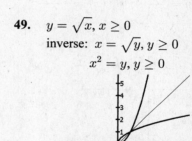

51. inverse $= \{(1,1),(4,2),(9,3),(16,4)\}$. Since each x-coordinate is paired with only one y-coordinate, the inverse relation **is a function**.

53.
$$y = -\sqrt[3]{x}$$
$$x = -\sqrt[3]{y}$$
$$-x = \sqrt[3]{y}$$
$$(-x)^3 = \left(\sqrt[3]{y}\right)^3$$
$$-x^3 = y, \text{ or } f^{-1}(x) = -x^3$$

55.
$$y = 2x^3 - 3$$
$$x = 2y^3 - 3$$
$$x + 3 = 2y^3$$
$$\frac{x+3}{2} = y^3$$
$$\sqrt[3]{\frac{x+3}{2}} = \sqrt[3]{y^3}, \text{ or } y = f^{-1}(x) = \sqrt[3]{\frac{x+3}{2}}$$

57.
$$f(x) = \frac{x-7}{3}$$
$$y = \frac{x-7}{3}$$
$$x = \frac{y-7}{3}$$
$$3x = y - 7$$
$$3x + 7 = y, \text{ or } y = f^{-1}(x) = 3x + 7$$

59.
$$4x - 5y = 20$$
$$5y = 4x - 20$$
$$y = f(x) = \frac{4}{5}x - 4$$
$$x = \frac{4}{5}y - 4$$
$$x + 4 = \frac{4}{5}y$$
$$\frac{5}{4}(x + 4) = y, \text{ or } y = f^{-1}(x) = \frac{5}{4}x + 5$$

61.
$$3x - y = 5$$
$$3y - x = 5$$
$$3y = x + 5$$
$$y = \frac{x+5}{3}; \; f^{-1}(x) = \frac{x+5}{2}$$

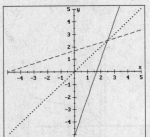

63.
$$3(x + y) = 2x + 4$$
$$3(y + x) = 2y + 4$$
$$3y + 3x = 2y + 4$$
$$y = 4 - 3x; \; f^{-1}(x) = -3x + 4$$

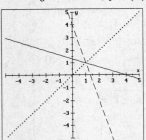

65. **Answers may vary.**

67.
$$y = \frac{x+1}{x-1}$$
$$x = \frac{y+1}{y-1}$$
$$x(y - 1) = y + 1$$
$$xy - x = y + 1$$
$$yx - y = x + 1$$
$$y(x - 1) = x + 1$$
$$y = \frac{x+1}{x-1}, \text{ or } f^{-1}(x) = \frac{x+1}{x-1}$$

Exercises 11.3 (page 831)

1. $3^x = 3^2 = 9$

3. $3(6^x) = 3(6^2) = 3(36) = 108$

5. $3^x = 3^{-2} = \frac{1}{3^2} = \frac{1}{9}$

7. $3(6^x) = 3(6^{-2}) = 3\left(\frac{1}{6^2}\right) = 3\left(\frac{1}{36}\right) = \frac{1}{12}$

9. $3x + 2x - 20 = 180$
$5x = 200$
$x = 40$

11. $m(\angle 2) = 3x = 3(40) = 120°$

13. exponential

15. $(0, \infty)$

17. increasing

19. $P\left(1 + \frac{r}{k}\right)^{kt}$

21. $(1, b)$

23. left

25. $y = f(x) = 3^x$
through $(0, 1)$ and $(1, 3)$

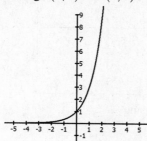

27. $y = f(x) = \left(\frac{5}{2}\right)^x$
through $(0, 1)$ and $\left(1, \frac{5}{2}\right)$

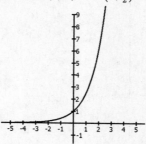

29. $y = f(x) = \left(\frac{1}{3}\right)^x$
through $(0, 1)$ and $\left(1, \frac{1}{3}\right)$

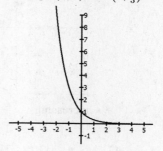

31. $y = f(x) = \left(\frac{2}{5}\right)^x$
through $(0, 1)$ and $\left(1, \frac{2}{5}\right)$

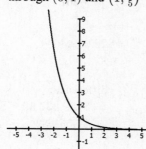

33. $y = b^x$
$\frac{1}{2} = b^1$
$\frac{1}{2} = b$

35. $y = b^x$
$3 = b^1$
$3 = b$

405

37. $f(x) = 3^x - 2$
Shift $y = 3^x$ down 2.

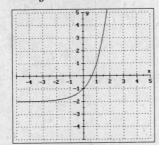

39. $f(x) = 3^{(x-1)}$
Shift $y = 3^x$ right 1.

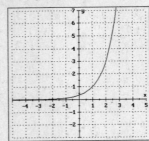

41. $y = b^x$
$2 = b^1$
$2 = b$

43. $y = f(x) = \dfrac{1}{2}\left(3^{x/2}\right)$

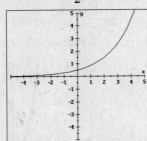

increasing

45. $y = f(x) = 2\left(3^{-x/2}\right)$

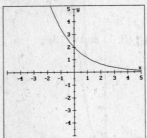

decreasing

47. $S(n) = 1.6(1.173)^n = 1.6(1.173)^1$
$\qquad\qquad\;\; = 1.8768$
There were about 1.9 billion users in 2005.

49. $A = A_0\left(\dfrac{2}{3}\right)^t$

$A = A_0\left(\dfrac{2}{3}\right)^8 = \dfrac{256}{6561}A_0$

51. $A = P\left(1 + \dfrac{r}{k}\right)^{kt} = 10{,}000\left(1 + \dfrac{0.08}{4}\right)^{4(10)} = 10{,}000(1.02)^{40} \approx \$22{,}080.40$

53. $A = P\left(1 + \dfrac{r}{k}\right)^{kt}$ $\qquad$ $A = P\left(1 + \dfrac{r}{k}\right)^{kt}$ $\qquad$ difference:
$\qquad\qquad\qquad\qquad\qquad\qquad\qquad\qquad\qquad\qquad\qquad\; = \$156{,}437.69 - \$148{,}886.37$
$\quad = 100000\left(1 + \dfrac{0.04}{4}\right)^{4(10)}$ $\quad = 100000\left(1 + \dfrac{0.045}{4}\right)^{4(10)}$ $\quad = \$7{,}551.32$
$\quad = 100000(1.01)^{40}$ $\qquad\qquad = 100000(1.01125)^{40}$
$\quad \approx \$148{,}886.37$ $\qquad\qquad\;\; \approx \$156{,}437.69$

55. $A = P\left(1 + \dfrac{r}{k}\right)^{kt} = 1\left(1 + \dfrac{0.05}{1}\right)^{1(300)} = 1(1.05)^{300} \approx \$2{,}273{,}996.13$

57. $C = (3 \times 10^{-4})(0.7)^t = (3 \times 10^{-4})(0.7)^5 \approx 5.0421 \times 10^{-5}$ coulombs

59. $A = P\left(1 + \dfrac{r}{k}\right)^{kt} = 4700\left(1 + \dfrac{-0.25}{1}\right)^{1(5)} = 4700(0.75)^5 \approx \$1115.33$

61. **Answers may vary.**

63. If the base were 0, then the function would not be defined for $x = 0 \Rightarrow y = 0^0$.

Exercises 11.4 (page 839)

Problems 1-5 are to be solved using a calculator. The keystrokes needed to solve each problem using a TI-84 graphing calculator appear in each solution. There may be other solutions. Keystrokes for other calculators may be slightly different.

1. $e^0 \Rightarrow$ [2nd] [LN] [0] [ENTER] $\{1\}$

3. $2e^3 \Rightarrow$ [2] [×] [2nd] [LN] [3] [ENTER] $\{40.17\}$

5. $e^{-1.5} \Rightarrow$ [2nd] [LN] [(−)] [1] [.] [5] [ENTER] $\{0.22\}$

7. $\sqrt{320x^7} = \sqrt{64x^6}\sqrt{5x} = 8x^3\sqrt{5x}$

9. $5\sqrt{98y^3} - 4y\sqrt{18y} = 5\sqrt{49y^2}\sqrt{2y} - 4y\sqrt{9}\sqrt{2y} = 5(7y)\sqrt{2y} - 4y(3)\sqrt{2y}$
$$= 35y\sqrt{2y} - 12y\sqrt{2y} = 23y\sqrt{2y}$$

11. 2.72

13. increasing

15. $A = Pe^{rt}$

17. $y = f(x) = e^x + 1$
Shift $y = e^x$ up 1.

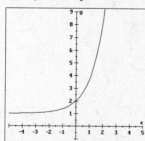

19. $y = f(x) = e^{x+3}$
Shift $y = e^x$ left 3.

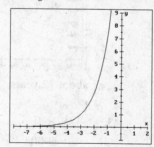

21. The graph should be increasing. The graph could not look like this.

407

23. The graph should go through the point $(0, 1)$. The graph could not look like this.

25. $y = f(x) = -e^x$; Reflect $y = e^x$ about the x-axis.

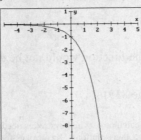

27. $y = f(x) = 2e^x$; Stretch $y = e^x$ vertically by a factor of 2.

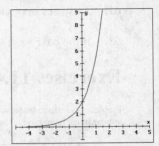

29.
$$A = Pe^{rt}$$
$$= 9000e^{0.04(20)}$$
$$= 9000e^{0.8}$$
$$\approx \$20{,}029.87$$

31.
$$A = Pe^{rt}$$
$$25000 = Pe^{0.05(15)}$$
$$25000 = Pe^{0.75}$$
$$25000 \approx P(2.117)$$
$$P \approx \frac{25000}{2.117}$$
$$P \approx \$11{,}809.16$$

33.
$$A = Pe^{rt}$$
$$= 6e^{0.019(30)}$$
$$= 6e^{0.57}$$
$$\approx 10.6 \text{ billion people}$$

35.
$$A = Pe^{rt}$$
$$= 6e^{0.019(50)}$$
$$= 6e^{0.95}$$
$$\approx 6(2.6)$$
It will increase by a factor of about 2.6.

37.
$$y = 1000e^{0.02x}$$
$$y = 31x + 2000$$

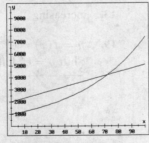

about 72 years

39.
$$A = A_0e^{-0.087t}$$
$$= 75e^{-0.087(40)}$$
$$= 75e^{-3.48}$$
$$\approx 2.31 \text{ grams}$$

41.
$$A = A_0e^{-0.00000693t}$$
$$= 2500e^{-0.00000693(100)}$$
$$= 2500e^{-0.000693}$$
$$\approx 2498.27 \text{ grams}$$

43.
$$A = Pe^{rt}$$
$$= 5000e^{0.085(5)}$$
$$= 5000e^{0.425}$$
$$\approx \$7{,}647.95 \quad \text{(continuous)}$$

$$A = P\left(1 + \frac{r}{k}\right)^{kt}$$
$$= 5000\left(1 + \frac{0.085}{1}\right)^{1(5)}$$
$$= 5000(1.085)^5$$
$$\approx \$7{,}518.28 \quad \text{(annual)}$$

45. $P = 8000e^{-0.008t}$
$= 8000e^{-0.008(50)}$
$= 8000e^{-0.4}$
≈ 5363

47. $x = 0.08\left(1 - e^{-0.1t}\right)$
$= 0.08\left(1 - e^{-0.1(30)}\right)$
$= 0.08\left(1 - e^{-3}\right)$
$\approx 0.08(1 - 0.049787)$
$\approx 0.08(0.950213)$
≈ 0.076

49. $v = 50\left(1 - e^{-0.2t}\right)$
$= 50\left(1 - e^{-0.2(0)}\right)$
$= 50\left(1 - e^{0}\right)$
$= 50(1 - 1)$
$= 50(0) = 0$ mps

51. $v = 50\left(1 - e^{-0.2t}\right)$
$= 50\left(1 - e^{-0.2(2)}\right)$
$= 50\left(1 - e^{-0.4}\right)$
$\approx 50(1 - 0.67032)$
$\approx 50(0.32968) \approx 16.5$ mps

$v = 50\left(1 - e^{-0.3t}\right)$
$= 50\left(1 - e^{-0.3(2)}\right)$
$= 50\left(1 - e^{-0.6}\right)$
$\approx 50(1 - 0.54881)$
$\approx 50(0.45119) \approx 22.6$ mps $\Rightarrow$ faster

53. $A = Pe^{rt}$
$= 4570e^{-0.06(6.5)}$
$\approx \$3094.15$

55. **Answers may vary.**

57. $e \approx 2.7182$; $1 + 1 + \frac{1}{2} + \frac{1}{2\cdot3} + \frac{1}{2\cdot3\cdot4} + \frac{1}{2\cdot3\cdot4\cdot5} \approx 2.7167$; Accurate to two decimal places.

59. $e^{t+10} = ke^t$
$e^t \cdot e^{10} = ke^t$
$e^{10}e^t = ke^t$
$k = e^{10}$

Exercises 11.5 (page 849)

1. $2^{\boxed{3}} = 8$

3. $\boxed{5}^3 = 125$

5. $5^{\boxed{2}} = 25$

7. $2^{\boxed{5}} = 32$

9. $\boxed{2}^{-2} = \frac{1}{4}$

11. $\sqrt{3x - 4} = \sqrt{-7x + 2}$
$3x - 4 = -7x + 2$
$10x = 6$
$x = \frac{6}{10} = \frac{3}{5}$
$\frac{3}{5}$ is extraneous $\Rightarrow \emptyset$

13. $3 - \sqrt{t - 3} = \sqrt{t}$
$\left(3 - \sqrt{t - 3}\right)^2 = t$
$9 - 6\sqrt{t - 3} + t - 3 = t$
$6 = 6\sqrt{t - 3}$
$1 = \sqrt{t - 3}$
$1 = t - 3$
$4 = t$

15. $(0, \infty)$

17. x

19. exponent

21. $(b, 1); (1, 0)$

409

23. $20 \log \dfrac{E_O}{E_I}$

25. $\log_4 64 = 3 \Rightarrow 4^3 = 64$

27. $\log_{1/2} \dfrac{1}{8} = 3 \Rightarrow \left(\dfrac{1}{2}\right)^3 = \dfrac{1}{8}$

29. $\log_4 \dfrac{1}{64} = -3 \Rightarrow 4^{-3} = \dfrac{1}{64}$

31. $\log_{1/2} \dfrac{1}{8} = 3 \Rightarrow \left(\dfrac{1}{2}\right)^3 = \dfrac{1}{8}$

33. $7^2 = 49 \Rightarrow \log_7 49 = 2$

35. $6^{-2} = \dfrac{1}{36} \Rightarrow \log_6 \dfrac{1}{36} = -2$

37. $\left(\dfrac{1}{2}\right)^{-5} = 32 \Rightarrow \log_{1/2} 32 = -5$

39. $x^y = z \Rightarrow \log_x z = y$

41. $\log_2 32 = x \Rightarrow 2^x = 32 \Rightarrow x = 5$

43. $\log_5 125 = x \Rightarrow 5^x = 125 \Rightarrow x = 3$

45. $\log_7 x = 2 \Rightarrow 7^2 = x \Rightarrow x = 49$

47. $\log_6 x = 1 \Rightarrow 6^1 = x \Rightarrow x = 6$

49. $\log_{25} x = \dfrac{1}{2} \Rightarrow 25^{1/2} = x \Rightarrow x = 5$

51. $\log_5 x = -2 \Rightarrow 5^{-2} = x \Rightarrow x = \dfrac{1}{25}$

53. $\log_x 5^3 = 3 \Rightarrow x^3 = 5^3 \Rightarrow x = 5$

55. $\log_x \dfrac{9}{4} = 2 \Rightarrow x^2 = \dfrac{9}{4} \Rightarrow x = \dfrac{3}{2}$

57. $\log_{1/2} \dfrac{1}{8} = x \Rightarrow \left(\dfrac{1}{2}\right)^x = \dfrac{1}{8} \Rightarrow x = 3$

59. $\log_{1/4} x = 3 \Rightarrow \left(\dfrac{1}{4}\right)^3 = x \Rightarrow x = \dfrac{1}{64}$

61. $y = f(x) = \log_3 x$
through $(1, 0)$ and $(3, 1)$

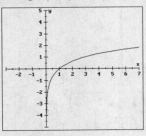

increasing

63. $y = f(x) = \log_{1/2} x$
through $(1, 0)$ and $\left(\dfrac{1}{2}, 1\right)$

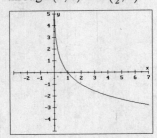

decreasing

65. $y = f(x) = 2^x$
$y = g(x) = \log_2 x$

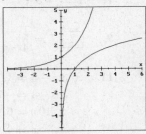

67. $y = f(x) = \left(\frac{1}{4}\right)^x$

$y = g(x) = \log_{1/4} x$

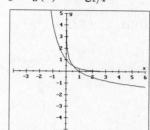

69. $y = f(x) = 3 + \log_3 x$

Shift $y = \log_3 x$ up 3.

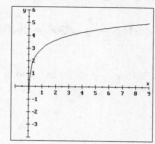

71. $y = f(x) = \log_{1/2}(x - 2)$

Shift $y = \log_{1/2} x$ right 2.

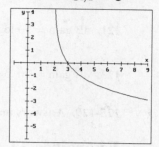

73. $\log 8.25 \approx 0.9165$

75. $\log 0.00867 \approx -2.0620$

77. $\log y = 4.24 \Rightarrow y = 17{,}378.01$

79. $\log y = -3.71 \Rightarrow y = 0.00$

81. $\log_{36} x = -\frac{1}{2} \Rightarrow 36^{-1/2} = x \Rightarrow x = \frac{1}{6}$

83. $\log_{1/2} 8 = x \Rightarrow \left(\frac{1}{2}\right)^x = 8 \Rightarrow x = -3$

85. $\log_{1/2} x = -3 \Rightarrow \left(\frac{1}{2}\right)^{-3} = x \Rightarrow x = 8$

87. $\log_4 2 = x \Rightarrow 4^x = 2 \Rightarrow x = \frac{1}{2}$

89. $\log_{100} \frac{1}{1000} = x \Rightarrow 100^x = \frac{1}{1000} \Rightarrow x = -\frac{3}{2}$

91. $\log_{27} 9 = x \Rightarrow 27^x = 9 \Rightarrow x = \frac{2}{3}$

93. $\log_{2\sqrt{2}} x = 2 \Rightarrow (2\sqrt{2})^2 = x \Rightarrow x = 8$

95. $\log_x \frac{1}{64} = -3 \Rightarrow x^{-3} = \frac{1}{64} \Rightarrow x = 4$

97. $2^{\log_2 4} = x \Rightarrow x = 4$

99. $x^{\log_4 6} = 6 \Rightarrow x = 4$

101. $\log 10^3 = x \Rightarrow 10^x = 10^3 \Rightarrow x = 3$

103. $10^{\log x} = 100 \Rightarrow \log x = 2 \Rightarrow x = 100$

105. $\log y = 1.4023 \Rightarrow y = 25.25$

107. $\log y = \log 8 \Rightarrow \log y = 0.9030 \Rightarrow y = 8$

109. $\log_b 9 = 2 \Rightarrow b^2 = 9 \Rightarrow b = 3$

111. $\log_b 2 = 0 \Rightarrow b^0 = 2$

No such value exists.

113. $R = \log \frac{A}{P} = \log \frac{5000}{0.2} = \log 25{,}000$
≈ 4.4

115. $R = \log \frac{A}{P} = \log \frac{2500}{0.25} = \log 10000$
$= 4$

117. $n = \dfrac{\log V - \log C}{\log \left(1 - \frac{2}{N}\right)} = \dfrac{\log 2000 - \log 17000}{\log \left(1 - \frac{2}{5}\right)} \approx \dfrac{-0.929419}{-0.221849} \approx 4.2$ years old

119. $n = \dfrac{\log\left[\frac{Ar}{P} + 1\right]}{\log\left(1 + r\right)} = \dfrac{\log\left[\frac{20{,}000(0.12)}{1000} + 1\right]}{\log\left(1 + 0.12\right)} = \dfrac{\log 3.4}{\log 1.12} \approx 10.8 \text{ years}$

121. dB gain $= 20 \log \dfrac{E_O}{E_I} = 20 \log \dfrac{20}{0.71} = 20 \log 28.169 \approx 29.0 \text{ dB}$

123. dB gain $= 20 \log \dfrac{E_O}{E_I} = 20 \log \dfrac{30}{0.1} = 20 \log 300 \approx 49.5 \text{ dB}$

125-129. Answers may vary.

Exercises 11.6 (page 856)

1. $\log_3 9 = 2 \Rightarrow 3^2 = 9$

3. $2^3 = 8 \Rightarrow \log_2 8 = 3$

5. $y = mx + b$
$y = -7x + 2$

7. $3x + 2y = 9$
$2y = -3x + 9$
$y = -\dfrac{3}{2}x + \dfrac{9}{2} \Rightarrow m = -\dfrac{3}{2}$

Use the parallel slope:
$y - y_1 = m(x - x_1)$
$y - 5 = -\dfrac{3}{2}(x - (-3))$
$y - 5 = -\dfrac{3}{2}x - \dfrac{9}{2}$
$y = -\dfrac{3}{2}x + \dfrac{1}{2}$

9. $y = 6$

11. $\dfrac{x+1}{x} + \dfrac{x-1}{x+1} = \dfrac{(x+1)(x+1)}{x(x+1)} + \dfrac{(x-1)x}{(x+1)x} = \dfrac{x^2+2x+1}{x(x+1)} + \dfrac{x^2-x}{x(x+1)} = \dfrac{2x^2+x+1}{x(x+1)}$

13. $\dfrac{1 + \frac{y}{x}}{\frac{y}{x} - 1} = \dfrac{\left(1 + \frac{y}{x}\right)x}{\left(\frac{y}{x} - 1\right)x} = \dfrac{x+y}{y-x}$

15. $(0, \infty); (-\infty, \infty)$

17. 10

19. $\dfrac{\ln 2}{r}$

21. $\ln 25.25 \approx 3.2288$

23. $\ln 9.89 \approx 2.2915$

25. $\ln y = 2.3015 \Rightarrow y = 9.9892$

27. $\ln y = 3.17 \Rightarrow y = 23.8075$

29. $y = -\ln x$

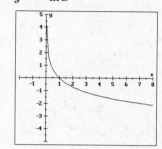

31. $y = \ln(-x)$

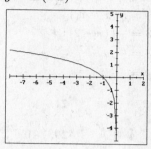

33. The graph must be increasing. The graph could not look like this.

35. The graph must go through $(1, 0)$. The graph could not look like this.

37. $\log(\ln 3) \approx \log(1.0986) \approx 0.0408$

39. $\ln(\log 0.7) = \ln(-0.1549) \Rightarrow$ no real value

41. $\ln y = -4.72 \Rightarrow y = 0.0089$

43. $\log y = \ln 4 \Rightarrow \log y \approx 1.3863 \Rightarrow y \approx 24.3385$
(The answer will vary if rounding is used on the calculator.)

45. $t = \dfrac{\ln 2}{r} = \dfrac{\ln 2}{0.12} \approx 5.8$ years

47. $t = \dfrac{\ln 2}{r} = \dfrac{\ln 2}{0.05} \approx 13.9$ years

49. $t = -\dfrac{1}{0.9} \ln \dfrac{50 - T_r}{212 - T_r} = -\dfrac{1}{0.9} \ln \dfrac{50 - 38}{212 - 38} = -\dfrac{1}{0.9} \ln \dfrac{12}{174} \approx -\dfrac{1}{0.9}(-2.6741) \approx 3.0$ hours

51. **Answers may vary.**

53. $P = P_0 e^{rt} = P_0 e^{r \frac{\ln 3}{r}} = P_0 e^{\ln 3} = 3P_0$

55. Let $t = \dfrac{\ln 5}{r} \Rightarrow P = P_0 e^{rt} = P_0 e^{r \frac{\ln 5}{r}} = P_0 e^{\ln 5} = 5P_0$

Exercises 11.7 (page 866)

1. $\log_5 25 = x \Rightarrow 5^x = 25 \Rightarrow x = 2$

3. $\log_7 x = 3 \Rightarrow 7^3 = x \Rightarrow x = 343$

5. $\log_9 x = \frac{1}{2} \Rightarrow 9^{1/2} = x \Rightarrow x = 3$

7. $\log_{1/3} x = 2 \Rightarrow \left(\frac{1}{3}\right)^2 = x \Rightarrow x = \frac{1}{9}$

9. $\log_x \frac{1}{4} = -2 \Rightarrow x^{-2} = \frac{1}{4} \Rightarrow x = 2$

11. $d = \sqrt{(-2 - 4)^2 + (3 - (-4))^2} = \sqrt{(-6)^2 + 7^2} = \sqrt{36 + 49} = \sqrt{85}$

13. Use the slope $m = -\frac{7}{6}$ from **#10**.
$$y - y_1 = m(x - x_1)$$
$$y - 3 = -\frac{7}{6}(x + 2)$$
$$y - 3 = -\frac{7}{6}x - \frac{7}{3}$$
$$y = -\frac{7}{6}x + \frac{2}{3}$$

15. 1

17. x

19. —

21. x

23. =

25. 0

27. 2

29. 10

31. 1

Problems 33-35 are to be solved using a calculator. The keystrokes needed to solve each problem using a TI-84 graphing calculator appear in each solution. There may be other solutions. Keystrokes for other calculators may be slightly different.

33.
| log | 2 | . | 5 | × | 3 | . | 7 | ENTER | {0.96614}
| log | 2 | . | 5 |) | + | log | 3 | . | 7 |) | ENTER | {0.96614}

35.
| ln | 2 | . | 2 | 5 | ^ | 4 | ENTER | {3.24372}
| 4 | ln | 2 | . | 2 | 5 | ENTER | {3.24372}

37. $\log_b 7xy = \log_b 7 + \log_b x + \log_b y$

39. $\log_b \dfrac{5x}{y} = \log_b 5x - \log_b y$
$$= \log_b 5 + \log_b x - \log_b y$$

41. $\log_b x^3 y^2 = \log_b x^3 + \log_b y^2 = 3\log_b x + 2\log_b y$

43. $\log_b \dfrac{x^3 \sqrt{z}}{y^5} = \log_b \dfrac{x^3 z^{1/2}}{y^5} = \log_b \left(x^3 z^{1/2}\right) - \log_b y^5 = \log_b x^3 + \log_b z^{1/2} - \log_b y^5$
$$= 3\log_b x + \frac{1}{2}\log_b z - 5\log_b y$$

45. $\log_b (x - 3) - 5\log_b x = \log_b (x - 3) - \log_b x^5 = \log_b \dfrac{x - 3}{x^5}$

47. $7\log_b x + \dfrac{1}{2}\log_b z = \log_b x^7 + \log_b z^{1/2} = \log_b x^7 z^{1/2} = \log_b \left(x^7 \sqrt{z}\right)$

49. $\log 28 = \log 4 \cdot 7 = \log 4 + \log 7 = 0.6021 + 0.8451 = 1.4472$

51. $\log 2.25 = \log \dfrac{9}{4} = \log 9 - \log 4 = 0.9542 - 0.6021 = 0.3521$

53. $\log \dfrac{49}{36} = \log \dfrac{7^2}{4 \cdot 9} = \log 7^2 - \log 4 \cdot 9 = 2\log 7 - (\log 4 + \log 9)$
$$= 2(0.8451) - 0.6021 - 0.9542 = 0.1339$$

55. $\log 252 = \log 4 \cdot 7 \cdot 9 = \log 4 + \log 7 + \log 9 = 0.6021 + 0.8451 + 0.9542 = 2.4014$

57. $\log_3 7 = \dfrac{\log 7}{\log 3} \approx 1.7712$

59. $\log_{1/3} 3 = \dfrac{\log 3}{\log \frac{1}{3}} \approx -1.0000$

61. $\log_3 8 = \dfrac{\log 8}{\log 3} \approx 1.8928$

63. $\log_{\sqrt{2}} \sqrt{5} = \dfrac{\log \sqrt{5}}{\log \sqrt{2}} \approx 2.3219$

65. 0 **67.** 7 **69.** 10 **71.** 1

73. $\log_b \left(\dfrac{xy}{z}\right)^{1/3} = \frac{1}{3} \log_b \left(\dfrac{xy}{z}\right) = \frac{1}{3}\left(\log_b x + \log_b y - \log_b z\right)$

75. $\log_b \dfrac{\sqrt[3]{x}}{\sqrt[4]{yz}} = \log_b \dfrac{x^{1/3}}{(yz)^{1/4}} = \log_b x^{1/3} - \log_b (yz)^{1/4} = \dfrac{1}{3}\log_b x - \dfrac{1}{4}\log_b yz$

$$= \dfrac{1}{3}\log_b x - \dfrac{1}{4}(\log_b y + \log_b z)$$

$$= \dfrac{1}{3}\log_b x - \dfrac{1}{4}\log_b y - \dfrac{1}{4}\log_b z$$

77. $3\log_b(x+2) - 4\log_b y + \dfrac{1}{2}\log_b(x+1) = \log_b(x+2)^3 + \log_b y^{-4} + \log_b(x+1)^{1/2}$

$$= \log_b(x+2)^3 y^{-4}(x+1)^{1/2} = \log_b \dfrac{(x+2)^3 \sqrt{x+1}}{y^4}$$

79. $\log_b \left(\dfrac{x}{z} + x\right) - \log_b \left(\dfrac{y}{z} + y\right) = \log_b \dfrac{\frac{x}{z} + x}{\frac{y}{z} + y} = \log_b \dfrac{x + xz}{y + yz} = \log_b \dfrac{x(1+z)}{y(1+z)} = \log_b \dfrac{x}{y}$

81. $\log 112 = \log 4^2 \cdot 7 = \log 4^2 + \log 7 = 2\log 4 + \log 7 = 2(0.6021) + 0.8451 = 2.0493$

83. $\log \dfrac{144}{49} = \log \dfrac{16 \cdot 9}{7^2} = \log \dfrac{4^2 \cdot 9}{7^2} = \log 4^2 + \log 9 - \log 7^2 = 2\log 4 + \log 9 - 2\log 7$

$$= 2(0.6021) + 0.9542 - 2(0.8451)$$

$$= 0.4682$$

85. $\boxed{\log}\ \boxed{\sqrt{}}\ \boxed{2}\ \boxed{4}\ \boxed{.}\ \boxed{3}\ \boxed{)}\ \boxed{\text{ENTER}}\ \{0.69280\}$

$\boxed{.}\ \boxed{5}\ \boxed{\log}\ \boxed{2}\ \boxed{4}\ \boxed{.}\ \boxed{3}\ \boxed{)}\ \boxed{\text{ENTER}}\ \{0.69280\}$

87. $\log_b 0 = 1 \Rightarrow b^1 = 0 \Rightarrow b = 0 \Rightarrow$ FALSE $(b^1 = b)$

89. $\log_b xy = (\log_b x)(\log_b y) \Rightarrow$ FALSE $(\log_b xy = \log_b x + \log_b y)$

91. $\log_7 7^7 = 7 \Rightarrow 7^7 = 7^7 \Rightarrow$ TRUE

93. $\dfrac{\log_b A}{\log_b B} = \log_b A - \log_b B \Rightarrow$ FALSE $\left(\log_b \dfrac{A}{B} = \log_b A - \log_b B\right)$

95. $3 \log_b \sqrt[3]{a} = 3 \log_b a^{1/3} = \dfrac{1}{3} \cdot 3 \log_b a = \log_b a \Rightarrow$ TRUE

97. $\log_b \dfrac{1}{a} = \log_b 1 - \log_b a = 0 - \log_b a = -\log_b a \Rightarrow$ TRUE

99. $\text{pH} = -\log [\text{H}^+] = -\log (1.7 \times 10^{-5}) \approx 4.77$

101. low pH: high pH:

    $\text{pH} = -\log [\text{H}^+]$ $\text{pH} = -\log [\text{H}^+]$

    $6.8 = -\log [\text{H}^+]$ $7.6 = -\log [\text{H}^+]$

    $-6.8 = \log [\text{H}^+]$ $-7.6 = \log [\text{H}^+]$

    $[\text{H}^+] = 1.5849 \times 10^{-7}$ $[\text{H}^+] = 2.5119 \times 10^{-8}$

103. $k \ln 2I = k(\ln 2 + \ln I)$

           $= k \ln 2 + k \ln I$

           $= k \ln 2 + L$

The loudness increases by $k \ln 2$.

105. $L = 3k \ln I = k \cdot 3 \ln I$

           $= k \ln I^3$

The intensity must be cubed.

107. **Answers may vary.**

109. $\ln(e^x) = \log_e(e^x) = x$

111. Let $\log_{b^2} x = y$. Then

    $(b^2)^y = x$

    $b^{2y} = x$

    $(b^{2y})^{1/2} = x^{1/2}$

    $b^y = x^{1/2}$

  $\log_b x^{1/2} = y$

  $\dfrac{1}{2} \log_b x = y$

Exercises 11.8   (page 877)

1. $\log x = 2$

      $x = 10^2$

3. $\log (x+1) = 1$

       $x+1 = 10^1$

5. $\boxed{\log}\ \boxed{4}\ \boxed{)}\ \boxed{\div}\ \boxed{\log}\ \boxed{3}\ \boxed{)}\ \boxed{\text{ENTER}}$ $\{1.2619\}$

7. $\boxed{(-)}\ \boxed{\log}\ \boxed{5}\ \boxed{\div}\ \boxed{7}\ \boxed{)}\ \boxed{\text{ENTER}}$ $\{0.1461\}$

9. $4x^2 - 64x = 0$
$4x(x - 16) = 0$
$4x = 0$ **or** $x - 16 = 0$
$x = 0$ $x = 16$

11. $3p^2 + 10p = 8$
$3p^2 + 10p - 8 = 0$
$(3p - 2)(p + 4) = 0$
$3p - 2 = 0$ **or** $p + 4 = 0$
$p = \frac{2}{3}$ $p = -4$

13. exponential

15. $A_0 e^{-kt}$

17. $3^x = 8$
$\log 3^x = \log 8$
$x \log 3 = \log 8$
$x = \dfrac{\log 8}{\log 3}$
$x \approx 1.8928$

19. $e^t = 50$
$\ln e^t = \ln 50$
$t \ln e = \ln 50$
$t = \ln 50$
$t \approx 3.9120$

21. $7 = 4.3(1.01)^t$
$\dfrac{7}{4.3} = (1.01)^t$
$\log \dfrac{7}{4.3} = \log (1.01)^t$
$\log \dfrac{7}{4.3} = t \log 1.01$
$\dfrac{\log \frac{7}{4.3}}{\log 1.01} = t$
$48.9728 \approx t$

23. $5 = 2.1(1.04)^t$
$\dfrac{5}{2.1} = (1.04)^t$
$\log \dfrac{5}{2.1} = \log (1.04)^t$
$\log \dfrac{5}{2.1} = t \log 1.04$
$\dfrac{\log \frac{5}{2.1}}{\log 1.04} = t$
$22.1184 \approx t$

25. $6^{x-2} = 4$
$\log 6^{x-2} = \log 4$
$(x - 2) \log 6 = \log 4$
$x - 2 = \dfrac{\log 4}{\log 6}$
$x = \dfrac{\log 4}{\log 6} + 2$
$x \approx 2.7737$

27. $4^{x+1} = 7^x$
$\log 4^{x+1} = \log 7^x$
$(x + 1) \log 4 = x \log 7$
$x \log 4 + \log 4 = x \log 7$
$\log 4 = x \log 7 - x \log 4$
$\log 4 = x(\log 7 - \log 4)$
$\dfrac{\log 4}{\log 7 - \log 4} = x$
$2.4772 \approx x$

29. $5^{x^2+2x} = 125$
$5^{x^2+2x} = 5^3$
$x^2 + 2x = 3$
$x^2 + 2x - 3 = 0$
$(x + 3)(x - 1) = 0$
$x + 3 = 0$ **or** $x - 1 = 0$
$x = -3$ $x = 1$

31. $3^{x^2+4x} = \frac{1}{81}$
$3^{x^2+4x} = 3^{-4}$
$x^2 + 4x = -4$
$x^2 + 4x + 4 = 0$
$(x + 2)(x + 2) = 0$
$x + 2 = 0$ **or** $x + 2 = 0$
$x = -2$ $x = -2$

33. $2^{x+1} = 7 \Rightarrow$ Graph $y = 2^{x+1}$ and $y = 7$.

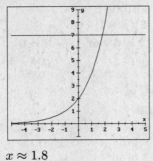

$x \approx 1.8$

35. $2^{x^2 - 2x} - 8 = 0 \Rightarrow$ Graph $y = 2^{x^2 - 2x} - 8$ and find any x-intercept(s).

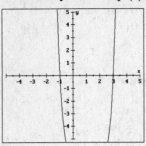

$x = 3$ or $x = -1$

37. $\log 9x = \log 27$
$$9x = 27$$
$$x = 3$$

39. $\log(4x + 3) = \log(x + 9)$
$$4x + 3 = x + 9$$
$$3x = 6$$
$$x = 2$$

41. $\log x^2 = 2$
$$10^2 = x^2$$
$$100 = x^2$$
$$\pm 10 = x$$

43. $\log x + \log(x - 3) = 1$
$$\log x(x - 3) = 1$$
$$10^1 = x(x - 3)$$
$$0 = x^2 - 3x - 10$$
$$0 = (x - 5)(x + 2)$$
$$x - 5 = 0 \quad \textbf{or} \quad x + 2 = 0$$
$$x = 5 \qquad\qquad x = -2$$
$$\text{Extraneous}$$

45. $\log x + \log(x - 15) = 2$
$$\log x(x - 15) = 2$$
$$10^2 = x(x - 15)$$
$$0 = x^2 - 15x - 100$$
$$0 = (x - 20)(x + 5)$$
$$x - 20 = 0 \quad \textbf{or} \quad x + 5 = 0$$
$$x = 20 \qquad\qquad x = -5$$
$$\text{Extraneous}$$

47. $\log(x + 90) = 3 - \log x$
$$\log x + \log(x + 90) = 3$$
$$\log x(x + 90) = 3$$
$$10^3 = x(x + 90)$$
$$0 = x^2 + 90x - 1000$$
$$0 = (x - 10)(x + 100)$$
$$x - 10 = 0 \quad \textbf{or} \quad x + 100 = 0$$
$$x = 10 \qquad\qquad x = -100$$
$$\text{Extraneous}$$

49.
$$\frac{\log(2x+1)}{\log(x-1)} = 2$$
$$\log(2x+1) = 2\log(x-1)$$
$$\log(2x+1) = \log(x-1)^2$$
$$2x+1 = (x-1)^2$$
$$2x+1 = x^2 - 2x + 1$$
$$0 = x^2 - 4x$$
$$0 = x(x-4)$$
$$x = 0 \quad \textbf{or} \quad x - 4 = 0$$
Extraneous $\qquad\qquad x = 4$

51.
$$\frac{\log(3x+4)}{\log x} = 2$$
$$\log(3x+4) = 2\log x$$
$$\log(3x+4) = \log x^2$$
$$3x+4 = x^2$$
$$0 = x^2 - 3x - 4$$
$$0 = (x-4)(x+1)$$
$$x - 4 = 0 \quad \textbf{or} \quad x + 1 = 0$$
$$x = 4 \qquad\qquad x = -1$$
$\qquad\qquad\qquad$ Extraneous

53. $\log x + \log(x-15) = 2$
Graph $y = \log x + \log(x-15)$ and $y = 2$.

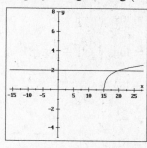

$x = 20$

55. $\ln(2x+5) - \ln 3 = \ln(x-1) \Rightarrow$ Graph
$y = \ln(2x+5) - \ln 3$ and $y = \ln(x-1)$.

$x = 8$

57.
$$5^x = 4^x$$
$$\log 5^x = \log 4^x$$
$$x \log 5 = x \log 4$$
$$0 = x \log 4 - x \log 5$$
$$0 = x(\log 4 - \log 5)$$
$$\frac{0}{\log 4 - \log 5} = x$$
$$0 = x$$

59.
$$7^{x^2} = 10$$
$$\log 7^{x^2} = \log 10$$
$$x^2 \log 7 = \log 10$$
$$x^2 = \frac{\log 10}{\log 7}$$
$$x^2 \approx 1.1833$$
$$x \approx \pm 1.0878$$

61.
$$8^{x^2} = 9^x$$
$$\log 8^{x^2} = \log 9^x$$
$$x^2 \log 8 = x \log 9$$
$$x^2 \log 8 - x \log 9 = 0$$
$$x(x \log 8 - \log 9) = 0$$
$$x = 0 \quad \textbf{or} \quad x \log 8 - \log 9 = 0$$
$$x \log 8 = \log 9$$
$$x = \frac{\log 9}{\log 8}$$
$$x \approx 1.0566$$

63.
$$\log(5-4x) - \log(x+25) = 0$$
$$\log(5-4x) = \log(x+25)$$
$$5 - 4x = x + 25$$
$$-20 = 5x$$
$$-4 = x$$

65. $\log (x - 6) - \log (x - 2) = \log \dfrac{5}{x}$

$\log \dfrac{x - 6}{x - 2} = \log \dfrac{5}{x}$

$\dfrac{x - 6}{x - 2} = \dfrac{5}{x}$

$x(x - 6) = 5(x - 2)$

$x^2 - 6x = 5x - 10$

$x^2 - 11x + 10 = 0$

$(x - 10)(x - 1) = 0$

$x - 10 = 0 \quad \textbf{or} \quad x - 1 = 0$

$x = 10 \qquad\qquad x = 1$

Extraneous

67. $4^{x+2} - 4^x = 15$

$4^x 4^2 - 4^x = 15$

$16 \cdot 4^x - 4^x = 15$

$15 \cdot 4^x = 15$

$4^x = 1$

$x = 0$

69. $\dfrac{\log (5x + 6)}{2} = \log x$

$\log (5x + 6) = 2 \log x$

$\log (5x + 6) = \log x^2$

$5x + 6 = x^2$

$0 = x^2 - 5x - 6$

$0 = (x - 6)(x + 1)$

$x - 6 = 0 \quad \textbf{or} \quad x + 1 = 0$

$x = 6 \qquad\qquad x = -1$

Extraneous

71. $\log_3 x = \log_3 \left(\dfrac{1}{x}\right) + 4$

$\log_3 x = \log_3 \left(\dfrac{1}{x}\right) + \log_3 81$

$\log_3 x = \log_3 \left(\dfrac{81}{x}\right)$

$x = \dfrac{81}{x}$

$x^2 = 81$

$x = 9 \ (-9 \text{ is extraneous.})$

73. $2(3^x) = 6^{2x}$

$\log 2(3^x) = \log 6^{2x}$

$\log 2 + \log 3^x = 2x \log 6$

$\log 2 + x \log 3 = 2x \log 6$

$\log 2 = 2x \log 6 - x \log 3$

$\log 2 = x(2 \log 6 - \log 3)$

$\dfrac{\log 2}{2 \log 6 - \log 3} = x$

$0.2789 \approx x$

75. $\log x^2 = (\log x)^2$

$2 \log x = (\log x)^2$

$0 = (\log x)^2 - 2 \log x$

$0 = \log x (\log x - 2)$

$\log x = 0 \quad \textbf{or} \quad \log x - 2 = 0$

$x = 1 \qquad\qquad \log x = 2$

$x = 100$

77. $2 \log_2 x = 3 + \log_2 (x - 2)$

$\log_2 x^2 - \log_2 (x - 2) = 3$

$\log_2 \dfrac{x^2}{x - 2} = 3$

$\dfrac{x^2}{x - 2} = 8$

$x^2 = 8(x - 2)$

$x^2 = 8(x - 2)$

$x^2 = 8x - 16$

$x^2 - 8x + 16 = 0$

$(x - 4)(x - 4) = 0$

$x - 4 = 0 \quad \textbf{or} \quad x - 4 = 0$

$x = 4 \qquad\qquad x = 4$

79.

$\log (7y + 1) = 2 \log (y + 3) - \log 2$
$\log (7y + 1) = \log (y + 3)^2 - \log 2$
$\log (7y + 1) = \log \dfrac{y^2 + 6y + 9}{2}$
$7y + 1 = \dfrac{y^2 + 6y + 9}{2}$
$2(7y + 1) = y^2 + 6y + 9$

$2(7y + 1) = y^2 + 6y + 9$
$14y + 2 = y^2 + 6y + 9$
$0 = y^2 - 8y + 7$
$0 = (y - 7)(y - 1)$
$y - 7 = 0$ **or** $y - 1 = 0$
$y = 7$ $\qquad y = 1$

81.

$\log \dfrac{4x + 1}{2x + 9} = 0$
$10^0 = \dfrac{4x + 1}{2x + 9}$
$1 = \dfrac{4x + 1}{2x + 9}$
$2x + 9 = 4x + 1$
$8 = 2x$
$4 = x$

83.

$A = A_0 e^{-0.013t}$
$0.5 A_0 = A_0 e^{-0.013t}$
$0.5 = e^{-0.013t}$
$\ln 0.5 = \ln e^{-0.013t}$
$\ln 0.5 = -0.013t$
$\dfrac{\ln 0.5}{-0.013} = t$
$53 \text{ days} \approx t$

85.

$A = A_0 2^{-t/h}$
$0.80 A_0 = A_0 2^{-2/h}$
$0.80 = 2^{-2/h}$
$\log 0.80 = \log 2^{-2/h}$
$\log 0.80 = -\dfrac{2}{h} \log 2$
$h \log 0.80 = -2 \log 2$
$h = \dfrac{-2 \log 2}{\log 0.80}$
$h \approx 6.2 \text{ years}$

87.

$A = A_0 2^{-t/h}$
$0.60 A_0 = A_0 2^{-t/5700}$
$0.60 = 2^{-t/5700}$
$\log 0.60 = \log 2^{-t/5700}$
$\log 0.60 = -\dfrac{t}{5700} \log 2$
$\dfrac{\log 0.60}{-\log 2} = \dfrac{t}{5700}$
$-5700 \dfrac{\log 0.60}{\log 2} = t$
$4200 \text{ years} \approx t$

89.

$P = P_0 e^{kt}$
$2P_0 = P_0 e^{5k}$
$2 = e^{5k}$
$\ln 2 = \ln e^{5k}$
$\ln 2 = 5k$
$\dfrac{\ln 2}{5} = k$

$P = P_0 e^{kt}$
$1{,}000{,}000 = 30{,}000 e^{\frac{\ln 2}{5} t}$
$33.333 \approx e^{\frac{\ln 2}{5} t}$
$\ln 33.333 \approx \ln e^{\frac{\ln 2}{5} t}$
$\ln 33.333 \approx \dfrac{\ln 2}{5} t$
$\dfrac{5 \ln 33.333}{\ln 2} \approx t$
$25.3 \text{ years} \approx t$

91.

$P = P_0 e^{kt}$
$2P_0 = P_0 e^{24k}$
$2 = e^{24k}$
$\ln 2 = \ln e^{24k}$
$\ln 2 = 24k$
$\dfrac{\ln 2}{24} = k$

$P = P_0 e^{kt}$
$P = P_0 e^{\frac{\ln 2}{24} (36)}$
$P = P_0 e^{\frac{3 \ln 2}{2}}$
$P = P_0 (2.828)$
It will be about 2.828 times larger.

93. $n = \dfrac{1}{\log 2} \left(\log \dfrac{B}{b} \right) = \dfrac{1}{\log 2} \log \dfrac{5 \times 10^6}{500} = \dfrac{1}{\log 2} \log 10{,}000 \approx 13.3 \text{ generations}$

95.
$$A = A_0 2^{-t/h}$$
$$0.20A_0 = A_0 2^{-t/18.4}$$
$$0.20 = 2^{-t/18.4}$$
$$\log 0.20 = \log 2^{-t/18.4}$$
$$\log 0.20 = -\frac{t}{18.4}\log 2$$
$$\frac{\log 0.20}{-\log 2} = \frac{t}{18.4}$$
$$-18.4\frac{\log 0.20}{\log 2} = t$$
$$42.7 \text{ days} \approx t$$

97.
$$A = P\left(1 + \frac{r}{k}\right)^{kt}$$
$$800 = 500\left(1 + \frac{0.085}{2}\right)^{2t}$$
$$1.6 = (1.0425)^{2t}$$
$$\log 1.6 = \log (1.0425)^{2t}$$
$$\log 1.6 = 2t \log (1.0425)$$
$$\frac{\log 1.6}{2 \log 1.0425} = t$$
$$5.6 \text{ years} \approx t$$

99.
$$A = P\left(1 + \frac{r}{k}\right)^{kt}$$
$$2100 = 1300\left(1 + \frac{0.09}{4}\right)^{4t}$$
$$\frac{2100}{1300} = (1.0225)^{4t}$$
$$\log\frac{2100}{1300} = \log (1.0225)^{4t}$$
$$\log\frac{21}{13} = 4t \log (1.0225)$$
$$\frac{\log\frac{21}{13}}{4 \log 1.0225} = t, \text{ so } t \approx 5.4 \text{ years}$$

101. doubling time $= t = \dfrac{\ln 2}{r}$
$$= \frac{100 \ln 2}{100r}$$
$$\approx \frac{70}{100r}$$
$$= \frac{70}{r, \text{ written as a \%}}$$

103. Answers may vary.

105. Since the logarithm of a negative number is not defined (as a real number), the values $x - 3$ and $x^2 + 2$ must be nonnegative. Since $x^2 + 2$ is always greater than 0, the only restriction is that $x - 3 > 0$, or $x > 3$. Thus, x cannot be a solution if $x \leq 3$.

Chapter 11 Review (page 883)

1. $(f + g)(x) = f(x) + g(x) = 2x + x + 1 = 3x + 1$; domain $= (-\infty, \infty)$

2. $(f - g)(x) = f(x) - g(x) = 2x - (x + 1) = x - 1$; domain $= (-\infty, \infty)$

3. $(f \cdot g)(x) = f(x)g(x) = 2x(x + 1) = 2x^2 + 2x$; domain $= (-\infty, \infty)$

4. $(f/g)(x) = \frac{f(x)}{g(x)} = \frac{2x}{x+1} \ (x \neq -1)$; domain $= (-\infty, -1) \cup (-1, \infty)$

5. $(f \circ g)(3) = f(g(3)) = f(3 + 1)$
$= f(4) = 2(4) = 8$

6. $(g \circ f)(-2) = g(f(-2)) = g(2(-2))$
$= g(-4)$
$= -4 + 1 = -3$

7. $(f \circ g)(x) = f(g(x)) = f(x + 1) = 2(x + 1)$

8. $(g \circ f)(x) = g(f(x)) = g(2x) = 2x + 1$

9. $f(x) = 2(x - 3)$

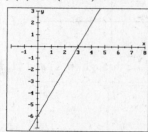

one-to-one

10. $f(x) = x(2x - 3)$

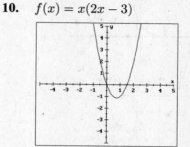

not one-to-one

11. $f(x) = -3(x - 2)^2 + 5$

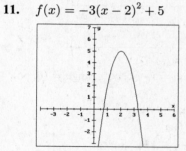

not one-to-one

12. $f(x) = |x|$

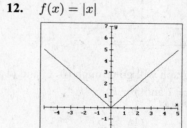

not one-to-one

13.
$$y = 7x - 2$$
$$x = 7y - 2$$
$$x + 2 = 7y$$
$$\frac{x+2}{7} = y, \text{ or } y = f^{-1}(x) = \frac{x+2}{7}$$

14.
$$y = 4x + 5$$
$$x = 4y + 5$$
$$x - 5 = 4y$$
$$\frac{x-5}{4} = y, \text{ or } y = f^{-1}(x) = \frac{x-5}{4}$$

15.
$$y = 2x^2 - 1$$
$$x = 2y^2 - 1$$
$$x + 1 = 2y^2$$
$$\frac{x+1}{2} = y^2$$
$$\sqrt{\frac{x+1}{2}} = y, \text{ or } y = f^{-1}(x) = \sqrt{\frac{x+1}{2}}$$

16.

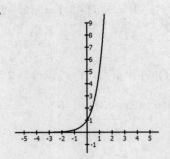

17.

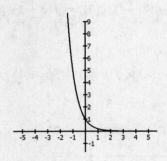

18. $y = 3^x$; through $(0, 1)$ and $(1, 3)$

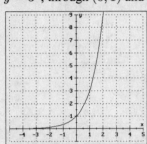

19. $y = \left(\frac{1}{3}\right)^x$; through $(0, 1)$ and $\left(1, \frac{1}{3}\right)$

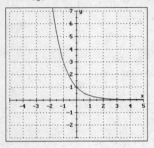

20. The graph will go through $(0, 1)$ and $(1, 6)$, so $x = 1$ and $y = 6$.

21. domain $= (-\infty, \infty)$; range $= (0, \infty)$

22. $y = f(x) = \left(\frac{1}{2}\right)^x - 2$
Shift $y = \left(\frac{1}{2}\right)^x$ down 2.

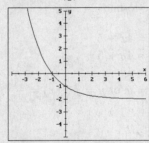

23. $y = f(x) = \left(\frac{1}{2}\right)^{x+2}$
Shift $y = \left(\frac{1}{2}\right)^x$ left 2.

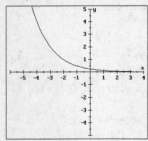

24. $A = P\left(1 + \dfrac{r}{k}\right)^{kt} = 25000\left(1 + \dfrac{0.05}{4}\right)^{4 \cdot 35} = 25000(1.0125)^{140} \approx \$142{,}312.97$

25. $A = Pe^{rt} = 25000e^{0.05(35)} = 25000e^{1.75} \approx \$143{,}865.07$

26. $y = f(x) = e^x + 1$; Shift $y = e^x$ up 1.

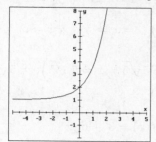

27. $y = f(x) = e^{x-3}$; Shift $y = e^x$ right 3.

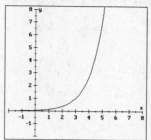

28. $P = P_0 e^{kt} = 275{,}000{,}000 e^{0.015(50)}$
$\approx 582{,}000{,}000$

29. $A = A_0 e^{-0.0244t} = 50 e^{-0.0244(20)} = 50 e^{-0.488}$
≈ 30.69 g

30. domain $= (0, \infty)$; range $= (-\infty, \infty)$

31. **Answers will vary.**

32. $\log_4 16 = x \Rightarrow 4^x = 16 \Rightarrow \log_4 16 = 2$

33. $\log_{16} \dfrac{1}{4} = x \Rightarrow 16^x = \dfrac{1}{4} \Rightarrow \log_{16} \dfrac{1}{4} = -\dfrac{1}{2}$

34. $\log_\pi 1 = x \Rightarrow \pi^x = 1 \Rightarrow \log_\pi 1 = 0$

35. $\log_5 0.04 = x \Rightarrow \log_5 \dfrac{1}{25} = x \Rightarrow 5^x = \dfrac{1}{25}$
$\log_5 0.04 = -2$

36. $\log_b \sqrt{b} = \log_b b^{1/2} = x \Rightarrow b^x = b^{1/2}$
$\log_b \sqrt{b} = \dfrac{1}{2}$

37. $\log_a \sqrt[3]{x} = \log_a a^{1/3} = x \Rightarrow a^x = a^{1/3}$
$\log_a \sqrt[3]{a} = \dfrac{1}{3}$

38. $\log_3 x = 4 \Rightarrow 3^4 = x \Rightarrow x = 81$

39. $\log_{\sqrt{3}} x = 4 \Rightarrow \left(\sqrt{3}\right)^4 = x \Rightarrow x = 9$

40. $\log_{\sqrt{3}} x = 6 \Rightarrow \left(\sqrt{3}\right)^6 = x \Rightarrow x = 27$

41. $\log_{0.1} 10 = x \Rightarrow (0.1)^x = 10$
$\left(\dfrac{1}{10}\right)^x = 10 \Rightarrow x = -1$

42. $\log_x 2 = -\dfrac{1}{3} \Rightarrow x^{-1/3} = 2$
$\left(x^{-1/3}\right)^{-3} = 2^{-3} \Rightarrow x = \dfrac{1}{8}$

43. $\log_x 16 = 4 \Rightarrow x^4 = 16 \Rightarrow x = 2$

44. $\log_{0.25} x = -1 \Rightarrow (0.25)^{-1} = x$
$\left(\dfrac{1}{4}\right)^{-1} = x \Rightarrow x = 4$

45. $\log_{0.125} x = -\dfrac{1}{3} \Rightarrow (0.125)^{-1/3} = x$
$\left(\dfrac{1}{8}\right)^{-1/3} = x \Rightarrow x = 2$

46. $\log_{\sqrt{2}} 32 = x \Rightarrow \left(\sqrt{2}\right)^x = 32$

$\left(2^{1/2}\right)^x = 2^5 \Rightarrow \frac{1}{2}x = 5 \Rightarrow x = 10$

47. $\log_{\sqrt{5}} x = -4 \Rightarrow \left(\sqrt{5}\right)^{-4} = x$

$\left(5^{1/2}\right)^{-4} = x \Rightarrow 5^{-2} = x \Rightarrow x = \frac{1}{25}$

48. $\log_{\sqrt{3}} 9\sqrt{3} = x \Rightarrow \left(\sqrt{3}\right)^x = 9\sqrt{3}$

$\left(3^{1/2}\right)^x = 3^{5/2} \Rightarrow x = 5$

49. $\log_{\sqrt{5}} 5\sqrt{5} = x \Rightarrow \left(\sqrt{5}\right)^x = 5\sqrt{5}$

$\left(5^{1/2}\right)^x = 5^{3/2} \Rightarrow x = 3$

50. $y = f(x) = \log(x - 2)$
Shift $y = \log x$ right 2.

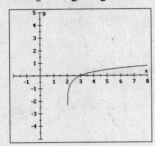

51. $y = f(x) = 3 + \log x$
Shift $y = \log x$ up 3.

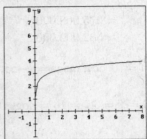

52. $y = 4^x$
$y = \log_4 x$

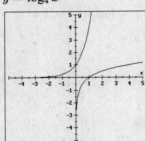

53. $y = \left(\frac{1}{3}\right)^x$
$y = \log_{1/3} x$

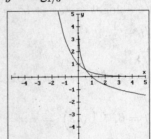

54. dB gain $= 20 \log \frac{E_O}{E_I} = 20 \log \frac{18}{0.04} = 20 \log 450 \approx 53$ dB

55. $R = \log \frac{A}{P} = \log \frac{7500}{0.3} = \log 25{,}000 \approx 4.4$

56. $\ln 362 \approx 5.8916$

57. $\ln(\log 7.85) \approx \ln 0.8949 \approx -0.1111$

58. $\ln x = 2.336 \Rightarrow x = 10.3398$

59. $\ln x = \log 8.8 \Rightarrow x = 2.5715$

60. $y = f(x) = 1 + \ln x$
Shift $y = \ln x$ up 1.

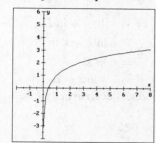

61. $y = f(x) = \ln(x + 1)$
Shift $y = \ln x$ left 1.

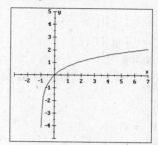

62. $t = \dfrac{\ln 2}{r} = \dfrac{\ln 2}{0.03} \approx 23$ years

63. $\log_5 1 = 0$ **64.** $\log_9 9 = 1$ **65.** $\log_7 7^3 = 3$ **66.** $8^{\log_8 5} = 5$

67. $\log 10^5 = 5$ **68.** $\ln 1 = 0$ **69.** $10^{\log_{10} 7} = 7$ **70.** $e^{\ln 2} = 2$

71. $\log_a a^6 = 6$ **72.** $\ln e^3 = 3$

73. $\log_b \dfrac{x^5 y^4}{z^2} = \log_b x^5 + \log_b y^4 - \log_b z^2 = 5 \log_b x + 4 \log_b y - 2 \log_b z$

74. $\log_b \sqrt{\dfrac{x}{yz^2}} = \log_b \left(\dfrac{x}{yz^2}\right)^{1/2} = \dfrac{1}{2} \log_b \dfrac{x}{yz^2} = \dfrac{1}{2}\left(\log_b x - \log_b y - \log_b z^2\right)$

$$= \dfrac{1}{2}\left(\log_b x - \log_b y - 2 \log_b z\right)$$

75. $3 \log_b x - 5 \log_b y + 7 \log_b z = \log_b x^3 - \log_b y^5 + \log_b z^7 = \log_b \dfrac{x^3 z^7}{y^5}$

76. $\dfrac{1}{2} \log_b x + 3 \log_b y - 7 \log_b z = \log_b x^{1/2} + \log_b y^3 - \log_b z^7 = \log_b \dfrac{y^3 \sqrt{x}}{z^7}$

77. $\log abc = \log a + \log b + \log c = 0.6 + 0.36 + 2.4 = 3.36$

78. $\log a^2 b = \log a^2 + \log b = 2 \log a + \log b = 2(0.6) + 0.36 = 1.56$

79. $\log \dfrac{ac}{b} = \log a + \log c - \log b = 0.6 + 2.4 - 0.36 = 2.64$

80. $\log \dfrac{a^2}{c^3 b^2} = \log a^2 - \log c^3 - \log b^2 = 2 \log a - 3 \log c - 2 \log b$

$$= 2(0.6) - 3(2.4) - 2(0.36) = -6.72$$

81. $\log_5 17 = \dfrac{\log 17}{\log 5} \approx 1.7604$

82.
$$\text{pH} = -\log[\text{H}^+]$$
$$3.1 = -\log[\text{H}^+]$$
$$-3.1 = \log[\text{H}^+]$$
$$[\text{H}^+] = 7.94 \times 10^{-4} \text{ gram-ions per liter}$$

83.
$$k\ln\left(\tfrac{1}{2}I\right) = k\left(\ln\tfrac{1}{2} + \ln I\right)$$
$$= k\ln\tfrac{1}{2} + k\ln I$$
$$= k\ln 2^{-1} + L$$
$$= -k\ln 2 + L$$
The loudness decreases by $k\ln 2$.

84.
$$3^x = 7$$
$$\log 3^x = \log 7$$
$$x\log 3 = \log 7$$
$$x = \frac{\log 7}{\log 3} \approx 1.7712$$

85.
$$5^{x+2} = 625$$
$$5^{x+2} = 5^4$$
$$x + 2 = 4$$
$$x = 2$$

86.
$$25 = 5.5(1.05)^t$$
$$\frac{25}{5.5} = (1.05)^t$$
$$\log\frac{25}{5.5} = \log(1.05)^t$$
$$\log\frac{25}{5.5} = t\log 1.05$$
$$\frac{\log\frac{25}{5.5}}{\log 1.05} = t$$
$$31.0335 \approx t$$

87.
$$4^{2t-1} = 64$$
$$4^{2t-1} = 4^3$$
$$2t - 1 = 3$$
$$2t = 4$$
$$t = 2$$

88.
$$2^x = 3^{x-1}$$
$$\log 2^x = \log 3^{x-1}$$
$$x\log 2 = (x-1)\log 3$$
$$x\log 2 = x\log 3 - \log 3$$
$$\log 3 = x\log 3 - x\log 2$$
$$\log 3 = x(\log 3 - \log 2)$$
$$\frac{\log 3}{\log 3 - \log 2} = x$$
$$2.7095 \approx x$$

89.
$$3^{x^2+4x} = \frac{1}{27}$$
$$3^{x^2+4x} = 3^{-3}$$
$$x^2 + 4x = -3$$
$$x^2 + 4x + 3 = 0$$
$$(x+3)(x+1) = 0$$
$$x + 3 = 0 \quad \textbf{or} \quad x + 1 = 0$$
$$x = -3 \qquad\qquad x = -1$$

90. $\log x + \log (29 - x) = 2$
$$\log x(29 - x) = 2$$
$$10^2 = x(29 - x)$$
$$100 = 29x - x^2$$
$$x^2 - 29x + 100 = 0$$
$$(x - 25)(x - 4) = 0$$
$$x - 25 = 0 \quad \textbf{or} \quad x - 4 = 0$$
$$x = 25 \qquad\qquad x = 4$$

91. $\log_2 x + \log_2 (x - 2) = 3$
$$\log_2 x(x - 2) = 3$$
$$x(x - 2) = 2^3$$
$$x^2 - 2x = 8$$
$$x^2 - 2x - 8 = 0$$
$$(x - 4)(x + 2) = 0$$
$$x - 4 = 0 \quad \textbf{or} \quad x + 2 = 0$$
$$x = 4 \qquad\qquad x = -2$$
$$\text{Extraneous}$$

92. $\log_2 (x + 2) + \log_2 (x - 1) = 2$
$$\log_2 (x + 2)(x - 1) = 2$$
$$(x + 2)(x - 1) = 2^2$$
$$x^2 + x - 2 = 4$$
$$x^2 + x - 6 = 0$$
$$(x - 2)(x + 3) = 0$$
$$x - 2 = 0 \quad \textbf{or} \quad x + 3 = 0$$
$$x = 2 \qquad\qquad x = -3$$
$$\text{Extraneous}$$

93. $\dfrac{\log (7x - 12)}{\log x} = 2$
$$\log (7x - 12) = 2 \log x$$
$$\log (7x - 12) = \log x^2$$
$$7x - 12 = x^2$$
$$0 = x^2 - 7x + 12$$
$$0 = (x - 4)(x - 3)$$
$$x - 4 = 0 \quad \textbf{or} \quad x - 3 = 0$$
$$x = 4 \qquad\qquad x = 3$$

94. $\log x + \log (x - 5) = \log 6$
$$\log x(x - 5) = \log 6$$
$$x(x - 5) = 6$$
$$x^2 - 5x - 6 = 0$$
$$(x - 6)(x + 1) = 0$$
$$x - 6 = 0 \quad \textbf{or} \quad x + 1 = 0$$
$$x = 6 \qquad\qquad x = -1$$
$$\text{Extraneous}$$

95. $\log 3 - \log (x - 1) = -1$
$$\log \frac{3}{x - 1} = -1$$
$$\frac{3}{x - 1} = 10^{-1}$$
$$\frac{3}{x - 1} = \frac{1}{10}$$
$$30 = x - 1$$
$$31 = x$$

96. $e^{x \ln 2} = 9$
$$e^{\ln 2^x} = 9$$
$$2^x = 9$$
$$\ln 2^x = \ln 9$$
$$x \ln 2 = \ln 9$$
$$x = \frac{\ln 9}{\ln 2} \approx 3.1699$$

97. $\ln x = \ln (x - 1)$
$$x = x - 1$$
$$0 = -1$$
There is no solution. $\emptyset$

98.
$$\ln x = \ln (x - 1) + 1$$
$$\ln x - \ln (x - 1) = 1$$
$$\ln \frac{x}{x - 1} = 1$$
$$\frac{x}{x - 1} = e^1$$
$$x = e(x - 1)$$
$$x = ex - e$$
$$e = ex - x$$
$$e = x(e - 1)$$
$$\frac{e}{e - 1} = x$$
$$1.5820 \approx x$$

99.
$$\ln x = \log_{10} x$$
$$\ln x = \frac{\ln x}{\ln 10}$$
$$\ln x \ln 10 = \ln x$$
$$\ln x \ln 10 - \ln x = 0$$
$$\ln x (\ln 10 - 1) = 0$$
$$\ln x = 0$$
$$x = 1$$

100.
$$A = A_0 2^{-t/h}$$
$$\frac{2}{3} A_0 = A_0 2^{-t/5730}$$
$$\frac{2}{3} = 2^{-t/5730}$$
$$\log \frac{2}{3} = \log 2^{-t/5730}$$
$$\log \frac{2}{3} = -\frac{t}{5730} \log 2$$
$$\frac{\log \frac{2}{3}}{-\log 2} = \frac{t}{5730}$$
$$-5730 \frac{\log \frac{2}{3}}{\log 2} = t$$
$$3352 \text{ years} \approx t \text{ [or about 3400 years]}$$

Chapter 11 Test   (page 890)

1. $(g + f)(x) = g(x) + f(x) = x - 1 + 4x = 5x - 1; \text{domain} = (-\infty, \infty)$

2. $(f - g)(x) = f(x) - g(x) = 4x - (x - 1) = 3x + 1; \text{domain} = (-\infty, \infty)$

3. $(g \cdot f)(x) = g(x)f(x) = (x - 1)4x = 4x^2 - 4x; \text{domain} = (-\infty, \infty)$

4. $(g/f)(x) = \dfrac{g(x)}{f(x)} = \dfrac{x - 1}{4x}; \text{domain} = (-\infty, 0) \cup (0, \infty)$

5. $(g \circ f)(1) = g(f(1)) = g(4(1))$
$= g(4) = 4 - 1 = 3$

6. $(f \circ g)(0) = f(g(0)) = f(0 - 1)$
$= f(-1) = 4(-1) = -4$

7. $(f \circ g)(-1) = f(g(-1)) = f(-1 - 1) = f(-2) = 4(-2) = -8$

8. $(g \circ f)(-2) = g(f(-2)) = g(4(-2)) = g(-8) = -8 - 1 = -9$

9. $(f \circ g)(x) = f(g(x)) = f(x - 1) = 4(x - 1)$

10. $(g \circ f)(x) = g(f(x)) = g(4x) = 4x - 1$

11. $3x + 2y = 12$
$3y + 2x = 12$
$\quad\quad 3y = -2x + 12$
$\quad\quad\quad y = \dfrac{-2x + 12}{3}; f^{-1}(x) = \dfrac{-2x + 12}{3}$

12. $f(x) = 2^x + 1$; Shift $y = 2^x$ up 1.

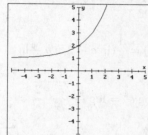

13. $f(x) = 2^{-x}$; Reflect $y = 2^x$ about y-axis.

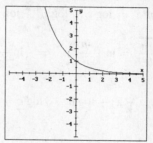

14. $A = A_0(2)^{-t} = 3(2)^{-6} = \dfrac{3}{2^6} = \dfrac{3}{64}$ gram

15. $A = A_0\left(1 + \dfrac{r}{k}\right)^{kt} = 1000\left(1 + \dfrac{0.06}{2}\right)^{2(1)} = 1000(1.03)^2 \approx \$1060.90$

16. $f(x) = e^x$

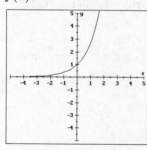

17. $A = A_0 e^{rt} = 2000e^{(0.08)10}$
$\quad\quad\quad = 2000e^{0.8}$
$\quad\quad\quad \approx \$4451.08$

18. $\log_9 81 = x \Rightarrow 9^x = 81 \Rightarrow x = 2$

19. $\log_x 81 = 4 \Rightarrow x^4 = 81 \Rightarrow x = 3$

20. $\log_3 x = -3 \Rightarrow 3^{-3} = x \Rightarrow x = \dfrac{1}{27}$

21. $\log_x 49 = 2 \Rightarrow x^2 = 49 \Rightarrow x = 7$

22. $\log_{3/2} \dfrac{9}{4} = x \Rightarrow \left(\dfrac{3}{2}\right)^x = \dfrac{9}{4} \Rightarrow x = 2$

23. $\log_{2/3} x = -3 \Rightarrow \left(\dfrac{2}{3}\right)^{-3} = x \Rightarrow x = \dfrac{27}{8}$

24. $f(x) = -\log_3 x$

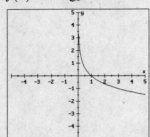

25. $f(x) = \ln x$

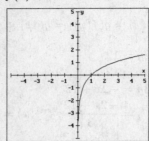

26. $\log xy^3 z^4 = \log x + \log y^3 + \log z^4 = \log x + 3\log y + 4\log z$

27. $\ln \sqrt{\dfrac{a}{b^2 c}} = \ln \left(\dfrac{a}{b^2 c}\right)^{1/2} = \dfrac{1}{2}\ln \dfrac{a}{b^2 c} = \dfrac{1}{2}(\ln a - \ln b^2 - \ln c) = \dfrac{1}{2}(\ln a - 2\ln b - \ln c)$

28. $\dfrac{1}{2}\log(a+2) + \log b - 3\log c = \log(a+2)^{1/2} + \log b - \log c^3 = \log \dfrac{b\sqrt{a+2}}{c^3}$

29. $\dfrac{1}{3}(\log a - 2\log b) - \log c = \dfrac{1}{3}(\log a - \log b^2) - \log c = \dfrac{1}{3}\log \dfrac{a}{b^2} - \log c = \log \sqrt[3]{\dfrac{a}{b^2}} - \log c$

$$= \log \dfrac{\sqrt[3]{a}}{c\sqrt[3]{b^2}}$$

30. $\log 18 = \log 2 \cdot 3^2 = \log 2 + \log 3^2 = \log 2 + 2\log 3 = 0.3010 + 2(0.4771) = 1.2552$

31. $\log \dfrac{8}{3} = \log \dfrac{2^3}{3} = \log 2^3 - \log 3 = 3\log 2 - \log 3 = 3(0.3010) - 0.4771 = 0.4259$

32. $\log_3 2 = \dfrac{\log 2}{\log 3}$ or $\dfrac{\ln 2}{\ln 3}$

33. $\log_\pi e = \dfrac{\log e}{\log \pi}$ or $\dfrac{\ln e}{\ln \pi} = \dfrac{1}{\ln \pi}$

34. $\log_a ab = \log_a a + \log_a b = 1 + \log_a b \Rightarrow$ TRUE

35. $\dfrac{\log a}{\log b} = \log a - \log b \Rightarrow$ FALSE $\left(\log \dfrac{a}{b} = \log a - \log b\right)$

36. $\log a^{-3} = -3\log a \neq \dfrac{1}{3\log a} \Rightarrow$ FALSE

37. $\ln(-x) = -\ln x \Rightarrow$ FALSE (You cannot take the logarithm of a negative number.)

38. pH $= -\log [\text{H}^+] = -\log(3.7 \times 10^{-7}) \approx 6.4$

39. dB gain $= 20\log \dfrac{E_O}{E_I} = 20\log \dfrac{60}{0.3} = 20\log 200 \approx 46$

40.
$$7^x = 4$$
$$\log 7^x = \log 4$$
$$x \log 7 = \log 4$$
$$x = \frac{\log 4}{\log 7}$$
$$x \approx 0.7124$$

41.
$$3^{x-1} = 100^x$$
$$\log 3^{x-1} = \log 100^x$$
$$(x-1)\log 3 = x \log 100$$
$$x \log 3 - \log 3 = 2x$$
$$x \log 3 - 2x = \log 3$$
$$x(\log 3 - 2) = \log 3$$
$$x = \frac{\log 3}{\log 3 - 2}$$
$$x \approx -0.3133$$

42.
$$\log(3x+1) = \log(5x-7)$$
$$3x + 1 = 5x - 7$$
$$-2x = -8$$
$$x = 4$$

43.
$$\log x + \log(x-9) = 1$$
$$\log x(x-9) = 1$$
$$x(x-9) = 10$$
$$x^2 - 9x - 10 = 0$$
$$(x-10)(x+1) = 0$$
$$x - 10 = 0 \quad \textbf{or} \quad x + 1 = 0$$
$$x = 10 \qquad\qquad x = -1$$
$$\text{Extraneous}$$

Exercises 12.1 (page 904)

1. $a = -5$: downward

3. $a = 1$: upward

5. $x^2 + 14x + 49 = (x+7)^2$

7. $x^2 - 20x + 100 = (x-10)^2$

9.
$$|2x - 5| = 7$$
$$2x - 5 = 7 \quad \textbf{or} \quad 2x - 5 = -7$$
$$2x = 12 \qquad\qquad 2x = -2$$
$$x = 6 \qquad\qquad x = -1$$

11.
$$|3x + 4| = |5x - 2|$$
$$3x + 4 = 5x - 2 \ \textbf{or} \ 3x + 4 = -(5x - 2)$$
$$-2x = -6 \qquad\qquad 3x + 4 = -5x + 2$$
$$x = 3 \qquad\qquad 8x = -2$$
$$x = -\frac{1}{4}$$

13. conic

15. standard; $(0, 3)$; 4

17. circle; general

19. parabola; $(3, 2)$; right

433

21. $x^2 + y^2 = 9$
C $(0,0)$; $r = \sqrt{9} = 3$

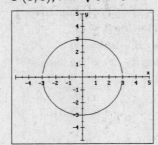

23. $(x-2)^2 + y^2 = 9$
C $(2,0)$; $r = \sqrt{9} = 3$

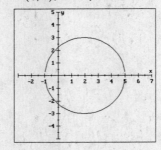

25. $(x-2)^2 + (y-4)^2 = 4$
C $(2,4)$; $r = \sqrt{4} = 2$

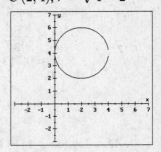

27. $(x+3)^2 + (y-1)^2 = 16$
C $(-3,1)$; $r = \sqrt{16} = 4$

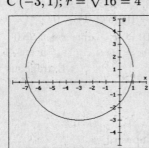

29. $x^2 + y^2 + 2x - 8 = 0$
$x^2 + 2x + y^2 = 8$
$x^2 + 2x + 1 + y^2 = 8 + 1$
$(x+1)^2 + y^2 = 9$
C $(-1,0)$; $r = \sqrt{9} = 3$

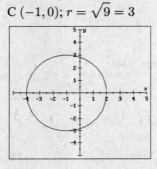

31. $9x^2 + 9y^2 - 12y = 5$
$x^2 + y^2 - \dfrac{4}{3}y = \dfrac{5}{9}$
$x^2 + y^2 - \dfrac{4}{3}y + \dfrac{4}{9} = \dfrac{5}{9} + \dfrac{4}{9}$
$x^2 + \left(y - \dfrac{2}{3}\right)^2 = 1$
C $\left(0, \dfrac{2}{3}\right)$; $r = \sqrt{1} = 1$

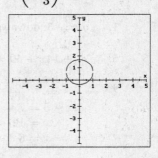

33.
$$x^2 + y^2 - 2x + 4y = -1$$
$$x^2 - 2x + y^2 + 4y = -1$$
$$x^2 - 2x + 1 + y^2 + 4y + 4 = -1 + 1 + 4$$
$$(x - 1)^2 + (y + 2)^2 = 4$$
C $(1, -2)$; $r = \sqrt{4} = 2$

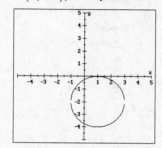

35.
$$x^2 + y^2 + 6x - 4y = -12$$
$$x^2 + 6x + y^2 - 4y = -12$$
$$x^2 + 6x + 9 + y^2 - 4y + 4 = -12 + 9 + 4$$
$$(x + 3)^2 + (y - 2)^2 = 1$$
C $(-3, 2)$; $r = \sqrt{1} = 1$

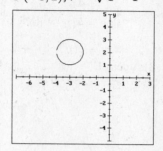

37.
$$(x - h)^2 + (y - k)^2 = r^2$$
$$(x - 0)^2 + (y - 0)^2 = 3^2$$
$$\boxed{x^2 + y^2 = 9}$$
$$\boxed{x^2 + y^2 - 9 = 0}$$

39.
$$(x - h)^2 + (y - k)^2 = r^2$$
$$(x - 2)^2 + (y - 4)^2 = 7^2$$
$$\boxed{(x - 2)^2 + (y - 4)^2 = 49}$$
$$x^2 - 4x + 4 + y^2 - 8y + 16 = 49$$
$$x^2 + y^2 - 4x - 8y + 20 = 49$$
$$\boxed{x^2 + y^2 - 4x - 8y - 29 = 0}$$

41.
$$(x - h)^2 + (y - k)^2 = r^2$$
$$(x - (-4))^2 + (y - 7)^2 = 10^2$$
$$\boxed{(x + 4)^2 + (y - 7)^2 = 100}$$
$$x^2 + 8x + 16 + y^2 - 14y + 49 = 100$$
$$x^2 + y^2 + 8x - 14y + 65 = 100$$
$$\boxed{x^2 + y^2 + 8x - 14y - 35 = 0}$$

43.
$$(x - h)^2 + (y - k)^2 = r^2$$
$$(x - 0)^2 + (y - 0)^2 = \left(4\sqrt{5}\right)^2$$
$$\boxed{x^2 + y^2 = 80}$$
$$\boxed{x^2 + y^2 - 80 = 0}$$

45. $x = y^2$
$x = (y - 0)^2 + 0$; V $(0, 0)$; opens R

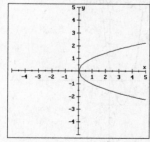

47. $x = -\frac{1}{4}y^2$
$x = -\frac{1}{4}(y - 0)^2 + 0$; V $(0, 0)$; opens L

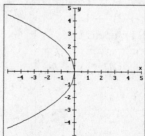

49. $y^2 + 4x - 6y = -1$

$$4x = -y^2 + 6y - 1$$

$$x = -\frac{1}{4}y^2 + \frac{3}{2}y - \frac{1}{4}$$

$$x = -\frac{1}{4}(y^2 - 6y) - \frac{1}{4}$$

$$x = -\frac{1}{4}(y^2 - 6y + 9) - \frac{1}{4} + \frac{9}{4}$$

$$x = -\frac{1}{4}(y - 3)^2 + 2$$

V $(2, 3)$; opens L

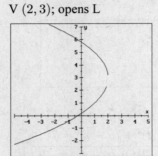

51. $y = -x^2 - x + 1$

$$y = -(x^2 + x) + 1$$

$$y = -\left(x^2 + x + \frac{1}{4}\right) + 1 + \frac{1}{4}$$

$$y = -\left(x + \frac{1}{2}\right)^2 + \frac{5}{4}$$

V $\left(-\frac{1}{2}, \frac{5}{4}\right)$; opens D

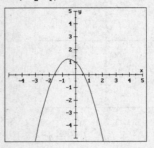

53. $x^2 + (y + 3)^2 = 1$

C $(0, -3)$; $r = \sqrt{1} = 1$

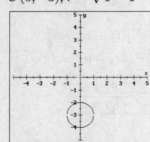

55. $y = 2(x - 1)^2 + 3$

V $(1, 3)$; opens U

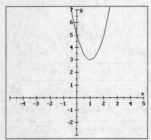

57. $y = x^2 + 4x + 5$

$$y = x^2 + 4x + 4 + 5 - 4$$

$$y = (x + 2)^2 + 1$$

V $(-2, 1)$; opens U

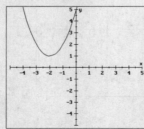

59. $y = 3(x + 1)^2 - 2$

V $(-1, -2)$; opens U

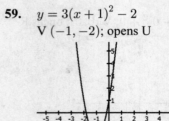

61. $3x^2 + 3y^2 = 16$

$\qquad 3y^2 = 16 - 3x^2$

$\qquad y^2 = \dfrac{16 - 3x^2}{3}$

$\qquad y = \pm\sqrt{\dfrac{16 - 3x^2}{3}}$

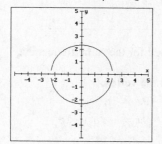

63. $(x+1)^2 + y^2 = 16$

$\qquad y^2 = 16 - (x+1)^2$

$\qquad y = \pm\sqrt{16 - (x+1)^2}$

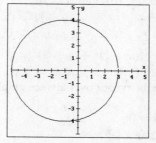

65. $x = 2y^2$

$\qquad y^2 = \dfrac{x}{2}$

$\qquad y = \pm\sqrt{\dfrac{x}{2}}$

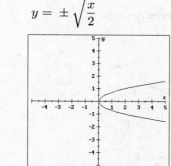

67. $x^2 - 2x + y = 6$

$\qquad y = 6 - x^2 + 2x$

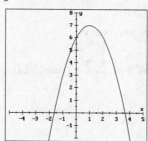

69. The radius of the larger gear is $\sqrt{16} = 4$. Centers: 7 units apart $\Rightarrow$ smaller gear $r = 3$.

$\qquad (x - h)^2 + (y - k)^2 = r^2$

$\qquad (x - 7)^2 + (y - 0)^2 = 3^2 \Rightarrow (x - 7)^2 + y^2 = 9$

71.
$$x^2 + y^2 - 8x - 20y + 16 = 0$$
$$x^2 - 8x + y^2 - 20y = -16$$
$$x^2 - 8x + 16 + y^2 - 20y + 100 = -16 + 16 + 100$$
$$(x - 4)^2 + (y - 10)^2 = 100$$
center: $(4, 10)$; radius $= 10$
$$x^2 + y^2 + 2x + 4y - 11 = 0$$
$$x^2 + 2x + 1 + y^2 + 4y + 4 = 11 + 1 + 4$$
$$(x + 1)^2 + (y + 2)^2 = 16$$
center: $(-1, -2)$; radius $= 4$
Since the ranges overlap (see graph), they can not be licensed for the same frequency.

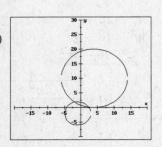

73. Set $y = 0$:
$$y = 30x - x^2$$
$$0 = 30x - x^2$$
$$0 = x(30 - x)$$
$$x = 0 \text{ or } x = 30$$
It lands 30 feet away.

75. Find the vertex:
$$2y^2 - 9x = 18$$
$$-9x = -2y^2 + 18$$
$$x = \frac{2}{9}y^2 - 2$$
$$x = \frac{2}{9}(y - 0)^2 - 2$$
vertex: $(-2, 0) \Rightarrow$ distance $= 2$ AU

77. Answers may vary.

79. Answers may vary.

Exercises 12.2 (page 915)

1.
$$\frac{x^2}{4} + \frac{y^2}{9} = 1$$
$$\frac{0^2}{4} + \frac{y^2}{9} = 1$$
$$\frac{y^2}{9} = 1$$
$$y^2 = 9$$
$$y = \pm 3$$
$$(0, 3), (0, -3)$$

3.
$$\frac{x^2}{4} + \frac{y^2}{9} = 1$$
$$\frac{x^2}{4} + \frac{0^2}{9} = 1$$
$$\frac{x^2}{4} = 1$$
$$x^2 = 4$$
$$x = \pm 2$$
$$(2, 0), (-2, 0)$$

5. $5x^{-3}y^4(7x^3 + 2y^{-4}) = 35x^0y^4 + 10x^{-3}y^0 = 35y^4 + \dfrac{10}{x^3}$

7. $\dfrac{x^{-2} + y^{-2}}{x^{-2} - y^{-2}} = \dfrac{x^{-2} + y^{-2}}{x^{-2} - y^{-2}} \cdot \dfrac{x^2 y^2}{x^2 y^2} = \dfrac{y^2 + x^2}{y^2 - x^2}$

9. ellipse; sum

11. center

13. $(0, 0)$; major axis; $2b$

438

15. $\dfrac{x^2}{4} + \dfrac{y^2}{9} = 1$

C $(0,0)$; move 2 horiz. and 3 vert.

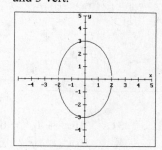

17. $\dfrac{(x-2)^2}{9} + \dfrac{(y-1)^2}{4} = 1$

C $(2,1)$; move 3 horiz. and 2 vert.

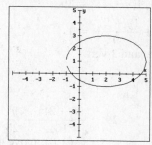

19. $x^2 + 9y^2 = 9$

$\dfrac{x^2}{9} + \dfrac{9y^2}{9} = \dfrac{9}{9}$

$\dfrac{x^2}{9} + \dfrac{y^2}{1} = 1$

C $(0,0)$; move 3 horiz. and 1 vert.

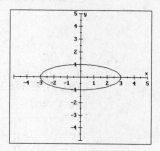

21. $(x+1)^2 + 4(y+2)^2 = 4$

$\dfrac{(x+1)^2}{4} + \dfrac{4(y+2)^2}{4} = \dfrac{4}{4}$

$\dfrac{(x+1)^2}{4} + \dfrac{(y+2)^2}{1} = 1$

C $(-1,-2)$; move 2 horiz. and 1 vert.

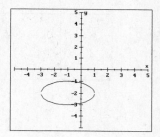

23. $\qquad x^2 + 4y^2 - 4x + 8y + 4 = 0$

$x^2 - 4x + 4\left(y^2 + 2y\right) = -4$

$x^2 - 4x + 4 + 4\left(y^2 + 2y + 1\right) = -4 + 4 + 4$

$(x-2)^2 + 4(y+1)^2 = 4$

$\dfrac{(x-2)^2}{4} + \dfrac{(y+1)^2}{1} = 1$

C $(2,-1)$; move 2 horiz. and 1 vert.

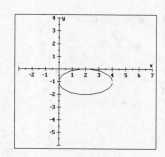

439

25.
$$9x^2 + 4y^2 - 18x + 16y = 11$$
$$9(x^2 - 2x) + 4(y^2 + 4y) = 11$$
$$9(x^2 - 2x + 1) + 4(y^2 + 4y + 4) = 11 + 9 + 16$$
$$9(x - 1)^2 + 4(y + 2)^2 = 36$$
$$\frac{(x - 1)^2}{4} + \frac{(y + 2)^2}{9} = 1$$
C $(1, -2)$; move 2 horiz. and 3 vert.

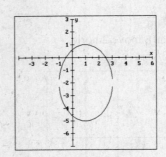

27. $\dfrac{x^2}{9} + \dfrac{y^2}{16} = 1$
C $(0, 0)$; move 3 horiz. and 4 vert.

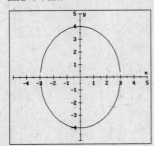

29. $\dfrac{(x - 2)^2}{16} + \dfrac{y^2}{25} = 1$
C $(2, 0)$; move 4 horiz. and 5 vert.

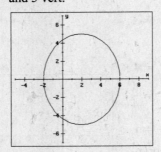

31.
$$16x^2 + 4y^2 = 64$$
$$\frac{16x^2}{64} + \frac{4y^2}{64} = \frac{64}{64}$$
$$\frac{x^2}{4} + \frac{y^2}{16} = 1$$
C $(0, 0)$; move 2 horiz. and 4 vert.

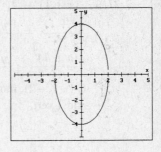

33.
$$25(x + 1)^2 + 9y^2 = 225$$
$$\frac{25(x + 1)^2}{225} + \frac{9y^2}{225} = \frac{225}{225}$$
$$\frac{(x + 1)^2}{9} + \frac{y^2}{25} = 1$$
C $(-1, 0)$; move 3 horiz. and 5 vert.

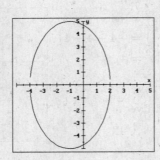

440

35. $\dfrac{x^2}{9} + \dfrac{y^2}{4} = 1$

$\dfrac{y^2}{4} = 1 - \dfrac{x^2}{9}$

$y^2 = 4\left(1 - \dfrac{x^2}{9}\right)$

$y = \pm\sqrt{4\left(1 - \dfrac{x^2}{9}\right)}$

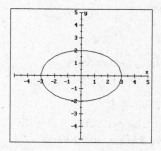

37. $\dfrac{x^2}{4} + \dfrac{(y-1)^2}{9} = 1$

$\dfrac{(y-1)^2}{9} = 1 - \dfrac{x^2}{4}$

$(y-1)^2 = 9\left(1 - \dfrac{x^2}{4}\right)$

$y - 1 = \pm\sqrt{9\left(1 - \dfrac{x^2}{4}\right)}$

$y = 1 \pm\sqrt{9\left(1 - \dfrac{x^2}{4}\right)}$

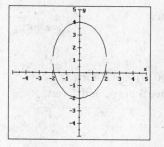

39. $a = 24/2 = 12, b = 10/2 = 5$

$\dfrac{x^2}{a^2} + \dfrac{y^2}{b^2} = 1$

$\dfrac{x^2}{12^2} + \dfrac{y^2}{5^2} = 1$

$\dfrac{x^2}{144} + \dfrac{y^2}{25} = 1$

41. Note: $a = 50/2 = 25, b = 20$

$\dfrac{x^2}{a^2} + \dfrac{y^2}{b^2} = 1$

$\dfrac{x^2}{625} + \dfrac{y^2}{400} = 1$

$10{,}000\left(\dfrac{x^2}{625} + \dfrac{y^2}{400}\right) = 10{,}000(1)$

$16x^2 + 25y^2 = 10{,}000$

$25y^2 = 10{,}000 - 16x^2$

$25y^2 = 16\left(625 - x^2\right)$

$y^2 = \tfrac{16}{25}\left(625 - x^2\right)$

$y = \tfrac{4}{5}\sqrt{625 - x^2}$

$\left[\text{just the top half of the ellipse, so no } \pm\,\right]$

43. $25x^2 + 4y^2 = 100$

$\dfrac{25x^2}{100} + \dfrac{4y^2}{100} = \dfrac{100}{100}$

$\dfrac{x^2}{4} + \dfrac{y^2}{25} = 1 \Rightarrow \text{area} = \pi ab = \pi(5)(2) = 10\pi \text{ square units}$

45. **Answers may vary.**

47. It is a circle.

Exercises 12.3 (page 926)

1.
$$\frac{x^2}{4} - \frac{y^2}{9} = 1$$
$$\frac{x^2}{4} - \frac{0^2}{9} = 1$$
$$\frac{x^2}{4} = 1$$
$$x^2 = 4$$
$$x = \pm 2$$
$$(2, 0), (-2, 0)$$

3. $-5x^4 + 10x^3 - 15x^2 = -5x^2(x^2 - 2x + 3)$

5. $14a^2 - 15ab - 9b^2 = (2a - 3b)(7a + 3b)$

7. hyperbola; difference

9. center

11. $(\pm a, 0)$; y-intercepts

13. $\frac{x^2}{9} - \frac{y^2}{4} = 1$
C $(0, 0)$; open horiz.;
move 3 horiz. and 2 vert.

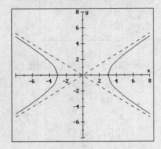

15. $\frac{y^2}{4} - \frac{x^2}{9} = 1$
C $(0, 0)$; open vert.;
move 3 horiz. and 2 vert.

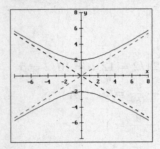

17. $\frac{(x-2)^2}{9} - \frac{y^2}{16} = 1$
C $(2, 0)$; open horiz.;
move 3 horiz. and 4 vert.

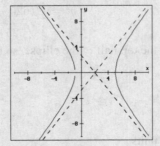

19. $\frac{(y+1)^2}{1} - \frac{(x-2)^2}{4} = 1$
C $(2, -1)$; open vert.;
move 2 horiz. and 1 vert.

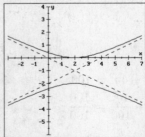

21.
$$4x^2 - y^2 + 8x - 4y = 4$$
$$4x^2 + 8x - y^2 - 4y = 4$$
$$4\left(x^2 + 2x\right) - \left(y^2 + 4y\right) = 4$$
$$4\left(x^2 + 2x + 1\right) - \left(y^2 + 4y + 4\right) = 4 + 4 - 4$$
$$4(x+1)^2 - (y+2)^2 = 4$$
$$\frac{(x+1)^2}{1} - \frac{(y+2)^2}{4} = 1$$
C $(-1, -2)$; opens horiz.; move 1 horiz. and 2 vert.

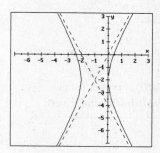

23.
$$4y^2 - x^2 + 8y + 4x = 4$$
$$4y^2 + 8y - x^2 + 4x = 4$$
$$4\left(y^2 + 2y\right) - \left(x^2 - 4x\right) = 4$$
$$4\left(y^2 + 2y + 1\right) - \left(x^2 - 4x + 4\right) = 4 + 4 - 4$$
$$4(y+1)^2 - (x-2)^2 = 4$$
$$\frac{(y+1)^2}{1} - \frac{(x-2)^2}{4} = 1$$
C $(2, -1)$; opens vert.; move 2 horiz. and 1 vert.

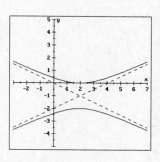

25. $xy = 10$

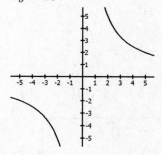

27. $xy = -12$

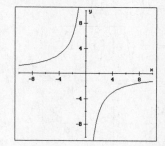

443

29. $25x^2 - y^2 = 25$

$\dfrac{25x^2}{25} - \dfrac{y^2}{25} = 1$

$\dfrac{x^2}{1} - \dfrac{y^2}{25} = 1$

C $(0, 0)$; open horiz.;
move 1 horiz. and 5 vert.

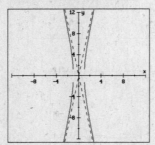

31. $4(x+3)^2 - (y-1)^2 = 4$

$\dfrac{4(x+3)^2}{4} - \dfrac{(y-1)^2}{4} = \dfrac{4}{4}$

$\dfrac{(x+3)^2}{1} - \dfrac{(y-1)^2}{4} = 1$

C $(-3, 1)$; open horiz.;
move 1 horiz. and 2 vert.

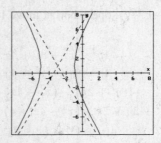

33. $\dfrac{x^2}{9} - \dfrac{y^2}{4} = 1$

$\dfrac{y^2}{4} = \dfrac{x^2}{9} - 1$

$y^2 = 4\left(\dfrac{x^2}{9} - 1\right)$

$y = \pm\sqrt{4\left(\dfrac{x^2}{9} - 1\right)}$

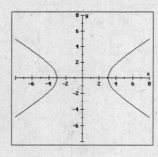

35. $\dfrac{x^2}{4} - \dfrac{(y-1)^2}{9} = 1$

$\dfrac{(y-1)^2}{9} = \dfrac{x^2}{4} - 1$

$(y-1)^2 = 9\left(\dfrac{x^2}{4} - 1\right)$

$y - 1 = \pm\sqrt{9\left(\dfrac{x^2}{4} - 1\right)}$

$y = 1 \pm\sqrt{9\left(\dfrac{x^2}{4} - 1\right)}$

444

37. $9y^2 - x^2 = 81$

$\dfrac{9y^2}{81} - \dfrac{x^2}{81} = \dfrac{81}{81}$

$\dfrac{y^2}{9} - \dfrac{x^2}{81} = 1$

distance $= \sqrt{9} = 3$ units

39. $x^2 - 4y^2 = 4$

$\dfrac{x^2}{4} - \dfrac{4y^2}{4} = \dfrac{4}{4}$

$\dfrac{x^2}{4} - \dfrac{y^2}{1} = 1$

$a = 2 \Rightarrow 2a = 4$

The smallest distance is 4 units.

41. **Answers may vary.**

43. If $a = b$, the rectangle is a square.

Exercises 12.4 (page 933)

1. 0, 1, 2

3. 0, 1, 2, 3, 4

5. $\sqrt{200y^2} - 7\sqrt{98y^2} = \sqrt{100y^2}\sqrt{2} - 7\sqrt{49y^2}\sqrt{2} = 10y\sqrt{2} - 7(7y)\sqrt{2} = -39y\sqrt{2}$

7. $\dfrac{3t\sqrt{2t} + 2\sqrt{2t^3}}{\sqrt{18t} + \sqrt{2t}} = \dfrac{3t\sqrt{2t} + 2\sqrt{t^2}\sqrt{2t}}{\sqrt{9}\sqrt{2t} + \sqrt{2t}} = \dfrac{3t\sqrt{2t} + 2t\sqrt{2t}}{3\sqrt{2t} + \sqrt{2t}} = \dfrac{5t\sqrt{2t}}{4\sqrt{2t}} = \dfrac{5t}{4}$

9. graphing; substitution

11. two

13. four

15. graphing

17. $\begin{cases} 8x^2 + 32y^2 = 256 \\ x = 2y \end{cases}$

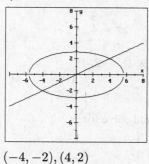

$(-4, -2), (4, 2)$

19. $\begin{cases} x^2 - 13 = -y^2 \\ y = 2x - 4 \end{cases}$

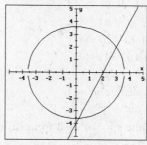

$\left(\dfrac{1}{5}, -\dfrac{18}{5}\right), (3, 2)$

21. $\begin{cases} (1) & 25x^2 + 9y^2 = 225 \\ (2) & 5x + 3y = 15 \end{cases}$

Substitute $x = -\frac{3}{5}y + 3$ from (2) into (1):
$$25x^2 + 9y^2 = 225$$
$$25\left(-\frac{3}{5}y + 3\right)^2 + 9y^2 = 225$$
$$25\left(\frac{9}{25}y^2 - \frac{18}{5}y + 9\right) + 9y^2 = 225$$
$$9y^2 - 90y + 225 + 9y^2 = 225$$
$$18y^2 - 90y = 0$$
$$18y(y - 5) = 0$$

$18y = 0 \quad$ **or** $\quad y - 5 = 0$
$\quad y = 0 \qquad\qquad\quad y = 5$

Substitute these and solve for x:

$5x + 3y = 15 \qquad\quad 5x + 3y = 15$
$5x + 3(0) = 15 \qquad 5x + 3(5) = 15$
$\quad 5x = 15 \qquad\qquad\quad 5x = 0$
$\quad\quad x = 3 \qquad\qquad\qquad x = 0$

Solutions: $(3, 0), (0, 5)$

23. $\begin{cases} (1) & x^2 + y^2 = 2 \\ (2) & x + y = 2 \end{cases}$

Substitute $x = 2 - y$ from (2) into (1):
$$x^2 + y^2 = 2$$
$$(2 - y)^2 + y^2 = 2$$
$$4 - 4y + y^2 + y^2 = 2$$
$$2y^2 - 4y + 2 = 0$$
$$2(y - 1)(y - 1) = 0$$

$y - 1 = 0 \quad$ **or** $\quad y - 1 = 0$
$\quad y = 1 \qquad\qquad\quad y = 1$

Substitute this and solve for x:
$$x = 2 - y$$
$$x = 2 - 1$$
$$x = 1$$

Solution: $(1, 1)$

25. $\begin{cases} (1) & x^2 + y^2 = 20 \\ (2) & y = x^2 \end{cases}$

Substitute $x^2 = y$ from (2) into (1):
$$x^2 + y^2 = 20$$
$$y + y^2 = 20$$
$$y^2 + y - 20 = 0$$
$$(y + 5)(y - 4) = 0$$

$y + 5 = 0 \quad$ **or** $\quad y - 4 = 0$
$\quad y = -5 \qquad\qquad\quad y = 4$

Substitute these and solve for x:

$x^2 = y \qquad\qquad x^2 = y$
$x^2 = -5 \qquad\quad x^2 = 4$
$\text{complex} \qquad\quad x = \pm 2$

Solutions: $(2, 4), (-2, 4)$

27. $\begin{cases} (1) & 2x^2 + y^2 = 6 \\ (2) & x^2 - y^2 = 3 \end{cases}$

Substitute $x^2 = y^2 + 3$ from (2) into (1):
$$2x^2 + y^2 = 6$$
$$2(y^2 + 3) + y^2 = 6$$
$$3y^2 + 6 = 6$$
$$3y^2 = 0$$
$$y^2 = 0$$

$y = 0$

Substitute and solve for x:
$$x^2 - y^2 = 3$$
$$x^2 - 0^2 = 3$$
$$x^2 = 3$$
$$x = \pm\sqrt{3}$$

Solutions: $\left(\sqrt{3}, 0\right), \left(-\sqrt{3}, 0\right)$

29.
$$x^2 + y^2 = 13$$
$$\underline{x^2 - y^2 = 5}$$
$$2x^2 = 18$$
$$x^2 = 9$$
$$x = \pm 3$$

Substitute and solve for y:

$$x^2 + y^2 = 13 \qquad\qquad x^2 + y^2 = 13$$
$$3^2 + y^2 = 13 \qquad\qquad (-3)^2 + y^2 = 13$$
$$y^2 = 4 \qquad\qquad\quad y^2 = 4$$
$$y = \pm 2 \qquad\qquad\quad y = \pm 2$$

Solutions: $(3, 2), (3, -2), (-3, 2), (-3, -2)$

31.
$$x^2 + y^2 = 25 \Rightarrow (\times 3) \quad 3x^2 + 3y^2 = 75$$
$$\underline{2x^2 - 3y^2 = 5 \Rightarrow} \qquad\qquad \underline{2x^2 - 3y^2 = 5}$$
$$\qquad\qquad\qquad\qquad\qquad 5x^2 = 80$$
$$\qquad\qquad\qquad\qquad\qquad x^2 = 16$$
$$\qquad\qquad\qquad\qquad\qquad x = \pm 4$$

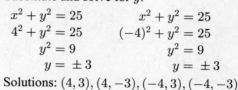

Substitute and solve for y:

$$x^2 + y^2 = 25 \qquad\qquad x^2 + y^2 = 25$$
$$4^2 + y^2 = 25 \qquad\qquad (-4)^2 + y^2 = 25$$
$$y^2 = 9 \qquad\qquad\quad y^2 = 9$$
$$y = \pm 3 \qquad\qquad\quad y = \pm 3$$

Solutions: $(4, 3), (4, -3), (-4, 3), (-4, -3)$

33. $\begin{cases} 2x - y > 4 \\ y < -x^2 + 2 \end{cases}$

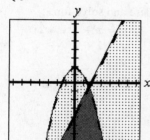

35. $\begin{cases} y > x^2 - 4 \\ y < -x^2 + 4 \end{cases}$

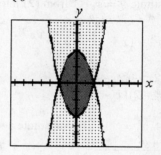

37.
$$9x^2 - 7y^2 = 81 \Rightarrow \qquad\quad 9x^2 - 7y^2 = 81$$
$$\underline{x^2 + y^2 = 9 \Rightarrow (\times 7)} \qquad \underline{7x^2 + 7y^2 = 63}$$
$$\qquad\qquad\qquad\qquad\qquad 16x^2 = 144$$
$$\qquad\qquad\qquad\qquad\qquad x^2 = 9$$
$$\qquad\qquad\qquad\qquad\qquad x = \pm 3$$

Substitute and solve for y:

$$x^2 + y^2 = 9 \qquad\qquad x^2 + y^2 = 9$$
$$3^2 + y^2 = 9 \qquad\qquad (-3)^2 + y^2 = 9$$
$$y^2 = 0 \qquad\qquad\quad y^2 = 0$$
$$y = 0 \qquad\qquad\quad y = 0$$

Solutions: $(3, 0), (-3, 0)$

39. $\begin{cases} (1) \quad x^2 + y^2 = 36 \\ (2) \quad 49x^2 + 36y^2 = 1764 \end{cases}$

Substitute $x^2 = 36 - y^2$ from (1) into (2):

$$49x^2 + 36y^2 = 1764$$
$$49(36 - y^2) + 36y^2 = 1764$$
$$1764 - 49y^2 + 36y^2 = 1764$$
$$-13y^2 = 0$$

$$y = 0$$

Substitute this and solve for x:

$$x^2 = 36 - y^2$$
$$x^2 = 36 - 0$$
$$x^2 = 36$$
$$x = \pm 6$$

Solutions: $(6, 0), (-6, 0)$

447

41. $\begin{cases} x^2 + y^2 = 10 \\ y = 3x^2 \end{cases}$

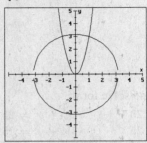

$(-1, 3), (1, 3)$

43. $\begin{cases} x^2 + y^2 = 25 \\ 12x^2 + 64y^2 = 768 \end{cases}$

$(-4, 3), (4, 3), (4, -3)$
$(-4, -3)$

45. $\begin{cases} (1) & x^2 + y^2 = 5 \\ (2) & y = x + 1 \end{cases}$

Substitute $y = x + 1$ from (2) into (1):
$$x^2 + y^2 = 5$$
$$x^2 + (x+1)^2 = 5$$
$$x^2 + x^2 + 2x + 1 = 5$$
$$2x^2 + 2x - 4 = 0$$
$$2(x + 2)(x - 1) = 0$$

$x + 2 = 0 \quad$ **or** $\quad x - 1 = 0$
$\quad x = -2 \qquad\qquad x = 1$

Substitute these and solve for x:
$y = x + 1 \qquad y = x + 1$
$y = -2 + 1 \qquad y = 1 + 1$
$y = -1 \qquad\quad y = 2$

Solutions: $(-2, -1), (1, 2)$

47. $\begin{aligned} x^2 + y^2 &= 15 \\ \underline{x^2 - y^2} &= \underline{-9} \\ 2x^2 &= 6 \\ x^2 &= 3 \\ x &= \pm\sqrt{3} \end{aligned}$

Substitute and solve for y:

$x^2 + y^2 = 15 \qquad\qquad x^2 + y^2 = 15$
$\left(\sqrt{3}\right)^2 + y^2 = 15 \qquad (-\sqrt{3})^2 + y^2 = 15$
$y^2 = 12 \qquad\qquad\qquad y^2 = 12$
$y = \pm\sqrt{12} \qquad\qquad y = \pm\sqrt{12}$
$y = \pm 2\sqrt{3} \qquad\qquad y = \pm 2\sqrt{3}$

Solutions: $\left(\sqrt{3}, 2\sqrt{3}\right), \left(\sqrt{3}, -2\sqrt{3}\right), \left(-\sqrt{3}, 2\sqrt{3}\right), \left(-\sqrt{3}, -2\sqrt{3}\right)$

49. $\begin{cases} (1) & y^2 = 40 - x^2 \\ (2) & y = x^2 - 10 \end{cases}$

Substitute $x^2 = 40 - y^2$ from (1) into (2):
$$y = x^2 - 10$$
$$y = 40 - y^2 - 10$$
$$y^2 + y - 30 = 0$$
$$(y + 6)(y - 5) = 0$$

$y + 6 = 0 \quad$ **or** $\quad y - 5 = 0$
$\quad y = -6 \qquad\qquad y = 5$

Substitute these and solve for x:
$x^2 = 40 - y^2 \qquad x^2 = 40 - y^2$
$x^2 = 40 - 36 \qquad x^2 = 40 - 25$
$x^2 = 4 \qquad\qquad x^2 = 15$
$x = \pm 2 \qquad\qquad x = \pm\sqrt{15}$

$(2, -6), (-2, -6), \left(\sqrt{15}, 5\right), \left(-\sqrt{15}, 5\right)$

51.

$\frac{1}{x} + \frac{2}{y} = 1 \Rightarrow$ $\frac{1}{x} + \frac{2}{y} = 1$ Substitute and solve for y:

$\frac{2}{x} - \frac{1}{y} = \frac{1}{3} \Rightarrow (\times 2)$ $\frac{4}{x} - \frac{2}{y} = \frac{2}{3}$ $\frac{1}{x} + \frac{2}{y} = 1$

$\overline{}$ $\overline{\frac{5}{x} = \frac{5}{3}}$ $\frac{1}{3} + \frac{2}{y} = 1$

$15 = 5x$ $\frac{2}{y} = \frac{2}{3}$

$3 = x$ $6 = 2y$

$3 = y$

Solution: $(3, 3)$

53. $\begin{cases} (1) & 3y^2 = xy \\ (2) & 2x^2 + xy - 84 = 0 \end{cases}$

From (1): $3y^2 - xy = 0$

$\qquad\qquad y(3y - x) = 0$

$y = 0 \quad$ or $\quad y = \frac{1}{3}x$

Substitute these into (2):

$2x^2 + xy - 84 = 0$ $2x^2 + xy - 84 = 0$

$2x^2 + x(0) - 84 = 0$ $2x^2 + x\left(\frac{1}{3}x\right) - 84 = 0$

$2x^2 = 84$ $2x^2 + \frac{1}{3}x^2 = 84$

$x^2 = 42$ $6x^2 + x^2 = 252$

$x = \pm\sqrt{42}$ $7x^2 = 252$

$x^2 = 36$

$x = \pm 6$

(substitute and solve for y)

Solutions: $\left(\sqrt{42}, 0\right), \left(-\sqrt{42}, 0\right), (6, 2), (-6, -2)$

55. $\begin{cases} (1) & xy = \frac{1}{6} \\ (2) & y + x = 5xy \end{cases}$

Substitute $x = \frac{1}{6y}$ from (1) into (2):

$y + \frac{1}{6y} = \frac{5y}{6y}$

$6y^2 + 1 = 5y$

$6y^2 - 5y + 1 = 0$

$(2y - 1)(3y - 1) = 0$

$2y - 1 = 0 \quad$ **or** $\quad 3y - 1 = 0$

$y = \frac{1}{2}$ $y = \frac{1}{3}$

Substitute these and solve for x:

$x = \frac{1}{6y}$ $x = \frac{1}{6y}$

$x = \frac{1}{6\left(\frac{1}{2}\right)}$ $x = \frac{1}{6\left(\frac{1}{3}\right)}$

$x = \frac{1}{3}$ $x = \frac{1}{2}$

Solutions: $\left(\frac{1}{3}, \frac{1}{2}\right), \left(\frac{1}{2}, \frac{1}{3}\right)$

57. Let the integers be x and y. Then the equations are

$\begin{cases} (1) & xy = 32 \\ (2) & x + y = 12 \end{cases}$

Substitute $x = \frac{32}{y}$ from (1) into (2):

$\frac{32}{y} + y = 12$

$32 + y^2 = 12y$

$y^2 - 12y + 32 = 0$

$(y - 4)(y - 8) = 0$

$y - 4 = 0 \quad$ **or** $\quad y - 8 = 0$

$y = 4$ $y = 8$

Substitute these and solve for x:

$x = \frac{32}{y} = \frac{32}{4} = 8$ $x = \frac{32}{y} = \frac{32}{8} = 4$

The integers are 8 and 4.

59. Let $l =$ the length of the rectangle, and $w =$ the width of the rectangle. Then the equations are:

$$\begin{cases} (1) & lw = 63 \\ (2) & 2l + 2w = 32 \end{cases}$$

Substitute $l = \frac{63}{w}$ from (1) into (2):

$$2\left(\tfrac{63}{w}\right) + 2w = 32$$
$$\tfrac{126}{w} + 2w = 32$$
$$126 + 2w^2 = 32w$$
$$2w^2 - 32y + 126 = 0$$
$$2(w - 7)(w - 9) = 0$$

$w - 7 = 0$ **or** $w - 9 = 0$
$w = 7$ $\qquad w = 9$

Substitute these and solve for l:

$l = \frac{63}{w} = \frac{63}{7} = 9$ $\qquad l = \frac{63}{w} = \frac{63}{9} = 7$

The dimensions are 7 cm by 9 cm.

61. Let $r =$ Carol's rate, and let $p =$ the amount Carol invested.

Then Juan invested $p + 150$ at a rate of $r + 0.015$. The equations are

$$\begin{cases} (1) & pr = 67.50 \Rightarrow p = \frac{67.5}{r} \\ (2) & (p + 150)(r + 0.015) = 94.5 \end{cases}$$

Substitute $p = \frac{67.5}{r}$ from (1) into (2):

$$\left(\tfrac{67.5}{r} + 150\right)(r + 0.015) = 94.5$$
$$67.5 + \tfrac{1.0125}{r} + 150r + 2.25 = 94.5$$
$$67.5r + 1.0125 + 150r^2 + 2.25r = 94.5r$$
$$150r^2 - 24.75r - 1.0125 = 0$$
$$12,000r^2 - 1980r + 81 = 0$$
$$(100r - 9)(120r - 9) = 0$$

$100r - 9 = 0$ **or** $120r - 9 = 0$
$r = 0.09$ $\qquad r = 0.075$

Substitute and solve for p:

$p = \frac{67.5}{r} = \frac{67.5}{0.09} = 750$ or $p = \frac{67.5}{r} = \frac{67.5}{0.075} = 900$

Carol invested \$750 at 9% or \$900 at 7.5%.

63. Let $r =$ Jim's rate and $t =$ Jim's time. Then his brother's rate was $r - 17$ and his time was $t + 1.5$.

$$\begin{cases} (1) & rt = 306 \Rightarrow t = \frac{306}{r} \\ (2) & (r - 17)(t + 1.5) = 306 \end{cases}$$

Substitute $t = \frac{306}{r}$ from (1) into (2):

$$(r - 17)\left(\tfrac{306}{r} + 1.5\right) = 306$$
$$306 + 1.5r - \tfrac{5202}{r} - 25.5 = 306$$
$$306r + 1.5r^2 - 5202 - 25.5r = 306r$$
$$1.5r^2 - 25.5r - 5202 = 0$$
$$3r^2 - 51r - 10,404 = 0$$
$$(3r + 153)(r - 68) = 0$$

$3r + 153 = 0$ **or** $r - 68 = 0$
$r = -153/3$ $\qquad r = 68$

Substitute and solve for t:

$t = \frac{306}{r} = \frac{306}{68} = 4.5$

Jim drove for 4.5 hours at 68 miles per hour.

65. **Answers may vary.**

67. $0, 1, 2, 3, 4$

69. $0, 1, 2,$ or infinitely many

Exercises 12.5 (page 940)

1. $f(1) = 1^2 = 1; f(2) = 2^2 = 4$
$f(3) = 3^2 = 9;$ increasing

3. $f(-3) = (-3)^2 = 9; f(-2) = (-2)^2 = 4$
$f(-1) = (-1)^2 = 1;$ decreasing

5. $(6x - 10) + (3x + 10) = 180$
$$9x = 180$$
$$x = 20$$

7. domains

9. constant; $f(x)$

11. step

13. increasing on $(-\infty, 0)$, decreasing on $(0, \infty)$

15. decreasing on $(-\infty, 0)$, constant on $(0, 2)$ increasing on $(2, \infty)$

17. $f(x) = \begin{cases} -1 \text{ when } x \leq 0 \\ x \text{ when } x > 0 \end{cases}$

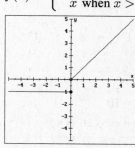

constant on $(-\infty, 0)$
increasing on $(0, \infty)$

19. $f(x) = \begin{cases} -x \text{ if } x \leq 0 \\ x \text{ if } 0 < x < 2 \\ -x \text{ if } x \geq 2 \end{cases}$

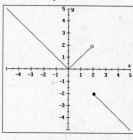

decreasing on $(-\infty, 0)$
increasing on $(0, 2)$
decreasing on $(2, \infty)$

21. $f(x) = -[[x]]$

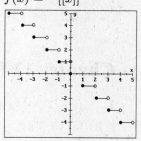

23. $f(x) = 2[[x]]$

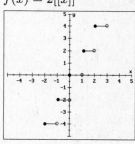

25. $f(x) = \begin{cases} -1 \text{ if } x < 0 \\ 0 \text{ if } x = 0 \\ 1 \text{ if } x > 0 \end{cases}$

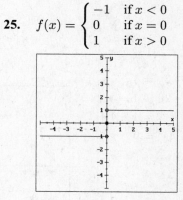

27. Find y when $x = 2.5$.
cost = \$30

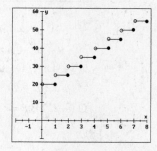

29. After 2 hours, B is cheaper.

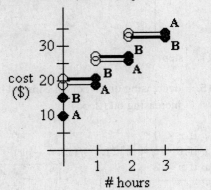

31. Answers may vary.

33. $f(x) = \begin{cases} x & \text{if } x < 6 \\ -x & \text{if } x > 6 \end{cases}$

Chapter 12 Review (page 945)

1. $(x-1)^2 + (y+2)^2 = 9$
C $(1, -2)$; $r = \sqrt{9} = 3$

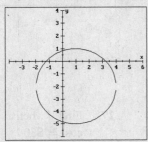

2. $x^2 + y^2 = 16$
C $(0,0)$; $r = \sqrt{16} = 4$

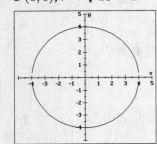

3.
$$x^2 + y^2 + 4x - 2y = 4$$
$$x^2 + 4x + y^2 - 2y = 4$$
$$x^2 + 4x + 4 + y^2 - 2y + 1 = 4 + 4 + 1$$
$$(x+2)^2 + (y-1)^2 = 9$$
C $(-2, 1)$; $r = \sqrt{9} = 3$

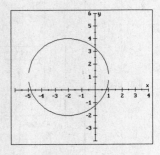

4. $x = -3(y - 2)^2 + 5$
V $(5, 2)$; opens L

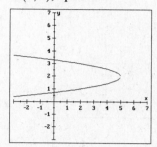

5. $x = 2(y + 1)^2 - 2$
V $(-2, -1)$; opens R

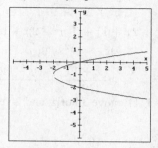

6. $y = -2(x - 1)^2 + 3$
V $(1, 3)$; opens D

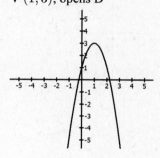

7. $y = (x + 2)^2 - 3$
V $(-2, -3)$; opens U

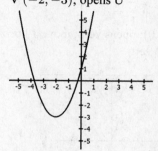

8. $9x^2 + 16y^2 = 144$

$$\frac{9x^2}{144} + \frac{16y^2}{144} = \frac{144}{144}$$

$$\frac{x^2}{16} + \frac{y^2}{9} = 1$$

C $(0, 0)$; move 4 horiz. and 3 vert.

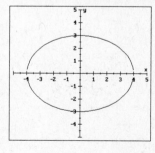

9. $\dfrac{(x - 2)^2}{4} + \dfrac{(y - 1)^2}{9} = 1$

C $(2, 1)$; move 2 horiz. and 3 vert.

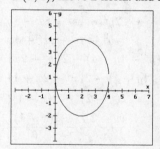

10.

$$4x^2 + 9y^2 + 8x - 18y = 23$$
$$4x^2 + 8x + 9y^2 - 18y = 23$$
$$4(x^2 + 2x) + 9(y^2 - 2y) = 23$$
$$4(x^2 + 2x + 1) + 9(y^2 - 2y + 1) = 23 + 4 + 9$$
$$4(x + 1)^2 + 9(y - 1)^2 = 36$$
$$\frac{(x + 1)^2}{9} + \frac{(y - 1)^2}{4} = 1$$

C $(-1, 1)$; move 3 horiz. and 2 vert.

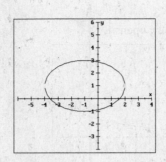

11.

$$9x^2 - y^2 = -9$$
$$\frac{9x^2}{-9} - \frac{y^2}{-9} = \frac{-9}{-9}$$
$$\frac{y^2}{9} - \frac{x^2}{1} = 1$$

C $(0, 0)$; opens vert.; move 1 horiz.
and 3 vert.

12. $xy = 9$

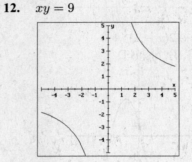

13.

$$2y^2 - 4x^2 + 8x - 8y = 8$$
$$y^2 - 4y - 2x^2 + 4x = 4$$
$$(y^2 - 4y + 4) - 2(x^2 - 2x) = 4 + 4$$
$$(y^2 - 4y + 4) - 2(x^2 - 2x + 1) = 8 - 2$$
$$(y - 2)^2 - 2(x - 1)^2 = 6$$
$$\frac{(y - 2)^2}{6} - \frac{2(x - 1)^2}{6} = 1$$
$$\frac{(y - 2)^2}{6} - \frac{(x - 1)^2}{3} = 1 \Rightarrow \text{hyperbola}$$

14.
$$9x^2 - 4y^2 - 18x - 8y = 31$$
$$9x^2 - 18x - 4y^2 - 8y = 31$$
$$9(x^2 - 2x) - 4(y^2 + 2y) = 31$$
$$9(x^2 - 2x + 1) - 4(y^2 + 2y + 1) = 31 + 9 - 4$$
$$9(x - 1)^2 - 4(y + 1)^2 = 36$$
$$\frac{(x-1)^2}{4} - \frac{(y+1)^2}{9} = 1$$

C $(1, -1)$; opens horiz.; move 2 horiz. and 3 vert.

15. $\begin{cases} (1) & x^2 + y^2 = 13 \\ (2) & y = x^2 - 1 \end{cases}$

Substitute $x^2 = 13 - y^2$ from (1) into (2):
$$y = x^2 - 1$$
$$y = 13 - y^2 - 1$$
$$y^2 + y - 12 = 0$$
$$(y + 4)(y - 3) = 0$$

$y + 4 = 0$ **or** $y - 3 = 0$
$\qquad y = -4 \qquad\qquad y = 3$

Substitute these and solve for x:

$x^2 = 13 - y^2$	$x^2 = 13 - y^2$
$x^2 = 13 - 16$	$x^2 = 13 - 9$
$x^2 = -3$	$x^2 = 4$
complex	$x = \pm 2$

Solutions: $(2, 3), (-2, 3)$

16.
$$\begin{array}{rcr} x^2 + y^2 &=& 20 \\ x^2 - y^2 &=& -12 \\ \hline 2x^2 &=& 8 \\ x^2 &=& 4 \\ x &=& \pm 2 \end{array}$$

Substitute and solve for y:

$x^2 + y^2 = 20$	$x^2 + y^2 = 20$
$2^2 + y^2 = 20$	$(-2)^2 + y^2 = 20$
$y^2 = 16$	$y^2 = 16$
$y = \pm 4$	$y = \pm 4$

Solutions: $(2, 4), (2, -4), (-2, 4), (-2, -4)$

17. $\begin{cases} y \geq x^2 - 4 \\ y < x + 3 \end{cases}$

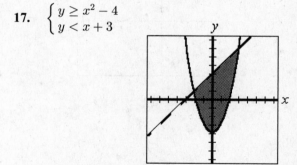

18. increasing on $(-\infty, -2)$; constant on $(-2, 1)$; decreasing on $(1, \infty)$

19. $f(x) = \begin{cases} x & \text{if } x \le 1 \\ -x^2 & \text{if } x > 1 \end{cases}$

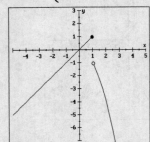

20. $f(x) = 3[[x]]$

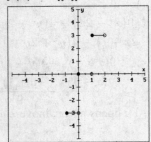

Chapter 12 Test (page 952)

1. $(x+5)^2 + (y-2)^2 = 9$

Center: $(-5, 2)$; radius $= \sqrt{9} = 3$

2.
$$x^2 + y^2 + 8x - 4y = 5$$
$$x^2 + 8x + y^2 - 4y = 5$$
$$x^2 + 8x + 16 + y^2 - 4y + 4 = 5 + 16 + 4$$
$$(x+4)^2 + (y-2)^2 = 25$$

Center: $(-4, 2)$; radius $= \sqrt{25} = 5$

3. $(x+1)^2 + (y-2)^2 = 9$

$C(-1, 2)$; $r = \sqrt{9} = 3$

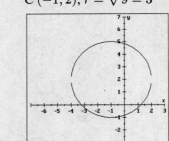

4. $x = (y-2)^2 - 1$

$V(-1, 2)$; opens R

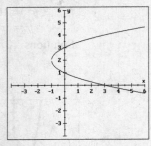

456

5.
$$9x^2 + 4y^2 = 36$$
$$\frac{9x^2}{36} + \frac{4y^2}{36} = \frac{36}{36}$$
$$\frac{x^2}{4} + \frac{y^2}{9} = 1$$
C $(0, 0)$; move 2 horiz. and 3 vert.

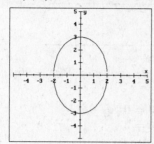

6.
$$\frac{(x-2)^2}{9} - y^2 = 1$$
C $(2, 0)$; opens horiz; move 3 horiz and 1 vert

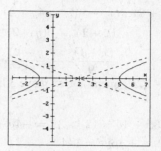

7.
$$4x^2 + y^2 - 24x + 2y = -33$$
$$4(x^2 - 6x) + (y^2 + 2y) = -33$$
$$4(x^2 - 6x + 9) + (y^2 + 2y + 1) = -33 + 36 + 1$$
$$4(x-3)^2 + (y+1)^2 = 4$$
$$\frac{(x-3)^2}{1} + \frac{(y+1)^2}{4} = 1$$
C $(3, -1)$; move 1 horiz. and 2 vert.

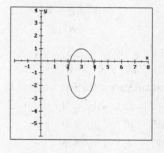

8.
$$x^2 + y^2 + 6x - 2y = -1$$
$$x^2 + 6x + y^2 - 2y = -1$$
$$x^2 + 6x + 9 + y^2 - 2y + 1 = -1 + 9 + 1$$
$$(x+3)^2 + (y-1)^2 = 9$$
Center: $(-3, 1)$; radius $= \sqrt{9} = 3$

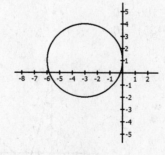

9.

$$x^2 - 9y^2 + 2x + 36y = 44$$
$$x^2 + 2x - 9y^2 + 36y = 44$$
$$\left(x^2 + 2x\right) - 9\left(y^2 - 4y\right) = 44$$
$$\left(x^2 + 2x + 1\right) - 9\left(y^2 - 4y + 4\right) = 44 + 1 - 36$$
$$(x+1)^2 - 9(y-2)^2 = 9$$
$$\frac{(x+1)^2}{9} - \frac{(y-2)^2}{1} = 1$$

C $(-1, 2)$; opens horiz.; move 3 horiz. and 1 vert.

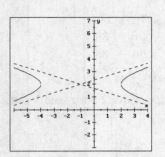

10. (1) $3x + 4y = 12 \Rightarrow 3x = -4y + 12 \Rightarrow (3x)^2 = (-4y + 12)^2 \Rightarrow 9x^2 = 16y^2 - 96y + 144$

(2) $9x^2 + 16y^2 = 144$

Substitute $9x^2 = 16y^2 - 96y + 144$:

$$9x^2 + 16y^2 = 144$$
$$16y^2 - 96y + 144 + 16y^2 = 144$$
$$32y^2 - 96y = 0$$
$$32y(y-3) = 0$$

$32y = 0$ **or** $y - 3 = 0$
$y = 0$ $\qquad\qquad y = 3$

Substitute these and solve for y:

$3x + 4y = 12$ $\qquad$ $3x + 4y = 12$ $\qquad$ Solutions:
$3x + 4(0) = 12$ $\quad$ $3x + 4(3) = 12$ $\qquad$ $(4, 0), (0, 3)$
$x = 4$ $\qquad\qquad$ $x = 0$

11. $\begin{cases} (1) & x^2 + y^2 = 25 \\ (2) & 4x^2 - 9y = 0 \end{cases}$

Substitute $x^2 = 25 - y^2$ from (1) into (2):

$$4x^2 - 9y = 0$$
$$4\left(25 - y^2\right) - 9y = 0$$
$$100 - 4y^2 - 9y = 0$$
$$-4y^2 - 9y + 100 = 0$$
$$4y^2 + 9y - 100 = 0$$
$$(y - 4)(4y + 25) = 0$$

$y - 4 = 0$ **or** $4y + 25 = 0$
$y = 4$ $\qquad\qquad y = -\frac{25}{4}$

Substitute these and solve for x:

$x^2 = 25 - y^2$ $\qquad$ $x^2 = 25 - y^2$
$x^2 = 25 - 4^2$ $\qquad$ $x^2 = 25 - \left(-\frac{25}{4}\right)^2$
$x^2 = 25 - 16 = 9$ $\quad$ $x^2 = 25 - \frac{625}{16} = -\frac{225}{16}$
$x = \pm 3$ $\qquad\qquad$ x is nonreal.

Solutions: $(3, 4), (-3, 4)$

12. $\begin{cases} x - y < 3 \\ y \leq x^2 - 6x + 7 \end{cases}$

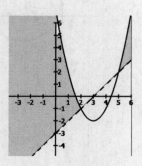

13. increasing: $(-3, 0)$; decreasing: $(0, 3)$

14. $f(x) = \begin{cases} -x^2 & \text{when } x < 0 \\ -x & \text{when } x \ge 0 \end{cases}$

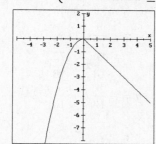

15. $f(x) = 2[[x]]$

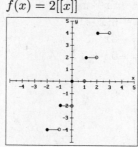

Cumulative Review Exercises (page 953)

1. $(2x - 5y)(4x + y) = 8x^2 + 2xy - 20xy - 5y^2 = 8x^2 - 18xy - 5y^2$

2. $(a^n + 1)(a^n - 3) = a^n a^n - 3a^n + a^n - 3 = a^{2n} - 2a^n - 3$

3. $\dfrac{7a - 14}{a^2 - 7a + 10} = \dfrac{7(a - 2)}{(a - 2)(a - 5)} = \dfrac{7}{a - 5}$

4. $\dfrac{a^4 - 5a^2 + 4}{a^2 + 3a + 2} = \dfrac{(a^2 - 4)(a^2 - 1)}{(a + 2)(a + 1)} = \dfrac{(a + 2)(a - 2)(a + 1)(a - 1)}{(a + 2)(a + 1)} = (a - 2)(a - 1)$

$$= a^2 - 3a + 2$$

5. $\dfrac{a^2 - a - 6}{a^2 - 4} \div \dfrac{a^2 - 9}{a^2 + a - 6} = \dfrac{a^2 - a - 6}{a^2 - 4} \cdot \dfrac{a^2 + a - 6}{a^2 - 9} = \dfrac{(a - 3)(a + 2)}{(a + 2)(a - 2)} \cdot \dfrac{(a + 3)(a - 2)}{(a + 3)(a - 3)} = 1$

6. $\dfrac{2}{a - 2} + \dfrac{3}{a + 2} - \dfrac{a - 1}{a^2 - 4} = \dfrac{2}{a - 2} + \dfrac{3}{a + 2} - \dfrac{a - 1}{(a + 2)(a - 2)}$

$$= \dfrac{2(a + 2)}{(a - 2)(a + 2)} + \dfrac{3(a - 2)}{(a + 2)(a - 2)} - \dfrac{a - 1}{(a + 2)(a - 2)}$$

$$= \dfrac{2(a + 2) + 3(a - 2) - (a - 1)}{(a + 2)(a - 2)}$$

$$= \dfrac{2a + 4 + 3a - 6 - a + 1}{(a + 2)(a - 2)} = \dfrac{4a - 1}{(a + 2)(a - 2)}$$

7. $3x - 4y = 12 \qquad\qquad y = \dfrac{3}{4}x - 5$

$-4y = -3x + 12 \qquad\quad m = \dfrac{3}{4}$

$y = \dfrac{3}{4}x - 3$

$m = \dfrac{3}{4}$

Parallel

8. $y = 3x + 4 \qquad x = -3y + 4$

$m = 3 \qquad\quad 3y = -x + 4$

$y = -\dfrac{1}{3}x + \dfrac{4}{3}$

$m = -\dfrac{1}{3}$

Perpendicular

9.
$$y - y_1 = m(x - x_1)$$
$$y - 3 = -6(x - 0)$$
$$y - 3 = -6x$$
$$y = -6x + 3$$

10.
$$m = \frac{y_2 - y_1}{x_2 - x_1} = \frac{4 - (-5)}{-5 - 8} = \frac{9}{-13} = -\frac{9}{13}$$
$$y - y_1 = m(x - x_1)$$
$$y - (-5) = -\frac{9}{13}(x - 8)$$
$$y + 5 = -\frac{9}{13}x + \frac{72}{13}$$
$$y = -\frac{9}{13}x + \frac{7}{13}$$

11. $2x - 3y < 6$

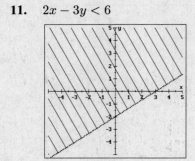

12. $y \geq x^2 - 4$

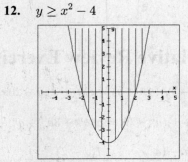

13. $\sqrt{98} + \sqrt{8} - \sqrt{32} = \sqrt{49}\sqrt{2} + \sqrt{4}\sqrt{2} - \sqrt{16}\sqrt{2} = 7\sqrt{2} + 2\sqrt{2} - 4\sqrt{2} = 5\sqrt{2}$

14. $12\sqrt[3]{648x^4} + 3\sqrt[3]{81x^4} = 12\sqrt[3]{216x^3}\sqrt[3]{3x} + 3\sqrt[3]{27x^3}\sqrt[3]{3x} = 12(6x)\sqrt[3]{3x} + 3(3x)\sqrt[3]{3x}$
$$= 72x\sqrt[3]{3x} + 9x\sqrt[3]{3x} = 81x\sqrt[3]{3x}$$

15.
$$\sqrt{3a + 1} = a - 1$$
$$\left(\sqrt{3a + 1}\right)^2 = (a - 1)^2$$
$$3a + 1 = a^2 - 2a + 1$$
$$0 = a^2 - 5a$$
$$0 = a(a - 5)$$
$$a = 0 \quad \textbf{or} \quad a - 5 = 0$$
$$\text{extraneous} \qquad a = 5$$

16.
$$x - \frac{2x}{x - 5} = 1 - \frac{10}{x - 5}$$
$$(x - 5)\left(x - \frac{2x}{x - 5}\right) = (x - 5)\left(1 - \frac{10}{x - 5}\right)$$
$$x(x - 5) - 2x = 1(x - 5) - 10$$
$$x^2 - 5x - 2x = x - 5 - 10$$
$$x^2 - 8x + 15 = 0$$
$$(x - 3)(x - 5) = 0$$
$$x - 3 = 0 \quad \textbf{or} \quad x - 5 = 0$$
$$x = 3 \qquad\qquad x = 5 \text{ (extraneous)}$$

CUMULATIVE REVIEW EXERCISES

17. $10a^2 + 21a - 10 = 0$

$(2a + 5)(5a - 2) = 0$

$2a + 5 = 0$ **or** $5a - 2 = 0$

$a = -\frac{5}{2}$ $\qquad a = \frac{2}{5}$

18. $2x^2 - 2x + 5 = 0$

$a = 2, b = -2, c = 5$

$x = \dfrac{-b \pm \sqrt{b^2 - 4ac}}{2a}$

$= \dfrac{-(-2) \pm \sqrt{(-2)^2 - 4(2)(5)}}{2(2)}$

$= \dfrac{2 \pm \sqrt{4 - 40}}{4}$

$= \dfrac{2 \pm \sqrt{-36}}{4}$

$= \dfrac{2}{4} \pm \dfrac{6i}{4} = \dfrac{1}{2} \pm \dfrac{3}{2}i$

19. $(f \circ g)(x) = f(g(x)) = f(4x + 1) = (4x + 1)^2 - 3 = 16x^2 + 8x + 1 - 3 = 16x^2 + 8x - 2$

20.

$y = 2x^3 - 1$

$x = 2y^3 - 1$

$x + 1 = 2y^3$

$\dfrac{x + 1}{2} = y^3$

$\sqrt[3]{\dfrac{x + 1}{2}} = y$

$y = f^{-1}(x) = \sqrt[3]{\dfrac{x + 1}{2}} = \dfrac{\sqrt[3]{4x + 4}}{2}$

21. $y = \left(\frac{1}{2}\right)^x$

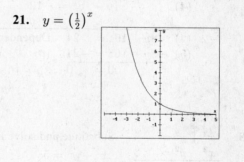

22. $y = \log_5 x \Rightarrow 5^y = x$

23. $2^{x+1} = 8^{x^2 - 1}$

$2^{x+1} = \left(2^3\right)^{x^2 - 1}$

$2^{x+1} = 2^{3x^2 - 3}$

$x + 1 = 3x^2 - 3$

$0 = 3x^2 - x - 4$

$0 = (3x - 4)(x + 1)$

$x = \frac{4}{3}$ or $x = -1$

24. $2\log 5 + \log x - \log 4 = 2$

$\log 5^2 + \log x - \log 4 = 2$

$\log \frac{25x}{4} = 2$

$10^2 = \frac{25x}{4}$

$400 = 25x$

$16 = x$

461

25. $x^2 + (y+1)^2 = 9$

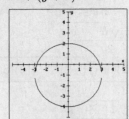

26. $x^2 - 9(y+1)^2 = 9$

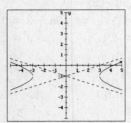

27.

(1) $x + 2y + 3z = 4$
(2) $5x + 6y + 7z = 8$
(3) $9x + 10y + 11z = 12$

$-5 \cdot (1)$ $-5x - 10y - 15z = -20$ $-9 \cdot (1)$ $-9x - 18y - 27z = -36$
  (2) $\underline{5x + 6y + 7z = 8}$ (3) $\underline{9x + 10y + 11z = 12}$
  (4) $- 4y - 8z = -12$ (5) $- 8y - 16z = -24$

$-2 \cdot (4)$ $8y + 16z = 24$ **Dependent equations** $x + 2y + 3z = 4$
  (5) $\underline{-8y - 16z = -24}$ If $z = z$, then $x + 2(-2z + 3) + 3z = 4$
   $0 = 0$ $8y + 16z = 24$ $x - 4z + 6 + 3z = 4$
      $8y = -16z + 24$ $x = z - 2$
       $y = -2z + 3$ **Solution:** $(z - 2, -2z + 3, z)$

28. $x^2 + y^2 = 5$ Substitute and solve for y:
   $\underline{x^2 - y^2 = 3}$
   $2x^2 = 8$ $x^2 + y^2 = 5$ $x^2 + y^2 = 5$
   $x^2 = 4$ $2^2 + y^2 = 5$ $(-2)^2 + y^2 = 5$
   $x = \pm 2$ $y^2 = 1$ $y^2 = 1$
       $y = \pm 1$ $y = \pm 1$

   Solutions: $(2, 1), (2, -1), (-2, 1), (-2, -1)$

29. $\begin{cases} (1) \quad x^2 + y^2 = 25 \\ (2) \quad 4x^2 - 9y = 0 \Rightarrow x^2 = \frac{9}{4}y \end{cases}$ $4y + 25 = 0$ **or** $y - 4 = 0$
   Substitute $x^2 = \frac{9}{4}y$ from (2) into (1): $y = -\frac{25}{4}$ $y = 4$

   $x^2 + y^2 = 25$ Substitute these and solve for x:

   $\frac{9}{4}y + y^2 = 25$ $x^2 = \frac{9}{4}y$ $x^2 = \frac{9}{4}y$

   $9y + 4y^2 = 100$ $x^2 = \frac{9}{4}\left(-\frac{25}{4}\right)$ $x^2 = \frac{9}{4}(4)$

   $4y^2 + 9y - 100 = 0$ $x^2 = \pm\sqrt{-\frac{225}{16}}$ $x^2 = 9$

   $(4y + 25)(y - 4) = 0$ complex $x = \pm 3$

               Solutions: $(3, 4), (-3, 4)$

29. $\begin{cases} (1) & x^2 + y^2 = 25 \\ (2) & 4x^2 - 9y = 0 \Rightarrow x^2 = \frac{9}{4}y \end{cases}$

Substitute $x^2 = \frac{9}{4}y$ from (2) into (1):

$$x^2 + y^2 = 25$$
$$\frac{9}{4}y + y^2 = 25$$
$$9y + 4y^2 = 100$$
$$4y^2 + 9y - 100 = 0$$
$$(4y + 25)(y - 4) = 0$$

$$4y + 25 = 0 \quad \textbf{or} \quad y - 4 = 0$$
$$y = -\tfrac{25}{4} \qquad\qquad y = 4$$

Substitute these and solve for x:

$$x^2 = \tfrac{9}{4}y \qquad\qquad x^2 = \tfrac{9}{4}y$$
$$x^2 = \tfrac{9}{4}\left(-\tfrac{25}{4}\right) \qquad x^2 = \tfrac{9}{4}(4)$$
$$x^2 = \pm\sqrt{-\tfrac{225}{16}} \qquad x^2 = 9$$
$$\text{complex} \qquad\qquad x = \pm 3$$

Solutions: $(3, 4), (-3, 4)$

30. $\begin{cases} y \geq x^2 \\ y < x + 3 \end{cases}$

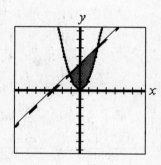

Exercises 13.1 (page 963)

1. $6 \cdot 5 \cdot 4 \cdot 3 \cdot 2 \cdot 1 = 720$

3. $\dfrac{6 \cdot 5 \cdot 4 \cdot 3 \cdot 2 \cdot 1}{4 \cdot 3 \cdot 2 \cdot 1} = \dfrac{720}{24} = 30$

5. $(x + 3y)^2 = (x + 3y)(x + 3y)$
$\qquad = x^2 + 6xy + 9y^2$

7. $(x - 3y)^2 = (x - 3y)(x - 3y)$
$\qquad = x^2 - 6xy + 9y^2$

9. $\log_9 81 = x \Rightarrow 9^x = 81 \Rightarrow x = 2$

11. $\log_{25} x = \frac{1}{2} \Rightarrow 25^{1/2} = x \Rightarrow x = 5$

13. one **15.** Pascal's **17.** $8!$ **19.** 1

21. $(a + b)^3 = 1a^3 + 3a^2b + 3ab^2 + 1b^3 = a^3 + 3a^2b + 3ab^2 + b^3$

23. $(a - b)^4 = 1a^4 + 4a^3(-b) + 6a^2(-b)^2 + 4a(-b)^3 + 1(-b)^4$
$\qquad = a^4 - 4a^3b + 6a^2b^2 - 4ab^3 + b^4$

25. $4! = 4 \cdot 3 \cdot 2 \cdot 1 = 24$

27. $-8! = -1 \cdot 8! = -8 \cdot 7 \cdot 6 \cdot 5 \cdot 4 \cdot 3 \cdot 2 \cdot 1$
$\qquad = -40,320$

29. $3! + 4! = 3 \cdot 2 \cdot 1 + 4 \cdot 3 \cdot 2 \cdot 1$
$\qquad = 6 + 24 = 30$

31. $3!(4!) = 3 \cdot 2 \cdot 1 \cdot 4 \cdot 3 \cdot 2 \cdot 1 = 144$

33. $\dfrac{9!}{11!} = \dfrac{9!}{11 \cdot 10 \cdot 9!} = \dfrac{1}{11 \cdot 10} = \dfrac{1}{110}$

35. $\dfrac{49!}{47!} = \dfrac{49 \cdot 48 \cdot 47!}{47!} = 49 \cdot 48 = 2352$

37. $\dfrac{9!}{7!\,0!} = \dfrac{9 \cdot 8 \cdot 7!}{7! \cdot 1} = 9 \cdot 8 = 72$

39. $\dfrac{5!}{3!(5-3)!} = \dfrac{5!}{3!\,2!} = \dfrac{5 \cdot 4 \cdot 3!}{3! \cdot 2 \cdot 1} = \dfrac{5 \cdot 4}{2 \cdot 1} = 10$

41. $(x+y)^3 = x^3 + \dfrac{3!}{1!(3-1)!}x^2 y + \dfrac{3!}{2!(3-2)!}xy^2 + y^3 = x^3 + \dfrac{3!}{1!2!}x^2 y + \dfrac{3!}{2!1!}xy^2 + y^3$

$$= x^3 + \dfrac{3 \cdot 2!}{1!2!}x^2 y + \dfrac{3 \cdot 2!}{2!1!}xy^2 + y^3$$

$$= x^3 + \dfrac{3}{1}x^2 y + \dfrac{3}{1}xy^2 + y^3$$

$$= x^3 + 3x^2 y + 3xy^2 + y^3$$

43. $(a+b)^6 = a^6 + \dfrac{6!}{1!(6-1)!}a^5 b + \dfrac{6!}{2!(6-2)!}a^4 b^2 + \dfrac{6!}{3!(6-3)!}a^3 b^3 + \dfrac{6!}{4!(6-4)!}a^2 b^4 + \cdots$

$$+ \dfrac{6!}{5!(6-5)!}ab^5 + \dfrac{6!}{6!(6-6)!}b^6$$

$$= a^6 + \dfrac{6!}{1!5!}a^5 b + \dfrac{6!}{2!4!}a^4 b^2 + \dfrac{6!}{3!3!}a^3 b^3 + \dfrac{6!}{4!2!}a^2 b^4 + \dfrac{6!}{5!1!}ab^5 + \dfrac{6!}{6!0!}b^6$$

$$= a^6 + \dfrac{6 \cdot 5!}{1!5!}a^5 b + \dfrac{6 \cdot 5 \cdot 4!}{2!4!}a^4 b^2 + \dfrac{6 \cdot 5 \cdot 4 \cdot 3!}{3!3!}a^3 b^3 + \dfrac{6 \cdot 5 \cdot 4!}{4!2!}a^2 b^4 + \dfrac{6 \cdot 5!}{5!1!}ab^5 + \dfrac{6!}{6!0!}b^6$$

$$= a^6 + \dfrac{6}{1}a^5 b + \dfrac{30}{2}a^4 b^2 + \dfrac{120}{6}a^3 b^3 + \dfrac{30}{2}a^2 b^4 + \dfrac{6}{1}ab^5 + b^6$$

$$= a^6 + 6a^5 b + 15a^4 b^2 + 20a^3 b^3 + 15a^2 b^4 + 6ab^5 + b^6$$

45. $(a-b)^4 = a^4 + \dfrac{4!}{1!(4-1)!}a^3(-b) + \dfrac{4!}{2!(4-2)!}a^2(-b)^2 + \dfrac{4!}{3!(4-3)!}a(-b)^3 + (-b)^4$

$$= a^4 + \dfrac{4!}{1!3!}(-a^3 b) + \dfrac{4!}{2!2!}a^2 b^2 + \dfrac{4!}{3!1!}(-ab^3) + b^4$$

$$= a^4 - \dfrac{4 \cdot 3!}{1!3!}a^3 b + \dfrac{4 \cdot 3 \cdot 2!}{2! \cdot 2 \cdot 1}a^2 b^2 - \dfrac{4 \cdot 3!}{3!1!}ab^3 + b^4$$

$$= a^4 - \dfrac{4}{1}a^3 b + \dfrac{12}{2}a^2 b^2 - \dfrac{4}{1}ab^3 + b^4$$

$$= a^4 - 4a^3 b + 6a^2 b^2 - 4ab^3 + b^4$$

47. $(x-2y)^3 = x^3 + \dfrac{3!}{1!(3-1)!}x^2(-2y) + \dfrac{3!}{2!(3-2)!}x(-2y)^2 + (-2y)^3$

$$= x^3 + \dfrac{3!}{1!2!} \cdot (-2x^2 y) + \dfrac{3!}{2!1!} \cdot 4xy^2 - 8y^3$$

$$= x^3 - \dfrac{3 \cdot 2!}{1!2!} \cdot 2x^2 y + \dfrac{3 \cdot 2!}{2!1!} \cdot 4xy^2 - 8y^3$$

$$= x^3 - \dfrac{3}{1} \cdot 2x^2 y + \dfrac{3}{1} \cdot 4xy^2 - 8y^3$$

$$= x^3 - 6x^2 y + 12xy^2 - 8y^3$$

49. $(2x + y)^3 = (2x)^3 + \dfrac{3!}{1!(3-1)!}(2x)^2 y + \dfrac{3!}{2!(3-2)!}2xy^2 + y^3$

$= 8x^3 + \dfrac{3!}{1!2!} \cdot 4x^2 y + \dfrac{3!}{2!1!} \cdot 2xy^2 + y^3$

$= 8x^3 + \dfrac{3 \cdot 2!}{1!2!} \cdot 4x^2 y + \dfrac{3 \cdot 2!}{2!1!} \cdot 2xy^2 + y^3$

$= 8x^3 + \dfrac{3}{1} \cdot 4x^2 y + \dfrac{3}{1} \cdot 2xy^2 + y^3$

$= 8x^3 + 12x^2 y + 6xy^2 + y^3$

51. $(2x + 3y)^3 = (2x)^3 + \dfrac{3!}{1!(3-1)!}(2x)^2(3y) + \dfrac{3!}{2!(3-2)!}2x(3y)^2 + (3y)^3$

$= 8x^3 + \dfrac{3!}{1!2!} \cdot 4x^2(3y) + \dfrac{3!}{2!1!} \cdot 2x(9y^2) + 27y^3$

$= 8x^3 + \dfrac{3 \cdot 2!}{1!2!} \cdot 12x^2 y + \dfrac{3 \cdot 2!}{2!1!} \cdot 18xy^2 + 27y^3$

$= 8x^3 + \dfrac{3}{1} \cdot 12x^2 y + \dfrac{3}{1} \cdot 18xy^2 + 27y^3$

$= 8x^3 + 36x^2 y + 54xy^2 + 27y^3$

53. In the 3rd term, the exponent on b is 2.
Variables: $a^2 b^2$
Coef. $= \dfrac{n!}{r!(n-r)!} = \dfrac{4!}{2!2!} = 6$
Term $= 6a^2 b^2$

55. In the 5th term, the exponent on y is 4.
Variables: $x^2 y^4$
Coef. $= \dfrac{n!}{r!(n-r)!} = \dfrac{6!}{4!2!} = 15$
Term $= 15x^2 y^4$

57. In the 4th term, the exponent on $-y$ is 3.
Variables: $x^1(-y)^3 = -xy^3$
Coef. $= \dfrac{n!}{r!(n-r)!} = \dfrac{4!}{3!1!} = 4$
Term $= 4(-xy^3) = -4xy^3$

59. In the 3rd term, the exponent on $-y$ is 2.
Variables: $x^6(-y)^2 = x^6 y^2$
Coef. $= \dfrac{n!}{r!(n-r)!} = \dfrac{8!}{2!6!} = 28$
Term $= 28x^6 y^2$

61. In the 3rd term, the exponent on y is 2.
Variables: $(4x)^3 y^2 = 64x^3 y^2$
Coef. $= \dfrac{n!}{r!(n-r)!} = \dfrac{5!}{2!3!} = 10$
Term $= 10(64x^3 y^2) = 640x^3 y^2$

63. In the 2nd term, the exponent on $-3y$ is 1.
Variables: $x^3(-3y)^1 = -3x^3 y$
Coef. $= \dfrac{n!}{r!(n-r)!} = \dfrac{4!}{1!3!} = 4$
Term $= 4(-3x^3 y) = -12x^3 y$

65. In the 4th term, the exponent on -5 is 3. Variables: $(2x)^4(-5)^3 = (16x^4)(-125) = -2000x^4$
Coef. $= \dfrac{n!}{r!(n-r)!} = \dfrac{7!}{3!4!} = 35;$ Term $= 35(-2000x^4) = -70{,}000x^4$

67. In the 5th term, the exponent on $-3y$ is 4. Variables: $(2x)^1(-3y)^4 = 2x(81y^4) = 162xy^4$

Coef. $= \dfrac{n!}{r!(n-r)!} = \dfrac{5!}{4!1!} = 5$; Term $= 5(162xy^4) = 810xy^4$

69. $8(7!) = 8 \cdot 7 \cdot 6 \cdot 5 \cdot 4 \cdot 3 \cdot 2 \cdot 1 = 40{,}320$

71. $\dfrac{7!}{5!(7-5)!} = \dfrac{7!}{5!2!} = \dfrac{7 \cdot 6 \cdot 5!}{5! \cdot 2 \cdot 1} = \dfrac{7 \cdot 6}{2 \cdot 1} = 21$

73. $\dfrac{5!(8-5)!}{4!7!} = \dfrac{5!3!}{4!7!} = \dfrac{5!}{7!} \cdot \dfrac{3!}{4!} = \dfrac{5!}{7 \cdot 6 \cdot 5!} \cdot \dfrac{3!}{4 \cdot 3!} = \dfrac{1}{7 \cdot 6} \cdot \dfrac{1}{4} = \dfrac{1}{168}$

75. $11! = 39{,}916{,}800$

77. $20! = 2.432902008 \times 10^{18}$

79. $(3 + 2y)^4 = 3^4 + \dfrac{4!}{1!(4-1)!}3^3(2y) + \dfrac{4!}{2!(4-2)!}3^2(2y)^2 + \dfrac{4!}{3!(4-3)!}3(2y)^3 + (2y)^4$

$= 81 + \dfrac{4!}{1!3!} \cdot 27(2y) + \dfrac{4!}{2!2!} \cdot 9(4y^2) + \dfrac{4!}{3!1!} \cdot 3(8y^3) + 16y^4$

$= 81 + \dfrac{4 \cdot 3!}{1!3!} \cdot 54y + \dfrac{4 \cdot 3 \cdot 2!}{2! \cdot 2 \cdot 1} \cdot 36y^2 + \dfrac{4 \cdot 3!}{3!1!} \cdot 24y^3 + 16y^4$

$= 81 + \dfrac{4}{1} \cdot 54y + \dfrac{12}{2} \cdot 36y^2 + \dfrac{4}{1} \cdot 24y^3 + 16y^4$

$= 81 + 216y + 216y^2 + 96y^3 + 16y^4$

81. $\left(\dfrac{x}{2} - \dfrac{y}{3}\right)^3 = \left(\dfrac{x}{2}\right)^3 + \dfrac{3!}{1!(3-1)!}\left(\dfrac{x}{2}\right)^2\left(-\dfrac{y}{3}\right) + \dfrac{3!}{2!(3-2)!}\left(\dfrac{x}{2}\right)\left(-\dfrac{y}{3}\right)^2 + \left(-\dfrac{y}{3}\right)^3$

$= \dfrac{x^3}{8} - \dfrac{3!}{1!2!} \cdot \dfrac{x^2}{4} \cdot \dfrac{y}{3} + \dfrac{3!}{2!1!} \cdot \dfrac{x}{2} \cdot \dfrac{y^2}{9} - \dfrac{y^3}{27}$

$= \dfrac{x^3}{8} - \dfrac{3 \cdot 2!}{1!2!} \cdot \dfrac{x^2y}{12} + \dfrac{3 \cdot 2!}{2!1!} \cdot \dfrac{xy^2}{18} - \dfrac{y^3}{27}$

$= \dfrac{x^3}{8} - \dfrac{3}{1} \cdot \dfrac{x^2y}{12} + \dfrac{3}{1} \cdot \dfrac{xy^2}{18} - \dfrac{y^3}{27}$

$= \dfrac{x^3}{8} - \dfrac{x^2y}{4} + \dfrac{xy^2}{6} - \dfrac{y^3}{27}$

83. $(x - y)^5 = x^5 + \dfrac{5!}{1!(5-1)!}x^4(-y) + \dfrac{5!}{2!(5-2)!}x^3(-y)^2 + \dfrac{5!}{3!(5-3)!}x^2(-y)^3$

$\qquad\qquad + \dfrac{5!}{4!(5-4)!}x(-y)^4 + \dfrac{5!}{5!(5-5)!}(-y)^5$

$= x^5 - \dfrac{5!}{1!4!}x^4y + \dfrac{5!}{2!3!}x^3y^2 - \dfrac{5!}{3!2!}x^2y^3 + \dfrac{5!}{4!1!}xy^4 - \dfrac{5!}{5!0!}y^5$

$= x^5 - \dfrac{5 \cdot 4!}{1!4!}x^4y + \dfrac{5 \cdot 4 \cdot 3!}{2!3!}x^3y^2 - \dfrac{5 \cdot 4 \cdot 3!}{3!2!}x^2y^3 + \dfrac{5 \cdot 4!}{4!1!}x^2y^4 - \dfrac{5!}{5!0!}y^5$

$= x^5 - \dfrac{5}{1}x^4y + \dfrac{20}{2}x^3y^2 - \dfrac{20}{2}x^2y^3 + \dfrac{5}{1}xy^4 - y^5$

$= x^5 - 5x^4y + 10x^3y^2 - 10x^2y^3 + 5xy^4 - y^5$

85. $\left(\dfrac{x}{2} + \dfrac{y}{3}\right)^4$

$$= \left(\frac{x}{2}\right)^4 + \frac{4!}{1!(4-1)!}\left(\frac{x}{2}\right)^3\left(\frac{y}{3}\right) + \frac{4!}{2!(4-2)!}\left(\frac{x}{2}\right)^2\left(\frac{y}{3}\right)^2 + \frac{4!}{3!(4-3)!}\left(\frac{x}{2}\right)\left(\frac{y}{3}\right)^3 + \left(\frac{y}{3}\right)^4$$

$$= \frac{x^4}{16} + \frac{4!}{1!3!}\cdot\frac{x^3}{8}\cdot\frac{y}{3} + \frac{4!}{2!2!}\cdot\frac{x^2}{4}\cdot\frac{y^2}{9} + \frac{4!}{3!1!}\cdot\frac{x}{2}\cdot\frac{y^3}{27} + \frac{y^4}{81}$$

$$= \frac{x^4}{16} + \frac{4\cdot 3!}{1!3!}\cdot\frac{x^3y}{24} + \frac{4\cdot 3\cdot 2!}{2!2!}\cdot\frac{x^2y^2}{36} + \frac{4\cdot 3!}{3!1!}\cdot\frac{xy^3}{54} + \frac{y^4}{81}$$

$$= \frac{x^4}{16} + \frac{4}{1}\cdot\frac{x^3y}{24} + \frac{12}{2}\cdot\frac{x^2y^2}{36} + \frac{4}{1}\cdot\frac{xy^3}{54} + \frac{y^4}{81}$$

$$= \frac{x^4}{16} + \frac{x^3y}{6} + \frac{x^2y^2}{6} + \frac{2xy^3}{27} + \frac{y^4}{81}$$

87.

```
                              1
                          1       1
                      1       2       1
                  1       3       3       1
              1       4       6       4       1
          1       5      10      10       5       1
      1       6      15      20      15       6       1
  1       7      21      35      35      21       7       1
1     8      28      56      70      56      28       8       1
1   9     36     84     126    126     84     36      9      1
```

89. In the 3rd term, the exponent on $\sqrt{3}y$ is 2. Variables: $(\sqrt{2}x)^4(\sqrt{3}y)^2 = (4x^4)(3y^2) = 12x^4y^2$

Coef. $= \dfrac{n!}{r!(n-r)!} = \dfrac{6!}{2!4!} = 15;$ Term $= 15(12x^4y^2) = 180x^4y^2$

91. In the 2nd term, the exponent on $-\dfrac{y}{3}$ is 1. Variables: $\left(\dfrac{x}{2}\right)^3\left(-\dfrac{y}{3}\right)^1 = \left(\dfrac{x^3}{8}\right)\left(-\dfrac{y}{3}\right) = -\dfrac{x^3y}{24}$

Coef. $= \dfrac{n!}{r!(n-r)!} = \dfrac{4!}{1!3!} = 4;$ Term $= 4\left(-\dfrac{x^3y}{24}\right) = -\dfrac{x^3y}{6} = -\dfrac{1}{6}x^3y$

93. In the 3rd term, the exponent on 5 is 2.
Variables: $x^4(5)^2 = 25x^4$

Coef. $= \dfrac{n!}{r!(n-r)!} = \dfrac{6!}{2!4!} = 15$

Term $= 15(25x^4) = 375x^4$

95. In the 4th term, the exponent on b is 3.
Variables: $a^{n-3}b^3$

Coef. $= \dfrac{n!}{r!(n-r)!} = \dfrac{n!}{3!(n-3)!}$

Term $= \dfrac{n!}{3!(n-3)!}a^{n-3}b^3$

97. In the 5th term, the exponent on $-b$ is 4. Variables: $a^{n-4}(-b)^4 = a^{n-4}b^4$

Coef. $= \dfrac{n!}{r!(n-r)!} = \dfrac{n!}{4!(n-4)!};$ Term $= \dfrac{n!}{4!(n-4)!}a^{n-4}b^4$

99. In the rth term, the coefficient on b is $r - 1$. Variables: $a^{n-(r-1)}b^{r-1} = a^{n-r+1}b^{r-1}$

Coef. $= \dfrac{n!}{r!(n-r)!} = \dfrac{n!}{(r-1)![n-(r-1)]!} = \dfrac{n!}{(r-1)!(n-r+1)!}$

Term $= \dfrac{n!}{(r-1)!(n-r+1)!}a^{n-r+1}b^{r-1}$

101. Answers may vary.   **103. Answers may vary.**

105. $\dfrac{n!}{0!(n-0)!} = \dfrac{n!}{0!n!} = \dfrac{n!}{1 \cdot n!} = \dfrac{n!}{n!} = 1$

107. $1, 1, 2, 3, 5, 8, 13, \ldots;$ Beginning with 2, each number is the sum of the previous two numbers.

109. $\left(a - \dfrac{1}{a}\right)^9 = \left(a - a^{-1}\right)^9$. The desired term occurs when the exponent is 5.

The $(r+1)$th term of $(a - a^{-1})^9$ is $\dfrac{9!}{r!(9-r)!}a^{9-r}\left(-a^{-1}\right)^r = \dfrac{9!}{r!(9-r)!}a^{9-r}(-a)^{-r}$.

But $\dfrac{9!}{r!(9-r)!}a^{9-r}(-a)^{-r} = (-1)^{-r}\dfrac{9!}{r!(9-r)!}a^{9-r}a^{-r} = (-1)^{-r}\dfrac{9!}{r!(9-r)!}a^{9-2r}$.

If $9 - 2r = 5$, then $r = 2$.

The term is $(-1)^{-2}\dfrac{9!}{2!(9-2)!}a^{9-2}a^{-2} = \dfrac{9!}{2!7!}a^5 = \dfrac{9 \cdot 8 \cdot 7!}{2 \cdot 1 \cdot 7!} = 36a^5$. Coefficient $= 36$

Exercises 13.2 (page 972)

1. $2(1) = 2, 2(2) = 4, 2(3) = 6, 2(4) = 8$ **3.** $1 + 4 = 5, 2 + 4 = 6, 3 + 4 = 7, 4 + 4 = 8$

5. $5 + (1-1)2 = 5 + 0(2) = 5, 5 + (2-1)2 = 5 + 1(2) = 7, 5 + (3-1)2 = 5 + 2(2) = 9,$
$5 + (4-1)2 = 5 + 3(2) = 11$

7. $5(3x^2 - 6x + 4) + 3(4x^2 + 6x - 2) = 15x^2 - 30x + 20 + 12x^2 + 18x - 6 = 27x^2 - 12x + 14$

9. $\dfrac{3a+4}{a-2} + \dfrac{3a-4}{a+2} = \dfrac{(3a+4)(a+2)}{(a-2)(a+2)} + \dfrac{(3a-4)(a-2)}{(a+2)(a-2)}$

$= \dfrac{3a^2 + 10a + 8 + 3a^2 - 10a + 8}{(a+2)(a-2)} = \dfrac{6a^2 + 16}{(a+2)(a-2)}$

11. sequence; finite; infinite   **13.** arithmetic; difference   **15.** arithmetic mean

17. series   **19.** $1 + 2 + 3 + 4 + 5 + 6 + 7$ **21.** $a_2 = 3(2) - 2 = 4$

23. $a_{30} = 3(30) - 2 = 88$ **25.** $3, 5, 7, 9, 11$ **27.** $-5, -8, -11, -14, -17$

29. $\quad$ 4, 7, 10, 13, 16
$$a_n = a_1 + (n-1)d$$
$$a_{15} = 4 + (15-1)3$$
$$a_{15} = 4 + (14)3 = 46$$

31. $\quad$ $-7, -9, -11, -13, -15$
$$a_n = a_1 + (n-1)d$$
$$a_{30} = -7 + (30-1)(-2)$$
$$a_{30} = -7 + (29)(-2) = -65$$

33. $\quad$ $a_1 = 2, d = 5$
$$a_n = a_1 + (n-1)d$$
$$a_{50} = 2 + (50-1)(5)$$
$$a_{50} = 2 + 49(5) = 247$$

35. $\quad$ $a_1 = 5, d = -6$
$$a_n = a_1 + (n-1)d$$
$$a_{49} = 5 + (49-1)(-6)$$
$$a_{49} = 5 + 48(-6) = -283$$

37. $\quad$
$$a_n = a_1 + (n-1)d$$
$$a_5 = a_1 + (5-1)d$$
$$29 = 5 + 4d$$
$$24 = 4d$$
$$6 = d; \ 5, 11, 17, 23, 29$$

39. $\quad$
$$a_n = a_1 + (n-1)d$$
$$a_6 = a_1 + (6-1)d$$
$$-39 = -4 + 5d$$
$$-35 = 5d$$
$$-7 = d$$
$$-4, -11, -18, -25, -32$$

41. Form an arithmetic sequence with a 1st term of 2 and a 5th term of 11:
$$a_n = a_1 + (n-1)d$$
$$a_5 = a_1 + (5-1)d$$
$$11 = 2 + 4d$$
$$9 = 4d$$
$$\frac{9}{4} = d$$
$$2, \boxed{\frac{17}{4}, \frac{13}{2}, \frac{35}{4}}, 11$$

43. Form an arithmetic sequence with a 1st term of 10 and a 6th term of 20:
$$a_n = a_1 + (n-1)d$$
$$a_6 = a_1 + (6-1)d$$
$$20 = 10 + 5d$$
$$10 = 5d$$
$$2 = d$$
$$10, \boxed{12, 14, 16, 18}, 20$$

45. $\quad$ $a_1 = 1, d = 3, n = 30$
$$a_n = a_1 + (n-1)d = 1 + 29(3) = 88$$
$$S_n = \frac{n(a_1 + a_n)}{2} = \frac{30(1+88)}{2} = 1335$$

47. $\quad$ $a_1 = -5, d = 4, n = 17$
$$a_n = a_1 + (n-1)d = -5 + 16(4) = 59$$
$$S_n = \frac{n(a_1 + a_n)}{2} = \frac{17(-5+59)}{2} = 459$$

49. $\quad$ $\displaystyle\sum_{k=1}^{4}(3k) = 3(1) + 3(2) + 3(3) + 3(4) = 3 + 6 + 9 + 12$

51. $\quad$ $\displaystyle\sum_{k=4}^{6} k^2 = 4^2 + 5^2 + 6^2 = 16 + 25 + 36$

53. $\quad$ $\displaystyle\sum_{k=1}^{4} 6k = 6(1) + 6(2) + 6(3) + 6(4) = 6 + 12 + 18 + 24 = 60$

55. $\quad$ $\displaystyle\sum_{k=3}^{4}(k^2 + 3) = (3^2 + 3) + (4^2 + 3) = 9 + 3 + 16 + 3 = 31$

57.
$$a_n = a_1 + (n-1)d$$
$$a_6 = a_1 + (6-1)d$$
$$-83 = a_1 + 5(7)$$
$$-83 = a_1 + 35$$
$$-118 = a_1$$
$$-118, -111, -104, -97, -90$$

59.
$$a_n = a_1 + (n-1)d$$
$$a_7 = a_1 + (7-1)d$$
$$16 = a_1 + 6(-3)$$
$$16 = a_1 - 18$$
$$34 = a_1$$
$$34, 31, 28, 25, 22$$

61.
$$a_n = a_1 + (n-1)d$$
$$a_{19} = a_1 + (19-1)d$$
$$131 = a_1 + 18d$$
$$a_n = a_1 + (n-1)d$$
$$a_{20} = a_1 + (20-1)d$$
$$138 = a_1 + 19d$$

$$a_1 + 18d = 131 \Rightarrow \times(-1) \quad -a_1 - 18d = -131$$
$$\underline{a_1 + 19d = 138} \Rightarrow \qquad \underline{a_1 + 19d = 138}$$
$$d = 7$$

Substitute and solve for a_1:
$$a_1 + 18d = 131$$
$$a_1 + 18(7) = 131$$
$$a_1 + 126 = 131$$
$$a_1 = 5 \Rightarrow 5, 12, 19, 26, 33$$

63. $a_n = a_1 + (n-1)d; \ a_{30} = a_1 + (30-1)d = 7 + 29(12) = 7 + 348 = 355$

65.
$$a_n = a_1 + (n-1)d$$
$$a_2 = a_1 + (2-1)d$$
$$-4 = a_1 + d$$
$$a_n = a_1 + (n-1)d$$
$$a_3 = a_1 + (3-1)d$$
$$-9 = a_1 + 2d$$

$$a_1 + d = -4 \Rightarrow \times(-1) \quad -a_1 - d = 4$$
$$\underline{a_1 + 2d = -9} \Rightarrow \qquad \underline{a_1 + 2d = -9}$$
$$d = -5$$

Substitute and solve for a_1:
$$a_1 + d = -4$$
$$a_1 + (-5) = -4$$
$$a_1 = 1$$

Find the desired term:
$$a_n = a_1 + (n-1)d$$
$$a_{37} = a_1 + (37-1)d$$
$$= 1 + 36(-5)$$
$$= 1 - 180 = \boxed{-179}$$

67.
$$a_n = a_1 + (n-1)d$$
$$a_{27} = a_1 + (27-1)d$$
$$263 = a_1 + 26(11)$$
$$263 = a_1 + 286$$
$$-23 = a_1$$

69.
$$a_n = a_1 + (n-1)d$$
$$a_{44} = a_1 + (44-1)d$$
$$556 = 40 + 43d$$
$$516 = 43d$$
$$12 = d$$

71. Form an arithmetic sequence with a 1st term of 10 and a 3rd term of 19:
$$a_n = a_1 + (n-1)d$$
$$a_3 = a_1 + (3-1)d$$
$$19 = 10 + 2d$$
$$9 = 2d$$
$$\frac{9}{2} = d$$
$$10, \boxed{\frac{29}{2}}, 19$$

73. Form an arithmetic sequence with a 1st term of -4.5 and a 3rd term of 7:
$$a_n = a_1 + (n-1)d$$
$$a_3 = a_1 + (3-1)d$$
$$7 = -4.5 + 2d$$
$$11.5 = 2d$$
$$5.75 = d$$
$$-4.5, \boxed{1.25}, 7$$

75.
$$a_n = a_1 + (n-1)d$$
$$a_2 = a_1 + (2-1)d$$
$$7 = a_1 + d$$
$$a_n = a_1 + (n-1)d$$
$$a_3 = a_1 + (3-1)d$$
$$12 = a_1 + 2d$$

$$\begin{aligned}a_1 + d &= 7 \Rightarrow \times(-1) \quad -a_1 - d = -7\\ a_1 + 2d &= 12 \Rightarrow \qquad\qquad \underline{a_1 + 2d = 12}\\ & \qquad\qquad\qquad\qquad\qquad d = 5\end{aligned}$$

Substitute and solve for a_1:
$$a_1 + d = 7$$
$$a_1 + 5 = 7$$
$$a_1 = 2, d = 5, n = 12$$
$$a_n = a_1 + (n-1)d = 2 + 11(5) = 57$$
$$S_n = \frac{n(a_1 + a_n)}{2} = \frac{12(2+57)}{2} = 354$$

77.
$$f(n) = 2n + 1 \Rightarrow f(1) = 3$$
$$f(n) = 2n + 1 = 31$$
$$2n = 30$$
$$n = 15$$
$$S_n = \frac{n(a_1 + a_n)}{2} = \frac{15(3+31)}{2} = 255$$

79.
$$a_1 = 1, d = 1, n = 50$$
$$a_n = a_1 + (n-1)d = 1 + 49(1) = 50$$
$$S_n = \frac{n(a_1 + a_n)}{2} = \frac{50(1+50)}{2} = 1275$$

81.
$$a_1 = 1, d = 2, n = 50$$
$$a_n = a_1 + (n-1)d = 1 + 49(2) = 99$$
$$S_n = \frac{n(a_1 + a_n)}{2} = \frac{50(1+99)}{2} = 2500$$

83. $\displaystyle\sum_{k=4}^{4}(2k+4) = 2(4) + 4 = 8 + 4 = 12$

85.
$$a_1 = 60, d = 50 \Rightarrow 60, 110, 160, 210, 260, 310; n = 121$$
$$a_n = a_1 + (n-1)d = 60 + (120-1)(50) = 60 + 119(50) = \$6010$$

87. $a_1 = 1, d = 1, n = 150, a_n = 150 \Rightarrow S_n = \frac{n(a_1 + a_n)}{2} = \frac{150(1+150)}{2} = 11{,}325$ bricks

89. After 1 sec.: $s = 16(1)^2 = 16$; After 2 sec.: $s = 16(2)^2 = 64$; After 3 sec.: $s = 16(3)^2 = 144$
During 2nd second $\Rightarrow$ falls $64 - 16 = 48$ ft; During 3rd second $\Rightarrow$ falls $144 - 64 = 80$ ft
The sequence of the amounts fallen during each second is $16, 48, 80 \Rightarrow a_1 = 16, d = 32$
$$a_n = a_1 + (n-1)d = 16 + (12-1)(32) = 16 + 11(32) = 368 \text{ ft}$$

91. Answers may vary.

93. $\displaystyle\sum_{n=1}^{6}\left(\tfrac{1}{2}n + 1\right)$: $\frac{3}{2}, 2, \frac{5}{2}, 3, \frac{7}{2}, 4$

95. Form an arithmetic sequence with a 1st term of a and a 3rd term of b:
$$a_n = a_1 + (n-1)d$$
$$b = a_1 + (3-1)d$$
$$b = a + 2d$$
$$b - a = 2d$$
$$\frac{b-a}{2} = d \Rightarrow \text{mean} = a_1 + \frac{b-a}{2} = a + \frac{b-a}{2} = \frac{2a}{2} + \frac{b-a}{2} = \frac{a+b}{2}$$

97. $\displaystyle\sum_{k=1}^{5} 5k = 5(1) + 5(2) + 5(3) + 5(4) + 5(5) = 5(1 + 2 + 3 + 4 + 5) = 5\sum_{k=1}^{5} k.$

99. $\displaystyle\sum_{k=1}^{n} 3 = \sum_{k=1}^{n} 3k^0 = 3(1)^0 + 3(2)^0 + \cdots + 3(n)^0 = 3 + 3 + \cdots + 3 = 3n$

Exercises 13.3 (page 980)

1. $2(4^2) = 2(16) = 32$

3. $4(3)^{5-1} = 4(3)^4 = 4(81) = 324$

5. $7\left(\frac{1}{3}\right)^{4-1} = 7\left(\frac{1}{3}\right)^3 = 7\left(\frac{1}{27}\right) = \frac{7}{27}$

7.
$$x^2 - 4x - 5 \le 0$$
$$(x - 5)(x + 1) \le 0$$
$x - 5 \quad -------- \ 0++++$
$x + 1 \quad --- \ 0++++++++++$
$$\xleftarrow{\qquad} \underset{-1}{[} \underset{5}{]} \xrightarrow{\qquad}$$
solution set: $[-1, 5]$

9.
$$\frac{x - 4}{x + 3} \ge 0$$
$x - 4 \quad -------- \ 0++++$
$x + 3 \quad ---0++++++++++$
$$\xleftarrow{\qquad} \underset{-3}{)} \underset{4}{[} \xrightarrow{\qquad}$$
solution set: $(-\infty, -3) \cup [4, \infty)$

11. geometric

13. common ratio

15. $S_n = \dfrac{a_1 - a_1 r^n}{1 - r}$

17. $9 \cdot 3 = 27$

19. $\dfrac{0.5}{0.2} = \dfrac{5}{2}$, or 2.5

21. $\dfrac{54}{18} = 3, 2 \cdot 3 = 6$

23. $4, 12, 36, 108, 324$
$$a_n = a_1 r^{n-1}$$
$$a_8 = 4(3)^{8-1} = 4\left(3^7\right) = 8748$$

25. $-5, -1, -\dfrac{1}{5}, -\dfrac{1}{25}, -\dfrac{1}{125}$
$$a_n = a_1 r^{n-1}$$
$$a_8 = -5\left(\frac{1}{5}\right)^{8-1} = -5\left(\frac{1}{5}\right)^7 = -\frac{1}{15{,}625}$$

27. $a_1 = 1, r = 3$
$$a_n = a_1 r^{n-1}$$
$$a_6 = 1(3)^{6-1} = 1(3)^5 = 243$$

29. $a_1 = 2, r = \dfrac{1}{2}$
$$a_n = a_1 r^{n-1}$$
$$a_7 = 2\left(\frac{1}{2}\right)^{7-1} = 2\left(\frac{1}{2}\right)^6 = \frac{1}{32}$$

31.
$$a_n = a_1 r^{n-1}$$
$$32 = 2r^{3-1}$$
$$32 = 2r^2$$
$$16 = r^2$$
$$\pm 4 = r, \text{ so } r = 4 \ (r > 0)$$
$$2, 8, 32, 128, 512$$

33.
$$a_n = a_1 r^{n-1}$$
$$50 = 2r^{3-1}$$
$$50 = 2r^2$$
$$25 = r^2$$
$$\pm 5 = r, \text{ so } r = -5 \ (r < 0)$$
$$2, -10, 50, -250, 1250$$

35. $a_1 = 2, a_5 = 162$
$a_n = a_1 r^{n-1}$
$a_5 = a_1 r^{5-1}$
$162 = 2r^4$
$81 = r^4$
$\pm 3 = r \Rightarrow$ choose $r = 3$
$2, \boxed{6, 18, 54}, 162$

37. $a_1 = -4, a_6 = -12500$
$a_n = a_1 r^{n-1}$
$a_6 = a_1 r^{6-1}$
$-12500 = -4r^5$
$3125 = r^5$
$5 = r$
$-4, \boxed{-20, -100, -500, -2500}, -12500$

39. $a_1 = 2, a_3 = 128$
$a_n = a_1 r^{n-1}$
$a_3 = a_1 r^{3-1}$
$128 = 2r^2$
$64 = r^2$
$\pm 8 = r \Rightarrow$ choose $r = -8$
$2, \boxed{-16}, 128$

41. $a_1 = 10, a_3 = 20$
$a_n = a_1 r^{n-1}$
$a_3 = a_1 r^{3-1}$
$20 = 10r^2$
$2 = r^2$
$\pm \sqrt{2} = r \Rightarrow$ choose $r = \sqrt{2}$
$10, \boxed{10\sqrt{2}}, 20$

43. $a_1 = 120, r = \frac{1}{2}, n = 4;$ $S_n = \dfrac{a_1 - a_1 r^n}{1 - r} = \dfrac{120 - 120\left(\frac{1}{2}\right)^4}{1 - \frac{1}{2}} = \dfrac{120 - 120\left(\frac{1}{16}\right)}{\frac{1}{2}} = \dfrac{112.5}{\frac{1}{2}} = 225$

45. $a_1 = 2, r = -3, n = 5;$ $S_n = \dfrac{a_1 - a_1 r^n}{1 - r} = \dfrac{2 - 2(-3)^5}{1 - (-3)} = \dfrac{2 - 2(-243)}{4} = \dfrac{488}{4} = 122$

47. If the 3rd term is 50 and the 2nd term is 10, then the common ratio $r = 50 \div 10 = 5$.
$a_1 = a_2 \div r = 10 \div 5 = 2$
$2, 10, 50, 250, 1250$

49. $a_n = a_1 r^{n-1}$
$-4 = -64r^{5-1}$
$-4 = -64r^4$
$\frac{1}{16} = r^4$
$\pm \frac{1}{2} = r$, so $r = -\frac{1}{2}$ $(r < 0)$
$-64, 32, -16, 8, -4$

51. $a_n = a_1 r^{n-1} = 4 \cdot 3^{9-1} = 4 \cdot 3^8 = 4 \cdot 6561 = 26{,}244$

53. $a_n = a_1 r^{n-1}$
$-81 = a_1(-3)^{8-1}$
$-81 = a_1(-3)^7$
$-81 = -2187 a_1$
$\frac{1}{27} = a_1$

55. $a_n = a_1 r^{n-1}$
$-1944 = -8r^{6-1}$
$-1944 = -8r^5$
$243 = r^5$
$3 = r$

473

57. $a_1 = -50, a_3 = 10$ $a_n = a_1 r^{n-1}$
$$a_3 = a_1 r^{3-1}$$
$$10 = -50r^2$$
$$-\tfrac{1}{5} = r^2 \text{: No such mean exists.}$$

59. $a_1 = 3, r = -2, n = 8;$ $S_n = \dfrac{a_1 - a_1 r^n}{1-r} = \dfrac{3 - 3(-2)^8}{1-(-2)} = \dfrac{3 - 3(256)}{3} = \dfrac{-765}{3} = -255$

61. $a_1 = 3, r = 2, n = 7;$ $S_n = \dfrac{a_1 - a_1 r^n}{1-r} = \dfrac{3 - 3(2)^7}{1-2} = \dfrac{3 - 3(128)}{-1} = \dfrac{-381}{-1} = 381$

63. If the 3rd term is $\tfrac{1}{5}$ and the 2nd term is 1, then the common ratio $r = \tfrac{1}{5} \div 1 = \tfrac{1}{5}$.

$a_1 = 1 \div \tfrac{1}{5} = 5, r = \tfrac{1}{5}, n = 4;$ $S_n = \dfrac{a_1 - a_1 r^n}{1-r} = \dfrac{5 - 5\left(\tfrac{1}{5}\right)^4}{1 - \tfrac{1}{5}} = \dfrac{5 - \tfrac{1}{125}}{\tfrac{4}{5}} = \dfrac{\tfrac{624}{125}}{\tfrac{4}{5}} = \dfrac{156}{25}$

65. If the 4th term is 1 and the 3rd term is -2, then the common ratio $r = 1 \div (-2) = -\tfrac{1}{2}$.

$a_2 = -2 \div \left(-\tfrac{1}{2}\right) = 4;$ $a_1 = 4 \div \left(-\tfrac{1}{2}\right) = -8;$ $a_1 = -8, r = -\tfrac{1}{2}, n = 6$

$S_n = \dfrac{a_1 - a_1 r^n}{1-r} = \dfrac{-8 - (-8)\left(-\tfrac{1}{2}\right)^6}{1 - \left(-\tfrac{1}{2}\right)} = \dfrac{-8 + \tfrac{1}{8}}{\tfrac{3}{2}} = \dfrac{-\tfrac{63}{8}}{\tfrac{3}{2}} = -\dfrac{21}{4}$

67. Sequence of population: $500, 500(1.06), 500(1.06)^2, \ldots$
$a_1 = 500, r = 1.06, n = 6 \Rightarrow a_n = a_1 r^{n-1} = 500(1.06)^5 \approx 669$

69. Sequence of amounts: $10000, 10000(0.88), 10000(0.88)^2, \ldots$
$a_1 = 10000, r = 0.88, n = 16 \Rightarrow a_n = a_1 r^{n-1} = 10000(0.88)^{15} \approx \$1{,}469.74$

71. Sequence of values: $70000, 70000(1.06), 70000(1.06)^2, \ldots$
$a_1 = 70000, r = 1.06, n = 13 \Rightarrow a_n = a_1 r^{n-1} = 70000(1.06)^{12} \approx \$140{,}853.75$

73. Sequence of areas: $1, \tfrac{1}{2}, \tfrac{1}{4}, \ldots$
$a_1 = 1, r = \tfrac{1}{2}, n = 12 \Rightarrow a_n = a_1 r^{n-1} = 1\left(\tfrac{1}{2}\right)^{11} = \left(\tfrac{1}{2}\right)^{11} \approx 0.0005$

75. Sequence of amounts: $1000(1.03), 1000(1.03)^2, 1000(1.03)^3, \ldots$
$a_1 = 1030, r = 1.03, n = 4 \Rightarrow S_n = \dfrac{a_1 - a_1 r^n}{1-r} = \dfrac{1030 - 1030(1.03)^4}{1 - 1.03} = \dfrac{-129.2740743}{-0.03}$
$$\approx \$4{,}309.14$$

77. **Answers may vary.**   **79.** **Answers may vary.**

81. arithmetic mean   **83.** **Answers may vary.**

Exercises 13.4 (page 986)

1. $r = \dfrac{6}{2} = 3$ **3.** $r = \dfrac{10}{50} = \dfrac{1}{5}$ **5.** $r = \dfrac{\frac{1}{5}}{1} = \dfrac{1}{5}$

7. $y = -2x^3 + 5$ function   **9.** $3x = y^2 + 4$ not a function   **11.** infinite   **13** $S_\infty = \dfrac{a_1}{1-r}$

15. $a_1 = 48, r = \dfrac{1}{4}$
$$S_\infty = \frac{a_1}{1-r} = \frac{48}{1-\frac{1}{4}} = \frac{48}{\frac{3}{4}} = 64$$

17. $a_1 = 200, r = \dfrac{1}{5}$
$$S_\infty = \frac{a_1}{1-r} = \frac{200}{1-\frac{1}{5}} = \frac{200}{\frac{4}{5}} = 250$$

19. $a_1 = 12, r = -\dfrac{1}{2}$
$$S_\infty = \frac{a_1}{1-r} = \frac{12}{1-\left(-\frac{1}{2}\right)} = \frac{12}{\frac{3}{2}} = 8$$

21. $a_1 = -81, r = -\dfrac{1}{3}$
$$S_\infty = \frac{a_1}{1-r} = \frac{-81}{1-\left(-\frac{1}{3}\right)} = \frac{-81}{\frac{4}{3}} = -\frac{243}{4}$$

23. $a_1 = -3, r = 2$
No sum since $|r| = 2 > 1$.

25. $a_1 = \frac{3}{4}, r = 2$
No sum since $|r| = 2 > 1$.

27. $0.\overline{1} = \frac{1}{10} + \frac{1}{100} + \frac{1}{1000} + \cdots \Rightarrow a_1 = \frac{1}{10}, r = \frac{1}{10} \Rightarrow S_\infty = \frac{a_1}{1-r} = \frac{\frac{1}{10}}{1-\frac{1}{10}} = \frac{\frac{1}{10}}{\frac{9}{10}} = \frac{1}{9}$

29. $-0.\overline{3} = -\frac{3}{10} - \frac{3}{100} - \frac{3}{1000} + \cdots \Rightarrow a_1 = -\frac{3}{10}, r = \frac{1}{10} \Rightarrow S_\infty = \frac{a_1}{1-r} = \frac{-\frac{3}{10}}{1-\frac{1}{10}} = \frac{-\frac{3}{10}}{\frac{9}{10}} = -\frac{1}{3}$

31. $0.\overline{12} = \frac{12}{100} + \frac{12}{10,000} + \frac{12}{1,000,000} + \cdots \Rightarrow a_1 = \frac{12}{100}, r = \frac{1}{100}$
$$S_\infty = \frac{a_1}{1-r} = \frac{\frac{12}{100}}{1-\frac{1}{100}} = \frac{\frac{12}{100}}{\frac{99}{100}} = \frac{12}{99} = \frac{4}{33}$$

33. $0.\overline{75} = \frac{75}{100} + \frac{75}{10,000} + \frac{75}{1,000,000} + \cdots \Rightarrow a_1 = \frac{75}{100}, r = \frac{1}{100}$
$$S_\infty = \frac{a_1}{1-r} = \frac{\frac{75}{100}}{1-\frac{1}{100}} = \frac{\frac{75}{100}}{\frac{99}{100}} = \frac{75}{99} = \frac{25}{33}$$

35. $a_1 = -54, r = -\dfrac{1}{3}$
$$S_\infty = \frac{a_1}{1-r} = \frac{-54}{1-\left(-\frac{1}{3}\right)} = \frac{-54}{\frac{4}{3}} = -\frac{81}{2}$$

37. $a_1 = -\dfrac{27}{2}, r = \dfrac{2}{3}$
$$S_\infty = \frac{a_1}{1-r} = \frac{-\frac{27}{2}}{1-\frac{2}{3}} = \frac{-\frac{27}{2}}{\frac{1}{3}} = -\frac{81}{2}$$

39. $S_\infty = \dfrac{a_1}{1-r} = \dfrac{1000}{1-0.8} = \dfrac{1000}{0.2} = 5{,}000$ moths

41. Distance ball travels down $= 10 + 5 + 2.5 + \cdots = \dfrac{a_1}{1 - r} = \dfrac{10}{1 - \frac{1}{2}} = \dfrac{10}{\frac{1}{2}} = 20$

Distance ball travels up $= 5 + 2.5 + 1.25 + \cdots = \dfrac{a_1}{1 - r} = \dfrac{5}{1 - \frac{1}{2}} = \dfrac{5}{\frac{1}{2}} = 10$

Total distance $= 20 + 10 = 30$ m

43. **Answers may vary.**

45.
$$S_\infty = \frac{a_1}{1 - r}$$
$$5 = \frac{1}{1 - r}$$
$$5(1 - r) = 1$$
$$5 - 5r = 1$$
$$4 = 5r$$
$$\frac{4}{5} = r$$

47. $0.\overline{9} = \dfrac{9}{10} + \dfrac{9}{100} + \dfrac{9}{1000} + \cdots$

$a = \dfrac{9}{10}, r = \dfrac{1}{10}$

$S_\infty = \dfrac{a_1}{1 - r} = \dfrac{\frac{9}{10}}{1 - \frac{1}{10}} = \dfrac{\frac{9}{10}}{\frac{9}{10}}$

$\qquad = \dfrac{9}{9} = 1$

49. No. $0.999999 = \dfrac{999,999}{1,000,000} < 1$

Exercises 13.5 (page 995)

1. $\dfrac{7!}{3!} = \dfrac{7 \cdot 6 \cdot 5 \cdot 4 \cdot 3!}{3!} = 840$

3. $\dfrac{5!}{1!} = \dfrac{5 \cdot 4 \cdot 3 \cdot 2 \cdot 1!}{1!} = 120$

5. $\dfrac{8!}{3!5!} = \dfrac{8 \cdot 7 \cdot 6 \cdot 5!}{3!5!} = \dfrac{336}{6} = 56$

7.
$$|2x - 7| = 17$$
$$2x - 7 = 17 \quad \textbf{or} \quad 2x - 7 = -17$$
$$2x = 24 \qquad\qquad 2x = -10$$
$$x = 12 \qquad\qquad x = -5$$

9.
$$\frac{3}{x - 5} = \frac{8}{x}$$
$$3x = 8(x - 5)$$
$$3x = 8x - 40$$
$$-5x = -40$$
$$x = 8$$

11. $p \cdot q$

13. permutation

15. $P(n, r) = \dfrac{n!}{(n - r)!}$

17. combination

19. $C(n, r) = \dfrac{n!}{r!(n - r)!}$

21. $P(5, 5) = \dfrac{5!}{(5 - 5)!} = \dfrac{5!}{0!} = \dfrac{120}{1} = 120$

23. $P(6, 2) = \dfrac{6!}{(6 - 2)!} = \dfrac{6!}{4!} = \dfrac{720}{24} = 30$

476

25. $P(4, 2) \cdot P(5, 2) = \dfrac{4!}{(4-2)!} \cdot \dfrac{5!}{(5-2)!} = \dfrac{4!}{2!} \cdot \dfrac{5!}{3!} = \dfrac{24}{2} \cdot \dfrac{120}{6} = 240$

27. $\dfrac{P(n, n)}{P(n, 0)} = \dfrac{\frac{n!}{(n-n)!}}{\frac{n!}{(n-0)!}} = \dfrac{\frac{n!}{0!}}{\frac{n!}{n!}} = \dfrac{\frac{n!}{1}}{\frac{1}{1}} = n!$

29. $C(4, 3) = \dfrac{4!}{3!(4-3)!} = \dfrac{4!}{3!1!} = \dfrac{24}{6 \cdot 1} = \dfrac{24}{6} = 4$

31. $\dbinom{8}{5} = \dfrac{8!}{5!(8-5)!} = \dfrac{8!}{5!3!} = \dfrac{40{,}320}{120 \cdot 6} = \dfrac{40{,}320}{720} = 56$

33. $\dbinom{7}{4}\dbinom{7}{5} = \dfrac{7!}{4!(7-4)!} \cdot \dfrac{7!}{5!(7-5)!} = \dfrac{7!}{4!3!} \cdot \dfrac{7!}{5!2!} = \dfrac{7 \cdot 6 \cdot 5 \cdot 4!}{4!3!} \cdot \dfrac{7 \cdot 6 \cdot 5!}{5!2!} = \dfrac{210}{6} \cdot \dfrac{42}{2}$
$= 35 \cdot 21 = 735$

35. $\dfrac{C(38, 37)}{C(19, 18)} = \dfrac{\frac{38!}{37!(38-37)!}}{\frac{19!}{18!(19-18)!}} = \dfrac{\frac{38 \cdot 37!}{37!1!}}{\frac{19 \cdot 18!}{18!1!}} = \dfrac{\frac{38}{1}}{\frac{19}{1}} = \dfrac{38}{19} = 2$

37. $(x+y)^4 = \dbinom{4}{0}x^4y^0 + \dbinom{4}{1}x^3y^1 + \dbinom{4}{2}x^2y^2 + \dbinom{4}{3}x^1y^3 + \dbinom{4}{4}x^0y^4$
$= \dfrac{4!}{0!4!}x^4 + \dfrac{4!}{1!3!}x^3y + \dfrac{4!}{2!2!}x^2y^2 + \dfrac{4!}{3!1!}xy^3 + \dfrac{4!}{4!0!}y^4 = x^4 + 4x^3y + 6x^2y^2 + 4xy^3 + y^4$

39. $(a+b)^5 = \dbinom{5}{0}a^5b^0 + \dbinom{5}{1}a^4b^1 + \dbinom{5}{2}a^3b^2 + \dbinom{5}{3}a^2b^3 + \dbinom{5}{4}a^1b^4 + \dbinom{5}{5}a^0b^5$
$= \dfrac{5!}{0!5!}a^5 + \dfrac{5!}{1!4!}a^4b + \dfrac{5!}{2!3!}a^3b^2 + \dfrac{5!}{3!2!}a^2b^3 + \dfrac{5!}{4!1!}a^1b^4 + \dfrac{5!}{5!0!}a^0b^5$
$= a^5 + 5a^4b + 10a^3b^2 + 10a^2b^3 + 5ab^4 + b^5$

41. $(2x+y)^3 = \dbinom{3}{0}(2x)^3y^0 + \dbinom{3}{1}(2x)^2y^1 + \dbinom{3}{2}(2x)^1y^2 + \dbinom{3}{3}(2x)^0y^3$
$= \dfrac{3!}{0!3!} \cdot 8x^3 + \dfrac{3!}{1!2!} \cdot 4x^2y + \dfrac{3!}{2!1!} \cdot 2xy^2 + \dfrac{3!}{3!0!}y^3$
$= 8x^3 + 12x^2y + 6xy^2 + y^3$

43. $(4x-y)^3 = \dbinom{3}{0}(4x)^3(-y)^0 + \dbinom{3}{1}(4x)^2(-y)^1 + \dbinom{3}{2}(4x)^1(-y)^2 + \dbinom{3}{3}(4x)^0(-y)^3$
$= \dfrac{3!}{0!3!} \cdot 64x^3 - \dfrac{3!}{1!2!} \cdot 16x^2y + \dfrac{3!}{2!1!} \cdot 4xy^2 - \dfrac{3!}{3!0!}y^3$
$= 64x^3 - 48x^2y + 12xy^2 - y^3$

45. $C(12, 0)C(12, 12) = \dfrac{12!}{0!(12-0)!} \cdot \dfrac{12!}{12!(12-12)!} = \dfrac{12!}{0!12!} \cdot \dfrac{12!}{12!0!} = \dfrac{12!}{12!} \cdot \dfrac{12!}{12!} = 1 \cdot 1 = 1$

47. $\dfrac{P(6,2) \cdot P(7,3)}{P(5,1)} = \dfrac{\frac{6!}{(6-2)!} \cdot \frac{7!}{(7-3)!}}{\frac{5!}{(5-1)!}} = \dfrac{\frac{6!}{4!} \cdot \frac{7!}{4!}}{\frac{5!}{4!}} = \dfrac{\frac{720}{24} \cdot \frac{5040}{24}}{\frac{120}{24}} = \dfrac{30 \cdot 210}{5} = 1{,}260$

49. $C(n,2) = \dfrac{n!}{2!(n-2)!}$

51. $(3x-2)^4$

$= \dbinom{4}{0}(3x)^4(-2)^0 + \dbinom{4}{1}(3x)^3(-2)^1 + \dbinom{4}{2}(3x)^2(-2)^2 + \dbinom{4}{3}(3x)^1(-2)^3$

$\hspace{6cm} + \dbinom{4}{4}(3x)^0(-2)^4$

$= \dfrac{4!}{0!4!} \cdot 81x^4(1) + \dfrac{4!}{1!3!} \cdot 27x^3(-2) + \dfrac{4!}{2!2!} \cdot 9x^2(4) + \dfrac{4!}{3!1!} \cdot 3x(-8) + \dfrac{4!}{4!0!} \cdot 1(16)$

$= 81x^4 - 216x^3 + 216x^2 - 96x + 16$

53. $\dbinom{5}{3}x^2(-5y)^3 = \dfrac{5!}{3!2!}x^2(-125y^3) = 10x^2(-125y^3) = -1{,}250x^2y^3$

55. $\dbinom{4}{1}(x^2)^3(-y^3)^1 = \dfrac{4!}{1!3!}x^6(-y^3) = -4x^6y^3$

57. $7 \cdot 5 = 35$

59. $7! = 5{,}040$

61. $10 \cdot 10 \cdot 10 \cdot 10 \cdot 10 \cdot 10 = 1{,}000{,}000$

63. $9 \cdot 9 \cdot 8 \cdot 7 \cdot 6 \cdot 5 = 136{,}080$

65. $8 \cdot 10 \cdot 10 \cdot 10 \cdot 10 \cdot 10 \cdot 10 = 8{,}000{,}000$

67. $4! \cdot 5! = 24 \cdot 120 = 2{,}880$

69. $25 \cdot 24 \cdot 23 = 13{,}800$

71. $P(10,3) = \dfrac{10!}{(10-3)!} = \dfrac{10!}{7!} = 720$

73. $9 \cdot 10 \cdot 10 \cdot 1 \cdot 1 = 900$

75. $C(14,3) = \dfrac{14!}{3!(14-3)!} = \dfrac{14!}{3!11!} = 364$

77. $C(5,3) = 10 \Rightarrow 5$ persons

79. $C(100,6) = \dfrac{100!}{6!(100-6)!} = \dfrac{100!}{6!94!} = \dfrac{100 \cdot 99 \cdot 98 \cdot 97 \cdot 96 \cdot 95 \cdot 94!}{6!94!} = 1{,}192{,}052{,}400$

81. $C(3,2) \cdot C(4,2) = \dfrac{3!}{2!1!} \cdot \dfrac{4!}{2!2!} = 3 \cdot 6 = 18$

83. $C(12,2) \cdot C(10,3) = \dfrac{12!}{2!10!} \cdot \dfrac{10!}{3!7!} = 66 \cdot 120 = 7{,}920$

85. Answers may vary.

87. Consider the two people who insist on standing together as one person. Then there are a total of 4 "persons" to be arranged. This can be done in $4! = 24$ ways. However, the two people who are standing together can be arranged in 2 different ways, so there are $24 \cdot 2 = 48$ arrangements.

Exercises 13.6 (page 1001)

1. $C(52, 26)$

3.
$$4^{3x} = \frac{1}{64}$$
$$4^{3x} = 4^{-3}$$
$$3x = -3$$
$$x = -1$$

5.
$$5^{x^2-2x} = 125$$
$$5^{x^2-2x} = 5^3$$
$$x^2 - 2x = 3$$
$$x^2 - 2x - 3 = 0$$
$$(x+1)(x-3) = 0$$
$$x + 1 = 0 \quad \text{or} \quad x - 3 = 0$$
$$x = -1 \qquad\qquad x = 3$$

7.
$$3^{x^2+4x} = \frac{1}{81}$$
$$3^{x^2+4x} = 3^{-4}$$
$$x^2 + 4x = -4$$
$$x^2 + 4x + 4 = 0$$
$$(x+2)(x+2) = 0$$
$$x + 2 = 0 \quad \text{or} \quad x + 2 = 0$$
$$x = -2 \qquad\qquad x = -2$$

9. experiment

11. $\dfrac{s}{n}$

13. 0

15.
  a. 6
  b. 52
  c. $\dfrac{6}{52}; \dfrac{3}{26}$

17. $\{(1, H), (2, H), (3, H), (4, H), (5, H), (6, H), (1, T), (2, T), (3, T), (4, T), (5, T), (6, T)\}$

19. $\{A, B, C, D, E, F, G, H, I, J, K, L, M, N, O, P, Q, R, S, T, U, V, W, X, Y, Z\}$

21. rolls of 5: $\{(1, 4), (2, 3), (3, 2), (4, 1)\}$; Probability $= \dfrac{4}{36} = \dfrac{1}{9}$

23. rolls < 6: $\{(1, 1), (1, 2), (1, 3), (1, 4), (2, 1), (2, 2), (2, 3), (3, 1), (3, 2), (4, 1)\}$
Probability $= \dfrac{10}{36} = \dfrac{5}{18}$

25. $\dfrac{\text{\# diamonds}}{\text{\# cards}} = \dfrac{13}{52} = \dfrac{1}{4}$

27. $\dfrac{\text{\# red face cards}}{\text{\# cards in deck}} = \dfrac{6}{52} = \dfrac{3}{26}$

29. $\dfrac{\substack{\text{\# ways to get 6}\\\text{diamonds}}}{\substack{\text{\# ways to get 6 cards}\\\text{from the deck of 52}}} = \dfrac{\binom{13}{6}}{\binom{52}{6}} = \dfrac{1716}{20{,}358{,}520}$
$= \dfrac{33}{391{,}510}$

31. $\dfrac{\text{\# ways to get 5 clubs}}{\substack{\text{\# ways to get 5 cards}\\\text{from the 26 black cards}}} = \dfrac{\binom{13}{5}}{\binom{26}{5}} = \dfrac{1287}{65780}$
$= \dfrac{9}{460}$

33. $\dfrac{1}{6}$ **35.** $\dfrac{4}{6} = \dfrac{2}{3}$ **37.** $\dfrac{19}{42}$ **39.** $\dfrac{13}{42}$

41. $\dfrac{3}{8}$ **43.** $\dfrac{0}{8} = 0$

45. $\dfrac{\#\text{ face} \cdot \#\text{ ten}}{\substack{\#\text{ ways to} \\ \text{get 2 cards}}} = \dfrac{12 \cdot 4}{52 \cdot 52} = \dfrac{48}{2704} = \dfrac{3}{169}$

47. rolls of 7: $\{(1, 6), (2, 5), (3, 4), (4, 3), (5, 2), (6, 1)\}$; Probability $= \dfrac{6}{36} = \dfrac{1}{6}$

49. $\dfrac{\#\text{ yellow}}{\#\text{ eggs}} = \dfrac{7}{12}$ **51.** $FFFF \Rightarrow \dfrac{1}{16}$

53. $PPFF, PFPF, PFFP, FPPF, FPFP, FFPP \Rightarrow \dfrac{6}{16} = \dfrac{3}{8}$

55. $SSSS \Rightarrow \dfrac{1}{16}$ **57.** $\dfrac{\binom{8}{4}}{\binom{10}{4}} = \dfrac{70}{210} = \dfrac{1}{3}$

59. $\dfrac{176}{282} = \dfrac{88}{141}$ **61.** $\dfrac{15}{71}$

63. Answers may vary.   **65.** $\begin{aligned} P\left(\text{lux and 2}^{\text{nd}}\text{ car}\right) &= P(\text{lux}) \cdot P\left(2^{\text{nd}} | \text{lux}\right) \\ &= 0.2(0.7) \\ &= 0.14 \end{aligned}$

Chapter 13 Review (page 1006)

1. $(5!)(2!) = 5 \cdot 4 \cdot 3 \cdot 2 \cdot 1 \cdot 2 \cdot 1 = 240$ **2.** $\dfrac{5!}{3!} = \dfrac{5 \cdot 4 \cdot 3!}{3!} = 5 \cdot 4 = 20$

3. $\dfrac{6!}{2!(6-2)!} = \dfrac{6!}{2!4!} = \dfrac{6 \cdot 5 \cdot 4!}{2 \cdot 1 \cdot 4!} = \dfrac{30}{2} = 15$ **4.** $\begin{aligned} \dfrac{12!}{3!(12-3)!} = \dfrac{12!}{3!9!} &= \dfrac{12 \cdot 11 \cdot 10 \cdot 9!}{3 \cdot 2 \cdot 1 \cdot 9!} \\ &= \dfrac{1320}{6} = 220 \end{aligned}$

5. $(n - n)! = 0! = 1$ **6.** $\dfrac{10!}{6!} = \dfrac{10 \cdot 9 \cdot 8 \cdot 7 \cdot 6!}{6!} = 5040$

7. $(x+y)^5 = x^5 + \dfrac{5!}{1!(5-1)!}x^4y + \dfrac{5!}{2!(5-2)!}x^3y^2 + \dfrac{5!}{3!(5-3)!}x^2y^3 + \dfrac{5!}{4!(5-4)!}xy^4 + y^5$

$\qquad = x^5 + \dfrac{5!}{1!4!}x^4y + \dfrac{5!}{2!3!}x^3y^2 + \dfrac{5!}{3!2!}x^2y^3 + \dfrac{5!}{4!1!}xy^4 + y^5$

$\qquad = x^5 + \dfrac{5\cdot4!}{1!4!}x^4y + \dfrac{5\cdot4\cdot3!}{2\cdot1\cdot3!}x^3y^2 + \dfrac{5\cdot4\cdot3!}{3!\cdot2\cdot1}x^2y^3 + \dfrac{5\cdot4!}{4!\cdot1}xy^4 + y^5$

$\qquad = x^5 + \dfrac{5}{1}x^4y + \dfrac{20}{2}x^3y^2 + \dfrac{20}{2}x^2y^3 + \dfrac{5}{1}xy^4 + y^5$

$\qquad = x^5 + 5x^4y + 10x^3y^2 + 10x^2y^3 + 5xy^4 + y^5$

8. $(x-y)^4 = x^4 + \dfrac{4!}{1!(4-1)!}x^3(-y) + \dfrac{4!}{2!(4-2)!}x^2(-y)^2 + \dfrac{4!}{3!(4-3)!}x(-y)^3 + (-y)^4$

$\qquad = x^4 + \dfrac{4!}{1!3!}(-x^3y) + \dfrac{4!}{2!2!}x^2y^2 + \dfrac{4!}{3!1!}(-xy^3) + y^4$

$\qquad = x^4 - \dfrac{4\cdot3!}{1!3!}x^3y + \dfrac{4\cdot3\cdot2!}{2!\cdot2\cdot1}x^2y^2 - \dfrac{4\cdot3!}{3!1!}xy^3 + y^4$

$\qquad = x^4 - \dfrac{4}{1}x^3y + \dfrac{12}{2}x^2y^2 - \dfrac{4}{1}xy^3 + y^4 = x^4 - 4x^3y + 6x^2y^2 - 4xy^3 + y^4$

9. $(4x-y)^3 = (4x)^3 + \dfrac{3!}{1!(3-1)!}(4x)^2(-y) + \dfrac{3!}{2!(3-2)!}(4x)(-y)^2 + (-y)^3$

$\qquad = 64x^3 + \dfrac{3!}{1!2!}\cdot(-16x^2y) + \dfrac{3!}{2!1!}\cdot4xy^2 - y^3$

$\qquad = 64x^3 - \dfrac{3\cdot2!}{1!2!}\cdot16x^2y + \dfrac{3\cdot2!}{2!1!}\cdot4xy^2 - y^3$

$\qquad = 64x^3 - \dfrac{3}{1}\cdot16x^2y + \dfrac{3}{1}\cdot4xy^2 - y^3 = 64x^3 - 48x^2y + 12xy^2 - y^3$

10. $(x+4y)^3 = x^3 + \dfrac{3!}{1!(3-1)!}x^2(4y) + \dfrac{3!}{2!(3-2)!}x(4y)^2 + (4y)^3$

$\qquad = x^3 + \dfrac{3!}{1!2!}\cdot4x^2y + \dfrac{3!}{2!1!}\cdot16xy^2 + 64y^3$

$\qquad = x^3 + \dfrac{3\cdot2!}{1!2!}\cdot4x^2y + \dfrac{3\cdot2!}{2!1!}\cdot16xy^2 + 64y^3$

$\qquad = x^3 + \dfrac{3}{1}\cdot4x^2y + \dfrac{3}{1}\cdot16xy^2 + 64y^3 = x^3 + 12x^2y + 48xy^2 + 64y^3$

11. In the 5th term, the exponent on y is 4.

Variables: x^1y^4

Coef. $= \dfrac{n!}{r!(n-r)!} = \dfrac{5!}{4!1!} = 5$

Term $= 5xy^4$

12. In the 4th term, the exponent on $-y$ is 3.

Variables: $x^2(-y)^3 = -x^2y^3$

Coef. $= \dfrac{n!}{r!(n-r)!} = \dfrac{5!}{3!2!} = 10$

Term $= 10(-x^2y^3) = -10x^2y^3$

13. 2nd term: The exponent on $-4y$ is 1. Variables: $(3x)^2(-4y)^1 = (9x^2)(-4y) = -36x^2y$

Coef. $= \dfrac{n!}{r!(n-r)!} = \dfrac{3!}{2!1!} = 3$; Term $= 3(-36x^2y) = -108x^2y$

14. 3rd term: The exponent on $3y$ is 2. Variables: $(4x)^2(3y)^2 = (16x^2)(9y^2) = 144x^2y^2$

Coef. $= \dfrac{n!}{r!(n-r)!} = \dfrac{4!}{2!2!} = 6$; Term $= 6(144x^2y^2) = 864x^2y^2$

15. $a_n = a_1 + (n-1)d = 7 + (8-1)5 = 7 + 7 \cdot 5 = 42$

16.
$a_n = a_1 + (n-1)d$ $a_1 + 8d = 242 \Rightarrow \times(-1)$ $-a_1 - 8d = -242$

$242 = a_1 + (9-1)d$ $\underline{a_1 + 6d = 212} \Rightarrow$ $\underline{a_1 + 6d = 212}$

$242 = a_1 + 8d$ $-2d = -30$

$a_n = a_1 + (n-1)d$ $d = 15$

$212 = a_1 + (7-1)d$ Substitute and solve for a_1:

$212 = a_1 + 6d$ $a_1 + 6d = 212$

                         $a_1 + 6(15) = 212$

                         $a_1 + 90 = 212$

                             $a_1 = 122 \Rightarrow 122, 137, 152, 167, 182$

17. Form an arithmetic sequence with a 1st term of 8 and a 4th term of 25:

$a_n = a_1 + (n-1)d$

$25 = 8 + (4-1)d$

$17 = 3d$

$\frac{17}{3} = d \Rightarrow 8, \boxed{\frac{41}{3}, \frac{58}{3}}, 25$

18. $a_1 = 11, d = 7, n = 20$

$a_n = a_1 + (n-1)d = 11 + 19(7) = 144$

$S_n = \dfrac{n(a_1 + a_n)}{2} = \dfrac{20(11 + 144)}{2} = 1,550$

19. $a_1 = 9, d = -\frac{5}{2}, n = 10$

$a_n = a_1 + (n-1)d = 9 + 9\left(-\frac{5}{2}\right) = -\frac{27}{2}$

$S_n = \dfrac{n(a_1 + a_n)}{2} = \dfrac{10\left(9 - \frac{27}{2}\right)}{2} = -\dfrac{45}{2}$

20. $\displaystyle\sum_{k=1}^{5} 4k = 4(1) + 4(2) + 4(3) + 4(4) + 4(5) = 4 + 8 + 12 + 16 + 20 = 60$

21. $\displaystyle\sum_{k=1}^{4} (k^2 + 1) = (1^2 + 1) + (2^2 + 1) + (3^2 + 1) + (4^2 + 1) = 2 + 5 + 10 + 17 = 34$

22. $\displaystyle\sum_{k=1}^{4} (3k - 4) = (3(1) - 4) + (3(2) - 4) + (3(3) - 4) + (3(4) - 4) = -1 + 2 + 5 + 8 = 14$

23. $\displaystyle\sum_{k=10}^{10} 36k = 36(10) = 360$

24. If the 5th term is $\frac{3}{2}$ and the 4th term is 3,

then the common ratio $r = \frac{3}{2} \div 3 = \frac{1}{2}$.

$a_3 = a_4 \div r = 3 \div \frac{1}{2} = 6$

$a_2 = a_3 \div r = 6 \div \frac{1}{2} = 12$

$a_1 = a_2 \div r = 12 \div \frac{1}{2} = 24$

$24, 12, 6, 3, \frac{3}{2}$

25. $a_n = a_1 r^{n-1}$

$\quad = \frac{1}{16}(2)^{5-1}$

$\quad = \frac{1}{16}(16) = 1$

26. $a_1 = -6, a_4 = 384$ $\qquad a_n = a_1 r^{n-1}$

$384 = -6r^{4-1}$

$-64 = r^3$

$-4 = r \Rightarrow -6, \boxed{24, -96}, 384$

27. $a_1 = 240, r = \frac{1}{2}, n = 6;\ S_n = \dfrac{a_1 - a_1 r^n}{1-r} = \dfrac{240 - 240\left(\frac{1}{2}\right)^6}{1 - \frac{1}{2}} = \dfrac{240 - 240\left(\frac{1}{64}\right)}{\frac{1}{2}} = \dfrac{\frac{945}{4}}{\frac{1}{2}} = \dfrac{945}{2}$

28. $a_1 = \frac{1}{8}, r = -2, n = 8;\ S_n = \dfrac{a_1 - a_1 r^n}{1-r} = \dfrac{\frac{1}{8} - \frac{1}{8}(-2)^8}{1-(-2)} = \dfrac{\frac{1}{8} - \frac{1}{8}(256)}{3} = \dfrac{-\frac{255}{8}}{3} = -\dfrac{85}{8}$

29. Sequence of amounts: $5000, 5000(0.8), 5000(0.8)^2, \ldots$

$a_1 = 5000, r = 0.8, n = 6 \Rightarrow a_n = a_1 r^{n-1} = 5000(0.8)^5 \approx \$1{,}638.40$

30. Sequence of amounts: $25700, 25700(1.18), 25700(1.18)^2, \ldots$

$a_1 = 25700, r = 1.18, n = 11 \Rightarrow a_n = a_1 r^{n-1} = 25700(1.18)^{10} \approx \$134{,}509.57$

31. $a_1 = 300, d = 75$

$\quad a_n = a_1 + (n-1)d$

$1200 = 300 + (n-1)75$

$1200 = 300 + 75n - 75$

$975 = 75n$

$13 = n$; In the 13th year, or after 12 years

of increases.

32. $16, 48, 80, \ldots;\ a_1 = 16, d = 32$

$a_n = a_1 + (n-1)d = 16 + (10-1)(32)$

$\qquad = 16 + 9(32) = 304\ \text{ft}$

$S_n = \dfrac{n(a_1 + a_n)}{2} = \dfrac{10(16 + 304)}{2} = 1600\ \text{ft}$

33. $a_1 = 100, r = \frac{1}{5};\ S_\infty = \dfrac{a_1}{1-r} = \dfrac{100}{1 - \frac{1}{5}} = \dfrac{100}{\frac{4}{5}} = 125$

34. $0.\overline{05} = \dfrac{5}{100} + \dfrac{5}{10{,}000} + \dfrac{5}{1{,}000{,}000} + \cdots \Rightarrow a_1 = \dfrac{5}{100}, r = \dfrac{1}{100}$

$S_\infty = \dfrac{a_1}{1-r} = \dfrac{\frac{5}{100}}{1 - \frac{1}{100}} = \dfrac{\frac{5}{100}}{\frac{99}{100}} = \dfrac{5}{99}$

35. $17 \cdot 8 = 136$

36. $P(7, 7) = \dfrac{7!}{(7-7)!} = \dfrac{7!}{0!} = 7! = 5{,}040$

37. $P(5,0) = \dfrac{5!}{(5-0)!} = \dfrac{5!}{5!} = 1$

38. $P(8,6) = \dfrac{8!}{(8-6)!} = \dfrac{8!}{2!} = \dfrac{40{,}320}{2} = 20{,}160$

39. $\dfrac{P(9,6)}{P(10,7)} = \dfrac{\frac{9!}{(9-6)!}}{\frac{10!}{(10-7)!}} = \dfrac{\frac{9!}{3!}}{\frac{10!}{3!}} = \dfrac{9!}{3!} \cdot \dfrac{3!}{10!} = \dfrac{9!}{10!} = \dfrac{9!}{10 \cdot 9!} = \dfrac{1}{10}$

40. $C(8,8) = \dfrac{8!}{8!(8-8)!} = \dfrac{8!}{8!0!} = \dfrac{8!}{8!} = 1$

41. $C(7,0) = \dfrac{7!}{0!(7-0)!} = \dfrac{7!}{0!7!} = \dfrac{7!}{7!} = 1$

42. $\dbinom{8}{6} = \dfrac{8!}{6!(8-6)!} = \dfrac{8 \cdot 7 \cdot 6!}{6!2!} = \dfrac{56}{2} = 28$

43. $\dbinom{10}{4} = \dfrac{10!}{4!(10-4)!} = \dfrac{10 \cdot 9 \cdot 8 \cdot 7 \cdot 6!}{4!6!} = \dfrac{5040}{24} = 210$

44. $C(6,3) \cdot C(7,3) = \dfrac{6!}{3!(6-3)!} \cdot \dfrac{7!}{3!(7-3)!} = \dfrac{6!}{3!3!} \cdot \dfrac{7!}{3!4!} = 20 \cdot 35 = 700$

45. $(x+2)^3 = \dbinom{3}{0}x^3 2^0 + \dbinom{3}{1}x^2 2^1 + \dbinom{3}{2}x^1 2^2 + \dbinom{3}{3}x^0 2^3$

$\qquad = \dfrac{3!}{0!3!} \cdot x^3 + \dfrac{3!}{1!2!} \cdot 2x^2 + \dfrac{3!}{2!1!} \cdot 4x + \dfrac{3!}{3!0!} \cdot 8 = x^3 + 6x^2 + 12x + 8$

46. $5! = 120$

47. $5! \cdot 3! = 720$

48. $\dbinom{10}{3} = \dfrac{10!}{3!7!} = 120$

49. $\dbinom{5}{2}\dbinom{6}{2} = 10 \cdot 15$
$\qquad\qquad = 150$

50. $\dfrac{6}{16} = \dfrac{3}{8}$

51. $\dfrac{8}{16} = \dfrac{1}{2}$

52. $\dfrac{14}{16} = \dfrac{7}{8}$

53. roll of 5 $\Rightarrow \{(1,4),(2,3),(3,2),(4,1)\}$

$\qquad \dfrac{\text{\# roll of 5}}{\text{total \# rolls}} = \dfrac{4}{36} = \dfrac{1}{9}$

54. impossible $\Rightarrow 0$

55. $\dfrac{\text{\# tens}}{\text{total \# cards}} = \dfrac{4}{52} = \dfrac{1}{13}$

56. $\dfrac{\overset{\text{\# ways to get}}{\underset{\text{3 aces}}{}} \cdot \overset{\text{\# ways to get 2}}{\underset{\text{other cards}}{}}}{\text{\# ways to draw 5 cards}} = \dfrac{\binom{4}{3}\binom{48}{2}}{\binom{52}{5}} = \dfrac{4 \cdot 1128}{2{,}598{,}960} = \dfrac{4512}{2{,}598{,}960} = \dfrac{94}{54145}$

57. $\dfrac{\text{\# ways to get 5 diamonds}}{\text{\# ways to get 5 cards}} = \dfrac{\binom{13}{5}}{\binom{52}{5}} = \dfrac{1287}{2{,}598{,}960} = \dfrac{33}{66{,}640}$

Chapter 13 Test (page 1011)

1. $\dfrac{9!}{7!} = \dfrac{9 \cdot 8 \cdot 7!}{7!} = 9 \cdot 8 = 72$

2. $0! = 1$

3. In the 2nd term, the exponent on $-y$ is 1.
Variables: $x^4(-y)^1 = -x^4y$
Coef. $= \dfrac{n!}{r!(n-r)!} = \dfrac{5!}{1!4!} = 5$
Term $= 5(-x^4y) = -5x^4y$

4. In the 3rd term, the exponent on $2y$ is 2.
Variables: $(x)^2(2y)^2 = 4x^2y^2$
Coef. $= \dfrac{n!}{r!(n-r)!} = \dfrac{4!}{2!2!} = 6$
Term $= 6(4x^2y^2) = 24x^2y^2$

5. $a_1 = 3, d = 7, n = 10; a_n = a_1 + (n-1)d = 3 + (10-1)(7) = 3 + 9(7) = 66$

6. $a_1 = -2, d = 5, n = 12; a_n = a_1 + (n-1)d = -2 + (12-1)(5) = -2 + 11(5) = 53$
$S_n = \dfrac{n(a_1 + a_n)}{2} = \dfrac{12(-2 + 53)}{2} = \dfrac{12(51)}{2} = 306$

7. Form an arithmetic sequence with a 1st term of 2 and a 4th term of 98:
$a_n = a_1 + (n-1)d$
$98 = 2 + (4-1)d$
$96 = 3d$
$32 = d \Rightarrow 2, \boxed{34, 66}, 98$

8. $\displaystyle\sum_{k=1}^{4}(3k-4) = (3(1) - 4) + (3(2) - 4) + (3(3) - 4) + (3(4) - 4) = -1 + 2 + 5 + 8 = 14$

9. $a_1 = -\frac{1}{9}, r = 3, n = 7; a_n = a_1 r^{n-1} = -\frac{1}{9}(3)^{7-1} = -\frac{1}{9}(3)^6 = -\frac{1}{9}(729) = -81$

10. $a_1 = \frac{1}{27}, r = 3, n = 6; S_n = \dfrac{a_1 - a_1 r^n}{1 - r} = \dfrac{\frac{1}{27} - \frac{1}{27}(3)^6}{1 - 3} = \dfrac{\frac{1}{27} - \frac{1}{27}(729)}{-2} = \dfrac{-\frac{728}{27}}{-2} = \dfrac{364}{27}$

11. $a_1 = 3, a_4 = 648$
$a_n = a_1 r^{n-1}$
$648 = 3r^{4-1}$
$216 = r^3$
$6 = r \Rightarrow 3, \boxed{18, 108}, 648$

12. $a_1 = 9, r = \dfrac{1}{3}$
$S_\infty = \dfrac{a_1}{1-r} = \dfrac{9}{1 - \frac{1}{3}} = \dfrac{9}{\frac{2}{3}} = \dfrac{27}{2}$

13. $r = \frac{10}{5} = 2$; Since $|r| > 1$, the sum does not exist.

14. $P(7, 3) = \dfrac{7!}{(7-3)!} = \dfrac{7!}{4!} = \dfrac{7 \cdot 6 \cdot 5 \cdot 4!}{4!}$
$= 210$

15. $P(8, 8) = \dfrac{8!}{(8-8)!} = \dfrac{8!}{0!} = 8! = 40{,}320$

16. $C(5,3) = \dfrac{5!}{3!2!} = \dfrac{5 \cdot 4 \cdot 3!}{3!2!} = \dfrac{20}{2} = 10$

17. $C(8,3) = \dfrac{8!}{3!5!} = \dfrac{8 \cdot 7 \cdot 6 \cdot 5!}{3 \cdot 2 \cdot 1 \cdot 5!} = \dfrac{56 \cdot 6}{6} = 56$

18. $C(6,0) \cdot P(6,5) = \dfrac{6!}{0!6!} \cdot \dfrac{6!}{1!} = 1 \cdot 6!$
$$= 720$$

19. $\dfrac{P(6,4)}{C(6,4)} = \dfrac{\frac{6!}{2!}}{\frac{6!}{4!2!}} = \dfrac{6!}{2!} \cdot \dfrac{4!2!}{6!} = 4! = 24$

20. $(x-3)^4 = \binom{4}{0}x^4(-3)^0 + \binom{4}{1}x^3(-3)^1 + \binom{4}{2}x^2(-3)^2 + \binom{4}{3}x^1(-3)^3 + \binom{4}{4}x^0(-3)^4$

$= \dfrac{4!}{0!4!} \cdot x^4(1) + \dfrac{4!}{1!3!} \cdot x^3(-3) + \dfrac{4!}{2!2!} \cdot x^2(9) + \dfrac{4!}{3!1!} \cdot x(-27) + \dfrac{4!}{4!0!} \cdot (81)$

$= x^4 - 12x^3 + 54x^2 - 108x + 81$

21. $\binom{8}{5} = \dfrac{8!}{5!3!} = 56$

22. $\underset{\substack{\text{2 women}}}{\text{\# ways to get}} \cdot \underset{\substack{\text{1 man}}}{\text{\# ways to get}} = \binom{4}{2}\binom{5}{1}$
$$= 6 \cdot 5 = 30$$

23. $\dfrac{1}{6}$

24. $\dfrac{8}{52} = \dfrac{2}{13}$

25. $\dfrac{\text{\# ways to get 5 hearts}}{\text{\# ways to get 5 cards}} = \dfrac{\binom{13}{5}}{\binom{52}{5}} = \dfrac{1287}{2,598,960} = \dfrac{33}{66,640}$

26. $\dfrac{\text{\# ways to get 2 heads}}{\text{\# ways to toss 5 times}} = \dfrac{\binom{5}{2}}{2^5} = \dfrac{10}{32} = \dfrac{5}{16}$

Cumulative Review Exercises (page 1011)

1. $\begin{cases} 2x + y = 5 \\ x - 2y = 0 \end{cases}$

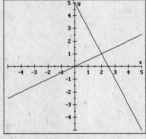

Solution: $(2,1)$

2. $\begin{cases} (1) & 3x + y = 4 \\ (2) & 2x - 3y = -1 \end{cases}$

Substitute $y = 4 - 3x$ from (1) into (2):
$$2x - 3y = -1$$
$$2x - 3(4 - 3x) = -1$$
$$2x - 12 + 9x = -1$$
$$11x = 11$$
$$x = 1$$

Substitute this and solve for y:
$$y = 4 - 3x = 4 - 3(1) = 1$$
Solution: $(1,1)$

CUMULATIVE REVIEW EXERCISES

23.
$$2^{x+5} = 3^x$$
$$\log 2^{x+5} = \log 3^x$$
$$(x+5)\log 2 = x\log 3$$
$$x\log 2 + 5\log 2 = x\log 3$$
$$5\log 2 = x\log 3 - x\log 2$$
$$5\log 2 = x(\log 3 - \log 2)$$
$$\frac{5\log 2}{\log 3 - \log 2} = x$$

24.
$$\log 5 + \log x - \log 4 = 1$$
$$\log\frac{5x}{4} = 1$$
$$10^1 = \frac{5x}{4}$$
$$40 = 5x$$
$$8 = x$$

25. $A = P\left(1 + \dfrac{r}{k}\right)^{kt} = 9000\left(1 + \dfrac{-0.12}{1}\right)^{1(9)} \approx \$2,848.31$

26. $\log_7 9 = \dfrac{\log 9}{\log 7} \approx 1.1292$

27. $\dfrac{5!7!}{6!} = \dfrac{5! \cdot 7 \cdot 6!}{6!} = 5! \cdot 7 = 840$

28. $(3a - b)^4$

$= (3a)^4 + \dfrac{4!}{1!(4-1)!}(3a)^3(-b) + \dfrac{4!}{2!(4-2)!}(3a)^2(-b)^2 + \dfrac{4!}{3!(4-3)!}(3a)(-b)^3 + (-b)^4$

$= 81a^4 + \dfrac{4!}{1!3!}(-27a^3b) + \dfrac{4!}{2!2!}(9a^2b^2) + \dfrac{4!}{3!1!}(-3ab^3) + b^4$

$= 81a^4 - \dfrac{4 \cdot 3!}{1!3!}(27a^3b) + \dfrac{4 \cdot 3 \cdot 2!}{2! \cdot 2 \cdot 1}(9a^2b^2) - \dfrac{4 \cdot 3!}{3!1!}(3ab^3) + b^4$

$= 81a^4 - \dfrac{4}{1}(27a^3b) + \dfrac{12}{2}(9a^2b^2) - \dfrac{4}{1}(3ab^3) + b^4$

$= 81a^4 - 108a^3b + 54a^2b^2 - 12ab^3 + b^4$

29. In the 7th term, the exponent on $-y$ is 6.
Variables: $(2x)^2(-y)^6 = 4x^2y^6$
Coef. $= \dfrac{n!}{r!(n-r)!} = \dfrac{8!}{6!2!} = 28$
Term $= 28(4x^2y^6) = 112x^2y^6$

30. $a_1 = -11, d = 6, n = 20$
$a_n = a_1 + (n-1)d$
$\quad = -11 + (19)(6)$
$\quad = -11 + 114 = 103$

31. $a_1 = 6, d = 3, n = 20$; $a_n = a_1 + (n-1)d = 6 + (20-1)(3) = 6 + 19(3) = 63$
$S_n = \dfrac{n(a_1 + a_n)}{2} = \dfrac{20(6+63)}{2} = \dfrac{20(69)}{2} = 690$

32. $a_1 = -3$; $a_4 = 30$:
$a_n = a_1 + (n-1)d$
$30 = -3 + (4-1)d$
$33 = 3d$
$11 = d \Rightarrow -3, \boxed{8, 19}, 30$

33. $\displaystyle\sum_{k=1}^{4} 2k^2 = 2(1)^2 + 2(2)^2 + 2(3)^2 + 2(4)^2$
$= 2 + 8 + 18 + 32 = 60$

34. $\sum_{k=3}^{6}(4k-5) = (4(3)-5) + (4(4)-5) + (4(5)-5) + (4(6)-5) = 7 + 11 + 15 + 19 = 52$

35. $a_1 = \frac{1}{27}, r = 3, n = 7; a_n = a_1 r^{n-1} = \frac{1}{27}(3)^{7-1} = \frac{1}{27}(3)^6 = \frac{1}{27}(729) = 27$

36. $a_1 = \frac{1}{64}, r = 2, n = 10; S_n = \frac{a_1 - a_1 r^n}{1-r} = \frac{\frac{1}{64} - \frac{1}{64}(2)^{10}}{1-2} = \frac{\frac{1}{64} - \frac{1}{64}(1024)}{-1} = \frac{-\frac{1023}{64}}{-1} = \frac{1023}{64}$

37. $a_1 = -3, a_4 = 192$
$$a_n = a_1 r^{n-1}$$
$$192 = -3r^{4-1}$$
$$-64 = r^3$$
$$-4 = r \Rightarrow -3, \boxed{12, -48}, 192$$

38. $a_1 = 9, r = \dfrac{1}{3}$
$$S_\infty = \frac{a_1}{1-r} = \frac{9}{1-\frac{1}{3}} = \frac{9}{\frac{2}{3}} = \frac{27}{2}$$

39. $P(9,3) = \dfrac{9!}{(9-3)!} = \dfrac{9!}{6!} = \dfrac{9 \cdot 8 \cdot 7 \cdot 6!}{6!} = 9 \cdot 8 \cdot 7 = 504$

40. $C(10,7) = \dfrac{10!}{7!(10-7)!} = \dfrac{10!}{7!3!} = \dfrac{10 \cdot 9 \cdot 8 \cdot 7!}{7! \cdot 3 \cdot 2 \cdot 1} = \dfrac{720}{6} = 120$

41. $\dfrac{C(8,4)C(8,0)}{P(6,2)} = \dfrac{\frac{8!}{4!4!} \cdot \frac{8!}{0!8!}}{\frac{6!}{4!}} = \dfrac{70 \cdot 1}{30} = \dfrac{7}{3}$

42. $C(n,n) = 1$ is smaller than $P(n,n) = n!$.

43. $7! = 5{,}040$

44. $\binom{9}{3} = \frac{9!}{3!6!} = 84$

45. $\dfrac{6}{52} = \dfrac{3}{26}$

Exercises A.1 (page A-11)

1. $27.3 \text{ yd} = \dfrac{27.3 \text{ yd}}{1} \cdot \dfrac{3 \text{ ft}}{1 \text{ yd}} = 81.9 \text{ ft}$

3. $20{,}320 \text{ ft} = \dfrac{20{,}320 \text{ ft}}{1} \cdot \dfrac{1 \text{ mi}}{5280 \text{ ft}} \approx 3.8 \text{ mi}$

5. $1{,}214{,}400 \text{ ft} = \dfrac{1{,}214{,}400 \text{ ft}}{1} \cdot \dfrac{1 \text{ mi}}{5280 \text{ ft}}$
$= 230 \text{ mi}$

7. $12{,}500 \text{ ft} = \dfrac{12{,}500 \text{ ft}}{1} \cdot \dfrac{1 \text{ mi}}{5280 \text{ ft}} \approx 2.4 \text{ mi}$

9. $\dfrac{1}{2} \text{ gal} = \dfrac{0.5 \text{ gal}}{1} \cdot \dfrac{4 \text{ qt}}{1 \text{ gal}} \cdot \dfrac{2 \text{ pt}}{1 \text{ qt}} \cdot \dfrac{2 \text{ C}}{1 \text{ pt}} = 8 \text{ C}$

11. $12 \text{ oz} \cdot 24 = 288 \text{ oz} = \dfrac{288 \text{ oz}}{1} \cdot \dfrac{1 \text{ bottle}}{32 \text{ oz}}$
$= 9 \text{ bottles}$

13. $5 \text{ gal} = \dfrac{5 \text{ gal}}{1} \cdot \dfrac{4 \text{ qt}}{1 \text{ gal}} \cdot \dfrac{2 \text{ pt}}{1 \text{ qt}} = 40 \text{ pt}$, so 40 bags

15. $4{,}500{,}000 \text{ lb} \div 2 = 2{,}250{,}000 \text{ lb} = \dfrac{2{,}250{,}000 \text{ lb}}{1} \cdot \dfrac{1 \text{ ton}}{2000 \text{ lb}} = 1125 \text{ tons}$

17. $1000 \text{ lb per acre} = \dfrac{1000 \text{ lb}}{1 \text{ acre}} \cdot \dfrac{3{,}500{,}000 \text{ acre}}{1 \text{ ranch}} \cdot \dfrac{1 \text{ ton}}{2000 \text{ lb}} = \dfrac{1{,}750{,}000 \text{ ton}}{\text{ranch}}$, or $1{,}750{,}000$ tons of cattle

$1{,}750{,}000 \text{ ton} = \dfrac{1{,}750{,}000 \text{ ton}}{1} \cdot \dfrac{2000 \text{ lb}}{1 \text{ ton}} \cdot \dfrac{1 \text{ cow}}{1650 \text{ lb}} \approx 2{,}121{,}212 \text{ head of cattle}$

19. $2 \text{ mm} = \dfrac{2 \text{ mm}}{1} \cdot \dfrac{1 \text{ m}}{1000 \text{ mm}} = 0.002 \text{ m}$ **21.** $15 \text{ L} = \dfrac{15 \text{ L}}{1} \cdot \dfrac{100 \text{ cl}}{1 \text{ L}} = 1500 \text{ cl}$

23. $5 \text{ T oz} = \dfrac{5 \text{ T oz}}{1} \cdot \dfrac{31.1034768 \text{ g}}{1 \text{ T oz}} \cdot \dfrac{1 \text{ kg}}{1000 \text{ g}} = 0.155517384 \text{ kg}$

25. $1 \text{ km} = \dfrac{1 \text{ km}}{1} \cdot \dfrac{0.62 \text{ mi}}{1 \text{ km}} \cdot \dfrac{5280 \text{ ft}}{1 \text{ mi}} \approx 3274 \text{ ft}$

27. $\dfrac{340.2 \text{ g}}{12 \text{ croissants}} \cdot \dfrac{0.035 \text{ oz}}{1 \text{ g}} = 0.99225 \text{ oz}$, or about 1 oz per croissant

Appendix A.2

1.

$$y = x^2 - 1$$

x-axis	y-axis	origin
$-y = x^2 - 1$	$y = (-x)^2 - 1$	$-y = (-x)^2 - 1$
not equivalent: no symmetry	$y = x^2 - 1$	$-y = x^2 - 1$
	equivalent: $\boxed{\text{symmetry}}$	not equivalent: no symmetry

3.

$$y = x^5$$

x-axis	y-axis	origin
$-y = x^5$	$y = (-x)^5$	$-y = (-x)^5$
not equivalent: no symmetry	$y = -x^5$	$-y = -x^5$
	not equivalent: no symmetry	$y = x^5$
		equivalent: $\boxed{\text{symmetry}}$

5.

$$y = -x^2 + 2$$

x-axis	y-axis	origin
$-y = -x^2 + 2$	$y = -(-x)^2 + 2$	$-y = (-x)^2 + 2$
not equivalent: no symmetry	$y = -x^2 + 2$	$-y = x^2 + 2$
	equivalent: $\boxed{\text{symmetry}}$	not equivalent: no symmetry

7.
$$y = x^2 - x$$

x-axis	y-axis	origin
$-y = x^2 - x$	$y = (-x)^2 - (-x)$	$-y = (-x)^2 - (-x)$
not equivalent: no symmetry	$y = x^2 + x$	$-y = x^2 + x$
	not equivalent: no symmetry	not equivalent: no symmetry

9.
$$y = -|x + 2|$$

x-axis	y-axis	origin
$-y = -|x + 2|$	$y = -|-x + 2|$	$-y = -|-x + 2|$
not equivalent: no symmetry	not equivalent: no symmetry	not equivalent: no symmetry

11.
$$|y| = x$$

x-axis	y-axis	origin
$|-y| = x$	$|y| = -x$	$|-y| = -x$
$|-1||y| = x$	not equivalent: no symmetry	$|-1||y| = -x$
$|y| = x$		$|y| = -x$
equivalent: $\boxed{\text{symmetry}}$		not equivalent: no symmetry

13. $y = x^4 - 4$

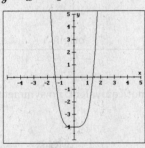

D $(-\infty, \infty)$; R $[-4, \infty)$

15. $y = -x^3$

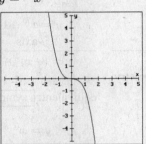

D $(-\infty, \infty)$; R $(-\infty, \infty)$

17. $y = x^4 + x^2$

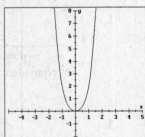

D $(-\infty, \infty)$; R $[0, \infty)$

19. $y = x^3 - x$

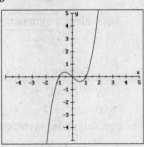

D $(-\infty, \infty)$; R $(-\infty, \infty)$

21. $y = \dfrac{1}{2}|x| - 1$

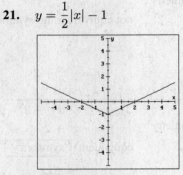

D $(-\infty, \infty)$; R $[-1, \infty)$

23. $y = -|x + 2|$

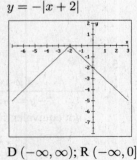

D $(-\infty, \infty)$; R $(-\infty, 0]$